Third Edition

TIME SERIES ANALYSIS

Forecasting and Control

GEORGE E. P. BOX
University of Wisconsin, U.S.A.

GWILYM M. JENKINS
University of Lancaster, U.K.

GREGORY C. REINSEL
University of Wisconsin, U.S.A.

PRENTICE HALL, Englewood Cliffs, New Jersey 07632

Library of Congress Cataloging-in-Publication Data

Box, George E. P.
Time series analysis : forecasting and control / George E. P. Box, Gwilym M. Jenkins, and Gregory C. Reinsel. -- 3rd ed.
p. cm.
Includes index.
ISBN 0-13-060774-6
1. Time-series analysis. 2. Prediction theory. 3. Transfer functions. 4. Feedback control systems—Mathematical models.
I. Jenkins, Gwilym M. II. Reinsel, Gregory C. III. Title.
QA280.B67 1994
003'.83—dc20 93-34620

To the memory of
Gwilym Meirion Jenkins

Acquisitions editor: *Jerome Grant*
Editorial/production supervision: *Edie Riker*
Cover design: *H. J. Salzbach*
Production coordinator: *Trudy Pisciotti*

Printed in the United States of America

10 9 8 7 6 5 4 3 2 1

ISBN 0-13-060774-6

Prentice-Hall International (UK) Limited, *London*
Prentice-Hall of Australia Pty. Limited, *Sydney*
Prentice-Hall Canada Inc., *Toronto*
Prentice-Hall Hispanoamericana, S.A., *Mexico*
Prentice-Hall of India Private Limited, *New Delhi*
Prentice-Hall of Japan, Inc., *Tokyo*
Simon & Schuster Asia Pte. Ltd., *Singapore*
Editora Prentice-Hall do Brasil, Ltda., *Rio de Janeiro*

CONTENTS

PREFACE

This book is concerned with the building of stochastic (statistical) models for time series and their use in important areas of application. This includes the topics of forecasting, model specification, estimation, and checking, transfer function modeling of dynamic relationships, modeling the effects of intervention events, and process control. Coincident with the first publication of *Time Series Analysis: Forecasting and Control*, there was a great upsurge in research in these topics. Thus, while the fundamental principles of the kind of time series analysis presented in that edition have remained the same, there has been a great influx of new ideas, modifications, and improvements provided by many authors.

The earlier editions of this book were written during a period in which Gwilym Jenkins was, with extraordinary courage, fighting a slowly debilitating illness. In the present revision, dedicated to his memory, we have preserved the general structure of the original book while revising, modifying, and omitting text where appropriate. In particular, Chapter 7 on estimation of ARMA models has been considerably modified. In addition, we have introduced entirely new sections on some important topics that have evolved since the first edition. These include presentations on various more recently developed methods for model specification, such as canonical correlation analysis and the use of model selection criteria, results on testing for unit root nonstationarity in ARIMA processes, the state space representation of ARMA models and its use for likelihood estimation and forecasting, score tests for model checking, structural components, and deterministic components in time series models and their estimation based on regression-time series model methods. A new chapter (12) has been developed on the important topic of *intervention* and *outlier* analysis, reflecting the substantial interest and research in this topic since the earlier editions.

Over the last few years, the new emphasis on industrial quality improvement has strongly focused attention on the role of control both in process *monitoring* as well as in process *adjustment*. The control section of this book has, therefore, been completely rewritten to serve as an introduction to these important topics and to provide a better understanding of their relationship.

The objective of this book is to provide practical techniques that will be available to most of the wide audience who could benefit from their use. While we have tried to remove the inadequacies of earlier editions, we have not attempted to produce here a rigorous mathematical treatment of the subject.

We wish to acknowledge our indebtedness to Meg (Margaret) Jenkins and to our wives, Claire and Sandy, for their continuing support and assistance throughout the long period of preparation of this revision.

Research on which the original book was based was supported by the Air Force Office of Scientific Research and by the British Science Research Council. Research incorporated in the third edition was partially supported by the Alfred P. Sloan Foundation and by the National Aeronautics and Space Administration. We are grateful to Professor E. S. Pearson and the Biometrika Trustees for permission to reprint condensed and adapted forms of Tables 1, 8, and 12 of *Biometrika Tables for Statisticians*, Vol. 1, edited by E. S. Pearson and H. O. Hartley, to Dr. Casimer Stralkowski for permission to reproduce and adapt three figures from his doctoral thesis, and to George Tiao, David Mayne, Emanuel Parzen, David Pierce, Granville Wilson, Donald Watts, John Hampton, Elaine Hodkinson, Patricia Blant, Dean Wichern, David Bacon, Paul Newbold, Hiro Kanemasu, Larry Haugh, John MacGregor, Bovas Abraham, Gina Chen, Johannes Ledolter, Greta Ljung, Carole Leigh, Mary Esser, and Meg Jenkins for their help, in many different ways, in preparing the earlier editions.

George Box and *Gregory Reinsel*

1

INTRODUCTION

A *time series* is a sequence of observations taken sequentially in time. Many sets of data appear as time series: a monthly sequence of the quantity of goods shipped from a factory, a weekly series of the number of road accidents, hourly observations made on the yield of a chemical process, and so on. Examples of time series abound in such fields as economics, business, engineering, the natural sciences (especially geophysics and meteorology), and the social sciences. Examples of data of the kind that we will be concerned with are displayed as time series plots in Figure 4.1 at the beginning of Chapter 4. An intrinsic feature of a time series is that, typically, adjacent observations are *dependent*. The nature of this dependence among observations of a time series is of considerable practical interest. *Time series analysis* is concerned with techniques for the analysis of this dependence. This requires the development of stochastic and dynamic models for time series data and the use of such models in important areas of application.

In the subsequent chapters of this book we present methods for building, identifying, fitting, and checking models for time series and dynamic systems. The methods discussed are appropriate for discrete (sampled-data) systems, where observation of the system occurs at equally spaced intervals of time.

We illustrate the use of these time series and dynamic models in four important areas of application.

1. The *forecasting* of future values of a time series from current and past values.
2. The determination of the *transfer function* of a system subject to

inertia—the determination of a dynamic input–output model that can show the effect on the output of a system of any given series of inputs.
3. The use of indicator input variables in transfer function models to represent and assess the effects of unusual *intervention* events on the behavior of a time series.
4. The design of simple *control schemes* by means of which potential deviations of the system output from a desired target may, so far as possible, be compensated by adjustment of the input series values.

1.1 FOUR IMPORTANT PRACTICAL PROBLEMS

1.1.1 Forecasting Time Series

The use at time t of available observations from a time series to forecast its value at some future time $t + l$ can provide a basis for (1) economic and business planning, (2) production planning, (3) inventory and production control, and (4) control and optimization of industrial processes. As originally described by Holt et al. [113], Brown [64], and the Imperial Chemical Industries monograph on short-term forecasting [178], forecasts are usually needed over a period known as the *lead time,* which varies with each problem. For example, the lead time in the inventory control problem was defined by Harrison [105] as a period that begins when an order to replenish stock is placed with the factory, and lasts until the order is delivered into stock.

We suppose that observations are available at *discrete,* equispaced intervals of time. For example, in a sales forecasting problem, the sales z_t in the current month t and the sales $z_{t-1}, z_{t-2}, z_{t-3}, \ldots$ in previous months might be used to forecast sales for lead times $l = 1, 2, 3, \ldots, 12$ months ahead. Denote by $\hat{z}_t(l)$ the forecast made at *origin* t of the sales z_{t+l} at some future time $t + l$, that is, at *lead time* l. The function $\hat{z}_t(l)$, $l = 1, 2, \ldots,$ which provides the forecasts at origin t for all future lead times, will be called the *forecast function* at origin t. Our objective is to obtain a forecast function which is such that the mean square of the deviations $z_{t+l} - \hat{z}_t(l)$ between the actual and forecasted values is as small as possible *for each lead time* l.

In addition to calculating the best forecasts, it is also necessary to specify their accuracy, so that, for example, the risks associated with decisions based upon the forecasts may be calculated. The accuracy of the forecasts may be expressed by calculating *probability limits* on either side of each forecast. These limits may be calculated for any convenient set of probabilities, for example 50% and 95%. They are such that the realized value of the time series, when it eventually occurs, will be included within these limits with the stated probability. To illustrate, Figure 1.1 shows the last 20 values of a time series culminating at time t. Also shown are fore-

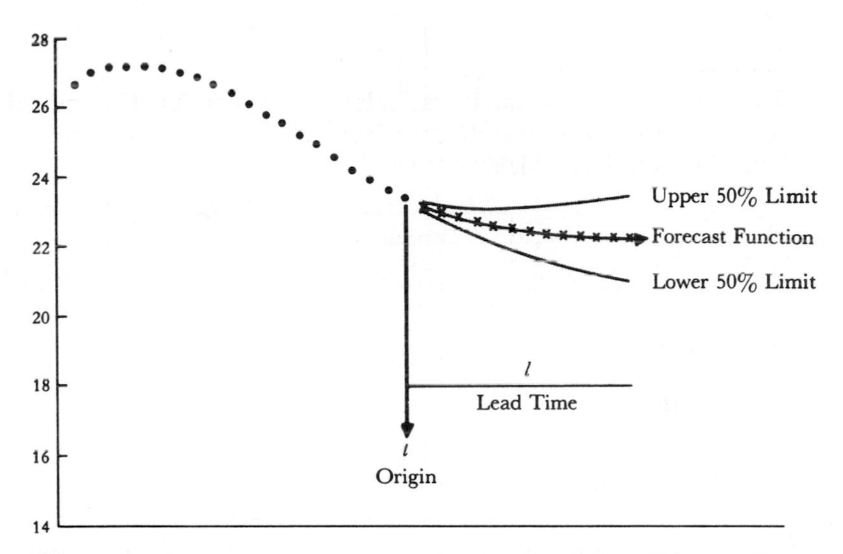

Figure 1.1 Values of a time series with forecast function and 50% probability limits.

casts made from origin t for lead times $l = 1, 2, \ldots, 13$, together with the 50% probability limits.

Methods for obtaining forecasts and estimating probability limits are discussed in detail in Chapter 5. These forecasting methods are developed based on the assumption that the time series z_t follows a *stochastic* model of a known form. Consequently, in Chapters 3 and 4 a useful class of such time series models that might be appropriate to represent the behavior of a series z_t, called autoregressive integrated moving average (ARIMA) models, are introduced and many of their properties are studied. Subsequently, in Chapters 6, 7, and 8 the practical matter of how these models may be fitted to actual time series data is explored and the methods are described through the three-stage procedure of tentative model identification or specification, estimation of model parameters, and model checking and diagnostics.

1.1.2 Estimation of Transfer Functions

A topic of considerable industrial interest is the study of process dynamics [15], [115]. Such a study is made (1) to achieve better control of existing plants, and (2) to improve the design of new plants. In particular, several methods have been proposed for estimating the transfer function of plant units from process records consisting of an input time series X_t and an output time series Y_t. Sections of such records are shown in Figure 1.2, where the input X_t is the rate of air supply and the output Y_t is the concentration of carbon dioxide produced in a furnace. The observations were made

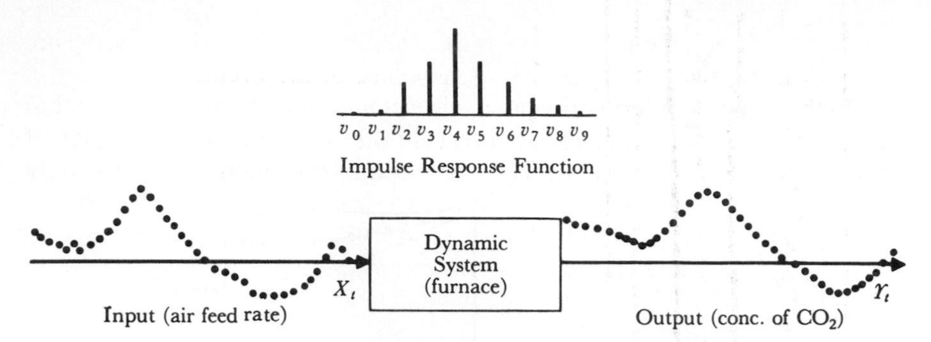

Figure 1.2 Input and output time series in relation to a dynamic system.

at 9-second intervals. A hypothetical impulse response function $v_j, j = 0, 1, 2, \ldots$, which determines the transfer function for the system through a dynamic linear relationship between input X_t and output Y_t of the form $Y_t = \Sigma_{j=0}^{\infty} v_j X_{t-j}$, is also shown in the figure as a bar chart. Transfer function models that relate an input process X_t to an output process Y_t are introduced in Chapter 10 and many of their properties are examined.

Methods for estimating transfer function models based on deterministic perturbations of the input, such as step, pulse, and sinusoidal changes, have not always been successful. This is because, for perturbations of a magnitude that are relevant and tolerable, the response of the system may be masked by uncontrollable disturbances referred to collectively as *noise*. Statistical methods for estimating transfer function models that make allowance for noise in the system are described in Chapter 11. The estimation of dynamic response is of considerable interest in economics, engineering, biology, and many other fields.

Another important application of transfer function models is in forecasting. If, for example, the dynamic relationship between two time series Y_t and X_t can be determined, past values of *both* series may be used in forecasting Y_t. In some situations this approach can lead to a considerable reduction in the errors of the forecasts.

1.1.3 Analysis of Effects of Unusual Intervention Events to a System

In some situations it may be known that certain exceptional external events, *intervention events,* could have affected the time series z_t under study. Examples of such intervention events include the incorporation of new environmental regulations, economic policy changes, strikes, and special promotion campaigns. Under such circumstances we may use transfer function models, as discussed in Section 1.1.2, to account for the effects of the intervention event on the series z_t, but where the "input" series will be in the form of a simple indicator variable taking only the values 1 and 0 to indicate (qualitatively) the presence or absence of the event.

In these cases, intervention analysis is undertaken to obtain a quantitative measure of the impact of the intervention event on the time series of interest. For example, Box and Tiao [59] used intervention models to study and quantify the impact of air pollution controls on smog-producing oxidant levels in the Los Angeles area and of economic controls on the consumer price index in the United States. Alternatively, intervention analysis may be undertaken to adjust for any unusual values in the series z_t that might have resulted as a consequence of the intervention event. This will ensure that the results of the time series analysis of the series, such as the structure of the fitted model, estimates of model parameters, and forecasts of future values, are not seriously distorted by the influence of these unusual values. Models for intervention analysis and their use, together with consideration of the related topic of detection of outlying or unusual values in a time series, are presented in Chapter 12.

1.1.4 Discrete Control Systems

In the past, to the statistician, the words "process control" have usually meant the *quality control techniques* developed originally by Shewhart [177] in the United States (see also Dudding and Jennet [85]). More recently, the sequential aspects of quality control have been emphasized, leading to the introduction of *cumulative sum charts* by Page [154], [155] and Barnard [21] and the *geometric moving average* charts of Roberts [170]. Such charts are frequently employed in industries concerned with the manufacture of discrete "parts" as one aspect of what is called *statistical process control* (SPC). In particular (see Deming [79]), they are used for continuous monitoring of a process. That is, they are used to supply a continuous screening mechanism for detecting assignable (or special) causes of variation. Appropriate display of plant data ensures that significant changes are quickly brought to the attention of those responsible for running the process. Knowing the answer to the question "*when* did a change of this particular kind occur?" we may be able to answer the question "*why* did it occur?" Hence a continuous incentive for process stabilization and improvement can be achieved.

By contrast, in the process and chemical industries, various forms of *feedback* and *feedforward* adjustment have been used in what we shall call *engineering process control* (EPC). Because the adjustments made by engineering process control are usually computed and applied automatically, this type of control is sometimes called *automatic process control* (APC). However, the *manner* in which these adjustments are made is a matter of convenience. This type of control is necessary when there are inherent *disturbances* or *noise* in the system inputs which are impossible or impractical to remove. When we can measure fluctuations in an input variable that can be observed but not changed, it may be possible to make appropriate compensatory changes in some other control variable. This is referred to as *feedforward control.* Alternatively, or in addition, we may be able to use the deviation from target or "error signal" of the output characteristic itself to calculate appropriate compensatory changes in the control variable. This is called *feedback control.* Unlike feedforward control, this mode of correction can be employed even when the source of the disturbances is not accurately known or the magnitude of the disturbance is not measured.

In Chapter 13 we draw on the earlier discussions in this book, on time series and transfer function models, to provide insight into the statistical aspects of these control methods and to appreciate better their relationships and different objectives. In particular, we show how some of the ideas of feedback control can be used to design simple charts for *manually adjusting* processes. For example, the upper chart of Figure 1.3 shows hourly measurements of the viscosity of a polymer made over a period of 42 hours. The viscosity is to be controlled about a target value of 90 units. As each viscosity measurement comes to hand, the process operator uses the nomogram shown in the middle of the figure to compute the adjustment to be made in

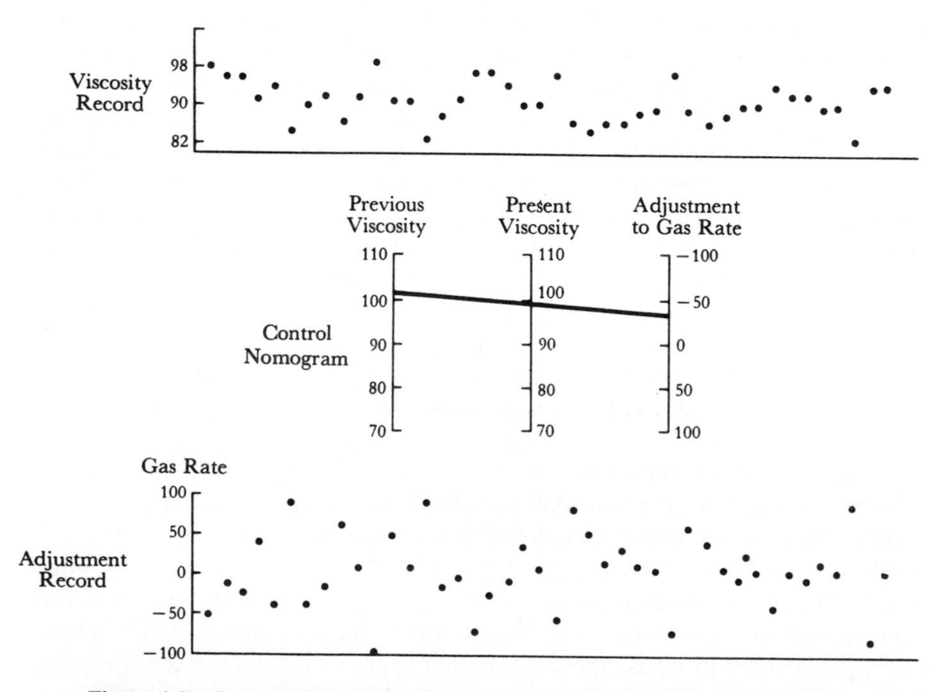

Figure 1.3 Control of viscosity. Record of observed viscosity and of adjustments in gas rate made using nomogram.

the manipulated variable (gas rate). The lower chart of Figure 1.3 shows the adjustments made in accordance with the nomogram.

1.2 STOCHASTIC AND DETERMINISTIC DYNAMIC MATHEMATICAL MODELS

The idea of using a mathematical model to describe the behavior of a physical phenomenon is well established. In particular, it is sometimes possible to derive a model based on physical laws, which enables us to calculate the value of some time-dependent quantity nearly exactly at any instant of time. Thus, we might calculate the trajectory of a missile launched in a known direction with known velocity. If exact calculation were possible, such a model would be entirely *deterministic*.

Probably no phenomenon is totally deterministic, however, because unknown factors can occur such as a variable wind velocity that can throw a missile slightly off course. In many problems we have to consider a time-dependent phenomenon, such as monthly sales of newsprint, in which there are many unknown factors and for which it is not possible to write a deterministic model that allows exact calculation of the future behavior of the phenomenon. Nevertheless, it may be possible to derive a model that can be used to calculate the *probability* of a future value lying between two specified limits. Such a model is called a probability model or a *stochastic model.* The models for time series that are needed, for example to achieve optimal forecasting and control, are in fact stochastic models. It is necessary in what follows to distinguish between the probability model or *stochastic process,* as it is sometimes called, and the observed time series. Thus a time series $z_1, z_2, \ldots, z_N$ of N successive observations is regarded as a sample realization from an infinite population of such time series that could have been generated by the stochastic process. Very often we shall omit the word stochastic from "stochastic process" and talk about the "process."

1.2.1 Stationary and Nonstationary Stochastic Models for Forecasting and Control

An important class of stochastic models for describing time series, which has received a great deal of attention, comprises what are called *stationary* models, which assume that the process remains in *equilibrium* about a *constant mean level.* However, forecasting has been of particular importance in industry, business, and economics, where many time series are often better represented as nonstationary and, in particular, as having no natural constant mean level over time. It is not surprising, therefore, that many of the economic forecasting methods originally proposed by Holt [112], [113], Winters [206], Brown [64], and the ICI Monograph [178] that used exponentially weighted moving averages can be shown to be appropriate for a particular type of *nonstationary* process. Although such methods are too narrow to deal efficiently with all time series, the fact that they give the right kind of forecast function supplies a clue to the *kind* of *nonstationary* model that might be useful in these problems.

The stochastic model for which the exponentially weighted moving average forecast yields minimum mean square error (Muth [147]) is a member of a class of *nonstationary* processes called autoregressive integrated moving average processes (ARIMA processes), which are discussed in Chapter 4. This wider class of processes provides a range of models, stationary and nonstationary, that adequately represent many of the time series met in practice. Our approach to forecasting has been first to derive an adequate stochastic model for the particular time series under study. As shown in Chapter 5, once an appropriate model has been determined for the series, the optimal forecasting procedure follows immediately. These forecasting procedures include the exponentially weighted moving average forecast as a special case.

Some simple operators. We shall employ extensively the *backward shift operator B*, which is defined by $Bz_t = z_{t-1}$; hence $B^m z_t = z_{t-m}$. The inverse operation is performed by the *forward shift operator* $F = B^{-1}$ given by $Fz_t = z_{t+1}$; hence $F^m z_t = z_{t+m}$. Another important operator is the *backward difference operator* ∇, which can be written in terms of B, since

$$\nabla z_t = z_t - z_{t-1} = (1 - B)z_t$$

Linear filter model. The stochastic models we employ are based on the idea (Yule [212]) that a time series in which successive values are highly dependent can frequently be regarded as generated from a series of *independent* "shocks" a_t. These shocks are *random* drawings from a fixed distribution, usually assumed Normal and having mean zero and variance σ_a^2. Such a sequence of random variables $a_t, a_{t-1}, a_{t-2}, \ldots$ is called a *white noise* process.

The white noise process a_t is supposed transformed to the process z_t by what is called a *linear filter*, as shown in Figure 1.4. The linear filtering operation simply takes a weighted sum of previous random shocks a_t, so that

$$\begin{aligned} z_t &= \mu + a_t + \psi_1 a_{t-1} + \psi_2 a_{t-2} + \cdots \\ &= \mu + \psi(B)a_t \end{aligned} \tag{1.2.1}$$

In general, μ is a parameter that determines the "level" of the process, and

$$\psi(B) = 1 + \psi_1 B + \psi_2 B^2 + \cdots$$

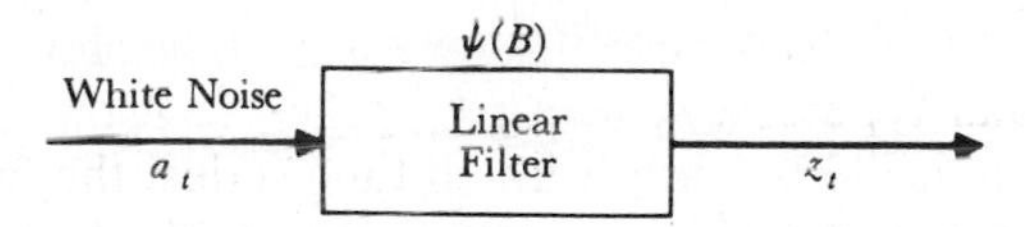

Figure 1.4 Representation of a time series as the output from a linear filter.

is the linear operator that transforms a_t into z_t and is called the *transfer function* of the filter.

The sequence $\psi_1, \psi_2, \ldots$ formed by the weights may, theoretically, be finite or infinite. If this sequence is finite, or infinite and absolutely summable in the sense that $\Sigma_{j=0}^{\infty} |\psi_j| < \infty$, the filter is said to be *stable* and the process z_t is stationary. The parameter μ is then the mean about which the process varies. Otherwise, z_t is nonstationary and μ has no specific meaning except as a reference point for the level of the process.

Autoregressive models. A stochastic model that can be extremely useful in the representation of certain practically occurring series is the *autoregressive* model. In this model, the current value of the process is expressed as a finite, linear aggregate of *previous values of the process* and a shock a_t. Let us denote the values of a process at equally spaced times t, $t-1$, $t-2$, ... by $z_t, z_{t-1}, z_{t-2}, \ldots$. Also let $\tilde{z}_t, \tilde{z}_{t-1}, \tilde{z}_{t-2}, \ldots$ be deviations from μ; for example, $\tilde{z}_t = z_t - \mu$. Then

$$\tilde{z}_t = \phi_1 \tilde{z}_{t-1} + \phi_2 \tilde{z}_{t-2} + \cdots + \phi_p \tilde{z}_{t-p} + a_t \tag{1.2.2}$$

is called an *autoregressive* (AR) *process of order p*. The reason for this name is that a linear model

$$\tilde{z} = \phi_1 \tilde{x}_1 + \phi_2 \tilde{x}_2 + \cdots + \phi_p \tilde{x}_p + a$$

relating a "dependent" variable z to a set of "independent" variables $x_1, x_2, \ldots, x_p$, plus an error term a, is often referred to as a *regression* model, and z is said to be "regressed" on $x_1, x_2, \ldots, x_p$. In (1.2.2) the variable z is regressed on previous values of itself; hence the model is *auto*regressive. If we define an *autoregressive operator* of order p by

$$\phi(B) = 1 - \phi_1 B - \phi_2 B^2 - \cdots - \phi_p B^p$$

the autoregressive model may be written economically as

$$\phi(B)\tilde{z}_t = a_t$$

The model contains $p + 2$ unknown parameters $\mu, \phi_1, \phi_2, \ldots, \phi_p, \sigma_a^2$, which in practice have to be estimated from the data. The additional parameter σ_a^2 is the variance of the white noise process a_t.

It is not difficult to see that the autoregressive model is a special case of the linear filter model of (1.2.1). For example, we can eliminate $\tilde{z}_{t-1}$ from the right-hand side of (1.2.2) by substituting

$$\tilde{z}_{t-1} = \phi_1 \tilde{z}_{t-2} + \phi_2 \tilde{z}_{t-3} + \cdots + \phi_p \tilde{z}_{t-p-1} + a_{t-1}$$

Similarly, we can substitute for $\tilde{z}_{t-2}$, and so on, to yield eventually an infinite series in the a's. Symbolically, we have that

$$\phi(B)\tilde{z}_t = a_t$$

is equivalent to

$$\tilde{z}_t = \psi(B)a_t$$

with

$$\psi(B) = \phi^{-1}(B)$$

Autoregressive processes can be stationary or nonstationary. For the process to be stationary, the ϕ's must be chosen so that the weights ψ_1, ψ_2, . . . in $\psi(B) = \phi^{-1}(B)$ form a convergent series. The necessary requirement for stationarity is that the autoregressive operator, $\phi(B) = 1 - \phi_1 B - \phi_2 B^2 - \cdots - \phi_p B^p$, considered as a polynomial in B of degree p, must have all roots of $\phi(B) = 0$ greater than 1 in absolute value; that is, all roots must lie outside the unit circle. We discuss these models in greater detail in Chapters 3 and 4.

Moving average models. The autoregressive model (1.2.2) expresses the deviation $\tilde{z}_t$ of the process as a *finite* weighted sum of p previous deviations $\tilde{z}_{t-1}, \tilde{z}_{t-2}, \ldots, \tilde{z}_{t-p}$ of the process, plus a random shock a_t. Equivalently, as we have just seen, it expresses $\tilde{z}_t$ as an *infinite* weighted sum of a's.

Another kind of model, of great practical importance in the representation of observed time series, is the finite *moving average* process. Here we make $\tilde{z}_t$ linearly dependent on a *finite* number q of previous a's. Thus

$$\tilde{z}_t = a_t - \theta_1 a_{t-1} - \theta_2 a_{t-2} - \cdots - \theta_q a_{t-q} \tag{1.2.3}$$

is called a *moving average* (MA) *process of order q*. The name "moving average" is somewhat misleading because the weights $1, -\theta_1, -\theta_2, \ldots, -\theta_q$, which multiply the a's, need not total unity nor need they be positive. However, this nomenclature is in common use, and therefore we employ it.

If we define a *moving average operator* of order q by

$$\theta(B) = 1 - \theta_1 B - \theta_2 B^2 - \cdots - \theta_q B^q$$

the moving average model may be written economically as

$$\tilde{z}_t = \theta(B)a_t$$

It contains $q + 2$ unknown parameters $\mu, \theta_1, \ldots, \theta_q, \sigma_a^2$, which in practice have to be estimated from the data.

Mixed autoregressive–moving average models. To achieve greater flexibility in fitting of actual time series, it is sometimes advantageous to include both autoregressive and moving average terms in the model. This leads to the mixed autoregressive–moving average model

$$\tilde{z}_t = \phi_1 \tilde{z}_{t-1} + \cdots + \phi_p \tilde{z}_{t-p} + a_t - \theta_1 a_{t-1} - \cdots - \theta_q a_{t-q} \tag{1.2.4}$$

or

$$\phi(B)\tilde{z}_t = \theta(B)a_t$$

which employs $p + q + 2$ unknown parameters μ; $\phi_1, \ldots, \phi_p$; $\theta_1, \ldots, \theta_q$; σ_a^2, that are estimated from the data. This model may also be written in the form of the linear filter (1.2.1) as $\tilde{z}_t = \phi^{-1}(B)\theta(B)a_t$. In practice, it is frequently true that adequate representation of actually occurring stationary time series can be obtained with autoregressive, moving average, or mixed models, in which p and q are not greater than 2 and often less than 2.

Nonstationary models. Many series actually encountered in industry or business (e.g., stock prices) exhibit nonstationary behavior and in particular do not vary about a fixed mean. Such series may nevertheless exhibit homogeneous behavior of a kind. In particular, although the general level about which fluctuations are occurring may be different at different times, the broad behavior of the series, when differences in level are allowed for, may be similar. We show in Chapter 4 and later chapters that such behavior may often be represented by a generalized autoregressive operator $\varphi(B)$, in which one or more of the zeros of the polynomial $\varphi(B)$ [i.e., one or more of the roots of the equation $\varphi(B) = 0$] lie on the unit circle. In particular, if there are d unit roots, the operator $\varphi(B)$ can be written

$$\varphi(B) = \phi(B)(1 - B)^d$$

where $\phi(B)$ is a stationary operator. Thus a model that can represent homogeneous nonstationary behavior is of the form

$$\varphi(B)z_t = \phi(B)(1 - B)^d z_t = \theta(B)a_t$$

that is,

$$\phi(B)w_t = \theta(B)a_t \tag{1.2.5}$$

where

$$w_t = \nabla^d z_t \tag{1.2.6}$$

Thus homogeneous nonstationary behavior can sometimes be represented by a model that calls for the dth difference of the process to be stationary. In practice, d is usually 0, 1, or at most 2.

The process defined by (1.2.5) and (1.2.6) provides a powerful model for describing stationary and nonstationary time series and is called an *autoregressive integrated moving average (ARIMA) process,* of order (p, d, q). The process is defined by

$$w_t = \phi_1 w_{t-1} + \cdots + \phi_p w_{t-p} + a_t - \theta_1 a_{t-1} - \cdots - \theta_q a_{t-q} \qquad (1.2.7)$$

with $w_t = \nabla^d z_t$. Note that if we replace w_t by $z_t - \mu$, when $d = 0$, the model (1.2.7) includes the stationary mixed model (1.2.4), as a special case, and also the pure autoregressive model (1.2.2) and the pure moving average model (1.2.3).

The reason for inclusion of the word "integrated" (which should perhaps more appropriately be "summed") in the ARIMA title is as follows. The relationship which is inverse to (1.2.6) is $z_t = S^d w_t$, where $S = \nabla^{-1} = (1 - B)^{-1}$ is the *summation operator* defined by

$$Sw_t = \sum_{j=0}^{\infty} w_{t-j} = w_t + w_{t-1} + w_{t-2} + \cdots$$

Thus the general autoregressive integrated moving average (ARIMA) process may be generated by summing or "integrating" the stationary ARMA process w_t, d times. In Chapter 9 we describe how a special form of the model (1.2.7) can be employed to represent seasonal time series.

1.2.2 Transfer Function Models

An important type of dynamic relationship between a continuous input and a continuous output, for which many physical examples can be found, is that in which the *deviations* of input X and output Y, from appropriate mean values, are related by a *linear* differential equation. In a similar way, for discrete data, in Chapter 10 we represent the transfer between an output Y and an input X, each measured at equispaced times, by the difference equation

$$(1 + \xi_1 \nabla + \cdots + \xi_r \nabla^r) Y_t = (\eta_0 + \eta_1 \nabla + \cdots + \eta_s \nabla^s) X_{t-b} \qquad (1.2.8)$$

in which the differential operator $D = d/dt$ is replaced by the difference operator ∇. An expression of the form (1.2.8), containing only a few parameters ($r \leqslant 2$, $s \leqslant 2$), may often be used as an approximation to a dynamic relationship whose true nature is more complex.

The linear model (1.2.8) may be written equivalently in terms of past values of the input and output by substituting $B = 1 - \nabla$ in (1.2.8), that is,

$$\begin{aligned}(1 - \delta_1 B - \cdots - \delta_r B^r) Y_t &= (\omega_0 - \omega_1 B - \cdots - \omega_s B^s) X_{t-b} \\ &= (\omega_0 B^b - \omega_1 B^{b+1} - \cdots - \omega_s B^{b+s}) X_t\end{aligned} \qquad (1.2.9)$$

or

$$\begin{aligned}\delta(B) Y_t &= \omega(B) B^b X_t \\ &= \Omega(B) X_t\end{aligned}$$

Alternatively, we can say that the output Y_t and input X_t are linked by a linear filter

$$\begin{aligned} Y_t &= v_0 X_t + v_1 X_{t-1} + v_2 X_{t-2} + \cdots \\ &= v(B) X_t \end{aligned} \tag{1.2.10}$$

for which the transfer function

$$v(B) = v_0 + v_1 B + v_2 B^2 + \cdots \tag{1.2.11}$$

can be expressed as a ratio of two polynomials,

$$v(B) = \frac{\Omega(B)}{\delta(B)} = \delta^{-1}(B)\Omega(B)$$

The linear filter (1.2.10) is said to be *stable* if the series (1.2.11) converges for $|B| \leq 1$. The series of weights $v_0, v_1, v_2, \ldots$, which appear in the transfer function (1.2.11), is called the *impulse response function*. We note that for the model (1.2.9), the first b weights $v_0, v_1, \ldots, v_{b-1}$ are zero. A hypothetical impulse response function for the system of Figure 1.2 is shown in the center of that diagram.

Models with superimposed noise. We have seen that the problem of estimating an appropriate model, linking an output Y_t and an input X_t, is equivalent to estimating the transfer function $v(B) = \delta^{-1}(B)\Omega(B)$. However, this problem is complicated in practice by the presence of noise N_t, which we assume corrupts the true relationship between input and output according to

$$Y_t = v(B)X_t + N_t$$

where N_t and X_t are independent. Suppose, as indicated by Figure 1.5, that the noise N_t can be described by a nonstationary stochastic model of the

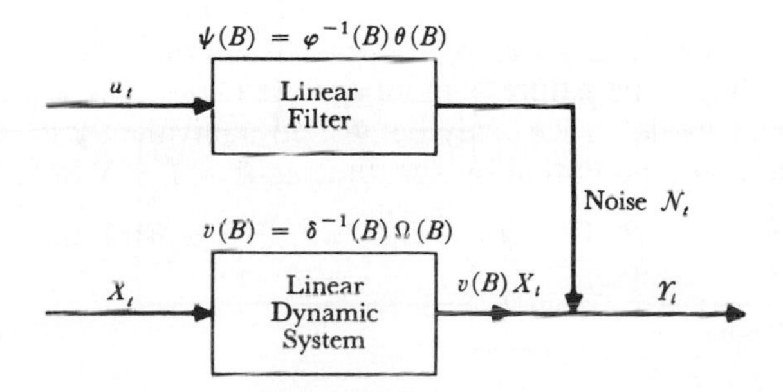

Figure 1.5 Transfer function model for dynamic system with superimposed noise model.

form (1.2.5) or (1.2.7), that is,

$$N_t = \psi(B)a_t = \varphi^{-1}(B)\theta(B)a_t$$

Then the observed relationship between output and input will be

$$\begin{aligned} Y_t &= v(B)X_t + \psi(B)a_t \\ &= \delta^{-1}(B)\Omega(B)X_t + \varphi^{-1}(B)\theta(B)a_t \end{aligned} \tag{1.2.12}$$

In practice, it is necessary to estimate the transfer function $\psi(B) = \varphi^{-1}(B)\theta(B)$ of the linear filter describing the noise, in addition to the transfer function $v(B) = \delta^{-1}(B)\Omega(B)$, which describes the dynamic relationship between the input and the output. Methods for doing this are discussed in Chapter 11.

1.2.3 Models for Discrete Control Systems

As stated in Section 1.1.4, control is an attempt to compensate for disturbances that infect a system. Some of these disturbances are measurable; others are not measurable and only manifest themselves as unexplained deviations from the target of the characteristic to be controlled. To illustrate the general principles involved, consider the special case where unmeasured disturbances affect the output Y_t of a system, and suppose that feedback control is employed to bring the output as close as possible to the desired target value by adjustments applied to an input variable X_t. This is illustrated in Figure 1.6. Suppose that N_t represents the effect at the output of various unidentified disturbances within the system, which in the absence

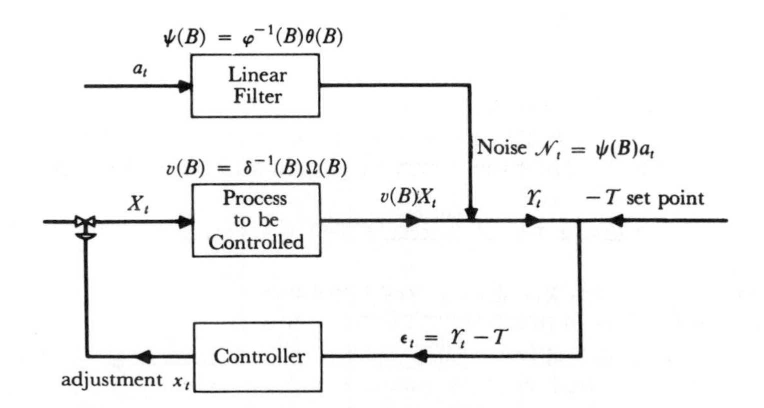

Figure 1.6 Feedback control scheme to compensate an unmeasured disturbance N_t.

of control could cause the output to drift away from the desired target value or *set point* T. Then, despite adjustments that have been made to the process, an error

$$\begin{aligned}\varepsilon_t &= Y_t - T \\ &= v(B)X_t + N_t - T\end{aligned}$$

will occur between the output and its target value T. The object is to so choose a control equation that the errors ε will have the smallest possible mean square. The control equation expresses the adjustment $x_t = X_t - X_{t-1}$ to be taken at time t, as a function of the present deviation ε_t, previous deviations $\varepsilon_{t-1}, \varepsilon_{t-2}, \ldots$, and previous adjustments $x_{t-1}, x_{t-2}, \ldots$. The mechanism (human, electrical, pneumatic, or electronic) that carries out the control action called for by the control equation is called the *controller*.

One procedure for designing a controller is equivalent to forecasting the deviation from target that would occur *if no control were applied*, and then calculating the adjustment that would be necessary to cancel out this deviation. It follows that the forecasting and control problems are closely linked. In particular, if a minimum mean square error forecast is used, the controller will produce minimum mean square error control. To forecast the deviation from target that could occur if no control were applied, it is necessary to build a model

$$N_t = \psi(B)a_t = \varphi^{-1}(B)\theta(B)a_t$$

for the disturbance. Calculation of the adjustment x_t that needs to be applied to the input variable at time t to cancel out a predicted change at the output requires the building of a dynamic model with transfer function

$$v(B) = \delta^{-1}(B)\Omega(B)$$

which links the input and output. The resulting adjustment x_t will consist, in general, of a linear aggregate of previous adjustments and current and previous control errors. Thus the control equation will be of the form

$$x_t = \zeta_1 x_{t-1} + \zeta_2 x_{t-2} + \cdots + \chi_0\varepsilon_t + \chi_1\varepsilon_{t-1} + \chi_2\varepsilon_{t-2} + \cdots \tag{1.2.13}$$

where $\zeta_1, \zeta_2, \ldots, \chi_0, \chi_1, \chi_2, \ldots$ are constants.

It turns out that in practice, minimum mean square error control sometimes results in unacceptably large adjustments x_t to the input variable. Consequently, modified control schemes are employed that restrict the amount of variation in the adjustments. Some of these issues are discussed in Chapter 13.

1.3 BASIC IDEAS IN MODEL BUILDING

1.3.1 Parsimony

We have seen that the mathematical models which we need to employ contain certain constants or parameters whose values must be estimated from the data. It is important, in practice, that we employ the *smallest possible* number of parameters for adequate representations. The central role played by this principle of *parsimony* [194] in the use of parameters will become clearer as we proceed. As a preliminary illustration, we consider the following simple example.

Suppose that we fitted a dynamic model (1.2.9) of the form

$$Y_t = (\omega_0 - \omega_1 B - \omega_2 B^2 - \cdots - \omega_s B^s)X_t \tag{1.3.1}$$

when dealing with a system that was adequately represented by

$$(1 - \delta B)Y_t = \omega_0 X_t \tag{1.3.2}$$

The model (1.3.2) contains only two parameters, δ and ω_0, but for s sufficiently large, it could be represented approximately by the model (1.3.1), through

$$Y_t = (1 - \delta B)^{-1}\omega_0 X_t = \omega_0(1 + \delta B + \delta^2 B^2 + \cdots)X_t$$

with $|\delta| < 1$. Because of experimental error, we could easily fail to recognize the relationship between the coefficients in the fitted equation. Thus we might needlessly fit a relationship like (1.3.1), containing $s + 1$ parameters, where the much simpler form (1.3.2), containing only two, would have been adequate. This could, for example, lead to unnecessarily poor estimation of the output Y_t for given values of the input $X_t, X_{t-1}, \ldots$.

Our objective, then, must be to obtain adequate but parsimonious models. Forecasting and control procedures could be seriously deficient if these models were either inadequate or unnecessarily prodigal in the use of parameters. Care and effort is needed in selecting the model. The process of selection is necessarily iterative; that is, it is a process of evolution, adaptation, or trial and error and is outlined briefly below.

1.3.2 Iterative Stages in the Selection of a Model

If the physical mechanism of a phenomenon were completely understood, it would be possible theoretically to write down a mathematical expression that described it exactly. We thus obtain a *mechanistic* or *theoretical* model. In most instances the complete knowledge or large experimental resources needed to produce a mechanistic model are not available, and we must resort to an empirical model. Of course, the exact mechanistic model and the exclusively empirical model represent extremes. Models actually employed usually lie somewhere in between. In particular, we may use incomplete theoretical knowledge to indicate a suitable class of mathematical functions, which will then be fitted empirically (e.g., [40]); that is, the number of terms needed in the model and the numerical values of the parameters are estimated from experimental data. This is the approach that we adopt in this book. As we have indicated previously, the stochastic and dynamic models we describe can be justified, at least partially, on theoretical grounds as having the right general properties.

It is normally supposed that successive values of the time series under consideration or of the input–output data are available for analysis. If possible, at least 50 and preferably 100 observations or more should be used. In those cases where a past history of 50 or more observations are not available, one proceeds by using experience and past information to yield a preliminary model. This model may be updated from time to time as more data become available.

In fitting dynamic models, a theoretical analysis can sometimes tell us not only the appropriate form for the model, but may also provide us with good estimates of the numerical values of its parameters. These values can then be checked later by analysis of data.

Figure 1.7 summarizes the iterative approach to model building for forecasting and control, which is employed in this book.

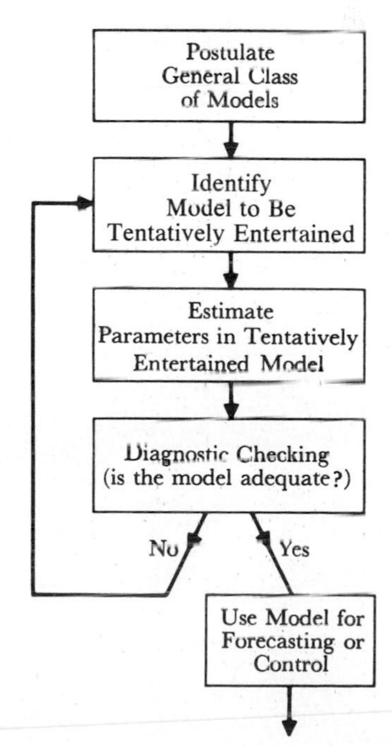

Figure 1.7 Stages in the iterative approach to model building.

1. From the interaction of theory and practice, a *useful class of models* for the purposes at hand is considered.
2. Because this class is too extensive to be conveniently fitted directly to data, rough methods for *identifying* subclasses of these models are developed. Such methods of model identification employ data and knowledge of the system to suggest an appropriate parsimonious subclass of models which may be tentatively entertained. In addition, the identification process can be used to yield rough preliminary estimates of the parameters in the model.
3. The tentatively entertained model is *fitted* to data and its parameters *estimated*. The rough estimates obtained during the identification stage can now be used as starting values in more refined iterative methods for estimating the parameters.
4. *Diagnostic checks* are applied with the object of uncovering possible lack of fit and diagnosing the cause. If no lack of fit is indicated, the model is ready to use. If any inadequacy is found, the iterative cycle of identification, estimation, and diagnostic checking is repeated until a suitable representation is found.

Identification, estimation, and diagnostic checking are discussed for univariate time series models in Chapters 6, 7, 8, and 9, and for transfer function models in Chapter 11.

Part One

STOCHASTIC MODELS AND THEIR FORECASTING

In the first part of this book, which includes Chapters 2, 3, 4, and 5, a valuable class of stochastic models is described and its use in forecasting discussed.

A model that describes the probability structure of a sequence of observations is called a *stochastic process*. A time series of N successive observations $\mathbf{z}' = (z_1, z_2, \ldots, z_N)$ is regarded as a sample realization, from an infinite population of such samples, which could have been generated by the process. A major objective of statistical investigation is to infer properties of the population from those of the sample. For example, to make a forecast is to infer the probability distribution of a *future observation* from the population, given a sample $\mathbf{z}$ of past values. To do this we need ways of describing stochastic processes and time series, and we also need classes of stochastic models that are capable of describing practically occurring situations. An important class of stochastic processes discussed in Chapter 2 is the *stationary* processes. They are assumed to be in a specific form of statistical equilibrium, and in particular, vary about a fixed mean. Useful devices for describing the behavior of stationary processes are the *autocorrelation function* and the *spectrum*.

Particular stationary stochastic processes of value in modeling time series are the autoregressive, moving average, and mixed autoregressive–moving average processes. The properties of these processes, in particular their correlation structures, are described in Chapter 3.

Because many practically occurring time series (e.g., stock prices and sales figures) have nonstationary characteristics, the stationary models introduced in Chapter 3 are developed further in Chapter 4 to give a useful class of nonstationary processes called autoregressive integrated moving average (ARIMA) models. The use of all these models in forecasting time series is discussed in Chapter 5 and is illustrated with examples.

2

AUTOCORRELATION FUNCTION AND SPECTRUM OF STATIONARY PROCESSES

A central feature in the development of time series models is an assumption of some form of *statistical equilibrium*. A particular assumption of this kind (an unduly restrictive one, as we shall see later) is that of *stationarity*. Usually, a stationary time series can be usefully described by its mean, variance, and *autocorrelation function*, or equivalently by its mean, variance, and *spectral density function*. In this chapter we consider the properties of these functions and, in particular, the properties of the autocorrelation function, which is used extensively in the chapters that follow.

2.1 AUTOCORRELATION PROPERTIES OF STATIONARY MODELS

2.1.1 Time Series and Stochastic Processes

Time series. A time series is a set of observations generated sequentially in time. If the set is continuous, the time series is said to be *continuous*. If the set is discrete, the time series is said to be *discrete*. Thus the observations from a discrete time series made at times $\tau_1, \tau_2, \ldots, \tau_t, \ldots, \tau_N$ may be denoted by $z(\tau_1), z(\tau_2), \ldots, z(\tau_t), \ldots, z(\tau_N)$. In this book we consider only discrete time series where observations are made at a fixed interval h. When we have N *successive* values of such a series available for analysis, we write $z_1, z_2, \ldots, z_t, \ldots, z_N$ to denote observations made at equidistant time intervals $\tau_0 + h, \tau_0 + 2h, \ldots, \tau_0 + th, \ldots, \tau_0 + Nh$. For many purposes the values of τ_0 and h are unimportant, but if the observation times need to be defined exactly, these two values can be specified. If we adopt τ_0 as the origin and h as the unit of time, we can regard z_t as the observation *at time t*.

Discrete time series may arise in two ways.

1. By *sampling* a continuous time series: for example, in the situation shown in Figure 1.2, where the continuous input and output from a gas furnace was sampled at intervals of 9 seconds.
2. By *accumulating* a variable over a period of time: examples are rainfall, which is usually accumulated over a period such as a day or a month, and the yield from a batch process, which is accumulated over the batch time. For example, Figure 2.1 shows a time series consisting of the yields from 70 consecutive batches of a chemical process.

Deterministic and statistical time series. If future values of a time series are exactly determined by some mathematical function such as

$$z_t = \cos(2\pi ft)$$

the time series is said to be *deterministic*. If the future values can be described only in terms of a probability distribution, the time series is said to be nondeterministic or simply a *statistical time series*. The batch data of Figure 2.1 provide an example of a statistical time series. Thus, although there is a well-defined high–low pattern in the series, it is impossible to forecast the exact yield for the next batch. It is with such statistical time series that we are concerned in this book.

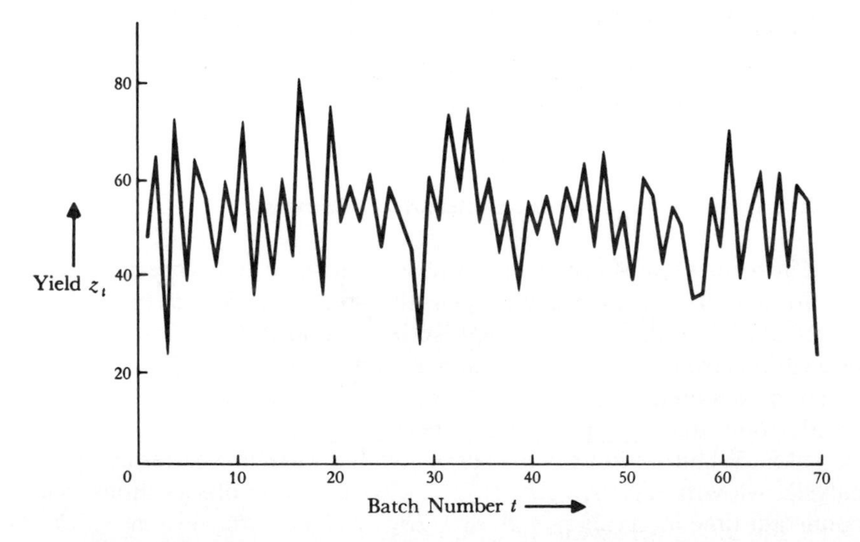

Figure 2.1 Yields of 70 consecutive batches from a chemical process.

Stochastic processes. A statistical phenomenon that evolves in time according to probabilistic laws is called a *stochastic process.* We shall often refer to it simply as a *process,* omitting the work "stochastic." The time series to be analyzed may then be thought of as a particular *realization,* produced by the underlying probability mechanism, of the system under study. In other words, *in analyzing a time series we regard it as a realization of a stochastic process.*

For example, to analyze the batch data in Figure 2.1, we can imagine other sets of observations (other realizations of the underlying stochastic process), which might have been generated by the same chemical system, in the same $N = 70$ batches. Thus Figure 2.2 shows the yields from batches $t = 21$ to $t = 30$ (thick line), together with other time series which *might* have been obtained from the population of time series defined by the underlying stochastic process. It follows that we can regard the observation z_t at a given time t, say $t = 25$, as a realization of a random variable z_t with probability density function $p(z_t)$. Similarly, the observations at any two times, say $t_1 = 25$ and $t_2 = 27$, may be regarded as realizations of two random variables z_{t_1} and z_{t_2} with joint probability density function $p(z_{t_1}, z_{t_2})$. For example, Figure 2.3 shows contours of constant density for such a joint distribution, together with the marginal distribution at time t_1. In general, the observations making up an equispaced time series can be described by an N-dimensional random variable $(z_1, z_2, \ldots, z_N)$ with probability distribution $p(z_1, z_2, \ldots, z_N)$.

2.1.2 Stationary Stochastic Processes

A very special class of stochastic processes, called *stationary processes,* is based on the assumption that the process is in a particular state of *statistical equilibrium.* A stochastic process is said to be *strictly stationary*

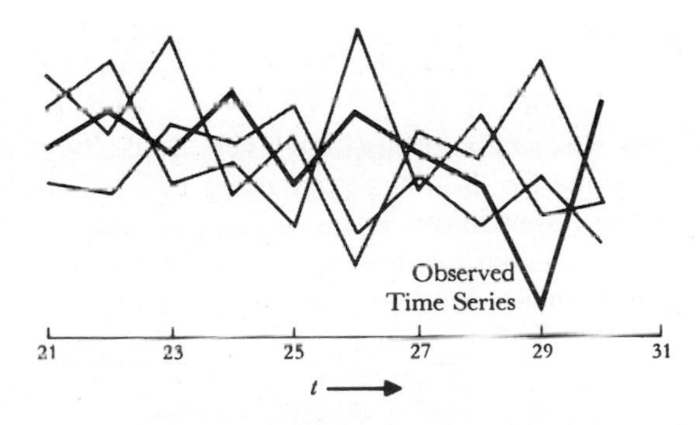

Figure 2.2 Observed time series (thick line), with other time series representing realizations of the same stochastic process.

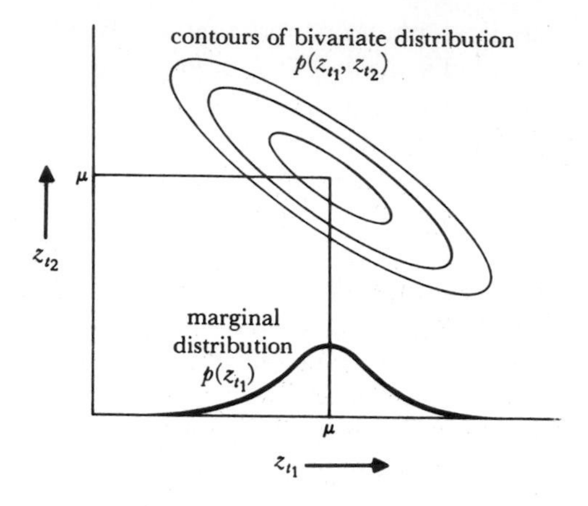

Figure 2.3 Contours of constant density of a bivariate probability distribution describing a stochastic process at two times t_1, t_2, together with the marginal distribution at time t_1.

if its properties are unaffected by a change of time origin, that is, if the joint probability distribution associated with m observations $z_{t_1}, z_{t_2}, \ldots, z_{t_m}$, made at *any* set of times $t_1, t_2, \ldots, t_m$, is the same as that associated with m observations $z_{t_1+k}, z_{t_2+k}, \ldots, z_{t_m+k}$, made at times $t_1 + k, t_2 + k, \ldots, t_m + k$. Thus for a discrete process to be strictly stationary, the joint distribution of any set of observations must be unaffected by shifting all the times of observation forward or backward by any integer amount k.

Mean and variance of a stationary process. When $m = 1$, the stationarity assumption implies that the probability distribution $p(z_t)$ is the same for all times t and may be written $p(z)$. Hence the stochastic process has a constant mean

$$\mu = E[z_t] = \int_{-\infty}^{\infty} zp(z)\, dz \tag{2.1.1}$$

which defines the level about which it fluctuates, and a constant variance

$$\sigma_z^2 = E[(z_t - \mu)^2] = \int_{-\infty}^{\infty} (z - \mu)^2 p(z)\, dz \tag{2.1.2}$$

which measures its *spread* about this time level. Since the probability distribution $p(z)$ is the same for all times t, its shape can be inferred by forming the histogram of the observations $z_1, z_2, \ldots, z_N$, making up the observed time series. In addition, the mean μ of the stochastic process can be estimated by the sample mean

$$\bar{z} = \frac{1}{N} \sum_{t=1}^{N} z_t \tag{2.1.3}$$

of the time series, and the variance σ_z^2 of the stochastic process can be estimated by the sample variance

$$\hat{\sigma}_z^2 = \frac{1}{N}\sum_{t=1}^{N}(z_t - \bar{z})^2 \tag{2.1.4}$$

of the time series.

Autocovariance and autocorrelation coefficients. The stationarity assumption also implies that the joint probability distribution $p(z_{t_1}, z_{t_2})$ is the same for all times t_1, t_2 which are a constant interval apart. It follows that

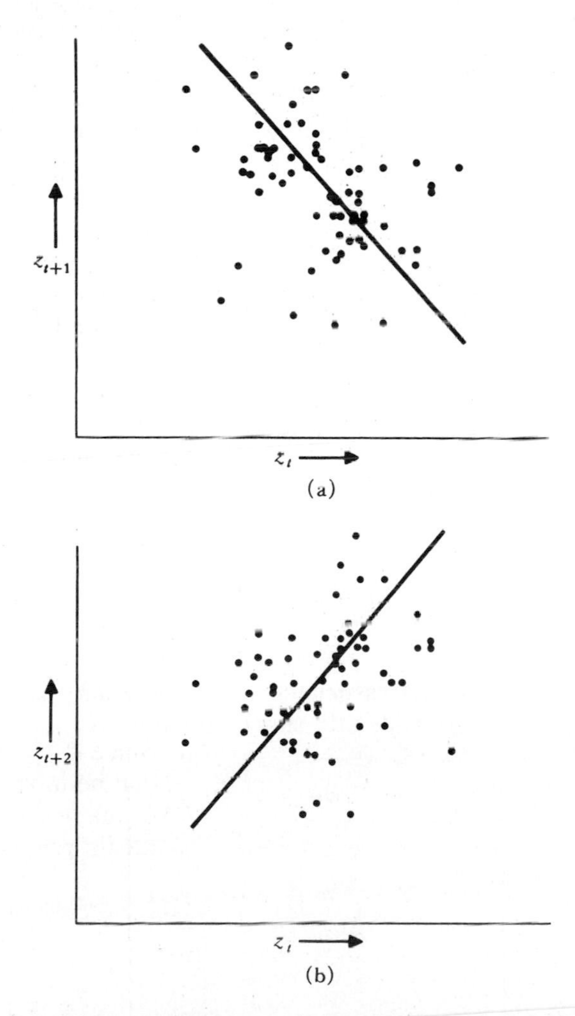

Figure 2.4 Scatter diagrams at lags 1 and 2 for the batch data of Figure 2.1.

the nature of this joint distribution can be inferred by plotting a scatter diagram using pairs of values (z_t, z_{t+k}) of the time series, separated by a constant interval or *lag* k. For the batch data, Figure 2.4(a) is a scatter diagram for lag $k = 1$, obtained by plotting z_{t+1} versus z_t, and Figure 2.4(b) is a scatter diagram for lag $k = 2$, obtained by plotting z_{t+2} versus z_t. We see that neighboring values of the time series are correlated, the correlation between z_t and z_{t+1} appearing to be negative and the correlation between z_t and z_{t+2} positive. The covariance between z_t and its value z_{t+k}, separated by k intervals of time, which under the stationarity assumption must be the same for all t, is called the *autocovariance* at lag k and is defined by

$$\gamma_k = \text{cov}[z_t, z_{t+k}] = E[(z_t - \mu)(z_{t+k} - \mu)] \tag{2.1.5}$$

Similarly, the *autocorrelation* at lag k is

$$\begin{aligned}\rho_k &= \frac{E[(z_t - \mu)(z_{t+k} - \mu)]}{\sqrt{E[(z_t - \mu)^2]E[(z_{t+k} - \mu)^2]}} \\ &= \frac{E[(z_t - \mu)(z_{t+k} - \mu)]}{\sigma_z^2}\end{aligned}$$

since, for a stationary process, the variance $\sigma_z^2 = \gamma_0$ is the same at time $t + k$ as at time t. Thus the autocorrelation at lag k, that is, the correlation between z_t and z_{t+k}, is

$$\rho_k = \frac{\gamma_k}{\gamma_0} \tag{2.1.6}$$

which implies that $\rho_0 = 1$.

2.1.3 Positive Definiteness and the Autocovariance Matrix

The covariance matrix associated with a stationary process for observations $(z_1, z_2, \ldots, z_n)$ made at n successive times is

$$\mathbf{\Gamma}_n = \begin{bmatrix} \gamma_0 & \gamma_1 & \gamma_2 & \cdots & \gamma_{n-1} \\ \gamma_1 & \gamma_0 & \gamma_1 & \cdots & \gamma_{n-2} \\ \gamma_2 & \gamma_1 & \gamma_0 & \cdots & \gamma_{n-3} \\ \vdots & \vdots & \vdots & \cdots & \vdots \\ \gamma_{n-1} & \gamma_{n-2} & \gamma_{n-3} & \cdots & \gamma_0 \end{bmatrix}$$

$$= \sigma_z^2 \begin{bmatrix} 1 & \rho_1 & \rho_2 & \cdots & \rho_{n-1} \\ \rho_1 & 1 & \rho_1 & \cdots & \rho_{n-2} \\ \rho_2 & \rho_1 & 1 & \cdots & \rho_{n-3} \\ \vdots & \vdots & \vdots & \cdots & \vdots \\ \rho_{n-1} & \rho_{n-2} & \rho_{n-3} & \cdots & 1 \end{bmatrix} = \sigma_z^2 \mathbf{P}_n \qquad (2.1.7)$$

A covariance matrix $\mathbf{\Gamma}_n$ of this form, which is symmetric with constant elements on any diagonal, will be called an *autocovariance matrix* and the corresponding correlation matrix $\mathbf{P}_n$, will be called an *autocorrelation matrix*. Now consider any linear function of the random variables z_t, $z_{t-1}, \ldots, z_{t-n+1}$:

$$L_t = l_1 z_t + l_2 z_{t-1} + \cdots + l_n z_{t-n+1} \qquad (2.1.8)$$

Since $\text{cov}[z_i, z_j] = \gamma_{|j-i|}$ for a stationary process, the variance of L_t is

$$\text{var}[L_t] = \sum_{i=1}^{n} \sum_{j=1}^{n} l_i l_j \gamma_{|j-i|}$$

which is necessarily greater than zero if the l's are not all zero. It follows that both an autocovariance matrix and an autocorrelation matrix are positive-definite for any stationary process. Correspondingly, it is seen that both the autocovariance function $\{\gamma_k\}$ and the autocorrelation function $\{\rho_k\}$, viewed as functions of the lag k, are positive-definite functions in the sense that $\Sigma_{i=1}^{n}\Sigma_{j=1}^{n} l_i l_j \gamma_{|j-i|} > 0$ for every positive integer n and all constants $l_1, \ldots, l_n$.

Conditions satisfied by the autocorrelations of a stationary process. The positive definiteness of the autocorrelation matrix (2.1.7) implies that its determinant and all principal minors are greater than zero. In particular, for $n = 2$,

$$\begin{vmatrix} 1 & \rho_1 \\ \rho_1 & 1 \end{vmatrix} > 0$$

so that

$$1 - \rho_1^2 > 0$$

and hence

$$-1 < \rho_1 < 1$$

Similarly, for $n = 3$, we must have

$$\begin{vmatrix} 1 & \rho_1 \\ \rho_1 & 1 \end{vmatrix} > 0, \qquad \begin{vmatrix} 1 & \rho_2 \\ \rho_2 & 1 \end{vmatrix} > 0$$

$$\begin{vmatrix} 1 & \rho_1 & \rho_2 \\ \rho_1 & 1 & \rho_1 \\ \rho_2 & \rho_1 & 1 \end{vmatrix} > 0$$

which implies that

$$-1 < \rho_1 < 1$$

$$-1 < \rho_2 < 1$$

$$-1 < \frac{\rho_2 - \rho_1^2}{1 - \rho_1^2} < 1$$

and so on. Since $\mathbf{P}_n$ must be positive-definite for *all* values of n, the autocorrelations of a stationary process must satisfy a very large number of conditions. As will be shown in Section 2.2.3, all of these conditions can be brought together in the definition of the spectrum.

Stationarity of linear functions. It follows from the definition of stationarity that the process L_t, obtained by performing the linear operation (2.1.8) on a stationary process z_t for fixed n and fixed coefficients $l_1, \ldots, l_n$, is also stationary. In particular, the first difference $\nabla z_t = z_t - z_{t-1}$, and higher differences $\nabla^d z_t$, are stationary. This result is of particular importance to the discussion of nonstationary time series presented in Chapter 4.

Gaussian processes. If the probability distribution of observations associated with *any* set of times is a multivariate Normal distribution, the process is called a *Normal* or *Gaussian* process. Since the multivariate Normal distribution is fully characterized by its moments of first and second order, the existence of a fixed mean μ and an autocovariance matrix $\mathbf{\Gamma}_n$ for all n would be sufficient to ensure the stationarity of a Gaussian process.

Weak stationarity. We have seen that for a process to be strictly stationary, the whole probability structure must depend only on time differences. A less restrictive requirement, called *weak stationarity* of order f, is that the moments up to some order f depend only on time differences. For example, the existence of a fixed mean μ and an autocovariance matrix $\mathbf{\Gamma}_n$ of the form (2.1.7) is sufficient to ensure stationarity up to second order. Thus second-order stationarity, plus an assumption of Normality, are sufficient to produce strict stationarity.

2.1.4 Autocovariance and Autocorrelation Functions

It was seen in Section 2.1.2 that the autocovariance coefficient γ_k, at lag k, measures the covariance between two values z_t and z_{t+k} a distance k apart. The plot of γ_k versus the lag k is called the *autocovariance function* $\{\gamma_k\}$ of the stochastic process. Similarly, the plot of the autocorrelation coefficient ρ_k as a function of the lag k is called the *autocorrelation function* $\{\rho_k\}$ of the process. Note that the autocorrelation function is dimensionless, that is, independent of the scale of measurement of the time series. Since $\gamma_k = \rho_k \sigma_z^2$, knowledge of the autocorrelation function $\{\rho_k\}$ and of the variance σ_z^2 is equivalent to knowledge of the autocovariance function $\{\gamma_k\}$.

The autocorrelation function, shown in Figure 2.5 as a plot of the diagonals of the autocorrelation matrix, reveals how the correlation between any two values of the series changes as their separation changes.

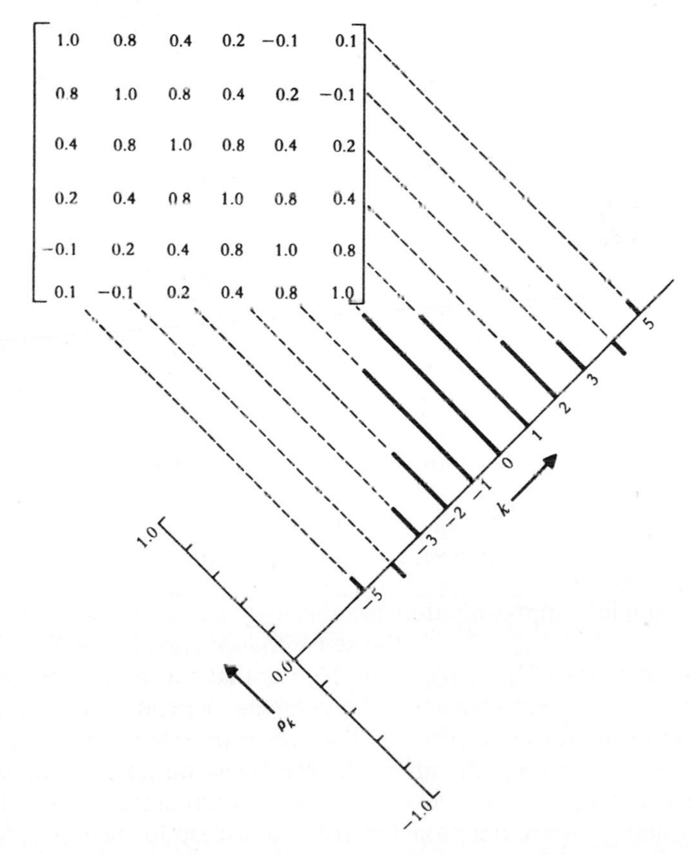

Figure 2.5 Autocorrelation matrix and resulting autocorrelation function of a process.

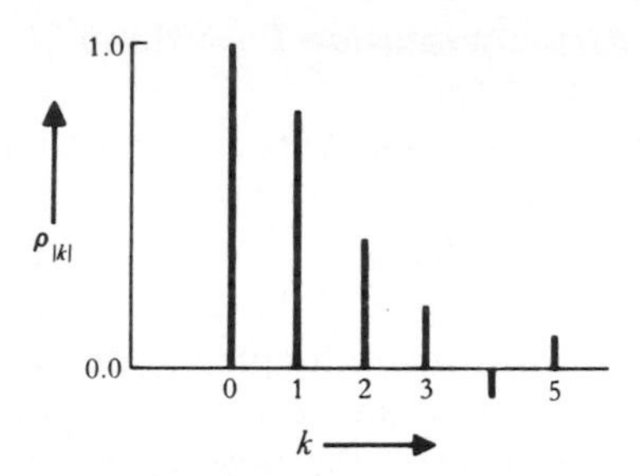

Figure 2.6 Positive half of the autocorrelation function of Figure 2.5.

Since $\rho_k = \rho_{-k}$, the autocorrelation function is necessarily symmetric about zero, and in practice it is only necessary to plot the positive half of the function. Figure 2.6 shows the positive half of the autocorrelation function given in Figure 2.5. Henceforth when we speak of the autocorrelation function, we shall often mean the positive half. In the past, the autocorrelation function has sometimes been called the *correlogram.*

From what has previously been shown, a *Normal* stationary process z_t is completely characterized by its mean μ and its autocovariance function $\{\gamma_k\}$, or equivalently by its mean μ, variance σ_z^2, and autocorrelation function $\{\rho_k\}$.

2.1.5 Estimation of Autocovariance and Autocorrelation Functions

Up to now we have only considered the theoretical autocorrelation function that describes a conceptual stochastic process. In practice, we have a finite time series $z_1, z_2, \ldots, z_N$, of N observations, from which we can only obtain *estimates* of the mean μ and the autocorrelations. The mean $\mu = E[z_t]$ is estimated as in (2.1.3), by the sample mean $\bar{z} = \Sigma_{t=1}^{N} z_t/N$. It is easy to see that $E[\bar{z}] = \mu$, so that $\bar{z}$ is an unbiased estimate of μ. As a measure of precision of $\bar{z}$ as an estimator of μ, we find that

$$\text{var}[\bar{z}] = \frac{1}{N^2}\sum_{t=1}^{N}\sum_{s=1}^{N}\gamma_{t-s} = \frac{\gamma_0}{N}\left[1 + 2\sum_{k=1}^{N-1}\left(1 - \frac{k}{N}\right)\rho_k\right]$$

A "large-sample" approximation for this variance expression is given by $\text{var}[\bar{z}] = (\gamma_0/N)(1 + 2\Sigma_{k=1}^{\infty}\rho_k)$, in the sense that $N\,\text{var}[\bar{z}] \rightarrow \gamma_0(1 + 2\Sigma_{k=1}^{\infty}\rho_k)$ as $N \rightarrow \infty$, assuming $\Sigma_{k=-\infty}^{\infty}|\rho_k| < \infty$. Notice that the first factor in $\text{var}[\bar{z}]$, γ_0/N, is the familiar expression for the variance of $\bar{z}$ obtained from independent random samples of size N, but the presense of autocorrelation among the z_t values can substantially affect the precision of $\bar{z}$. For example, in the case where a stationary process has autocorrelations $\rho_k = \phi^{|k|}$, $|\phi| < 1$, the large-sample approximation for the variance of $\bar{z}$ becomes $\text{var}[\bar{z}] = (\gamma_0/N)[(1 + \phi)/(1 - \phi)]$, and the second factor can obviously differ substantially from 1.

A number of estimates of the autocorrelation function have been suggested by statisticians and their properties are discussed in particular in [122]. It is concluded that the most satisfactory estimate of the kth lag autocorrelation ρ_k is

$$r_k = \frac{c_k}{c_0} \tag{2.1.9}$$

where

$$c_k = \frac{1}{N} \sum_{t=1}^{N-k} (z_t - \bar{z})(z_{t+k} - \bar{z}) \qquad k = 0, 1, 2, \ldots, K \tag{2.1.10}$$

is the estimate of the autocovariance γ_k, and $\bar{z}$ is the sample mean of the time series. The values r_k in (2.1.9) may be called the *sample* autocorrelation function.

We now illustrate (2.1.10) by calculating r_1 for the first 10 values of the batch data of Figure 2.1, given in Table 2.1. The mean $\bar{z}$ of the first 10 values in the table is 51, so the deviations about the mean are -4, 13, -28, 20, -13, 13, 4, -10, 8, and -3. Thus

$$\begin{aligned}\sum_{t=1}^{9} (z_t - \bar{z})(z_{t+1} - \bar{z}) &= (-4)(13) + (13)(-28) + \cdots + (8)(-3) \\ &= -1497\end{aligned}$$

TABLE 2.1 Series of 70 Consecutive Yields from a Batch Chemical Process[a]

1–15	16–30	31–45	46–60	61–70
47	44	50	62	68
64	80	71	44	38
23	55	56	64	50
71	37	74	43	60
38	74	50	52	39
64	51	58	38	59
55	57	45	59	40
41	50	54	55	57
59	60	36	41	54
48	45	54	53	23
71	57	48	49	
35	50	55	34	
57	45	45	35	
40	25	57	54	
58	59	50	45	

[a] This series also appears as series F in the "Collection of Time Series" in Part Five.

Hence $c_1 = -1497/10 = -149.7$. Similarly, we find that $c_0 = 189.6$. Hence

$$r_1 = \frac{c_1}{c_0} = \frac{-149.7}{189.6} = -0.79$$

it being sufficient for most practical purposes to round off the autocorrelation to two decimal places. The calculation above is made for illustration only. In practice, to obtain a useful estimate of the autocorrelation function, we would need at least 50 observations, and the estimated autocorrelations r_k would be calculated for $k = 0, 1, \ldots, K$, where K was not larger than, say, $N/4$.

The first 15 values of r_k, based on the entire series of 70 observations, are given in Table 2.2 and plotted in Figure 2.7. The estimated autocorrelation function is characterized by correlations that alternate in sign and tend to damp out with increasing lag. Autocorrelation functions of this kind are not uncommon in production data and can arise because of "carryover" effects. In this particular example, a high-yielding batch tended to produce tarry residues which were not entirely removed from the vessel and adversely affected the yield of the next batch.

2.1.6 Standard Error of Autocorrelation Estimates

To identify a model for a time series, using methods to be described in Chapter 6, it is useful to have a rough check on whether ρ_k is effectively zero beyond a certain lag. For this purpose, use can be made of the following approximate expression for the variance of the estimated autocorrelation coefficient of a stationary Normal process given by Bartlett [25]:

$$\text{var}[r_k] \simeq \frac{1}{N} \sum_{v=-\infty}^{+\infty} (\rho_v^2 + \rho_{v+k}\rho_{v-k} - 4\rho_k\rho_v\rho_{v-k} + 2\rho_v^2\rho_k^2) \qquad (2.1.11)$$

For example, if $\rho_k = \phi^{|k|}$ $(-1 < \phi < 1)$, that is, the autocorrelation function damps out exponentially, (2.1.11) gives

$$\text{var}[r_k] \simeq \frac{1}{N}\left[\frac{(1+\phi^2)(1-\phi^{2k})}{1-\phi^2} - 2k\phi^{2k}\right] \qquad (2.1.12)$$

TABLE 2.2 Estimated Autocorrelation Function of Batch Data

k	r_k	k	r_k	k	r_k
1	−0.39	6	−0.05	11	0.11
2	0.30	7	0.04	12	−0.07
3	−0.17	8	−0.04	13	0.15
4	0.07	9	−0.01	14	0.04
5	−0.10	10	0.01	15	−0.01

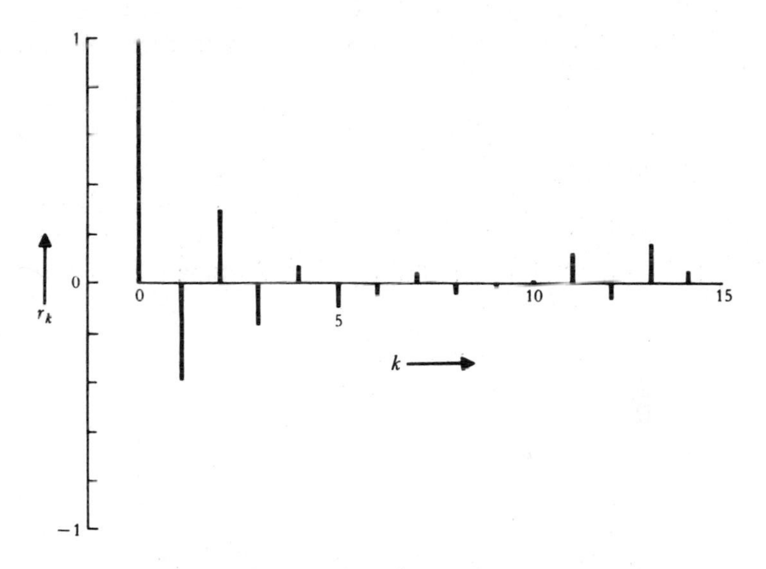

Figure 2.7 Estimated autocorrelation function of batch data.

and in particular

$$\operatorname{var}[r_1] \simeq \frac{1}{N}(1 - \phi^2)$$

For any process for which all the autocorrelations ρ_v are zero for $v > q$, all terms except the first appearing in the right-hand side of (2.1.11) are zero when $k > q$. Thus for the variance of the estimated autocorrelation r_k, at lags k greater than some value q *beyond which the theoretical autocorrelation function may be deemed to have "died out,"* Bartlett's approximation gives

$$\operatorname{var}[r_k] \simeq \frac{1}{N}\left(1 + 2\sum_{v=1}^{q} \rho_v^2\right) \qquad k > q \tag{2.1.13}$$

To use (2.1.13) in practice, the estimated autocorrelations r_k ($k = 1, 2, \ldots, q$) are substituted for the theoretical autocorrelations ρ_k, and when this is done we shall refer to the square root of (2.1.13) as the *large-lag standard error*. On the assumption that the theoretical autocorrelations ρ_k are all essentially zero beyond some hypothesized lag $k = q$, the large lag standard error approximates the standard deviation of r_k for suitably large lags ($k > q$).

Similar approximate expressions for the covariance between the estimated correlations r_k and r_{k+s} at two different lags k and $k + s$ have been given by Bartlett [25]. In particular, the large-lag approximation reduces to

$$\text{cov}[r_k, r_{k+s}] \simeq \frac{1}{N} \sum_{v=-q}^{q} \rho_v \rho_{v+s} \qquad k > q \tag{2.1.14}$$

Bartlett's result (2.1.14) shows that care is required in the interpretation of individual autocorrelations because large covariances can exist between neighboring values. This effect can sometimes distort the visual appearance of the autocorrelation function, which may fail to damp out according to expectation [122], [126].

A special case of particular interest occurs for $q = 0$, that is, when the ρ_k are taken to be zero for all lags (other than lag 0) and hence the series is completely random. Then the standard errors from (2.1.13) for estimated autocorrelations r_k take the simple form $1/N^{1/2}$. In addition, in this case the result in (2.1.14) indicates that estimated autocorrelations r_k and r_{k+s} at two different lags are not correlated, and since the r_k are also known to be approximately normally distributed for large N, a collection of estimated autocorrelations for different lags will tend to be independently and normally distributed with mean 0 and variance $1/N$.

Example. The following estimated autocorrelations were obtained from a time series of length $N = 200$ observations, generated from a stochastic process for which it was *known* that $\rho_1 = -0.4$ and $\rho_k = 0$ for $k \geqslant 2$:

k	r_k	k	r_k
1	−0.38	6	0.00
2	−0.08	7	0.00
3	0.11	8	0.00
4	−0.08	9	0.07
5	0.02	10	−0.08

On the assumption that the series is completely random, we have that $q = 0$. Then for *all lags,* (2.1.13) yields

$$\text{var}[r_k] \simeq \frac{1}{N} = \frac{1}{200} = 0.005$$

The corresponding standard error is $0.07 = (0.005)^{1/2}$. Since the value of -0.38 for r_1 is over five times this standard error, it can be concluded that ρ_1 is nonzero. Moreover, the estimated autocorrelations for lags greater than 1 are small. Therefore, it might be reasonable to ask next whether the series was compatible with a hypothesis (whose relevance will be discussed later) whereby ρ_1 was nonzero but $\rho_k = 0$ $(k \geqslant 2)$. Using (2.1.13) with $q = 1$ and substituting r_1 for ρ_1, the estimated large-lag variance under this assumption is

$$\text{var}[r_k] \simeq \frac{1}{200}[1 + 2(-0.38)^2] = 0.0064 \qquad k > 1$$

yielding a standard error of 0.08. Since the estimated autocorrelations for lags greater than 1 are small compared with this standard error, there is no reason to doubt the adequacy of the model $\rho_1 \neq 0$, $\rho_k = 0$ $(k \geqslant 2)$.

2.2 SPECTRAL PROPERTIES OF STATIONARY MODELS

2.2.1 Periodogram of a Time Series

Another way of analyzing a time series is based on the assumption that it is made up of sine and cosine waves with different frequencies. A device that uses this idea, introduced by Schuster [174] in 1898, is the *periodogram* (see also [182]). The periodogram was originally used to detect and estimate the amplitude of a sine component, of known frequency, buried in noise. We shall use it later to provide a check on the randomness of a series (usually, a series of residuals after fitting a particular model), where we consider the possibility that periodic components of unknown frequency may remain in the series.

To illustrate the calculation of the periodogram, suppose that the number of observations $N = 2q + 1$ is odd. If we fit the Fourier series model

$$z_t = \alpha_0 + \sum_{i=1}^{q} (\alpha_i c_{it} + \beta_i s_{it}) + e_t \tag{2.2.1}$$

where $c_{it} = \cos(2\pi f_i t)$, $s_{it} = \sin(2\pi f_i t)$, and $f_i = i/N$ is the ith harmonic of the fundamental frequency $1/N$, the least squares estimates of the coefficients α_0 and (α_i, β_i) will be

$$a_0 = \bar{z} \tag{2.2.2}$$

$$\left.\begin{aligned} a_i &= \frac{2}{N}\sum_{t=1}^{N} z_t c_{it} && (2.2.3) \\ b_i &= \frac{2}{N}\sum_{t=1}^{N} z_t s_{it} && (2.2.4) \end{aligned}\right\} i = 1, 2, \ldots, q$$

The periodogram then consists of the $q = (N - 1)/2$ values

$$I(f_i) = \frac{N}{2}(a_i^2 + b_i^2) \qquad i = 1, 2, \ldots, q \tag{2.2.5}$$

where $I(f_i)$ is called the *intensity* at frequency f_i.

When N is even, we set $N = 2q$ and (2.2.2), (2.2.3), (2.2.4), and (2.2.5) apply for $i = 1, 2, \ldots, (q - 1)$, but

$$a_q = \frac{1}{N} \sum_{t=1}^{N} (-1)^t z_t$$

$$b_q = 0$$

and

$$I(f_q) = I(0.5) = N a_q^2$$

Note that the highest frequency is 0.5 cycle per time interval because the smallest period is 2 intervals.

2.2.2 Analysis of Variance

In an analysis of variance table associated with the fitted regression (2.2.1), when N is odd, we can isolate $(N - 1)/2$ pairs of degrees of freedom, after eliminating the mean. These are associated with the pairs of coefficients $(a_1, b_1), (a_2, b_2), \ldots, (a_q, b_q)$, and hence with the frequencies $1/N$, $2/N, \ldots, q/N$. The periodogram $I(f_i) = (N/2)(a_i^2 + b_i^2)$ is seen to be simply the "sum of squares" associated with the pair of coefficients (a_i, b_i), and hence with the frequency $f_i = i/N$. Thus

$$\sum_{t=1}^{N} (z_t - \bar{z})^2 = \sum_{i=1}^{q} I(f_i) \tag{2.2.6}$$

When N is even, there are $(N - 2)/2$ pairs of degrees of freedom, and a further single degree of freedom associated with the coefficient a_q.

If the series were truly random, containing no systematic sinusoidal component, that is,

$$z_t = \alpha_0 + e_t$$

with α_0 the fixed mean, and the e's independent and Normal, with mean zero and variance σ^2, each component $I(f_i)$ would have expectation $2\sigma^2$ and would be distributed* as $\sigma^2\chi^2(2)$, independently of all the other components. By contrast, if the series contained a systematic sine component having frequency f_i, amplitude A, and phase angle F, so that

$$z_t = \alpha_0 + \alpha \cos(2\pi f_i t) + \beta \sin(2\pi f_i t) + e_t$$

with $A \sin F = \alpha$ and $A \cos F = \beta$, the sum of squares $I(f_i)$ would tend to be inflated, since its expected value would be $2\sigma^2 + N(\alpha^2 + \beta^2)/2 = 2\sigma^2 + NA^2/2$.

In practice, it is unlikely that the frequency f of an unknown systematic sine component would exactly match any of the frequencies f_i for which

* It is to be understood that $\chi^2(m)$ refers to a random variable having a chi-square distribution with m degrees of freedom, defined explicitly, for example, in Appendix A7.1.

TABLE 2.3 Mean Monthly Temperatures for Central England in 1964

t	z_t	c_{1t}	t	z_t	c_{1t}
1	3.4	0.87	7	16.1	−0.87
2	4.5	0.50	8	15.5	−0.50
3	4.3	0.00	9	14.1	0.00
4	8.7	−0.50	10	8.9	0.50
5	13.3	−0.87	11	7.4	0.87
6	13.8	−1.00	12	3.6	1.00

intensities have been calculated. In this case the periodogram would show an increase in the intensities in the immediate vicinity of f.

Example. A large number of observations would generally be used in calculation of the periodogram. However, to illustrate the process of calculation we use the set of 12 mean monthly temperatures (in degrees Celsius) for central England during 1964, given in Table 2.3. Table 2.3 gives $c_{1t} = \cos(2\pi t/12)$, which is required in the calculation of a_1, obtained from

$$a_1 = \tfrac{1}{6}[(3.4)(0.87) + \cdots + (3.6)(1.00)]$$
$$= -5.30$$

The values of the a_i, b_i, $i = 1, 2, \ldots, 6$, are given in Table 2.4 and yield the analysis of variance of Table 2.5. As would be expected, the major component of this temperature data has a period of 12 months, that is, a frequency of 1/12 cycle per month.

2.2.3 Spectrum and Spectral Density Function

For completeness we add here a brief discussion of the spectrum and spectral density function. The use of these important tools is described more fully in [122]. We do not apply them to the analysis of time series in this book, and this section can be omitted on first reading.

TABLE 2.4 Amplitudes of Sines and Cosines at Different Harmonics for Temperature Data

i	a_i	b_i
1	−5.30	−3.82
2	0.05	0.17
3	0.10	0.50
4	0.52	−0.52
5	0.09	−0.58
6	−0.30	

TABLE 2.5 Analysis of Variance Table for Temperature Series

i	Frequency f_i	Period	Periodogram $I(f_i)$	Degrees of Freedom	Mean Square
1	1/12	12	254.96	2	127.48
2	1/6	6	0.19	2	0.10
3	1/4	4	1.56	2	0.78
4	1/3	3	3.22	2	1.61
5	5/12	12/5	2.09	2	1.05
6	1/2	2	1.08	1	1.08
			263.10	11	23.92

Sample spectrum. The definition (2.2.5) of the periodogram assumes that the frequencies $f_i = i/N$ are harmonics of the fundamental frequency $1/N$. By way of introduction to the spectrum, we relax this assumption and allow the frequency f to vary continuously in the range 0 to 0.5 cycle. The definition (2.2.5) of the periodogram may be modified to

$$I(f) = \frac{2}{N}(a_f^2 + b_f^2) \qquad 0 \leqslant f \leqslant \tfrac{1}{2} \tag{2.2.7}$$

and $I(f)$ is then referred to as the *sample spectrum* [122]. Like the periodogram, it can be used to detect and estimate the amplitude of a sinusoidal component of unknown frequency f buried in noise and is, indeed, a more appropriate tool for this purpose if it is known that the frequency f is not harmonically related to the length of the series. Moreover, it provides a starting point for the theory of spectral analysis, using a result given in Appendix A2.1. This result shows that the sample spectrum $I(f)$ and the estimate c_k of the autocovariance function are linked by the important relation

$$I(f) = 2\left[c_0 + 2\sum_{k=1}^{N-1} c_k \cos(2\pi f k)\right] \qquad 0 \leqslant f \leqslant \tfrac{1}{2} \tag{2.2.8}$$

That is, the sample spectrum is the Fourier cosine transform of the estimate of the autocovariance function.

Spectrum. The periodogram and sample spectrum are appropriate tools for analyzing time series made up of mixtures of sine and cosine waves, at *fixed* frequencies buried in noise. However, stationary time series of the kind described in Section 2.1 are characterized by random changes of frequency, amplitude, and phase. For this type of series, the sample spectrum $I(f)$ fluctuates wildly and is not capable of any meaningful interpretation [122].

However, suppose that the sample spectrum was calculated for a time series of N observations, which is a realization of a stationary Normal process. As we have said above, such a process would not have any cosine or sine deterministic components, but we could formally carry through the Fourier analysis and obtain values of (a_f, b_f) for any given frequency f. If repeated realizations of N observations were taken from the stochastic process, we could build up a population of values for a_f, b_f, and $I(f)$. Thus we could calculate the mean value of $I(f)$ in repeated realizations of size N, namely,

$$E[I(f)] = 2\left[E[c_0] + 2\sum_{k=1}^{N-1} E[c_k]\cos(2\pi fk)\right] \tag{2.2.9}$$

For large N it may be shown, for example in [122], that the average value of the estimate of the autocovariance coefficient c_k in repeated realizations tends to the theoretical autocovariance γ_k, that is,

$$\lim_{N\to\infty} E[c_k] = \gamma_k$$

On taking the limit of (2.2.9) as N tends to infinity, the *power spectrum* $p(f)$ is defined by

$$p(f) = \lim_{N\to\infty} E[I(f)] = 2\left[\gamma_0 + 2\sum_{k=1}^{\infty} \gamma_k\cos(2\pi fk)\right] \qquad 0 \leqslant f \leqslant \tfrac{1}{2} \tag{2.2.10}$$

We note that since

$$\begin{aligned} |p(f)| &\leqslant 2\left[|\gamma_0| + 2\sum_{k=1}^{\infty} |\gamma_k||\cos(2\pi fk)|\right] \\ &\leqslant 2\left(|\gamma_0| + 2\sum_{k=1}^{\infty} |\gamma_k|\right) \end{aligned} \tag{2.2.11}$$

a sufficient condition for the spectrum to converge is that γ_k damps out rapidly enough for the series (2.2.11) to converge. *Since the power spectrum is the Fourier cosine transform of the autocovariance function,* knowledge of the autocovariance function is mathematically equivalent to knowledge of the spectrum, and vice versa. From now on we refer to the power spectrum as simply the spectrum.

On integrating (2.2.10) between the limits 0 and $\frac{1}{2}$, the variance of the process z_t is

$$\gamma_0 = \sigma_z^2 = \int_0^{1/2} p(f)\,df \tag{2.2.12}$$

Hence in the same way that the periodogram $I(f)$ shows how the variance (2.2.6) of a series, consisting of mixtures of sines and cosines, is distributed between the various distinct harmonic frequencies, the spectrum $p(f)$ shows how the variance of a stochastic process is distributed between a continuous range of frequencies. One can interpret $p(f)\,df$ as measuring approximately the variance of the process in the frequency range f to $f + df$. In addition, from the definition in (2.2.10), the spectral representation for the autocovariance function $\{\gamma_k\}$ can be obtained as $\gamma_k = \int_0^{1/2} \cos(2\pi fk)p(f)\,df$, which together with (2.2.10) directly exhibits the one-to-one correspondence between the spectrum and the autocovariance function of a process. Conversely, since the $\{\gamma_k\}$ form a positive-definite sequence, provided that the series (2.2.11) converges, it follows from Herglotz's theorem (see [140]) that a unique function $p(f)$ exists such that the γ_k have the spectral representation $\gamma_k = \frac{1}{2}\int_{-1/2}^{1/2} e^{i2\pi fk}p(f)\,df$. Consequently, the spectrum $p(f)$ of a stationary process, for which (2.2.11) converges, can be defined as this unique function, which is guaranteed to exist and must have the form of the right-hand side of (2.2.10) by the spectral representation.

Spectral density function. It is sometimes more convenient to base the definition (2.2.10) of the spectrum on the autocorrelations ρ_k rather than the autocovariances γ_k. The resulting function

$$
\begin{aligned}
g(f) &= \frac{p(f)}{\sigma_z^2} \\
&= 2\left[1 + 2\sum_{k=1}^{\infty} \rho_k \cos(2\pi fk)\right] \qquad 0 \leqslant f \leqslant \tfrac{1}{2}
\end{aligned}
\tag{2.2.13}
$$

is called the *spectral density function*. Using (2.2.12), it is seen that the spectral density function has the property

$$\int_0^{1/2} g(f)\,df = 1$$

Since $g(f)$ is also positive, it has the same properties as an ordinary probability density function. This analogy extends to the estimation properties of these two functions, as we discuss next.

Estimation of the spectrum. One would expect that an estimate of the spectrum could be obtained from (2.2.10), by replacing the theoretical autocovariances γ_k with their estimates c_k. Because of (2.2.8), this corresponds to taking the sample spectrum as an estimate of $p(f)$. However, it may be shown (see [122]) that the sample spectrum of a stationary time series fluctuates violently about the theoretical spectrum. An intuitive explanation of this fact is that the sample spectrum corresponds to using an interval, in the frequency domain, whose width is too small. This is analogous to using too small a group interval for the histogram when estimating an ordinary probability distribution. By using a modified or *smoothed* estimate

$$\hat{p}(f) = 2\left[c_0 + 2\sum_{k=1}^{N-1} \lambda_k c_k \cos(2\pi f k)\right] \tag{2.2.14}$$

where the λ_k are suitably chosen weights called a *lag window*, it is possible to increase the *bandwidth* of the estimate and to obtain a smoother estimate of the spectrum. The weights λ_k in (2.2.14) are typically chosen so that they die out to zero for lags $k > M$, where M is known as the *truncation point* and $M < N$ is moderately small in relation to series length N. As an alternative computational form, one can also obtain an estimate of the spectrum smoother than the sample spectrum $I(f)$ by forming a weighted average of a number of periodogram values $I(f_{i+j})$ in a small neighborhood of frequencies around a given frequency f_i. Specifically, a smoothed periodogram estimator of $p(f_i)$ takes the form

$$\hat{p}(f_i) = \sum_{j=-m}^{m} W(f_j) I\left(f_i + \frac{j}{N}\right)$$

where $\Sigma_{j=-m}^{m} W(f_j) = 1$, the symmetric weighting function $W(f_j)$ is referred to as the *spectral window*, and m is chosen to be much smaller than $N/2$.

Figure 2.8 shows an estimate of the spectrum of the batch data. It is seen that most of the variance of the series is concentrated at high frequencies. This is due to the rapid oscillations in the original series, shown in Figure 2.1.

2.2.4 Simple Examples of Autocorrelation and Spectral Density Functions

For illustration, we show below equivalent representations of two simple stochastic processes by:

1. Their theoretical models
2. Their theoretical autocorrelation functions
3. Their theoretical spectra

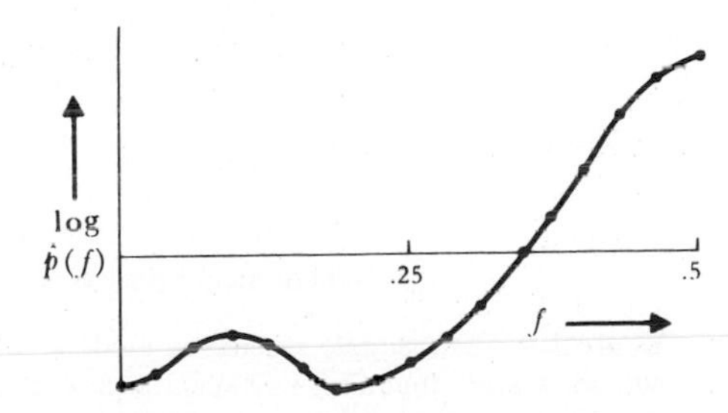

Figure 2.8 Estimated power spectrum of batch data.

Consider the two processes

$$z_t = 10 + a_t + a_{t-1} \qquad z_t = 10 + a_t - a_{t-1}$$

where $a_t, a_{t-1}, \ldots$, are a sequence of uncorrelated random Normal variables with mean zero and variance 1, that is, white noise. Using the definition (2.1.5),

$$\gamma_k = \text{cov}[z_t, z_{t+k}] = E[(z_t - \mu)(z_{t+k} - \mu)]$$

where $E[z_t] = E[z_{t+k}] = \mu = 10$, the autocovariances of these two stochastic processes are

$$\gamma_k = \begin{cases} 2.0 & k = 0 \\ 1.0 & k = 1 \\ 0 & k \geq 2 \end{cases} \qquad \gamma_k = \begin{cases} 2.0 & k = 0 \\ -1.0 & k = 1 \\ 0 & k \geq 2 \end{cases}$$

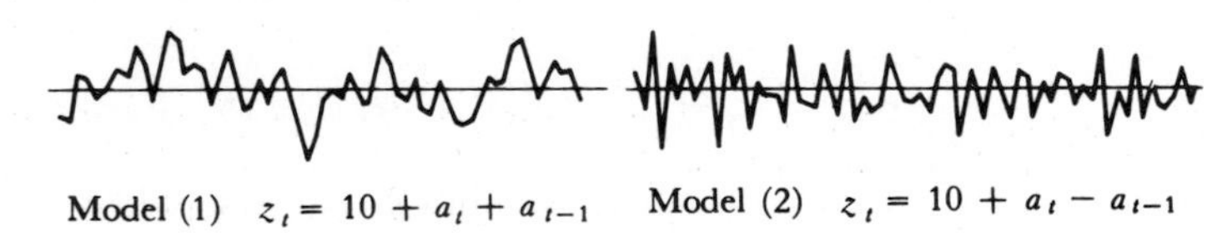

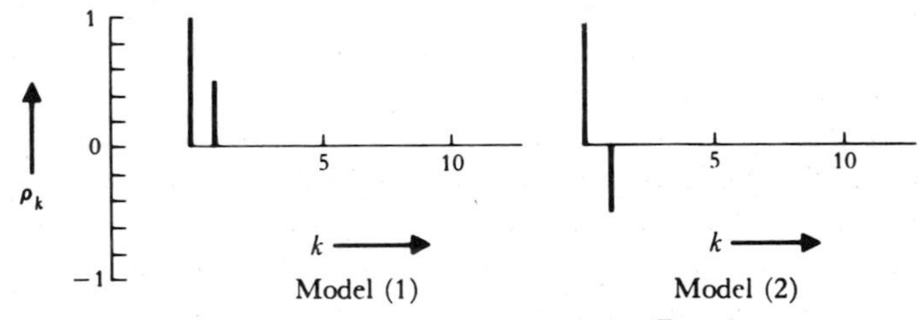

Theoretical Autocorrelation Functions

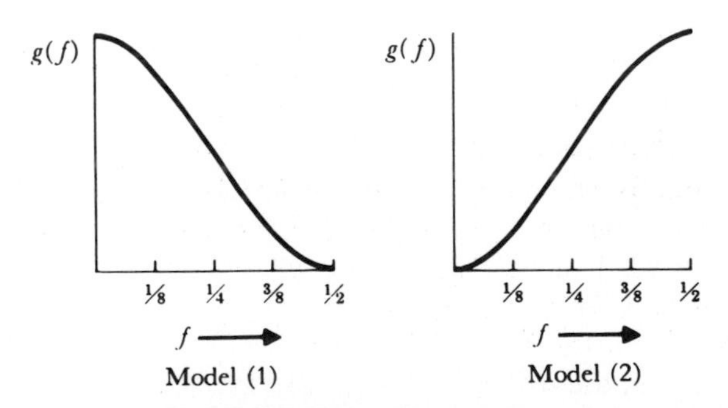

Theoretical Spectral Density Functions

Figure 2.9 Two simple stochastic models with their corresponding theoretical autocorrelation functions and spectral density functions.

Thus the theoretical autocorrelation functions are

$$\rho_k = \begin{cases} 0.5 & k = 1 \\ 0.0 & k \geqslant 2 \end{cases} \qquad \rho_k = \begin{cases} -0.5 & k = 1 \\ 0.0 & k \geqslant 2 \end{cases}$$

and using (2.2.13), the theoretical spectral density functions are

$$g(f) = 2[1 + \cos(2\pi f)] \qquad g(f) = 2[1 - \cos(2\pi f)]$$

The autocorrelation functions and spectral density functions are plotted in Figure 2.9 together with a sample time series from each process.

1. It should be noted that for these two stationary processes, knowledge of either the autocorrelation function or the spectral density function, with the mean and variance of the process, is equivalent to knowledge of the model (given the Normality assumption).
2. It will be seen that the autocorrelation function reflects one aspect of the behavior of the series. The comparatively smooth nature of the first series is accounted for by the positive association between successive values. The alternating tendency of the second series, in which positive deviations usually follow negative ones, is accounted for by the negative association between successive values.
3. The spectral density throws light on a different, but equivalent aspect. The predominance of low frequencies in the first series and high frequencies in the second is shown up by the spectra.

2.2.5 Advantages and Disadvantages of the Autocorrelation and Spectral Density Functions

Because the autocorrelation function and the spectrum are transforms of each other, they are mathematically equivalent, and therefore any discussion of their advantages and disadvantages turns not on mathematical questions, but on their representational value. Because, as we have seen, each sheds light on a different aspect of the data, they should be regarded not as rivals but as allies. Each contributes something to an understanding of the stochastic process in question.

The obtaining of sample estimates of the autocorrelation function and of the spectrum are nonstructural approaches, analogous to the representation of an empirical distribution function by a histogram. They are both ways of letting data from stationary series "speak for themselves" and provide a first step in the analysis of time series, just as a histogram can provide a first step in the distributional analysis of data, pointing the way to some parametric model on which subsequent analysis will be based.

Parametric time series models such as those of Section 2.2.4 are not necessarily associated with a simple autocorrelation function or a simple spectrum. Working with either of these nonstructural methods, we may be involved in the estimation of many lag correlations and many spectral ordinates, even when a parametric model containing only one or two parameters could represent the data. Each correlation and each spectral ordinate is a parameter to be estimated, so that these nonstructural approaches might be very prodigal with parameters, when the approach via the model could be parsimonious. On the other hand, initially, we probably do not know what type of model may be appropriate and initial use of one or other of these nonstructural approaches is necessary to *identify* the type of model that is needed (in the same way that plotting a histogram helps to indicate which family of distributions may be appropriate). The choice between the spectrum and the autocorrelation function as a tool in model building depends upon the nature of the models that turn out to be practically useful. The models that we have found useful, which we consider in later chapters of this book, are simply described in terms of the autocorrelation function, and it is this tool that we employ for identification of series.

APPENDIX A2.1 LINK BETWEEN THE SAMPLE SPECTRUM AND AUTOCOVARIANCE FUNCTION ESTIMATE

Here we derive the result (2.2.8):

$$I(f) = 2\left[c_0 + 2\sum_{k=1}^{N-1} c_k \cos(2\pi fk)\right] \qquad 0 \leq f \leq \tfrac{1}{2}$$

which links the sample spectrum $I(f)$ and the estimate c_k of the autocovariance function. Suppose that the least squares estimates a_f and b_f of the cosine and sine components, at frequency f, in a series are combined according to $d_f = a_f - ib_f$, where $i = \sqrt{-1}$; then

$$\begin{aligned} I(f) &= \frac{N}{2}(a_f - ib_f)(a_f + ib_f) \\ &= \frac{N}{2} d_f d_f^* \end{aligned} \qquad \text{(A2.1.1)}$$

where d_f^* is the complex conjugate of d_f. Then, using (2.2.3) and (2.2.4), we obtain

$$d_f = \frac{2}{N} \sum_{t=1}^{N} z_t[\cos(2\pi f t) - i \sin(2\pi f t)]$$

$$= \frac{2}{N} \sum_{t=1}^{N} z_t e^{-i2\pi f t} \tag{A2.1.2}$$

$$= \frac{2}{N} \sum_{t=1}^{N} (z_t - \bar{z}) e^{-i2\pi f t}$$

Substituting (A2.1.2) in (A2.1.1) yields

$$I(f) = \frac{2}{N} \sum_{t=1}^{N} \sum_{t'=1}^{N} (z_t - \bar{z})(z_{t'} - \bar{z}) e^{-i2\pi f(t-t')} \tag{A2.1.3}$$

Since

$$c_k = \frac{1}{N} \sum_{t=1}^{N-k} (z_t - \bar{z})(z_{t+k} - \bar{z})$$

the transformation $k = t - t'$ transforms (A2.1.3) into

$$I(f) = 2 \sum_{k=-N+1}^{N-1} c_k e^{-i2\pi f k}$$

$$= 2\left[c_0 + 2 \sum_{k=1}^{N-1} c_k \cos(2\pi f k)\right] \qquad 0 \leqslant f \leqslant \tfrac{1}{2}$$

which is the required result.

3

LINEAR STATIONARY MODELS

A general linear stochastic model is described that supposes a time series to be generated by a linear aggregation of random shocks. For practical representation it is desirable to employ models that use parameters parsimoniously. Parsimony may often be achieved by representation of the linear process in terms of a small number of autoregressive and moving average terms. The properties of the resulting autoregressive-moving average (ARMA) models are discussed in preparation for their use in model building.

3.1 GENERAL LINEAR PROCESS

3.1.1 Two Equivalent Forms for the Linear Process

In Section 1.2.1 we discussed the representation of a stochastic process as the output from a linear filter, whose input is white noise a_t, that is,

$$\begin{aligned}\tilde{z}_t &= a_t + \psi_1 a_{t-1} + \psi_2 a_{t-2} + \cdots \\ &= a_t + \sum_{j=1}^{\infty} \psi_j a_{t-j}\end{aligned} \tag{3.1.1}$$

where $\tilde{z}_t = z_t - \mu$ is the deviation of the process from some origin, or from its mean, if the process is stationary. The *general linear process* (3.1.1) allows us to represent $\tilde{z}_t$ as a weighted sum of present and past values of the "white noise" process a_t. Important references in the development of linear stochastic models are [25], [84], [98], [100], [126], [162], [163], [171], [180], [195], [207], and [212]. The white noise process a_t may be regarded as a *series of shocks* which drive the system. It consists of a sequence of uncorrelated random variables with mean zero and constant variance, that is

$$E[a_t] = 0 \qquad \text{var}[a_t] = \sigma_a^2$$

Since the random variables a_t are uncorrelated, it follows that their autocovariance function is

$$\gamma_k = E[a_t a_{t+k}] = \begin{cases} \sigma_a^2 & k = 0 \\ 0 & k \neq 0 \end{cases} \tag{3.1.2}$$

Thus the autocorrelation function of white noise has the particularly simple form

$$\rho_k = \begin{cases} 1 & k = 0 \\ 0 & k \neq 0 \end{cases} \tag{3.1.3}$$

A fundamental result in the development of stationary processes is that of Wold [207], who established that any zero-mean purely nondeterministic stationary process $\tilde{z}_t$ possesses a linear representation as in (3.1.1) with $\sum_{j=0}^{\infty} \psi_j^2 < \infty$; the a_t are uncorrelated with common variance σ_a^2 but *need not be independent*. We will reserve the term *linear processes* for processes z_t of the form of (3.1.1) in which the a_t are independent random variables.

For $\tilde{z}_t$ defined by (3.1.1) to represent a valid stationary process, it is necessary for the coefficients ψ_j to be *absolutely summable*, that is, for $\sum_{j=0}^{\infty} |\psi_j| < \infty$. The model (3.1.1) implies that, under suitable conditions (see [133, p. 254]), $\tilde{z}_t$ is also a weighted sum of past values of the $\tilde{z}$'s, plus an added shock a_t, that is,

$$\begin{aligned} \tilde{z}_t &= \pi_1 \tilde{z}_{t-1} + \pi_2 \tilde{z}_{t-2} + \cdots + a_t \\ &= \sum_{j=1}^{\infty} \pi_j \tilde{z}_{t-j} + a_t \end{aligned} \tag{3.1.4}$$

The alternative form (3.1.4) may be thought of as one where the current deviation $\tilde{z}_t$ from the level μ is "regressed" on past deviations $\tilde{z}_{t-1}, \tilde{z}_{t-2}, \ldots$ of the process.

Relationships between the ψ weights and π weights. The relationships between the ψ weights and π weights may be obtained using the previously defined *backward shift operator*

$$Bz_t = z_{t-1} \qquad B^j z_t = z_{t-j}$$

Later we shall also need to use the forward shift operator $F = B^{-1}$, such that

$$Fz_t = z_{t+1} \qquad F^j z_t = z_{t+j}$$

As an example of the use of the operator B, consider the model

$$\tilde{z}_t = a_t - \theta a_{t-1} = (1 - \theta B)a_t$$

in which $\psi_1 = -\theta$, $\psi_j = 0$ for $j > 1$. Expressing a_t in terms of the $\tilde{z}$'s we obtain

$$(1 - \theta B)^{-1}\tilde{z}_t = a_t$$

Hence for $|\theta| < 1$,

$$(1 + \theta B + \theta^2 B^2 + \theta^3 B^3 + \cdots)\tilde{z}_t = a_t$$

and the deviation $\tilde{z}_t$ expressed in terms of previous deviations, as in (3.1.4), is

$$\tilde{z}_t = -\theta\tilde{z}_{t-1} - \theta^2\tilde{z}_{t-2} - \theta^3\tilde{z}_{t-3} - \cdots + a_t$$

so that for this model, $\pi_j = -\theta^j$.

In general, (3.1.1) may be written

$$\tilde{z}_t = \left(1 + \sum_{j=1}^{\infty} \psi_j B^j\right) a_t$$

or

$$\tilde{z}_t = \psi(B)a_t \tag{3.1.5}$$

where

$$\psi(B) = 1 + \sum_{j=1}^{\infty} \psi_j B^j = \sum_{j=0}^{\infty} \psi_j B^j$$

with $\psi_0 = 1$. As mentioned in Section 1.2.1, $\psi(B)$ is called the *transfer function* of the linear filter relating $\tilde{z}_t$ to a_t. It can also be regarded as the *generating function* of the ψ weights, with B now treated simply as a variable whose jth power is the coefficient of ψ_j.

Similarly, (3.1.4) may be written

$$\left(1 - \sum_{j=1}^{\infty} \pi_j B^j\right) \tilde{z}_t = a_t$$

or

$$\pi(B)\tilde{z}_t = a_t \tag{3.1.6}$$

Thus

$$\pi(B) = 1 - \sum_{j=1}^{\infty} \pi_j B^j$$

is the generating function of the π weights. After operating on both sides of (3.1.6) by $\psi(B)$, we obtain

$$\psi(B)\pi(B)\tilde{z}_t = \psi(B)a_t = \tilde{z}_t$$

Hence $\psi(B)\pi(B) = 1$, that is,

$$\pi(B) = \psi^{-1}(B) \tag{3.1.7}$$

The relationship (3.1.7) may be used to derive the π weights, knowing the ψ weights, and vice versa.

3.1.2 Autocovariance Generating Function of a Linear Process

A basic data analysis tool for identifying models in Chapter 6 will be the autocorrelation function. Therefore, it is important to know the autocorrelation function of a linear process. It is shown in Appendix A3.1 that the autocovariance function of the linear process (3.1.1) is given by

$$\gamma_k = \sigma_a^2 \sum_{j=0}^{\infty} \psi_j \psi_{j+k} \tag{3.1.8}$$

In particular, by setting $k = 0$ in (3.1.8), we find that its variance is

$$\gamma_0 = \sigma_z^2 = \sigma_a^2 \sum_{j=0}^{\infty} \psi_j^2 \tag{3.1.9}$$

It follows that the stationarity condition of absolute summability of the coefficients ψ_j, $\Sigma_{j=0}^{\infty} |\psi_j| < \infty$, implies that the series on the right of (3.1.9) converges, and hence guarantees that the process will have a finite variance.

Another way of obtaining the autocovariances of a linear process is via the *autocovariance generating function*

$$\gamma(B) = \sum_{k=-\infty}^{\infty} \gamma_k B^k \tag{3.1.10}$$

in which it is noted that γ_0, the variance of the process, is the coefficient of $B^0 = 1$, while γ_k, the autocovariance of lag k, is the coefficient of both B^j and of $B^{-j} = F^j$. It is shown in Appendix A3.1 that

$$\gamma(B) = \sigma_a^2 \psi(B)\psi(B^{-1}) = \sigma_a^2 \psi(B)\psi(F) \tag{3.1.11}$$

For example, suppose that

$$\tilde{z}_t = a_t - \theta a_{t-1} = (1 - \theta B)a_t$$

so that $\psi(B) = 1 - \theta B$. Then, substituting in (3.1.11) gives

$$\begin{aligned}\gamma(B) &= \sigma_a^2(1 - \theta B)(1 - \theta B^{-1}) \\ &= \sigma_a^2[-\theta B^{-1} + (1 + \theta^2) - \theta B]\end{aligned}$$

Comparing with (3.1.10), the autocovariances are

$$\gamma_0 = (1 + \theta^2)\sigma_a^2$$

$$\gamma_1 = -\theta\sigma_a^2$$

$$\gamma_k = 0 \qquad k \geqslant 2$$

In the development that follows, when treated as a variable in a generating function, B will be supposed capable of taking complex values. In particular, it will often be necessary to consider the different situations occurring when $|B| < 1$, $|B| = 1$, or $|B| > 1$, that is, when the complex number B lies inside, on, or outside the unit circle.

3.1.3 Stationarity and Invertibility Conditions for a Linear Process

Stationarity. The convergence of the series (3.1.9) ensures that the process has a finite variance. Also, we have seen in Section 2.1.3 that the autocovariances and autocorrelations must satisfy a set of conditions to ensure stationarity. For a linear process (3.1.1) these conditions are guaranteed by the single condition that $\Sigma_{j=0}^{\infty} |\psi_j| < \infty$. This condition can also be embodied in the condition that the series $\psi(B)$, which is the generating function of the ψ weights, must converge for $|B| \leqslant 1$, that is, on or within the unit circle. This result is discussed in Appendix A3.1.

Spectrum of a linear stationary process. It is shown in Appendix A3.1 that if we substitute $B = e^{-i2\pi f}$, where $i = \sqrt{-1}$, in the autocovariance generating function (3.1.11), we obtain one half of the power spectrum. Thus the spectrum of a linear process is

$$\begin{aligned}p(f) &= 2\sigma_a^2\psi(e^{-i2\pi f})\psi(e^{i2\pi f}) \\ &= 2\sigma_a^2|\psi(e^{-i2\pi f})|^2 \qquad 0 \leqslant f \leqslant \tfrac{1}{2}\end{aligned} \qquad (3.1.12)$$

In fact, (3.1.12) is the well-known expression [122], which relates the spectrum $p(f)$ of the output from a linear system to the uniform spectrum $2\sigma_a^2$ of a white noise input by multiplying by the squared gain $G^2(f) = |\psi(e^{-i2\pi f})|^2$ of the system.

Invertibility. We have seen above that the ψ weights of a linear process must satisfy the condition that $\psi(B)$ converges on or within the unit circle if the process is to be stationary. We now consider a restriction applied to the π weights to ensure what is called *invertibility*. The invertibility condition is independent of the stationarity condition and is also applicable to the nonstationary linear models, which we introduce in Chapter 4.

To illustrate the basic idea of invertibility, consider again the model

$$\tilde{z}_t = (1 - \theta B)a_t \tag{3.1.13}$$

Expressing the a's in terms of the present and past $\tilde{z}$'s, (3.1.13) becomes

$$a_t = (1 - \theta B)^{-1}\tilde{z}_t = (1 + \theta B + \theta^2 B^2 + \cdots + \theta^k B^k)(1 - \theta^{k+1}B^{k+1})^{-1}\tilde{z}_t$$

that is,

$$\tilde{z}_t = -\theta\tilde{z}_{t-1} - \theta^2\tilde{z}_{t-2} - \cdots - \theta^k z_{t-k} + a_t - \theta^{k+1}a_{t-k-1} \tag{3.1.14}$$

If $|\theta| < 1$, on letting k tend to infinity, we obtain the infinite series

$$\tilde{z}_t = -\theta\tilde{z}_{t-1} - \theta^2\tilde{z}_{t-2} - \cdots + a_t \tag{3.1.15}$$

and the π weights of the model in the form of (3.1.4) are $\pi_j = -\theta^j$. Whatever the value of θ, (3.1.13) defines a perfectly proper stationary process. However, if $|\theta| \geqslant 1$, the current deviation $\tilde{z}_t$ in (3.1.14) depends on $\tilde{z}_{t-1}, \tilde{z}_{t-2}, \ldots, \tilde{z}_{t-k}$, with weights that increase as k increases. We avoid this situation by requiring that $|\theta| < 1$. We shall then say that the series is *invertible*. We see that this condition is equivalent to $\sum_{j=0}^{\infty} |\theta|^j \equiv \sum_{j=0}^{\infty} |\pi_j| < \infty$, so that the series

$$\pi(B) = (1 - \theta B)^{-1} = \sum_{j=0}^{\infty} \theta^j B^j$$

converges for all $|B| \leqslant 1$, that is, on or within the unit circle.

In Chapter 6, where we consider questions of uniqueness of these models, we shall see that a convergent expansion for a_t is possible when $|\theta| \geqslant 1$, but only in terms of $z_t, z_{t+1}, z_{t+2}, \ldots$ (i.e., in terms of present and *future* values of the process). The requirement of invertibility is needed if we are interested in associating present events with *past* happenings in a sensible manner.

In general, the linear process (3.1.1) is invertible and has the representation

$$\pi(B)\tilde{z}_t = a_t$$

if the weights π_j are absolutely summable, that is, if $\sum_{j=0}^{\infty} |\pi_j| < \infty$, which implies that the series $\pi(B)$ converges on or within the unit circle.

To sum up, a linear process (3.1.1) is *stationary* if $\sum_{j=0}^{\infty} |\psi_j| < \infty$ and is *invertible* if $\sum_{j=0}^{\infty} |\pi_j| < \infty$, where $\pi(B) = \psi^{-1}(B) = 1 - \sum_{j=1}^{\infty} \pi_j B^j$.

3.1.4 Autoregressive and Moving Average Processes

The representations (3.1.1) and (3.1.4) of the general linear process would not be very useful in practice if they contained an infinite number of parameters ψ_j and π_j. We now consider how to introduce parsimony and yet retain models that are representationally useful.

Autoregressive processes. Consider the special case of (3.1.4), in which only the first p of the weights are nonzero. The model may be written

$$\tilde{z}_t = \phi_1 \tilde{z}_{t-1} + \phi_2 \tilde{z}_{t-2} + \cdots + \phi_p \tilde{z}_{t-p} + a_t \tag{3.1.16}$$

where we now use the symbols $\phi_1, \phi_2, \ldots, \phi_p$ for the *finite* set of weight parameters. The process defined by (3.1.16) is called an *autoregressive* process of order p, or more succinctly, an AR(p) process. In particular, the autoregressive processes of first order ($p = 1$), and of second order ($p = 2$),

$$\tilde{z}_t = \phi_1 \tilde{z}_{t-1} + a_t$$

$$\tilde{z}_t = \phi_1 \tilde{z}_{t-1} + \phi_2 \tilde{z}_{t-2} + a_t$$

are of considerable practical importance.

Now we can write (3.1.16) in the equivalent form

$$(1 - \phi_1 B - \phi_2 B^2 - \cdots - \phi_p B^p)\tilde{z}_t = a_t$$

or

$$\phi(B)\tilde{z}_t = a_t \tag{3.1.17}$$

Since (3.1.17) implies that

$$\tilde{z}_t = \frac{1}{\phi(B)} a_t = \phi^{-1}(B) a_t$$

the autoregressive process can be thought of as the output $\tilde{z}_t$ from a linear filter with transfer function $\phi^{-1}(B)$ when the input is white noise a_t.

Moving average processes. Consider the special case of (3.1.1) when only the first q of the ψ weights are nonzero. The process may be written as

$$\tilde{z}_t = a_t - \theta_1 a_{t-1} - \theta_2 a_{t-2} - \cdots - \theta_q a_{t-q} \tag{3.1.18}$$

where we now use the symbols $-\theta_1, -\theta_2, \ldots, -\theta_q$ for the *finite* set of weight parameters. The process defined by (3.1.18) is called a *moving average* process* of order q, which we sometimes abbreviate as MA(q). In

* As we remarked in Chapter 1, the term "moving average" is somewhat misleading, since the weights do not sum to unity. However, this nomenclature is now well established and we shall use it.

particular, the processes of first order ($q = 1$), and second order ($q = 2$),

$$\tilde{z}_t = a_t - \theta_1 a_{t-1}$$

$$\tilde{z}_t = a_t - \theta_1 a_{t-1} - \theta_2 a_{t-2}$$

are particularly important in practice.

We can also write (3.1.18) in the equivalent form

$$\tilde{z}_t = (1 - \theta_1 B - \theta_2 B^2 - \cdots - \theta_q B^q) a_t$$

or

$$\tilde{z}_t = \theta(B) a_t \tag{3.1.19}$$

Hence the moving average process can be thought of as the output $\tilde{z}_t$ from a linear filter with transfer function $\theta(B)$ when the input is white noise a_t.

Mixed autoregressive–moving average processes. We have seen in Section 3.1.1 that the *finite* moving average process

$$\tilde{z}_t = a_t - \theta_1 a_{t-1} = (1 - \theta_1 B) a_t \qquad |\theta_1| < 1$$

can be written as an *infinite* autoregressive process

$$\tilde{z}_t = -\theta_1 \tilde{z}_{t-1} - \theta_1^2 \tilde{z}_{t-2} - \cdots + a_t$$

Hence, if the process were really MA(1), we would obtain a nonparsimonious representation in terms of an autoregressive model. Conversely, an AR(1) could not be parsimoniously represented using a moving average process. In practice, to obtain a parsimonious parameterization, it will sometimes be necessary to include both autoregressive and moving average terms in the model. Thus

$$\tilde{z}_t = \phi_1 \tilde{z}_{t-1} + \cdots + \phi_p \tilde{z}_{t-p} + a_t - \theta_1 a_{t-1} - \cdots - \theta_q a_{t-q}$$

or

$$\phi(B) \tilde{z}_t = \theta(B) a_t \tag{3.1.20}$$

is called the *mixed autoregressive–moving average* process of order (p, q), which we sometimes abbreviate as ARMA(p, q). For example, the ARMA(1, 1) process is

$$\tilde{z}_t - \phi_1 \tilde{z}_{t-1} = a_t - \theta_1 a_{t-1}$$

Since (3.1.20) may be written

$$\begin{aligned} \tilde{z}_t &= \phi^{-1}(B)\theta(B) a_t \\ &= \frac{\theta(B)}{\phi(B)} a_t = \frac{1 - \theta_1 B - \cdots - \theta_q B^q}{1 - \phi_1 B - \cdots - \phi_p B^p} a_t \end{aligned}$$

the mixed autoregressive–moving average process can be thought of as the output $\tilde{z}_t$ from a linear filter, whose transfer function is the ratio of two polynomials $\theta(B)$ and $\phi(B)$, when the input is white noise a_t. Notice that since $\tilde{z}_t = z_t - \mu$, where $\mu = E[z_t]$ is the mean of the process in the stationary case, the general ARMA(p, q) process in (3.1.20) may be expressed in the equivalent form in terms of the original process z_t as

$$\phi(B) z_t = \theta_0 + \theta(B) a_t \tag{3.1.21}$$

where the constant term θ_0 in the model is

$$\theta_0 = (1 - \phi_1 - \phi_2 - \cdots - \phi_p)\mu \tag{3.1.22}$$

In the following sections we discuss important characteristics of autoregressive, moving average, and mixed models. We study their variances, autocorrelation functions, spectra, and the stationarity and invertibility conditions that must be imposed upon their parameters.

3.2 AUTOREGRESSIVE PROCESSES

3.2.1 Stationarity Conditions for Autoregressive Processes

The set of adjustable parameters $\phi_1, \phi_2, \ldots, \phi_p$ of an AR(p) process

$$\tilde{z}_t = \phi_1 \tilde{z}_{t-1} + \cdots + \phi_p \tilde{z}_{t-p} + a_t$$

or

$$(1 - \phi_1 B - \cdots - \phi_p B^p)\tilde{z}_t = \phi(B)\tilde{z}_t = a_t$$

must satisfy certain conditions for the process to be stationary.

For illustration, the first-order autoregressive process

$$(1 - \phi_1 B)\tilde{z}_t = a_t$$

may be written

$$\tilde{z}_t = (1 - \phi_1 B)^{-1} a_t = \sum_{j=0}^{\infty} \phi_1^j a_{t-j}$$

Hence

$$\psi(B) = (1 - \phi_1 B)^{-1} = \sum_{j=0}^{\infty} \phi_1^j B^j \tag{3.2.1}$$

We have seen in Section 3.1.3 that for stationarity, $\psi(B)$ must converge for $|B| \leqslant 1$. From (3.2.1) we see that this implies that the parameter ϕ_1 of an AR(1) process must satisfy the condition $|\phi_1| < 1$ to ensure stationarity. Since the root of $1 - \phi_1 B = 0$ is $B = \phi_1^{-1}$, this condition is equivalent to saying that the root of $1 - \phi_1 B = 0$ must lie *outside* the unit circle.

For the general AR(p) process $\tilde{z}_t = \phi^{-1}(B)a_t$, we obtain

$$\phi(B) = (1 - G_1 B)(1 - G_2 B) \cdots (1 - G_p B)$$

where $G_1^{-1}, \ldots, G_p^{-1}$ are the roots of $\phi(B) = 0$, and expanding $\phi^{-1}(B)$ in partial fractions yields

$$\tilde{z}_t = \phi^{-1}(B)a_t = \sum_{i=1}^{p} \frac{K_i}{1 - G_i B} a_t$$

Hence, if $\psi(B) = \phi^{-1}(B)$ is to be a convergent series for $|B| \leq 1$, that is, if the weights $\psi_j = \sum_{i=1}^{p} K_i G_i^j$ are to be absolutely summable so that the AR(p) will represent a stationary process, we must have $|G_i| < 1$ for $i = 1, 2, \ldots, p$. Equivalently, the roots of $\phi(B) = 0$ must lie *outside* the unit circle. The roots of the equation $\phi(B) = 0$ may be referred to as the zeros of the polynomial $\phi(B)$. Thus the stationarity condition may be expressed by saying that the zeros of $\phi(B)$ must lie *outside* the unit circle. A similar argument may be applied when the zeros of $\phi(B)$ are not all distinct. The equation $\phi(B) = 0$ is called the *characteristic equation* for the process. In addition, from the relation $\phi(B)\psi(B) = 1$ it readily follows that the weights ψ_j for the AR(p) process satisfy the difference equation

$$\psi_j = \phi_1 \psi_{j-1} + \phi_2 \psi_{j-2} + \cdots + \phi_p \psi_{j-p} \qquad j > 0$$

with $\psi_0 = 1$ and $\psi_j = 0$ for $j < 0$, from which the weights ψ_j can easily be computed recursively in terms of the ϕ_i. In fact, as seen from the principles of linear difference equations as discussed in Appendix A4.1, the fact that the weights ψ_j satisfy the difference equation above implies that they have an explicit representation in the form $\psi_j = \sum_{i=1}^{p} K_i G_i^j$ for the case of distinct roots.

Since the series

$$\pi(B) = \phi(B) = 1 - \phi_1 B - \phi_2 B^2 - \cdots - \phi_p B^p$$

is finite, no restrictions are required on the parameters of an autoregressive process to ensure invertibility.

3.2.2 Autocorrelation Function and Spectrum of Autoregressive Processes

Autocorrelation function. An important recurrence relation for the autocorrelation function of a stationary autoregressive process is found by multiplying throughout in

$$\tilde{z}_t = \phi_1 \tilde{z}_{t-1} + \phi_2 \tilde{z}_{t-2} + \cdots + \phi_p \tilde{z}_{t-p} + a_t$$

by $\tilde{z}_{t-k}$, to obtain

$$\tilde{z}_{t-k}\tilde{z}_t = \phi_1 \tilde{z}_{t-k}\tilde{z}_{t-1} + \phi_2 \tilde{z}_{t-k}\tilde{z}_{t-2} + \cdots + \phi_p \tilde{z}_{t-k}\tilde{z}_{t-p} + \tilde{z}_{t-k}a_t \qquad (3.2.2)$$

On taking expected values in (3.2.2), we obtain the difference equation

$$\gamma_k = \phi_1\gamma_{k-1} + \phi_2\gamma_{k-2} + \cdots + \phi_p\gamma_{k-p} \qquad k > 0 \tag{3.2.3}$$

Note that the expectation $E[\tilde{z}_{t-k}a_t]$ vanishes when $k > 0$, since $\tilde{z}_{t-k}$ can only involve the shocks a_j up to time $t - k$, which are uncorrelated with a_t. On dividing throughout in (3.2.3) by γ_0, it is seen that the autocorrelation function satisfies the same form of difference equation

$$\rho_k = \phi_1\rho_{k-1} + \phi_2\rho_{k-2} + \cdots + \phi_p\rho_{k-p} \qquad k > 0 \tag{3.2.4}$$

We note that this is analogous to the difference equation satisfied by the process $\tilde{z}_t$ itself.

Now suppose that (3.2.4) is written

$$\phi(B)\rho_k = 0$$

where $\phi(B) = 1 - \phi_1 B - \cdots - \phi_p B^p$ and B now operates on k and not t. Then, writing

$$\phi(B) = \prod_{i=1}^{p} (1 - G_i B)$$

the general solution (see, e.g., Appendix A4.1) of (3.2.4) is

$$\rho_k = A_1 G_1^k + A_2 G_2^k + \cdots + A_p G_p^k \tag{3.2.5}$$

where $G_1^{-1}, G_2^{-1}, \ldots, G_p^{-1}$ are the roots of the *characteristic equation*

$$\phi(B) = 1 - \phi_1 B - \phi_2 B^2 - \cdots - \phi_p B^p = 0$$

For stationarity we require that $|G_i| < 1$. Thus two situations can arise in practice if we assume that the roots G_i are distinct.

1. A root G_i is real, in which case a term $A_i G_i^k$ in (3.2.5) decays to zero geometrically as k increases. We shall frequently refer to this as a damped exponential.
2. A pair of roots G_i, G_j are complex conjugates, in which case they contribute a term

$$D^k \sin(2\pi f k + F)$$

to the autocorrelation function (3.2.5), which follows a damped sine wave, with damping factor $D = |G_i| = |G_j|$ and frequency f such that $2\pi f = \cos^{-1}[|\text{Re}(G_i)|/D]$.

In general, the autocorrelation function of a stationary autoregressive process will consist of a mixture of damped exponentials and damped sine waves.

Autoregressive parameters in terms of the autocorrelations: Yule–Walker equations. If we substitute $k = 1, 2, \ldots, p$ in (3.2.4), we obtain a set of linear equations for $\phi_1, \phi_2, \ldots, \phi_p$ in terms of $\rho_1, \rho_2, \ldots, \rho_p$, that is,

$$\begin{aligned} \rho_1 &= \phi_1 &+ \phi_2\rho_1 &+ \cdots + \phi_p\rho_{p-1} \\ \rho_2 &= \phi_1\rho_1 &+ \phi_2 &+ \cdots + \phi_p\rho_{p-2} \\ \vdots & \quad \vdots & \vdots & \quad \cdots \quad \vdots \\ \rho_p &= \phi_1\rho_{p-1} &+ \phi_2\rho_{p-2} &+ \cdots + \phi_p \end{aligned} \tag{3.2.6}$$

These are usually called the *Yule–Walker* equations [195], [212]. We obtain *Yule–Walker estimates* of the parameters by replacing the theoretical autocorrelations ρ_k by the estimated autocorrelations r_k. Note that if we write

$$\boldsymbol{\phi} = \begin{bmatrix} \phi_1 \\ \phi_2 \\ \vdots \\ \phi_p \end{bmatrix} \qquad \boldsymbol{\rho}_p = \begin{bmatrix} \rho_1 \\ \rho_2 \\ \vdots \\ \rho_p \end{bmatrix} \qquad \mathbf{P}_p = \begin{bmatrix} 1 & \rho_1 & \rho_2 & \cdots & \rho_{p-1} \\ \rho_1 & 1 & \rho_1 & \cdots & \rho_{p-2} \\ \vdots & \vdots & \vdots & \cdots & \vdots \\ \rho_{p-1} & \rho_{p-2} & \rho_{p-3} & \cdots & 1 \end{bmatrix}$$

the solution of (3.2.6) for the parameters $\boldsymbol{\phi}$ in terms of the autocorrelations may be written

$$\boldsymbol{\phi} = \mathbf{P}_p^{-1}\boldsymbol{\rho}_p \tag{3.2.7}$$

Variance. When $k = 0$, the contribution from the term $E[z_{t-k}a_t]$, on taking expectations in (3.2.2), is $E[a_t^2] = \sigma_a^2$, since the only part of z_t that will be correlated with a_t is the most recent shock, a_t. Hence, when $k = 0$,

$$\gamma_0 = \phi_1\gamma_{-1} + \phi_2\gamma_{-2} + \cdots + \phi_p\gamma_{-p} + \sigma_a^2$$

On dividing throughout by $\gamma_0 = \sigma_z^2$ and substituting $\gamma_k = \gamma_{-k}$, the variance σ_z^2 may be written

$$\sigma_z^2 = \frac{\sigma_a^2}{1 - \rho_1\phi_1 - \rho_2\phi_2 - \cdots - \rho_p\phi_p} \tag{3.2.8}$$

Spectrum. For the AR(p) process, $\psi(B) = \phi^{-1}(B)$ and

$$\phi(B) = 1 - \phi_1 B - \phi_2 B^2 - \cdots - \phi_p B^p$$

Therefore, using (3.1.12), the spectrum of an autoregressive process is

$$p(f) = \frac{2\sigma_a^2}{|1 - \phi_1 e^{-i2\pi f} - \phi_2 e^{-i4\pi f} - \cdots - \phi_p e^{-i2\pi p f}|^2} \qquad 0 \leqslant f \leqslant \tfrac{1}{2} \tag{3.2.9}$$

We now discuss two particularly important autoregressive processes, those of first and second order.

3.2.3 First-Order Autoregressive (Markov) Process

The first-order autoregressive process is

$$\begin{aligned}\tilde{z}_t &= \phi_1 \tilde{z}_{t-1} + a_t \\ &= a_t + \phi_1 a_{t-1} + \phi_1^2 a_{t-2} + \cdots\end{aligned} \tag{3.2.10}$$

where it has been shown in Section 3.2.1 that ϕ_1 must satisfy the condition $-1 < \phi_1 < 1$ for the process to be stationary.

Autocorrelation function. Using (3.2.4), the autocorrelation function satisfies the first-order difference equation

$$\rho_k = \phi_1 \rho_{k-1} \qquad k > 0 \tag{3.2.11}$$

which, with $\rho_0 = 1$, has the solution

$$\rho_k = \phi_1^k \qquad k \geqslant 0 \tag{3.2.12}$$

As shown in Figure 3.1, the autocorrelation function decays exponentially to zero when ϕ_1 is positive, but decays exponentially to zero and oscillates in sign when ϕ_1 is negative. In particular, it will be noted that

$$\rho_1 = \phi_1 \tag{3.2.13}$$

Variance. Using (3.2.8), the variance of the process is

$$\begin{aligned}\sigma_z^2 &= \frac{\sigma_a^2}{1 - \rho_1 \phi_1} \\ &= \frac{\sigma_a^2}{1 - \phi_1^2}\end{aligned} \tag{3.2.14}$$

on substituting $\rho_1 = \phi_1$.

Spectrum. Finally, using (3.2.9), the spectrum is

$$\begin{aligned}p(f) &= \frac{2\sigma_a^2}{|1 - \phi_1 e^{-i2\pi f}|^2} \\ &= \frac{2\sigma_a^2}{1 + \phi_1^2 - 2\phi_1 \cos(2\pi f)} \qquad 0 \leqslant f \leqslant \tfrac{1}{2}\end{aligned} \tag{3.2.15}$$

Figure 3.1 shows realizations from processes with $\phi_1 = 0.8$, $\phi_1 = -0.8$, and the corresponding theoretical autocorrelation functions and spectra.

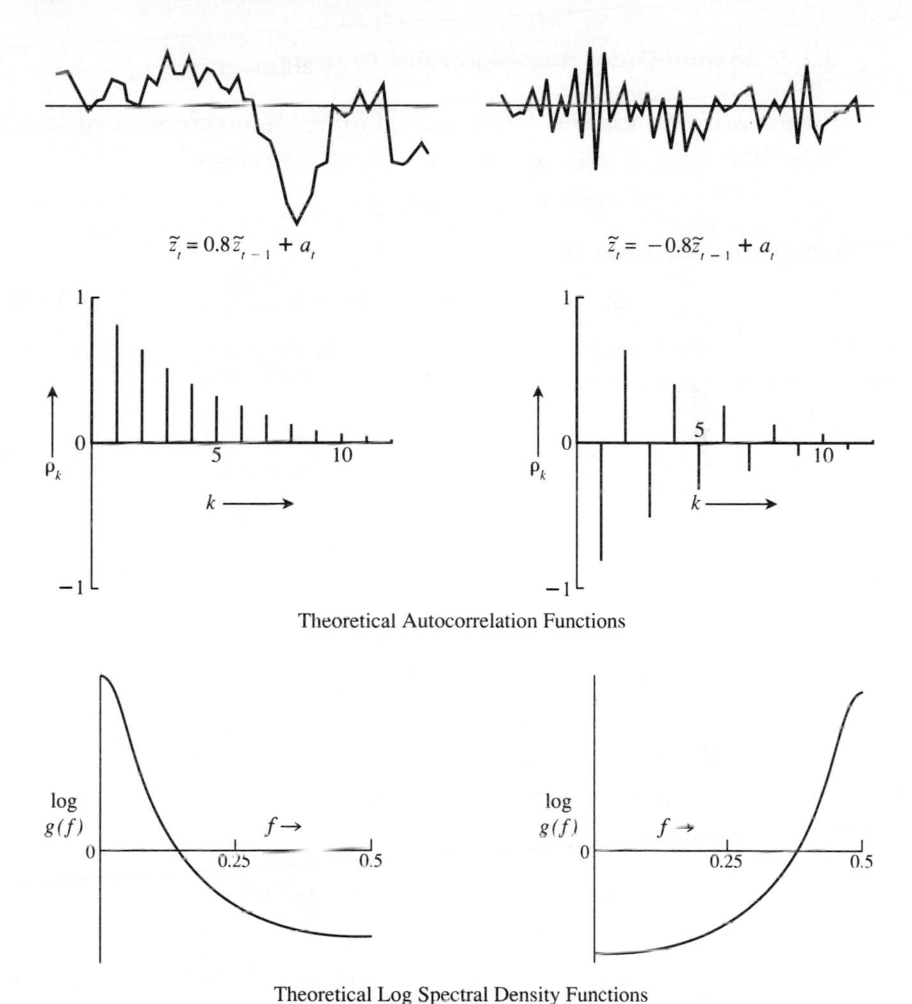

Figure 3.1 Realizations from first order autoregressive processes and their corresponding theoretical autocorrelation functions and spectral density functions.

Thus, when the parameter has the large positive value $\phi_1 = 0.8$, neighboring values in the series are similar and the series exhibits marked trends. This is reflected in the autocorrelation function, which slowly decays exponentially to zero, and in the spectrum, which is dominated by low frequencies. When, on the other hand, the parameter has the large negative value $\phi_1 = -0.8$, the series tends to oscillate rapidly, and this is reflected in the autocorrelation function, which alternates in sign as it decays to zero, and in the spectrum, which is dominated by high frequencies.

3.2.4 Second-Order Autoregressive Process

Stationarity condition. The second-order autoregressive process may be written

$$\tilde{z}_t = \phi_1 \tilde{z}_{t-1} + \phi_2 \tilde{z}_{t-2} + a_t \tag{3.2.16}$$

For stationarity, the roots of

$$\phi(B) = 1 - \phi_1 B - \phi_2 B^2 = 0 \tag{3.2.17}$$

must lie outside the unit circle, which implies that the parameters ϕ_1 and ϕ_2 must lie in the triangular region

$$\begin{aligned} \phi_2 + \phi_1 &< 1 \\ \phi_2 - \phi_1 &< 1 \\ -1 < \phi_2 &< 1 \end{aligned} \tag{3.2.18}$$

shown in Figure 3.2.

Autocorrelation function. Using (3.2.4), the autocorrelation function satisfies the second-order difference equation

$$\rho_k = \phi_1 \rho_{k-1} + \phi_2 \rho_{k-2} \qquad k > 0 \tag{3.2.19}$$

with starting values $\rho_0 = 1$ and $\rho_1 = \phi_1/(1 - \phi_2)$. From (3.2.5), the general solution of the difference equation (3.2.19) is

$$\begin{aligned} \rho_k &= A_1 G_1^k + A_2 G_2^k \\ &= \frac{G_1(1 - G_2^2)G_1^k - G_2(1 - G_1^2)G_2^k}{(G_1 - G_2)(1 + G_1 G_2)} \end{aligned} \tag{3.2.20}$$

where G_1^{-1} and G_2^{-1} are the roots of the characteristic equation (3.2.17). When the roots are real, the autocorrelation function consists of a mixture of damped exponentials. This occurs when $\phi_1^2 + 4\phi_2 \geqslant 0$ and corresponds to regions 1 and 2, which lie above the parabolic boundary in Figure 3.2. Specifically, in region 1, the autocorrelation function remains positive as it damps out, corresponding to a positive dominant root in (3.2.20). In region 2, the autocorrelation function alternates in sign as it damps out, corresponding to a negative dominant root.

If the roots G_1 and G_2 are complex ($\phi_1^2 + 4\phi_2 < 0$), a second-order autoregressive process displays *pseudo-periodic behavior*. This behavior is reflected in the autocorrelation function, for on substituting $G_1 = De^{i2\pi f_0}$ and $G_2 = De^{-i2\pi f_0}$ in (3.2.20), we obtain

$$\rho_k = \frac{[\text{sgn}(\phi_1)]^k D^k \sin(2\pi f_0 k + F)}{\sin F} \tag{3.2.21}$$

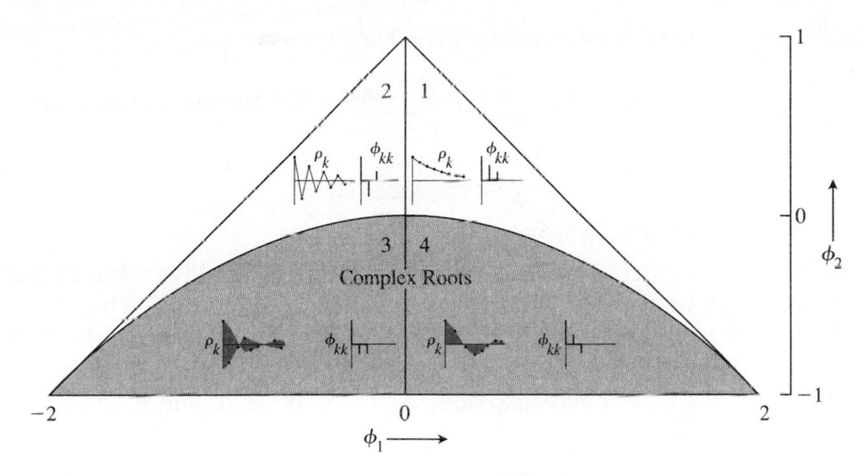

Figure 3.2 Typical autocorrelation and partial autocorrelation functions ρ_k and ϕ_{kk} for various stationary AR(2) models. (From [183].)

where $\text{sgn}(\phi_1)$ is $+1$ if ϕ_1 is positive and -1 if ϕ_1 is negative. In either case, we refer to (3.2.21) as a *damped sine wave* with damping factor D, *frequency* f_0, and *phase* F. These factors are related to the process parameters as follows:

$$D = |G_i| = \sqrt{-\phi_2} \tag{3.2.22}$$

where the positive square root is taken,

$$\cos(2\pi f_0) = \frac{|\text{Re}(G_i)|}{D} = \frac{|\phi_1|}{2\sqrt{-\phi_2}} \tag{3.2.23}$$

$$\tan F = \frac{1 + D^2}{1 - D^2} \tan(2\pi f_0) \tag{3.2.24}$$

Again referring to Figure 3.2, the autocorrelation function is a damped sine wave in regions 3 and 4, the phase angle F being less than 90° in region 4 and lying between 90° and 180° in region 3. This means that the autocorrelation function starts with a positive value throughout region 4, but always switches sign from lag zero to lag 1 in region 3.

Yule–Walker equations Substituting $p = 2$ in (3.2.6), the Yule–Walker equations are

$$\begin{aligned} \rho_1 &= \phi_1 + \phi_2 \rho_1 \\ \rho_2 &= \phi_1 \rho_1 + \phi_2 \end{aligned} \tag{3.2.25}$$

which, when solved for ϕ_1 and ϕ_2, give

$$\phi_1 = \frac{\rho_1(1 - \rho_2)}{1 - \rho_1^2}$$
$$\phi_2 = \frac{\rho_2 - \rho_1^2}{1 - \rho_1^2} \tag{3.2.26}$$

Chart B in the "Collection of Tables and Charts" in Part Five allows values of ϕ_1 and ϕ_2 to be read off for any given values of ρ_1 and ρ_2. The chart is used in Chapters 6 and 7 to obtain estimates of the ϕ's from values of the estimated autocorrelations r_1 and r_2.

Equations (3.2.25) may also be solved to express ρ_1 and ρ_2 in terms of ϕ_1 and ϕ_2, to give

$$\rho_1 = \frac{\phi_1}{1 - \phi_2}$$
$$\rho_2 = \phi_2 + \frac{\phi_1^2}{1 - \phi_2} \tag{3.2.27}$$

which explains the starting values for (3.2.19) quoted above. The forms (3.2.20) and (3.2.21) for the autocorrelation function are useful for explaining the different types that may arise in practice. However, for computing the autocorrelations of an AR(2) process, with given values of ϕ_1 and ϕ_2, it is simplest to make direct use of the difference equation (3.2.19).

Using the stationarity conditions (3.2.18) and the expressions (3.2.27) for ρ_1 and ρ_2, it is found that the admissible values of ρ_1 and ρ_2, for a stationary AR(2) process, must lie in the region

$$-1 < \rho_1 < 1$$
$$-1 < \rho_2 < 1$$
$$\rho_1^2 < \tfrac{1}{2}(\rho_2 + 1)$$

Figure 3.3(a) shows the admissible region for the parameters ϕ_1 and ϕ_2, and Figure 3.3(b) shows the corresponding admissible region for ρ_1 and ρ_2.

Variance. From (3.2.8), the variance of the process is

$$\sigma_z^2 = \frac{\sigma_a^2}{1 - \rho_1\phi_1 - \rho_2\phi_2}$$
$$= \left(\frac{1 - \phi_2}{1 + \phi_2}\right) \frac{\sigma_a^2}{\{(1 + \phi_2)^2 - \phi_1^2\}} \tag{3.2.28}$$

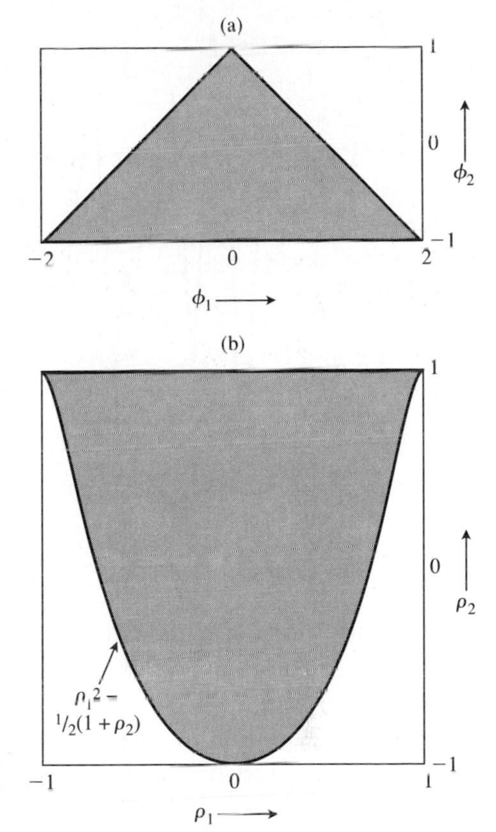

Figure 3.3 Admissible regions for (a) ϕ_1, ϕ_2 and (b) ρ_1, ρ_2, for a stationary AR(2) process.

Spectrum. From (3.2.9), the spectrum is

$$
\begin{aligned}
p(f) &= \frac{2\sigma_a^2}{|1 - \phi_1 e^{-i2\pi f} - \phi_2 e^{-i4\pi f}|^2} \\
&= \frac{2\sigma_a^2}{1 + \phi_1^2 + \phi_2^2 - 2\phi_1(1 - \phi_2)\cos(2\pi f) - 2\phi_2 \cos(4\pi f)} \qquad (3.2.29) \\
& \qquad 0 \leq f \leq \tfrac{1}{2}
\end{aligned}
$$

The spectrum also reflects the pseudo-periodic behavior that the series exhibits when the roots of the characteristic equation are complex. For illustration, Figure 3.4 shows 70 terms of a series generated by the second-order autoregressive model

$$\tilde{z}_t = 0.75\tilde{z}_{t-1} - 0.50\tilde{z}_{t-2} + a_t$$

obtained by setting $\phi_1 = 0.75$ and $\phi_2 = -0.50$ in (3.2.16). Figure 3.5 shows the corresponding theoretical autocorrelation function calculated from

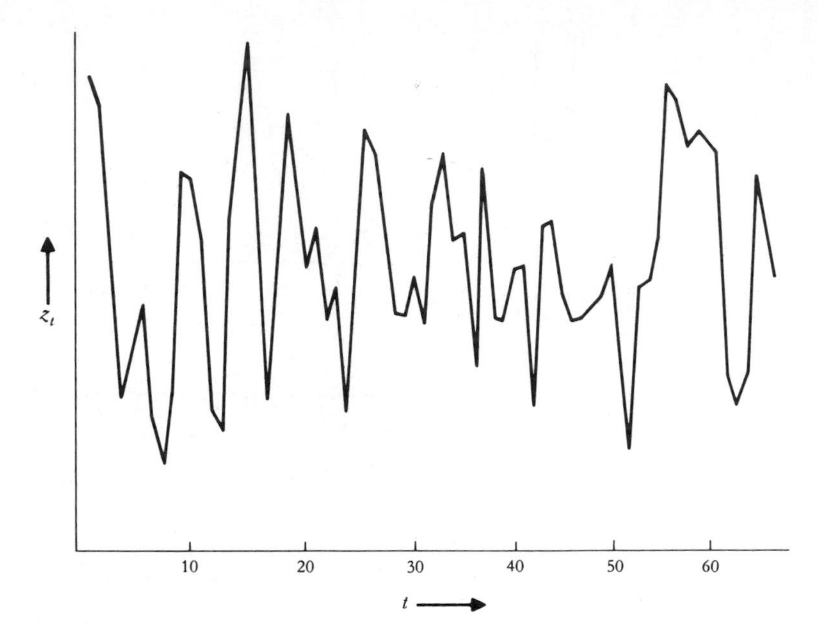

Figure 3.4 Time series generated from a second-order autoregressive process $\tilde{z}_t = 0.75\tilde{z}_{t-1} - 0.50\tilde{z}_{t-2} + a_t$.

(3.2.19), with starting values of $\rho_0 = 1$ and $\rho_1 = 0.75/[1 - (-0.5)] = 0.5$. The roots of the characteristic equation

$$1 - 0.75B + 0.5B^2 = 0$$

are complex, so that pseudo-periodic behavior which may be observed in the series is to be expected. We clearly see this behavior reflected in the theoretical autocorrelation function of Figure 3.5, the average apparent period being about 6.

The damping factor D and frequency f_0, from (3.2.22) and (3.2.23), are

$$D = \sqrt{0.50} = 0.71 \qquad f_0 = \frac{\cos^{-1}(0.5303)}{2\pi} = \frac{1}{6.2}$$

Thus the fundamental period of the autocorrelation function is 6.2.

Finally, the theoretical spectral density function in Figure 3.6, obtained from (3.2.29), shows that a large proportion of the variance of the series is accounted for by frequencies in the neighborhood of f_0.

3.2.5 Partial Autocorrelation Function

Initially, we may not know which order of autoregressive process to fit to an observed time series. This problem is analogous to deciding on the number of independent variables to be included in a multiple regression.

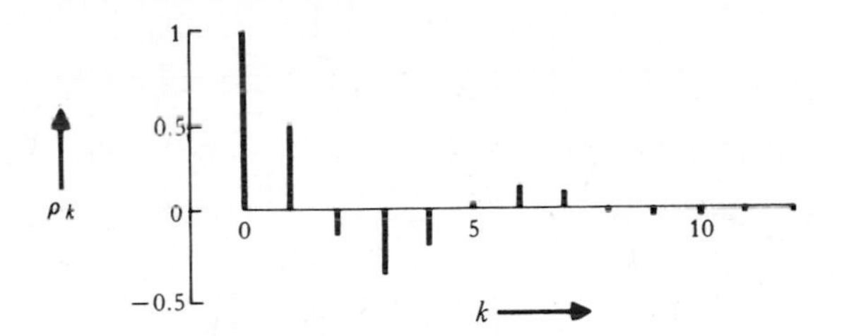

Figure 3.5 Theoretical autocorrelation function of second-order autoregressive process $\tilde{z}_t = 0.75\tilde{z}_{t-1} - 0.50\tilde{z}_{t-2} + a_t$.

The partial autocorrelation function is a device which exploits the fact that whereas an AR(p) process has an autocorrelation function which is infinite in extent, it can by its very nature be described in terms of p nonzero *functions* of the autocorrelations. Denote by ϕ_{kj}, the jth coefficient in an autoregressive representation of order k, so that ϕ_{kk} is the last coefficient. From (3.2.4), the ϕ_{kj} satisfy the set of equations

$$\rho_j = \phi_{k1}\rho_{j-1} + \cdots + \phi_{k(k-1)}\rho_{j-k+1} + \phi_{kk}\rho_{j-k} \qquad j = 1, 2, \ldots, k \tag{3.2.30}$$

leading to the Yule–Walker equations (3.2.6), which may be written

$$\begin{bmatrix} 1 & \rho_1 & \rho_2 & \cdots & \rho_{k-1} \\ \rho_1 & 1 & \rho_1 & \cdots & \rho_{k-2} \\ \vdots & \vdots & \vdots & \vdots\vdots\vdots & \vdots \\ \rho_{k-1} & \rho_{k-2} & \rho_{k-3} & \cdots & 1 \end{bmatrix} \begin{bmatrix} \phi_{k1} \\ \phi_{k2} \\ \vdots \\ \phi_{kk} \end{bmatrix} = \begin{bmatrix} \rho_1 \\ \rho_2 \\ \vdots \\ \rho_k \end{bmatrix} \tag{3.2.31}$$

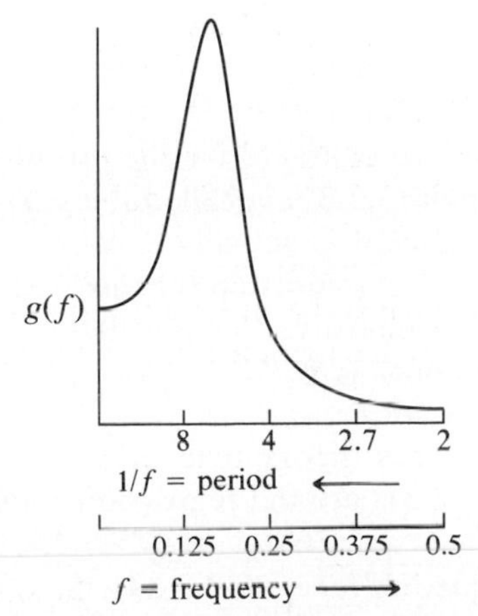

Figure 3.6 Theoretical spectral density function of second-order autoregressive process $\tilde{z}_t = 0.75\tilde{z}_{t-1} - 0.50\tilde{z}_{t-2} + a_t$.

or

$$\mathbf{P}_k \boldsymbol{\phi}_k = \boldsymbol{\rho}_k \tag{3.2.32}$$

Solving these equations for $k = 1, 2, 3, \ldots$, successively, we obtain

$$\phi_{11} = \rho_1$$

$$\phi_{22} = \frac{\begin{vmatrix} 1 & \rho_1 \\ \rho_1 & \rho_2 \end{vmatrix}}{\begin{vmatrix} 1 & \rho_1 \\ \rho_1 & 1 \end{vmatrix}} = \frac{\rho_2 - \rho_1^2}{1 - \rho_1^2}$$

$$\phi_{33} = \frac{\begin{vmatrix} 1 & \rho_1 & \rho_1 \\ \rho_1 & 1 & \rho_2 \\ \rho_2 & \rho_1 & \rho_3 \end{vmatrix}}{\begin{vmatrix} 1 & \rho_1 & \rho_2 \\ \rho_1 & 1 & \rho_1 \\ \rho_2 & \rho_1 & 1 \end{vmatrix}} \tag{3.2.33}$$

In general, for ϕ_{kk}, the determinant in the numerator has the same elements as that in the denominator, but with the last column replaced by $\boldsymbol{\rho}_k$. The quantity ϕ_{kk}, regarded as a function of the lag k, is called the *partial autocorrelation* function.

For an autoregressive process of order p, the partial autocorrelation function ϕ_{kk} will be nonzero for k less than or equal to p and *zero for k greater than p*. In other words, the partial autocorrelation function of a pth-order autoregressive process has a *cutoff* after lag p. For the second-order autoregressive process, partial autocorrelation functions ϕ_{kk} are shown in each of the four regions of Figure 3.2.

The partial autocorrelation function ϕ_{kk} given through (3.2.31) is defined for any stationary process as a function of the autocorrelations ρ_k of the process, but the distinctive feature that $\phi_{kk} = 0$ for all $k > p$ in an AR(p) process serves to characterize the pth-order AR process. The quantity ϕ_{kk} defined through (3.2.31) is called the *partial autocorrelation* of the process $\{z_t\}$ at lag k, since it is actually equal to the partial correlation between the variables z_t and z_{t-k} adjusted for the intermediate variables $z_{t-1}, z_{t-2}, \ldots, z_{t-k+1}$, and ϕ_{kk} measures the correlation between z_t and z_{t-k} after adjusting for the effects of $z_{t-1}, z_{t-2}, \ldots, z_{t-k+1}$ (or the correlation between z_t and z_{t-k} not accounted for by $z_{t-1}, z_{t-2}, \ldots, z_{t-k+1}$). Now it is easy to establish from least squares theory that the values $\phi_{k1}, \phi_{k2}, \ldots, \phi_{kk}$ which are the solution to (3.2.31) are the regression coefficients in the linear regression of z_t on $z_{t-1}, \ldots, z_{t-k}$; that is, they are the values of coefficients $b_1, \ldots, b_k$ which minimize $E[(z_t - b_0 - \Sigma_{i=1}^k b_i z_{t-i})^2]$. Hence, assuming for convenience that the process $\{z_t\}$ has mean zero, the best linear predictor (or linear regression), in the mean square error sense, of z_t based on $z_{t-1}, z_{t-2}, \ldots, z_{t-k+1}$ is

$$\hat{z}_t = \phi_{k-1,1}z_{t-1} + \phi_{k-1,2}z_{t-2} + \cdots + \phi_{k-1,k-1}z_{t-k+1}$$

whether the process is an AR or not. Similarly, the best linear predictor or regression of z_{t-k} based on the (future) values $z_{t-1}, z_{t-2}, \ldots, z_{t-k+1}$ is

$$\hat{z}_{t-k} = \phi_{k-1,1}z_{t-k+1} + \phi_{k-1,2}z_{t-k+2} + \cdots + \phi_{k-1,k-1}z_{t-1}$$

Then the lag k partial autocorrelation of $\{z_t\}$, ϕ_{kk}, can be defined as the correlation between the "adjusted values" of z_t and z_{t-k}, that is,

$$\phi_{kk} = \text{corr}[z_t - \hat{z}_t, z_{t-k} - \hat{z}_{t-k}] \tag{3.2.34}$$

As examples, we find that $\phi_{11} = \text{corr}[z_t, z_{t-1}] = \rho_1$, while

$$\begin{aligned}\phi_{22} &= \text{corr}[z_t - \rho_1 z_{t-1}, z_{t-2} - \rho_1 z_{t-1}] \\ &= \frac{\gamma_2 - 2\rho_1\gamma_1 + \rho_1^2\gamma_0}{[(\gamma_0 + \rho_1^2\gamma_0 - 2\rho_1\gamma_1)^2]^{1/2}} = \frac{\rho_2 - \rho_1^2}{1 - \rho_1^2}\end{aligned}$$

which is seen to agree with the solution to the equations (3.2.33) with $k = 2$ given earlier. Higher-order partial autocorrelations ϕ_{kk} as defined through (3.2.34) can similarly be shown to be the solution to the appropriate set of Yule–Walker equations (3.2.31) (see e.g., [63, p. 164]). Because ϕ_{kk} is the solution to the "normal" equations for the least squares regression of z_t on $z_{t-1}, \ldots, z_{t-k}$, it can also be viewed as a partial regression coefficient of z_t on z_{t-k}, adjusted for $z_{t-1}, \ldots, z_{t-k+1}$. That is,

$$\phi_{kk} = \frac{\text{cov}[z_t - \hat{z}_t, z_{t-k} - \hat{z}_{t-k}]}{\text{var}[z_{t-k} - \hat{z}_{t-k}]} = \text{corr}[z_t - \hat{z}_t, z_{t-k} - \hat{z}_{t-k}]$$

which equals $(\rho_k - \Sigma_{i=1}^{k-1} \phi_{k-1,i}\rho_{k-i})/(1 - \Sigma_{i=1}^{k-1} \phi_{k-1,i}\rho_i)$, since

$$\sigma_{k-1}^2 = \text{var}[z_t - \hat{z}_t] = \text{var}[z_{t-k} - \hat{z}_{t-k}] = \gamma_0\left(1 - \sum_{i=1}^{k-1} \phi_{k-1,i}\rho_i\right)$$

because the process $\{z_t\}$ has the same correlation structure in both forward and backward time directions.

3.2.6 Estimation of the Partial Autocorrelation Function

The partial autocorrelations may be estimated by fitting successively autoregressive models of orders 1, 2, 3, . . . by least squares, as will be described in Chapter 7, and picking out the estimates $\hat{\phi}_{11}, \hat{\phi}_{22}, \hat{\phi}_{33}, \ldots$ of the last coefficient fitted at each stage. Alternatively, if the values of the parameters are not too close to the nonstationary boundaries, approximate Yule–Walker estimates of the successive autoregressive models may be employed. The estimated partial autocorrelations can then be obtained by substituting estimates r_j for the theoretical autocorrelations in (3.2.30), to yield

$$r_j = \hat{\phi}_{k1} r_{j-1} + \hat{\phi}_{k2} r_{j-2} + \cdots + \hat{\phi}_{k(k-1)} r_{j-k+1} + \hat{\phi}_{kk} r_{j-k} \qquad j = 1, 2, \ldots, k \tag{3.2.35}$$

and solving the resultant equations for $k = 1, 2, \ldots$. A simple recursive method for doing this, due to Durbin [87], is given in Appendix A3.2. However, these estimates obtained from (3.2.35) become very sensitive to rounding errors and should not be used if the values of the parameters are close to the nonstationary boundaries.

3.2.7 Standard Errors of Partial Autocorrelation Estimates

It was shown by Quenouille [161] (see also [77] and [118]) that on the hypothesis that the process is autoregressive of order p, the estimated partial autocorrelations of order $p + 1$, and higher, are approximately independently and normally distributed with zero mean. Also, if n is the number of observations used in fitting,

$$\text{var}[\hat{\phi}_{kk}] \simeq \frac{1}{n} \qquad k \geqslant p + 1$$

Thus the standard error (S.E.) of the estimated partial autocorrelation $\hat{\phi}_{kk}$ is

$$\text{S.E.}[\hat{\phi}_{kk}] = \hat{\sigma}[\hat{\phi}_{kk}] \simeq \frac{1}{\sqrt{n}} \qquad k \geqslant p + 1 \tag{3.2.36}$$

Table 3.1 shows the first 15 estimated partial autocorrelations for the batch data of Table 2.1, obtained by direct fitting* of autoregressive models of increasing order. These partial autocorrelations are plotted in Figure 3.7

TABLE 3.1 Estimated Partial Autocorrelation Function for Batch Data of Table 2.1

k	$\hat{\phi}_{kk}$	k	$\hat{\phi}_{kk}$	k	$\hat{\phi}_{kk}$
1	−0.40	6	−0.15	11	0.18
2	0.19	7	0.05	12	−0.05
3	0.01	8	0.00	13	0.09
4	−0.07	9	−0.10	14	0.18
5	−0.07	10	0.05	15	0.01

* Approximate values agreeing to the first decimal may be obtained by solving equations (3.2.35).

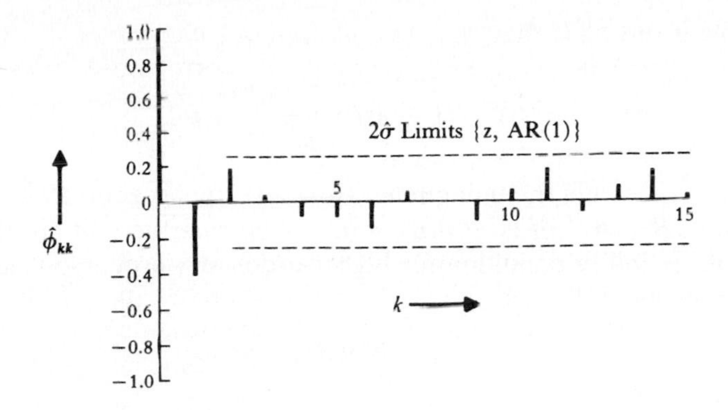

Figure 3.7 Estimated partial autocorrelation function for batch data of Figure 2.1, together with two standard error limits calculated on the assumption that the model is AR(1).

and may be compared with the autocorrelations of Figure 2.7. The behavior of these functions resembles that associated with an AR(1) process with a negative value of ϕ_1 (see Figure 3.1), or possibly an AR(2) process with a dominant negative root (see region 2 of Figure 3.2). Also shown in Figure 3.7 by dashed lines are the 2 S.E. limits for $\hat{\phi}_{22}, \hat{\phi}_{33}, \ldots$, calculated from (3.2.36) on the assumption that the process is AR(1). Since $\hat{\phi}_{22}$ is the second biggest partial autocorrelation of those considered, the possibility that the process is AR(2) ought to be kept in mind.

The use of the partial autocorrelation function for identifying models is discussed more fully in Chapter 6.

3.3 MOVING AVERAGE PROCESSES

3.3.1 Invertibility Conditions for Moving Average Processes

We now derive the conditions that the parameters $\theta_1, \theta_2, \ldots, \theta_q$ must satisfy to ensure the invertibility of the MA(q) process:

$$\begin{aligned} \tilde{z}_t &= a_t - \theta_1 a_{t-1} - \cdots - \theta_q a_{t-q} \\ &= (1 - \theta_1 B - \cdots - \theta_q B^q) a_t \qquad (3.3.1) \\ &= \theta(B) a_t \end{aligned}$$

We have already seen in Section 3.1.3 that the first-order moving average process

$$\tilde{z}_t = (1 - \theta_1 B) a_t$$

is invertible if $|\theta_1| < 1$; that is,

$$\pi(B) = (1 - \theta_1 B)^{-1} = \sum_{j=0}^{\infty} \theta_1^j B^j$$

converges on or within the unit circle. However, this is equivalent to saying that the root, $B = \theta_1^{-1}$ of $(1 - \theta_1 B) = 0$, lies *outside* the unit circle.

The invertibility condition for higher-order MA processes may be obtained by writing (3.3.1) as

$$a_t = \theta^{-1}(B)\tilde{z}_t$$

Hence, if

$$\theta(B) = \prod_{i=1}^{q} (1 - H_i B)$$

then, on expanding in partial fractions, we obtain

$$\pi(B) = \theta^{-1}(B) = \sum_{i=1}^{q} \left(\frac{M_i}{1 - H_i B}\right)$$

which converges, or equivalently, the weights $\pi_j = -\Sigma_{i=1}^q M_i H_i^j$ are absolutely summable, if $|H_i| < 1$, for $i = 1, 2, \ldots, q$. Since the roots of $\theta(B) = 0$ are H_i^{-1}, it follows that the invertibility condition for a MA(q) process is that the roots of the characteristic equation

$$\theta(B) = 1 - \theta_1 B - \theta_2 B^2 - \cdots - \theta_q B^q = 0 \qquad (3.3.2)$$

lie *outside* the unit circle. From the relation $\theta(B)\pi(B) = 1$ it follows that the weights π_j satisfy the difference equation

$$\pi_j = \theta_1 \pi_{j-1} + \theta_2 \pi_{j-2} + \cdots + \theta_q \pi_{j-q} \qquad j > 0$$

with the convention that $\pi_0 = -1$ and $\pi_j = 0$ for $j < 0$, from which the weights π_j can easily be computed recursively in terms of the θ_i.

Note that since the series

$$\psi(B) = \theta(B) = 1 - \theta_1 B - \theta_2 B^2 - \cdots - \theta_q B^q$$

is finite, no restrictions are needed on the parameters of the moving average process to ensure stationarity.

3.3.2 Autocorrelation Function and Spectrum of Moving Average Processes

Autocorrelation function. Using (3.3.1), the autocovariance function of a MA(q) process is

$$\gamma_k = E[(a_t - \theta_1 a_{t-1} - \cdots - \theta_q a_{t-q})(a_{t-k} - \theta_1 a_{t-k-1} - \cdots - \theta_q a_{t-k-q})]$$

Hence the variance of the process is

$$\gamma_0 = (1 + \theta_1^2 + \theta_2^2 + \cdots + \theta_q^2)\sigma_a^2 \tag{3.3.3}$$

and

$$\gamma_k = \begin{cases} (-\theta_k + \theta_1\theta_{k+1} + \theta_2\theta_{k+2} + \cdots + \theta_{q-k}\theta_q)\sigma_a^2 & k = 1, 2, \ldots, q \\ 0 & k > q \end{cases}$$

Thus the autocorrelation function is

$$\rho_k = \begin{cases} \dfrac{-\theta_k + \theta_1\theta_{k+1} + \cdots + \theta_{q-k}\theta_q}{1 + \theta_1^2 + \cdots + \theta_q^2} & k = 1, 2, \ldots, q \\ 0 & k > q \end{cases} \tag{3.3.4}$$

We see that the autocorrelation function of a MA(q) process is zero, beyond the order q of the process. In other words, the autocorrelation function of a moving average process has a *cutoff* after lag q.

Moving average parameters in terms of the autocorrelations. If $\rho_1, \rho_2, \ldots, \rho_q$ are known, the q equations (3.3.4) may be solved for the parameters $\theta_1, \theta_2, \ldots, \theta_q$. However, unlike the Yule–Walker equations (3.2.6) for an autoregressive process, which are linear, the equations (3.3.4) are nonlinear. Hence, except in the simple case where $q = 1$, which is discussed shortly, these equations have to be solved iteratively as described in Appendix A6.2. By substituting estimates r_k for ρ_k in (3.3.4) and solving the resulting equations, initial estimates of the moving average parameters may be obtained. Unlike the corresponding autoregressive estimates, obtained by a corresponding substitution in the Yule–Walker equations, the resulting estimates may not have high statistical efficiency. However, they can provide useful rough estimates at the identification stage discussed in Chapter 6. Furthermore, they provide useful starting values for an iterative procedure, discussed in Chapter 7, which converges to the efficient maximum likelihood estimates.

Spectrum. For the MA(q) process,

$$\psi(B) = \theta(B) = 1 - \theta_1 B - \theta_2 B^2 - \cdots - \theta_q B^q$$

Therefore, using (3.1.12), the spectrum of a MA(q) process is

$$p(f) = 2\sigma_a^2|1 - \theta_1 e^{-i2\pi f} - \theta_2 e^{-i4\pi f} - \cdots - \theta_q e^{-i2\pi q f}|^2 \qquad 0 \leq f \leq \tfrac{1}{2} \tag{3.3.5}$$

We now discuss in greater detail the moving average processes of first and second order, which are of considerable practical importance.

3.3.3 First-Order Moving Average Process

We have already met this process in the form

$$\begin{aligned}\tilde{z}_t &= a_t - \theta_1 a_{t-1} \\ &= (1 - \theta_1 B)a_t\end{aligned}$$

and it has been shown in Section 3.1.3 that θ_1 must lie in the range $-1 < \theta_1 < 1$ for the process to be invertible. However, the process is of course stationary for all values of θ_1.

Autocorrelation function. Using (3.3.3), the variance of the process is

$$\gamma_0 = (1 + \theta_1^2)\sigma_a^2$$

and using (3.3.4), the autocorrelation function is

$$\rho_k = \begin{cases} \dfrac{-\theta_1}{1 + \theta_1^2} & k = 1 \\ 0 & k \geqslant 2 \end{cases} \tag{3.3.6}$$

from which it is noted that ρ_1 for an MA(1) must satisfy $|\rho_1| \leq \frac{1}{2}$. From (3.3.6), with $k = 1$, we find that

$$\theta_1^2 + \frac{\theta_1}{\rho_1} + 1 = 0 \tag{3.3.7}$$

with roots for θ_1 equal to $\theta_1 = (-1 \pm \sqrt{1 - 4\rho_1^2})/(2\rho_1)$. Since the product of the roots is unity, we see that if θ_1 is a solution, so is θ_1^{-1}. Furthermore, if θ_1 satisfies the invertibility condition $|\theta_1| < 1$, the other root θ_1^{-1} will be greater than unity and will not satisfy the condition. For example, if $\rho_1 = -0.4$, (3.3.7) has two solutions, $\theta_1 = 0.5$ and $\theta_1 = 2.0$. However, only the solution $\theta_1 = 0.5$ corresponds to an invertible process. Table A in the "Collection of Tables and Charts" in Part Five allows such solutions for θ_1 to be read off over the entire range of possible values $-0.5 < \rho_1 < 0.5$.

Spectrum. Using (3.3.5), the spectrum is

$$\begin{aligned}p(f) &= 2\sigma_a^2|1 - \theta_1 e^{-i2\pi f}|^2 \\ &= 2\sigma_a^2[1 + \theta_1^2 - 2\theta_1 \cos(2\pi f)] \qquad 0 \leqslant f \leqslant \tfrac{1}{2}\end{aligned} \tag{3.3.8}$$

In general when θ_1 is negative, ρ_1 is positive, and the spectrum is dominated by low frequencies. Conversely, when θ_1 is positive, ρ_1 is negative, and the spectrum is dominated by high frequencies.

Partial autocorrelation function. Using (3.2.31) with $\rho_1 = -\theta_1/(1 + \theta_1^2)$ and $\rho_k = 0$, for $k > 1$, we obtain after some algebraic manipulation

$$\phi_{kk} = \frac{-\theta_1^k(1 - \theta_1^2)}{(1 - \theta_1^{2(k+1)})}$$

Thus $|\phi_{kk}| < \theta_1^k$, and the partial autocorrelation function is dominated by a damped exponential. If ρ_1 is positive, so that θ_1 is negative, the partial autocorrelations alternate in sign. If, however, ρ_1 is negative, so that θ_1 is positive, the partial autocorrelations are negative. From (3.1.15) it has been seen that the weights π_j for the MA(1) process are $\pi_j = -\theta_1^j$, and hence since these are coefficients in the infinite autoregressive form of the process it is sensible that the partial autocorrelation function ϕ_{kk} for the MA(1) essentially mimics the exponential decay feature of the weights π_j.

We now note a duality between the AR(1) and MA(1) processes. Thus, whereas the autocorrelation function of a MA(1) process has a cutoff after lag 1, the autocorrelation function of an AR(1) process tails off exponentially. Conversely, whereas the partial autocorrelation function of a MA(1) process tails off and is dominated by a damped exponential, the partial autocorrelation function of an AR(1) process has a cutoff after lag 1. It turns out that a corresponding approximate duality of this kind occurs in general in the autocorrelation and partial autocorrelation functions between AR and MA processes.

3.3.4 Second-Order Moving Average Process

Invertibility conditions. The second-order moving average process is defined by

$$\tilde{z}_t = a_t - \theta_1 a_{t-1} - \theta_2 a_{t-2}$$

and is stationary for all values of θ_1 and θ_2. However, it is invertible only if the roots of the characteristic equation

$$1 - \theta_1 B - \theta_2 B^2 = 0 \tag{3.3.9}$$

lie outside the unit circle, that is,

$$\begin{aligned} \theta_2 + \theta_1 &< 1 \\ \theta_2 - \theta_1 &< 1 \\ -1 < \theta_2 &< 1 \end{aligned} \tag{3.3.10}$$

These are parallel to the conditions (3.2.18) required for the *stationarity* of an AR(2) process.

Autocorrelation function. Using (3.3.3), the variance of the process is

$$\gamma_0 = \sigma_a^2(1 + \theta_1^2 + \theta_2^2)$$

and using (3.3.4), the autocorrelation function is

$$\rho_1 = \frac{-\theta_1(1 - \theta_2)}{1 + \theta_1^2 + \theta_2^2}$$

$$\rho_2 = \frac{-\theta_2}{1 + \theta_1^2 + \theta_2^2} \tag{3.3.11}$$

$$\rho_k = 0 \qquad k \geqslant 3$$

Thus the autocorrelation function has a cutoff after lag 2.

It follows from (3.3.10) and (3.3.11) that the first two autocorrelations of an invertible MA(2) process must lie within the area bounded by segments of the curves

$$\rho_2 + \rho_1 = -0.5$$

$$\rho_2 - \rho_1 = -0.5 \tag{3.3.12}$$

$$\rho_1^2 = 4\rho_2(1 - 2\rho_2)$$

The invertibility region (3.3.10) for the parameters is shown in Figure 3.8(a) and the corresponding region (3.3.12) for the autocorrelations in Figure 3.8(b). The latter shows whether a given pair of autocorrelations ρ_1 and ρ_2 is consistent with the hypothesis that the model is a MA(2) process. If they are consistent, the values of the parameters θ_1 and θ_2 can be obtained by solving the nonlinear equations (3.3.11). To facilitate this calculation, Chart C in the "Collection of Tables and Charts" in Part Five has been prepared so that the values of θ_1 and θ_2 can be read off directly, given ρ_1 and ρ_2.

Spectrum. Using (3.3.5), the spectrum is

$$\begin{aligned} p(f) &= 2\sigma_a^2|1 - \theta_1 e^{-i2\pi f} - \theta_2 e^{-i4\pi f}|^2 \\ &= 2\sigma_a^2[1 + \theta_1^2 + \theta_2^2 - 2\theta_1(1 - \theta_2)\cos(2\pi f) - 2\theta_2 \cos(4\pi f)] \\ & \qquad 0 \leqslant f \leqslant \tfrac{1}{2} \end{aligned} \tag{3.3.13}$$

and is the reciprocal of the spectrum of a second-order autoregressive process (3.2.29), apart from the constant $2\sigma_a^2$.

Partial autocorrelation function. The exact expression for the partial autocorrelation function of an MA(2) process is complicated, but it is dominated by the sum of two exponentials if the roots of the characteristic

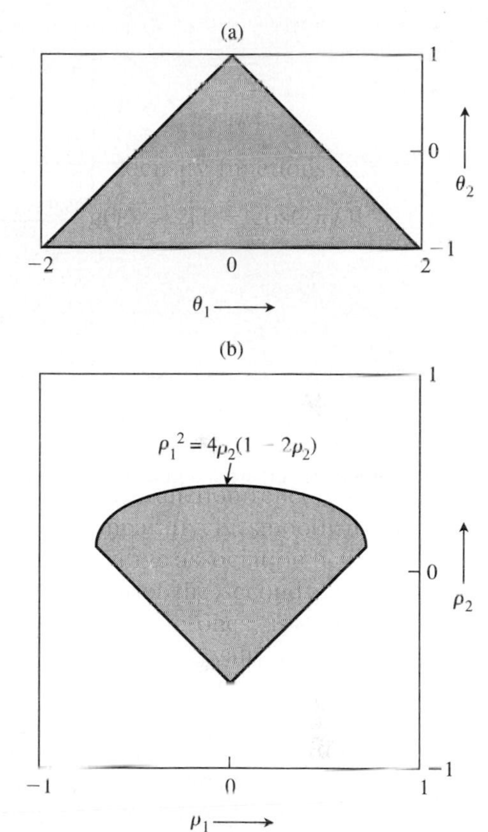

Figure 3.8 Admissible regions for (a) θ_1, θ_2 and (b) ρ_1, ρ_2 for an invertible MA(2) process.

equation (3.3.9) are real, and by a damped sine wave if the roots of (3.3.9) are complex. Thus it behaves like the autocorrelation function of an AR(2) process. The autocorrelation functions (left-hand curves) and partial autocorrelation functions (right-hand curves) for various values of the parameters within the invertible region are shown in Figure 3.9, which is taken from Stralkowski [183]. Comparison of Figure 3.9 with Figure 3.2, which shows the corresponding autocorrelations and partial autocorrelations for an AR(2) process, illustrates the duality between the MA(2) and the AR(2) processes.

3.3.5 Duality between Autoregressive and Moving Average Processes

The results of the previous sections have shown further aspects of the *duality* between autoregressive and finite moving average processes. As illustrated in Table 3.3 at the end of this chapter, this duality has the following consequences:

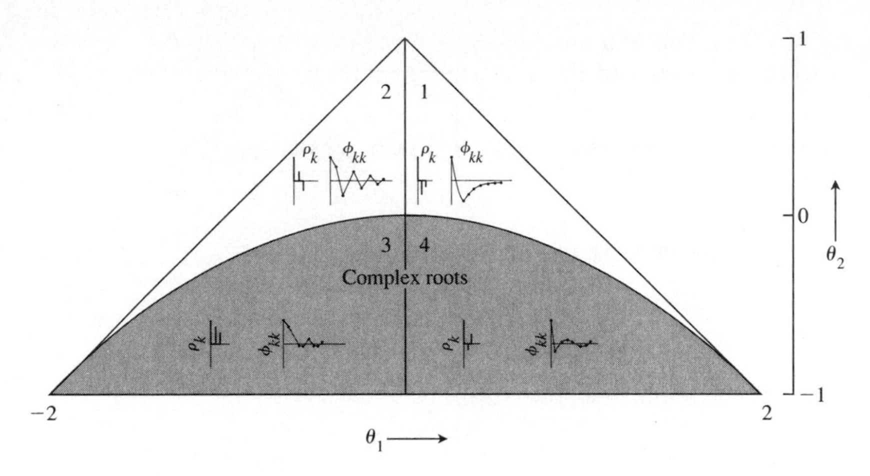

Figure 3.9 Autocorrelation and partial autocorrelation functions ρ_k and ϕ_{kk} for various MA(2) models. (From [183].)

1. In a stationary autoregressive process of order p, a_t can be represented as a *finite* weighted sum of previous $\tilde{z}$'s, or $\tilde{z}_t$ as an infinite weighted sum

$$\tilde{z}_t = \phi^{-1}(B)a_t$$

of previous a's. Also, in an invertible moving average process of order q, $\tilde{z}_t$ can be represented as a finite weighted sum of previous a's, or a_t as an infinite weighted sum

$$\theta^{-1}(B)\tilde{z}_t = a_t$$

of previous $\tilde{z}$'s.

2. The finite MA process has an autocorrelation function that is zero beyond a certain point, but since it is equivalent to an infinite AR process, its partial autocorrelation function is infinite in extent and is dominated by damped exponentials and/or damped sine waves. Conversely, the AR process has a partial autocorrelation function that is zero beyond a certain point, but its autocorrelation function is infinite in extent and consists of a mixture of damped exponentials and/or damped sine waves.
3. For an autoregressive process of finite order p, the parameters are not required to satisfy any conditions to ensure invertibility. However, for stationarity, the roots of $\phi(B) = 0$ must lie outside the unit circle. Conversely, the parameters of the MA process are not required to satisfy any conditions to ensure stationarity. However, for invertibility of the MA process, the roots of $\theta(B) = 0$ must lie outside the unit circle.

4. The spectrum of a moving average process has an inverse relationship to the spectrum of the corresponding autoregressive process.

3.4 MIXED AUTOREGRESSIVE–MOVING AVERAGE PROCESSES

3.4.1 Stationarity and Invertibility Properties

We have noted in Section 3.1.4 that to achieve parsimony it may be necessary to include both autoregressive and moving average terms. Thus we may need to employ the mixed autoregressive–moving average (ARMA) model

$$\tilde{z}_t = \phi_1 \tilde{z}_{t-1} + \cdots + \phi_p \tilde{z}_{t-p} + a_t - \theta_1 a_{t-1} - \cdots - \theta_q a_{t-q} \tag{3.4.1}$$

that is,

$$(1 - \phi_1 B - \phi_2 B^2 - \cdots - \phi_p B^p)\tilde{z}_t = (1 - \theta_1 B - \theta_2 B^2 - \cdots - \theta_q B^q)a_t$$

or

$$\phi(B)\tilde{z}_t = \theta(B)a_t$$

where $\phi(B)$ and $\theta(B)$ are polynomials of degree p and q in B.

We subsequently refer to this process as an ARMA(p, q) process. It may be thought of in two ways:

1. As a pth-order autoregressive process

   $$\phi(B)\tilde{z}_t = e_t$$

   with e_t following the qth-order moving average process

   $$e_t = \theta(B)a_t$$

2. As a qth-order moving average process

   $$\tilde{z}_t = \theta(B)b_t$$

   with b_t following the pth-order autoregressive process

   $$\phi(B)b_t = a_t$$

   so that

   $$\phi(B)\tilde{z}_t = \theta(B)\phi(B)b_t = \theta(B)a_t$$

It is obvious that moving average terms on the right of (3.4.1) will not affect the argument of Section 3.2.1, which establishes conditions for stationarity of an autoregressive process. Thus $\phi(B)\tilde{z}_t = \theta(B)a_t$ will define a stationary process provided that the characteristic equation $\phi(B) = 0$ has all its roots lying outside the unit circle. Similarly, the roots of $\theta(B) = 0$ must lie outside the unit circle if the process is to be invertible.

Thus the stationary and invertible ARMA(p, q) process (3.4.1) has both the infinite moving average representation

$$\tilde{z}_t = \psi(B)a_t = \sum_{j=0}^{\infty} \psi_j a_{t-j}$$

where $\psi(B) = \phi^{-1}(B)\theta(B)$, and the infinite autoregressive representation

$$\pi(B)\tilde{z}_t = \tilde{z}_t - \sum_{j=1}^{\infty} \pi_j \tilde{z}_{t-j} = a_t$$

where $\pi(B) = \theta^{-1}(B)\phi(B)$, with both the ψ_j weights and the π_j weights being absolutely summable. The weights ψ_j are determined from the relation $\phi(B)\psi(B) = \theta(B)$ to satisfy

$$\psi_j = \phi_1\psi_{j-1} + \phi_2\psi_{j-2} + \cdots + \phi_p\psi_{j-p} - \theta_j \qquad j > 0$$

with $\psi_0 = 1$, $\psi_j = 0$ for $j < 0$, and $\theta_j = 0$ for $j > q$, while from the relation $\theta(B)\pi(B) = \phi(B)$ the π_j are determined to satisfy

$$\pi_j = \theta_1\pi_{j-1} + \theta_2\pi_{j-2} + \cdots + \theta_q\pi_{j-q} + \phi_j \qquad j > 0$$

with $\pi_0 = -1$, $\pi_j = 0$ for $j < 0$ and $\phi_j = 0$ for $j > p$. From these relations the ψ_j and π_j weights can readily be computed recursively in terms of the ϕ_i and θ_i coefficients.

3.4.2 Autocorrelation Function and Spectrum of Mixed Processes

Autocorrelation function. The autocorrelation function of the mixed process may be derived by a method similar to that used for autoregressive processes in Section 3.2.2. On multiplying throughout in (3.4.1) by $\tilde{z}_{t-k}$ and taking expectations, we see that the autocovariance function satisfies the difference equation

$$\gamma_k = \phi_1\gamma_{k-1} + \cdots + \phi_p\gamma_{k-p} + \gamma_{za}(k) - \theta_1\gamma_{za}(k-1) - \cdots - \theta_q\gamma_{za}(k-q)$$

where $\gamma_{za}(k)$ is the cross covariance function between z and a and is defined by $\gamma_{za}(k) = E[\tilde{z}_{t-k}a_t]$. Since z_{t-k} depends only on shocks that have occurred up to time $t - k$ through the infinite moving average representation $\tilde{z}_{t-k} = \psi(B)a_{t-k} = \sum_{j=0}^{\infty} \psi_j a_{t-k-j}$, it follows that

$$\gamma_{za}(k) = \begin{cases} 0 & k > 0 \\ \psi_{-k}\sigma_a^2 & k \leqslant 0 \end{cases}$$

Hence the preceding equation for γ_k may be expressed as

$$\gamma_k = \phi_1\gamma_{k-1} + \cdots + \phi_p\gamma_{k-p} - \sigma_a^2(\theta_k\psi_0 + \theta_{k+1}\psi_1 + \cdots + \theta_q\psi_{q-k}) \qquad (3.4.2)$$

with the convention that $\theta_0 = -1$.

We see that (3.4.2) implies

$$\gamma_k = \phi_1\gamma_{k-1} + \phi_2\gamma_{k-2} + \cdots + \phi_p\gamma_{k-p} \qquad k \geqslant q + 1$$

and hence

$$\rho_k = \phi_1\rho_{k-1} + \phi_2\rho_{k-2} + \cdots + \phi_p\rho_{k-p} \qquad k \geqslant q + 1 \qquad (3.4.3)$$

or

$$\phi(B)\rho_k = 0 \qquad k \geqslant q + 1$$

Thus, for the ARMA(p, q) process, there will be q autocorrelations ρ_q, $\rho_{q-1}, \ldots, \rho_1$ whose values depend directly, through (3.4.2), on the choice of the q moving average parameters $\boldsymbol{\theta}$, as well as on the p autoregressive parameters $\boldsymbol{\phi}$. Also, the p values $\rho_q, \rho_{q-1}, \ldots, \rho_{q-p+1}$ provide the necessary starting values for the difference equation $\phi(B)\rho_k = 0$, where $k \geqslant q + 1$, which then entirely determines the autocorrelations at higher lags. If $q - p < 0$, the whole autocorrelation function ρ_j, for $j = 0, 1, 2, \ldots$, will consist of a mixture of damped exponentials and/or damped sine waves, whose nature is dictated by the polynomial $\phi(B)$ and the starting values. If, however, $q - p \geqslant 0$ there will be $q - p + 1$ initial values $\rho_0, \rho_1, \ldots, \rho_{q-p}$, which do not follow this general pattern. These facts are useful in identifying mixed series.

Variance. When $k = 0$, we have

$$\gamma_0 = \phi_1\gamma_1 + \cdots + \phi_p\gamma_p + \sigma_a^2(1 - \theta_1\psi_1 - \cdots - \theta_q\psi_q) \qquad (3.4.4)$$

which has to be solved along with the p equations (3.4.2) for $k = 1, 2, \ldots, p$ to obtain $\gamma_0, \gamma_1, \ldots, \gamma_p$.

Spectrum. Using (3.1.12), the spectrum of a mixed process is

$$\begin{aligned} p(f) &= 2\sigma_a^2 \frac{|\theta(e^{-i2\pi f})|^2}{|\phi(e^{-i2\pi f})|^2} \\ &= 2\sigma_a^2 \frac{|1 - \theta_1 e^{-i2\pi f} - \cdots - \theta_q e^{-i2\pi q f}|^2}{|1 - \phi_1 e^{-i2\pi f} - \cdots - \phi_p e^{-i2\pi p f}|^2} \qquad 0 \leqslant f \leqslant \tfrac{1}{2} \end{aligned} \qquad (3.4.5)$$

Partial autocorrelation function. The process (3.4.1) may be written

$$a_t = \theta^{-1}(B)\phi(B)\tilde{z}_t$$

and $\theta^{-1}(B)$ is an infinite series in B. Hence the partial autocorrelation function of a mixed process is infinite in extent. It behaves eventually like the partial autocorrelation function of a pure moving average process, being dominated by a mixture of damped exponentials and/or damped sine waves, depending on the order of the moving average and the values of the parameters it contains.

3.4.3 First-Order Autoregressive–First-Order Moving Average Process

A mixed process of considerable practical importance is the first-order autoregressive–first-order moving average ARMA(1, 1) process

$$\tilde{z}_t - \phi_1 \tilde{z}_{t-1} = a_t - \theta_1 a_{t-1} \tag{3.4.6}$$

that is,

$$(1 - \phi_1 B)\tilde{z}_t = (1 - \theta_1 B)a_t$$

We now derive some of its more important properties.

Stationarity and invertibility conditions. First, we note that the process is stationary if $-1 < \phi_1 < 1$, and invertible if $-1 < \theta_1 < 1$. Hence the admissible parameter space is the square shown in Figure 3.10(a). In addition, from the relations $\psi_1 = \phi_1\psi_0 - \theta_1 = \phi_1 - \theta_1$ and $\psi_j = \phi_1\psi_{j-1}$ for $j > 1$, we find that the ψ_j weights are given by $\psi_j = (\phi_1 - \theta_1)\phi_1^{j-1}, j \geqslant 1$, and similarly it is easily seen that $\pi_j = (\phi_1 - \theta_1)\theta_1^{j-1}, j \geqslant 1$, for the stationary and invertible ARMA(1, 1) process.

Autocorrelation function. From (3.4.2) and (3.4.4) we obtain

$$\gamma_0 = \phi_1\gamma_1 + \sigma_a^2(1 - \theta_1\psi_1)$$

$$\gamma_1 = \phi_1\gamma_0 - \theta_1\sigma_a^2$$

$$\gamma_k = \phi_1\gamma_{k-1} \qquad k \geqslant 2$$

with $\psi_1 = \phi_1 - \theta_1$. Hence, solving the first two equations for γ_0 and γ_1, the autocovariance function of the process is

$$\gamma_0 = \frac{1 + \theta_1^2 - 2\phi_1\theta_1}{1 - \phi_1^2}\sigma_a^2$$

$$\gamma_1 = \frac{(1 - \phi_1\theta_1)(\phi_1 - \theta_1)}{1 - \phi_1^2}\sigma_a^2 \tag{3.4.7}$$

$$\gamma_k = \phi_1\gamma_{k-1} \qquad k \geqslant 2$$

From the last equations we have that $\rho_k = \phi_1\rho_{k-1}$, $k \geqslant 2$, so $\rho_k = \rho_1\phi_1^{k-1}$, $k > 1$. Thus the autocorrelation function decays exponentially from the

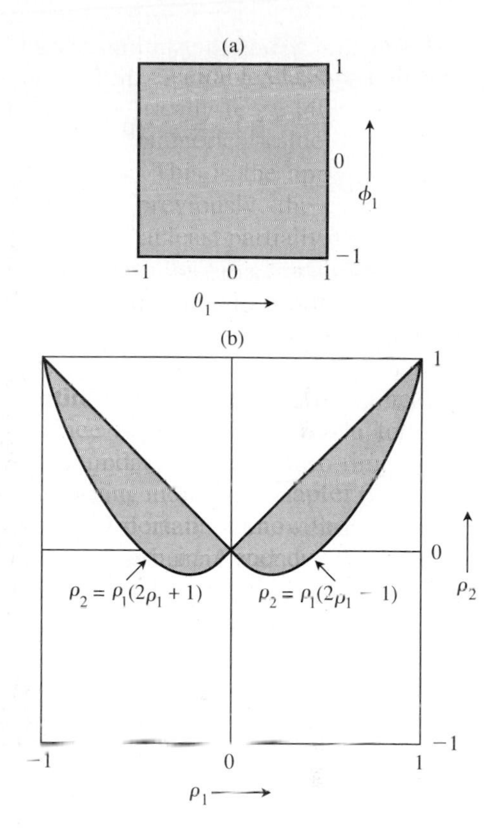

Figure 3.10 Admissible regions for (a) θ_1, ϕ_1 and (b) ρ_1, ρ_2, for a stationary and invertible ARMA(1, 1) process.

starting value ρ_1, which depends on θ_1 as well as on ϕ_1.* As shown in Figure 3.11, this exponential decay is smooth if ϕ_1 is positive and alternates if ϕ_1 is negative. Furthermore, the sign of ρ_1 is determined by the sign of $(\phi_1 - \theta_1)$ and dictates from which side of zero the exponential decay takes place.

From (3.4.7), the first two autocorrelations may be expressed in terms of the parameters of the process, as follows:

$$\rho_1 = \frac{(1 - \phi_1\theta_1)(\phi_1 - \theta_1)}{1 + \theta_1^2 - 2\phi_1\theta_1}$$

$$\rho_2 = \phi_1\rho_1 \tag{3.4.8}$$

Chart D in the "Collection of Tables and Charts" is so constructed that the solution of the equations (3.4.8) for ϕ_1 and θ_1 can be read off, knowing ρ_1 and ρ_2. By substituting estimates r_1 and r_2 for ρ_1 and ρ_2, initial estimates for the parameters ϕ_1 and θ_1 can be obtained.

* By contrast the autocorrelation function for the AR(1) process decays exponentially from the starting value $\rho_0 = 1$.

Using (3.4.8) and the stationarity and invertibility conditions, it may be shown that ρ_1 and ρ_2 must lie in the region

$$
\begin{aligned}
&|\rho_2| < |\rho_1| \\
&\rho_2 > \rho_1(2\rho_1 + 1) \qquad \rho_1 < 0 \\
&\rho_2 > \rho_1(2\rho_1 - 1) \qquad \rho_1 > 0
\end{aligned} \tag{3.4.9}
$$

Figure 3.10(b) shows the admissible space for ρ_1 and ρ_2; that is, it indicates which combinations of ρ_1 and ρ_2 are possible for the mixed (1, 1) stationary, invertible process.

Partial autocorrelation function. The partial autocorrelation function of the mixed ARMA(1, 1) process (3.4.6) consists of a single initial value $\phi_{11} = \rho_1$. Thereafter it behaves like the partial autocorrelation function of a pure MA(1) process and is dominated by a damped exponential. Thus, as shown in Figure 3.11, when θ_1 is positive it is dominated by a smoothly damped exponential which decays from a value of ρ_1, with sign determined by the sign of $(\phi_1 - \theta_1)$. Similarly, when θ_1 is negative, it is dominated by an exponential which oscillates as it decays from a value of ρ_1, with sign determined by the sign of $(\phi_1 - \theta_1)$.

As a numerical example to illustrate the behavior of the partial autocorrelation function and other features of an ARMA(1, 1) process, consider the

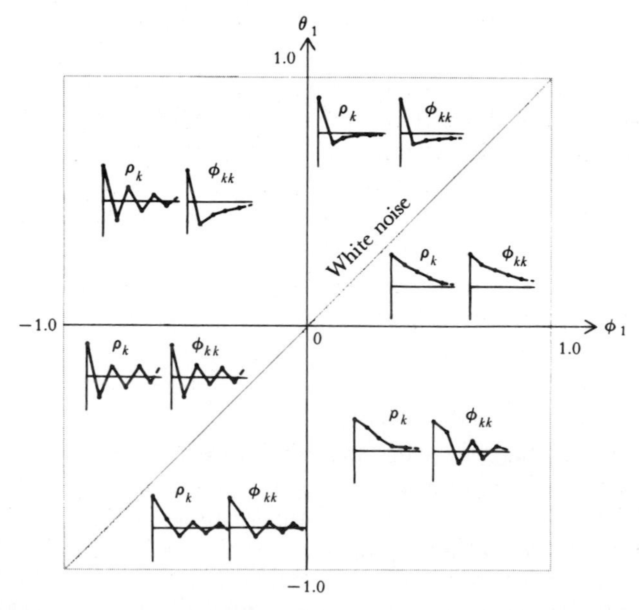

Figure 3.11 Autocorrelation and partial autocorrelation functions ρ_k and ϕ_{kk} for various ARMA(1, 1) models.

TABLE 3.2 Autoregressive Coefficients ϕ_{kj} and Values of $\sigma_k^2 = \gamma_0(1 - \sum_{j=1}^{k} \phi_{kj}\rho_j)$ for kth-Order Autoregressive "Approximations" for the ARMA(1, 1) Model Example, $(1 - 0.8B)\tilde{z}_t = (1 + 0.6B)a_t$, $\sigma_a^2 = 1$[a]

k	σ_k^2	"Autoregressive Coefficients" ϕ_{kj}, $j = 1, \ldots, k$						
1	1.304	0.893						
2	1.084	1.260	−0.411					
3	1.028	1.354	−0.697	0.227				
4	1.010	1.384	−0.790	0.407	−0.133			
5	1.003	1.394	−0.822	0.469	−0.242	0.079		
.	.	.	.	.	.	.		
.	.	.	.	.	.	.		
.	.	.	.	.	.	.		
∞	1.000	1.400	−0.840	0.504	−0.302	0.181	−0.109	. . .

[a] Diagonal elements are values ϕ_{kk} of the partial autocorrelation function.

model $(1 - 0.8B)\tilde{z}_t = (1 + 0.6B)a_t$, that is, $\phi_1 = 0.8$ and $\theta_1 = -0.6$, with $\sigma_a^2 = 1.0$. From (3.4.7) and (3.4.8), we find that the variance of the process is $\gamma_0 = 6.444$ and $\rho_1 = 0.893$. Also, the autocorrelation function satisfies $\rho_j = 0.8\rho_{j-1}$, $j \geqslant 2$, so $\rho_j = 0.893(0.8)^{j-1}$, $j \geqslant 2$. The autoregressive coefficients $\phi_{k1}, \ldots, \phi_{kk}$, obtained by solving the Yule–Walker equations (3.2.31) for this process, are presented in Table 3.2 for AR orders $k = 1$ to $k = 5$, together with the residual variances $\sigma_k^2 = \gamma_0(1 - \sum_{i=1}^{k} \phi_{ki}\rho_i)$, which are the variances of $\tilde{z}_t - \sum_{i=1}^{k} \phi_{ki}\tilde{z}_{t-i}$. Notice that the diagonal elements in Table 3.2, which are the values ϕ_{kk} of the partial autocorrelation function of this process, exhibit an exponential decaying pattern that oscillates in sign due to the negative value of θ_1. Also, the last row in the table contains the π_j weights for this process, obtained as $\pi_1 = \phi_1 - \theta_1 = 1.4$ and $\pi_j = \theta_1\pi_{j-1} = -0.6\pi_{j-1}$, $j > 1$, and the values of the coefficients ϕ_{kj} in the AR "approximations" of order k of the model approach the values π_j as k increases. Finally, note that the values σ_k^2 are very close to σ_a^2 for $k \geqslant 4$, suggesting that an autoregressive model representation of order 4 or greater would provide a good approximation to the ARMA(1, 1) model, but at the expense of additional parameters.

3.4.4 Summary

Figure 3.12 brings together the admissible regions for the parameters and for the correlations ρ_1, ρ_2 for AR(2), MA(2), and ARMA(1, 1) processes which are restricted to being both stationary and invertible. Table 3.3 summarizes the properties of mixed autoregressive–moving average processes, and brings together all the important results for autoregressive, moving average, and mixed processes, which will be needed in Chapter 6 to identify models for observed time series. In Chapter 4 we extend the mixed ARMA

TABLE 3.3 Summary of Properties of Autoregressive, Moving Average, and Mixed ARMA Processes

	Autoregressive Processes	Moving Average Processes	Mixed Processes
Model in terms of previous $\tilde{z}$'s	$\phi(B)\tilde{z}_t = a_t$	$\theta^{-1}(B)\tilde{z}_t = a_t$	$\theta^{-1}(B)\phi(B)\tilde{z}_t = a_t$
Model in terms of previous a's	$\tilde{z}_t = \phi^{-1}(B)a_t$	$\tilde{z}_t = \theta(B)a_t$	$\tilde{z}_t = \phi^{-1}(B)\theta(B)a_t$
π weights	Finite series	Infinite series	Infinite series
ψ weights	Infinite series	Finite series	Infinite series
Stationarity condition	Roots of $\phi(B) = 0$ lie outside the unit circle	Always stationary	Roots of $\phi(B) = 0$ lie outside the unit circle
Invertibility condition	Always invertible	Roots of $\theta(B) = 0$ lie outside unit circle	Roots of $\theta(B) = 0$ lie outside unit circle
Autocorrelation function	Infinite (damped exponentials and/or damped sine waves)	Finite	Infinite (damped exponentials and/or damped sine waves after first $q - p$ lags)
	Tails off	Cuts off after lag q	Tails off
Partial autocorrelation function	Finite	Infinite (dominated by damped exponentials and/or sine waves)	Infinite (dominated by damped exponentials and/or sine waves after first $p - q$ lags)
	Cuts off after lag p	Tails off	Tails off

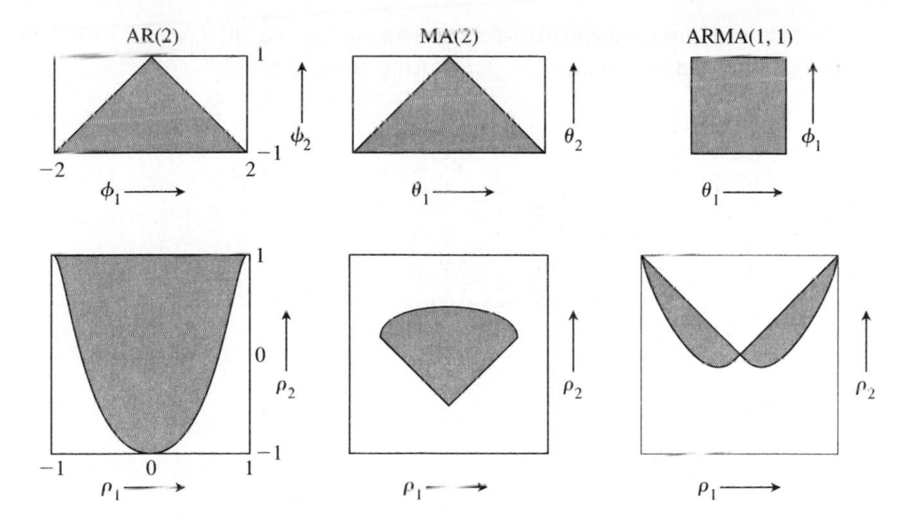

Figure 3.12 Admissible regions for the parameters and ρ_1, ρ_2 for AR(2), MA(2), and ARMA(1, 1) processes that are restricted to being both stationary and invertible.

model to produce models which can describe nonstationary behavior of the kind that is frequently met in practice.

APPENDIX A3.1 AUTOCOVARIANCES, AUTOCOVARIANCE GENERATING FUNCTION, AND STATIONARITY CONDITIONS FOR A GENERAL LINEAR PROCESS

Autocovariances. The autocovariance at lag k of the linear process

$$\tilde{z}_t = \sum_{j=0}^{\infty} \psi_j a_{t-j}$$

with $\psi_0 = 1$ is clearly

$$\begin{aligned}\gamma_k &= E[\tilde{z}_t \tilde{z}_{t+k}] \\ &= E\left[\sum_{j=0}^{\infty} \sum_{h=0}^{\infty} \psi_j \psi_h a_{t-j} a_{t+k-h}\right] \qquad \text{(A3.1.1)} \\ &= \sigma_a^2 \sum_{j=0}^{\infty} \psi_j \psi_{j+k}\end{aligned}$$

using the property (3.1.2) for the autocovariance function of white noise.

Autocovariance generating function. The result (A3.1.1) may be substituted in the autocovariance generating function

$$\gamma(B) = \sum_{k=-\infty}^{\infty} \gamma_k B^k \tag{A3.1.2}$$

to give

$$\gamma(B) = \sigma_a^2 \sum_{k=-\infty}^{\infty} \sum_{j=0}^{\infty} \psi_j \psi_{j+k} B^k$$

$$= \sigma_a^2 \sum_{j=0}^{\infty} \sum_{k=-j}^{\infty} \psi_j \psi_{j+k} B^k$$

since $\psi_h = 0$ for $h < 0$. Writing $j + k = h$, so that $k = h - j$, we have

$$\gamma(B) = \sigma_a^2 \sum_{j=0}^{\infty} \sum_{h=0}^{\infty} \psi_j \psi_h B^{h-j}$$

$$= \sigma_a^2 \sum_{h=0}^{\infty} \psi_h B^h \sum_{j=0}^{\infty} \psi_j B^{-j}$$

that is,

$$\gamma(B) = \sigma_a^2 \psi(B)\psi(B^{-1}) = \sigma_a^2 \psi(B)\psi(F) \tag{A3.1.3}$$

which is the result (3.1.11) quoted in the text.

Stationarity conditions. If we substitute $B = e^{-i2\pi f}$ and $F = B^{-1} = e^{i2\pi f}$ in the autocovariance generating function (A3.1.2), we obtain half the power spectrum. Hence the power spectrum of a linear process is

$$\begin{aligned} p(f) &= 2\sigma_a^2 \psi(e^{-i2\pi f})\psi(e^{i2\pi f}) \\ &= 2\sigma_a^2 |\psi(e^{-i2\pi f})|^2 \qquad 0 \leqslant f \leqslant \tfrac{1}{2} \end{aligned} \tag{A3.1.4}$$

It follows that the variance of the process is

$$\sigma_z^2 = \int_0^{1/2} p(f)\, df = 2\sigma_a^2 \int_0^{1/2} \psi(e^{-i2\pi f})\psi(e^{i2\pi f})\, df \tag{A3.1.5}$$

Now if the integral (A3.1.5) is to converge, it may be shown [98] that the infinite series $\psi(B)$ must converge for B on or within the unit circle. More directly, for the linear process $\tilde{z}_t = \sum_{j=0}^{\infty} \psi_j a_{t-j}$, the condition $\sum_{j=0}^{\infty} |\psi_j| < \infty$ of absolute summability of the coefficients ψ_j implies (see [63], [93]) that the sum $\sum_{j=0}^{\infty} \psi_j a_{t-j}$ converges with probability 1 and hence represents a valid stationary process.

APPENDIX A3.2 RECURSIVE METHOD FOR CALCULATING ESTIMATES OF AUTOREGRESSIVE PARAMETERS

We now show how Yule–Walker estimates for the parameters of an AR($p + 1$) model may be obtained when the estimates for an AR(p) model, fitted to the same time series, are known. This recursive method of calculation can be used to approximate the partial autocorrelation function, as described in Section 3.2.6.

To illustrate the recursion, consider the equations (3.2.35). Yule–Walker estimates are obtained for $k = 2, 3$ from

$$\begin{aligned} r_2 &= \hat{\phi}_{21} r_1 + \hat{\phi}_{22} \\ r_1 &= \hat{\phi}_{21} + \hat{\phi}_{22} r_1 \end{aligned} \tag{A3.2.1}$$

and

$$\begin{aligned} r_3 &= \hat{\phi}_{31} r_2 + \hat{\phi}_{32} r_1 + \hat{\phi}_{33} \\ r_2 &= \hat{\phi}_{31} r_1 + \hat{\phi}_{32} + \hat{\phi}_{33} r_1 \\ r_1 &= \hat{\phi}_{31} + \hat{\phi}_{32} r_1 + \hat{\phi}_{33} r_2 \end{aligned} \tag{A3.2.2}$$

The coefficients $\hat{\phi}_{31}$ and $\hat{\phi}_{32}$ may be expressed in terms of $\hat{\phi}_{33}$ using the last two equations of (A3.2.2). The solution may be written in matrix form as

$$\begin{pmatrix} \hat{\phi}_{31} \\ \hat{\phi}_{32} \end{pmatrix} = \mathbf{R}_2^{-1} \begin{pmatrix} r_2 - \hat{\phi}_{33} r_1 \\ r_1 - \hat{\phi}_{33} r_2 \end{pmatrix} \tag{A3.2.3}$$

where

$$\mathbf{R}_2 = \begin{bmatrix} r_1 & 1 \\ 1 & r_1 \end{bmatrix}$$

Now (A3.2.3) may be rewritten as

$$\begin{bmatrix} \hat{\phi}_{31} \\ \hat{\phi}_{32} \end{bmatrix} = \mathbf{R}_2^{-1} \begin{bmatrix} r_2 \\ r_1 \end{bmatrix} - \hat{\phi}_{33} \mathbf{R}_2^{-1} \begin{bmatrix} r_1 \\ r_2 \end{bmatrix} \tag{A3.2.4}$$

Using the fact that (A3.2.1) may be rewritten as

$$\begin{bmatrix} \hat{\phi}_{21} \\ \hat{\phi}_{22} \end{bmatrix} = \mathbf{R}_2^{-1} \begin{bmatrix} r_2 \\ r_1 \end{bmatrix}$$

it follows that (A3.2.4) becomes

$$\begin{bmatrix} \hat{\phi}_{31} \\ \hat{\phi}_{32} \end{bmatrix} = \begin{bmatrix} \hat{\phi}_{21} \\ \hat{\phi}_{22} \end{bmatrix} - \hat{\phi}_{33} \begin{bmatrix} \hat{\phi}_{22} \\ \hat{\phi}_{21} \end{bmatrix}$$

that is,

$$\begin{aligned}\hat{\phi}_{31} &= \hat{\phi}_{21} - \hat{\phi}_{33}\hat{\phi}_{22} \\ \hat{\phi}_{32} &= \hat{\phi}_{22} - \hat{\phi}_{33}\hat{\phi}_{21}\end{aligned} \tag{A3.2.5}$$

To complete the calculation of $\hat{\phi}_{31}$ and $\hat{\phi}_{32}$, we need an expression for $\hat{\phi}_{33}$. On substituting (A3.2.5) in the first of the equations (A3.2.2), we obtain

$$\hat{\phi}_{33} = \frac{r_3 - \hat{\phi}_{21}r_2 - \hat{\phi}_{22}r_1}{1 - \hat{\phi}_{21}r_1 - \hat{\phi}_{22}r_2} \tag{A3.2.6}$$

Thus the partial autocorrelation $\hat{\phi}_{33}$ is first calculated from $\hat{\phi}_{21}$ and $\hat{\phi}_{22}$, using (A3.2.6), and then the other two coefficients, $\hat{\phi}_{31}$ and $\hat{\phi}_{32}$, may be obtained from (A3.2.5).

In general, the recursive formulas, which are due to Durbin [87], are

$$\hat{\phi}_{p+1,j} = \hat{\phi}_{pj} - \hat{\phi}_{p+1,p+1}\hat{\phi}_{p,p-j+1} \qquad j = 1, 2, \ldots, p \tag{A3.2.7}$$

$$\hat{\phi}_{p+1,p+1} = \frac{r_{p+1} - \sum_{j=1}^{p} \hat{\phi}_{pj}r_{p+1-j}}{1 - \sum_{j=1}^{p} \hat{\phi}_{pj}r_j} \tag{A3.2.8}$$

Example. As an illustration, consider the calculations of the estimates $\hat{\phi}_{31}$, $\hat{\phi}_{32}$, and $\hat{\phi}_{33}$ of the parameters of an AR(3) model, fitted to Wölfer's sunspot numbers. The estimated autocorrelations to three-decimal accuracy are $r_1 = 0.806$, $r_2 = 0.428$, and $r_3 = 0.070$. Then

$$\hat{\phi}_{21} = \frac{r_1(1 - r_2)}{1 - r_1^2} = 1.316$$

$$\hat{\phi}_{22} = \frac{r_2 - r_1^2}{1 - r_1^2} = -0.632$$

Using (A3.2.6), we obtain

$$\hat{\phi}_{33} = \frac{0.070 - (1.316)(0.428) + (0.632)(0.806)}{1 - (1.316)(0.806) + (0.632)(0.428)} = 0.077$$

Substituting the values for $\hat{\phi}_{21}$, $\hat{\phi}_{22}$, and $\hat{\phi}_{33}$ in (A3.2.5) yields

$$\hat{\phi}_{31} = 1.316 + (0.077)(0.632) = 1.365$$

$$\hat{\phi}_{32} = -0.632 - (0.077)(1.316) = -0.733$$

We remind the reader that these estimates of the partial autocorrelations differ somewhat from the maximum likelihood values obtained by fitting autoregressive models of successively higher order. They are very sensitive to rounding errors, particularly when the process approaches nonstationarity.

4

LINEAR NONSTATIONARY MODELS

Many empirical time series (e.g., stock prices) behave as through they had no fixed mean. Even so, they exhibit homogeneity in the sense that apart from local level, or perhaps local level and trend, one part of the series behaves much like any other part. Models that describe such homogeneous nonstationary behavior can be obtained by supposing some suitable *difference* of the process to be stationary. We now consider the properties of the important class of models for which the dth difference is a stationary mixed autoregressive–moving average process. These models are called autoregressive integrated moving average (ARIMA) processes.

4.1 AUTOREGRESSIVE INTEGRATED MOVING AVERAGE PROCESSES

4.1.1 Nonstationary First-Order Autoregressive Process

Figure 4.1 shows sections of four time series encountered in practice. These series have arisen in forecasting and control problems, and all of them exhibit behavior suggestive of nonstationarity. Series A, C, and D represent "uncontrolled" outputs (concentration, temperature, and viscosity, respectively) from three different chemical processes. These series were collected to show the effect on these outputs of uncontrolled and unmeasured disturbances such as variations in feedstock and ambient temperature. The tem-

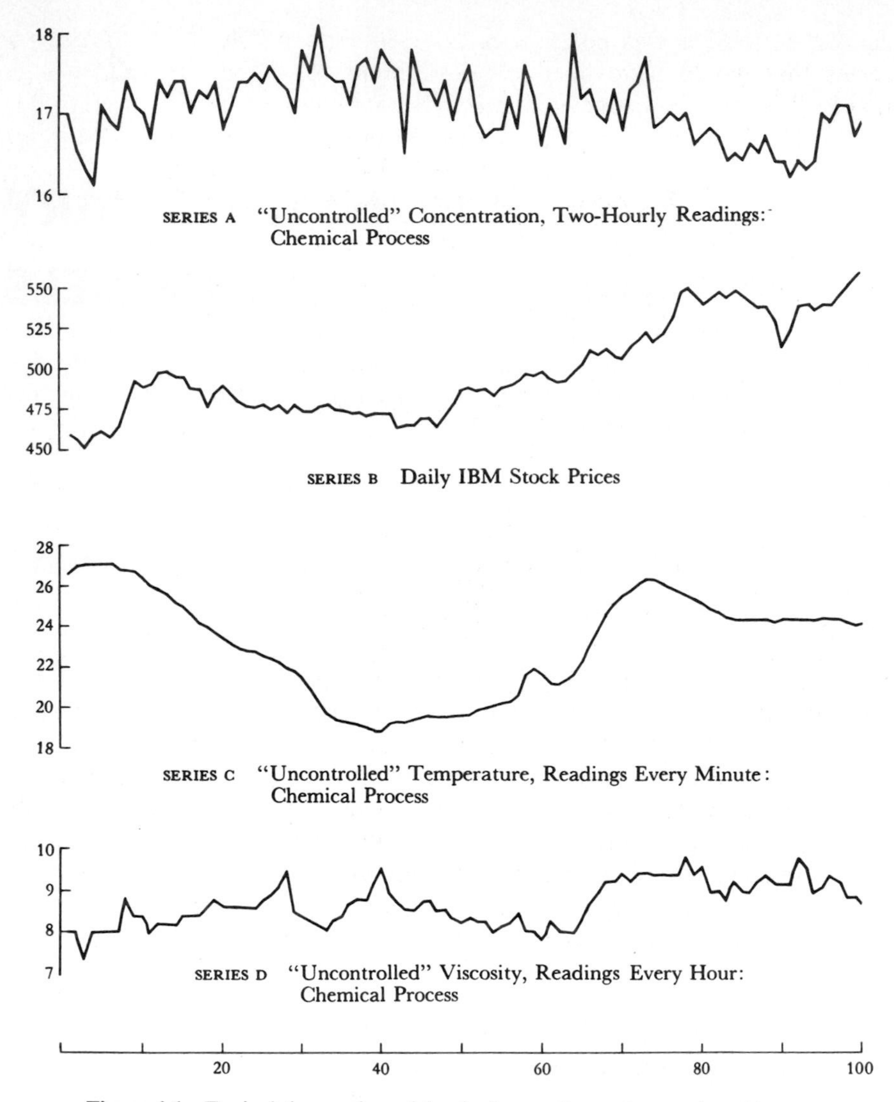

Figure 4.1 Typical time series arising in forecasting and control problems.

perature series C was obtained by temporarily disconnecting the controllers on the pilot plant involved and recording the subsequent temperature fluctuations. Both A and D were collected on full-scale processes, where it was necessary to maintain some output quality characteristic as close as possible to a fixed level. To achieve this control, another variable had been manipulated to approximately cancel out variations in the output. However, the effect of these manipulations on the output was in each case accurately known, so that it was possible to compensate numerically for the control action. That is, it was possible to calculate very nearly the values of the series that would have been obtained if no corrective action had been taken. It is these compensated values that are recorded here and referred to as the "uncontrolled" series. Series B consists of the daily IBM stock prices during a period beginning in May 1961. A complete listing of all the series is given in the collection of time series at the end of this volume. In Figure 4.1, 100 successive observations have been plotted from each series and the points joined by straight lines.

There are an unlimited number of ways in which a process can be nonstationary. However, the types of economic and industrial series that we wish to analyze frequently exhibit a particular kind of homogeneous nonstationary behavior that can be represented by a stochastic model, which is a modified form of the ARMA model. In Chapter 3 we considered the mixed autoregressive–moving average model

$$\phi(B)\tilde{z}_t = \theta(B)a_t \tag{4.1.1}$$

with $\phi(B)$ and $\theta(B)$ polynomials in B, of degree p and q, respectively. To ensure stationarity, the roots of $\phi(B) = 0$ must lie outside the unit circle. A natural way of obtaining nonstationary processes would be to relax this restriction.

To gain some insight into the possibilities, consider the first-order autoregressive model

$$(1 - \phi B)\tilde{z}_t = a_t \tag{4.1.2}$$

which is stationary for $|\phi| < 1$. Let us study the behavior of this process for $\phi = 2$, a value outside the stationary range.

Table 4.1 shows a set of unit random Normal deviates a_t and the corresponding values of the series $\tilde{z}_t$, generated by the model $\tilde{z}_t = 2\tilde{z}_{t-1} + a_t$, with $\tilde{z}_0 = 0.7$. The series is plotted in Figure 4.2. It is seen that after a short induction period, the series "breaks loose" and essentially follows an exponential curve, with the generating a's playing almost no further part. The behavior of series generated by processes of higher order, which violate the stationarity condition, is similar. Furthermore, this behavior is essentially

TABLE 4.1 First 11 Values of a Nonstationary First-Order Autoregressive Process

t	a_t	$\tilde{z}_t$	t	a_t	$\tilde{z}_t$
0		0.7	6	−0.8	22.8
1	0.1	1.5	7	0.8	46.4
2	−1.1	1.9	8	0.1	92.9
3	0.2	4.0	9	0.1	185.9
4	−2.0	6.0	10	−0.9	370.9
5	−0.2	11.8			

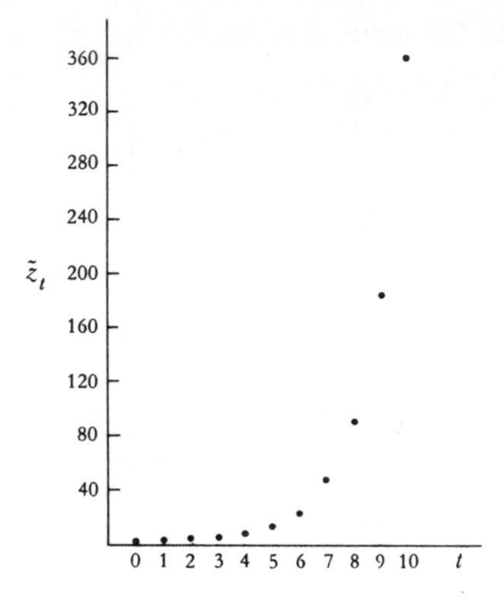

Figure 4.2 Realization of the nonstationary first-order autoregressive process $\tilde{z}_t = 2\tilde{z}_{t-1} + a_t$ with $\sigma_a^2 = 1$.

the same whether or not moving average terms are introduced on the right of the model.

4.1.2 General Model for a Nonstationary Process Exhibiting Homogeneity

Autoregressive integrated moving average model. Although nonstationary models of the kind described above are of value to represent explosive or evolutionary behavior (such as bacterial growth), the situations that we describe in this book are not of this type. So far we have seen that an ARMA process is stationary if the roots of $\phi(B) = 0$ lie *outside* the unit circle, and exhibits explosive nonstationary behavior if the roots lie *inside* the unit circle. The only other case open to us is that for which the roots of $\phi(B) = 0$ lie *on* the unit circle. It turns out that the resulting models are of great value in representing homogeneous nonstationary time series. In particular, nonseasonal series are often well represented by models in which one or more of these roots are *unity* and these are considered in the present chapter.*

Let us consider the model

$$\varphi(B)\tilde{z}_t = \theta(B)a_t \tag{4.1.3}$$

where $\varphi(B)$ is a nonstationary autoregressive operator such that d of the roots of $\varphi(B) = 0$ are unity and the remainder lie outside the unit circle. Then we can express the model (4.1.3) in the form

* In Chapter 9 we consider models, capable of representing seasonality of period s, for which the characteristic equation has roots lying on the unit circle that are the sth roots of unity.

$$\varphi(B)z_t = \phi(B)(1 - B)^d \tilde{z}_t = \theta(B)a_t \tag{4.1.4}$$

where $\phi(B)$ is a *stationary* autoregressive operator. Since $\nabla^d \tilde{z}_t = \nabla^d z_t$, for $d \geqslant 1$, where $\nabla = 1 - B$ is the differencing operator, we can write the model as

$$\phi(B)\nabla^d z_t = \theta(B)a_t \tag{4.1.5}$$

Equivalently, the process is defined by the two equations

$$\phi(B)w_t = \theta(B)a_t \tag{4.1.6}$$

and

$$w_t = \nabla^d z_t \tag{4.1.7}$$

Thus we see that the model corresponds to assuming that the dth difference of the series can be represented by a stationary, invertible ARMA process. An alternative way of looking at the process for $d \geqslant 1$ results from inverting (4.1.7) to give

$$z_t = S^d w_t \tag{4.1.8}$$

where S is the infinite summation operator defined by

$$\begin{aligned} Sx_t &= \sum_{h=-\infty}^{t} x_h = (1 + B + B^2 + \cdots)x_t \\ &= (1 - B)^{-1}x_t = \nabla^{-1}x_t \end{aligned}$$

Thus

$$S = (1 - B)^{-1} = \nabla^{-1}$$

The operator S^2 is similarly defined as

$$\begin{aligned} S^2 x_t &= Sx_t + Sx_{t-1} + Sx_{t-2} + \cdots \\ &= \sum_{i=-\infty}^{t} \sum_{h=-\infty}^{i} x_h = (1 + 2B + 3B^2 + \cdots)x_t \end{aligned}$$

and so on for higher-order d. Equation (4.1.8) implies that the process (4.1.5) can be obtained by summing (or "integrating") the stationary process (4.1.6) d times. Therefore, we call the process (4.1.5) an *autoregressive integrated moving average (ARIMA) process*. The ARIMA models for nonstationary time series, which have also been considered by Yaglom [209], are of fundamental importance to problems of forecasting and control [42], [43], [44], [45], [46], [47], [49]. For a further discussion of nonstationary processes, see also Zadeh and Ragazzini [213], Kalman [124], and Kalman and Bucy [125]. An earlier procedure for time series analysis that employed differencing was the "variate difference method." (See Tintner [187] and

Rao and Tintner [165].) However, the motivation, methods, and objectives of this procedure were quite different from those discussed here.

Technically, the infinite summation operator $S = (1 - B)^{-1}$ cannot actually be used in defining the nonstationary ARIMA processes, since the infinite sums involved will not be convergent. Instead, we can consider the finite summation operator S_m for any positive integer m, given by

$$S_m = (1 + B + B^2 + \cdots + B^{m-1}) \equiv \frac{1 - B^m}{1 - B}$$

Similarly, the finite double summation operator can be defined as

$$\begin{aligned} S_m^{(2)} &= \sum_{j=0}^{m-1} \sum_{i=j}^{m-1} B^i = (1 + 2B + 3B^2 + \cdots + mB^{m-1}) \\ &\equiv \frac{1 - B^m - mB^m(1 - B)}{(1 - B)^2} \end{aligned}$$

since $(1 - B)S_m^{(2)} = S_m - mB^m$, and so on. Then the relation between an integrated ARMA process z_t with $d = 1$, for example, and the corresponding stationary ARMA process $w_t = (1 - B)z_t$, in terms of values back to some earlier time origin $k < t$, can be expressed as

$$z_t = \frac{S_{t-k}}{1 - B^{t-k}} w_t = \frac{1}{1 - B^{t-k}} (w_t + w_{t-1} + \cdots + w_{k+1})$$

so that $z_t = w_t + w_{t-1} + \cdots + w_{k+1} + z_k$ can be thought of as the sum of a finite number of terms from the stationary process w plus an initializing value of the process z at time k. Hence, in the formal definition of the stochastic properties of a nonstationary ARIMA process as generated in (4.1.3), it would typically be necessary to specify initializing conditions for the process at some time point k in the finite (but possibly remote) past. However, these initial condition specifications will have little effect on most of the important characteristics of the process, and such specifications will for the most part not be emphasized in this book.

As mentioned in Chapter 1, the model (4.1.5) is equivalent to representing the process z_t as the output from a linear filter (unless $d = 0$ this is an *unstable* linear filter), whose input is white noise a_t. Alternatively, we can regard it as a device *for transforming the highly dependent, and possibly nonstationary process* z_t*, to a sequence of uncorrelated random variables* a_t, that is, for transforming the process to white noise.

If in (4.1.5), the autoregressive operator $\phi(B)$ is of order p, the dth difference is taken, and the moving average operator $\theta(B)$ is of order q, we say that we have an ARIMA model of order (p, d, q), or simply an ARIMA(p, d, q) process.

Two interpretations of the ARIMA model. We now show that (4.1.5) is an intuitively reasonable model for the time series we wish to describe in practice. First, we note that a basic characteristic of the first-order autoregressive process (4.1.2), for $|\phi| < 1$ and for $|\phi| > 1$, is that the local behavior of a series generated from the model is heavily dependent on the *level* of $\tilde{z}_t$. This is to be contrasted with the behavior of series such as those in Figure 4.1, where the local behavior of the series appears to be independent of its level.

If we are to use models for which the behavior of the process is independent of its level, we must choose the autoregressive operator $\varphi(B)$ such that

$$\varphi(B)(\tilde{z}_t + c) = \varphi(B)\tilde{z}_t$$

where c is any constant. Thus $\varphi(B)$ must be of the form

$$\varphi(B) = \phi_1(B)(1 - B) = \phi_1(B)\nabla$$

Therefore, a class of processes having the desired property will be of the form

$$\phi_1(B)w_t = \theta(B)a_t$$

where $w_t = \nabla\tilde{z}_t = \nabla z_t$. Required homogeneity excludes the possibility that w_t should increase explosively. This means that either $\phi_1(B)$ is a stationary autoregressive operator, or $\phi_1(B) = \phi_2(B)(1 - B)$, so that $\phi_2(B)w_t = \theta(B)a_t$, where now $w_t = \nabla^2 z_t$. In the latter case the same argument can be applied to the second difference, and so on.

Eventually, we arrive at the conclusion that for the representation of time series which are nonstationary but nevertheless exhibit homogeneity, the operator on the left of (4.1.3) should be of the form $\phi(B)\nabla^d$, where $\phi(B)$ is a *stationary* autoregressive operator. Thus we are led back to the model (4.1.5).

To approach the model from a somewhat different viewpoint, consider the situation where $d = 0$ in (4.1.4), so that $\phi(B)\tilde{z}_t = \theta(B)a_t$. The requirement that the zeros of $\phi(B)$ lie outside the unit circle would ensure not only that the process $\tilde{z}_t$ was stationary with mean zero, but also that ∇z_t, $\nabla^2 z_t$, $\nabla^3 z_t$, . . . were each stationary with mean zero. Figure 4.3(a) shows one kind of nonstationary series we would like to represent. This series is homogeneous except in level, in that except for a vertical translation, one part of it looks much the same as another. We can represent such behavior by retaining the requirement that each of the differences be stationary with zero mean, but letting the level "go free." This we do by using the model

$$\phi(B)\nabla z_t = \theta(B)a_t$$

Figure 4.3(b) shows a second kind of nonstationarity of fairly common occurrence. The series has neither a fixed level nor a fixed slope, but its

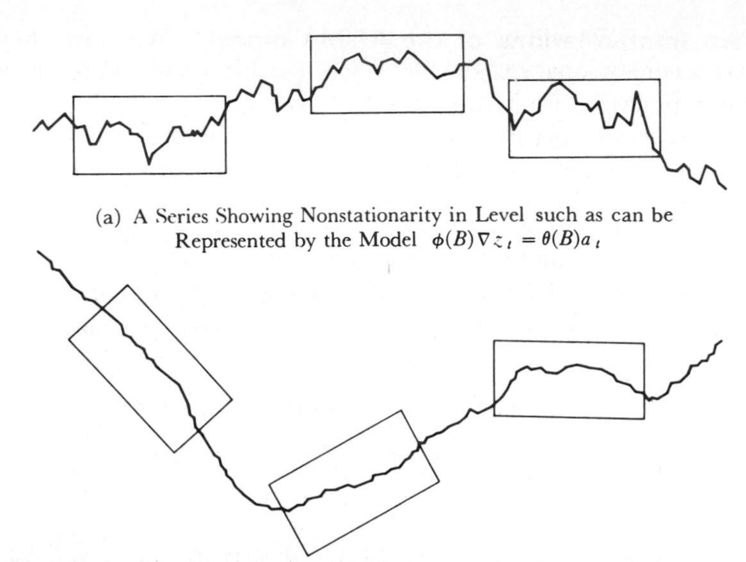

(a) A Series Showing Nonstationarity in Level such as can be Represented by the Model $\phi(B)\nabla z_t = \theta(B)a_t$

(b) A Series Showing Nonstationarity in Level and in Slope such as can be Represented by the Model $\phi(B)\nabla^2 z_t = \theta(B)a_t$

Figure 4.3 Two kinds of homogeneous nonstationary behavior.

behavior is homogeneous if we allow for differences in these characteristics. We can represent such behavior by the model

$$\phi(B)\nabla^2 z_t = \theta(B)a_t$$

which ensures stationarity and zero mean for all differences after the first and second but allows the level and the slope to "go free."

4.1.3 General Form of the Autoregressive Integrated Moving Average Process

For reasons to be given below, it is sometimes useful to consider a slight extension of the ARIMA model (4.1.5), by adding a constant term θ_0. Thus a rather general form of the model that we shall use to describe time series is the autoregressive integrated moving average process

$$\varphi(B)z_t = \phi(B)\nabla^d z_t = \theta_0 + \theta(B)a_t \tag{4.1.9}$$

where

$$\phi(B) = 1 - \phi_1 B - \phi_2 B^2 - \cdots - \phi_p B^p$$

$$\theta(B) = 1 - \theta_1 B - \theta_2 B^2 - \cdots - \theta_q B^q$$

In what follows:

1. $\phi(B)$ will be called the *autoregressive operator*; it is assumed to be stationary, that is, the roots of $\phi(B) = 0$ lie outside the unit circle.

2. $\varphi(B) = \phi(B)\nabla^d$ will be called the *generalized autoregressive operator*; it is a nonstationary operator with d of the roots of $\varphi(B) = 0$ equal to unity.
3. $\theta(B)$ will be called the *moving average operator*; it is assumed to be invertible, that is, the roots of $\theta(B) = 0$ lie outside the unit circle.

When $d = 0$, the model (4.1.9) represents a stationary process. The requirements of stationarity and invertibility apply independently, and in general, the operators $\phi(B)$ and $\theta(B)$ will not be of the same order. Examples of the stationarity regions for the simple cases of $p = 1, 2$ and the identical invertibility regions for $q = 1, 2$ have been given in Chapter 3.

Stochastic and deterministic trends. We have seen in Section 4.1.2 that when the constant term θ_0 is omitted, the model (4.1.9) is capable of representing series which have *stochastic* trends, as typified for example, by random changes in the level and slope of the series. In general, however, we may wish to include a *deterministic* function of time $f(t)$ in the model. In particular, automatic allowance for a deterministic polynomial trend, of degree d, can be made by permitting θ_0 to be nonzero. For example, when $d = 1$, we may use the model with $\theta_0 \neq 0$ to estimate a possible deterministic linear trend in the presence of nonstationary noise. Since, from (3.1.22), to allow θ_0 to be nonzero is equivalent to permitting

$$E[w_t] = E[\nabla^d z_t] = \mu_w = \frac{\theta_0}{1 - \phi_1 - \phi_2 - \cdots - \phi_p}$$

to be nonzero, an alternative way of expressing this more general model (4.1.9) is in the form of a stationary invertible ARMA process in $\tilde{w}_t = w_t - \mu_w$. That is,

$$\phi(B)\tilde{w}_t = \theta(B)a_t \tag{4.1.10}$$

In many applications, where no physical reason for a deterministic component exists, the mean of w can be assumed to be zero unless such an assumption proves contrary to facts presented by the data. It is clear that for many applications, the assumption of a stochastic trend is often more realistic than the assumption of a deterministic trend. This is of special importance in forecasting a time series, since a stochastic trend does not necessitate the series to follow the identical pattern that it has developed in the past. In what follows, when $d > 0$, we shall often assume that $\mu_w = 0$, or equivalently, that $\theta_0 = 0$, unless it is clear from the data or from the nature of the problem that a nonzero mean, or more generally a deterministic component of known form, is needed.

Some important special cases of the ARIMA model. In Chapter 3 we have become acquainted with some important special cases of the model (4.1.9), corresponding to the stationary situation, $d = 0$. The following models represent some special cases of the nonstationary model ($d \geqslant 1$), which seem to be of frequent occurrence.

1. *The (0, 1, 1) process*:

$$\nabla z_t = a_t - \theta_1 a_{t-1}$$
$$= (1 - \theta_1 B)a_t$$

corresponding to $p = 0$, $d = 1$, $q = 1$, $\phi(B) = 1$, $\theta(B) = 1 - \theta_1 B$.

2. *The (0, 2, 2) process*:

$$\nabla^2 z_t = a_t - \theta_1 a_{t-1} - \theta_2 a_{t-2}$$
$$= (1 - \theta_1 B - \theta_2 B^2)a_t$$

corresponding to $p = 0$, $d = 2$, $q = 2$, $\phi(B) = 1$, $\theta(B) = 1 - \theta_1 B - \theta_2 B^2$.

3. *The (1, 1, 1) process*:

$$\nabla z_t - \phi_1 \nabla z_{t-1} = a_t - \theta_1 a_{t-1}$$

or

$$(1 - \phi_1 B)\nabla z_t = (1 - \theta_1 B)a_t$$

corresponding to $p = 1$, $d = 1$, $q = 1$, $\phi(B) = 1 - \phi_1 B$, $\theta(B) = 1 - \theta_1 B$.

For the representation of nonseasonal time series (seasonal models are considered in Chapter 9), we rarely seem to meet situations for which either p, d, or q need be greater than 2. Frequently, values of zero or unity will be appropriate for one or more of these coefficients. For example, we show later that series A, B, C, D given in Figure 4.1 are reasonably well fitted* by the simple models shown in Table 4.2.

Nonlinear transformation of *z*. A considerable widening of the range of useful application of the model (4.1.9) is achieved if we allow the possibility of transformation. Thus we may substitute $z_t^{(\lambda)}$ for z_t in (4.1.9),

TABLE 4.2 Summary of Simple Nonstationary Models Fitted to Time Series of Figure 4.1

Series	Model	Order of Model
A	$\nabla z_t = (1 - 0.7B)a_t$	(0, 1, 1)
B	$\nabla z_t = (1 + 0.1B)a_t$	(0, 1, 1)
C	$(1 - 0.8B)\nabla z_t = a_t$	(1, 1, 0)
D	$\nabla z_t = (1 - 0.1B)a_t$	(0, 1, 1)

* As is discussed more fully later, there are certain advantages in using a nonstationary rather than a stationary model in cases of doubt. In particular, none of the fitted models above assume that z_t has a fixed mean. However, we show in Chapter 7 that it is possible in certain cases to obtain stationary models of slightly better fit.

where $z_t^{(\lambda)}$ is some nonlinear transformation of z_t involving one or more transformation parameters λ. A suitable transformation may be suggested by the situation, or in some cases it can be estimated from the data. For example, if we were interested in the sales of a recently introduced commodity, we might find that sales volume was increasing at a rapid rate and that it was the *percentage* fluctuation which showed nonstationary stability rather than the absolute fluctuation. This would support the analysis of the logarithm of sales since

$$\nabla \log(z_t) = \log(z_t/z_{t-1}) = \log\left(1 + \frac{\nabla z_t}{z_{t-1}}\right) \simeq \frac{\nabla z_t}{z_{t-1}}$$

where $\nabla z_t/z_{t-1}$ are the relative or percentage changes, the approximation holding if the relative changes are not excessively large. When the data cover a wide range and especially for seasonal data, estimation of the transformation using the approach of Box and Cox [38] may be helpful (Section 9.4).

4.2 THREE EXPLICIT FORMS FOR THE AUTOREGRESSIVE INTEGRATED MOVING AVERAGE MODEL

We now consider three different "explicit" forms for the general model (4.1.9). Each of these allows some special aspect to be appreciated. Thus the current value z_t of the process can be expressed:

1. In terms of previous values of the z's and current and previous values of the a's, by direct use of the *difference equation*
2. In terms of *current and previous shocks* a_{t-j} only
3. In terms of a weighted sum of *previous values* z_{t-j} of the process and the current shock a_t

In this chapter we are concerned primarily with *nonstationary* models in which $\nabla^d z_t$ is a stationary process and d is greater than zero. For such models we can, without loss of generality, omit μ from the specification or equivalently replace $\tilde{z}_t$ by z_t. The results of this chapter and the next will, however, apply to stationary models for which $d = 0$, provided that z_t is then interpreted as the *deviation* from the mean μ.

4.2.1 Difference Equation Form of the Model

Direct use of the difference equation permits us to express the current value z_t of the process in terms of previous values of the z's and of current and previous values of the a's. Thus if

$$\varphi(B) = \phi(B)(1 - B)^d = 1 - \varphi_1 B - \varphi_2 B^2 - \cdots - \varphi_{p+d} B^{p+d}$$

the general model (4.1.9), with $\theta_0 = 0$, may be written

$$z_t = \varphi_1 z_{t-1} + \cdots + \varphi_{p+d} z_{t-p-d} - \theta_1 a_{t-1} \cdots - \theta_q a_{t-q} + a_t \tag{4.2.1}$$

For example, consider the process represented by the model of order (1, 1, 1)

$$(1 - \phi B)(1 - B)z_t = (1 - \theta B)a_t$$

where for convenience we drop the subscript 1 on ϕ_1 and θ_1. Then this process may be written

$$[1 - (1 + \phi)B + \phi B^2]z_t = (1 - \theta B)a_t$$

that is,

$$z_t = (1 + \phi)z_{t-1} - \phi z_{t-2} + a_t - \theta a_{t-1} \tag{4.2.2}$$

For many purposes, and in particular for calculating the forecasts in Chapter 5, the difference equation (4.2.1) is the most convenient form to employ.

4.2.2 Random Shock Form of the Model

Model in terms of current and previous shocks. We have seen in Section 3.1.1 that a linear model can be written as the output z_t from the linear filter

$$\begin{aligned} z_t &= a_t + \psi_1 a_{t-1} + \psi_2 a_{t-2} + \cdots \\ &= a_t + \sum_{j=1}^{\infty} \psi_j a_{t-j} \qquad (4.2.3) \\ &= \psi(B) a_t \end{aligned}$$

whose input is white noise, or a sequence of uncorrelated shocks a_t. It is sometimes useful to express the ARIMA model in the form (4.2.3), and in particular, the ψ weights will be needed in Chapter 5 to calculate the variance of the forecasts. However, since the nonstationary ARIMA processes are not in statistical equilibrium over time, they cannot be assumed to extend infinitely into the past, and hence an infinite representation as in (4.2.3) will not be possible. But a related finite truncated form, which will be discussed subsequently, always exists. We now show that the ψ weights for an ARIMA process may be obtained directly from the difference equation form of the model.

General expression for the ψ weights. If we operate on both sides of (4.2.3) with the generalized autoregressive operator $\varphi(B)$, we obtain

$$\varphi(B)z_t = \varphi(B)\psi(B)a_t$$

However, since

$$\varphi(B)z_t = \theta(B)a_t$$

it follows that

$$\varphi(B)\psi(B) = \theta(B) \tag{4.2.4}$$

Therefore, the ψ weights may be obtained by equating coefficients of B in the expansion

$$\begin{aligned}(1 - \varphi_1 B - \cdots - \varphi_{p+d}B^{p+d})(1 + \psi_1 B + \psi_2 B^2 + \cdots) \\ = (1 - \theta_1 B - \cdots - \theta_q B^q)\end{aligned} \tag{4.2.5}$$

Thus we find that the ψ_j weights of the ARIMA process can be determined recursively through the equations

$$\psi_j = \varphi_1\psi_{j-1} + \varphi_2\psi_{j-2} + \cdots + \varphi_{p+d}\psi_{j-p-d} - \theta_j \qquad j > 0$$

with $\psi_0 = 1$, $\psi_j = 0$ for $j < 0$, and $\theta_j = 0$ for $j > q$. We note that for j greater than the larger of $p + d - 1$ and q, the ψ weights satisfy the difference equation defined by the generalized autoregressive operator, that is,

$$\varphi(B)\psi_j = \phi(B)(1 - B)^d\psi_j = 0 \tag{4.2.6}$$

where B now operates on the subscript j. Thus, for sufficiently large j, the weights ψ_j are represented by a mixture of polynomials, damped exponentials, and damped sinusoids in the argument j.

Example. To illustrate the use of (4.2.5), consider the (1, 1, 1) process (4.2.2), for which

$$\begin{aligned}\varphi(B) &= (1 - \phi B)(1 - B) \\ &= 1 - (1 + \phi)B + \phi B^2\end{aligned}$$

and

$$\theta(B) = 1 - \theta B$$

Substituting in (4.2.5) gives

$$[1 - (1 + \phi)B + \phi B^2](1 + \psi_1 B + \psi_2 B^2 + \cdots) = 1 - \theta B$$

and hence the ψ_j satisfy the recursion $\psi_j = (1 + \phi)\psi_{j-1} - \phi\psi_{j-2}$, $j \geq 2$, with $\psi_0 = 1$ and $\psi_1 = (1 + \phi) - \theta$. Thus since the roots of $\varphi(B) = (1 - \phi B)(1 - B) = 0$ are $G_1^{-1} = 1$ and $G_2^{-1} = \phi^{-1}$, we have, in general,

$$\psi_j = A_0 + A_1\phi^j \tag{4.2.7}$$

where the constants A_0 and A_1 are determined from the initial values $\psi_0 = A_0 + A_1 = 1$ and $\psi_1 = A_0 + A_1\phi = 1 + \phi - \theta$ as

$$A_0 = \frac{1-\theta}{1-\phi} \qquad A_1 = \frac{\theta - \phi}{1-\phi}$$

Thus, informally, we may wish to express model (4.2.2) in the equivalent form

$$z_t = \sum_{j=0}^{\infty} (A_0 + A_1 \phi^j) a_{t-j} \tag{4.2.8}$$

Since $|\phi| < 1$, the weights ψ_j tend to A_0 for large j, so that shocks a_{t-j}, which entered in the remote past receive a constant weight A_0. However, the representation in (4.2.8) is strictly not valid because the infinite sum on the right does not converge in any sense; that is, the weights ψ_j are not absolutely summable as in the case of a stationary process. A related truncated version of the random shock form of the model is always valid, as we discuss in detail shortly. Nevertheless, for notational convenience we will often refer to the infinite random shock form (4.2.3) of an ARIMA process, even though this form is strictly not convergent, as a simple notational device to represent the valid truncated form in (4.2.14) below, in situations where the distinction between the two forms is not important.

Truncated form of the random shock model. For technical purposes it is necessary and also sometimes convenient to consider the model in a slightly different form from (4.2.3). Suppose that we wish to express the current value z_t of the process in terms of the $t - k$ shocks $a_t, a_{t-1}, \ldots, a_{k+1}$, which have entered the system since some time origin $k < t$. This time origin k might, for example, be the time at which the process was first observed.

The general model

$$\varphi(B) z_t = \theta(B) a_t \tag{4.2.9}$$

is a difference equation with the solution

$$z_t = C_k(t-k) + I_k(t-k) \tag{4.2.10}$$

A short discussion of linear difference equations is given in Appendix A4.1. We remind the reader that the solution of such equations closely parallels the solution of linear differential equations. The *complementary function* $C_k(t-k)$ is the general solution of the difference equation

$$\varphi(B) C_k(t-k) = 0 \tag{4.2.11}$$

In general, this solution will consist of a *linear* combination of certain functions of time. These functions are powers t^j, real geometric (exponential) terms G^t, and complex geometric (exponential) terms $D^t \sin(2\pi f_0 t + F)$, where the constants G, f_0, and F are functions of the parameters $(\boldsymbol{\phi}, \boldsymbol{\theta})$ of the model. The coefficients that form the linear combinations of these terms can be determined so as to satisfy a set of initial conditions defined by the values of the process before time $k + 1$. The *particular integral* $I_k(t - k)$ is *any* function that satisfies

$$\varphi(B)I_k(t - k) = \theta(B)a_t \tag{4.2.12}$$

It should be carefully noted that in this expression B operates on t and *not on* k. It is shown in Appendix A4.1 that this equation is satisfied for $t - k > q$ by

$$I_k(t - k) = \sum_{j=0}^{t-k-1} \psi_j a_{t-j} = a_t + \psi_1 a_{t-1} + \cdots + \psi_{t-k-1} a_{k+1} \qquad t > k \tag{4.2.13}$$

with $I_k(t - k) = 0, t \leq k$. This particular integral $I_k(t - k)$ thus represents the finite truncated form of the infinite random shock form (4.2.3), while the complementary function $C_k(t - k)$ embodies the "initializing" features of the process z in the sense that $C_k(t - k)$ is already determined or specified by the time $k + 1$. Hence the truncated form of the random shock model for the ARIMA process (4.1.3) is given by

$$z_t = \sum_{j=0}^{t-k-1} \psi_j a_{t-j} + C_k(t - k) \tag{4.2.14}$$

For illustration, consider Figure 4.4. The discussion above implies that any observation z_t can be considered in relation to any previous time k and can be divided up into two additive parts. The first part $C_k(t - k)$ is the component of z_t, *already determined at time k*, and indicates what the observations prior to time $k + 1$ had to tell us about the value of the series at time

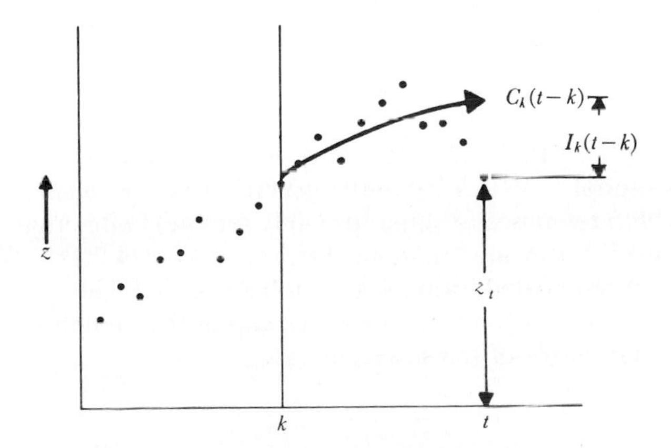

Figure 4.4 Role of the complementary function $C_k(t - k)$ and of the particular integral $I_k(t - k)$ in describing the behavior of a time series.

t. It represents the course that the process would take if at time k, the source of shocks a_t had been "switched off." The second part, $I_k(t - k)$, represents an additional component, *unpredictable at time* k, which embodies the entire effect of shocks entering the system after time k. Hence, to specify an ARIMA process, one must specify the initializing component $C_k(t - k)$ in (4.2.14) for some time origin k in the finite (but possibly remote) past, with the remaining course of the process being determined through the truncated random shock terms in (4.2.14).

Example. For illustration, consider again the example

$$(1 - \phi B)(1 - B)z_t = (1 - \theta B)a_t$$

The complementary function is the solution of the difference equation

$$(1 - \phi B)(1 - B)C_k(t - k) = 0$$

that is,

$$C_k(t - k) = b_0^{(k)} + b_1^{(k)}\phi^{t-k}$$

where $b_0^{(k)}$, $b_1^{(k)}$ are coefficients which depend on the past history of the process and, it will be noted, *change with the origin k*.

Making use of the ψ weights (4.2.7), a particular integral (4.2.13) is

$$I_k(t - k) = \sum_{j=0}^{t-k-1} (A_0 + A_1\phi^j)a_{t-j}$$

so that finally we can write the model (4.2.8) in the equivalent form

$$z_t = b_0^{(k)} + b_1^{(k)}\phi^{t-k} + \sum_{j=0}^{t-k-1} (A_0 + A_1\phi^j)a_{t-j} \tag{4.2.15}$$

Note that since $|\phi| < 1$, when $t - k$ is sufficiently large, the term involving ϕ^{t-k} in (4.2.15) is negligible and may be ignored.

Link between the truncated and nontruncated forms of the random shock model. Returning to the general case, we can always think of the process with reference to some (possibly remote) finite origin k, with the process having the truncated random shock form as in (4.2.14). By comparison with the nontruncated form in (4.2.3), one can see that we might, informally, make the correspondence of representing the complementary function $C_k(t - k)$ in terms of the ψ weights as

$$C_k(t - k) = \sum_{j=t-k}^{\infty} \psi_j a_{t-j} \tag{4.2.16}$$

even though, formally, the infinite sum on the right of (4.2.16) does not converge. As mentioned earlier, for notational simplicity, we will often use this correspondence.

In summary, then, for the general model (4.2.9):

1. We can express the value z_t of the process, informally, as an infinite weighted sum of current and previous shocks a_{t-j}, according to

$$z_t = \sum_{j=0}^{\infty} \psi_j a_{t-j} = \psi(B)a_t$$

2. The value of z_t can be expressed, more formally, as a weighted finite sum of the $t - k$ current and previous shocks occurring after some origin k, plus a complementary function $C_k(t - k)$. This finite sum consists of the first $t - k$ terms of the infinite sum, so that

$$z_t = C_k(t - k) + \sum_{j=0}^{t-k-1} \psi_j a_{t-j} \tag{4.2.17}$$

Finally, the complementary function $C_k(t - k)$ can be taken, for notational convenience, to be represented as the truncated infinite sum, so that

$$C_k(t - k) = \sum_{j=t-k}^{\infty} \psi_j a_{t-j} \tag{4.2.18}$$

For illustration, consider once more the model

$$(1 - \phi B)(1 - B)z_t = (1 - \theta B)a_t$$

We can write z_t either, informally, as an infinite weighted sum of the a_{t-j}'s

$$z_t = \sum_{j=0}^{\infty} (A_0 + A_1\phi^j)a_{t-j}$$

or, more formally, in terms of the weighted finite sum as

$$z_t = C_k(t - k) + \sum_{j=0}^{t-k-1} (A_0 + A_1\phi^j)a_{t-j}$$

Furthermore, the complementary function can be written as

$$C_k(t - k) = b_0^{(k)} + b_1^{(k)}\phi^{t-k}$$

and it can be determined that $b_0^{(k)}$ and $b_1^{(k)}$, which satisfy the initial conditions through time k, are

$$b_0^{(k)} = \frac{z_k - \phi z_{k-1} - \theta a_k}{1 - \phi} \qquad b_1^{(k)} = \frac{-\phi(z_k - z_{k-1}) + \theta a_k}{1 - \phi}$$

The complementary function can also be represented, informally, as the truncated infinite sum

$$C_k(t-k) = \sum_{j=t-k}^{\infty} (A_0 + A_1\phi^j)a_{t-j}$$

from which it can be seen that $b_0^{(k)}$ and $b_1^{(k)}$ may be represented as

$$b_0^{(k)} = A_0 \sum_{j=t-k}^{\infty} a_{t-j} = \frac{1-\theta}{1-\phi} \sum_{j=t-k}^{\infty} a_{t-j}$$

$$b_1^{(k)} = A_1 \sum_{j=t-k}^{\infty} \phi^{j-(t-k)}a_{t-j} = \frac{\theta-\phi}{1-\phi} \sum_{j=t-k}^{\infty} \phi^{j-(t-k)}a_{t-j}$$

Complementary function as a conditional expectation. One consequence of the truncated form (4.2.14) is that for $m > 0$,

$$C_k(t-k) = C_{k-m}(t-k+m) + \psi_{t-k}a_k + \psi_{t-k+1}a_{k-1} + \cdots + \psi_{t-k+m-1}a_{k-m+1} \qquad (4.2.19)$$

which shows how the complementary function changes as the origin k is changed. Now denote by $E_k[z_t]$ the *conditional expectation of* z_t *at time* k. That is the expectation given complete historical knowledge up to, but not beyond time k. To calculate this expectation, note that

$$E_k[a_j] = \begin{cases} 0 & j > k \\ a_j & j \leq k \end{cases}$$

That is, *standing at time* k, the expected values of a's that have yet to happen is zero and the expectation of those which have happened already is the value they have actually realized.

By taking conditional expectations at time k on both sides of (4.2.17), we obtain $E_k[z_t] = C_k(t-k)$. Thus, for $(t-k) > q$, the complementary function provides the expected value of the future value z_t of the process, *viewed from time* k and based on knowledge of the past. The particular integral shows how that expectation is modified by *subsequent* events represented by the shocks $a_{k+1}, a_{k+2}, \ldots, a_t$. In the problem of forecasting, which we discuss in Chapter 5, it will turn out that $C_k(t-k)$ is the minimum mean square error forecast of z_t made at time k. Equation (4.2.19) may be used in "updating" this forecast.

4.2.3 Inverted Form of the Model

Model in terms of previous z's and the current shock a_t. We have seen in Section 3.1.1 that the model

$$z_t = \psi(B)a_t$$

may also be written in the inverted form

$$\psi^{-1}(B)z_t = a_t$$

or

$$\pi(B)z_t = \left(1 - \sum_{j=1}^{\infty} \pi_j B^j\right) z_t = a_t \tag{4.2.20}$$

Thus z_t is an infinite weighted sum of previous values of z, plus a random shock

$$z_t = \pi_1 z_{t-1} + \pi_2 z_{t-2} + \cdots + a_t$$

Because of the invertibility condition, the π weights in (4.2.20) must form a convergent series; that is, $\pi(B)$ must converge on or within the unit circle.

General expression for the π weights. To derive the π weights for the general ARIMA model, we can substitute (4.2.20) in

$$\varphi(B)z_t = \theta(B)a_t$$

to obtain

$$\varphi(B)z_t = \theta(B)\pi(B)z_t$$

Hence the π weights can be obtained explicitly by equating coefficients of B in

$$\varphi(B) = \theta(B)\pi(B) \tag{4.2.21}$$

that is,

$$\begin{aligned}(1 - \varphi_1 B - \cdots - \varphi_{p+d}B^{p+d}) = (1 - \theta_1 B - \cdots - \theta_q B^q) \\ \times (1 - \pi_1 B - \pi_2 B^2 - \cdots)\end{aligned} \tag{4.2.22}$$

Thus we find that the π_j weights of the ARIMA process can be determined recursively through

$$\pi_j = \theta_1 \pi_{j-1} + \theta_2 \pi_{j-2} + \cdots + \theta_q \pi_{j-q} + \varphi_j \qquad j > 0$$

with the convention $\pi_0 = -1$, $\pi_j = 0$ for $j < 0$, and $\varphi_j = 0$ for $j > p + d$. It will be noted that for j greater than the larger of $p + d$ and q, the π weights satisfy the difference equation defined by the *moving average operator*

$$\theta(B)\pi_j = 0$$

where B now operates on j. Hence for sufficiently large j, the π weights will exhibit similar behavior as the autocorrelation function (3.2.5) of an autoregressive process; that is, they follow a mixture of damped exponentials and damped sine waves.

Another interesting fact is that if $d \geqslant 1$, the π weights in (4.2.20) sum to unity. This may be verified by substituting $B = 1$ in (4.2.21). Thus $\varphi(B) = \phi(B)(1 - B)^d$ is zero when $B = 1$ and $\theta(1) \neq 0$, because the roots of $\theta(B) = 0$ lie outside the unit circle. Hence it follows from (4.2.21) that $\pi(1) = 0$, that is,

$$\sum_{j=1}^{\infty} \pi_j = 1 \tag{4.2.23}$$

Therefore, if $d \geqslant 1$, the process may be written in the form

$$z_t = \bar{z}_{t-1}(\pi) + a_t \tag{4.2.24}$$

where

$$\bar{z}_{t-1}(\pi) = \sum_{j=1}^{\infty} \pi_j z_{t-j}$$

is a *weighted average* of previous values of the process.

Example. We again consider, for illustration, the ARIMA(1, 1, 1) process

$$(1 - \phi B)(1 - B)z_t = (1 - \theta B)a_t$$

Then, using (4.2.21),

$$\pi(B) = \varphi(B)\theta^{-1}(B) = [1 - (1 + \phi)B + \phi B^2](1 + \theta B + \theta^2 B^2 + \cdots)$$

so that

$$\pi_1 = \phi + (1 - \theta), \quad \pi_2 = (\theta - \phi)(1 - \theta), \ \pi_j = (\theta - \phi)(1 - \theta)\theta^{j-2} \qquad j \geqslant 3$$

For example, the first seven π weights corresponding to $\phi = -0.3$ and $\theta = 0.5$ are given in Table 4.3. Thus the tth value of the process would be generated by a weighted average of previous values, plus an additional shock, according to

$$z_t = (0.2z_{t-1} + 0.4z_{t-2} + 0.2z_{t-3} + 0.1z_{t-4} + \cdots) + a_t$$

We notice in particular that the π weights die out as more and more remote values of z_{t-j} are involved. This is the property that results from requiring that the series be invertible (in this case by requiring that $-1 < \theta < 1$).

We mention in passing that, for statistical models representing practically occurring time series, the convergent π weights usually die out rather quickly. Thus although z_t may be theoretically dependent on the remote past, the representation

$$z_t = \sum_{j=1}^{\infty} \pi_j z_{t-j} + a_t$$

TABLE 4.3 First Seven π Weights for an ARIMA(1, 1, 1) Process with $\phi = -0.3$, $\theta = 0.5$

j	1	2	3	4	5	6	7
π_j	0.2	0.4	0.2	0.1	0.05	0.025	0.0125

will usually show that z_t is dependent *to an important extent* only on recent past values z_{t-j} of the time series. This is still true even though for nonstationary models with $d > 0$, the ψ weights in the "weighted shock" representation

$$z_t = \sum_{j=0}^{\infty} \psi_j a_{t-j}$$

do not converge. What happens, of course, is that all the information that remote values of the shocks a_{t-j} supply about z_t is contained in recent values $z_{t-1}, z_{t-2}, \ldots$ of the series. In particular the expectation $E_k[z_t]$, in theory conditional on complete historical knowledge up to time k, can usually be computed to sufficient accuracy from *recent* values of the time series. This fact is particularly important in forecasting applications.

4.3 INTEGRATED MOVING AVERAGE PROCESSES

A nonstationary model that is useful in representing some commonly occurring series is the (0, 1, 1) process,

$$\nabla z_t = a_t - \theta a_{t-1}$$

The model contains only two parameters, θ and σ_a^2. Figure 4.5 shows two

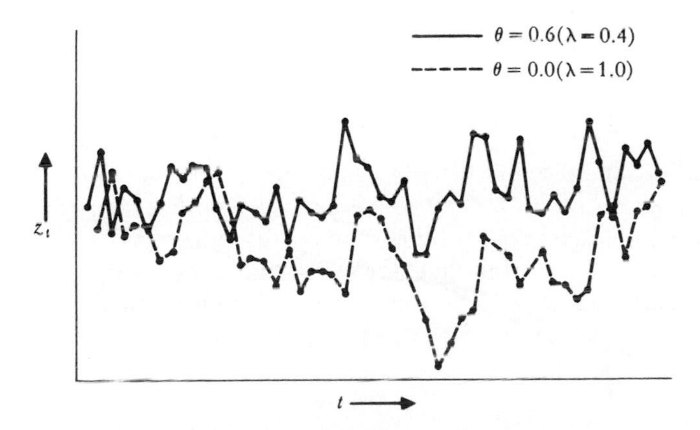

Figure 4.5 Two time series generated from IMA(0, 1, 1) models.

time series generated by this model from the same sequence of random Normal deviates a_t. For the first series, $\theta = 0.6$, and for the second, $\theta = 0$. Models of this kind have often been found useful in inventory control problems, in representing certain kinds of disturbance occurring in industrial processes, and in econometrics. As we noted in Section 4.1.3, it will be shown in Chapter 7 that this simple process can, with suitable parameter values, supply useful representations of series A, B, and D shown in Figure 4.1. Another valuable model is the (0, 2, 2) process

$$\nabla^2 z_t = a_t - \theta_1 a_{t-1} - \theta_2 a_{t-2}$$

which contains three parameters, θ_1, θ_2, and σ_a^2. Figure 4.6 shows two series generated from the model using the same set of Normal deviates. For the first series the parameters $(\theta_1, \theta_2) = (0, 0)$, and for the second $(\theta_1, \theta_2) = (1.5, -0.8)$. The series tend to be much smoother than those generated by the (0, 1, 1) process. The (0, 2, 2) models are useful in representing disturbances (such as series C) in systems with a large degree of inertia. Both the (0, 1, 1) and (0, 2, 2) models are special cases of the class

$$\nabla^d z_t = \theta(B) a_t \tag{4.3.1}$$

We call such models as (4.3.1) *integrated moving average* (IMA) processes, of order (0, d, q), and consider their properties in the following section.

4.3.1 Integrated Moving Average Process of Order (0, 1, 1)

Difference equation form. The (0, 1, 1) process

$$\nabla z_t = (1 - \theta B) a_t \qquad -1 < \theta < 1$$

possesses useful representational capability and we now study its properties in more detail. The model can be written in terms of the z's and the a's in the

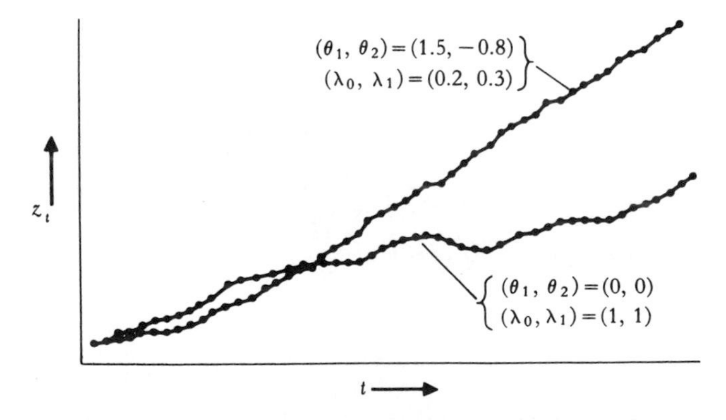

Figure 4.6 Two time series generated from IMA(0, 2, 2) models.

form

$$z_t = z_{t-1} + a_t - \theta a_{t-1} \tag{4.3.2}$$

Random shock form of model. Alternatively, we can obtain z_t in terms of the a's alone by summing on both sides of (4.3.2). Before doing this, there is some advantage in expressing the right-hand operator in terms of ∇ rather than B. Thus we can write

$$1 - \theta B = (1 - \theta)B + (1 - B) = (1 - \theta)B + \nabla = \lambda B + \nabla$$

where $\lambda = 1 - \theta$, and the invertibility region in terms of λ is defined by $0 < \lambda < 2$. Hence

$$\nabla z_t = \lambda a_{t-1} + \nabla a_t$$

Relative to some time origin $k < t$, applying the finite summation operator $S_{t-k} = 1 + B + \cdots + B^{t-k-1} = (1 - B^{t-k})/(1 - B)$, we obtain

$$(1 - B^{t-k})z_t = \lambda S_{t-k}a_{t-1} + (1 - B^{t-k})a_t \tag{4.3.3}$$

so that

$$z_t = a_t + \lambda(a_{t-1} + a_{t-2} + \cdots + a_{k+1}) + (z_k - \theta a_k) \tag{4.3.4}$$

In comparison to $z_t = \sum_{j=0}^{t-k-1} \psi_j a_{t-j} + C_k(t - k)$, the weights are $\psi_0 = 1$, $\psi_j = \lambda$ for $j \geq 1$. Also, the complementary function is $C_k(t - k) = z_k - \theta a_k = b_0^{(k)}$ (a "constant" b_0 for each k), which is the solution of the difference equation $(1 - B)C_k(t - k) = 0$. Moreover, in the infinite form $z_t = a_t + \lambda \sum_{j=1}^{\infty} a_{t-j}$, we may identify $b_0^{(k)}$ with $\lambda \sum_{j=t-k}^{\infty} a_{t-j}$. For this model, then, the complementary function is simply a constant (i.e., a polynomial in t of degree zero) representing the current "level" of the process and associated with the particular origin of reference k. If the origin is changed from $k - 1$ to k, then b_0 is "updated" according to

$$b_0^{(k)} = b_0^{(k-1)} + \lambda a_k$$

since using (4.3.2), $b_0^{(k)} = z_k + (\lambda - 1)a_k = z_{k-1} - \theta a_{k-1} + \lambda a_k$.

Inverted form of model. Finally, we can consider the model in the form

$$\pi(B)z_t = a_t$$

or equivalently, in the form

$$z_t = \sum_{j=1}^{\infty} \pi_j z_{t-j} + a_t = \bar{z}_{t-1}(\pi) + a_t$$

where $\bar{z}_{t-1}(\pi)$ is a weighted moving average of previous values of the process.

Using (4.2.21), the π weights for the IMA(0, 1, 1) process are given by

$$(1 - \theta B)\pi(B) = 1 - B$$

that is,

$$\pi(B) = \frac{1 - B}{1 - \theta B} = \frac{1 - \theta B - (1 - \theta)B}{1 - \theta B}$$
$$= 1 - (1 - \theta)(B + \theta B^2 + \theta^2 B^3 + \cdots)$$

whence

$$\pi_j = (1 - \theta)\theta^{j-1} = \lambda(1 - \lambda)^{j-1} \qquad j \geqslant 1$$

Thus the process may be written

$$z_t = \bar{z}_{t-1}(\lambda) + a_t \tag{4.3.5}$$

The weighted moving average of previous values of the process

$$\bar{z}_{t-1}(\lambda) = \lambda \sum_{j=1}^{\infty} (1 - \lambda)^{j-1} z_{t-j} \tag{4.3.6}$$

is in this case an *exponentially weighted moving average* (EWMA).

This moving average (4.3.6) is said to be exponentially (or geometrically) weighted because the weights

$$\lambda \qquad \lambda(1 - \lambda) \qquad \lambda(1 - \lambda)^2 \qquad \lambda(1 - \lambda)^3 \cdots$$

fall off exponentially (i.e., as a geometric progression). The weight function for an IMA process of order (0, 1, 1), with $\lambda = 0.4$ ($\theta = 0.6$), is shown in Figure 4.7.

Although the invertibility condition is satisfied for $0 < \lambda < 2$, we are in practice most often concerned with values of λ between zero and 1. We note that if λ had a value equal to 1, the weight function would consist of a single spike ($\pi_1 = 1$, $\pi_j = 0$ for $j > 1$). As the value λ approaches zero, the exponential weights die out more and more slowly and the EWMA stretches back further into the process. Finally, with $\lambda = 0$ and $\theta = 1$, the model $(1 - B)z_t = (1 - B)a_t$ is equivalent to $z_t = \theta_0 + a_t$, with θ_0 being given by the mean of all past values.

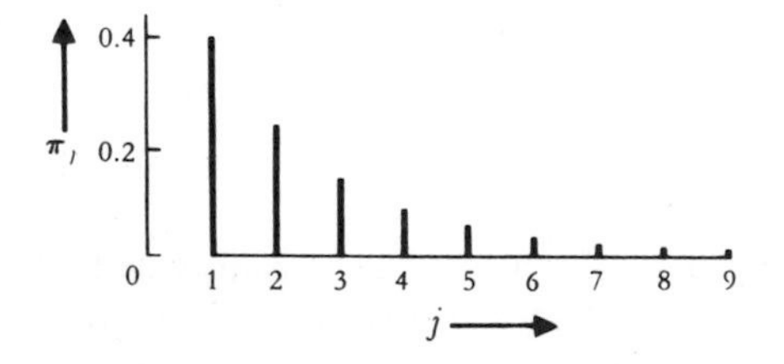

Figure 4.7 π weights for an IMA process of order (0, 1, 1) with $\lambda = 1 - \theta = 0.4$.

Since $b_0^{(k)} = z_k - \theta a_k = z_{k+1} - a_{k+1}$, or $z_{k+1} = b_0^{(k)} + a_{k+1}$, on comparison with (4.3.5) it follows that for this process, the complementary function $b_0^{(k)} = C_k(t - k)$ in (4.3.4) is

$$b_0^{(k)} = \bar{z}_k(\lambda) \tag{4.3.7}$$

an exponentially weighted average of values up to the origin k. In fact, (4.3.4) may be written

$$z_t = \bar{z}_k(\lambda) + \lambda \sum_{j=1}^{t-k-1} a_{t-j} + a_t$$

We have seen that the complementary function $C_k(t - k)$ can be thought of as telling us what is known of the future value of the process at time t, based on knowledge of the past when *we are standing at time k*. For the IMA(0, 1, 1) process, this takes the form of information about the "level" or location of the process $b_0^{(k)} = \bar{z}_k(\lambda)$. At time k, our knowledge of the future behavior of the process is that it will diverge from this level in accordance with the "random walk" represented by $\lambda \sum_{j=1}^{t-k-1} a_{t-j} + a_t$, whose expectation is zero and whose behavior we cannot predict. As soon as a new observation is available, that is, as soon as we move our origin to time $k + 1$, the level will be updated to $b_0^{(k+1)} = \bar{z}_{k+1}(\lambda)$.

Important properties of the IMA(0, 1, 1) process. Since the process is nonstationary, it does not vary in a stable manner about a fixed mean. However, the exponentially weighted moving average $\bar{z}_t(\lambda)$ can be regarded as measuring the location or "level" of the process at time t. From its definition (4.3.6), we obtain the well-known recursion formula for the EWMA,

$$\bar{z}_t(\lambda) = \lambda z_t + (1 - \lambda)\bar{z}_{t-1}(\lambda) \tag{4.3.8}$$

This expression shows that for the IMA(0, 1, 1) model, each new level is arrived at by interpolating between the new observation and the previous level. If λ is equal to unity, $\bar{z}_t(\lambda) = z_t$, which would ignore all evidence concerning location coming from previous observations. On the other hand, if λ had some value close to zero, $\bar{z}_t(\lambda)$ would rely heavily on the previous value $\bar{z}_{t-1}(\lambda)$, which would have weight $1 - \lambda$. Only the small weight λ would be given to the new observation.

Now consider the two equations

$$\begin{aligned} z_t &= \bar{z}_{t-1}(\lambda) + a_t \\ \bar{z}_t(\lambda) &= \bar{z}_{t-1}(\lambda) + \lambda a_t \end{aligned} \tag{4.3.9}$$

the latter being obtained by substituting (4.3.5) in (4.3.8) and is also directly derivable from (4.3.7).

It has been pointed out by Muth [147] that the two equations (4.3.9) provide a useful way of thinking about the generation of the process. The first equation shows how, with the "level" of the system at $\bar{z}_{t-1}(\lambda)$, a shock a_t is added at time t and produces the value z_t. However, the second equation shows that only a proportion λ of the shock is actually absorbed into the level and has a lasting influence, the remaining proportion $\theta = 1 - \lambda$ of the shock being dissipated. Now a new level $\bar{z}_t(\lambda)$ having been established by the absorption of a_t, a new shock a_{t+1} enters the system at time $t + 1$. The equations (4.3.9), with subscripts increased by unity, will then show how this shock produces z_{t+1} and how a proportion λ of it is absorbed into the system to produce the new level $\bar{z}_{t+1}(\lambda)$, and so on.

Equation (4.3.4) can be used to obtain variance and correlation features of the IMA(0, 1, 1) process directly. For example, with reference to the origin k and treating the initializing function $b_0^{(k)}$ as constant, we find that

$$\text{var}[z_t] = \sigma_a^2[1 + (t - k - 1)\lambda^2] \tag{4.3.10}$$

which does not converge as t increases. We might also view this variance as, essentially, the variance of the difference $z_t - z_k$ [treating $a_k = 0$ in (4.3.4)]. In particular, in the case of a random walk process, $z_t = z_{t-1} + a_t$, we have $\lambda = 1$ and this variance function grows proportionally with $t - k$, whereas for more common situations with $0 < \lambda < 1$ (i.e., $0 < \theta < 1$) and especially for λ close to zero, the variance function of $z_t - z_k$ grows much more slowly with $t - k$. In addition, for $s > 0$, $\text{cov}[z_t, z_{t+s}] = \sigma_a^2[\lambda + (t - k - 1)\lambda^2]$, which implies that $\text{corr}[z_t, z_{t+s}]$ will be close to 1 for $t - k$ large relative to s (and λ not close to zero). Hence it follows that adjacent values of the process will be highly positively correlated, so the process will tend to exhibit rather smooth behavior (unless λ is close to zero).

The properties of the IMA(0, 1, 1) process with deterministic drift

$$\nabla z_t = \theta_0 + (1 - \theta_1 B)a_t$$

are discussed in Appendix A4.2.

4.3.2 Integrated Moving Average Process of Order (0, 2, 2)

Difference equation form. The process

$$\nabla^2 z_t = (1 - \theta_1 B - \theta_2 B^2)a_t \tag{4.3.11}$$

possesses representational possibilities for series possessing stochastic trends (e.g., see Figure 4.6) and we now study its general properties within the invertibility region

$$-1 < \theta_2 < 1 \qquad \theta_2 + \theta_1 < 1 \qquad \theta_2 - \theta_1 < 1$$

Proceeding as before, z_t can be written explicitly in terms of z's and a's as

$$z_t = 2z_{t-1} - z_{t-2} + a_t - \theta_1 a_{t-1} \quad \theta_2 a_{t-2}$$

Alternatively, before we obtain z_t in terms of the a's, we first rewrite the right-hand operator in terms of differences

$$1 - \theta_1 B - \theta_2 B^2 = (\lambda_0 \nabla + \lambda_1)B + \nabla^2$$

and on equating coefficients, we find expressions for the θ's in terms of the λ's, and vice versa, as follows:

$$\begin{aligned} \theta_1 &= 2 - \lambda_0 - \lambda_1 \qquad & \lambda_0 &= 1 + \theta_2 \\ \theta_2 &= \lambda_0 - 1 \qquad & \lambda_1 &= 1 - \theta_1 - \theta_2 \end{aligned} \tag{4.3.12}$$

The model (4.3.11) may then be rewritten as

$$\nabla^2 z_t = (\lambda_0 \nabla + \lambda_1)a_{t-1} + \nabla^2 a_t \tag{4.3.13}$$

There is an important advantage in using the form (4.3.13) of the model, as compared with (4.3.11). This stems from the fact that if we set $\lambda_1 = 0$ in (4.3.13), we obtain

$$\nabla z_t = [1 - (1 - \lambda_0)B]a_t$$

which corresponds to a (0, 1, 1) process, with $\theta = 1 - \lambda_0$. However, if we set $\theta_2 = 0$ in (4.3.11), we obtain

$$\nabla^2 z_t = (1 - \theta_1 B)a_t$$

It is shown later in Chapter 5 that for a series generated by the (0, 2, 2) process, the optimal forecasts lie along a straight line, the level and *slope* of which are continually updated as new data become available. By contrast, a series generated by a (0, 1, 1) process can supply no information about slope, but only about a continually updated level. It can be an important question whether a linear trend, as well as the level, can be forecasted and updated. When the choice is between these two models, this question turns on whether or not λ_1 in (4.3.13) is zero.

The invertibility region for an IMA (0, 2, 2) process is the same as that given for a MA(2) process in Chapter 3. It may be written in terms of the θ's and λ's as follows:

$$\begin{aligned} \theta_2 + \theta_1 &< 1 \qquad & 0 < 2\lambda_0 + \lambda_1 &< 4 \\ \theta_2 - \theta_1 &< 1 \qquad & \lambda_1 &> 0 \\ -1 < \theta_2 &< 1 \qquad & \lambda_0 &> 0 \end{aligned} \tag{4.3.14}$$

The triangular region for the θ's was shown in Figure 3.8 and the corresponding region for the λ's is shown in Figure 4.8.

Truncated and infinite random shock forms of model. On applying the finite double summation operator $S_{t-k}^{(2)}$, relative to a time origin k, to

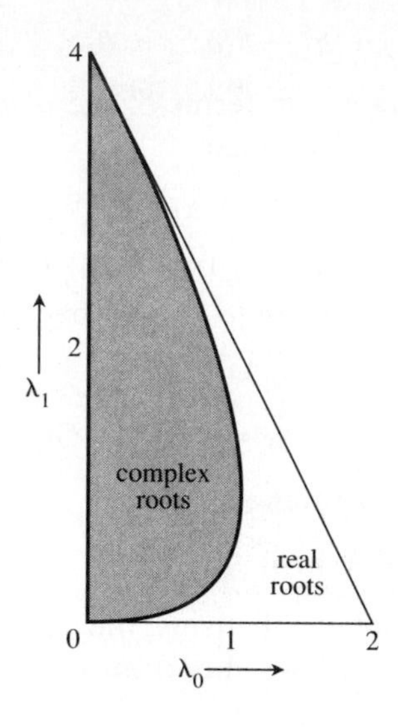

Figure 4.8 Invertibility region for parameters λ_0 and λ_1 of an IMA(0, 2, 2) process.

(4.3.13), we find that

$$[1 - B^{t-k} - (t - k)B^{t-k}(1 - B)]z_t = [\lambda_0(S_{t-k} - (t - k)B^{t-k}) + \lambda_1 S^{(2)}_{t-k}]a_{t-1} + [1 - B^{t-k} - (t - k)B^{t-k}(1 - B)]a_t$$

Hence we obtain the truncated form of the random shock model as

$$\begin{aligned} z_t &= \lambda_0 S_{t-k-1} a_{t-1} + \lambda_1 S^{(2)}_{t-k-1} a_{t-1} + a_t + b_0^{(k)} + b_1^{(k)}(t - k) \\ &= \lambda_0 \sum_{j=1}^{t-k-1} a_{t-j} + \lambda_1 \sum_{j=1}^{t-k-1} j a_{t-j} + a_t + C_k(t - k) \end{aligned} \tag{4.3.15}$$

So, for this process, the ψ weights are

$$\psi_0 = 1, \quad \psi_1 = (\lambda_0 + \lambda_1), \quad \ldots, \quad \psi_j = (\lambda_0 + j\lambda_1), \quad \ldots$$

The complementary function is the solution of

$$(1 - B)^2 C_k(t - k) = 0$$

that is,

$$C_k(t - k) = b_0^{(k)} + b_1^{(k)}(t - k) \tag{4.3.16}$$

which is a polynomial in $(t - k)$ of degree 1 whose coefficients depend on the location of the origin k. From (4.3.15) we find that these coefficients are given explicitly as

$$\begin{aligned} b_0^{(k)} &= z_k - (1 - \lambda_0)a_k \\ b_1^{(k)} &= z_k - z_{k-1} - (1 - \lambda_1)a_k + (1 - \lambda_0)a_{k-1} \end{aligned}$$

Also, by considering the differences $b_0^{(k)} - b_0^{(k-1)}$ and $b_1^{(k)} - b_1^{(k-1)}$, it follows that if the origin is updated from $k - 1$ to k, then b_0 and b_1 are updated according to

$$\begin{aligned} b_0^{(k)} &= b_0^{(k-1)} + b_1^{(k-1)} + \lambda_0 a_k \\ b_1^{(k)} &= b_1^{(k-1)} + \lambda_1 a_k \end{aligned} \tag{4.3.17}$$

We see that when this model is appropriate, our expectation of the future behavior of the series, judged from origin k, would be represented by the straight line (4.3.16), having location $b_0^{(k)}$ and slope $b_1^{(k)}$. In practice, the process will, by time t, have diverged from this line because of the influence of the random component

$$\lambda_0 \sum_{j=1}^{t-k-1} a_{t-j} + \lambda_1 \sum_{j=1}^{t-k-1} j a_{t-j} + a_t$$

which at time k is unpredictable. Moreover, on moving from origin $k - 1$ to origin k, the intercept and slope are updated according to (4.3.17).

Informally, through (4.3.15) we may also obtain the infinite random shock form as

$$z_t = \lambda_0 \sum_{j=1}^{\infty} a_{t-j} + \lambda_1 \sum_{j=1}^{\infty} j a_{t-j} + a_t = \lambda_0 S a_{t-1} + \lambda_1 S^2 a_{t-1} + a_t \tag{4.3.18}$$

So by comparison with (4.3.15), the complementary function can be represented informally as

$$C_k(t - k) = \lambda_0 \sum_{j=t-k}^{\infty} a_{t-j} + \lambda_1 \sum_{j=t-k}^{\infty} j a_{t-j} = b_0^{(k)} + b_1^{(k)}(t - k)$$

By writing the second infinite sum above in the form $\sum_{j=t-k}^{\infty} j a_{t-j} = (t - k) \sum_{j=t-k}^{\infty} a_{t-j} + \sum_{j=t-k}^{\infty} [j - (t - k)] a_{t-j}$, we see that the coefficients $b_0^{(k)}$ and $b_1^{(k)}$ can be associated with

$$\begin{aligned} b_0^{(k)} &= \lambda_0 S a_k + \lambda_1 S^2 a_{k-1} = (\lambda_0 - \lambda_1) S a_k + \lambda_1 S^2 a_k \\ b_1^{(k)} &= \lambda_1 S a_k \end{aligned}$$

Inverted form of model. Finally, we consider the model in the inverted form

$$z_t = \sum_{j=1}^{\infty} \pi_j z_{t-j} + a_t = \bar{z}_{t-1}(\pi) + a_t$$

Using (4.2.22), we find on equating coefficients in

$$1 - 2B + B^2 = (1 - \theta_1 B - \theta_2 B^2)(1 - \pi_1 B - \pi_2 B^2 - \cdots)$$

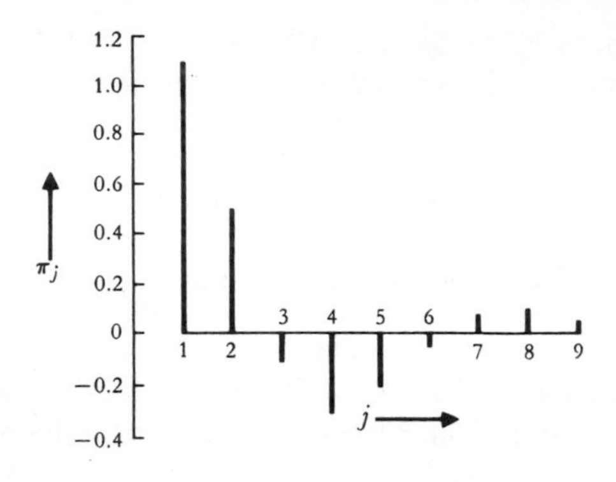

Figure 4.9 π weights for an IMA process of order (0, 2, 2) with $\lambda_0 = 0.5$, $\lambda_1 = 0.6$.

that the π weights of the IMA(0, 2, 2) process are

$$\begin{aligned}\pi_1 &= 2 - \theta_1 = \lambda_0 + \lambda_1 \\ \pi_2 &= \theta_1(2 - \theta_1) - (1 + \theta_2) = \lambda_0 + 2\lambda_1 - (\lambda_0 + \lambda_1)^2 \\ (1 - \theta_1 B &- \theta_2 B^2)\pi_j = 0 \qquad j \geqslant 3\end{aligned} \tag{4.3.19}$$

where B now operates on j.

If the roots of the characteristic equation $1 - \theta_1 B - \theta_2 B^2 = 0$ are real, the π weights applied to previous z's are a mixture of two damped exponentials. If the roots are complex, the weights follow a damped sine wave. Figure 4.9 shows the weights for a process with $\theta_1 = 0.9$ and $\theta_2 = -0.5$, that is, $\lambda_0 = 0.5$ and $\lambda_1 = 0.6$. We see from Figures 3.9 and 4.8 that for these values the characteristic equation has complex roots (the discriminant $\theta_1^2 + 4\theta_2 = -1.19$ is less than zero). Hence the weights in Figure 4.9 would be expected to follow a damped sine wave, as they do.

4.3.3 General Integrated Moving Average Process of Order (0, d, q)

Difference equation form. The general integrated moving average process of order (0, d, q) is

$$\nabla^d z_t = (1 - \theta_1 B - \theta_2 B^2 - \cdots - \theta_q B^q)a_t = \theta(B)a_t \tag{4.3.20}$$

where the zeros of $\theta(B)$ must lie outside the unit circle for the process to be invertible. The model (4.3.20) may be written explicitly in terms of past z's and a's in the form

$$z_t = dz_{t-1} - \tfrac{1}{2}d(d - 1)z_{t-2} + \cdots + (-1)^{d+1}z_{t-d} + a_t - \theta_1 a_{t-1} - \cdots - \theta_q a_{t-q}$$

Random shock form of model. To obtain z_t in terms of the a's, we write the right-hand operator in (4.3.20) in terms of $\nabla = 1 - B$. In this way we obtain

$$(1 - \theta_1 B - \cdots - \theta_q B^q) - (\lambda_{d-q}\nabla^{q-1} + \cdots + \lambda_0 \nabla^{d-1} + \cdots + \lambda_{d-1})B + \nabla^d \tag{4.3.21}$$

where, as before, the λ's may be written explicitly in terms of the θ's, by equating coefficients of B.

On substituting (4.3.21) in (4.3.20) and summing d times, informally, we obtain

$$z_t = (\lambda_{d-q}\nabla^{q-d-1} + \cdots + \lambda_0 S + \cdots + \lambda_{d-1}S^d)a_{t-1} + a_t \tag{4.3.22}$$

From (4.3.22), if $q > d$, we notice that in addition to the d sums, we pick up $q - d$ additional terms $\nabla^{q-d-1}a_{t-1}, \ldots$ involving $a_{t-1}, a_{t-2}, \ldots, a_{t+d-q}$.

If we write this solution in terms of finite sums of a's entering the system after some origin k, we obtain the same form of equation, but with an added complementary function, which is the solution of

$$\nabla^d C_k(t - k) = 0$$

that is, the polynomial

$$C_k(t-k) = b_0^{(k)} + b_1^{(k)}(t-k) + b_2^{(k)}(t-k)^2 + \cdots + b_{d-1}^{(k)}(t-k)^{d-1}$$

As before, the complementary function $C_k(t-k)$ represents that aspect of the finite behavior of the process, which is predictable at time k. Similarly, the coefficients $b_j^{(k)}$ may be expressed, informally, in terms of the infinite sums up to origin k, that is, $Sa_k, S^2a_k, \ldots, S^d a_k$. Accordingly, we can discover how the coefficients $b_j^{(k)}$ change as the origin is changed, from $k - 1$ to k.

Inverted form of model. Finally, the model can be expressed in the inverted form

$$\pi(B)z_t = a_t$$

or

$$z_t = \bar{z}_{t-1}(\pi) + a_t$$

The π weights may be obtained by equating coefficients in (4.2.22), that is,

$$(1 - B)^d = (1 - \theta_1 B - \theta_2 B^2 - \cdots - \theta_q B^q)(1 - \pi_1 B - \pi_2 B^2 - \cdots) \tag{4.3.23}$$

For a given model, they are best obtained by substituting numerical values in (4.3.23), rather than by deriving a general formula. We note that (4.3.23) implies that for j greater than the larger of d and q, the π weights satisfy the difference equation

$$\theta(B)\pi_j = 0$$

defined by the moving average operator. Hence, for sufficiently large j, the weights π_j follow a mixture of damped exponentials and sine waves.

IMA process of order (0, 2, 3). One final special case of sufficient interest to merit comment is the IMA process of order (0, 2, 3),

$$\nabla^2 z_t = (1 - \theta_1 B - \theta_2 B^2 - \theta_3 B^3)a_t$$

Proceeding as before, if we apply the finite double summation operator, this model can be written in truncated random shock form as

$$z_t = \lambda_{-1} a_{t-1} + \lambda_0 \sum_{j=1}^{t-k-1} a_{t-j} + \lambda_1 \sum_{j=1}^{t-k-1} j a_{t-j} + a_t + b_0^{(k)} + b_1^{(k)}(t - k)$$

where the relations between the λ's and θ's are

$$\begin{aligned} \theta_1 &= 2 - \lambda_{-1} - \lambda_0 - \lambda_1 & \lambda_{-1} &= -\theta_3 \\ \theta_2 &= \lambda_0 - 1 + 2\lambda_{-1} & \lambda_0 &= 1 + \theta_2 + 2\theta_3 \\ \theta_3 &= -\lambda_{-1} & \lambda_1 &= 1 - \theta_1 - \theta_2 - \theta_3 \end{aligned}$$

Alternatively, it can be written, informally, in the infinite integrated form as

$$z_t = \lambda_{-1} a_{t-1} + \lambda_0 S a_{t-1} + \lambda_1 S^2 a_{t-1} + a_t$$

Finally, the invertibility region is defined by

$$\begin{aligned} \theta_1 + \theta_2 + \theta_3 &< 1 & \lambda_1 &> 0 \\ -\theta_1 + \theta_2 - \theta_3 &< 1 & 2\lambda_0 + \lambda_1 &< 4(1 - \lambda_{-1}) \\ \theta_3(\theta_3 - \theta_1) - \theta_2 &< 1 & \lambda_0(1 + \lambda_{-1}) &> -\lambda_1 \lambda_{-1} \\ |\theta_3| &< 1 & |\lambda_{-1}| &< 1 \end{aligned}$$

and is shown in Figure 4.10.

In Chapter 5 we show how future values of a time series can be forecast in an optimal manner when the model is an ARIMA process. In studying these forecasts we make considerable use of the various model forms discussed in this chapter.

APPENDIX A4.1 LINEAR DIFFERENCE EQUATIONS

In this book we are often concerned with linear difference equations. In particular, the ARIMA model relates an output z_t to an input a_t in terms of the difference equation

$$\begin{aligned} z_t - \varphi_1 z_{t-1} - \varphi_2 z_{t-2} - \cdots - \varphi_{p'} z_{t-p'} \\ = a_t - \theta_1 a_{t-1} - \theta_2 a_{t-2} - \cdots - \theta_q a_{t-q} \end{aligned} \tag{A4.1.1}$$

where $p' = p + d$.

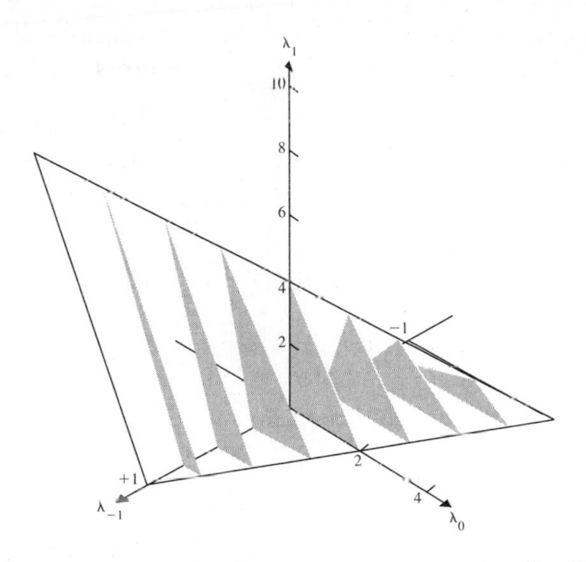

Figure 4.10 Invertibility region for parameters λ_{-1}, λ_0, and λ_1 of an IMA(0, 2, 3) process.

Alternatively, we may write (A4.1.1) as

$$\varphi(B)z_t = \theta(B)u_t$$

where

$$\varphi(B) = 1 - \varphi_1 B - \varphi_2 B^2 - \cdots - \varphi_{p'} B^{p'}$$
$$\theta(B) = 1 - \theta_1 B - \theta_2 B^2 - \cdots - \theta_q B^q$$

We now derive an expression for the general solution of the difference equation (A4.1.1) relative to an origin $k < t$.

1. We show that the general solution may be written as
$$z_t = C_k(t-k) + I_k(t-k)$$
where $C_k(t-k)$ is the complementary function and $I_k(t-k)$ a "particular integral."
2. We then derive a general expression for the complementary function $C_k(t-k)$.
3. Finally, we derive a general expression for a particular integral $I_k(t-k)$.

General solution. The argument is identical to that for the solution of linear differential or linear algebraic equations. Suppose that z_t' is any particular solution of

$$\varphi(B)z_t = \theta(B)a_t \tag{A4.1.2}$$

that is, it satisfies

$$\varphi(B)z_t' = \theta(B)a_t \tag{A4.1.3}$$

On subtracting (A4.1.3) from (A4.1.2), we obtain

$$\varphi(B)(z_t - z_t') = 0$$

Thus $z_t'' = z_t - z_t'$ satisfies

$$\varphi(B)z_t'' = 0 \tag{A4.1.4}$$

Now

$$z_t = z_t' + z_t''$$

and hence the general solution of (A4.1.2) is the sum of the complementary function z_t'', which is the general solution of the homogeneous difference equation (A4.1.4), and a particular integral z_t', which is any particular solution of (A4.1.2). Relative to any origin $k < t$ we denote the complementary function z_t'' by $C_k(t-k)$ and the particular integral z_t' by $I_k(t-k)$.

Evaluation of the complementary function

Distinct roots. Consider the homogeneous difference equation

$$\varphi(B)z_t = 0 \tag{A4.1.5}$$

where

$$\varphi(B) = (1 - G_1B)(1 - G_2B)\cdots(1 - G_{p'}B) \tag{A4.1.6}$$

and where we assume in the first instance that $G_1, G_2, \ldots, G_{p'}$ are *distinct*. Then it is shown below that the general solution of (A4.1.5) at time t, when the series is referred to an origin at time k, is

$$z_t = A_1G_1^{t-k} + A_2G_2^{t-k} + \cdots + A_{p'}G_{p'}^{t-k} \tag{A4.1.7}$$

where the A_i's are constants. Thus a real root of $\varphi(B) = 0$ contributes a damped exponential term G^{t-k} to the complementary function. A pair of complex roots contributes a damped sine wave term $D^{t-k}\sin(2\pi f_0 t + F)$.

To see that (A4.1.7) does satisfy (A4.1.5), we can substitute (A4.1.7) in (A4.1.5) to give

$$\varphi(B)(A_1G_1^{t-k} + A_2G_2^{t-k} + \cdots + A_{p'}G_{p'}^{t-k}) = 0 \tag{A4.1.8}$$

Now consider

$$\begin{aligned}\varphi(B)G_i^{t-k} &= (1 - \varphi_1B - \varphi_2B^2 - \cdots - \varphi_{p'}B^{p'})G_i^{t-k}\\ &= G_i^{t-k-p'}(G_i^{p'} - \varphi_1G_i^{p'-1} - \cdots - \varphi_{p'})\end{aligned}$$

We see that $\varphi(B)G_i^{t-k}$ vanishes for each value of i if

$$G_i^{p'} - \varphi_1 G_i^{p'-1} - \cdots - \varphi_{p'} = 0$$

that is, if $B = 1/G_i$ is a root of $\varphi(B) = 0$. Now, since (A4.1.6) implies that the roots of $\varphi(B) = 0$ are $B = 1/G_i$, it follows that $\varphi(B)G_i^{t-k}$ is zero for all i and hence (A4.1.8) holds, confirming that (A4.1.7) is a general solution of (A4.1.5).

To prove (A4.1.7) directly, consider the special case of the second-order equation

$$(1 - G_1 B)(1 - G_2 B)z_t = 0$$

which we can write as

$$(1 - G_1 B)y_t = 0 \tag{A4.1.9}$$

where

$$y_t = (1 - G_2 B)z_t \tag{A4.1.10}$$

Now (A4.1.9) implies that

$$y_t = G_1 y_{t-1} = G_1^2 y_{t-2} = \cdots = G_1^{t-k} y_k$$

and hence

$$y_t = D_1 G_1^{t-k}$$

where $D_1 = y_k$ is a constant determined by the starting value y_k. Hence (A4.1.10) may be written

$$\begin{aligned} z_t &= G_2 z_{t-1} + D_1 G_1^{t-k} \\ &= G_2(G_2 z_{t-2} + D_1 G_1^{t-k-1}) + D_1 G_1^{t-k} \\ &\vdots \\ &= G_2^{t-k} z_k + D_1(G_1^{t-k} + G_2 G_1^{t-k-1} + \cdots + G_2^{t-k-1} G_1) \qquad \text{(A4.1.11)} \\ &= G_2^{t-k} z_k + \frac{D_1}{1 - G_2/G_1}(G_1^{t-k} - G_2^{t-k}) \\ &= A_1 G_1^{t-k} + A_2 G_2^{t-k} \end{aligned}$$

where A_1, A_2 are constants determined by the starting values of the series. By an extension of the argument above, it may be shown that the general solution of (A4.1.5), when the roots of $\varphi(B) = 0$ are distinct, is given by (A4.1.7).

Equal roots. Suppose that $\varphi(B) = 0$ has d equal roots G_0^{-1}, so that $\varphi(B)$ contains a factor $(1 - G_0 B)^d$. In particular, consider the solution (A4.1.11) for the second-order equation when both G_1 and G_2 are equal to G_0. Then (A4.1.11) reduces to

$$z_t = G_0^{t-k} z_k + D_1 G_0^{t-k}(t-k)$$

or

$$z_t = [A_0 + A_1(t-k)]G_0^{t-k}$$

In general, if there are d equal roots G_0, it may be verified by direct substitution in (A4.1.5) that the general solution is

$$z_t = [A_0 + A_1(t-k) + A_2(t-k)^2 + \cdots + A_{d-1}(t-k)^{d-1}]G_0^{t-k} \tag{A4.1.12}$$

In particular, when the equal roots G_0 are all equal to unity as in the IMA(0, d, q) process, the solution is

$$z_t = A_0 + A_1(t-k) + \cdots + A_{d-1}(t-k)^{d-1} \tag{A4.1.13}$$

that is, a polynomial in $t - k$ of degree $d - 1$.

In general, when $\varphi(B)$ factors according to

$$(1 - G_1B)(1 - G_2B)\cdots(1 - G_pB)(1 - G_0B)^d$$

the complementary function is

$$C_k(t-k) = G_0^{t-k}\sum_{j=0}^{d-1} A_j(t-k)^j + \sum_{i=1}^{p} D_i G_i^{t-k} \tag{A4.1.14}$$

Thus, in general, the complementary function consists of a mixture of damped exponential terms G^{t-k}, polynomial terms $(t-k)^j$, damped sine wave terms $D^{t-k}\sin(2\pi f_0 t + F)$, and combinations of these functions.

Evaluation of the "particular integral." We now show that a particular integral $I_k(s-k)$, satisfying

$$\varphi(B)I_k(t-k) = \theta(B)a_t \qquad t - k > q \tag{A4.1.15}$$

is a function defined as follows:

$$\begin{aligned} I_k(s-k) &= 0 \qquad s \leqslant k \\ I_k(1) &= a_{k+1} \\ I_k(2) &= a_{k+2} + \psi_1 a_{k+1} \\ &\vdots \\ I_k(t-k) &= a_t + \psi_1 a_{t-1} + \psi_2 a_{t-2} + \cdots + \psi_{t-k-1}a_{k+1} \qquad t > k \end{aligned} \tag{A4.1.16}$$

where the ψ weights are those appearing in the form (4.2.3) of the model. Thus the ψ weights satisfy

$$\varphi(B)\psi(B)a_t = \theta(B)a_t \tag{A4.1.17}$$

Now the terms on the left-hand side of (A4.1.17) may be set out as follows:

$$
\begin{array}{lcl|l}
 & a_t + \psi_1 a_{t-1} + \psi_2 a_{t-2} + \cdots + \psi_{t-k-1} a_{k+1} & & + \psi_{t-k} a_k + \cdots \\
-\varphi_1(& a_{t-1} + \psi_1 a_{t-2} + \cdots + \psi_{t-k-2} a_{k+1} & & + \psi_{t-k-1} a_k + \cdots \\
-\varphi_2(& \cdots \quad \cdots \quad \cdots \quad \cdots & & \cdots \quad \cdots \quad \cdots \\
\vdots & \vdots \quad \vdots \quad \vdots \quad \vdots & & \vdots \quad \vdots \quad \vdots \\
-\varphi_{p'}(& a_{t-p'} + \cdots + \psi_{t-k-p'-1} a_{k+1} & & + \psi_{t-k-p'} a_k + \cdots
\end{array}
\tag{A4.1.18}
$$

Since the right-hand side of (A4.1.17) is

$$a_t - \theta_1 a_{t-1} - \cdots - \theta_q a_{t-q}$$

it follows that the first $q + 1$ columns in this array sum to a_t, $-\theta_1 a_{t-1}$, $\ldots, -\theta_q a_{t-q}$. Now the left-hand term in (A4.1.15), where $I_k(s - k)$ is given by (A4.1.16), is equal to the sum of the terms in the first $(t - k)$ columns of the array, that is, those to the left of the vertical line. Therefore, if $t - k < q$, that is, the vertical line is drawn after $q + 1$ columns, the sum of all terms up to the vertical line is equal to $\theta(B)a_t$. This shows that (A4.1.16) is a "particular integral" of the difference equation.

Example. Consider the IMA(0, 1, 1) process

$$z_t - z_{t-1} = a_t - \theta a_{t-1} \tag{A4.1.19}$$

for which $\psi_j = 1 - \theta$ for $j \geq 1$. Then

$$
\begin{aligned}
I_k(0) &= 0 \\
I_k(1) &= a_{k+1} \\
\vdots \quad & \quad \vdots \\
I_k(t-k) &= a_t + (1-\theta) \sum_{j=1}^{t-k-1} a_{t-j} \qquad t - k > 1
\end{aligned}
\tag{A4.1.20}
$$

Now if $z_t = I_k(t - k)$ is a solution of (A4.1.19), then

$$I_k(t - k) - I_k(t - k - 1) = a_t - \theta a_{t-1}$$

and as is easily verified, while this is not satisfied by (A4.1.20) for $t - k = 1$, it is satisfied by (A4.1.20) for $t - k > 1$, that is, for $t - k > q$.

APPENDIX A4.2 IMA(0, 1, 1) PROCESS WITH DETERMINISTIC DRIFT

The general model $\phi(B)\nabla^d z_t = \theta_0 + \theta(B)a_t$ can also be written

$$\phi(B)\nabla^d z_t = \theta(B)\varepsilon_t$$

with the shocks ε_t having a nonzero mean $\xi = \theta_0/(1 - \theta_1 - \cdots - \theta_q)$. For example, the IMA(0, 1, 1) model is then

$$\nabla z_t = (1 - \theta B)\varepsilon_t$$

with $E[\varepsilon_t] = \xi = \theta_0/(1 - \theta)$. In this form, z_t could represent, for example, the outlet temperature from a reactor when heat was being supplied from a heating element at a fixed rate. Now if

$$\varepsilon_t = \xi + a_t \tag{A4.2.1}$$

where a_t is white noise with zero mean, then with reference to a time origin k, the integrated form for the model is

$$z_t = b_0^{(k)} + \lambda \sum_{j=1}^{t-k-1} \varepsilon_{t-j} + \varepsilon_t \tag{A4.2.2}$$

with $\lambda = 1 - \theta$. Substituting for (A4.2.1) in (A4.2.2), the model written in terms of the a's is

$$z_t = b_0^{(k)} + \lambda\xi(t - k - 1) + \xi + \lambda \sum_{j=1}^{t-k-1} a_{t-j} + a_t \tag{A4.2.3}$$

Thus we see that z_t contains a deterministic slope or drift due to the term $\lambda\xi(t - k - 1)$, with the slope of the deterministic linear trend equal to $\lambda\xi = \theta_0$. Moreover, if we denote the "level" of the process at time $t - 1$ by l_{t-1}, where

$$z_t = l_{t-1} + a_t$$

we see that the level is changed from time $t - 1$ to time t, according to

$$l_t = l_{t-1} + \lambda\xi + \lambda a_t$$

The change in the level thus contains a deterministic component $\lambda\xi = \theta_0$, as well as a stochastic component λa_t.

APPENDIX A4.3 ARIMA PROCESSES WITH ADDED NOISE

In this appendix we consider the effect of adding noise (e.g., measurement error) to a general ARIMA(p, d, q) process.

A4.3.1 Sum of Two Independent Moving Average Processes

As a necessary preliminary to what follows, consider a stochastic process w_t, which is the sum of two *independent* moving average processes of orders q_1 and q_2, respectively. That is,

$$w_t = \theta_1(B)a_t + \theta_2(B)b_t \tag{A4.3.1}$$

where $\theta_1(B)$ and $\theta_2(B)$ are polynomials in B, of order q_1 and q_2, and the white noise processes a_t and b_t have zero means and are mutually independent. Suppose that $q = \max(q_1, q_2)$; then it is clear that the autocovariance function γ_j for w_t must be zero for $j > q$. It follows that there exists a representation of w_t as a single moving average process of order q:

$$w_t = \theta_3(B)u_t \tag{A4.3.2}$$

where u_t is a white noise process with mean zero. Thus the sum of two independent moving average processes is another moving average process, whose order is the same as that of the component process of higher order.

A4.3.2 Effect of Added Noise on the General Model

Correlated noise. Consider the general nonstationary model of order (p, d, q)

$$\phi(B)\nabla^d z_t = \theta(B)a_t \tag{A4.3.3}$$

Suppose that we cannot observe z_t itself, but only $Z_t = z_t + b_t$, where b_t represents some extraneous noise (e.g., measurement error) and may be correlated. We wish to determine the nature of the observed process Z_t. In general, we have

$$\phi(B)\nabla^d Z_t = \theta(B)a_t + \phi(B)\nabla^d b_t$$

If the noise follows a stationary ARMA process of order $(p_1, 0, q_1)$,

$$\phi_1(B)b_t = \theta_1(B)\alpha_t \tag{A4.3.4}$$

where α_t is a white noise process independent of the a_t process, then

$$\underbrace{\phi_1(B)\phi(B)\nabla^d Z_t}_{p_1 + p + d} = \underbrace{\phi_1(B)\theta(B)a_t}_{p_1 + q} + \underbrace{\phi(B)\theta_1(B)\nabla^d \alpha_t}_{p + q_1 + d} \tag{A4.3.5}$$

where the values below the braces indicate the degrees of the various polynomials in B. Now the right-hand side of (A4.3.5) is of the form (A4.3.1). Let $P = p_1 + p$ and Q be equal to whichever of $(p_1 + q)$ and $(p + q_1 + d)$ is larger. Then we can write

$$\phi_2(B)\nabla^d Z_t = \theta_2(B)u_t$$

with u_t a white noise process, and the Z_t process is seen to be of order (P, d, Q).

Added white noise. If, as might be true in some applications, the added noise is white, then $\phi_1(B) = \theta_1(B) = 1$ in (A4.3.4), and we obtain

$$\phi(B)\nabla^d Z_t = \theta_2(B)u_t \tag{A4.3.6}$$

with

$$\theta_2(B)u_t = \theta(B)a_t + \phi(B)\nabla^d b_t$$

which is of order (p, d, Q) where Q is the larger of q and $(p + d)$. If $p + d \leqslant q$, the order of the process with error is the same as that of the original process. The only effect of the added white noise is to change the values of the θ's (but not the ϕ's).

Effect of added white noise on an integrated moving average process. In particular, an IMA process of order $(0, d, q)$, with white noise added, remains an IMA of order $(0, d, q)$ if $d \leqslant q$; otherwise, it becomes an IMA of order $(0, d, d)$. In either case the parameters of the process are changed by the addition of noise. The nature of these changes can be determined by equating the autocovariances of the dth differences of the process, with added noise, to those of the dth differences of a simple IMA process. The procedure will now be illustrated with an example.

A4.3.3 Example for an IMA(0, 1, 1) Process with Added White Noise

Consider the properties of the process $Z_t = z_t + b_t$ when

$$z_t = z_{t-1} - (1 - \lambda)a_{t-1} + a_t \tag{A4.3.7}$$

and the b_t and a_t are mutually independent white noise processes. The Z_t process has first difference $W_t = Z_t - Z_{t-1}$, given by

$$W_t = [1 - (1 - \lambda)B]a_t + (1 - B)b_t \tag{A4.3.8}$$

The autocovariances for the first differences W_t are

$$\begin{aligned} \gamma_0 &= \sigma_a^2[1 + (1 - \lambda)^2] + 2\sigma_b^2 \\ \gamma_1 &= -\sigma_a^2(1 - \lambda) - \sigma_b^2 \\ \gamma_j &= 0 \qquad j \geqslant 2 \end{aligned} \tag{A4.3.9}$$

The fact that the γ_j are zero beyond the first confirms that the process with added noise is, as expected, an IMA process of order (0, 1, 1). To obtain explicitly the parameters of the IMA that represents the noisy process, we suppose that it can be written

$$Z_t = Z_{t-1} - (1 - \Lambda)u_{t-1} + u_t \tag{A4.3.10}$$

with u_t a white noise process. The process (A4.3.10) has first differences $W_t = Z_t - Z_{t-1}$ with autocovariances

$$\begin{aligned} \gamma_0 &= \sigma_u^2[1 + (1 - \Lambda)^2] \\ \gamma_1 &= -\sigma_u^2(1 - \Lambda) \\ \gamma_j &= 0 \qquad j \geqslant 2 \end{aligned} \tag{A4.3.11}$$

Equating (A4.3.9) and (A4.3.11), we can solve for Λ and for σ_u^2 explicitly. Thus

$$\frac{\Lambda^2}{1-\Lambda} = \frac{\lambda^2}{1-\lambda+\sigma_b^2/\sigma_a^2} \tag{A4.3.12}$$

$$\sigma_u^2 = \sigma_a^2 \frac{\lambda^2}{\Lambda^2}$$

Suppose, for example, that the original series has $\lambda = 0.5$ and $\sigma_b^2 = \sigma_a^2$; then $\Lambda = 0.33$ and $\sigma_u^2 = 2.25\sigma_a^2$.

A4.3.4 Relation between the IMA(0, 1, 1) Process and a Random Walk

The process

$$z_t = z_{t-1} + a_t \tag{A4.3.13}$$

which is an IMA(0, 1, 1) process, with $\lambda = 1$, is sometimes called a *random walk*. If the a_t are steps taken forward or backward at time t, then z_t will represent the position of the walker at time t.

Any IMA(0, 1, 1) process can be thought of as a random walk buried in white noise b_t, uncorrelated with the shocks a_t associated with the random walk process. If the noisy process is $Z_t = z_t + b_t$, where z_t is defined by (A4.3.13), then using (A4.3.12), we have

$$Z_t - Z_{t-1} = (1-\Lambda)u_{t-1} + u_t$$

with

$$\frac{\Lambda^2}{1-\Lambda} = \frac{\sigma_a^2}{\sigma_b^2} \qquad \sigma_u^2 = \frac{\sigma_a^2}{\Lambda^2} \tag{A4.3.14}$$

A4.3.5 Autocovariance Function of the General Model with Added Correlated Noise

Suppose that the basic process is an ARIMA process of order (p, d, q)

$$\phi(B)\nabla^d z_t = \theta(B)a_t$$

and that $Z_t = z_t + b_t$ is observed, where the stationary b_t process, which has autocovariance function $\gamma_j(b)$, is independent of the a_t process, and hence of z_t. Suppose that $\gamma_j(w)$ is the autocovariance function for $w_t = \nabla^d z_t = \phi^{-1}(B)\theta(B)a_t$ and that $W_t = \nabla^d Z_t$. We require the autocovariance function for W_t. Now

$$\nabla^d(Z_t - b_t) = \phi^{-1}(B)\theta(B)a_t$$

$$W_t = w_t + v_t$$

where

$$v_t = \nabla^d b_t = (1 - B)^d b_t$$

Hence

$$\gamma_j(W) = \gamma_j(w) + \gamma_j(v)$$
$$\gamma_j(v) = (1 - B)^d(1 - F)^d\gamma_j(b)$$
$$= (-1)^d(1 - B)^{2d}\gamma_{j+d}(b)$$

and

$$\gamma_j(W) = \gamma_j(w) + (-1)^d(1 - B)^{2d}\gamma_{j+d}(b) \qquad \text{(A4.3.15)}$$

For example, suppose that correlated noise b_t is added to an IMA(0, 1, 1) process defined by $w_t = \nabla z_t = (1 - \theta B)a_t$. Then the autocovariances of the first difference W_t of the "noisy" process will be

$$\gamma_0(W) = \sigma_a^2(1 + \theta^2) + 2[\gamma_0(b) - \gamma_1(b)]$$
$$\gamma_1(W) = -\sigma_a^2\theta + [2\gamma_1(b) - \gamma_0(b) - \gamma_2(b)]$$
$$\gamma_j(W) = [2\gamma_j(b) - \gamma_{j-1}(b) - \gamma_{j+1}(b)] \qquad j \geqslant 2$$

In particular, if b_t was first-order autoregressive, so that

$$b_t = \phi b_{t-1} + \alpha_t$$
$$\gamma_0(W) = \sigma_a^2(1 + \theta^2) + 2\sigma_b^2(1 - \phi)$$
$$\gamma_1(W) = -\sigma_a^2\theta - \sigma_b^2(1 - \phi)^2$$
$$\gamma_j(W) = -\sigma_b^2\phi^{j-1}(1 - \phi)^2 \qquad j \geqslant 2$$

In fact, from (A4.3.5), the resulting noisy process $Z_t = z_t + b_t$ is in this case defined by

$$(1 - \phi B)\nabla Z_t = (1 - \phi B)(1 - \theta B)a_t + (1 - B)\alpha_t$$

which is of order (1, 1, 2).

5

FORECASTING

Having considered in Chapter 4 some of the properties of ARIMA models, we now show how they may be used to forecast future values of an observed time series. In Part Two we consider the problem of fitting the model to actual data. For the present, however, we proceed as if the model were known *exactly*, bearing in mind that estimation errors in the parameters will not seriously affect the forecasts unless the number of data points, used to fit the model, is small.

In this chapter we consider nonseasonal time series. The forecasting of seasonal time series is described in Chapter 9. We show how minimum mean square error forecasts may be generated directly from the *difference equation* form of the model. A further recursive calculation yields the probability limits for the forecasts. It is to be emphasized that for practical computation of the forecasts, this approach via the difference equation is the simplest and most elegant. However, to provide insight into the nature of the forecasts, we also consider them from other viewpoints.

5.1 MINIMUM MEAN SQUARE ERROR FORECASTS AND THEIR PROPERTIES

In Section 4.2 we discussed three explicit forms for the general ARIMA model

$$\varphi(B)z_t = \theta(B)a_t \tag{5.1.1}$$

where $\varphi(B) = \phi(B)\nabla^d$. We begin by recalling these three forms, since each one throws light on a different aspect of the forecasting problem.

We shall be concerned with forecasting a value z_{t+l}, $l \geqslant 1$, when we are currently standing at time t. This forecast is said to be made at *origin t* for *lead time l*. We now summarize the results of Section 4.2, but writing $t + l$ for t and t for k.

Three explicit forms for the model. An observation z_{t+l} generated by the process (5.1.1) may be expressed as follows:

1. Directly in terms of the difference equation by

$$z_{t+l} = \varphi_1 z_{t+l-1} + \cdots + \varphi_{p+d} z_{t+l-p-d} - \theta_1 a_{t+l-1} - \cdots - \theta_q a_{t+l-q} + a_{t+l} \tag{5.1.2}$$

2. As an infinite weighted sum of current and previous shocks a_j,

$$z_{t+l} = \sum_{j=0}^{\infty} \psi_j a_{t+l-j} \tag{5.1.3}$$

where $\psi_0 = 1$ and, as in (4.2.5), the ψ weights may be obtained by equating coefficients in

$$\varphi(B)(1 + \psi_1 B + \psi_2 B^2 + \cdots) = \theta(B) \tag{5.1.4}$$

Equivalently, for positive l, with reference to origin $k < t$, the model may be written in the truncated form

$$\begin{aligned} z_{t+l} &= a_{t+l} + \psi_1 a_{t+l-1} + \cdots + \psi_{l-1} a_{t+1} \\ &\quad + \psi_l a_t + \cdots + \psi_{t+l-k-1} a_{k+1} + C_k(t + l - k) \\ &= a_{t+l} + \psi_1 a_{t+l-1} + \cdots + \psi_{l-1} a_{t+1} + C_t(l) \end{aligned} \tag{5.1.5}$$

where $C_k(t + l - k)$ is the complementary function relative to the finite origin k of the process, and from (4.2.19) we recall that the complementary function relative to the forecast origin t can be expressed as $C_t(l) = C_k(t + l - k) + \psi_l a_t + \psi_{l+1} a_{t-1} + \cdots + \psi_{t+l-k-1} a_{k+1}$. Informally, $C_t(l)$ is associated with the truncated infinite sum

$$C_t(l) = \sum_{j=l}^{\infty} \psi_j a_{t+l-j} \tag{5.1.6}$$

3. As an infinite weighted sum of previous observations, plus a random shock,

$$z_{t+l} = \sum_{j=1}^{\infty} \pi_j z_{t+l-j} + a_{t+l} \tag{5.1.7}$$

Also, if $d \geq 1$,

$$\bar{z}_{t+l-1}(\pi) = \sum_{j=1}^{\infty} \pi_j z_{t+l-j} \tag{5.1.8}$$

will be a weighted average, since then $\sum_{j=1}^{\infty} \pi_j = 1$.

As in (4.2.22), the π weights may be obtained from

$$\varphi(B) = (1 - \pi_1 B - \pi_2 B^2 - \cdots)\theta(B) \tag{5.1.9}$$

5.1.1 Derivation of the Minimum Mean Square Error Forecasts

Now suppose, standing at origin t, that we are to make a forecast $\hat{z}_t(l)$ of z_{t+l} which is to be a linear function of current and previous observations $z_t, z_{t-1}, z_{t-2}, \ldots$. Then it will also be a linear function of current and previous shocks $a_t, a_{t-1}, a_{t-2}, \ldots$.

Suppose, then, that the best forecast is

$$\hat{z}_t(l) = \psi_l^* a_t + \psi_{l+1}^* a_{t-1} + \psi_{l+2}^* a_{t-2} + \cdots$$

where the weights $\psi_l^*, \psi_{l+1}^*, \ldots$ are to be determined. Then, using (5.1.3), the mean square error of the forecast is

$$E[z_{t+l} - \hat{z}_t(l)]^2 = (1 + \psi_1^2 + \cdots + \psi_{l-1}^2)\sigma_a^2 + \sum_{j=0}^{\infty} (\psi_{l+j} - \psi_{l+j}^*)^2 \sigma_a^2 \tag{5.1.10}$$

which is minimized by setting $\psi_{l+j}^* = \psi_{l+j}$, a conclusion that is a special case of more general results in prediction theory due to Wold [207], Kolmogoroff [130], [131], [132], Wiener [200], and Whittle [198]. We then have

$$z_{t+l} = (a_{t+l} + \psi_1 a_{t+l-1} + \cdots + \psi_{l-1} a_{t+1}) + (\psi_l a_t + \psi_{l+1} a_{t-1} + \cdots) \tag{5.1.11}$$

$$= e_t(l) + \hat{z}_t(l) \tag{5.1.12}$$

where $e_t(l)$ is the error of the forecast $\hat{z}_t(l)$ at lead time l.

Certain important facts emerge. As before, denote $E[z_{t+l}|z_t, z_{t-1}, \ldots]$, the conditional expectation of z_{t+l} given knowledge of all the z's up to time t, by $E_t[z_{t+l}]$. We will assume that the a_t are a sequence of independent random variables.

1. Then $E[a_{t+j}|z_t, z_{t-1}, \ldots] = 0, j > 0$, and so from (5.1.3),

$$\hat{z}_t(l) = \psi_l a_t + \psi_{l+1} a_{t-1} + \cdots = E_t[z_{t+l}] \tag{5.1.13}$$

Thus the minimum mean square error forecast at origin t, for lead time l, is the conditional expectation of z_{t+l} at time t. When $\hat{z}_t(l)$ is regarded as a function of l for fixed t, it will be called the *forecast function* for origin t. We note that a minimum requirement on the random shocks a_t in the model (5.1.1) in order for the conditional expectation $E_t[z_{t+l}]$, which always equals the minimum mean square error forecast, to coincide with the minimum mean square error *linear* forecast is that $E_t[a_{t+j}] = 0, j > 0$. This situation may not hold for certain types of intrinsically nonlinear processes, whose study has recently been receiving considerable attention (see, e.g., [160], [188]). Such processes may, in fact, possess a linear representation as in (5.1.1), but the shocks a_t will not be independent, only uncorrelated, and the best forecast $E_t[z_{t+l}]$ may not coincide with the best linear forecast $\hat{z}_t(l)$ as obtained in (5.1.11).

The preceding development, which yields (5.1.13), based only informally on the infinite random shock form (5.1.3) for nonstationary cases when $d \geq 1$, can also be derived formally from the truncated form of the model (5.1.5). For this we assume that the complementary function relative to the initial fixed origin $k < t$, $C_k(t - k)$, which is already determined at time k, is independent of all subsequent random shocks a_{k+i}, $i > 0$. Then, taking conditional expectations of the right side of the first equation in (5.1.5), we obtain

$$E_t[z_{t+l}] = \hat{z}_t(l) = \psi_l a_t + \psi_{l+1} a_{t-1} + \cdots + \psi_{t+l-k-1} a_{k+1} + C_k(t + l - k)$$

as the minimum mean square error forecast at origin t. This expression is thus informally represented by the second term on the right side of (5.1.11).

2. The forecast error for lead time l is

$$e_t(l) = a_{t+l} + \psi_1 a_{t+l-1} + \cdots + \psi_{l-1} a_{t+1} \tag{5.1.14}$$

Since

$$E_t[e_t(l)] = 0 \tag{5.1.15}$$

the forecast is unbiased. Also, the variance of the forecast error is

$$V(l) = \text{var}[e_t(l)] = (1 + \psi_1^2 + \psi_2^2 + \cdots + \psi_{l-1}^2)\sigma_a^2 \tag{5.1.16}$$

3. It is readily shown that not only is $\hat{z}_t(l)$ the minimum mean square error forecast of z_{t+l}, but that any linear function $\sum_{l=1}^{L} w_l \hat{z}_t(l)$ of the forecasts is a minimum mean square error forecast of the corresponding linear function $\sum_{l=1}^{L} w_l z_{t+l}$ of the future observations. For example, suppose that using (5.1.13), we have obtained, from monthly data, minimum mean square error forecasts $\hat{z}_t(1)$, $\hat{z}_t(2)$, and $\hat{z}_t(3)$ of the sales of a product 1, 2, and 3 months ahead. Then it is true that $\hat{z}_t(1) + \hat{z}_t(2) + \hat{z}_t(3)$ is the minimum mean square error forecast of the sales $z_{t+1} + z_{t+2} + z_{t+3}$, during the next quarter.

4. *The shocks as one-step-ahead forecast errors.* Using (5.1.14), the one-step-ahead forecast error is

$$e_t(1) = z_{t+1} - \hat{z}_t(1) = a_{t+1} \tag{5.1.17}$$

Hence the shocks a_t, which generate the process, and which so far we have introduced merely as a set of independent random variables or shocks, turn out to be the *one-step-ahead forecast errors*.

It follows that for a minimum mean square error forecast, the one-step-ahead forecast errors must be uncorrelated. This is eminently sensible, for if one-step-ahead errors were correlated, the forecast error a_{t+1} could, to some extent, be predicted from available forecast errors $a_t, a_{t-1}, a_{t-2}, \ldots$. If the prediction so obtained was $\hat{a}_{t+1}$, then $\hat{z}_t(1) + \hat{a}_{t+1}$ would be a better forecast of z_{t+1} than was $\hat{z}_t(1)$.

5. *Correlation between the forecast errors.* Although the optimal forecast errors at lead time 1 will be uncorrelated, the forecast errors for longer lead times in general will be correlated. In Section A5.1.1 we derive a general expression for the correlation between the forecast errors $e_t(l)$ and $e_{t-j}(l)$, made at the *same* lead time l from *different* origins t and $t - j$.

Now it is also true that the forecast errors $e_t(l)$ and $e_t(l + j)$, made at different lead times from the same origin t, are correlated. One consequence of this is that there will often be a tendency for the forecast function to lie either wholly above or below the values of the series when they eventually come to hand. In Section A5.1.2 we give a general expression for the correlation between the forecast errors $e_t(l)$ and $e_t(l + j)$, made from the same origin.

5.1.2 Three Basic Forms for the Forecast

We have seen that the minimum mean square error forecast $\hat{z}_t(l)$ for lead time l is the conditional expectation $E_t[z_{t+l}]$, of z_{t+l}, at origin t. Using this fact, we can write down expressions for the forecast in any one of three different ways, corresponding to the three ways of expressing the model summarized earlier in this section. For simplicity in notation, we will temporarily adopt the convention that square brackets imply that the conditional expectation, at time t, is to be taken. Thus

$$[a_{t+l}] = E_t[a_{t+l}] \qquad [z_{t+l}] = E_t[z_{t+l}]$$

For $l > 0$, the three different ways of expressing the forecasts are:

Forecasts from difference equation. Taking conditional expectations at time t in (5.1.2), we obtain

$$\begin{aligned}[z_{t+l}] = \hat{z}_t(l) = \varphi_1[z_{t+l-1}] + \cdots + \varphi_{p+d}[z_{t+l-p-d}] - \theta_1[a_{t+l-1}] \cdots \\ - \theta_q[a_{t+l-q}] + [a_{t+l}]\end{aligned} \qquad (5.1.18)$$

Forecasts in integrated form. Using (5.1.3) gives us

$$\begin{aligned}[z_{t+l}] = \hat{z}_t(l) = [a_{t+l}] + \psi_1[a_{t+l-1}] + \cdots + \psi_{l-1}[a_{t+1}] \\ \psi_l[a_t] + \psi_{l+1}[a_{t-1}] + \cdots\end{aligned} \qquad (5.1.19)$$

yielding the form (5.1.13) that we have met already. Alternatively, using the truncated form of the model (5.1.5), we have

$$\begin{aligned}[z_{t+l}] = \hat{z}_t(l) &= [a_{t+l}] + \psi_1[a_{t+l-1}] + \cdots \\ &\qquad + \psi_{t+l-k-1}[a_{k+1}] + C_k(t+l-k) \qquad (5.1.20)\\ &= [a_{t+l}] + \psi_1[a_{t+l-1}] + \cdots + \psi_{l-1}[a_{t+1}] + C_t(l)\end{aligned}$$

where $C_t(l)$ is the complementary function at origin t.

Forecasts as a weighted average of previous observations and forecasts made at previous lead times from the same origin. Finally, taking conditional expectations in (5.1.7) yields

$$[z_{t+l}] = \hat{z}_t(l) = \sum_{j=1}^{\infty} \pi_j[z_{t+l-j}] + [a_{t+l}] \qquad (5.1.21)$$

It is to be noted that the minimum mean square error forecast is defined in terms of the conditional expectation

$$[z_{t+l}] = E_t[z_{t+l}] = E[z_{t+l}|z_t, z_{t-1}, \ldots]$$

which theoretically requires knowledge of the z's stretching back into the infinite past. However, the requirement of invertibility, which we have imposed on the general ARIMA model, ensures that the π weights in (5.1.21) form a convergent series. Hence, for the computation of a forecast to a given degree of accuracy, for some k, the dependence on z_{t-j} for $j > k$ can be ignored. In practice, the π weights usually decay rather quickly, so that whatever form of the model is employed in the computation, only a moderate length of series $z_t, z_{t-1}, \ldots, z_{t-k}$ is needed to calculate the forecasts to sufficient accuracy. The methods we discuss are easily modified to calculate the exact finite sample forecasts, $E[z_{t+l}|z_t, z_{t-1}, \ldots, z_1]$, based on the finite length of data $z_t, z_{t-1}, \ldots, z_1$.

To calculate the conditional expectations that occur in the expressions (5.1.18) to (5.1.21), we note that if j is a nonnegative integer,

$$\begin{aligned}[z_{t-j}] &= E_t[z_{t-j}] = z_{t-j} && j = 0, 1, 2, \ldots\\ [z_{t+j}] &= E_t[z_{t+j}] = \hat{z}_t(j) && j = 1, 2, \ldots\\ [a_{t-j}] &= E_t[a_{t-j}] = a_{t-j} = z_{t-j} - \hat{z}_{t-j-1}(1) && j = 0, 1, 2, \ldots\\ [a_{t+j}] &= E_t[a_{t+j}] = 0 && j = 1, 2, \ldots\end{aligned} \qquad (5.1.22)$$

Therefore, to obtain the forecast $\hat{z}_t(l)$, one writes down the model for z_{t+l} in any one of the three explicit forms above and treats the terms on the right according to the following rules:

1. The z_{t-j} $(j = 0, 1, 2, \ldots)$, which have already happened at origin t, are left unchanged.
2. The z_{t+j} $(j = 1, 2, \ldots)$, which have not yet happened, are replaced by their forecasts $\hat{z}_t(j)$ at origin t.
3. The a_{t-j} $(j = 0, 1, 2, \ldots)$, which have happened, are available from $z_{t-j} - \hat{z}_{t-j-1}(1)$.
4. The a_{t+j} $(j = 1, 2, \ldots)$, which have not yet happened, are replaced by zeros.

For routine calculation it is by far the simplest to work directly with the difference equation form (5.1.18). Hence the forecasts are explicitly calculated, for $l = 1, 2, \ldots,$ recursively as

$$\hat{z}_t(l) = \sum_{j=1}^{p+d} \varphi_j \hat{z}_t(l-j) - \sum_{j=l}^{q} \theta_j a_{t+l-j}$$

where $\hat{z}_t(-j) = [z_{t-j}]$ denotes the observed value z_{t-j} for $j \geq 0$, and the moving average terms are not present for lead times $l > q$.

Example: forecasting using the difference equation form. It will be shown in Chapter 7 that series C is closely represented by the model

$$(1 - 0.8B)(1 - B)z_{t+1} = a_{t+1}$$

that is,

$$(1 - 1.8B + 0.8B^2)z_{t+1} = a_{t+1}$$

or

$$z_{t+l} = 1.8z_{t+l-1} - 0.8z_{t+l-2} + a_{t+l}$$

The forecasts at origin t are given by

$$\begin{aligned} \hat{z}_t(1) &= 1.8z_t - 0.8z_{t-1} \\ \hat{z}_t(2) &= 1.8\hat{z}_t(1) - 0.8z_t \\ \hat{z}_t(l) &= 1.8\hat{z}_t(l-1) - 0.8\hat{z}_t(l-2) \qquad l = 3, 4, \ldots \end{aligned} \tag{5.1.23}$$

It is seen that the forecasts are readily generated recursively in the order $\hat{z}_t(1), \hat{z}_t(2), \ldots$.

In the example above, there happen to be no moving average terms in the model; such terms produce no added difficulties. Thus we consider later in this chapter a series arising in a control problem, for which the model at time $t + l$ is

$$\nabla^2 z_{t+l} = (1 - 0.9B + 0.5B^2)a_{t+l}$$

Then

$$z_{t+l} = 2z_{t+l-1} - z_{t+l-2} + a_{t+l} - 0.9a_{t+l-1} + 0.5a_{t+l-2}$$
$$\hat{z}_t(1) = 2z_t - z_{t-1} - 0.9a_t + 0.5a_{t-1}$$
$$\hat{z}_t(2) = 2\hat{z}_t(1) - z_t + 0.5a_t$$
$$\hat{z}_t(l) = 2\hat{z}_t(l-1) - \hat{z}_t(l-2) \qquad l = 3, 4, \ldots$$

In these expressions we remember that $a_t = z_t - \hat{z}_{t-1}(1)$, $a_{t-1} = z_{t-1} - \hat{z}_{t-2}(1)$, and the forecasting process may be started off initially by setting unknown a's equal to their unconditional expected values of zero.

That is, assuming by convention that data are available starting from time $s = 1$, the necessary a_s's are computed recursively from the difference equation form (5.1.2) of the model, as

$$a_s = z_s - \hat{z}_{s-1}(1) = z_s - \left(\sum_{j=1}^{p+d} \varphi_j z_{s-j} - \sum_{j=1}^{q} \theta_j a_{s-j}\right) \qquad s = p + d + 1, \ldots, t$$

setting initial a_s's equal to zero, for $s < p + d + 1$. A more exact form of forecast can be obtained by using a method in which the necessary initial a_s's, as well as initial z_s's, which occur just prior to the start of the data period, are determined by the technique of *back-forecasting*. This technique, which essentially determines the conditional expectations of the initial a_s's and z_s's prior to the data period based on the available data for the *finite* data period starting at time $s = 1$, is discussed in Chapter 7 in regard to parameter estimation of ARIMA models. Using this back-forecasting technique to obtain the necessary initial a's and z's will ultimately produce the minimum mean square error forecasts $E[z_{t+l}|z_t, z_{t-1}, \ldots, z_1]$ based on the finite past knowledge available from the sample data $z_t, z_{t-1}, \ldots, z_1$. As noted earlier, provided that a sufficient length of data series $z_t, z_{t-1}, \ldots, z_1$ is available, the two different treatments of the initial values will have a negligible effect on the forecasts $\hat{z}_t(l)$.

Note that in the special case of an ARI(p, d) process, that is, an ARIMA process with $q = 0$, there is no dependence of the forecasts on initial values of the a's or the z's. In fact, from (5.1.18) it is seen that the forecasts in this case are

$$\hat{z}_t(l) = \varphi_1 \hat{z}_t(l-1) + \varphi_2 \hat{z}_t(l-2) + \cdots + \varphi_{p+d} \hat{z}_t(l-p-d) \qquad l = 1, 2, \ldots$$

starting with $\hat{z}_t(1) = \varphi_1 z_t + \varphi_2 z_{t-1} + \cdots + \varphi_{p+d} z_{t+1-p-d}$. Hence we see that forecasts $\hat{z}_t(l)$ for all lead times l for this model depend only on the $p + d$ most recent observed values of the process, $z_t, z_{t-1}, \ldots, z_{t+1-p-d}$.

In general, if the moving average operator $\theta(B)$ is of degree q, the forecast equations for $\hat{z}_t(1), \hat{z}_t(2), \ldots, \hat{z}_t(q)$ will depend directly on the a's, but forecasts at longer lead times will not. It should not, of course, be thought that the influence of these a's is not contained in forecasts at longer lead times. In the example above, for instance, $\hat{z}_t(3)$ depends on $\hat{z}_t(2)$ and $\hat{z}_t(1)$, which in turn depend on a_t and a_{t-1}.

We finally note that for nonstationary ARIMA processes with $d \geq 1$, there is a further way to view the forecasts that might be useful. For example, for $d = 1$ we have $w_t = (1 - B)z_t$ or $z_t = w_t + z_{t-1}$, where w_t is a stationary ARMA(p, q) process. Hence forecasts for the process z can be represented as

$$\hat{z}_t(l) = \hat{w}_t(l) + \hat{z}_t(l-1) = \hat{w}_t(l) + \hat{w}_t(l-1) + \cdots + \hat{w}_t(1) + z_t$$

where the forecasts $\hat{w}_t(l)$ are obtained from the ARMA(p, q) model for the process w in the usual way based on the data $w_t, w_{t-1}, \ldots$. Similarly, for $d = 2$, from $w_t = (1 - B)^2 z_t = z_t - 2z_{t-1} + z_{t-2}$ we obtain $\hat{z}_t(l) = \hat{w}_t(l) + 2\hat{z}_t(l-1) - \hat{z}_t(l-2) = \sum_{j=0}^{l-1} (j+1)\hat{w}_t(l-j) + z_t + l(z_t - z_{t-1})$, and so on.

5.2 CALCULATING AND UPDATING FORECASTS

5.2.1 Convenient Format for the Forecasts

It is frequently the case that forecasts are needed for several lead times, say at 1, 2, 3, . . . , L steps ahead. The forecasts are very easily built up one from the other in the calculation scheme illustrated in Table 5.1, which shows the forecasts made at origin $t = 20$, for lead times $l = 1, 2, 3, \ldots, 14$ for series C.

The diagonal arrangement allows each forecast to appear opposite the value it forecasts. Thus $\hat{z}_{20}(6) = 22.51$, and this forecast made at origin 20 is a forecast of z_{26} and so appears opposite that value. The actual values of z_t at $t = 21, 22$, and so on, are shown in italic type as a reminder that these values would not actually be available when the forecast was made.

The calculations are easily performed using (5.1.23). For example,

$$\hat{z}_{20}(1) = (1.8 \times 23.4) - (0.8 \times 23.7) = 23.16$$

$$\hat{z}_{20}(2) = (1.8 \times 23.16) - (0.8 \times 23.4) = 22.97$$

and so on. As soon as the new piece of data z_{21} became available, we could immediately generate a new set of forecasts which would fill a diagonal immediately below that shown. Using the fact that $a_t = z_t - \hat{z}_{t-1}(1)$, each a_t is computed, as each new piece of data z_t comes to hand, as the difference of entries on its immediate right and immediate left. Thus, as soon as $z_{21} = 23.1$ is available, we could insert the entry $-0.06 = 23.1 - 23.16$ for a_{21}.

5.2.2 Calculation of the ψ Weights

Suppose that forecasts at lead times 1, 2, . . . , L are required. To obtain probability limits for these forecasts and to allow new forecasts to be calculated by a process of updating the old, it is necessary to calculate the

TABLE 5.1 Convenient Format for the Forecasts

Lead Time $l \rightarrow$			1	2	3	4	5	6	7	8	9	10	11	12	13	14
Coefficient $\psi_l \rightarrow$			1.80	2.44	2.95	3.36	3.69	3.95	4.16	4.33	4.46	4.57	4.65	4.72	4.78	4.82
Nature of Forecast $\rightarrow$			$\hat{z}_{t-1}(1)$	$\hat{z}_{t-2}(2)$	$\hat{z}_{t-3}(3)$	$\hat{z}_{t-4}(4)$	$\hat{z}_{t-5}(5)$	$\hat{z}_{t-6}(6)$	$\hat{z}_{t-7}(7)$	$\hat{z}_{t-8}(8)$	$\hat{z}_{t-9}(9)$	$\hat{z}_{t-10}(10)$	$\hat{z}_{t-11}(11)$	$\hat{z}_{t-12}(12)$	$\hat{z}_{t-13}(13)$	$\hat{z}_{t-14}(14)$
95% Limit ±			0.26	0.55	0.84	1.15	1.46	1.75	2.04	2.32	2.59	2.84	3.09	3.32	3.58	3.77
50% Limit ±			0.09	0.19	0.29	0.39	0.50	0.60	0.70	0.79	0.88	0.97	1.05	1.13	1.22	1.29
t	z_t	a_t														
19	**23.7**															
Origin 20	**23.4**															
21	*23.1*	*−0.06*	**23.16**													
22	*22.9*			**22.97**												
23	*22.8*				**22.81**											
24	*22.7*					**22.69**										
25	*22.6*						**22.59**									
26	*22.4*							**22.51**								
27	*22.2*								**22.45**							
28	*22.0*									**22.40**						
29	*21.8*										**22.36**					
30	*21.4*											**22.32**				
31	*20.9*												**22.30**			
32	*20.3*													**22.28**		
33	*19.7*														**22.27**	
34	*19.4*															**22.25**
35	*19.3*															

weights $\psi_1, \psi_2, \ldots, \psi_{L-1}$. This is accomplished using (5.1.4), namely

$$\varphi(B)\psi(B) = \theta(B) \tag{5.2.1}$$

That is, equating coefficients of powers of B in

$$\begin{aligned}(1 - \varphi_1 B - \cdots - \varphi_{p+d}B^{p+d})(1 + \psi_1 B + \psi_2 B^2 + \cdots)\\ = (1 - \theta_1 B - \theta_2 B^2 - \cdots - \theta_q B^q)\end{aligned} \tag{5.2.2}$$

Knowing the values of the φ's and the θ's, the ψ's may be obtained by equating coefficients of B as follows:

$$\begin{aligned}\psi_1 &= \varphi_1 - \theta_1\\ \psi_2 &= \varphi_1\psi_1 + \varphi_2 - \theta_2\\ &\vdots\\ \psi_j &= \varphi_1\psi_{j-1} + \cdots + \varphi_{p+d}\psi_{j-p-d} - \theta_j\end{aligned} \tag{5.2.3}$$

where $\psi_0 = 1$, $\psi_j = 0$ for $j < 0$ and $\theta_j = 0$ for $j > q$. If K is the greater of the integers $p + d - 1$ and q, then for $j > K$ the ψ's satisfy the difference equation

$$\psi_j = \varphi_1\psi_{j-1} + \varphi_2\psi_{j-2} + \cdots + \varphi_{p+d}\psi_{j-p-d} \tag{5.2.4}$$

Thus the ψ's are easily calculated recursively. For example, for the model $(1 - 1.8B + 0.8B^2)z_t = a_t$, appropriate to series C, we have

$$(1 - 1.8B + 0.8B^2)(1 + \psi_1 B + \psi_2 B^2 + \cdots) = 1$$

Either by directly equating coefficients of B^j or by using (5.2.3) and (5.2.4) with $\varphi_1 = 1.8$ and $\varphi_2 = -0.8$, we obtain

$$\begin{aligned}\psi_0 &= 1\\ \psi_1 &= 1.8\\ \psi_j &= 1.8\psi_{j-1} - 0.8\psi_{j-2} \qquad j = 2, 3, 4, \ldots\end{aligned}$$

Thus

$$\begin{aligned}\psi_2 &= (1.8 \times 1.8) - (0.8 \times 1.0) = 2.44\\ \psi_3 &= (1.8 \times 2.44) - (0.8 \times 1.8) = 2.95\end{aligned}$$

and so on. The ψ's for this example are displayed in the second row of Table 5.1.

5.2.3 Use of the ψ Weights in Updating the Forecasts

It is interesting to consider yet another way of generating the forecasts. Using (5.1.13), we can express the forecasts $\hat{z}_{t+1}(l)$ and $\hat{z}_t(l + 1)$, of the future observation z_{t+l+1}, made at origins $t + 1$ and t, as

$$\hat{z}_{t+1}(l) = \psi_l a_{t+1} + \psi_{l+1} a_t + \psi_{l+2} a_{t-1} + \cdots$$
$$\hat{z}_t(l + 1) = \psi_{l+1} a_t + \psi_{l+2} a_{t-1} + \cdots$$

On subtraction, it follows that

$$\hat{z}_{t+1}(l) = \hat{z}_t(l + 1) + \psi_l a_{t+1} \tag{5.2.5}$$

Explicitly, the t-origin forecast of z_{t+l+1} can be updated to become the $t + 1$ origin forecast of the same z_{t+l+1}, by adding a constant multiple of the one-step-ahead forecast error a_{t+1}, with multiplier ψ_l.

This leads to a rather remarkable conclusion. Suppose that we currently have forecasts at origin t for lead times $1, 2, \ldots, L$. Then, as soon as z_{t+1} becomes available, we can calculate $a_{t+1} = z_{t+1} - \hat{z}_t(1)$ and proportionally update to obtain forecasts $\hat{z}_{t+1}(l) = \hat{z}_t(l + 1) + \psi_l a_{t+1}$ at origin $t + 1$, for lead times $1, 2, \ldots, L - 1$. The new forecast $\hat{z}_{t+1}(L)$, for lead time L, cannot be calculated by this means, but is easily obtained from the forecasts at shorter lead times, using the difference equation.

Referring once more to the forecasting of series C, in Table 5.2, the forecasts at origin $t = 21$ have been added to those shown previously in Table 5.1. These can either be obtained directly, as was done to obtain the forecasts at origin $t = 20$, or we can employ the updating equation (5.2.5). The values of the ψ's are located in the second row of the table so that this can be done conveniently.

Specifically, as soon as we know $z_{21} = 23.1$, we can calculate $a_{21} = 23.1 - 23.16 = -0.06$. Then

$$\hat{z}_{21}(1) = 22.86 = 22.97 + (1.8)(-0.06)$$

$$\hat{z}_{21}(2) = 22.67 = 22.81 + (2.44)(-0.06)$$

$$\hat{z}_{21}(3) = 22.51 = 22.69 + (2.95)(-0.06)$$

and so on.

5.2.4 Calculation of the Probability Limits of the Forecasts at Any Lead Time

The expression (5.1.16) shows that, in general, the variance of the l-steps-ahead forecast error for any origin t is the expected value of

$$e_t^2(l) = [z_{t+l} - \hat{z}_t(l)]^2$$

and is given by

$$V(l) = \left(1 + \sum_{j=1}^{l-1} \psi_j^2\right) \sigma_a^2$$

TABLE 5.2 Updating of Forecasts

Lead Time $l \rightarrow$			1	2	3	4	5	6	7	8	9	10	11	12	13	14
Coefficient $\psi_l \rightarrow$			1.8	2.44	2.95	3.36	3.69	3.95	4.16	4.33	4.46	4.57	4.65	4.72	4.78	4.82
Nature of Forecast $\rightarrow$			$\hat{z}_{t-1}(1)$	$\hat{z}_{t-2}(2)$	$\hat{z}_{t-3}(3)$	$\hat{z}_{t-4}(4)$	$\hat{z}_{t-5}(5)$	$\hat{z}_{t-6}(6)$	$\hat{z}_{t-7}(7)$	$\hat{z}_{t-8}(8)$	$\hat{z}_{t-9}(9)$	$\hat{z}_{t-10}(10)$	$\hat{z}_{t-11}(11)$	$\hat{z}_{t-12}(12)$	$\hat{z}_{t-13}(13)$	$\hat{z}_{t-14}(14)$
95% Limit $\pm$			0.26	0.55	0.84	1.15	1.46	1.75	2.04	2.32	2.59	2.84	3.09	3.32	3.58	3.77
50% Limit $\pm$			0.09	0.19	0.29	0.39	0.50	0.60	0.70	0.79	0.88	0.97	1.05	1.13	1.22	1.29
t	z_t	a_t														
19	**23.7**															
20	**23.4**															
21	**23.1**	**−0.06**	*23.16*													
22	*22.9*		**22.86**	*22.97*												
23	*22.8*			**22.67**	*22.81*											
24	*22.7*				**22.51**	*22.69*										
25	*22.6*					**22.39**	*22.59*									
26	*22.4*						**22.29**	*22.51*								
27	*22.2*							**22.21**	*22.45*							
28	*22.0*								**22.15**	*22.40*						
29	*21.8*									**22.10**	*22.36*					
30	*21.4*										**22.06**	*22.32*				
31	*20.9*											**22.03**	*22.30*			
32	*20.3*												**22.00**	*22.28*		
33	*19.7*													**21.99**	*22.27*	
34	*19.4*														**21.96**	*22.25*
35	*19.3*															**21.95**

TABLE 5.3 Variance Function for Series C

l	1	2	3	4	5	6	7	8	9	10
$V(l)/\sigma_a^2$	1.00	4.24	10.19	18.96	30.24	43.86	59.46	76.79	95.52	115.41

For example, using the weights given in Table 5.1, the function $V(l)/\sigma_a^2$ for series C is shown in Table 5.3.

Assuming that the a's are Normal, it follows that given information up to time t, the conditional probability distribution $p(z_{t+l}|z_t, z_{t-1}, \ldots)$ of a future value z_{t+l} of the process will be Normal with mean $\hat{z}_t(l)$ and standard deviation $(1 + \sum_{j=1}^{l-1} \psi_j^2)^{1/2}\sigma_a$. Figure 5.1 shows the conditional probability distributions of future values z_{21}, z_{22}, z_{23} for series C, given information up to origin $t = 20$.

We show in Chapter 7 how an estimate s_a^2, of the variance σ_a^2, may be obtained from time series data. When the number of observations on which such an estimate is based is, say, at least 50, s_a may be substituted for σ_a and approximate $1 - \varepsilon$ probability limits $z_{t+l}(-)$ and $z_{t+l}(+)$ for z_{t+l} will be given by

$$z_{t+l}(\pm) = \hat{z}_t(l) \pm u_{\varepsilon/2} \left(1 + \sum_{j=1}^{l-1} \psi_j^2\right)^{1/2} s_a \tag{5.2.6}$$

where $u_{\varepsilon/2}$ is the deviate exceeded by a proportion $\varepsilon/2$ of the unit Normal distribution.

It is shown in Table 7.13 that for series C, $s_a = 0.134$; hence the 50% and 95% limits, for $\hat{z}_t(2)$, for example, are given by

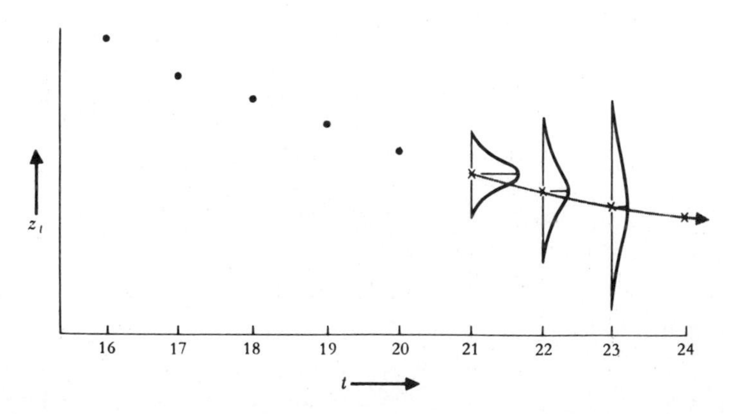

Figure 5.1 Conditional probability distributions of future values z_{21}, z_{22}, and z_{23} for series C, given information up to origin $t = 20$.

$$\text{50\% limits:} \quad \hat{z}_t(2) \pm (0.674)(1 + 1.8^2)^{1/2}(0.134) = \hat{z}_t(2) \pm 0.19$$

$$\text{95\% limits:} \quad \hat{z}_t(2) \pm 1.96(1 + 1.8^2)^{1/2}(0.134) = \hat{z}_t(2) \pm 0.55$$

The quantities to be added and subtracted from the forecast to obtain the 50% and 95% limits are shown in the fourth and fifth rows of the headings of Tables 5.1 and 5.2. They apply to the forecasts that are immediately below them.

In Figure 5.2, a section of series C is shown together with the several-steps-ahead forecasts (indicated by crosses) from origins $t = 20$ and $t = 67$. Also shown are the 50% and 95% probability limits for z_{20+l}, for $l = 1$ to 14. The interpretation of the limits $z_{t+l}(-)$ and $z_{t+l}(+)$ should be noted carefully. These limits are such that *given the information available at origin t*, there is a probability of $1 - \varepsilon$ that the actual value z_{t+l}, when it occurs, will be within them, that is,

$$\Pr\{z_{t+l}(-) < z_{t+l} < z_{t+l}(+)\} = 1 - \varepsilon$$

It should also be explained that the probabilities quoted apply to *individual* forecasts and not jointly to the forecasts at all the different lead times. For example, it is true that with 95% probability, the limits for lead time 10 will include the value z_{t+10} when it occurs. It is not true that the series can be expected to remain within *all* the limits simultaneously at this level of probability.

5.3 FORECAST FUNCTION AND FORECAST WEIGHTS

Forecasts are calculated most simply in the manner just described—by direct use of the difference equation. From the purely *computational* stand-

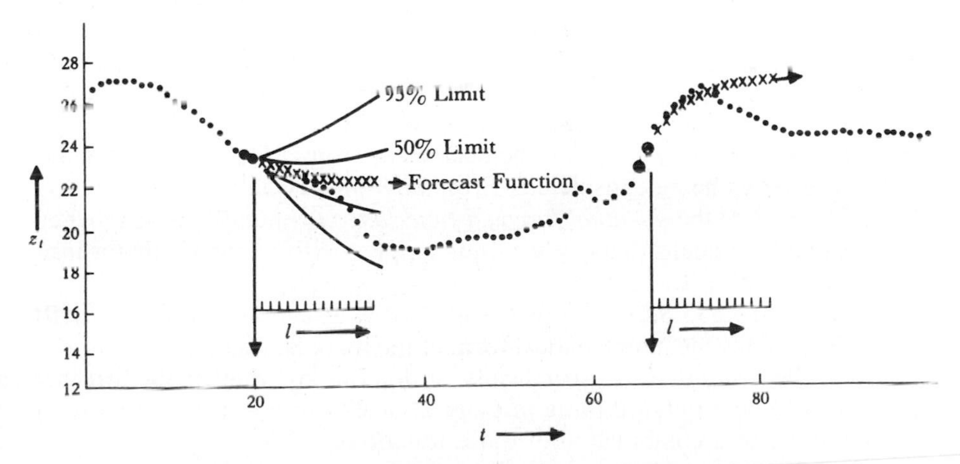

Figure 5.2 Forecasts for series C and probability limits.

point, the other model forms are less convenient. However, from the point of view of studying the nature of the forecasts, it is profitable to consider in greater detail the alternative forms discussed in Section 5.1.2 and, in particular, to consider the explicit form of the forecast function.

5.3.1 Eventual Forecast Function Determined by the Autoregressive Operator

At time $t + l$ the ARIMA model may be written

$$z_{t+l} - \varphi_1 z_{t+l-1} - \cdots - \varphi_{p+d} z_{t+l-p-d} = a_{t+l} - \theta_1 a_{t+l-1} - \cdots - \theta_q a_{t+l-q} \tag{5.3.1}$$

Taking conditional expectations at time t in (5.3.1), we have, for $l > q$,

$$\hat{z}_t(l) - \varphi_1 \hat{z}_t(l-1) - \cdots - \varphi_{p+d}\hat{z}_t(l-p-d) = 0 \qquad l > q \tag{5.3.2}$$

where it is understood that $\hat{z}_t(-j) = z_{t-j}$ for $j \geqslant 0$. The difference equation (5.3.2) has the solution

$$\hat{z}_t(l) = b_0^{(t)} f_0(l) + b_1^{(t)} f_1(l) + \cdots + b_{p+d-1}^{(t)} f_{p+d-1}(l) \tag{5.3.3}$$

for $l > q - p - d$. Note that the forecast $\hat{z}_t(l)$ is the complementary function introduced in Chapter 4. In (5.3.3), $f_0(l), f_1(l), \ldots, f_{p+d-1}(l)$ are functions of the lead time l. In general, they could include polynomials, exponentials, sines and cosines, and products of these functions. The functions $f_0(l), f_1(l), \ldots, f_{p+d-1}(l)$ consist of p damped exponential and damped sinusoidal terms of the form G^l and $D^l \sin(2\pi fl + F)$, respectively, associated with the roots of $\phi(B) = 0$ for the stationary autoregressive operator, and d polynomial terms l^i, $i = 0, \ldots, d-1$, of degree $d - 1$, associated with the nonstationary differencing operator $\nabla^d = (1 - B)^d$. For a *given origin* t, the coefficients $b_j^{(t)}$ are constants applying for all lead times l, but they change from one origin to the next, *adapting* themselves appropriately to the particular part of the series being considered. From now on we shall call the function defined by (5.3.3) the *eventual forecast function*; "eventual" because when it occasionally happens that $q > p + d$, it supplies the forecasts only for lead times $l > q - p - d$.

We see from (5.3.2) that it is the general autoregressive operator $\varphi(B)$ which determines the mathematical form of the forecast function, that is, the nature of the f's in (5.3.3). Specifically, it determines whether the forecast function is to be a polynomial, a mixture of sines and cosines, a mixture of exponentials, or a combination of these functions.

5.3.2 Role of the Moving Average Operator in Fixing the Initial Values

While the autoregressive operator decides the nature of the eventual forecast function, the moving average operator is influential in determining how that function is to be "fitted" to the data and hence how the coefficients $b_0^{(t)}, b_1^{(t)}, \ldots, b_{p+d-1}^{(t)}$ are to be calculated and updated.

For example, consider the IMA(0, 2, 3) process

$$z_{t+l} - 2z_{t+l-1} + z_{t+l-2} = a_{t+l} - \theta_1 a_{t+l-1} - \theta_2 a_{t+l-2} - \theta_3 a_{t+l-3}$$

Using the conditional expectation argument of Section 5.1.2, the forecast function is defined by

$$\hat{z}_t(1) = 2z_t - z_{t-1} - \theta_1 a_t - \theta_2 a_{t-1} - \theta_3 a_{t-2}$$
$$\hat{z}_t(2) = 2\hat{z}_t(1) - z_t - \theta_2 a_t - \theta_3 a_{t-1}$$
$$\hat{z}_t(3) = 2\hat{z}_t(2) - \hat{z}_t(1) - \theta_3 a_t$$
$$\hat{z}_t(l) = 2\hat{z}_t(l-1) - \hat{z}_t(l-2) \qquad l > 3$$

Therefore, since $\varphi(B) = (1 - B)^2$ in this model, the eventual forecast function is the unique straight line

$$\hat{z}_t(l) = b_0^{(t)} + b_1^{(t)} l \qquad l > 1$$

which passes through $\hat{z}_t(2)$ and $\hat{z}_t(3)$ as shown in Figure 5.3. However, note that if the θ_3 term had been omitted, then $q - p - d = 0$, and the forecast would have been given at *all lead times* by the straight line passing through $\hat{z}_t(1)$ and $\hat{z}_t(2)$.

In general, since only one function of the form (5.3.3) can pass through $p + d$ points, the eventual forecast function is that unique curve of the form required by $\varphi(B)$, which passes through the $p + d$ "pivotal" values $\hat{z}_t(q)$,

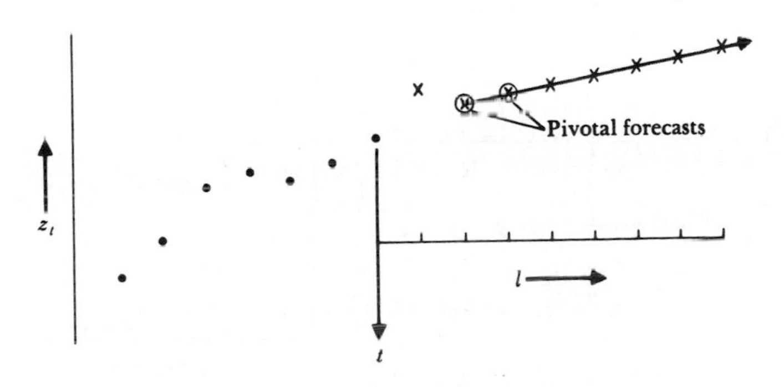

Figure 5.3 Eventual forecast function for an IMA(0, 2, 3) process.

$\hat{z}_t(q-1), \ldots, \hat{z}_t(q-p-d+1)$, where $\hat{z}_t(-j) = z_{t-j}(j = 0, 1, 2, \ldots)$. In the extreme case where $q = 0$, so that the model is of the purely autoregressive form $\varphi(B)z_t = a_t$, the curve passes through the points $z_t, z_{t-1}, \ldots, z_{t-p-d+1}$. Thus the pivotal values can consist of forecasts or of actual values of the series; they are indicated in the figures by circled points.

The moving average terms, which appear in the model form, help to decide the way in which we "reach back" into the series to fit the forecast function determined by the autoregressive operator $\varphi(B)$. Figure 5.4 illustrates the situation for the model of order (1, 1, 3) given by $(1 - \phi B)\nabla z_t = (1 - \theta_1 B - \theta_2 B^2 - \theta_3 B^3)a_t$. The (hypothetical) weight functions indicate the linear functional dependence of the three forecasts, $\hat{z}_t(1)$, $\hat{z}_t(2)$, and $\hat{z}_t(3)$, on $z_t, z_{t-1}, z_{t-2}, \ldots$. Since the forecast function contains $p + d = 2$ coefficients, it is uniquely determined by the forecasts $\hat{z}_t(3)$ and $\hat{z}_t(2)$, that is, by $\hat{z}_t(q)$ and $\hat{z}_t(q-1)$. We next consider how the forecast weight functions, referred to above, are determined.

5.3.3 Lead / Forecast Weights

The fact that the general model may also be written in inverted form,

$$a_t = \pi(B)z_t = (1 - \pi_1 B - \pi_2 B^2 - \pi_3 B^3 - \cdots)z_t \tag{5.3.4}$$

results in our being able to write the forecast as in (5.1.21). On substituting for the conditional expectations in (5.1.21), we obtain

$$\hat{z}_t(l) = \sum_{j=1}^{\infty} \pi_j \hat{z}_t(l-j) \tag{5.3.5}$$

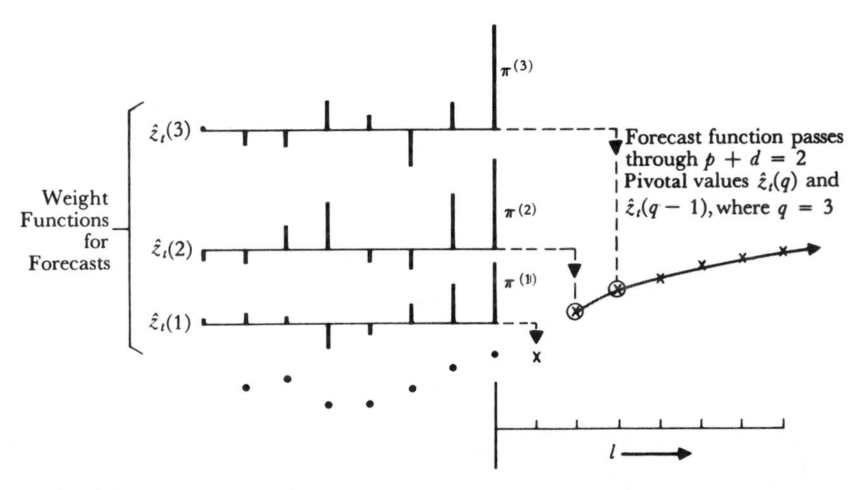

Figure 5.4 Dependence of forecast function on observations for (1, 1, 3) process $(1 - \phi B)\nabla z_t = (1 - \theta_1 B - \theta_2 B^2 - \theta_3 B^3)a_t$.

where, as before, $\hat{z}_t(-h) = z_{t-h}$ for $h = 0, 1, 2, \ldots$. Thus, in general,

$$\hat{z}_t(l) = \pi_1 \hat{z}_t(l-1) + \cdots + \pi_{l-1}\hat{z}_t(1) + \pi_l z_t + \pi_{l+1} z_{t-1} + \cdots \tag{5.3.6}$$

and in particular,

$$\hat{z}_t(1) = \pi_1 z_t + \pi_2 z_{t-1} + \pi_3 z_{t-2} + \cdots$$

The forecasts for higher lead times may also be expressed directly as linear functions of the observations $z_t, z_{t-1}, z_{t-2}, \ldots$. For example, the lead 2 forecast at origin t is

$$\begin{aligned}\hat{z}_t(2) &= \pi_1 \hat{z}_t(1) + \pi_2 z_t + \pi_3 z_{t-1} + \cdots \\ &= \pi_1 \sum_{j=1}^{\infty} \pi_j z_{t-j+1} + \sum_{j=1}^{\infty} \pi_{j+1} z_{t-j+1} \\ &= \sum_{j=1}^{\infty} \pi_j^{(2)} z_{t-j+1}\end{aligned}$$

where

$$\pi_j^{(2)} = \pi_1 \pi_j + \pi_{j+1} \qquad j = 1, 2, \ldots \tag{5.3.7}$$

Proceeding in this way, it is readily shown that

$$\hat{z}_t(l) = \sum_{j=1}^{\infty} \pi_j^{(l)} z_{t-j+1} \tag{5.3.8}$$

where

$$\pi_j^{(l)} = \pi_{j+l-1} + \sum_{h=1}^{l-1} \pi_h \pi_j^{(l-h)} \qquad j = 1, 2, \ldots \tag{5.3.9}$$

and $\pi_j^{(1)} = \pi_j$. Alternative methods for computing these weights are given in Appendix A5.2.

As we have seen in (4.2.22) and (5.1.9), the π_j's themselves may be obtained explicitly by equating coefficients in

$$\theta(B)(1 - \pi_1 B - \pi_2 B - \cdots) = \varphi(B)$$

Given these values, the $\pi_j^{(l)}$'s may readily be obtained, if so desired, using (5.3.9) or the results of Appendix A5.2.

As an example, consider again the model

$$\nabla^2 z_t = (1 - 0.9B + 0.5B^2)a_t$$

which was fitted to a series, a part of which is shown in Figure 5.5. Equating coefficients in

$$(1 - 0.9B + 0.5B^2)(1 - \pi_1 B - \pi_2 B^2 \cdots) = 1 - 2B + B^2$$

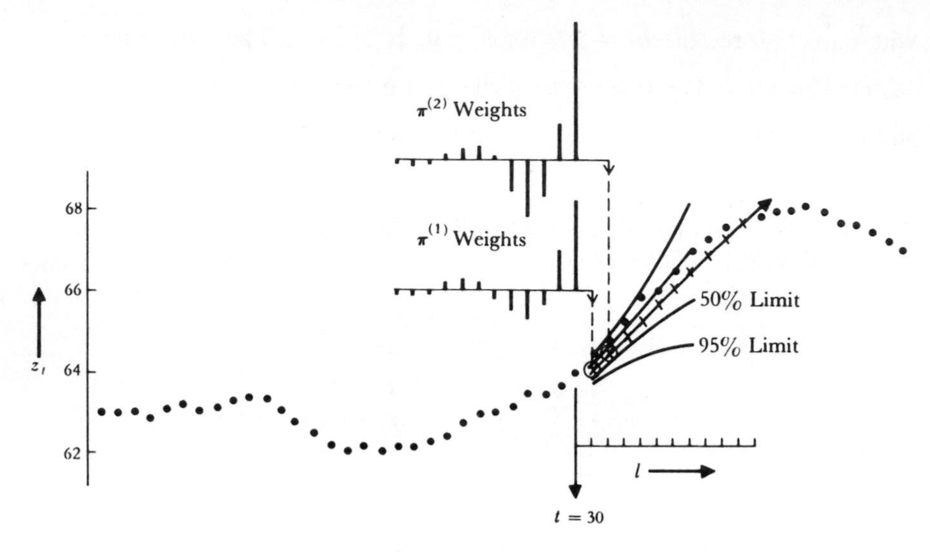

Figure 5.5 Part of a series fitted by $\nabla^2 z_t = (1 - 0.9B + 0.5B^2)a_t$ with forecast function for origin $t = 30$, forecast weights, and probability limits.

yields the weights $\pi_j = \pi_j^{(1)}$, from which the weights $\pi_j^{(2)}$ may be computed using (5.3.7). The two sets of weights are given for $j = 1, 2, \ldots, 12$ in Table 5.4.

In this example, the lead 1 and lead 2 forecasts, expressed in terms of the observations $z_t, z_{t-1}, \ldots$, are

$$\hat{z}_t(1) = 1.10z_t + 0.49z_{t-1} - 0.11z_{t-2} - 0.34z_{t-3} - 0.25z_{t-4} - \cdots$$

TABLE 5.4 π Weights for the Model $\nabla^2 z_t = (1 - 0.9B + 0.5B^2)a_t$

j	$\pi_j^{(1)}$	$\pi_j^{(2)}$
1	1.100	1.700
2	0.490	0.430
3	−0.109	−0.463
4	−0.343	−0.632
5	−0.254	−0.336
6	−0.057	0.013
7	0.076	0.181
8	0.097	0.156
9	0.049	0.050
10	−0.004	−0.032
11	−0.028	−0.054
12	−0.023	−0.026

and

$$\hat{z}_t(2) = 1.70z_t + 0.43z_{t-1} - 0.46z_{t-2} - 0.63z_{t-3} - 0.34z_{t-4} + \cdots$$

In fact, the weights follow damped sine waves as shown in Figure 5.5.

5.4 EXAMPLES OF FORECAST FUNCTIONS AND THEIR UPDATING

The forecast functions for some special cases of the general ARIMA model will now be considered. We shall exhibit these in the three different forms discussed in Section 5.1.2. As mentioned earlier, the forecasts are most easily computed from the difference equation itself. The other forms are useful because they provide insight into the nature of the forecast function in particular cases.

5.4.1 Forecasting an IMA(0, 1, 1) Process

The model is $\nabla z_t = (1 - \theta B)a_t$.

Difference equation approach. At time $t + l$ the model may be written

$$z_{t+l} = z_{t+l-1} + a_{t+l} - \theta a_{t+l-1}$$

Taking conditional expectations at origin t yields

$$\begin{aligned} \hat{z}_t(1) &= z_t - \theta a_t \\ \hat{z}_t(l) &= \hat{z}_t(l-1) \qquad l \geqslant 2 \end{aligned} \tag{5.4.1}$$

Hence for all lead times, the forecasts at origin t will follow a straight line parallel to the time axis. Using the fact that $z_t = \hat{z}_{t-1}(1) + a_t$, we can write (5.4.1) in either of two useful forms.

The first of these is

$$\hat{z}_t(l) = \hat{z}_{t-1}(l) + \lambda a_t \tag{5.4.2}$$

where $\lambda = 1 - \theta$. This implies that having seen that our previous forecast $\hat{z}_{t-1}(l)$ falls short of the realized value by a_t, we adjust it by an amount λa_t. It will be recalled from Section 4.3.1 that λ measures the proportion of any given shock a_t, which is permanently absorbed by the "level" of the process. Therefore, it is reasonable to increase the forecast by that part λa_t of a_t, which we expect to be absorbed.

The second way of rewriting (5.4.1) is

$$\hat{z}_t(l) = \lambda z_t + (1 - \lambda)\hat{z}_{t-1}(l) \tag{5.4.3}$$

This implies that the new forecast is a linear interpolation at argument λ between old forecast and new observation. The form (5.4.3) makes it clear that if λ is very small, we shall be relying principally on a weighted average of past data and heavily discounting the new observation z_t. By contrast, if $\lambda = 1$, the evidence of past data is completely ignored, $\hat{z}_t(1) = z_t$, and the forecast for all future time is the current value. With $\lambda > 1$, we induce an extrapolation rather than an interpolation between $\hat{z}_{t-1}(l)$ and z_t. The forecast error must now be *magnified* in (5.4.2) to indicate the change in the forecast.

Forecast function in integrated form. The eventual forecast function is the solution of $(1 - B)\hat{z}_t(l) = 0$. Thus $\hat{z}_t(l) = b_0^{(t)}$, and since $q - p - d = 0$, it provides the forecast for all lead times, that is

$$\hat{z}_t(l) = b_0^{(t)} \qquad l > 0 \tag{5.4.4}$$

For *any fixed origin*, $b_0^{(t)}$ is a constant, and the forecasts for all lead times will follow a straight line parallel to the time axis. However, the coefficient $b_0^{(t)}$ will be updated as a new observation becomes available and the origin advances. Thus the forecast function can be thought of as a polynomial of degree zero in the lead time l, with a coefficient that is adaptive with respect to the origin t.

By comparison of (5.4.4) with (5.4.1), we see that

$$b_0^{(t)} = \hat{z}_t(l) = z_t - \theta a_t$$

Equivalently, by referring to (4.3.4), since the truncated integrated form of the model, relative to an initial origin k, is

$$\begin{aligned} z_t &= \lambda S_{t-k-1} a_{t-1} + a_t + (z_k - \theta a_k) \\ &= \lambda(a_{t-1} + \cdots + a_{k+1}) + a_t + (z_k - \theta a_k) \end{aligned}$$

it follows that

$$\hat{z}_t(l) = b_0^{(t)} = \lambda S_{t-k} a_t + (z_k - \theta a_k) = \lambda(a_t + \cdots + a_{k+1}) + (z_k - \theta a_k)$$

Also, $\psi_j = \lambda (j = 1, 2, \ldots)$ and hence the adaptive coefficient $b_0^{(t)}$ can be updated from origin t to origin $t + 1$ according to

$$b_0^{(t+1)} = b_0^{(t)} + \lambda a_{t+1} \tag{5.4.5}$$

similar to (5.4.2).

Forecast as a weighted average of previous observations. Since, for this process, the $\pi_j^{(l)}$ weights of (5.3.8) are also the weights for the one-step-ahead forecast, we can also write, using (4.3.6),

$$\hat{z}_t(l) = b_0^{(t)} = \lambda z_t + \lambda(1 - \lambda)z_{t-1} + \lambda(1 - \lambda)^2 z_{t-2} + \cdots \tag{5.4.6}$$

Thus, for the IMA(0, 1, 1) model, the forecast for all future time is an *exponentially weighted moving average* of current and past z's.

Example: forecasting series A. It will be shown in Chapter 7 that series A is closely fitted by the model

$$(1 - B)z_t = (1 - 0.7B)a_t$$

In Figure 5.6 the forecasts at origins $t = 39, 40, 41, 42$, and 43 and also at origin $t = 79$ are shown for lead times $1, 2, \ldots, 20$. The weights π_j, which for this model are forecast weights for any lead time, are given in Table 5.5. These weights are shown diagrammatically in their appropriate positions for the forecast $\hat{z}_{39}(l)$ in Figure 5.6.

Variance functions. Since for this model, $\psi_j = \lambda (j = 1, 2, \ldots)$, the expression (5.1.16) for the variance of the lead l forecasts is

$$V(l) = \sigma_a^2[1 + (l - 1)\lambda^2] \tag{5.4.7}$$

TABLE 5.5 Forecast Weights Applied to Previous z's for Any Lead Time Used in Forecasting Series A with Model $\nabla z_t = (1 - 0.7B)a_t$

j	π_j	j	π_j
1	0.300	7	0.035
2	0.210	8	0.025
3	0.147	9	0.017
4	0.103	10	0.012
5	0.072	11	0.008
6	0.050	12	0.006

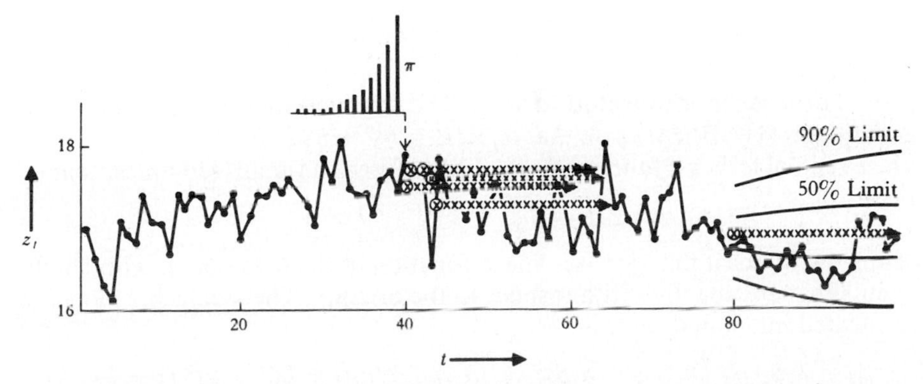

Figure 5.6 Part of series A with forecasts at origins $t = 39, 40, 41, 42, 43$ and at $t = 79$.

Using the estimate $s_a^2 = 0.101$, appropriate for series A, in (5.4.7), 50% and 95% probability limits were calculated and are shown in Figure 5.6 for origin $t = 79$.

5.4.2 Forecasting an IMA(0, 2, 2) Process

The model is $\nabla^2 z_t = (1 - \theta_1 B - \theta_2 B^2)a_t$.

Difference equation approach. At time $t + l$, the model may be written

$$z_{t+l} = 2z_{t+l-1} - z_{t+l-2} + a_{t+l} - \theta_1 a_{t+l-1} - \theta_2 a_{t+l-2}$$

On taking conditional expectations at time t we obtain

$$\hat{z}_t(1) = 2z_t - z_{t-1} - \theta_1 a_t - \theta_2 a_{t-1}$$

$$\hat{z}_t(2) = 2\hat{z}_t(1) - z_t - \theta_2 a_t$$

$$\hat{z}_t(l) = 2\hat{z}_t(l-1) - \hat{z}_t(l-2) \qquad l \geqslant 3$$

from which the forecasts may be calculated. Forecasting of the series of Figure 5.5 in this way was illustrated in Section 5.1.2. An alternative way of generating the first $L - 1$ of L forecasts is via the updating formula (5.2.5),

$$\hat{z}_{t+1}(l) = \hat{z}_t(l+1) + \psi_l a_{t+1} \tag{5.4.8}$$

The truncated integrated model, as in (4.3.15), is

$$z_t = \lambda_0 S_{t-k-1} a_{t-1} + \lambda_1 S^{(2)}_{t-k-1} a_{t-1} + a_t + b_0^{(k)} + b_1^{(k)}(t-k)$$
$$\lambda_0 = 1 + \theta_2, \quad \lambda_1 = 1 - \theta_1 - \theta_2 \tag{5.4.9}$$

so that $\psi_j = \lambda_0 + j\lambda_1$ $(j = 1, 2, \ldots)$. Therefore, the updating function for this model is

$$\hat{z}_{t+1}(l) = \hat{z}_t(l+1) + (\lambda_0 + l\lambda_1)a_{t+1} \tag{5.4.10}$$

Forecast in integrated form. The eventual forecast function is the solution of $(1 - B)^2\hat{z}_t(l) = 0$, that is, $\hat{z}_t(l) = b_0^{(t)} + b_1^{(t)}l$. Since $q - p - d = 0$, the eventual forecast function provides the forecast for all lead times, that is,

$$\hat{z}_t(l) = b_0^{(t)} + b_1^{(t)}l \qquad l > 0 \tag{5.4.11}$$

Thus the forecast function is a linear function of the lead time l, with coefficients that are adaptive with respect to the origin t. The stochastic model in truncated integrated form is

$$z_{t+l} = \lambda_0 S_{t+l-k-1} a_{t+l-1} + \lambda_1 S^{(2)}_{t+l-k-1} a_{t+l-1} + a_{t+l} + b_0^{(k)} + b_1^{(k)}(t + l - k)$$

and taking expectations at origin t, we obtain

$$\hat{z}_t(l) = \lambda_0 S_{t-k} a_t + \lambda_1(l a_t + (l+1)a_{t-1} + \cdots + (l+t-k-1)a_{k+1}) + b_0^{(k)} + b_1^{(k)}(t+l-k)$$

$$= [\lambda_0 S_{t-k} a_t + \lambda_1 S^{(2)}_{t-k-1} a_{t-1} + b_0^{(k)} + b_1^{(k)}(t-k)] + (\lambda_1 S_{t-k} a_t + b_1^{(k)})l$$

The adaptive coefficients may thus be identified as

$$\begin{aligned} b_0^{(t)} &= \lambda_0 S_{t-k} a_t + \lambda_1 S^{(2)}_{t-k-1} a_{t-1} + b_0^{(k)} + b_1^{(k)}(t-k) \\ b_1^{(t)} &= \lambda_1 S_{t-k} a_t + b_1^{(k)} \end{aligned} \tag{5.4.12}$$

or informally based on the infinite integrated form as $b_0^{(t)} = \lambda_0 S a_t + \lambda_1 S^2 a_{t-1}$ and $b_1^{(t)} = \lambda_1 S a_t$. Hence their updating formulas are

$$\begin{aligned} b_0^{(t)} &= b_0^{(t-1)} + b_1^{(t-1)} + \lambda_0 a_t \\ b_1^{(t)} &= b_1^{(t-1)} + \lambda_1 a_t \end{aligned} \tag{5.4.13}$$

similar to relations (4.3.17). The additional slope term $b_1^{(t-1)}$, which occurs in the updating formula for $b_0^{(t)}$, is an adjustment to change the location parameter b_0 to a value appropriate to the new origin. It will also be noted that λ_0 and λ_1 are the fractions of the shock a_t which are transmitted to the location parameter and the slope parameter, respectively.

Forecasts as a weighted average of previous observations. For this model, then, the forecast function is the straight line that passes through the forecasts $\hat{z}_t(1)$ and $\hat{z}_t(2)$. This is illustrated for the series in Figure 5.5, which shows the forecasts made at origin $t = 30$, with appropriate weight functions. It will be seen how dependence of the entire forecast function on previous z's in the series is a reflection of the dependence of $\hat{z}_t(1)$ and $\hat{z}_t(2)$ on these values. The weight functions for $\hat{z}_t(1)$ and $\hat{z}_t(2)$, plotted in the figure, have been given in Table 5.4.

The example illustrates once more that while the AR operator $\varphi(B)$ determines the form of function to be used (a straight line in this case), the MA operator is of importance in determining the way in which that function is "fitted" to previous data.

Dependence of the adaptive coefficients in the forecast function on previous z's. Since, for the general model, the values of the adaptive coefficients in the forecast function are determined by $\hat{z}_t(q), \hat{z}_t(q-1), \ldots, \hat{z}_t(q-p-d+1)$, which can be expressed as functions of the observations, it follows that the adaptive coefficients themselves may be so expressed.

For instance, in the case of the model $\nabla^2 z_t = (1 - 0.9B + 0.5B^2)a_t$ of Figure 5.5,

$$\hat{z}_t(1) = b_0^{(t)} + b_1^{(t)} = \sum_{j=1}^{\infty} \pi_j^{(1)} z_{t-j+1}$$

$$\hat{z}_t(2) = b_0^{(t)} + 2b_1^{(t)} = \sum_{j=1}^{\infty} \pi_j^{(2)} z_{t-j+1}$$

so that

$$b_0^{(t)} = 2\hat{z}_t(1) - \hat{z}_t(2) = \sum_{j=1}^{\infty} (2\pi_j^{(1)} - \pi_j^{(2)})z_{t-j+1}$$

and

$$b_1^{(t)} = \hat{z}_t(2) - \hat{z}_t(1) = \sum_{j=1}^{\infty} (\pi_j^{(2)} - \pi_j^{(1)})z_{t-j+1}$$

These weight functions are plotted in Figure 5.7.

Variance of the forecast error. Using (5.1.16) and the fact that $\psi_j = \lambda_0 + j\lambda_1$, the variance of the lead l forecast error is

$$V(l) = \sigma_a^2[1 + (l-1)\lambda_0^2 + \tfrac{1}{6}l(l-1)(2l-1)\lambda_1^2 + \lambda_0\lambda_1 l(l-1)] \qquad (5.4.14)$$

Using the estimate $s_a^2 = 0.032$, $\lambda_0 = 0.5$, and $\lambda_1 = 0.6$, the 50% and 95% limits are shown in Figure 5.5 for the forecast at origin $t = 30$.

5.4.3 Forecasting a General IMA(0, d, q) Process

As an example, consider the process of order (0, 1, 3),

$$(1 - B)z_{t+1} = (1 - \theta_1 B - \theta_2 B^2 - \theta_3 B^3)a_{t+1}$$

Taking conditional expectations at time t, we obtain

$$\begin{aligned}
\hat{z}_t(1) - z_t &= -\theta_1 a_t - \theta_2 a_{t-1} - \theta_3 a_{t-2} \\
\hat{z}_t(2) - \hat{z}_t(1) &= -\theta_2 a_t - \theta_3 a_{t-1} \\
\hat{z}_t(3) - \hat{z}_t(2) &= -\theta_3 a_t \\
\hat{z}_t(l) - \hat{z}_t(l-1) &= 0 \qquad l = 4, 5, 6, \ldots
\end{aligned}$$

Hence $\hat{z}_t(l) = \hat{z}_t(3) = b_0^{(t)}$ for all $l > 2$, as expected, since $q - p - d = 2$. As shown in Figure 5.8, the forecast function makes two initial "jumps," de-

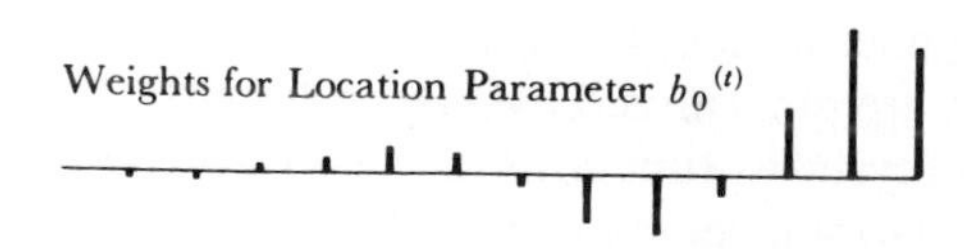

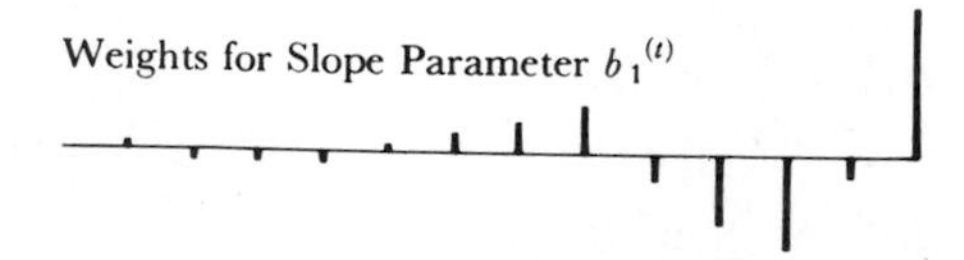

Figure 5.7 Weights applied to previous z's determining location and slope for the model $\nabla^2 z_t = (1 - 0.9B + 0.5B^2)a_t$.

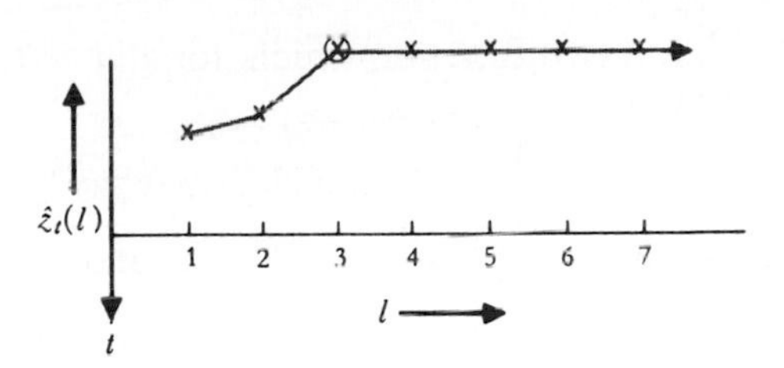

Figure 5.8 Forecast function for an IMA(0, 1, 3) process.

pending on previous a's, before leveling out to the eventual forecast function.

For the IMA(0, d, q) process, the eventual forecast function satisfies $(1 - B)^d\hat{z}_t(l) = 0$, and has for its solution, a polynomial in l of degree $d - 1$,

$$\hat{z}_t(l) = b_0^{(t)} + b_1^{(t)}l + b_2^{(t)}l^2 + \cdots + b_{d-1}^{(t)}l^{d-1}$$

This will provide the forecasts $\hat{z}_t(l)$ for $l > q - d$. The coefficients $b_0^{(t)}, b_1^{(t)}, \ldots, b_{d-1}^{(t)}$, must be updated progressively as the origin advances. The forecast for origin t will make $q - d$ initial "jumps," which depend on $a_t, a_{t-1}, \ldots, a_{t-q+1}$, and after this, will follow the polynomial above.

5.4.4 Forecasting Autoregressive Processes

Consider a process of order (p, d, 0)

$$\varphi(B)z_t = a_t$$

The eventual forecast function is the solution of $\varphi(B)\hat{z}_t(l) = 0$. It applies for all lead times and passes through the last $p + d$ available values of the series. For example, the model for the IBM stock series (series B) is very nearly

$$(1 - B)z_t = a_t$$

so that

$$\hat{z}_t(l) \approx z_t$$

The best forecast for all future time is very nearly the current value of the stock. The weight function for $\hat{z}_t(l)$ is a spike at time t and there is no averaging over past history.

Stationary autoregressive models. The process $\phi(B)\tilde{z}_t = a_t$, of order (p, 0, 0), where $\phi(B)$ is a stationary operator and $\tilde{z}_t = z_t - \mu$, with $E[z_t] = \mu$, will in general produce a forecast function that is a mixture of exponentials and damped sines. In particular, for $p = 1$ the model of order (1, 0, 0)

$$(1 - \phi B)\tilde{z}_t = a_t \qquad 1 < \phi < 1$$

has a forecast function which, for all $l > 0$, is the solution of $(1 - \phi B)\hat{\tilde{z}}_t(l) = 0$. Thus

$$\hat{\tilde{z}}_t(l) = b_0^{(t)}\phi^l \qquad l > 0 \tag{5.4.15}$$

Also, $\hat{\tilde{z}}_t(1) = \phi\tilde{z}_t$, so that $b_0^{(t)} = \tilde{z}_t$ and

$$\hat{\tilde{z}}_t(l) = \tilde{z}_t\phi^l$$

So the forecasts for the original process z_t are $\hat{z}_t(l) = \mu + \phi^l(z_t - \mu)$.

Hence the minimum mean square error forecast predicts the current deviation from the mean decaying exponentially to zero. In Figure 5.9(a) a time series is shown that is generated from the process $(1 - 0.5B)\tilde{z}_t = a_t$, with the forecast function at origin $t = 14$. The course of this function is seen to be determined entirely by the single deviation $\tilde{z}_{14}$. Similarly, the minimum mean square error forecast for a second-order autoregressive process is such that the current deviation from the mean decays to zero via a damped sine wave or a mixture of two exponentials. Figure 5.9(b) shows a time series generated from the process $(1 - 0.75B + 0.50B^2)\tilde{z}_t = a_t$ and the forecast at origin $t = 14$. Here the course of the forecast function at origin t is determined entirely by the last two deviations, $\tilde{z}_{14}$ and $\tilde{z}_{13}$.

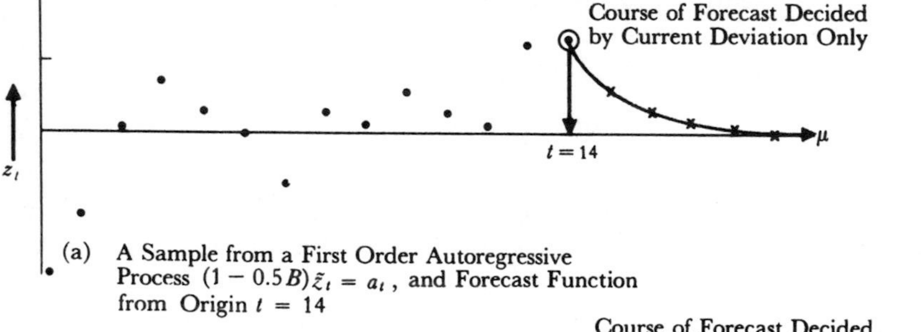

(a) A Sample from a First Order Autoregressive Process $(1 - 0.5B)\tilde{z}_t = a_t$, and Forecast Function from Origin $t = 14$

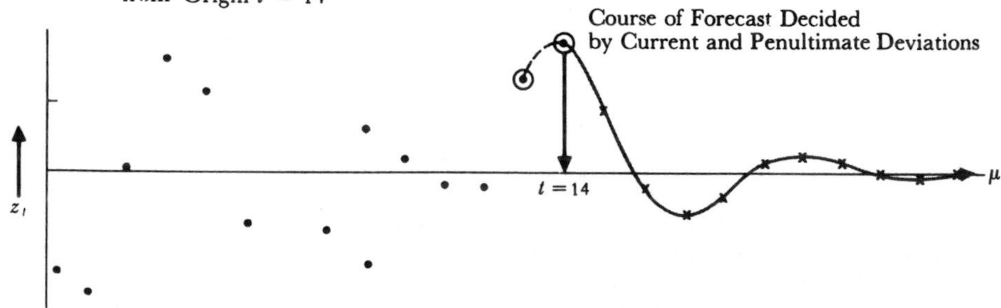

(b) A Sample from a Second Order Autoregressive Process $(1 - 0.75B + 0.5B^2)\tilde{z}_t = a_t$, and Forecast Function from Origin $t = 14$

Figure 5.9 Forecast functions for first- and second-order autoregressive processes.

Variance function for the forecast from a (1, 0, 0) process. As a further illustration of the use of (5.1.16), we derive the variance function for a first-order autoregressive process. Since the model at time $t + l$ may be written

$$\tilde{z}_{t+l} = a_{t+l} + \phi a_{t+l-1} + \cdots + \phi^{l-1} a_{t+1} + \phi^l \tilde{z}_t$$

it follows from (5.4.15) that

$$e_t(l) = \tilde{z}_{t+l} - \hat{\tilde{z}}_t(l) = a_{t+1} + \phi a_{t+l-1} + \cdots + \phi^{l-1} a_{t+1}$$

Hence

$$\begin{aligned} V(l) = \text{var}[e_t(l)] &= \sigma_a^2(1 + \phi^2 + \cdots + \phi^{2l-2}) \\ &= \frac{\sigma_a^2(1 - \phi^{2l})}{1 - \phi^2} \end{aligned} \qquad (5.4.16)$$

We see that for this stationary process, as l tends to infinity the variance increases to a constant value $\gamma_0 = \sigma_a^2/(1 - \phi^2)$, associated with the variation of the process about the ultimate forecast μ. This is in contrast to the behavior of forecast variance functions for nonstationary models that "blow up" for large lead times.

Nonstationary autoregressive models of order (p, d, 0). For the model

$$\phi(B)\nabla^d z_t = a_t$$

it will be the dth difference of the process that decays back to its mean when projected several steps ahead. The mean of $\nabla^d z_t$ will usually be assumed to be zero unless contrary evidence is available. When needed, it is possible, as discussed in Chapter 4, to introduce a nonzero mean by replacing $\nabla^d z_t$ by $\nabla^d z_t - \mu_w$ in the model. For example, consider the model

$$(1 - \phi B)(\nabla z_t - \mu_w) = a_t \qquad (5.4.17)$$

After substituting $t + j$ for t and taking conditional expectations at origin t, we readily obtain [compare with (5.4.15) et seq.]

$$\hat{z}_t(j) - \hat{z}_t(j - 1) - \mu_w = \phi^j(z_t - z_{t-1} - \mu_w)$$

or $\hat{w}_t(j) - \mu_w = \phi^j(w_t - \mu_w)$, where $w_t = \nabla z_t$, which shows how the forecasted *difference* decays exponentially from the initial value $w_t = z_t - z_{t-1}$ to its mean value μ_w. On summing this expression from $j = 1$ to $j = l$, that is, using $\hat{z}_t(l) = \hat{w}_t(l) + \cdots + \hat{w}_t(1) + z_t$, we obtain the forecast function

$$\hat{z}_t(l) = z_t + \mu_w l + (z_t - z_{t-1} - \mu_w)\frac{\phi(1 - \phi^l)}{1 - \phi} \qquad l \geqslant 1$$

which approaches asymptotically the straight line

$$f(l) = z_t + \mu_w l + (z_t - z_{t-1} - \mu_w)\frac{\phi}{1-\phi}$$

with deterministic slope μ_w. Figure 5.10 shows forecasts for the two cases $\phi = 0.8$, $\mu_w = 0$ and $\phi = 0.8$, $\mu_w = 0.2$. We show in Chapter 7 that the model (5.4.17), with $\phi = 0.8$, $\mu_w = 0$, closely represents series C and forecasts based on this model have already been illustrated in Figures 5.1 and 5.2. We now consider the forecasting of some important mixed models.

5.4.5 Forecasting a (1, 0, 1) Process

Difference equation approach. Consider the stationary model

$$(1 - \phi B)\tilde{z}_t = (1 - \theta B)a_t$$

The forecasts are readily obtained from

$$\begin{aligned} \hat{\tilde{z}}_t(1) &= \phi\tilde{z}_t - \theta a_t \\ \hat{\tilde{z}}_t(l) &= \phi\hat{\tilde{z}}_t(l-1) \qquad l \geqslant 2 \end{aligned} \tag{5.4.18}$$

The forecasts decay geometrically to the mean, as in the first-order autoregressive process, but with a lead 1 forecast modified by a factor depending

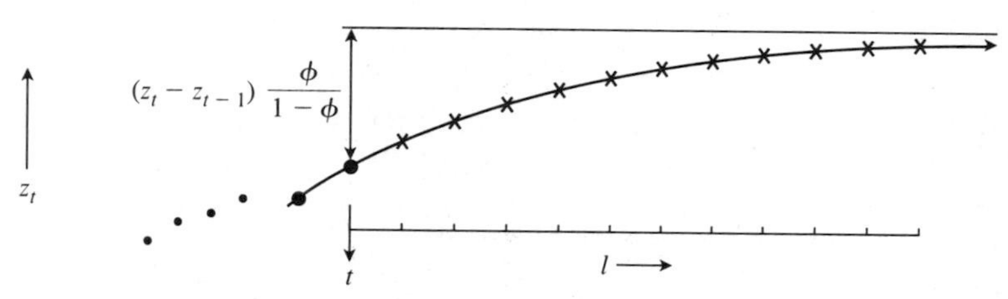

(a) Forecasts of a (1, 1, 0) Process $(1 - 0.8B)\nabla z_t = a_t$

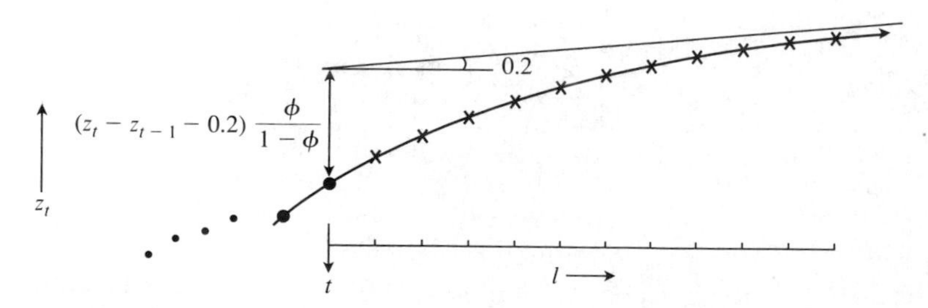

(b) Forecasts of a (1, 1, 0) Process $(1 - 0.8B)(\nabla z_t - 0.2) = a_t$

Figure 5.10 Forecast functions for two (1, 1, 0) processes.

on $a_t = z_t - \hat{z}_{t-1}(1)$. The ψ weights are

$$\psi_j = (\phi - \theta)\phi^{j-1} \qquad j = 1, 2, \ldots$$

and hence, using (5.2.5), the updated forecasts for lead times 1, 2, . . . , $L - 1$ could be obtained from previous forecasts for lead times 2, 3, . . . , L according to

$$\hat{\tilde{z}}_{t+1}(l) = \hat{\tilde{z}}_t(l + 1) + (\phi - \theta)\phi^{l-1}a_{t+1}$$

Integrated form. The eventual forecast function for all $l > 0$ is the solution of $(1 - \phi B)\hat{\tilde{z}}_t(l) = 0$, that is,

$$\hat{\tilde{z}}_t(l) = b_0^{(t)}\phi^l \qquad l > 0$$

However,

$$\hat{\tilde{z}}_t(1) = b_0^{(t)}\phi = \phi\tilde{z}_t - \theta a_t = \left[\left(1 - \frac{\theta}{\phi}\right)\tilde{z}_t + \frac{\theta}{\phi}\hat{\tilde{z}}_{t-1}(1)\right]\phi$$

Thus

$$\hat{\tilde{z}}_t(l) = \left[\left(1 - \frac{\theta}{\phi}\right)\tilde{z}_t + \frac{\theta}{\phi}\hat{\tilde{z}}_{t-1}(1)\right]\phi^l \tag{5.4.19}$$

Hence, the forecasted deviation at lead l decays exponentially from an initial value, which is a linear interpolation between the previous lead 1 forecasted deviation and the current deviation. When ϕ is equal to unity, the forecast for all lead times becomes the familiar exponentially weighted moving average and (5.4.19) becomes equal to (5.4.3).

Weights applied to previous observations. The π weights, and hence the weights applied to previous observations to obtain the lead 1 forecasts, are

$$\pi_j = (\phi - \theta)\theta^{j-1} \qquad j = 1, 2, \ldots$$

Note that the weights for this stationary process sum to $(\phi - \theta)/(1 - \theta)$ and not to unity. If ϕ were equal to 1, the process would become a nonstationary IMA(0, 1, 1) process, the weights would then sum to unity, and the behavior of the generated series would be independent of the level of z_t.

For example, series A is later fitted to a (1, 0, 1) model with $\phi = 0.9$ and $\theta = 0.6$, and hence the weights are $\pi_1 = 0.30$, $\pi_2 = 0.18$, $\pi_3 = 0.11$, $\pi_4 = 0.07$, . . . , which sum to 0.75. The forecasts (5.4.19) decay very slowly to the mean, and for short lead times, are practically indistinguishable from the forecasts obtained from the alternative IMA(0, 1, 1) model $\nabla z_t = a_t - 0.7a_{t-1}$, for which the weights are $\pi_1 = 0.30$, $\pi_2 = 0.21$, $\pi_3 = 0.15$, $\pi_4 = 0.10$, and so on, and sum to unity. The latter model has the advantage that it does not tie the process to a fixed mean.

Variance function. Since the ψ weights are given by

$$\psi_j = (\phi - \theta)\phi^{j-1} \qquad j = 1, 2, \ldots$$

it follows that the variance function is

$$V(l) = \sigma_a^2 \left[1 + (\phi - \theta)^2 \frac{1 - \phi^{2l-2}}{1 - \phi^2} \right] \tag{5.4.20}$$

which increases asymptotically to the value $\sigma_a^2(1 - 2\phi\theta + \theta^2)/(1 - \phi^2)$, the variance γ_0 of the process.

5.4.6 Forecasting a (1, 1, 1) Process

Another important mixed model is the nonstationary (1, 1, 1) process

$$(1 - \phi B)\nabla z_t = (1 - \theta B)a_t$$

Difference equation approach. At time $t + l$, the model may be written

$$z_{t+l} = (1 + \phi)z_{t+l-1} - \phi z_{t+l-2} + a_{t+l} - \theta a_{t+l-1}$$

On taking conditional expectations, we obtain

$$\begin{aligned} \hat{z}_t(1) &= (1 + \phi)z_t - \phi z_{t-1} - \theta a_t \\ \hat{z}_t(l) &= (1 + \phi)\hat{z}_t(l - 1) - \phi \hat{z}_t(l - 2) \qquad l > 1 \end{aligned} \tag{5.4.21}$$

Integrated form. Since $q < p + d$, the eventual forecast function for all $l > 0$ is the solution of $(1 - \phi B)(1 - B)\hat{z}_t(l) = 0$, which is

$$\hat{z}_t(l) = b_0^{(t)} + b_1^{(t)}\phi^l$$

Substituting for $\hat{z}_t(1)$ and $\hat{z}_t(2)$ in (5.4.21), we find explicitly that

$$b_0^{(t)} = z_t + \frac{\phi}{1 - \phi}(z_t - z_{t-1}) - \frac{\theta}{1 - \phi} a_t$$

$$b_1^{(t)} = \frac{\theta a_t - \phi(z_t - z_{t-1})}{1 - \phi}$$

Thus, finally,

$$\hat{z}_t(l) = z_t + \phi \frac{1 - \phi^l}{1 - \phi}(z_t - z_{t-1}) - \theta \frac{1 - \phi^l}{1 - \phi} a_t \tag{5.4.22}$$

Evidently for large l, the forecast tends to $b_0^{(t)}$.

Weights applied to previous observations. Eliminating a_t from (5.4.22), we obtain the alternative form for the forecast in terms of previous z's:

$$\hat{z}_t(l) = \left[1 - \frac{\theta - \phi}{1 - \phi}(1 - \phi^l)\right] z_t + \left[\frac{\theta - \phi}{1 - \phi}(1 - \phi^l)\right] \bar{z}_{t-1}(\theta) \qquad (5.4.23)$$

where $\bar{z}_{t-1}(\theta)$ is an exponentially weighted moving average with parameter θ, that is, $\bar{z}_{t-1}(\theta) = (1 - \theta)\sum_{j=1}^{\infty}\theta^{j-1}z_{t-j}$. Thus the π weights for the process consist of a "spike" at time t and an EWMA starting at time $t - 1$. If we refer to $(1 - \alpha)x + \alpha y$ as a linear interpolation between x and y at argument α, the forecast (5.4.23) is a linear interpolation between z_t and $\bar{z}_{t-1}(\theta)$. The argument for lead time 1 is $\theta - \phi$, but as the lead time is increased, the argument approaches $(\theta - \phi)/(1 - \phi)$. For example, when $\theta = 0.9$ and $\phi = 0.5$, the lead 1 forecast is

$$\hat{z}_t(1) = 0.6z_t + 0.4\bar{z}_{t-1}(\theta)$$

and for long lead times, the forecast approaches

$$\hat{z}_t(\infty) = 0.2z_t + 0.8\bar{z}_{t-1}(\theta)$$

5.5 USE OF STATE SPACE MODEL FORMULATION FOR EXACT FORECASTING

5.5.1 State Space Model Representation for the ARIMA Process

There has been much recent interest in the representation of ARIMA models in the state space form, for purposes of forecasting, as well as for model specification and maximum likelihood estimation of parameters. We briefly consider the state space form of an ARIMA model in this section and discuss its uses in exact finite sample forecasting. For an ARIMA(p, d, q) process $\varphi(B)z_t = \theta(B)a_t$, define the forecasts $\hat{z}_t(j) = E_t[z_{t+j}]$ as in Section 5.1, for $j = 0, 1, \ldots, r$, with $r = \max(p + d, q + 1)$, and $\hat{z}_t(0) = z_t$.

From the updating equations (5.2.5), we have that $\hat{z}_t(j - 1) = \hat{z}_{t-1}(j) + \psi_{j-1}a_t$, $j = 1, 2, \ldots, r - 1$. Also for $j = r > q$, recall from (5.3.2) that $\hat{z}_t(j - 1) = \hat{z}_{t-1}(j) + \psi_{j-1}a_t = \sum_{i=1}^{p+d}\varphi_i\hat{z}_{t-1}(j - i) + \psi_{j-1}a_t$. So we define the "state" vector at time t, $\boldsymbol{Y}_t$, with r components as $\boldsymbol{Y}_t = (z_t, \hat{z}_t(1), \ldots, \hat{z}_t(r - 1))'$. Then from the relations above we find that $\boldsymbol{Y}_t$ satisfies the equations

$$\boldsymbol{Y}_t = \begin{bmatrix} 0 & 1 & 0 & \cdots & 0 \\ 0 & 0 & 1 & \cdots & 0 \\ \cdots & \cdots & \cdots & \cdots & \cdots \\ 0 & 0 & & \cdots & 1 \\ \varphi_r & \varphi_{r-1} & & \cdots & \varphi_1 \end{bmatrix} \boldsymbol{Y}_{t-1} + \begin{bmatrix} 1 \\ \psi_1 \\ \vdots \\ \psi_{r-1} \end{bmatrix} a_t$$

where $\varphi_i = 0$ if $i > p + d$. So we have

$$\boldsymbol{Y}_t = \boldsymbol{\Phi} \boldsymbol{Y}_{t-1} + \boldsymbol{\Psi} a_t \tag{5.5.1}$$

together with the observation equation

$$Z_t = z_t + N_t = [1, 0, \ldots, 0]\boldsymbol{Y}_t + N_t = \mathbf{H}\boldsymbol{Y}_t + N_t \tag{5.5.2}$$

where the additional noise N_t would be present only if the process z_t is observed subject to additional white noise; otherwise, we simply have $z_t = \mathbf{H}\boldsymbol{Y}_t$. The two equations above constitute what is known as the state space representation of the model, which consists of a state or transition equation (5.5.1) and an observation equation (5.5.2), and $\boldsymbol{Y}_t$ is known as the state vector. We note that there are many other constructions of the state vector $\boldsymbol{Y}_t$ that will give rise to state space equations of the general form of (5.5.1)–(5.5.2); that is, the state space form of an ARIMA model is not unique. The two equations of the form above, in general, represent what is known as a state space model, with unobservable state vector $\boldsymbol{Y}_t$ and observations Z_t, and can arise in time series settings more general than the context of ARIMA models.

Consider a state space model of a slightly more general form, with state equation

$$\boldsymbol{Y}_t = \boldsymbol{\Phi}_t \boldsymbol{Y}_{t-1} + \boldsymbol{a}_t \tag{5.5.3}$$

and observation equation

$$Z_t = \boldsymbol{H}_t \boldsymbol{Y}_t + N_t \tag{5.5.4}$$

where it is assumed that $\boldsymbol{a}_t$ and N_t are independent white noise processes, $\boldsymbol{a}_t$ is a vector white noise process with covariance matrix Σ_a, and N_t has variance σ_N^2. In this model, the (unobservable) state vector $\boldsymbol{Y}_t$ summarizes the state of the dynamic system through time t and the state equation (5.5.3) describes the evolution of the dynamic system in time, while the measurement equation (5.5.4) indicates that the observations Z_t consist of linear combinations of the state variables corrupted by additive white noise. The matrix $\boldsymbol{\Phi}_t$ in (5.5.3) is an $r \times r$ transition matrix and $\mathbf{H}_t$ in (5.5.4) is a $1 \times r$ vector, which are allowed to vary with time t. Often, in applications these are constant matrices, $\boldsymbol{\Phi}_t \equiv \boldsymbol{\Phi}$ and $\mathbf{H}_t \equiv \mathbf{H}$ for all t, that do not depend on t, as in the state space form (5.5.1)–(5.5.2) of the ARIMA model. In this case the system or model is said to be *time invariant*. The minimal dimension r of the state vector $\boldsymbol{Y}_t$ in a state space model needs to be sufficiently large so that the dynamics of the system can be represented by the simple Markovian (first-order) structure as in (5.5.3).

5.5.2 Kalman Filtering Relations for Use in Prediction

For the state space model (5.5.3)–(5.5.4), define the finite sample optimal (minimum mean square error matrix) estimate of the state vector $\boldsymbol{Y}_{t+l}$

based on observations $Z_t, \ldots, Z_1$ over the finite past time period, as $\hat{Y}_{t+l|t} = E[Y_{t+l}|Z_t, \ldots, Z_1]$, with $\mathbf{V}_{t+l|t} = E[(Y_{t+l} - \hat{Y}_{t+l|t})(Y_{t+l} - \hat{Y}_{t+l|t})']$ equal to the error covariance matrix. A convenient computational procedure, known as the *Kalman filter* equations, is then available to obtain the current estimate $\hat{Y}_{t|t}$, in particular. It is known that, starting from some appropriate initial values $Y_0 \equiv \hat{Y}_{0|0}$ and $\mathbf{V}_0 \equiv \mathbf{V}_{0|0}$, the optimal filtered estimate, $\hat{Y}_{t|t}$, is given through the following recursive relations:

$$\hat{Y}_{t|t} = \hat{Y}_{t|t-1} + \mathbf{K}_t(Z_t - \mathbf{H}_t\hat{Y}_{t|t-1}) \tag{5.5.5}$$

where

$$\mathbf{K}_t = \mathbf{V}_{t|t-1}\mathbf{H}_t'[\mathbf{H}_t\mathbf{V}_{t|t-1}\mathbf{H}_t' + \sigma_N^2]^{-1} \tag{5.5.6}$$

with

$$\hat{Y}_{t|t-1} = \mathbf{\Phi}_t\hat{Y}_{t-1|t-1} \qquad \mathbf{V}_{t|t-1} = \mathbf{\Phi}_t V_{t-1|t-1}\mathbf{\Phi}_t' + \Sigma_a \tag{5.5.7}$$

and

$$\begin{aligned} V_{t|t} &= [I - \mathbf{K}_t\mathbf{H}_t]\mathbf{V}_{t|t-1} \\ &= \mathbf{V}_{t|t-1} - \mathbf{V}_{t|t-1}\mathbf{H}_t'[\mathbf{H}_t\mathbf{V}_{t|t-1}\mathbf{H}_t' + \sigma_N^2]^{-1}\mathbf{H}_t\mathbf{V}_{t|t-1} \end{aligned} \tag{5.5.8}$$

for $t = 1, 2, \ldots$.

In (5.5.5), the quantity $a_{t|t-1} = Z_t - \mathbf{H}_t\hat{Y}_{t|t-1} \equiv Z_t - \hat{Z}_{t|t-1}$ is called the (finite sample) innovation at time t, because it is the new information provided by the measurement Z_t that was not available from the previous observed (finite) history of the system. The factor $\mathbf{K}_t$ is called the "Kalman gain" matrix. The filtering procedure in (5.5.5) has the recursive "prediction–correction" or "updating" form, and the validity of these equations as representing the minimum mean square error predictor can readily be verified through the principles of "updating." For example, verification of (5.5.5) follows from the principle, for linear prediction, that

$$\begin{aligned} E[Y_t|Z_t, \ldots, Z_1] &= E[Y_t|Z_t - \hat{Z}_{t|t-1}, Z_{t-1}, \ldots, Z_1] \\ &= E[Y_t|Z_{t-1}, \ldots, Z_1] + E[Y_t|Z_t - \hat{Z}_{t|t-1}] \end{aligned}$$

since $a_{t|t-1} = Z_t - \hat{Z}_{t|t-1}$ is independent of $Z_{t-1}, \ldots, Z_1$. From (5.5.5) it is seen that the estimate of Y_t based on observations through time t equals the prediction of Y_t from observations through time $t - 1$ updated by the factor $\mathbf{K}_t$ times the innovation $a_{t|t-1}$. The equation in (5.5.6) indicates that $\mathbf{K}_t$ can be interpreted as the regression coefficients of Y_t on the innovation $a_{t|t-1}$, with $\text{var}[a_{t|t-1}] = \mathbf{H}_t\mathbf{V}_{t|t-1}\mathbf{H}_t' + \sigma_N^2$ and $\text{cov}[Y_t, a_{t|t-1}] = \mathbf{V}_{t|t-1}\mathbf{H}_t'$ following directly from (5.5.4) since $a_{t|t-1} = \mathbf{H}_t(Z_t - \hat{Z}_{t|t-1}) + N_t$. Thus the general *updating relation* is

$$\hat{Y}_{t|t} = \hat{Y}_{t|t-1} + \text{cov}[Y_t, a_{t|t-1}]\{\text{var}[a_{t|t-1}]\}^{-1}a_{t|t-1}$$

where $a_{t|t-1} = Z_t - \hat{Z}_{t|t-1}$, and the relation in (5.5.8) is the usual updating of the error covariance matrix to account for the new information available from the innovation $a_{t|t-1}$, while the *prediction relations* (5.5.7) follow directly from (5.5.3).

In general, forecasts of future state values are available directly as $\hat{\boldsymbol{Y}}_{t+l|t} = \boldsymbol{\Phi}_{t+l}\hat{\boldsymbol{Y}}_{t+l-1|t}$ for $l = 1, 2, \ldots$, with the covariance matrix of the forecast errors generated recursively essentially through (5.5.7) as $\mathbf{V}_{t+l|t} = \boldsymbol{\Phi}_{t+l}\mathbf{V}_{t+l-1|t}\boldsymbol{\Phi}'_{t+l} + \Sigma_a$. Finally, forecasts of future observations Z_{t+l} are then available as $\hat{Z}_{t+l|t} = \mathbf{H}_{t+l}\hat{\boldsymbol{Y}}_{t+l|t}$, since $Z_{t+l} = \mathbf{H}_{t+l}\boldsymbol{Y}_{t+l} + N_{t+l}$, with forecast error variance $v_{t+l|t} = E[(Z_{t+l} - \hat{Z}_{t+l|t})^2] = \mathbf{H}_{t+l}\mathbf{V}_{t+l|t}\mathbf{H}'_{t+l} + \sigma_N^2$.

For ARIMA models, with state space representation (5.5.1)–(5.5.2) and $Z_t = z_t = \mathbf{H}\boldsymbol{Y}_t$ with $\mathbf{H} = [1, 0, \ldots, 0]$, this Kalman filtering procedure constitutes an alternative method to obtain exact finite sample forecasts, based on data $z_t, z_{t-1}, \ldots, z_1$, for future values in the ARIMA process, subject to specification of appropriate initial conditions to use in (5.5.5)–(5.5.8). For stationary zero-mean processes z_t, the appropriate initial values are $\hat{\boldsymbol{Y}}_{0|0} = \mathbf{0}$, a vector of zeros, and $\mathbf{V}_{0|0} = \text{cov}[\boldsymbol{Y}_0] \equiv \mathbf{V}_*$, the covariance matrix of $\boldsymbol{Y}_0$, which can easily be determined under stationarity through the definition of $\boldsymbol{Y}_t$. Specifically, since the state vector $\boldsymbol{Y}_t$ follows the stationary vector AR(1) model $\boldsymbol{Y}_t = \boldsymbol{\Phi}\boldsymbol{Y}_{t-1} + \boldsymbol{\Psi}a_t$, its covariance matrix $\mathbf{V}_* = \text{cov}[\boldsymbol{Y}_t]$ satisfies $\mathbf{V}_* = \boldsymbol{\Phi}\mathbf{V}_*\boldsymbol{\Phi}' + \sigma_a^2\boldsymbol{\Psi}\boldsymbol{\Psi}'$, which can be readily solved for $\mathbf{V}_*$. For nonstationary ARIMA processes, additional assumptions need to be specified (see, e.g., Ansley and Kohn [13], Bell and Hillmer [29]). The "steady-state" values of the Kalman-filtering procedure l-step-ahead forecasts $\hat{z}_{t+l|t}$ and their forecast error variances $v_{t+l|t}$, which are rapidly approached as t increases, will be identical to the expressions given in Sections 5.1 and 5.2, $\hat{z}_t(l)$ and $V(l) = \sigma_a^2(1 + \Sigma_{j=1}^{l-1} \psi_j^2)$.

In particular, for the ARIMA process in state space form, we can obtain the exact (finite sample) one-step-ahead forecasts $\hat{z}_{t-1}(1) = E[z_t|z_{t-1}, \ldots, z_1] = \mathbf{H}\hat{\boldsymbol{Y}}_{t|t-1}$, and their error variances $v_t = \mathbf{H}\mathbf{V}_{t|t-1}\mathbf{H}'$, conveniently through the Kalman-filtering equations (5.5.5)–(5.5.8). This can be particularly useful for evaluation of the likelihood function, based on n observations $z_1, \ldots, z_n$ from the ARIMA process, applied to the problem of maximum likelihood estimation of model parameters (see, e.g., Jones [123] and Gardner, Harvey, and Phillips [94]). This will be discussed again briefly in Section 7.5.

5.6 SUMMARY

The results of this chapter may be summarized as follows: Let $\tilde{z}_t$ be the deviation of an observed time series from any known deterministic function of time $f(t)$. In particular, for a stationary series, $f(t)$ could be equal to μ, the mean of the series, or it could be equal to zero, so that $\tilde{z}_t$ was the observed series. Then consider the general ARIMA model

$$\phi(B)\nabla^d \tilde{z}_t = \theta(B) a_t$$

or

$$\varphi(B)\tilde{z}_t = \theta(B) a_t$$

Minimum mean square error forecast. Given knowledge of the series up to some origin t, the minimum mean square error forecast $\hat{\tilde{z}}_t(l)$ $(l > 0)$ of $\tilde{z}_{t+l}$, is the conditional expectation

$$\hat{\tilde{z}}_t(l) = [\tilde{z}_{t+l}] = E[\tilde{z}_{t+l} | \tilde{z}_t, \tilde{z}_{t-1}, \ldots]$$

Lead 1 forecast errors. A necessary consequence is that the lead 1 forecast errors are the generating a's in the model and are uncorrelated.

Calculation of the forecasts. It is usually simplest in practice to compute the forecasts directly from the difference equation to give

$$\hat{\tilde{z}}_t(l) = \varphi_1[\tilde{z}_{t+l-1}] + \cdots + \varphi_{p+d}[\tilde{z}_{t+l-p-d}] + [a_{t+l}] - \theta_1[a_{t+l-1}] - \cdots - \theta_q[a_{t+l-q}] \tag{5.6.1}$$

The conditional expectations in (5.6.1) are evaluated by inserting actual $\tilde{z}$'s when these are known, forecasted $\tilde{z}$'s for future values, actual a's when these are known, and zeros for future a's. The forecasting process may be initiated by approximating a's by zeros and, in practice, the appropriate form for the model and suitable estimates for the parameters are obtained by methods set out in Chapters 6, 7, and 8.

Probability limits for forecasts. These may be obtained:

1. By first calculating the ψ weights from

$$\begin{aligned} \psi_0 &= 1 \\ \psi_1 &= \varphi_1 - \theta_1 \\ \psi_2 &= \varphi_1\psi_1 + \varphi_2 - \theta_2 \\ &\vdots \\ \psi_j &= \varphi_1\psi_{j-1} + \cdots + \varphi_{p+d}\psi_{j-p-d} - \theta_j \end{aligned} \tag{5.6.2}$$

   where, $\theta_j = 0$, $j > q$.

2. For each desired level of probability ε, and for each lead time l, substituting in

$$\tilde{z}_{t+l}(\pm) = \hat{\tilde{z}}_t(l) \pm u_{\varepsilon/2}\left(1 + \sum_{j=1}^{l-1} \psi_j^2\right)^{1/2} \sigma_a \tag{5.6.3}$$

where in practice σ_a is replaced by an estimate s_a, of the standard deviation of the white noise process a_t, and $u_{\varepsilon/2}$ is the deviate exceeded by a proportion $\varepsilon/2$ of the unit Normal distribution.

Updating the forecasts. When a new deviation $\tilde{z}_{t+1}$ comes to hand, the forecasts may be updated to origin $t + 1$, by calculating the new forecast error $a_{t+1} = \tilde{z}_{t+1} - \hat{\tilde{z}}_t(1)$ and using the difference equation (5.6.1) with $t + 1$ replacing t. However, an *alternative* method is to use the forecasts $\hat{\tilde{z}}_t(1)$, $\hat{\tilde{z}}_t(2), \ldots, \hat{\tilde{z}}_t(L)$ at origin t, to obtain the first $L - 1$ forecasts $\hat{\tilde{z}}_{t+1}(1)$, $\hat{\tilde{z}}_{t+1}(2), \ldots, \hat{\tilde{z}}_{t+1}(L - 1)$ at origin $t + 1$, from

$$\hat{\tilde{z}}_{t+1}(l) = \hat{\tilde{z}}_t(l + 1) + \psi_l a_{t+1} \tag{5.6.4}$$

and then generate the last forecast $\hat{\tilde{z}}_{t+1}(L)$ using the difference equation (5.6.1).

Other ways of expressing the forecasts. The above is all that is needed for *practical* utilization of the forecasts. However, the following alternative forms provide theoretical insight into the nature of the forecasts generated by different models:

1. *Forecasts in integrated form.* For $l > q - p - d$, the forecasts lie on the unique curve

$$\hat{\tilde{z}}(l) = b_0^{(t)} f_0(l) + b_1^{(t)} f_1(l) + \cdots + b_{p+d-1}^{(t)} f_{p+d-1}(l) \tag{5.6.5}$$

determined by the "pivotal" values $\hat{\tilde{z}}_t(q), \hat{\tilde{z}}_t(q - 1), \ldots, \hat{\tilde{z}}_t(q - p - d + 1)$, where $\hat{\tilde{z}}_t(-j) = \tilde{z}_{t-j} (j = 0, 1, 2, \ldots)$. If $q > p + d$, the first $q - p - d$ forecasts do not lie on this curve. In general, the stationary autoregressive operator contributes damped exponential and damped sine wave terms to (5.6.5) and the nonstationary operator ∇^d contributes polynomial terms up to degree $d - 1$.

The adaptive coefficients $b_j^{(t)}$ in (5.6.5) may be updated from origin t to $t + 1$ by amounts depending on the last lead 1 forecast error a_{t+1}, according to the general formula

$$\mathbf{b}^{(t+1)} = \mathbf{L}'\mathbf{b}^{(t)} + \mathbf{g}a_{t+1} \tag{5.6.6}$$

given in Appendix A5.3. Specific examples of the updating are given in (5.4.5) and (5.4.13) for the IMA(0, 1, 1) and IMA(0, 2, 2) processes, respectively.

2. *Forecasts as a weighted sum of past observations.* It is instructive from a theoretical point of view to express the forecasts as a weighted sum of past observations. Thus if the model is written in inverted form,

$$a_t = \pi(B)\tilde{z}_t = (1 - \pi_1 B - \pi_2 B^2 - \cdots)\tilde{z}_t$$

the lead 1 forecast is

$$\hat{\tilde{z}}_t(1) = \pi_1 \tilde{z}_t + \pi_2 \tilde{z}_{t-1} + \cdots \tag{5.6.7}$$

and the forecasts for longer lead times may be obtained from

$$\hat{z}_t(l) = \pi_1[\tilde{z}_{t+l-1}] + \pi_2[\tilde{z}_{t+l-2}] + \cdots \tag{5.6.8}$$

where the conditional expectations in (5.6.8) are evaluated by replacing $\tilde{z}$'s by actual values when known, and by forecasted values when unknown.

Alternatively, the forecast for any lead time may be written as a linear function of the available observations. Thus

$$\hat{z}_t(l) = \sum_{j=1}^{\infty} \pi_j^{(l)} \tilde{z}_{t+j-1}$$

where the $\pi_j^{(l)}$ are functions of the π_j's.

Role of constant term in forecasts. We also note the impact on forecasts of allowance of a nonzero constant term θ_0 in the ARIMA(p, d, q) model, $\varphi(B)z_t = \theta_0 + \theta(B)a_t$, where $\varphi(B) = \phi(B)\nabla^d$. Then, in (5.3.3) an additional deterministic polynomial term of degree d, $(\mu_w/d!)l^d$ with $w_t = \nabla^d z_t$ and $\mu_w = E[w_t] = \theta_0/(1 - \phi_1 - \phi_2 - \cdots - \phi_p)$, will be present. This follows since in place of the relation $\varphi(B)\hat{z}_t(l) = 0$ in (5.3.2), the forecasts now satisfy $\varphi(B)\hat{z}_t(l) = \theta_0$, $l > q$, and the deterministic polynomial term of degree d represents a particular solution to this nonhomogeneous difference equation. Hence, in the instance of a nonzero constant term θ_0, the ARIMA model is also expressible as $\phi(B)(\nabla^d z_t - \mu_w) = \theta(B)a_t$, $\mu_w \neq 0$, and the forecast in the form (5.6.5) may be viewed as representing the forecast value of $\tilde{z}_{t+l} = z_{t+l} - f(t + l)$, where $f(t + l) = (\mu_w/d!)(t + l)^d + g(t + l)$ and $g(t)$ is any fixed deterministic polynomial in t of degree less than or equal to $d - 1$ [including the possibility $g(t) = 0$]. For example, in an ARIMA model with $d = 1$ such as the ARIMA(1, 1, 1) model example of Section 5.4.6, but with $\theta_0 \neq 0$, the eventual forecast function of the form $\hat{z}_t(l) = b_0^{(t)} + b_1^{(t)}\phi^l$ will now contain the additional deterministic linear trend term $\mu_w l$, where $\mu_w = \theta_0/(1 - \phi)$, similar to the result in the example for the ARIMA(1, 1, 0) model in (5.4.17). Note that in the special case of a stationary process z_t, with $d = 0$, the additional deterministic term in (5.3.3) reduces to the mean of the process z_t, $\mu = E[z_t]$.

APPENDIX A5.1 CORRELATIONS BETWEEN FORECAST ERRORS

A5.1.1 Autocorrelation Function of Forecast Errors at Different Origins

Although it is true that for an optimal forecast, the forecast errors for lead time 1 will be uncorrelated, this will not generally be true of forecasts at longer lead times. Consider forecasts for lead times l, made at origins t and

$t - j$, respectively, where j is a positive integer. Then if $j = l, l + 1, l + 2, \ldots$, the forecast errors will contain no common component, but for $j = 1, 2, \ldots, l - 1$, certain of the a's will be included in both forecast errors. Specifically,

$$e_t(l) = z_{t+l} - \hat{z}_t(l) = a_{t+l} + \psi_1 a_{t+l-1} + \cdots + \psi_{l-1} a_{t+1}$$

$$e_{t-j}(l) = z_{t+l-j} - \hat{z}_{t-j}(l) = a_{t-j+l} + \psi_1 a_{t-j+l-1} + \cdots + \psi_{l-1} a_{t-j+1}$$

and for $j < l$, the lag j autocovariance of the forecast errors for lead time l is

$$E[e_t(l)e_{t-j}(l)] = \sigma_a^2 \sum_{i=j}^{l-1} \psi_i \psi_{i-j} \tag{A5.1.1}$$

where $\psi_0 = 1$. The corresponding autocorrelations are

$$\rho[e_t(l), e_{t-j}(l)] = \begin{cases} \dfrac{\sum_{i=j}^{l-1} \psi_i \psi_{i-j}}{\sum_{i=0}^{l-1} \psi_i^2} & 0 \leqslant j < l \\ 0 & j \geqslant l \end{cases} \tag{A5.1.2}$$

We show in Chapter 7 that series C of Figure 4.1 is well fitted by the (1, 1, 0) model $(1 - 0.8B)\nabla z_t = a_t$. To illustrate (A5.1.2), we calculate the autocorrelation function of the forecast errors at lead time 6 for this model. It is shown in Section 5.2.2 that the ψ weights $\psi_1, \psi_2, \ldots, \psi_5$ for this model, are 1.80, 2.44, 2.95, 3.36, 3.69. Thus, for example, the lag 1 autocovariance is

$$\begin{aligned} E[e_t(6)e_{t-1}(6)] &= \sigma_a^2[(1.80 \times 1.00) + (2.44 \times 1.80) + \cdots + (3.69 \times 3.36)] \\ &= 35.70\sigma_a^2 \end{aligned}$$

On dividing by $E[e_t^2(6)] = 43.86\sigma_a^2$, we obtain $\rho[e_t(6), e_{t-1}(6)] = 0.81$. The first six autocorrelations are shown in Table A5.1 and plotted in Figure A5.1(a). As expected, the autocorrelations beyond the fifth are zero.

A5.1.2 Correlation between Forecast Errors at the Same Origin with Different Lead Times

Suppose that we make a series of forecasts for different lead times from the *same* fixed origin t. Then the errors for these forecasts will be correlated. We have for $j = 1, 2, 3, \ldots$,

$$e_t(l) = z_{t+l} - \hat{z}_t(l) = a_{t+l} + \psi_1 a_{t+l-1} + \cdots + \psi_{l-1} a_{t+1}$$

$$\begin{aligned} e_t(l + j) = z_{t+l+j} - \hat{z}_t(l + j) &= a_{t+l+j} + \psi_1 a_{t+l+j-1} + \cdots + \psi_j a_{t+l} \\ &\quad + \psi_{j+1} a_{t+l-1} + \cdots + \psi_{l+j-1} a_{t+1} \end{aligned}$$

so that the covariance between the t-origin forecasts at lead times l and $l + j$ is $\sigma_a^2 \sum_{i=0}^{l-1} \psi_i \psi_{j+i}$, where $\psi_0 = 1$.

TABLE A5.1 Autocorrelations of Forecast Errors at Lead 6 for Series C

j	0	1	2	3	4	5	6
$\rho[e_t(6), e_{t-j}(6)]$	1.00	0.81	0.61	0.41	0.23	0.08	0.00

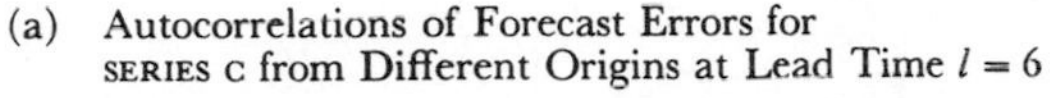

(a) Autocorrelations of Forecast Errors for
SERIES C from Different Origins at Lead Time $l = 6$

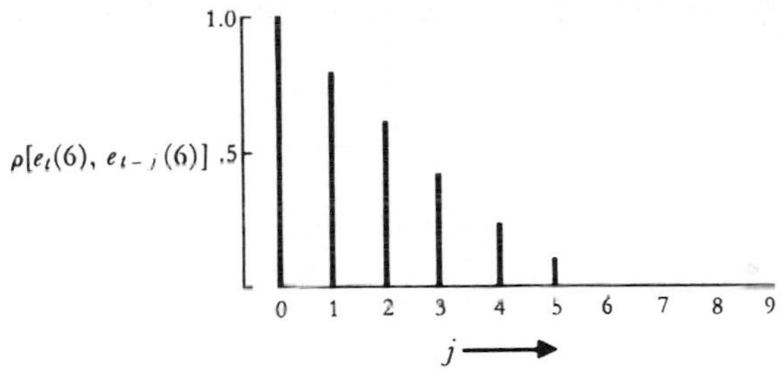

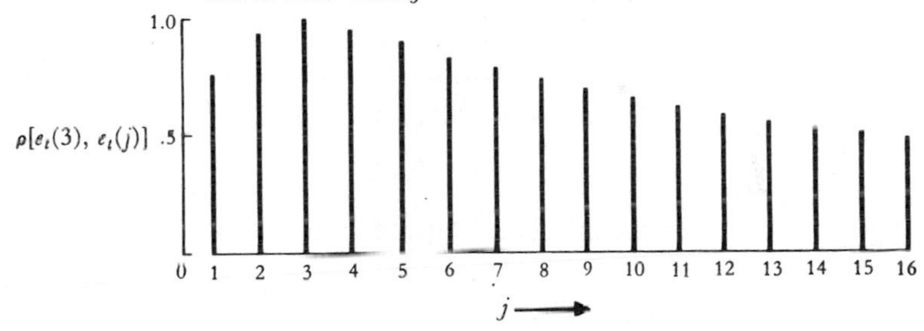

Figure A5.1 Correlations between various forecast errors for series C.

Thus the correlation coefficient between the t-origin forecast errors at lead times l and $l + j$ is

$$\rho[e_t(l), e_t(l + j)] = \frac{\sum_{i=0}^{l-1} \psi_i \psi_{j+i}}{(\sum_{h=0}^{l-1} \psi_h^2 \sum_{g=0}^{l+j-1} \psi_g^2)^{1/2}} \tag{A5.1.3}$$

To illustrate (A5.1.3), we compute, for forecasts made from the same origin, the correlation between the forecast error at lead time 3 and the forecast errors at lead times $j = 1, 2, 3, 4, \ldots, 16$ for series C. For example, using (A5.1.3) and the ψ weights given in Section 5.2.2,

$$\begin{aligned} E[e_t(3)e_t(5)] &= \sigma_a^2[(1.00 \times 2.44) + (1.80 \times 2.95) + (2.44 \times 3.36)] \\ &= 15.94\sigma_a^2 \end{aligned}$$

The correlations for lead times $j = 1, 2, \ldots, 16$ are shown in Table A5.2 and

TABLE A5.2 Correlation between Forecast Errors at Lead 3 and at Lead j Made from a Fixed Origin (Series C)

j	$\rho[e_t(3), e_t(j)]$	j	$\rho[e_t(3), e_t(j)]$
1	0.76	9	0.71
2	0.94	10	0.67
3	1.00	11	0.63
4	0.96	12	0.60
5	0.91	13	0.57
6	0.85	14	0.54
7	0.80	15	0.52
8	0.75	16	0.50

plotted in Figure A5.1(b). As is to be expected, forecasts made from the same origin at different lead times are highly correlated.

APPENDIX A5.2 FORECAST WEIGHTS FOR ANY LEAD TIME

In this appendix we consider an alternative procedure for calculating the forecast weights $\pi_j^{(l)}$ applied to previous z's for any lead time l. To derive this result, we make use of the identity (3.1.7), namely

$$(1 + \psi_1 B + \psi_2 B^2 + \cdots)(1 - \pi_1 B - \pi_2 B^2 - \cdots) = 1$$

from which the π weights may be obtained in terms of the ψ weights, and vice versa.

On equating coefficients, we find, for $j \geqslant 1$,

$$\psi_j = \sum_{i=1}^{j} \psi_{j-i}\pi_i \qquad (\psi_0 = 1) \tag{A5.2.1}$$

Thus, for example,

$$\begin{aligned} \psi_1 &= \pi_1 & \pi_1 &= \psi_1 \\ \psi_2 &= \psi_1\pi_1 + \pi_2 & \pi_2 &= \psi_2 - \psi_1\pi_1 \\ \psi_3 &= \psi_2\pi_1 + \psi_1\pi_2 + \pi_3 & \pi_3 &= \psi_3 - \psi_2\pi_1 - \psi_1\pi_2 \end{aligned}$$

Now from (5.3.6),

$$\hat{z}_t(l) = \pi_1\hat{z}_t(l-1) + \pi_2\hat{z}_t(l-2) + \cdots + \pi_{l-1}\hat{z}_t(1) + \pi_l z_t + \pi_{l+1}z_{t-1} + \cdots \tag{A5.2.2}$$

Since each of the forecasts in (A5.2.2) is itself a function of the observations $z_t, z_{t-1}, z_{t-2}, \ldots$, we can write

$$\hat{z}_t(l) = \pi_1^{(l)} z_t + \pi_2^{(l)} z_{t-1} + \pi_3^{(l)} z_{t-2} + \cdots$$

where the lead l forecast weights may be calculated from the lead 1 forecast weights $\pi_j^{(1)} = \pi_j$. We now show that the weights $\pi_j^{(l)}$ can be obtained using the identity

$$\pi_j^{(l)} = \sum_{i=1}^{l} \psi_{l-i}\pi_{i+j-1} \tag{A5.2.3}$$

For example, the weights for the forecast at lead time 3 are

$$\pi_1^{(3)} = \pi_3 + \psi_1\pi_2 + \psi_2\pi_1$$

$$\pi_2^{(3)} = \pi_4 + \psi_1\pi_3 + \psi_2\pi_2$$

$$\pi_3^{(3)} = \pi_5 + \psi_1\pi_4 + \psi_2\pi_3$$

and so on. To derive (A5.2.3), we write

$$\hat{z}_t(l) = \psi_l a_t + \psi_{l+1}a_{t-1} + \cdots$$

$$\hat{z}_{t+l-1}(1) = \psi_1 a_{t+l-1} + \cdots + \psi_l a_t + \psi_{l+1}a_{t-1} + \cdots$$

On subtraction we obtain

$$\hat{z}_t(l) = \hat{z}_{t+l-1}(1) - \psi_1 a_{t+l-1} - \psi_2 a_{t+l-2} - \cdots - \psi_{l-1}a_{t+1}$$

Hence

$$\begin{aligned}\hat{z}_t(l) = {} & \pi_1 z_{t+l-1} + \pi_2 z_{t+l-2} + \cdots + \pi_{l-1} z_{t+1} + \pi_l z_t + \pi_{l+1} z_{t-1} + \cdots \\ & + \psi_1(-z_{t+l-1} + \pi_1 z_{t+l-2} + \cdots + \pi_{l-2}z_{t+1} + \pi_{l-1}z_t + \pi_l z_{t-1} + \cdots) \\ & + \psi_2(-z_{t+l-2} + \cdots + \pi_{l-3}z_{t+1} + \pi_{l-2}z_t + \pi_{l-1}z_{t-1} + \cdots) \\ & + \cdots \\ & + \psi_{l-1}(-z_{t+1} + \pi_1 z_t + \pi_2 z_{t-1} + \cdots)\end{aligned}$$

Using the relations (A5.2.1), each one of the coefficients of $z_{t+l-1}, \ldots, z_{t+1}$ is seen to vanish, as they should, and on collecting terms, we obtain the required result (A5.2.3). Alternatively, we may use the formula in the recursive form

$$\pi_j^{(l)} = \pi_{j+1}^{(l-1)} + \psi_{l-1}\pi_j \tag{A5.2.4}$$

Using the model $\nabla^2 z_t = (1 - 0.9B + 0.5B^2)a_t$ for illustration, we calculate the weights for lead time 2. Equation (A5.2.4) gives

$$\pi_j^{(2)} = \pi_{j+1} + \psi_1\pi_j$$

and using the weights in Table 5.4 with $\psi_1 = 1.1$ we have, for example,

$$\pi_1^{(2)} = \pi_2 + \psi_1\pi_1 = 0.490 + (1.1)(1.1) = 1.700$$

$$\pi_2^{(2)} = \pi_3 + \psi_1\pi_2 = -0.109 + (1.1)(0.49) = 0.430$$

and so on. The first 12 weights have been given in Table 5.4.

APPENDIX A5.3 FORECASTING IN TERMS OF THE GENERAL INTEGRATED FORM

A5.3.1 General Method of Obtaining the Integrated Form

We emphasize once more that for practical computation of the forecasts, the difference equation procedure is by far the simplest. The following general treatment of the integrated form is given only to elaborate further on the forecasts obtained. In this treatment, rather than solve explicitly for the forecast function as we did in the examples given in Section 5.4, it will be appropriate to write down the general form of the eventual forecast function involving $p + d$ adaptive coefficients. We then show how the eventual forecast function needs to be modified to deal with the first $q - p - d$ forecasts if $q > p + d$. Finally, we show how to update the adaptive coefficients from origin t to origin $t + 1$.

If it is understood that $\hat{z}_t(-j) = z_{t-j}$ for $j = 0, 1, 2, \ldots$, then using the conditional expectation argument of Section 5.1.2, the forecasts satisfy the difference equation

$$\begin{aligned}
\hat{z}_t(1) - \varphi_1\hat{z}_t(0) - \cdots - \varphi_{p+d}\hat{z}_t(1 - p - d) &= -\theta_1 a_t - \cdots - \theta_q a_{t-q+1} \\
\hat{z}_t(2) - \varphi_1\hat{z}_t(1) - \cdots - \varphi_{p+d}\hat{z}_t(2 - p - d) &= -\theta_2 a_t - \cdots - \theta_q a_{t-q+2} \\
\vdots \qquad\qquad \vdots \qquad\qquad \vdots \qquad\qquad \vdots \qquad\quad &\qquad \vdots \qquad\qquad \vdots \qquad\qquad (A5.3.1) \\
\hat{z}_t(q) - \varphi_1\hat{z}_t(q - 1) - \cdots - \varphi_{p+d}\hat{z}_t(q - p - d) &= -\theta_q a_t \\
\hat{z}_t(l) - \varphi_1\hat{z}_t(l - 1) - \cdots - \varphi_{p+d}\hat{z}_t(l - p - d) &= 0 \qquad l > q
\end{aligned}$$

The eventual forecast function is the solution of the last equation and may be written

$$\hat{z}_t(l) = b_0^{(t)}f_0(l) + b_1^{(t)}f_1(l) + \cdots + b_{p+d-1}^{(t)}f_{p+d-1}(l) = \sum_{i=0}^{p+d-1} b_i^{(t)}f_i(l) \qquad l > q - p - d \qquad \text{(A5.3.2)}$$

When q is less than, or equal to $p + d$, the eventual forecast function will provide forecasts $\hat{z}_t(1), \hat{z}_t(2), \hat{z}_t(3), \ldots$ for all lead times $l \geqslant 1$.

As an example of such a model with $q \leqslant p + d$, suppose that

$$(1 - B)(1 - \sqrt{3}B + B^2)^2 z_t = (1 - 0.5B)a_t$$

so that $p + d = 5$ and $q = 1$. Then

$$(1 - B)(1 - \sqrt{3}B + B^2)^2\hat{z}_t(l) = 0 \qquad l = 2, 3, 4, \ldots,$$

where B now operates on l and not on t. Solution of this difference equation yields the forecast function

$$\hat{z}_t(l) = b_0^{(t)} + b_1^{(t)} \cos\left(\frac{2\pi l}{12}\right) + b_2^{(t)} l \cos\left(\frac{2\pi l}{12}\right) + b_3^{(t)} \sin\left(\frac{2\pi l}{12}\right) + b_4^{(t)} l \sin\left(\frac{2\pi l}{12}\right) \qquad l = 1, 2, \ldots.$$

If q is greater than $p + d$, then for lead times $l \leq q - p - d$, the forecast function will have additional terms containing a's. Thus

$$\hat{z}_t(l) = \sum_{i=0}^{p+d-1} b_i^{(t)} f_i(l) + \sum_{i=0}^{j} d_{li} a_{t-i} \qquad l \leq q - p - d \tag{A5.3.3}$$

where $j = q - p - d - l$ and the d's may be obtained explicitly by substituting (A5.3.3) in (A5.3.1). For example, consider the stochastic model

$$\nabla^2 z_t = (1 - 0.8B + 0.5B^2 - 0.4B^3 + 0.1B^4)a_t$$

in which $p + d = 2$, $q = 4$, $q - p - d = 2$ and $\varphi_1 = 2$, $\varphi_2 = -1$, $\theta_1 = 0.8$, $\theta_2 = -0.5$, $\theta_3 = 0.4$, and $\theta_4 = -0.1$. Using the recurrence relation (5.2.3), we obtain $\psi_1 = 1.2$, $\psi_2 = 1.9$, $\psi_3 = 2.2$, $\psi_4 = 2.6$. Now, from (A5.3.3),

$$\begin{aligned} \hat{z}_t(1) &= b_0^{(t)} + b_1^{(t)} + d_{10}a_t + d_{11}a_{t-1} \\ \hat{z}_t(2) &= b_0^{(t)} + 2b_1^{(t)} + d_{20}a_t \\ \hat{z}_t(l) &= b_0^{(t)} + b_1^{(t)} l \qquad l > 2 \end{aligned} \tag{A5.3.4}$$

Using (A5.3.1) gives

$$\hat{z}_t(4) - 2\hat{z}_t(3) + \hat{z}_t(2) = 0.1a_t$$

so that from (A5.3.4)

$$d_{20}a_t = 0.1a_t$$

and hence $d_{20} = 0.1$. Similarly, from (A5.3.1),

$$\hat{z}_t(3) - 2\hat{z}_t(2) + \hat{z}_t(1) = -0.4a_t + 0.1a_{t-1}$$

and hence, using (A5.3.4),

$$-0.2a_t + d_{10}a_t + d_{11}a_{t-1} = -0.4a_t + 0.1a_{t-1}$$

yielding

$$d_{10} = -0.2 \qquad d_{11} = 0.1$$

Hence the forecast function is

$$\begin{aligned} \hat{z}_t(1) &= b_0^{(t)} + b_1^{(t)} - 0.2a_t + 0.1a_{t-1} \\ \hat{z}_t(2) &= b_0^{(t)} + 2b_1^{(t)} + 0.1a_t \\ \hat{z}_t(l) &= b_0^{(t)} + b_1^{(t)} l \qquad l > 2 \end{aligned}$$

A5.3.2 Updating the General Integrated Form

Updating formulas for the coefficients may be obtained using the identity (5.2.5) with $t + 1$ replaced by t:

$$\hat{z}_t(l) = \hat{z}_{t-1}(l + 1) + \psi_l a_t$$

Then, for $l > q - p - d$,

$$\sum_{i=0}^{p+d-1} b_i^{(t)} f_i(l) = \sum_{i=0}^{p+d-1} b_i^{(t-1)} f_i(l + 1) + \psi_l a_t \tag{A5.3.5}$$

By solving $p + d$ such equations for different values of l, we obtain the required updating formula for the individual coefficients, in the form

$$b_i^{(t)} = \sum_{j=0}^{p+d-1} L_{ij} b_j^{(t-1)} + g_i a_t$$

Note that the updating of each of the coefficients of the forecast function depends only on the lead one forecast error $a_t = z_t - \hat{z}_{t-1}(1)$.

A5.3.3 Comparison with the Discounted Least Squares Method

Although to work with the integrated form is an unnecessarily complicated way of computing forecasts, it allows us to compare the present mean square error forecast with another type of forecast which has received considerable attention. Let us write

$$\mathbf{F}_l = \begin{bmatrix} f_0(l) & f_1(l) & \cdots & f_{p+d-1}(l) \\ f_0(l + 1) & f_1(l + 1) & \cdots & f_{p+d-1}(l + 1) \\ f_0(l + p + d - 1) & f_1(l + p + d - 1) & \cdots & f_{p+d-1}(l + p + d - 1) \end{bmatrix}$$

$$\mathbf{b}^{(t)} = \begin{bmatrix} b_0^{(t)} \\ b_1^{(t)} \\ b_{p+d-1}^{(t)} \end{bmatrix} \qquad \boldsymbol{\psi}_l = \begin{bmatrix} \psi_l \\ \psi_{l+1} \\ \psi_{l+p+d} \end{bmatrix}$$

Then, using (A5.3.5) for $l, l + 1, \ldots, l + p + d$, we obtain for $l > q - p - d$,

$$\mathbf{F}_l \mathbf{b}^{(t)} = \mathbf{F}_{l+1} \mathbf{b}^{(t-1)} + \boldsymbol{\psi}_l a_t$$

yielding

$$\mathbf{b}^{(t)} = (\mathbf{F}_l^{-1} \mathbf{F}_{l+1}) \mathbf{b}^{(t-1)} + (\mathbf{F}_l^{-1} \boldsymbol{\psi}_l) a_t$$

or

$$\mathbf{b}^{(t)} = \mathbf{L}' \mathbf{b}^{(t-1)} + \mathbf{g} a_t \tag{A5.3.6}$$

(A5.3.6) is of the same algebraic *form* as the updating function given by the "discounted least squares" procedure of Brown [64], [65]. For comparison, if we denote the forecast error given by that method by e_t, then Brown's updating formula may be written

$$\boldsymbol{\beta}^{(t)} = \mathbf{L}'\boldsymbol{\beta}^{(t-1)} + \mathbf{h}e_t \tag{A5.3.7}$$

where $\boldsymbol{\beta}^{(t)}$ is his vector of adaptive coefficients. The same matrix $\mathbf{L}$ appears in (A5.3.6) and (A5.3.7). This is inevitable, for this first factor merely allows for changes in the coefficients arising from translation to the new origin and would have to occur in any such formula. For example, consider the straight-line forecast function

$$\hat{z}_{t-1}(l) = b_0^{(t-1)} + b_1^{(t-1)}l$$

where $b_0^{(t-1)}$ is the ordinate at time $t-1$, the origin of the forecast. This can equally well be written

$$\hat{z}_{t-1}(l) = (b_0^{(t-1)} + b_1^{(t-1)}) + b_1^{(t-1)}(l-1)$$

where now $(b_0^{(t-1)} + b_1^{(t-1)})$ is the ordinate at time t. Obviously, if we update the forecast to origin t, the coefficient b_0 must be suitably adjusted even if the forecast function were to remain unchanged.

In general, the matrix $\mathbf{L}$ does not change the forecast function, it merely relocates it. The actual updating is done by the vector of coefficients $\mathbf{g}$ and $\mathbf{h}$. We shall see that the coefficients $\mathbf{g}$, which yield the minimum mean square error forecasts, and the coefficients $\mathbf{h}$ given by Brown are in general completely different.

Brown's method of forecasting

1. A forecast function is selected from the general class of linear combinations and products of polynomials, exponentials, and sines and cosines.
2. The selected forecast function is fitted to past values by a "discounted least squares" procedure. In this procedure, the coefficients are estimated and updated so that the sum of squares of weighted discrepancies

$$S_\omega = \sum_{j=0}^{\infty} \omega_j[z_{t-j} - \hat{z}_t(-j)]^2 \tag{A5.3.8}$$

between past values of the series and the value given by the forecast function at the corresponding past time are minimized. The weight function ω_j is chosen arbitrarily to fall off geometrically, so that $\omega_j = (1-\alpha)^j$, where the constant α, usually called the *smoothing constant*, is (again arbitrarily) set equal to a value in the range 0.1 to 0.3.

Difference between the minimum mean square error forecasts and those of Brown. To illustrate these comments, consider the forecasting of IBM stock prices, discussed by Brown [64, p. 141]. In this study he used a quadratic model which would be, in the present notation,

$$\hat{z}_t(l) = \beta_0^{(t)} + \beta_1^{(t)}l + \tfrac{1}{2}\beta_2^{(t)}l^2$$

With this model he employed his method of discounted least squares to forecast stock prices 3 days ahead. The results obtained from this method are shown for a section of the IBM series in Figure A5.2, where they are compared with the minimum mean square error forecasts.

The discounted least squares method can be criticized on the following grounds:

1. The nature of the forecast function ought to be decided by the autoregressive operator $\varphi(B)$ in the stochastic model, and not arbitrarily. In particular, it cannot be safely chosen by visual inspection of the time series itself. For example, consider the IBM stock prices plotted in Figure A5.2. It will be seen that a quadratic function might well be used to *fit* short pieces of this series to values already available. If such fitting were relevant to forecasting, we might conclude, as did Brown, that a polynomial forecast function of degree 2 was indicated. The

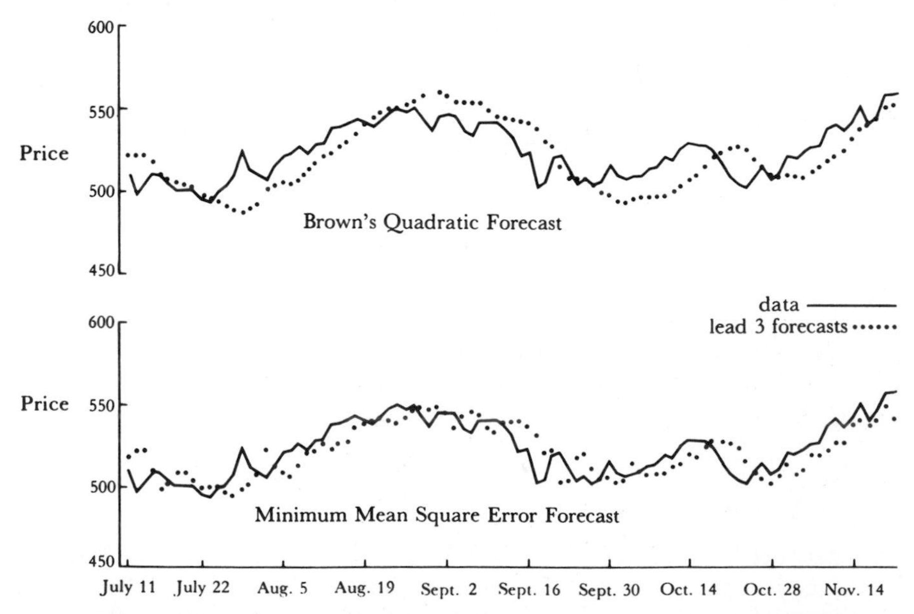

Figure A5.2 IBM stock price series with comparison of lead 3 forecasts obtained from best IMA(0, 1, 1) process and Brown's quadratic forecast for a period beginning July 11, 1960.

most general linear process for which a quadratic function would produce *minimum mean square error* forecasts at every lead time $l = 1, 2, \ldots$ is defined by the (0, 3, 3) model

$$\nabla^3 z_t = (1 - \theta_1 B - \theta_2 B^2 - \theta_3 B^3)a_t$$

which, arguing as in Section 4.3.3, can be written

$$\nabla^3 z_t = \nabla^3 a_t + \lambda_0 \nabla^2 a_{t-1} + \lambda_1 \nabla a_{t-1} + \lambda_2 a_{t-1}$$

However, we shall show in Chapter 7 that if this model is correctly fitted, the least squares estimates of the parameters are $\lambda_1 = \lambda_2 = 0$, and $\lambda_0 \simeq 1.0$. Thus $\nabla z_t = (1 - \theta B)a_t$, with $\theta = 1 - \lambda_0$ close to zero, is the appropriate stochastic model, and the appropriate forecasting polynomial is $\hat{z}_t(l) = \beta_0^{(t)}$, which is of degree zero in l and not of degree 2.

2. The choice of the weight function ω_j in (A5.3.8) must correspondingly be decided by the stochastic model, and not arbitrarily. The use of the discounted least squares fitting procedure would produce minimum mean square error forecasts in the very restricted case where:
   (a) The process was of order (0, 1, 1), so that $\nabla z_t = (1 - \theta B)a_t$.
   (b) A polynomial of degree zero was fitted.
   (c) The smoothing constant α was set equal to our $\lambda = 1 - \theta$.
   In the present example, even if the correct polynomial model of degree zero had been chosen, the value $\alpha = \lambda = 0.1$, actually used by Brown, would have been quite inappropriate. The correct value λ for this series is close to unity.
3. The exponentially discounted weighted least squares procedure forces all the $p + d$ coefficients in the updating vector $\mathbf{h}$ to be functions of the single smoothing parameter α. In fact, they should be functions of the $p + q$ independent parameters $(\boldsymbol{\phi}, \boldsymbol{\theta})$.

Thus the differences between the two methods are not trivial, and it is interesting to compare their performances on the IBM data. The minimum mean square error forecast is $\hat{z}_t(l) = b_0^{(t)}$, with updating $b_0^{(t)} = b_0^{(t-1)} + \lambda a_t$, where $\lambda \simeq 1.0$. If λ is taken to be exactly equal to unity, this is equivalent to using

$$\hat{z}_t(l) = z_t$$

which implies that the best forecast of the stock price for all future time is the present price.* The suggestion that stock prices behave in this way is of course not new and goes back to Bachelier [17]. Since $z_t = Sa_t$ when $\lambda = 1$, this implies that z_t is a random walk.

* This result is approximately true supposing that no relevant information except past values of the series itself is available and that fairly short forecasting periods are being considered. For longer periods, growth and inflationary factors would become important.

TABLE A5.3 Comparison of Mean Square Error of Forecasts Obtained at Various Lead Times Using Best IMA(0, 1, 1) Process and Brown's Quadratic Forecasts

	Lead Time l									
	1	2	3	4	5	6	7	8	9	10
MSE (Brown)	102	158	218	256	363	452	554	669	799	944
MSE ($\lambda = 0.9$)	42	91	136	180	282	266	317	371	427	483

To compare the minimum mean square error (MSE) forecast with Brown's quadratic forecasts, a direct comparison was made using the IBM stock price series from July 11, 1960 to February 10, 1961 (150 observations). For this stretch of the series the minimum MSE forecast is obtained using the model $\nabla z_t = a_t - \theta a_{t-1}$, with $\theta = 0.1$, or $\lambda = 1 - \theta = 0.9$. Figure A5.2 shows the minimum MSE forecasts for lead time 3 and the corresponding values of Brown's quadratic forecasts. It is seen that the minimum MSE forecasts, which are virtually equivalent to using today's price to predict that 3 days ahead, are considerably better than those obtained using Brown's much more complicated procedure.

The mean square errors for the forecast at various lead times, computed by direct comparison of the value of the series and their lead l forecasts, are shown in Table A5.3 for the two types of forecasts. It is seen that Brown's quadratic forecasts have mean square errors which are much larger than those obtained by the minimum mean square error method.

Part Two

STOCHASTIC MODEL BUILDING

We have seen that an ARIMA process of order (p, d, q) provides a class of models capable of representing time series which, although not necessarily stationary, are homogeneous and in statistical equilibrium.

The ARIMA process is defined by the equation

$$\phi(B)(1 - B)^d z_t = \theta_0 + \theta(B)a_t$$

where $\phi(B)$ and $\theta(B)$ are operators in B of degree p and q, respectively, whose zeros lie outside the unit circle. We have noted that the model is very general, subsuming autoregressive models, moving average models, mixed autoregressive–moving average models, and the integrated forms of all three.

Iterative Approach to Model Building

The relating of a model of this kind to data is usually best achieved by a three-stage iterative procedure based on identification, estimation, and diagnostic checking.

1. By *identification* we mean the use of the data, and of any information on how the series was generated, to suggest a subclass of parsimonious models worthy to be entertained.
2. By *estimation* we mean efficient use of the data to make inferences about parameters conditional on the adequacy of the model entertained.

3. By *diagnostic checking* we mean checking the fitted model in its relation to the data with intent to reveal model inadequacies and so to achieve model improvement.

In Chapter 6, which follows, we discuss identification, in Chapter 7 estimation, and in Chapter 8 diagnostic checking. In Chapter 9 all these techniques are illustrated by applying them to modeling seasonal time series.

6

MODEL IDENTIFICATION

In this chapter we discuss methods for identifying nonseasonal time series models. Identification methods are rough procedures applied to a set of data to indicate the kind of representational model that is worthy of further investigation. The specific aim here is to obtain some idea of the values of p, d, and q needed in the general linear ARIMA model and to obtain initial estimates for the parameters. The tentative model so obtained provides a starting point for the application of the more formal and efficient estimation methods described in Chapter 7.

6.1 OBJECTIVES OF IDENTIFICATION

It should first be said that identification and estimation necessarily overlap. Thus we may estimate the parameters in a model, which is more elaborate than that which we expect to find, so as to decide *at what point* simplification is possible. Here we employ the estimation procedure to carry out part of the identification. It should also be explained that identification is necessarily inexact. It is inexact because the question of what types of models occur in practice and in what circumstances, is a property of the behavior of the physical world and therefore cannot be decided by purely mathematical argument. Furthermore, because at the identification stage no precise formulation of the problem is available, statistically "inefficient" methods must necessarily be used. It is a stage at which graphical methods are particularly useful and judgment must be exercised. However, it should be borne in mind that preliminary identification commits us to nothing except tentative consideration of a class of models that will later be efficiently fitted and checked.

6.1.1 Stages In the Identification Procedure

Our task, then, is to identify an appropriate subclass of models from the general ARIMA family

$$\phi(B)\nabla^d z_t = \theta_0 + \theta(B)a_t \tag{6.1.1}$$

which may be used to represent a given time series. Our approach will be as follows:

1. To difference z_t as many times as is needed to produce stationarity, hopefully reducing the process under study to the mixed autoregressive–moving average process

$$\phi(B)w_t = \theta_0 + \theta(B)a_t$$

where

$$w_t = (1 - B)^d z_t = \nabla^d z_t$$

2. To identify the resulting ARMA process.

Our principal tools for putting 1 and 2 into effect will be the sample autocorrelation function and the sample partial autocorrelation function. They are used not only to help guess the form of the model, but also to obtain approximate estimates of the parameters. Such approximations are often useful at the estimation stage to provide starting values for iterative procedures employed at that stage.

6.2 IDENTIFICATION TECHNIQUES

6.2.1 Use of the Autocorrelation and Partial Autocorrelation Functions in Identification

Identifying the degree of differencing. We have seen in Section 3.4.2 that for a stationary mixed autoregressive–moving average process of order $(p, 0, q)$, $\phi(B)\tilde{z}_t = \theta(B)a_t$, the autocorrelation function satisfies the difference equation

$$\phi(B)\rho_k = 0 \qquad k > q$$

Also, if $\phi(B) = \prod_{i=1}^{p} (1 - G_i B)$, the solution of this difference equation for the kth autocorrelation is, assuming distinct roots, of the form

$$\rho_k = A_1 G_1^k + A_2 G_2^k + \cdots + A_p G_p^k \qquad k > q - p \tag{6.2.1}$$

The stationarity requirement that the zeros of $\phi(B)$ lie outside the unit circle implies that the roots $G_1, G_2, \ldots, G_p$ lie inside the unit circle.

Inspection of (6.2.1) shows that in the case of a stationary model in which none of the roots lie close to the boundary of the unit circle, the autocorrelation function will quickly "die out" for moderate and large k. Suppose now that a single real root, say G_1, approaches unity, so that

$$G_1 = 1 - \delta$$

where δ is some small positive quantity. Then, since for k large

$$\rho_k \simeq A_1(1 - k\delta)$$

the autocorrelation function will not die out quickly and will fall off slowly and very nearly linearly. A similar argument may be applied if more than one of the roots approaches unity.

Therefore, a tendency for the autocorrelation function not to die out quickly is taken as an indication that a root close to unity may exist. The estimated autocorrelation function tends to follow the behavior of the theoretical autocorrelation function. Therefore, failure of the estimated autocorrelation function to die out rapidly might logically suggest that we should treat the underlying stochastic process as nonstationary in z_t, but possibly as stationary in ∇z_t, or in some higher difference.

It turns out that it is failure of the estimated autocorrelation function to die out rapidly that suggests nonstationarity. It need not happen that the estimated correlations are extremely high even at low lags. This is illustrated in Appendix A6.1, where the expected behavior of the estimated autocorrelation function is considered for the nonstationary (0, 1, 1) process $\nabla z_t = (1 - \theta B)a_t$. The ratio $E[c_k]/E[c_0]$ of expected values falls off only slowly, but depends initially on the value of θ and on the number of observations in the series, and need not be close to unity if θ is close to 1. We shall illustrate this point again in Section 6.3.4 for series A of Figure 4.1.

For the reasons given, it is assumed that the degree of differencing d, necessary to achieve stationarity, has been reached when the autocorrelation function of $w_t = \nabla^d z_t$ dies out fairly quickly. In practice, d is normally either 0, 1, or 2, and it is usually sufficient to inspect the first 20 or so estimated autocorrelations of the original series, and of its first and second differences if necessary.

Identifying the resultant stationary ARMA process. Having tentatively decided what d should be, we next study the general appearance of the estimated autocorrelation and partial autocorrelation functions of the appropriately differenced series, $w_t = (1 - B)^d z_t$, to provide clues about the choice of the orders p and q for the autoregressive and moving average operators. In doing so, we recall the characteristic behavior of the theoretical autocorrelation function and of the theoretical partial autocorrelation function for moving average, autoregressive, and mixed processes, discussed in Chapter 3.

Briefly, whereas the autocorrelation function of an autoregressive process of order p tails off, its partial autocorrelation function has a cutoff after lag p. Conversely, the autocorrelation function of a moving average process of order q has a cutoff after lag q, while its partial autocorrelation tails off. If both the autocorrelations and partial autocorrelations tail off, a mixed process is suggested. Furthermore, the autocorrelation function for a mixed process, containing a pth-order autoregressive component and a qth-order moving average component, is a mixture of exponentials and damped sine waves after the first $q - p$ lags. Conversely, the partial autocorrelation function for a mixed process is dominated by a mixture of exponentials and damped sine waves after the first $p - q$ lags (see Table 3.3).

In general, autoregressive (moving average) behavior, as measured by the autocorrelation function, tends to mimic moving average (autoregressive) behavior as measured by the partial autocorrelation function. For example, the autocorrelation function of a first-order autoregressive process decays exponentially, while the partial autocorrelation function cuts off after the first lag. Correspondingly, for a first-order moving average process, the autocorrelation function cuts off after the first lag. Although not precisely exponential, the partial autocorrelation function is dominated by exponential terms and has the general appearance of an exponential.

Of particular importance are the autoregressive and moving average processes of first and second order and the simple mixed $(1, d, 1)$ process. The properties of the theoretical autocorrelation and partial autocorrelation functions for these processes are summarized in Table 6.1, which requires careful study and provides a convenient reference table. The reader should also refer again to Figures 3.2, 3.9, and 3.11, which show typical behavior of the autocorrelation function and the partial autocorrelation function for the second-order autoregressive process, the second-order moving average process, and the simple mixed ARMA(1,1) process.

Relation between estimated and theoretical autocorrelations. Estimated autocorrelations can have rather large variances and can be highly autocorrelated with each other. For this reason, as was emphasized by Kendall [126], *detailed* adherence to the theoretical autocorrelation function cannot be expected in the estimated function. In particular, moderately large estimated autocorrelations can occur after the theoretical autocorrelation function has damped out, and apparent ripples and trends can occur in the estimated function which have no basis in the theoretical function. In employing the estimated autocorrelation function as a tool for identification, it is usually possible to be fairly sure about broad characteristics, but more subtle indications may or may not represent real effects, and two or more

TABLE 6.1 Behavior of the Autocorrelation Functions for the *d*th *Difference* of an ARIMA Process of Order (p, d, q)[a]

	Order				
	$(1, d, 0)$	$(0, d, 1)$	$(2, d, 0)$	$(0, d, 2)$	$(1, d, 1)$
Behavior of ρ_k	Decays exponentially	Only ρ_1 nonzero	Mixture of exponentials or damped sine wave	Only ρ_1 and ρ_2 nonzero	Decays exponentially from first lag
Behavior of ϕ_{kk}	Only ϕ_{11} nonzero	Exponential dominates decay	Only ϕ_{11} and ϕ_{22} nonzero	Dominated by mixture of exponentials or damped sine wave	Dominated by exponential decay from first lag
Preliminary estimates from:	$\phi_1 = \rho$	$\rho_1 = \dfrac{-\theta_1}{1 + \theta_1^2}$	$\phi_1 = \dfrac{\rho_1(1 - \rho_2)}{1 - \rho_1^2}$ $\phi_2 = \dfrac{\rho_2 - \rho_1^2}{1 - \rho_1^2}$	$\rho_1 = \dfrac{-\theta_1(1 - \theta_2)}{1 + \theta_1^2 + \theta_2^2}$ $\rho_2 = \dfrac{-\theta_2}{1 + \theta_1^2 + \theta_2^2}$	$\rho_1 = \dfrac{(1 - \theta_1\phi_1)(\phi_1 - \theta_1)}{1 + \theta_1^2 - 2\phi_1\theta_1}$ $\rho_2 = \rho_1\phi_1$
Admissible region	$-1 < \phi_1 < 1$	$-1 < \theta_1 < 1$	$-1 < \phi_2 < 1$ $\phi_2 + \phi_1 < 1$ $\phi_2 - \phi_1 < 1$	$-1 < \theta_2 < 1$ $\theta_2 + \theta_1 < 1$ $\theta_2 - \theta_1 < 1$	$-1 < \phi_1 < 1$ $-1 < \theta_1 < 1$

[a] Table A and charts B, C, and D are included at the end of this volume to facilitate the calculation of approximate estimates of the parameters for first-order moving average, second-order autoregressive, second-order moving average, and for the mixed ARMA(1, 1) process.

related models may need to be entertained and investigated further at the estimation and diagnostic checking stages of model building.

6.2.2 Standard Errors for Estimated Autocorrelations and Partial Autocorrelations

Since we do not know the theoretical correlations and since the estimated values that we compute will differ somewhat from their theoretical counterparts, it is important to have some indication of how far an estimated value may differ from the corresponding theoretical value. In particular, we need some means for judging whether the autocorrelations and partial autocorrelations are effectively zero after some specific lag q or p, respectively. For *larger lags*, on the hypothesis that the process is moving average of order q, we can compute standard errors of estimated autocorrelations from the simplified form of Bartlett's formula (2.1.13), with sample estimates replacing theoretical autocorrelations. Thus

$$\hat{\sigma}[r_k] \simeq \frac{1}{n^{1/2}}[1 + 2(r_1^2 + r_2^2 + \cdots + r_q^2)]^{1/2} \qquad k > q \qquad (6.2.2)$$

For the partial correlations we use the result quoted in (3.2.36) that, on the hypothesis that the process is autoregressive of order p, the standard error for estimated partial correlations of order $p + 1$ and higher is

$$\hat{\sigma}[\hat{\phi}_{kk}] \simeq \frac{1}{n^{1/2}} \qquad k > p \qquad (6.2.3)$$

It was shown by Anderson [6] that for moderate n, the distribution of an estimated autocorrelation coefficient, whose theoretical value is zero, is approximately Normal. Thus, on the hypothesis that the theoretical autocorrelation ρ_k is zero, the estimate r_k divided by its standard error will be approximately distributed as a unit Normal deviate. A similar result is true for the partial autocorrelations. These facts may be used to provide an informal guide as to whether theoretical autocorrelations and partial autocorrelations beyond a particular lag are essentially zero. It is usually sufficient to remember that for the Normal distribution, deviations exceeding one standard error in either direction have a probability of about one third, while deviations in either direction exceeding two standard deviations have a probability of about one twentieth.

6.2.3 Identification of Some Actual Time Series

In this section the techniques referred to above are applied to six time series, designated A, B, . . . , F. Series A to D are plotted in Figure 4.1, series E in Figure 6.1, and series F in Figure 2.1. The data for all these series are listed in the "Collection of Time Series" in Part Five. Series A, B, C,

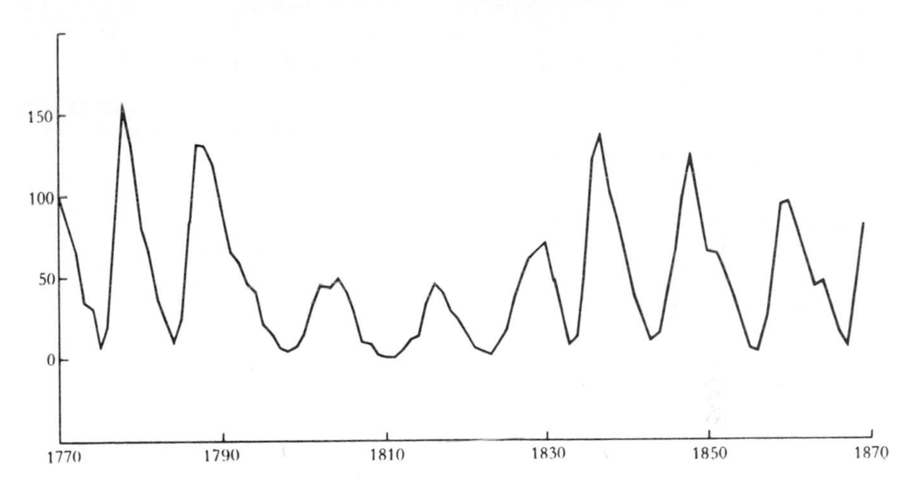

Figure 6.1 Series E: Wölfer annual sunspot numbers (1770–1869).

and D have been described in Chapter 4, and series F in Chapter 2. Series E is the series of annual Wölfer sunspot numbers and measures the average number of sunspots on the sun during each year. As we have remarked in Chapter 4, we might expect series A, C, and D to possess nonstationary characteristics, since they represent the "uncontrolled" behavior of certain process outputs. Similarly, we would expect the stock price series B to have no fixed level. On the other hand, we would expect series F to be stationary, because it represents the variation in the yields of batches processed under uniformly controlled conditions. Similarly, we would expect the sunspot series to remain in equilibrium over long periods.

The estimated autocorrelations of z, ∇z, and $\nabla^2 z$ for series A to F are shown in Table 6.2. Table 6.3 shows the corresponding estimated partial autocorrelations. Plotting of the correlation functions greatly assists their understanding, and for illustration, autocorrelations and partial autocorrelations are plotted in Figures 6.2 and 6.3 for series A, and in Figures 6.4 and 6.5 for series C.

For series A the autocorrelations for ∇z are small after the first lag. This suggests that this time series might be described by an IMA(0, 1, 1) process. However, from the autocorrelation function of z it is seen that *after lag 1* the correlations do decrease fairly regularly. Therefore, an alternative possibility is that the series is a mixed ARMA of order (1, 0, 1). The partial autocorrelation function for z tends to support this possibility. We shall see later that the two possibilities result in virtually the same model.

The autcorrelations of series C, shown in Figure 6.4, suggest that at least one differencing is necessary. The roughly exponential falloff in the correlations for the first difference suggests a process of order (1, 1, 0), with an autoregressive parameter ϕ of about 0.8. Alternatively, it will be noticed

TABLE 6.2 Estimated Autocorrelations of Series A to F

	Lag	*Autocorrelation* 1	2	3	4	5	6	7	8	9	10
		Series A: chemical process concentration readings: every 2 hours, 197 observations									
z	1–10	0.57	0.50	0.40	0.36	0.33	0.35	0.39	0.32	0.30	0.26
	11–20	0.19	0.16	0.20	0.24	0.14	0.18	0.20	0.20	0.14	0.18
∇z	1–10	−0.41	0.02	−0.07	−0.01	−0.07	−0.02	0.15	−0.07	0.04	0.02
	11–20	−0.05	−0.06	−0.01	0.16	−0.17	0.03	0.01	0.08	−0.12	0.15
$\nabla^2 z$	1–10	−0.65	0.18	−0.04	0.04	−0.04	−0.04	0.13	−0.11	0.04	0.02
	11–20	−0.02	−0.02	−0.04	0.18	−0.19	0.08	−0.03	0.09	−0.17	0.20
		Series B: IBM common stock closing prices: daily, May 17, 1961–November 2, 1962, 369 observations									
z	1–10	0.99	0.99	0.98	0.97	0.96	0.96	0.95	0.94	0.93	0.92
	11–20	0.91	0.91	0.90	0.89	0.88	0.87	0.86	0.85	0.84	0.83
∇z	1–10	0.09	0.00	−0.05	−0.04	−0.02	0.12	0.07	0.04	−0.07	0.02
	11–20	0.08	0.05	−0.05	0.07	−0.07	0.12	0.12	0.05	0.05	0.07
$\nabla^2 z$	1–10	−0.45	−0.02	−0.04	0.00	−0.07	0.11	−0.01	0.04	−0.10	0.02
	11–20	0.04	0.04	−0.12	0.13	−0.17	0.10	0.05	−0.04	−0.01	0.09
		Series C: chemical process temperature readings: every minute, 226 observations									
z	1–10	0.98	0.94	0.90	0.85	0.80	0.75	0.69	0.64	0.58	0.52
	11–20	0.47	0.41	0.36	0.30	0.25	0.20	0.15	0.10	0.05	0.00
∇z	1–10	0.80	0.65	0.53	0.44	0.38	0.32	0.26	0.19	0.14	0.14
	11–20	0.10	0.09	0.07	0.07	0.07	0.07	0.09	0.05	0.04	0.04
$\nabla^2 z$	1–10	−0.08	−0.07	−0.12	−0.06	0.01	−0.02	0.05	−0.05	−0.12	0.12
	11–20	−0.12	0.07	−0.08	0.03	−0.01	−0.06	0.17	−0.10	−0.01	−0.02

Series D: chemical process viscosity readings: every hour, 310 observations

z	1–10	0.86	0.74	0.62	0.53	0.46	0.41	0.35	0.31	0.27	0.24
	11–20	0.22	0.20	0.18	0.15	0.14	0.13	0.16	0.19	0.21	0.23
∇z	1–10	−0.05	−0.06	−0.07	−0.08	−0.06	0.00	−0.02	−0.02	−0.03	−0.06
	11–20	−0.01	0.04	0.02	−0.07	−0.03	−0.09	−0.02	0.05	−0.01	0.06
$\nabla^2 z$	1–10	−0.50	0.00	0.00	−0.01	−0.02	0.04	−0.01	0.00	0.01	−0.04
	11–20	0.00	0.04	0.03	−0.06	0.04	−0.06	0.00	0.06	−0.06	0.06

Series E: Wölfer sunspot numbers: yearly, 100 observations

z	1–10	0.81	0.43	0.07	−0.17	−0.27	−0.21	−0.04	0.16	0.33	0.41
	11–20	0.39	0.29	0.14	0.02	−0.06	−0.10	−0.14	−0.18	−0.17	−0.10
∇z	1–10	0.55	−0.02	−0.30	−0.40	−0.40	−0.33	−0.20	0.04	0.26	0.31
	11–20	0.29	0.15	−0.03	−0.12	−0.10	−0.09	−0.09	−0.12	−0.14	−0.05
$\nabla^2 z$	1–10	0.15	−0.31	−0.20	−0.11	−0.09	−0.02	−0.11	−0.04	0.19	0.05
	11–20	0.13	0.09	−0.10	−0.11	0.04	0.01	0.00	−0.03	−0.10	−0.04

Series F: yields from a batch chemical process, 70 observations

z	1–10	−0.39	0.30	−0.17	0.07	−0.10	−0.05	0.04	−0.04	−0.01	0.01
	11–20	0.11	−0.07	0.15	0.04	−0.01	0.17	−0.11	0.02	−0.05	0.02
∇z	1–10	−0.74	0.43	−0.27	0.16	−0.10	0.01	0.05	−0.05	0.04	−0.05
	11–20	0.11	−0.16	0.12	−0.01	−0.08	0.16	−0.14	0.08	−0.07	0.03
$\nabla^2 z$	1–10	−0.83	0.54	−0.33	0.21	−0.12	0.03	0.04	−0.06	0.06	−0.07
	11–20	0.12	−0.16	0.11	0.00	−0.10	0.16	−0.15	0.10	−0.07	0.03

TABLE 6.3 Estimated Partial Autocorrelations of Series A to F[a]

	Lag	*Partial Autocorrelation* 1	2	3	4	5	6	7	8	9	10
		Series A: chemical process concentration readings: every 2 hours, 197 observations									
z	1–10	0.57	0.25	0.08	0.09	0.07	0.15	0.19	−0.03	0.01	−0.01
	11–20	−0.09	−0.04	0.04	0.08	−0.15	0.06	0.13	0.09	−0.06	0.07
∇z	1–10	−0.41	−0.19	−0.17	−0.14	−0.20	−0.23	0.01	−0.04	−0.01	0.06
	11–20	0.02	−0.07	−0.10	0.13	−0.09	−0.15	−0.11	0.04	−0.08	0.12
$\nabla^2 z$	1–10	−0.66	−0.43	−0.33	−0.23	−0.20	−0.36	−0.23	−0.21	−0.23	−0.16
	11–20	−0.07	−0.04	−0.25	0.00	0.04	−0.02	−0.16	−0.03	−0.22	−0.03
		Series B: IBM common stock closing prices: daily, May 17, 1961–November 2, 1962, 369 observations									
z	1–10	0.996	−0.09	0.01	0.05	0.02	0.02	−0.12	−0.05	−0.02	0.06
	11–20	−0.05	−0.09	−0.03	0.07	−0.08	0.06	−0.14	−0.10	−0.01	−0.08
∇z	1–10	0.09	−0.01	−0.05	−0.03	−0.02	0.13	0.05	0.02	−0.06	0.05
	11–20	0.09	0.03	−0.08	0.08	−0.06	0.14	0.10	0.00	0.07	0.08
$\nabla^2 z$	1–10	−0.45	−0.28	−0.24	−0.20	−0.29	−0.17	−0.13	−0.03	−0.14	−0.16
	11–20	−0.09	0.02	−0.13	0.01	−0.19	−0.13	−0.03	−0.10	−0.10	0.06
		Series C: chemical process temperature readings: every minute, 226 observations									
z	1–10	0.99	−0.81	−0.03	−0.02	−0.10	−0.07	−0.01	−0.03	0.04	−0.04
	11–20	−0.15	0.10	−0.14	0.01	−0.10	−0.02	−0.07	−0.11	0.11	−0.13
∇z	1–10	0.81	−0.01	−0.01	0.06	0.03	−0.03	−0.01	−0.08	0.00	0.10
	11–20	−0.14	0.10	−0.05	0.05	0.02	0.06	0.06	−0.17	0.09	0.00
$\nabla^2 z$	1–10	−0.08	−0.08	−0.14	−0.10	−0.03	−0.05	0.02	−0.06	−0.16	0.09
	11–20	−0.14	0.01	−0.09	−0.02	−0.05	−0.09	0.13	−0.13	−0.03	−0.05

		Series D: chemical process viscosity readings: every hour, 310 observations									
z	1–10	0.86	−0.02	0.00	0.01	0.03	0.03	−0.02	0.01	0.00	0.01
	11–20	0.05	0.01	−0.04	−0.03	0.07	0.04	0.10	0.06	0.00	0.06
∇z	1–10	−0.05	−0.06	−0.07	−0.09	−0.08	−0.03	−0.05	−0.05	−0.05	−0.09
	11–20	−0.05	0.01	−0.01	−0.10	−0.07	−0.13	−0.09	−0.02	−0.08	0.00
$\nabla^2 z$	1–10	−0.50	−0.32	−0.24	−0.20	−0.22	−0.16	−0.14	−0.11	−0.07	−0.12
	11–20	−0.15	−0.12	−0.02	−0.06	−0.01	−0.07	−0.12	−0.06	−0.13	−0.07
		Series E: Wölfer sunspot numbers: yearly, 100 observations									
z	1–10	0.81	−0.71	0.21	−0.15	0.10	0.10	0.18	0.23	0.01	0.00
	11–20	0.14	−0.16	0.12	0.03	−0.08	−0.14	−0.06	−0.12	0.00	0.05
∇z	1–10	0.57	−0.48	−0.06	−0.27	−0.22	−0.26	−0.29	−0.05	−0.02	−0.16
	11–20	0.13	−0.15	−0.04	0.06	0.12	0.02	0.07	−0.06	−0.09	−0.06
$\nabla^2 z$	1–10	0.15	−0.35	−0.10	−0.21	−0.16	−0.17	−0.36	−0.26	−0.09	−0.33
	11–20	−0.02	−0.13	−0.20	−0.21	−0.10	−0.13	0.00	0.03	−0.01	−0.08
		Series F: yields from a batch chemical process, 70 observations									
z	1–10	−0.40	0.19	0.01	−0.07	−0.07	−0.15	0.05	0.00	−0.10	0.05
	11–20	0.18	−0.05	0.09	0.18	0.01	0.43	0.01	−0.14	0.11	0.18
∇z	1–10	−0.76	−0.32	−0.19	−0.16	−0.09	−0.24	−0.15	−0.06	−0.18	−0.28
	11–20	−0.02	−0.16	−0.24	−0.06	−0.44	−0.02	0.12	−0.12	−0.17	−0.24
$\nabla^2 z$	1–10	−0.83	−0.52	−0.38	−0.33	−0.15	−0.24	−0.26	−0.14	−0.09	−0.31
	11–20	−0.12	−0.09	−0.26	0.08	−0.38	−0.39	−0.07	−0.05	−0.03	−0.30

[a] Obtained by fitting autoregressive processes of increasing order, using least squares.

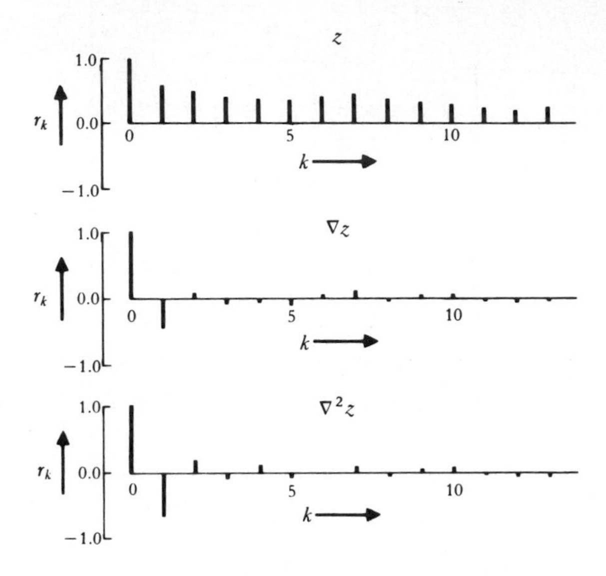

Figure 6.2 Estimated autocorrelations of various differences of series A.

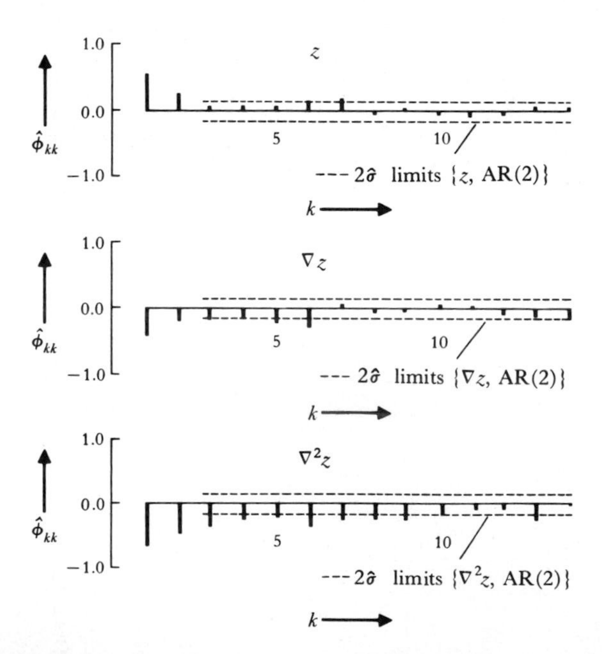

Figure 6.3 Estimated partial autocorrelations of various differences of series A.

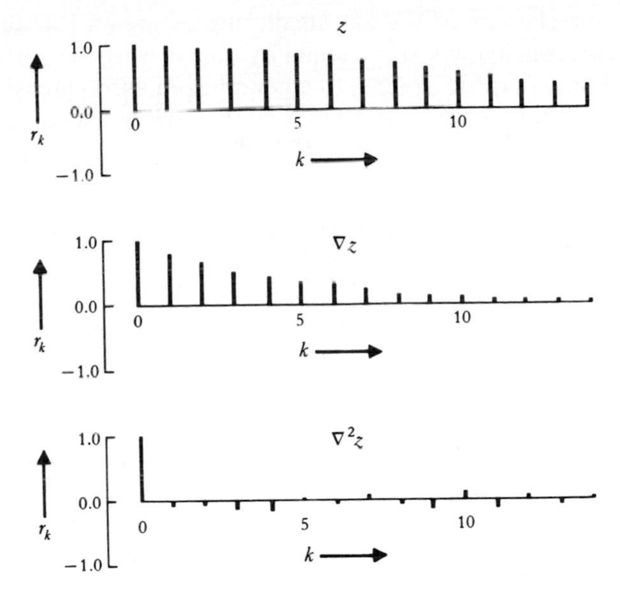

Figure 6.4 Estimated autocorrelations of various differences of series C.

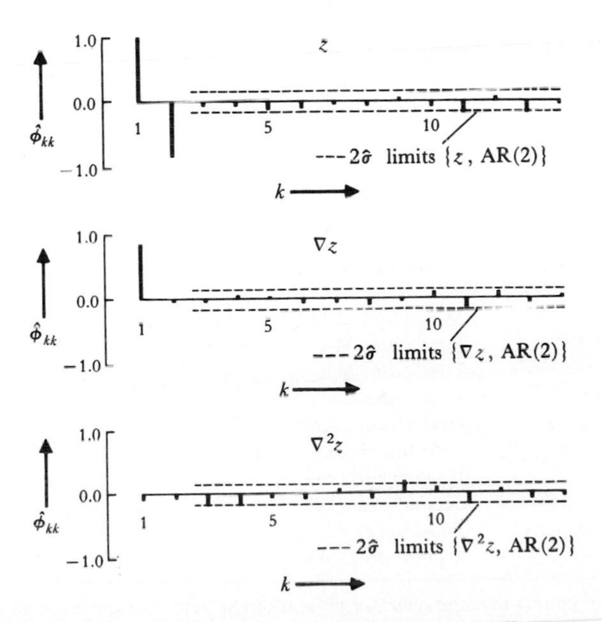

Figure 6.5 Estimated partial autocorrelations of various differences of series C.

that the autocorrelations of $\nabla^2 z$ are small, suggesting an IMA(0, 2, 0) process. The same conclusions are reached by considering the partial autocorrelation function. That for z points to a generalized autoregressive operator of degree 2. That for ∇z points to a first-order autoregressive process in ∇z_t, with ϕ equal to about 0.8, and that for $\nabla^2 z_t$ to uncorrelated noise. Thus the possibilities are

$$(1 - 0.8B)(1 - B)z_t = a_t$$

$$(1 - B)^2 z_t = a_t$$

The second model is very similar to the first, differing only in the choice of 0.8 rather than 1 for one of the autoregressive coefficients.

In assessing the estimated correlation functions, it is very helpful to plot "control" lines about zero at $\pm\hat{\sigma}$ or $\pm 2\hat{\sigma}$. An indication of the hypothesis in mind should accompany the limits. Thus, in Figure 6.5, the annotation "$2\hat{\sigma}$ limits $\{z, \text{AR}(2)\}$" on the limit lines indicates that the value $\hat{\sigma}$ employed is approximately correct on the hypothesis that the process is second-order autoregressive. Tentative identification for each of the series A to F is given in Table 6.4.

Three other points concerning this identification procedure need to be mentioned:

1. Simple differencing of the kind we have used will not produce stationarity in series containing seasonal components. In Chapter 9 we discuss the appropriate modifications for such seasonal time series.
2. As discussed in Chapter 4, a nonzero value for θ_0 in (6.1.1) implies the existence of a systematic polynomial trend of degree d. For the nonstationary models in Table 6.4, a value of $\theta_0 = 0$ can perfectly well

TABLE 6.4 Tentative Identification of Models for Series A to F

Series	Degree of Differencing	Apparent Nature of Differenced Series	Identification for z_t
A	Either 0	Mixed first-order AR with first-order MA	(1, 0, 1)
	or 1	First-order MA	(0, 1, 1)
B	1	First-order MA	(0, 1, 1)
C	Either 1	First-order AR	(1, 1, 0)
	or 2	Uncorrelated noise	(0, 2, 2)[a]
D	Either 0	First-order AR	(1, 0, 0)
	or 1	Uncorrelated noise	(0, 1, 1)[a]
E	Either 0	Second-order AR	(2, 0, 0)
	or 0	Third-order AR	(3, 0, 0)
F	0	Second-order AR	(2, 0, 0)

[a]The order of the moving average operator appears to be zero, but the more general form is retained for subsequent consideration.

account for the behavior of the series. Occasionally, however, there will be some real physical phenomenon requiring the provision of such a component. In other cases it might be uncertain whether or not such a provision should be made. Some indication of the evidence supplied by the data, for the inclusion of θ_0 in the model, can be obtained at the identification stage by comparing the mean $\overline{w}$ of $w_t = \nabla^d z_t$ with its approximate standard error, using $\sigma^2(\overline{w}) = n^{-1}\sigma_w^2 \, [1 + 2r_1(w) + 2r_2(w) + \cdots]$.

3. It was noted in Section 3.4.2 that, for any ARMA(p, q) process with $p - q > 0$, the *whole positive half* of the autocorrelation function will be a mixture of damped sine waves and exponentials. This does not, of course, prevent us from tentatively identifying q, because (a) the partial autocorrelation function will show $p - q$ "anomalous" values before behaving like that of an MA(q) process, and (b) q must be such that the autocorrelation function could take, as starting values following the general pattern, ρ_q back to $\rho_{-(p-q-1)}$.

6.2.4 Some Additional Model Identification Tools

Although the sample autocorrelation and partial autocorrelation functions are extremely useful in model identification, there are sometimes cases involving mixed models where they will not provide unambiguous results. This might not be considered a serious problem, since as has been emphasized, model specification is always tentative and subject to further examination, diagnostic checking, and modification if necessary. Nevertheless, there has been considerable interest in developing additional tools for use at the model identification stage. These include the R and S array approach proposed by Gray, Kelley, and McIntire [97], the generalized partial autocorrelation function studied by Woodward and Gray [208], the inverse autocorrelation function considered by Cleveland [71] and Chatfield [70], the extended sample autocorrelation function of Tsay and Tiao [192], and the use of canonical correlation analysis as examined by Akaike [4], Cooper and Wood [72], and Tsay and Tiao [193].

Canonical correlation methods. For illustration, we discuss aspects of the use of canonical correlation analysis for model identification. In general, for two sets of random variables, $\boldsymbol{Y}_1 = (y_{11}, y_{12}, \ldots, y_{1k})'$ and $\boldsymbol{Y}_2 = (y_{21}, y_{22}, \ldots, y_{2l})'$, of dimensions k and l (assume that $k \leq l$), canonical correlation analysis involves determining linear combinations $U_i = \boldsymbol{a}_i'\boldsymbol{Y}_1$ and $V_i = \boldsymbol{b}_i'\boldsymbol{Y}_2$, $i = 1, \ldots, k$, and corresponding correlations $\rho(i) = \text{corr}[U_i, V_i]$, with $\rho(1) \geq \rho(2) \geq \cdots \geq \rho(k) \geq 0$, such that the U_i and V_j are mutually uncorrelated for $i \neq j$, U_1 and V_1 have the maximum possible correlation $\rho(1)$ among all linear combinations of $\boldsymbol{Y}_1$ and $\boldsymbol{Y}_2$, U_2 and V_2 have the maximum possible correlation $\rho(2)$ among all linear combinations of $\boldsymbol{Y}_1$ and $\boldsymbol{Y}_2$ that are uncorrelated with U_1 and V_1, and so on. The resulting correlations $\rho(i)$ are called the *canonical correlations between* $\boldsymbol{Y}_1$ and $\boldsymbol{Y}_2$, and the variables U_i and V_i are the corresponding canonical variates. If $\boldsymbol{\Omega} = \text{cov}[\boldsymbol{Y}]$ denotes the covariance matrix of $\boldsymbol{Y} = (\boldsymbol{Y}_1', \boldsymbol{Y}_2')'$, with $\boldsymbol{\Omega}_{ij} = \text{cov}[\boldsymbol{Y}_i, \boldsymbol{Y}_j]$, it is known that the values $\rho^2(i)$ are the ordered eigenvalues of the matrix $\boldsymbol{\Omega}_{11}^{-1}\boldsymbol{\Omega}_{12}\boldsymbol{\Omega}_{22}^{-1}\boldsymbol{\Omega}_{21}$ and the vectors $\boldsymbol{a}_i$, such that $U_i = \boldsymbol{a}_i'\boldsymbol{Y}_1$, are the corresponding (normalized) eigenvectors; that is, the $\rho^2(i)$ and $\boldsymbol{a}_i$ satisfy

$$[\rho^2(i)\mathbf{I} - \boldsymbol{\Omega}_{11}^{-1}\boldsymbol{\Omega}_{12}\boldsymbol{\Omega}_{22}^{-1}\boldsymbol{\Omega}_{21}]\boldsymbol{a}_i = 0 \qquad i = 1, \ldots, k \tag{6.2.4}$$

with $\rho^2(1) \geq \rho^2(2) \geq \cdots \geq \rho^2(k) \geq 0$ (e.g., Anderson [7, p. 490]). Similarly, one can define the notion of *partial canonical correlations* between $\boldsymbol{Y}_1$ and $\boldsymbol{Y}_2$, given another set of variables $\boldsymbol{Y}_3$, as the canonical correlations between $\boldsymbol{Y}_1$ and $\boldsymbol{Y}_2$ after they have been "adjusted" for the effects of $\boldsymbol{Y}_3$ by linear regression on $\boldsymbol{Y}_3$, analogous to the definition of partial correlations as discussed in Section 3.2.5. A useful property to note is that if there exist (at least) $s \leq k$ linearly independent linear combinations of $\boldsymbol{Y}_1$ which are completely uncorrelated with $\boldsymbol{Y}_2$, say $\boldsymbol{U} = \mathbf{A}'\boldsymbol{Y}_1$ such that $\text{cov}[\boldsymbol{Y}_2, \boldsymbol{U}] = \boldsymbol{\Omega}_{21}\mathbf{A} = 0$, there are (at least) s zero canonical correlations between $\boldsymbol{Y}_1$ and $\boldsymbol{Y}_2$, which follows easily from (6.2.4) since there will be (at least) s linearly independent eigenvectors satisfying (6.2.4) with corresponding $\rho(i) = 0$. In effect, then, the number s of *zero* canonical correlations is equal to $s = k - r$, where $r = \text{rank}(\boldsymbol{\Omega}_{21})$.

In the ARMA time series model context, following the approach of Tsay and Tiao [193], we consider $\boldsymbol{Y}_{m,t} = (\tilde{z}_t, \tilde{z}_{t-1}, \ldots, \tilde{z}_{t-m})'$ and examine the canonical correlation structure between the variables $\mathbf{Y}_{m,t}$ and $\boldsymbol{Y}_{m,t-j-1} = (\tilde{z}_{t-j-1}, \tilde{z}_{t-j-2}, \ldots, \tilde{z}_{t-j-1-m})'$ for various combinations of $m = 0, 1, \ldots$ and $j = 0, 1, \ldots$. A key feature to recall is that the autocovariance function γ_k of an ARMA(p, q) process $\tilde{z}_t$ satisfies (3.4.2), and, in particular,

$$\gamma_k - \sum_{i=1}^{p} \phi_i \gamma_{k-i} = 0 \qquad k > q$$

Thus, for example, if $m \geq p$, there is (at least) one linear combination of $\boldsymbol{Y}_{m,t}$,

$$\tilde{z}_t - \sum_{i=1}^{p} \phi_i \tilde{z}_{t-i} = (1, -\phi_1, \ldots, -\phi_p, 0, \ldots, 0)\boldsymbol{Y}_{m,t} \equiv \boldsymbol{a}'\boldsymbol{Y}_{m,t} \tag{6.2.5}$$

such that

$$\boldsymbol{a}'\boldsymbol{Y}_{m,t} = a_t - \sum_{i=1}^{q} \theta_i a_{t-i}$$

which is uncorrelated with $\boldsymbol{Y}_{m,t-j-1}$ for $j \geq q$. In particular, then, for $m = p$ and $j = q$ there is one canonical correlation between $\mathbf{Y}_{p,t}$ and $\mathbf{Y}_{p,t-q-1}$, as well as between $\boldsymbol{Y}_{p,t}$ and $\boldsymbol{Y}_{p,t-j-1}$, $j > q$, and between $\mathbf{Y}_{m,t}$ and $\boldsymbol{Y}_{m,t-q-1}$, $m > p$, while in general it is not difficult to establish that there are $s = \min(m + 1 - p, j + 1 - q)$ zero canonical correlations between $\boldsymbol{Y}_{m,t}$ and $\boldsymbol{Y}_{m,t-j-1}$ for $m > p$ and $j > q$. Hence one can see that determination of the structure of the zero canonical correlations between $\boldsymbol{Y}_{m,t}$ and $\boldsymbol{Y}_{m,t-j-1}$ for various values of m and j will serve to characterize the orders p and q of the ARMA model, so the canonical correlations will be useful in model identification. We note that the special cases of these canonical correlations are as follows. First, when $m = 0$ we are simply examining the autocorrelations ρ_{j+1} between z_t and z_{t-j-1}, which will all equal zero in an MA(q) process for $j \geq q$. Second, when $j = 0$ we are examining the partial autocorrelations $\phi_{m+1,m+1}$ between z_t and z_{t-m-1}, given $z_{t-1}, \ldots, z_{t-m}$, and these will all equal zero in an AR(p) process for $m \geq p$. Hence the canonical correlation analysis can be viewed as an extension of the analysis of the autocorrelation and partial autocorrelation functions of the process.

In practice, based on (6.2.4), one is led to consider the sample canonical correlations $\hat{\rho}(i)$ which are determined from the eigenvalues of the matrix

$$\left(\sum_t \boldsymbol{Y}_{m,t}\boldsymbol{Y}'_{m,t}\right)^{-1}\left(\sum_t \boldsymbol{Y}_{m,t}\boldsymbol{Y}'_{m,t-j-1}\right)\left(\sum_t \boldsymbol{Y}_{m,t-j-1}\boldsymbol{Y}'_{m,t-j-1}\right)^{-1}\left(\sum_t \boldsymbol{Y}_{m,t-j-1}\boldsymbol{Y}'_{m,t}\right) \quad (6.2.6)$$

for various values of lag $j = 0, 1, \ldots$ and $m = 0, 1, \ldots$. Tsay and Tiao [193] use a chi-squared test statistic approach based on the *smallest* eigenvalue (squared sample canonical correlation) $\hat{\lambda}(m, j)$ of (6.2.6). They propose the statistic $c(m, j) = -(n - m - j)\ \ln[1 - \hat{\lambda}(m, j)/d(m, j)]$, where $d(m,j) = 1 + 2\sum_{i=1}^{j} r_i^2(w'), j > 0$, $r_i(w')$ denotes the sample autocorrelation at lag i of $w'_t = z_t - \hat{\phi}_1^{(j)} z_{t-1} - \cdots - \hat{\phi}_m^{(j)} z_{t-m}$, and the $\hat{\phi}_i^{(j)}$ are estimates of the ϕ_i's obtained from the eigenvector [see, e.g., equation (6.2.5)] corresponding to $\hat{\lambda}(m, j)$, and establish that $c(m, j)$ has an asymptotic χ_1^2 distribution when $m = p$ and $j \geq q$ or when $m \geq p$ and $j = q$. This statistic is used to test whether there exists a zero canonical correlation in theory, and hence if the sample statistics exhibit a pattern such that they are all insignificant, relative to a χ_1^2 distribution, for $m \geq p$ and $j \geq q$ for some p and q values, the model might reasonably be identified as an ARMA(p, q) for the smallest values (p, q) such that this pattern holds. An additional feature of this approach is that Tsay and Tiao also establish that the procedure is valid for nonstationary ARIMA models $\varphi(B)z_t = \theta(B)a_t$, in the sense that the overall order $p + d$ of the generalized AR operator $\varphi(B)$ can be determined by the procedure, without deciding initially on differencing of the original series z_t.

Canonical correlation methods were also proposed for ARMA modeling by Akaike [4] and Cooper and Wood [72]. Their approach is to perform a canonical correlation analysis between the vector of present and past values, $\boldsymbol{P}_t \equiv Y_{m,t} = (\tilde{z}_t, \tilde{z}_{t-1}, \ldots, \tilde{z}_{t-m})'$, and the vector of future values, $\boldsymbol{F}_{t+1} = (\tilde{z}_{t+1}, \tilde{z}_{t+2}, \ldots)'$. In practice, the finite lag m used to construct the vector of present and past values $\boldsymbol{P}_t$ may be fixed by use of an order determination criterion such as AIC, to be discussed a little later in this section, applied to fitting of AR models of various orders. The canonical correlation analysis is performed sequentially by adding elements to $\boldsymbol{F}_{t+1}$ one at a time [starting with $\boldsymbol{F}_{t+1}^* = (\tilde{z}_{t+1})$] until the first zero canonical correlation between $\boldsymbol{P}_t$ and the $\boldsymbol{F}_{t+1}$ is determined. Akaike [4] uses an AIC-type criterion called DIC to judge whether the smallest sample canonical correlation can be taken to be zero, while Cooper and Wood [72] use a traditional chi-squared statistic approach to assess the significance of the smallest canonical correlation, although as pointed out by Tsay [191], to be valid in the presence of a moving average component this statistic needs to be modified.

At a given stage in the procedure, when the smallest sample canonical correlation between $\boldsymbol{P}_t$ and $\boldsymbol{F}_{t+1}^*$ is first judged to be 0 and $\tilde{z}_{t+K+1}$ is the most recent variable to be included in $\boldsymbol{F}_{t+1}^*$, a linear combination of $\tilde{z}_{t+K+1}$ in terms of the remaining elements of $\boldsymbol{F}_{t+1}^*$ is identified which is uncorrelated with the past. Specifically, the linear combination $\tilde{z}_{t+K+1} - \sum_{j=1}^K \phi_j \tilde{z}_{t+K+1-j}$ of the elements in the vector $\boldsymbol{F}_{t+1}^*$ of future values is (in theory) determined to be uncorrelated with the past $\boldsymbol{P}_t$. Hence this canonical correlation analysis procedure determines that the forecasts $\hat{z}_t(K+1)$ of the process satisfy

$$\hat{z}_t(K+1) - \sum_{j=1}^{K} \phi_j \hat{z}_t(K+1-j) = \theta_0$$

By reference to the relation (5.3.2) in Section 5.3, for a stationary process, this implies that an ARMA model is identified for the process, with $K = \max\{p, q\}$.

As can be seen in the notation of Tsay and Tiao, the methods of Akaike and Cooper and Wood represent canonical correlation analysis between $\boldsymbol{Y}_{m,t}$ and $\boldsymbol{Y}_{n-1,t+n}$ for various $n = 1, 2, \ldots$. Since the Tsay and Tiao method considers canonical correlation analysis between $\boldsymbol{Y}_{m,t}$ and $\boldsymbol{Y}_{m,t-j-1}$ for various combinations of $m = 0, 1, \ldots$ and $j = 0, 1, \ldots$, it is more general and, in principle, it is capable of providing information on the orders p and q of the AR and MA parts of the model separately, rather than just the maximum of these two values. In practice, when using the methods of Akaike and Cooper and Wood, the more detailed information on the individual orders p and q would be determined at the stage of maximum likelihood estimation of the parameters of the ARMA(K, K) model.

Use of model selection criteria. Another approach to model selection is the use of information criteria such as AIC proposed by Akaike [3] or the BIC of Schwarz [176]. In the implementation of this approach, a range of potential ARMA models is estimated by maximum likelihood methods to be discussed in Chapter 7, and for each, a criterion such as AIC (normalized by sample size n), given by

$$\mathrm{AIC}_{p,q} = \frac{-2 \ln(\text{maximized likelihood}) + 2r}{n} \approx \ln(\hat{\sigma}_a^2) + r\frac{2}{n} + \text{constant}$$

or the related BIC given by

$$\mathrm{BIC}_{p,q} = \ln(\hat{\sigma}_a^2) + r\frac{\ln(n)}{n}$$

is evaluated, where $\hat{\sigma}_a^2$ denotes the maximum likelihood estimate of σ_a^2, and $r = p + q + 1$ denotes the number of parameters estimated in the model, including a constant term. In the criteria above, the first term essentially corresponds to minus $2/n$ times the log of the maximized likelihood, while the second term is a "penalty factor" for inclusion of additional parameters in the model. In the information criteria approach, models that yield a minimum value for the criterion are to be preferred, and the AIC or BIC values are compared among various models as the basis for selection of the model. Hence since the BIC criterion imposes a greater penalty for the number of estimated model parameters than does AIC, use of minimum BIC for model selection would always result in a chosen model whose number of parameters is no greater than that chosen under AIC.

One immediate disadvantage of this approach is that several models may have to be estimated by maximum likelihood, which is computationally time consuming and expensive. For this reason, Hannan and Rissanen [103] proposed the following model selection procedure. At the first stage of their procedure, one obtains estimates of the random shock series a_t by approximation of the unknown ARMA model by a (sufficiently high order) AR model of order m^*. The order m^* of the approximating AR model might itself be chosen by use of the AIC model selection criterion above. From the AR(m^*) model selected, one obtains residuals $\tilde{a}_t = \tilde{w}_t - \sum_{j=1}^{m^*} \hat{\phi}_{m^*j}\tilde{w}_{t-j}$. In the second stage of the procedure, one regresses $\tilde{w}_t$ on $\tilde{w}_{t-1}, \ldots, \tilde{w}_{t-p}$ and $\tilde{a}_{t-1}, \ldots, \tilde{a}_{t-q}$, for various values of p and q. That is, one estimates approximate models of the form

$$\tilde{w}_t = \sum_{j=1}^{p} \phi_j \tilde{w}_{t-j} - \sum_{j=1}^{q} \theta_j \tilde{a}_{t-j} + a_t \tag{6.2.7}$$

by ordinary least squares regression, and let $\hat{\sigma}_{p,q}^2$ denote the estimated error variance (uncorrected for degrees of freedom). Then by application of the BIC criterion above, the order (p, q) of the ARMA model is chosen as the one that minimizes $\ln(\hat{\sigma}_{p,q}^2) + (p + q)\ln(n)/n$. The appeal of this procedure is that computation of maximum likelihood estimates over a wide range of possible ARMA models is avoided. Such procedures and use of information criteria in model selection have been useful, but they should be viewed as supplementary guidelines to assist in the model selection process. In particular, they should not be used as a substitute for careful examination of characteristics of the estimated autocorrelation and partial autocorrelation functions of the series, and critical examination of the residuals $\hat{a}_t$ for model inadequacies should always be included as a major aspect of the overall model selection process.

6.3 INITIAL ESTIMATES FOR THE PARAMETERS

6.3.1 Uniqueness of Estimates Obtained from the Autocovariance Function

While it is true that a given ARMA model has a unique covariance structure, the converse is not true. At first sight this would seem to rule out the use of the estimated autocovariances as a means of identification. We show later in Section 6.4 that the estimated autocovariance function may be so used. The reason is that, although there exists a multiplicity of ARMA models possessing the same autocovariance function, there exists only one which expresses the current value of $\nabla^d z_t = w_t$, exclusively in terms of previous history and in stationary invertible form.

6.3.2 Initial Estimates for Moving Average Processes

It has been shown [see (3.3.4)] that the first q autocorrelations of a MA (q) process are nonzero and can be written in terms of the parameters of the model as

$$\rho_k = \frac{-\theta_k + \theta_1\theta_{k+1} + \theta_2\theta_{k+2} + \cdots + \theta_{q-k}\theta_q}{1 + \theta_1^2 + \theta_2^2 + \cdots + \theta_q^2} \qquad k = 1, 2, \ldots, q \tag{6.3.1}$$

The expression (6.3.1) for $\rho_1, \rho_2, \ldots, \rho_q$, in terms of $\theta_1, \theta_2, \ldots, \theta_q$, supplies q equations in q unknowns. Preliminary estimates of the θ's can be obtained by substituting the estimates r_k for ρ_k in (6.3.1) and solving the resulting nonlinear equations. A preliminary estimate of σ_a^2 may then be obtained from

$$\gamma_0 = \sigma_a^2(1 + \theta_1^2 + \cdots + \theta_q^2)$$

by substituting the preliminary estimates of the θ's and replacing $\gamma_0 = \sigma_w^2$ by its estimate c_0.

Preliminary estimates for a (0, *d*, 1) process. Table A in Part Five relates ρ_1 to θ_1, and by substituting $r_1(w)$ for ρ_1 can be used to provide initial estimates for any (0, d, 1) process $w_t = (1 - \theta_1 B)a_t$, where $w_t = \nabla^d z_t$.

Preliminary estimates for a (0, *d*, 2) process. Chart C in Part Five relates ρ_1 and ρ_2 to θ_1 and θ_2, and by substituting $r_1(w)$ and $r_2(w)$ for ρ_1 and ρ_2 can be used to provide initial estimates for any $(0, d, 2)$ process.

In obtaining preliminary estimates in this way, the following points should be borne in mind:

1. The autocovariances are second moments of the joint distribution of the w's. Thus the parameter estimates are obtained by equating sample moments to their theoretical values. It is well known that the method of moments is not necessarily efficient, and it can be demonstrated that it lacks efficiency in these particular cases. However, the rough estimates obtained can be useful in obtaining fully efficient estimates, because they supply an approximate idea of "where in the parameter space to look" for the most efficient estimates.
2. In general, the equations (6.3.1), obtained by equating moments, will have multiple solutions. For instance, when $q = 1$,

$$\rho_1 = \frac{-\theta_1}{1 + \theta_1^2} \tag{6.3.2}$$

and hence both

$$\theta_1 = -\frac{1}{2\rho_1} + \left[\frac{1}{(2\rho_1)^2} - 1\right]^{1/2}$$

and

$$\theta_1 = -\frac{1}{2\rho_1} - \left[\frac{1}{(2\rho_1)^2} - 1\right]^{1/2} \tag{6.3.3}$$

are possible solutions. Thus, from Table 6.2, the first lag autocorrelation, of the first difference of series A, is about -0.4. Substitution in (6.3.3) yields the pair of solutions $\theta_1 \simeq 0.5$ and $\theta_1' \simeq 2.0$. However, the chosen value $\theta_1 \simeq 0.5$ is the only value that lies within the invertibility interval $-1 < \theta_1 < 1$. In fact, it is shown in Section 6.4.1 that it is always true that only one of the multiple solutions can satisfy the invertibility condition.

Examples. Series A, B, and D have all been identified in Table 6.4 as possible IMA processes of order (0, 1, 1). We have seen in Section 4.3.1 that this model may be written in the alternative forms

$$\nabla z_t = (1 - \theta_1 B)a_t$$

$$\nabla z_t = \lambda_0 a_{t-1} + \nabla a_t \qquad (\lambda_0 = 1 - \theta_1)$$

$$z_t = \lambda_0 \sum_{j-1}^{\infty} (1 - \lambda_0)^{j-1} z_{t-j} + a_t$$

TABLE 6.5 Initial Estimates of Parameters for Series A, B, and D

Series	r_1	$\hat{\theta}_1$	$\hat{\lambda}_0 = 1 - \hat{\theta}_1$
A	−0.41	0.5	0.5
B	0.09	−0.1	1.1
D	−0.05	0.1	0.9

Using Table A in Part Five, the approximate estimates of the parameters shown in Table 6.5 were obtained.

Series C has been tentatively identified in Table 6.4 as an IMA(0, 2, 2) process

$$\nabla^2 z_t = (1 - \theta_1 B - \theta_2 B^2) a_t$$

or equivalently,

$$\nabla^2 z_t = (\lambda_0 \nabla + \lambda_1) a_{t-1} + \nabla^2 a_t$$

Since the first two autocorrelations of $\nabla^2 z_t$, given in Table 6.2, are approximately zero, then using Chart C in Part Five, $\hat{\theta}_1 = 0$, $\hat{\theta}_2 = 0$, so that $\hat{\lambda}_0 = 1 + \hat{\theta}_2 = 1$ and $\hat{\lambda}_1 = 1 - \hat{\theta}_1 - \hat{\theta}_2 = 1$. On this basis the series would be represented by

$$\nabla^2 z_t = a_t \tag{6.3.4}$$

This would mean that the second difference, $\nabla^2 z_t$, was very nearly a random series.

6.3.3 Initial Estimates for Autoregressive Processes

For an assumed AR process of order 1 or 2, initial estimates for ϕ_1 and ϕ_2 can be calculated by substituting estimates r_j for the theoretical autocorrelations ρ_j in the formulas of Table 6.1, which are obtained from the Yule–Walker equations (3.2.6). In particular, for an AR(1), $\hat{\phi}_{11} = r_1$, and for an AR(2),

$$\begin{aligned} \hat{\phi}_{21} &= \frac{r_1(1 - r_2)}{1 - r_1^2} \\ \hat{\phi}_{22} &= \frac{r_2 - r_1^2}{1 - r_1^2} \end{aligned} \tag{6.3.5}$$

where ϕ_{pj} denotes the jth autoregressive parameter in a process of order p. The corresponding formulas given by the Yule–Walker equations for higher-order schemes may be obtained by substituting the r_j for the ρ_j in (3.2.7). Thus

$$\hat{\boldsymbol{\phi}} = \mathbf{R}_p^{-1} \mathbf{r}_p \tag{6.3.6}$$

where $\mathbf{R}_p$ is an estimate of the $p \times p$ matrix of correlations up to order $p - 1$, and $\mathbf{r}_p'$ is the vector $(r_1, r_2, \ldots, r_p)$. For example, if $p = 3$, (6.3.6) becomes

$$\begin{bmatrix} \hat{\phi}_{31} \\ \hat{\phi}_{32} \\ \hat{\phi}_{33} \end{bmatrix} = \begin{bmatrix} 1 & r_1 & r_2 \\ r_1 & 1 & r_1 \\ r_2 & r_1 & 1 \end{bmatrix}^{-1} \begin{bmatrix} r_1 \\ r_2 \\ r_3 \end{bmatrix} \tag{6.3.7}$$

A recursive method of obtaining the estimates for an AR(p) from those of an AR($p - 1$) has been given in Appendix A3.2.

It will be shown in Chapter 7 that by contrast to the situation for MA processes, the autoregressive parameters obtained from (6.3.6) approximate the fully efficient maximum likelihood estimates.

Example. Series E behaves in its undifferenced form like* an autoregressive process of second order,

$$(1 - \phi_1 B - \phi_2 B^2)\tilde{z}_t = a_t$$

Substituting the estimates $r_1 = 0.81$ and $r_2 = 0.43$, obtained from Table 6.2, in (6.3.5), we have $\hat{\phi}_1 = 1.32$ and $\hat{\phi}_2 = -0.63$.

As a second example, consider again series C identified as either of order (1, 1, 0) or possibly (0, 2, 2). The first possibility would give

$$(1 - \phi_1 B)\nabla z_t = a_t$$

with $\hat{\phi}_1 = 0.8$, since r_1 for ∇z is 0.81.

This example is especially interesting because it makes clear that the two alternative models that have been identified for this series are closely related. On the supposition that the series is of order (0, 2, 2), we have suggested a model

$$(1 - B)(1 - B)z_t = a_t \tag{6.3.8}$$

The alternative

$$(1 - 0.8B)(1 - B)z_t = a_t \tag{6.3.9}$$

is very similar.

* The sunspot series has been the subject of much investigation (see, e.g., Schuster [175], Yule [212], and Moran [146]). The series is almost certainly not adequately fitted by a second-order autoregressive process. A model related to the actual mechanism at work would, of course, be the most satisfactory. Recent unpublished work has suggested that empirically, a second-order autoregressive model would provide a better fit if a suitable transformation were first applied to z. Inclusion of a higher-order term, at lag 9, in the AR model also improves the fit. Other possibilities include the use of nonlinear time series models, such as bilinear or threshold autoregressive models, as has been investigated by Subba Rao and Gabr [184], Tong and Lim [189], and Tong [188].

6.3.4 Initial Estimates for Mixed Autoregressive–Moving Average Processes

It will often be found, either initially or after suitable differencing, that $w_t = \nabla^d z_t$ is most economically represented by a mixed ARMA process

$$\phi(B)w_t = \theta(B)a_t$$

It was noted in Section 6.2.1 that a mixed process is indicated if both the autocorrelation and partial autocorrelation functions tail off. Another fact [see (3.4.3)] of help in identifying the mixed process is that after lag $q - p$ the theoretical autocorrelations of the mixed process behave like the autocorrelations of the pure autoregressive process $\phi(B)w_t = a_t$. In particular, if the autocorrelation function of the dth difference appeared to be falling off exponentially from an aberrant first value r_1 we would suspect that we had a process of order $(1, d, 1)$, that is,

$$(1 - \phi_1 B)w_t = (1 - \theta_1 B)a_t \tag{6.3.10}$$

where $w_t = \nabla^d z_t$.

Approximate values for the parameters of the process (6.3.10) are obtained by substituting the estimates $r_1(w)$ and $r_2(w)$ for ρ_1 and ρ_2 in the expressions (3.4.8) and also given in Table 6.1. Thus

$$r_1 = \frac{(1 - \hat{\theta}_1\hat{\phi}_1)(\hat{\phi}_1 - \hat{\theta}_1)}{1 + \hat{\theta}_1^2 - 2\hat{\phi}_1\hat{\theta}_1}$$

$$r_2 = r_1\hat{\phi}_1$$

Chart D in Part Five relates ρ_1 and ρ_2 to ϕ_1 and θ_1, and can be used to provide initial estimates of the parameters for any $(1, d, 1)$ process.

For example, using Figure 6.2, series A was identified as of order $(0, 1, 1)$, with θ_1 about 0.5. Looking at the autocorrelation function of z_t rather than that of $w_t = \nabla z_t$, we see that from r_1 onward the autocorrelations decay roughly exponentially, although slowly. Thus an alternative identification of series A is that it is generated by a stationary process of order $(1, 0, 1)$. The estimated autocorrelations and the corresponding initial estimates of the parameters are then

$$r_1 = 0.57 \qquad r_2 = 0.50 \qquad \hat{\phi}_1 \simeq 0.87 \qquad \hat{\theta}_1 \simeq 0.48$$

This identification yields the approximate model of order $(1, 0, 1)$

$$(1 - 0.9B)\tilde{z}_t = (1 - 0.5B)a_t$$

whereas the previously identified model of order $(0, 1, 1)$, given in Table 6.5, is

$$(1 - B)z_t = (1 - 0.5B)a_t$$

Again we see that the "alternative" models are nearly the same. A more general method for obtaining initial estimates of the parameters for a mixed autoregressive–moving average process is given in Appendix A6.2.

Compensation between autoregressive and moving average operators. The alternative models identified above are even more alike than they appear. This is because small changes in the autoregressive operator of a mixed model can be nearly compensated by corresponding changes in the moving average operator. In particular, if we have a model

$$[1 - (1 - \delta)B]\tilde{z}_t = (1 - \theta B)a_t$$

where δ is small and positive, we can write

$$\begin{aligned}(1 - B)\tilde{z}_t &= [1 - (1 - \delta)B]^{-1}(1 - B)(1 - \theta B)a_t \\ &= \{1 - \delta B[1 + (1 - \delta)B + (1 - \delta)^2B^2 + \cdots]\}(1 - \theta B)a_t \\ &= [1 - (\theta + \delta)B]a_t + \text{terms in } a_{t-2}, a_{t-3}, \ldots, \text{ of order } \delta\end{aligned}$$

6.3.5 Choice between Stationary and Nonstationary Models in Doubtful Cases

The apparent ambiguity displayed in Table 6.4 in identifying models for series A, C, and D is, of course, more apparent than real. It arises whenever the roots of $\phi(B) = 0$ approach unity. When this happens it becomes less and less important whether a root near unity is included in $\phi(B)$ or an additional difference is included corresponding to a unit root. A more precise evaluation is possible using the estimation procedures discussed in Chapter 7, but the following should be borne in mind:

1. From time series that are necessarily of finite length, it is never possible to *prove* that a zero of the autoregressive operator is exactly equal to unity.
2. There is, of course, no sudden transition from stationary behavior to nonstationary behavior. This can be understood by considering the behavior of the simple mixed model

$$(1 - \phi_1 B)(z_t - \mu) = (1 - \theta_1 B)a_t$$

Series generated by such a model behave in a more and more nonstationary manner as ϕ_1 increases toward unity. For example, a series with $\phi_1 = 0.99$ can wander away from its mean μ and not return for very long periods. It is as if the attraction that the mean exerts in the series becomes less and less as ϕ_1 approaches unity, and finally, when ϕ_1 is equal to unity, the behavior of the series is completely independent of μ.

In doubtful cases there may be an advantage in employing the nonstationary model rather than the stationary alternative (e.g., in treating a ϕ_1, whose estimate is close to unity, as being *equal* to unity). This is particularly true in forecasting and control problems. Where ϕ_1 is close to unity, we do not really know whether the mean of the series has meaning or not. Therefore, it may be advantageous to employ the nonstationary model, which does not include a mean μ. If we use such a model, forecasts of future behavior will not in any way depend on an estimated mean, calculated from a previous period, which may have no relevance to the future level of the series.

6.3.6 More Formal Tests for Unit Roots in ARIMA Models

As noted, initially, the decision concerning the need for differencing is based, informally, on characteristics of the time series plot of z_t and of its sample autocorrelation function (e.g., failure of the r_k to dampen out sufficiently quickly). This can be evaluated further based on efficient estimation of model parameters as discussed in Chapter 7. However, it must be borne in mind that the distribution theory for estimates of parameters differs markedly between stationary and nonstationary models. This has led to an interest in more formal inference procedures concerning the appropriateness of a differencing operator (or a unit root in the AR operator) in the model, including the estimation, testing, and distribution theory for parameter estimates in unit root nonstationary time series models.

Consider first the simple AR(1) model $z_t = \phi z_{t-1} + a_t$, $t = 1, 2, \ldots, n$, $z_0 = 0$, and consider testing for a random walk model, that is, $\phi = 1$. The least squares (LS) estimator of ϕ is given by

$$\hat{\phi} = \frac{\sum_{t=2}^{n} z_{t-1} z_t}{\sum_{t=2}^{n} z_{t-1}^2} = \phi + \frac{\sum_{t=2}^{n} z_{t-1} a_t}{\sum_{t=2}^{n} z_{t-1}^2}$$

In the stationary case with $|\phi| < 1$, it is indicated in Chapter 7 that $n^{1/2}(\hat{\phi} - \phi)$ has an approximate standard Normal distribution with zero mean and variance $(1 - \phi^2)$. However, when $\phi = 1$, so that $z_t = \sum_{j=0}^{t-1} a_{t-j} + z_0$ in the integrated form, it can be shown that

$$n(\hat{\phi} - 1) = n^{-1} \sum_{t=2}^{n} z_{t-1} a_t / n^{-2} \sum_{t=2}^{n} z_{t-1}^2 = O_p(1)$$

bounded in probability as $n \to \infty$, with both the numerator and denominator possessing nondegenerate and nonnormal limiting distributions. Hence in the nonstationary case the estimator $\hat{\phi}$ approaches its true value $\phi = 1$ with increasing sample size n at a faster rate than in the stationary case.

A representation for the limiting distribution of $n(\hat{\phi} - 1)$ has been given by Dickey and Fuller [82], such that

$$n(\hat{\phi} - 1) \xrightarrow{D} \frac{\frac{1}{2}(\Lambda^2 - 1)}{\Gamma} \tag{6.3.11}$$

where $(\Gamma, \Lambda) = (\Sigma_{i=1}^{\infty} \gamma_i^2 Z_i^2, \Sigma_{i=1}^{\infty} 2^{1/2}\gamma_i Z_i)$, with $\gamma_i = 2(-1)^{i+1}/[(2i-1)\pi]$, and the Z_i are i.i.d. $N(0, 1)$ distributed random variables. An equivalent representation for the distribution (see Chan and Wei [68]) is given by

$$n(\hat{\phi} - 1) \xrightarrow{D} \frac{\int_0^1 B(u)\, dB(u)}{\int_0^1 B(u)^2\, du} = \frac{\frac{1}{2}(B(1)^2 - 1)}{\int_0^1 B(u)^2\, du} \tag{6.3.12}$$

where $B(u)$ is a (continuous-parameter) standard Brownian motion process on [0, 1]. Tables for the percentiles of the limiting distribution of $n(\hat{\phi} - 1)$ have been given by Fuller [93]. Also, the "Studentized" statistic

$$\hat{\tau} = \frac{\hat{\phi} - 1}{s_a(\Sigma_{t=2}^{n} z_{t-1}^2)^{-1/2}} \tag{6.3.13}$$

where $s_a^2 = (n-2)^{-1}(\Sigma_{t=2}^{n} z_t^2 - \hat{\phi}\, \Sigma_{t=2}^{n} z_{t-1}z_t)$ is the residual mean square, has been considered. The limiting distribution of the statistic $\hat{\tau}$ has been derived, and tables of the percentiles of this distribution under $\phi = 1$ available in [93] can be used to test the random walk hypothesis that $\phi = 1$. The test rejects $\phi = 1$ when $\hat{\tau}$ is "too negative."

For higher-order (generalized) AR processes, $\varphi(B)z_t = a_t$, we consider the case where $\varphi(B) = \phi(B)(1 - B)$ and $\phi(B) = 1 - \Sigma_{j=1}^{p} \phi_j B^j$ is a pth-order *stationary* AR operator. Hence,

$$\varphi(B)z_t = \phi(B)(1 - B)z_t = z_t - z_{t-1} - \sum_{j=1}^{p} \phi_j(z_{t-j} - z_{t-j-1})$$

and testing for a unit root in $\varphi(B)$ is equivalent to testing $\rho = 1$ in the model $z_t = \rho z_{t-1} + \Sigma_{j=1}^{p} \phi_j(z_{t-j} - z_{t-j-1}) + a_t$, or testing $\rho - 1 = 0$ in the model

$$z_t - z_{t-1} = (\rho - 1)z_{t-1} + \sum_{j=1}^{p} \phi_j(z_{t-j} - z_{t-j-1}) + a_t \tag{6.3.14}$$

In fact, for any (generalized) AR($p + 1$) model $z_t = \Sigma_{j=1}^{p+1} \varphi_j z_{t-j} + a_t$, it is seen that the model can be expressed in an equivalent form as $w_t = (\rho - 1)z_{t-1} + \Sigma_{j=1}^{p} \phi_j w_{t-j} + a_t$, where $w_t = z_t - z_{t-1}$, $\rho - 1 = -\varphi(1) = \Sigma_{j=1}^{p+1} \varphi_j - 1$, and $\phi_j = -\Sigma_{i=j+1}^{p+1} \varphi_i$. Hence the existence of a unit root in the AR operator $\varphi(B)$ is equivalent to $\rho = \Sigma_{j=1}^{p+1} \varphi_j = 1$. So, based on this last form of the model, let $(\hat{\rho} - 1, \hat{\phi}_1, \ldots, \hat{\phi}_p)$ denote the usual least squares regression estimates for the model in (6.3.14), obtained by regressing $w_t = z_t - z_{t-1}$, on $z_{t-1}, w_{t-1}, \ldots, w_{t-p}$, where $w_t = z_t - z_{t-1}$. Then, under the unit root model where $\rho = 1$ and $\phi(B)$ is stationary, it has been shown [93, Th. 8.5.1 and Cor. 8.5.1] that $(\hat{\rho} - 1)/s_a(\Sigma_{t=p+2}^{n} z_{t-1}^2)^{-1/2}$ has the same limiting distribution as the statistic $\hat{\tau}$

in (6.3.13) for the AR(1) case, while $(n - p - 1)(\hat{\rho} - 1)\, c$, where $c = \Sigma_{j=0}^{\infty} \psi_j$, with $\psi(B) = \phi^{-1}(B)$, has approximately the same distribution as the statistic $n(\hat{\phi} - 1)$ for the AR(1) case. Also, it follows that the statistic, denoted as $\hat{\tau}$, formed by dividing $(\hat{\rho} - 1)$ by its "usual estimated standard error" from the least squares regression will be asymptotically equivalent to the statistic $(\hat{\rho} - 1)/s_a(\Sigma_{t=p+2}^{n} z_{t-1}^2)^{-1/2}$, and hence will have the same limiting distribution as the statistic $\hat{\tau}$ for the AR(1) case.

The test statistic $\hat{\tau}$ formed as described above can be used to test the hypothesis that $\rho = 1$ in the AR$(p + 1)$ model [i.e., to test for a unit root in the AR$(p + 1)$ model $\varphi(B)z_t = a_t$]. Furthermore, it has been shown [93, Th. 8.5.1] that the limiting distribution of the LSEs $(\hat{\phi}_1, \ldots, \hat{\phi}_p)$ for the parameters of the "stationary operator" $\phi(B)$ of the model is the same as the standard asymptotic distribution for LSEs $\tilde{\phi}_1, \ldots, \tilde{\phi}_p$ obtained by regressing the stationary differenced series $w_t = z_t - z_{t-1}$ on $w_{t-1}, \ldots, w_{t-p}$. (Estimation results for the stationary AR model are discussed in Chapter 7.) It is also noted that the results above extend to the case where a constant term is included in the least squares regression estimation, with the statistic analogous to $\hat{\tau}$ denoted as $\hat{\tau}_\mu$, although the limiting distribution for $\hat{\tau}_\mu$ is derived when the "true" value of the constant term in the model, $\theta_0 = (1 - \varphi_1 - \cdots - \varphi_{p+1})\mu = (1 - \rho)\mu$, is equal to zero under the hypothesis that $\rho = 1$. For example, in the AR(1) model $z_t = \phi z_{t-1} + \theta_0 + a_t$, one obtains the LSE

$$\hat{\phi}_\mu = \frac{\Sigma_{t=2}^{n} (z_{t-1} - \bar{z}_{(1)})(z_t - \bar{z}_{(0)})}{\Sigma_{t=2}^{n} (z_{t-1} - \bar{z}_{(1)})^2}$$

where $\bar{z}_{(i)} = (n - 1)^{-1}\Sigma_{t=2}^{n} z_{t-i}$, $i = 0, 1$, and the representation similar to (6.3.12) for its limiting distribution is given by

$$n(\hat{\phi}_\mu - 1) \xrightarrow{D} \frac{\int_0^1 B(u)\, dB(u) - \xi B(1)}{\int_0^1 B(u)^2\, du - \xi^2}$$

where $\xi = \int_0^1 B(u)\, du$, and it is assumed that $\theta_0 = (1 - \phi)\mu = 0$ when $\phi = 1$. The corresponding test statistic for $\phi = 1$ in the AR(1) case is

$$\hat{\tau}_\mu = \frac{\hat{\phi}_\mu - 1}{s_a[\Sigma_{t=2}^{n} (z_{t-1} - \bar{z}_{(1)})^2]^{-1/2}} \qquad (6.3.15)$$

and tables of percentiles of the distribution of $\hat{\tau}_\mu$ when $\phi = 1$ are also available in [93]. Notice that under $\phi = 1$, since $z_t = \Sigma_{j=0}^{t-1} a_{t-j} + z_0$ in the truncated random shock or integrated form, the terms $z_t - \bar{z}_{(0)}$ and $z_{t-1} - \bar{z}_{(1)}$ do not involve the initial value z_0, and hence it is seen that the distribution theory for the LSE $\hat{\phi}_\mu$ does not depend on any assumption concerning z_0.

These test procedures and similar ones have also been extended for use in testing for a unit root in mixed ARIMA$(p, 1, q)$ models (e.g., see Said and

Dickey [172] and Solo [181]), as well as for higher-order differencing (e.g., Dickey and Pantula [83]). Also, a review of various results concerning testing for unit roots in univariate ARIMA models has been given by Dickey, Bell, and Miller [81]. These formal test procedures could be used to supplement other information in cases where the use of differencing is in doubt, and generally, only in cases where sufficient evidence is found against a unit root hypothesis would the choice of a stationary model alternative be clearly preferable.

Example. To illustrate, consider series C, for which two tentatively identified models were the ARIMA(1, 1, 0) and the ARIMA(0, 2, 0). Since there is some doubt about the need for the second differencing in the ARIMA(0, 2, 0) model, with the alternative model being a stationary AR(1) for the first differences, we investigate this more formally. The AR(1) model $\nabla z_t = \phi \nabla z_{t-1} + a_t$ for the first differences can be written as $\nabla^2 z_t = (\phi - 1)\nabla z_{t-1} + a_t$, and in this form the least squares regression estimate $\hat{\phi} - 1 = -0.187$ is obtained, with estimated standard error of 0.038 and $\hat{\sigma}_a^2 = 0.018$. Note that this implies that $\hat{\phi} = 0.813$, similar to Table 6.7. The "Studentized" statistic to test $\phi = 1$ is $\hat{\tau} = -4.87$, which is far more negative than the lower one percentage point of -2.58 in the tables of [93]. Hence these estimation results do not support the need for second differencing, and point to a preference for the ARIMA(1, 1, 0) model.

For comparison with the more efficient methods of estimation to be described in Chapter 7, it is interesting to see how much additional information about the model can be extracted at the identification stage. We have already shown how to obtain initial estimates of the parameters $(\hat{\boldsymbol{\phi}}, \hat{\boldsymbol{\theta}})$ in the ARMA model, identified for an appropriate difference $w_t = \nabla^d z_t$ of the series. To complete the picture, we now show how to obtain preliminary estimates of the residual variance σ_a^2 and an approximate standard error for the mean of the appropriately differenced series.

6.3.7 Initial Estimate of Residual Variance

An initial estimate of the residual variance may be obtained by substituting an estimate c_0 in the expressions for the variance γ_0 given in Chapter 3. Thus, substituting in (3.2.8), an initial estimate of σ_a^2 for an AR process may be obtained from

$$\hat{\sigma}_a^2 = c_0(1 - \hat{\phi}_1 r_1 - \hat{\phi}_2 r_2 - \cdots - \hat{\phi}_p r_p) \tag{6.3.16}$$

Similarly, from (3.3.3), an initial estimate for a MA process may be obtained from

$$\hat{\sigma}_a^2 = \frac{c_0}{1 + \hat{\theta}_1^2 + \cdots + \hat{\theta}_q^2} \tag{6.3.17}$$

The form of the estimate for a mixed process is more complicated and is most easily obtained as described in Appendix A6.2. For the important ARMA(1, 1) process, it takes the form [see (3.4.7)]

$$\hat{\sigma}_a^2 = \frac{1 - \hat{\phi}_1^2}{1 + \hat{\theta}_1^2 - 2\hat{\phi}_1\hat{\theta}_1} c_0 \tag{6.3.18}$$

For example, consider the (1, 0, 1) model identified for series A. Using (6.3.18) with $\hat{\phi}_1 = 0.87$, $\hat{\theta}_1 = 0.48$, and $c_0 = 0.1586$, we obtain the estimate

$$\hat{\sigma}_a^2 = \frac{1 - (0.87)^2}{1 + (0.48)^2 - 2(0.87)(0.48)} 0.1586 = 0.098$$

6.3.8 Approximate Standard Error for $\bar{w}$

The general ARIMA model, for which the mean μ_w of $w_t = \nabla^d z_t$ is not necessarily zero, may be written in any one of the three forms

$$\phi(B)(w_t - \mu_w) = \theta(B)a_t \tag{6.3.19}$$

$$\phi(B)w_t = \theta_0 + \theta(B)a_t \tag{6.3.20}$$

$$\phi(B)w_t = \theta(B)(a_t + \xi) \tag{6.3.21}$$

where

$$\mu_w = \frac{\theta_0}{1 - \phi_1 - \phi_2 - \cdots - \phi_p} = \frac{(1 - \theta_1 - \theta_2 - \cdots - \theta_q)\xi}{1 - \phi_1 - \phi_2 - \cdots - \phi_p}$$

Hence, if $1 - \phi_1 - \phi_2 - \cdots - \phi_p \neq 0$ and $1 - \theta_1 - \theta_2 - \cdots - \theta_q \neq 0$, $\mu_w = 0$ implies that $\theta_0 = 0$ and that $\xi = 0$. Now, in general, when $d = 0$, μ_z will not be zero. However, consider the eventual forecast function associated with the general model (6.3.19) when $d > 0$. With $\mu_w = 0$, this forecast function already contains an adaptive polynomial component of degree $d - 1$. The effect of allowing μ_w to be nonzero is to introduce a *fixed* polynomial term into this function, of degree d. For example, if $d = 2$ and μ_w is nonzero, the forecast function $\hat{z}_t(l)$ includes a quadratic component in l, in which the coefficient of the quadratic term is fixed and does not adapt to the series. Because models of this kind are often inapplicable when $d > 0$, the hypothesis that $\mu_w = 0$ will frequently not be contradicted by the data. Indeed, as we have indicated, we usually assume that $\mu_w = 0$ unless evidence to the contrary presents itself.

At this, the identification stage of model building, an indication of whether or not a nonzero value for μ_w is needed, may be obtained by comparison of $\bar{w} = \Sigma w_t/n$ with its approximate standard error (see Section 2.1.5). With $n = N - d$ differences available,

$$\sigma^2(\bar{w}) = n^{-1}\gamma_0 \sum_{-\infty}^{+\infty} \rho_j = n^{-1} \sum_{-\infty}^{+\infty} \gamma_j$$

that is,

$$\sigma^2(\overline{w}) = n^{-1}\gamma(1) \tag{6.3.22}$$

where $\gamma(B)$ is the autocovariance generating function defined in (3.1.10) and $\gamma(1)$ is its value when $B = B^{-1} = 1$ is substituted.

For illustration, consider the process of order $(1, d, 0)$,

$$(1 - \phi B)(w_t - \mu_w) = a_t$$

with $w_t = \nabla^d z_t$. From (3.1.11) we obtain

$$\gamma(B) = \frac{\sigma_a^2}{(1 - \phi B)(1 - \phi F)}$$

so that

$$\sigma^2(\overline{w}) = n^{-1}(1 - \phi)^{-2}\sigma_a^2$$

But $\sigma_a^2 = \sigma_w^2(1 - \phi^2)$, so that

$$\sigma^2(\overline{w}) = \frac{\sigma_w^2}{n}\frac{1 - \phi^2}{(1 - \phi)^2} = \frac{\sigma_w^2}{n}\frac{1 + \phi}{1 - \phi}$$

and

$$\sigma(\overline{w}) = \sigma_w\left[\frac{1 + \phi}{n(1 - \phi)}\right]^{1/2}$$

Now σ_w^2 and ϕ are estimated by c_0 and r_1, respectively, as defined in (2.1.9) and (2.1.10). Thus for a $(1, d, 0)$ process, the required standard error is given by

$$\hat{\sigma}(\overline{w}) = \left[\frac{c_0(1 + r_1)}{n(1 - r_1)}\right]^{1/2}$$

Proceeding in this way, the expressions for $\sigma(\overline{w})$ given in Table 6.6 may be obtained.

TABLE 6.6 Approximate Standard Error for $\overline{w}$, Where $w_t = \nabla^d z_t$ and z_t Is an ARIMA Process of Order (p, d, q)

$(1, d, 0)$	$(0, d, 1)$
$\left[\frac{c_0(1 + r_1)}{n(1 - r_1)}\right]^{1/2}$	$\left[\frac{c_0(1 + 2r_1)}{n}\right]^{1/2}$
$(2, d, 0)$	$(0, d, 2)$
$\left[\frac{c_0(1 + r_1)(1 - 2r_1^2 + r_2)}{n(1 - r_1)(1 - r_2)}\right]^{1/2}$	$\left[\frac{c_0(1 + 2r_1 + 2r_2)}{n}\right]^{1/2}$
$(1, d, 1)$	
$\left[\frac{c_0}{n}\left(1 + \frac{2r_1^2}{r_1 - r_2}\right)\right]^{1/2}$	

TABLE 6.7 Summary of Models Identified for Series A to F, with Initial Estimates Inserted

Series	Differ-encing	$\bar{w} \pm \hat{\sigma}(\bar{w})$[a]	$\hat{\sigma}_w^2 = c_0$	Identified Model	$\hat{\sigma}_a^2$
A	Either 0	17.06 ± 0.10	0.1586	$z_t - 0.87z_{t-1} = 2.45 + a_t - 0.48a_{t-1}$	0.098
	or 1	0.002 ± 0.011	0.1364	$\nabla z_t = a_t - 0.53a_{t-1}$	0.107
B	1	−0.28 ± 0.41	52.54	$\nabla z_t = a_t + 0.09a_{t-1}$	52.2
C	Either 1	−0.035 ± 0.047	0.0532	$\nabla z_t - 0.81\nabla z_{t-1} = a_t$	0.019
	or 2	−0.003 ± 0.008	0.0198	$\nabla^2 z_t = a_t - 0.09a_{t-1} - 0.07a_{t-2}$	0.020
D	Either 0	9.13 ± 0.04	0.3620	$z_t - 0.86z_{t-1} = 1.32 + a_t$	0.093
	or 1	0.004 ± 0.017	0.0965	$\nabla z_t = a_t - 0.05a_{t-1}$	0.096
E	Either 0	46.9 ± 5.4	1382.2	$z_t - 1.32z_{t-1} + 0.63z_{t-2} = 14.9 + a_t$	289
	or 0	46.9 ± 5.4	1382.2	$z_t - 1.37z_{t-1} + 0.74z_{t-2} - 0.08z_{t-3} = 13.7 + a_t$	287
F	0	51.1 ± 1.1	139.80	$z_t + 0.32z_{t-1} - 0.18z_{t-2} = 58.3 + a_t$	115

[a] When $d = 0$, read z for w.

Tentative identification of models A to F. Table 6.7 summarizes the models tentatively identified for series A to F, with the preliminary parameter estimates inserted. These models are used as initial guesses for the more efficient estimation methods to be described in Chapter 7.

6.4 MODEL MULTIPLICITY

6.4.1 Multiplicity of Autoregressive–Moving Average Models

With the Normal assumption, knowledge of the first and second moments of a probability distribution implies complete knowledge of the distribution. In particular, knowledge of the mean of $w_t = \nabla^d z_t$ and of its autocovariance function uniquely determines the probability structure for w_t. We now show that although this unique probability structure can be represented by a *multiplicity* of linear models, nevertheless, uniqueness is achieved in the model when we introduce the appropriate stationarity and invertibility restrictions.

Suppose that w_t, having covariance generating function $\gamma(B)$, is represented by the linear model

$$\phi(B)w_t = \theta(B)a_t \tag{6.4.1}$$

where the zeros of $\phi(B)$ and of $\theta(B)$ lie outside the unit circle. Then this linear model may also be written

$$\prod_{i=1}^{p}(1 - G_i B)w_t = \prod_{j=1}^{q}(1 - H_j B)a_t \tag{6.4.2}$$

where the G_i^{-1} are the roots of $\phi(B) = 0$ and H_j^{-1} are the roots of $\theta(B) = 0$, and G_i, H_j lie inside the unit circle. Using (3.1.11), the covariance generating function for w is

$$\gamma(B) = \prod_{i-1}^{p}(1 - G_i B)^{-1}(1 - G_i F)^{-1}\prod_{j=1}^{q}(1 - H_j B)(1 - H_j F)\sigma_a^2$$

Multiple choice of moving average parameters. Since

$$(1 - H_j B)(1 - H_j F) = H_j^2(1 - H_j^{-1}B)(1 - H_j^{-1}F)$$

it follows that any one of the stochastic models

$$\prod_{i=1}^{p}(1 - G_i B)w_t = \prod_{j=1}^{q}(1 - H_j^{\pm 1}B)ka_t$$

can have the same covariance generating function if the constant k is chosen appropriately. In the above, it is understood that for complex roots, reciprocals of both members of the congugate pair will be taken. However, if a real root H is inside the unit circle, H^{-1} will lie outside, or if a complex pair, say H_1 and H_2, are inside, then the pair H_1^{-1} and H_2^{-1} will lie outside. It follows that there will be only *one stationary invertible* model of the form (6.4.2), which has a given autocovariance function.

Backward representations. Now $\gamma(B)$ also remains unchanged if in (6.4.2) we replace $1 - G_i B$ by $1 - G_i F$ or $1 - H_j B$ by $1 - H_j F$. Thus all the stochastic models

$$\prod_{i=1}^{p}(1 - G_i B^{\pm 1})w_t = \prod_{j=1}^{q}(1 - H_j B^{\pm 1})a_t$$

have identical covariance structure. However, representations containing the operator $B^{-1} = F$ refer to future w's and/or future a's, so that although stationary and invertible representations exist in which w_t is expanded in terms of future w's and a's, only one such representation, (6.4.2), exists that relates w_t entirely to *past* history.

A model form which, somewhat surprisingly, is of practical interest is that in which *all* B's are replaced by F's in (6.4.1), so that

$$\phi(F)w_t = \theta(F)e_t$$

where e_t is a sequence of independently distributed random variables having mean zero and variance $\sigma_e^2 = \sigma_a^2$. This then is a stationary invertible representation in which w_t is expressed *entirely* in terms of future w's and e's. We refer to it as the *backward* form of the process, or more simply as the *backward process*.

Equation (6.4.2) is not the most general form of stationary invertible linear model having the covariance generating function $\gamma(B)$. For example, the model (6.4.2) may be multiplied on both sides by any factor $1 - QB$. Thus the process

$$(1 - QB) \prod_{i=1}^{p} (1 - G_i B) w_t = (1 - QB) \prod_{j=1}^{q} (1 - H_j B) a_t$$

has the same covariance structure as (6.4.2). This fact will present no particular difficulty where identification is concerned, since we will be naturally led to choose the simplest representation. However, we shall find in Chapter 7 that we will need to be alert to the possibility of factoring of the operators when fitting the process.

Finally, we reach the conclusion that a stationary-invertible model, in which a current value w_t is expressed in terms only of *previous* history, and which contains no common factors, is uniquely determined by the covariance structure.

Proper understanding of model multiplicity is of importance for a number of reasons:

1. We are reassured by the foregoing argument that the covariance function can logically be used to identify a linear stationary-invertible model which expresses w_t in terms of previous history.
2. The nature of the multiple solutions for moving average parameters obtained by equating moments is clarified.
3. The backward process

$$\phi(F) w_t = \theta(F) e_t$$

obtained by replacing B by F in the linear model, is useful in estimating values of the series that have occurred before the first observation was made.

Now we consider (2) and (3) in greater detail.

6.4.2 Multiple Moment Solutions for Moving Average Parameters

In estimating the q parameters $\theta_1, \theta_2, \ldots, \theta_q$ in the MA model by equating covariances, we have seen in Section 3.3 that multiple solutions are obtained. To each combination of roots there will correspond a linear representation, but to only one such combination will there be an invertible representation in terms of past history.

For example, consider the MA(1) process in w_t,

$$w_t = (1 - \theta_1 B)a_t$$

and suppose that $\gamma_0(w)$ and $\gamma_1(w)$ are known and we want to deduce the values of θ_1 and σ_a^2. Since

$$\gamma_0 = (1 + \theta_1^2)\sigma_a^2 \qquad \gamma_1 = -\theta_1\sigma_a^2 \qquad \gamma_k = 0 \qquad k > 1 \tag{6.4.3}$$

then

$$-\frac{\gamma_0}{\gamma_1} = \theta_1^{-1} + \theta_1$$

and if $(\theta_1 = \theta, \sigma_a^2 = \sigma^2)$ is a solution for given γ_0 and γ_1, so is $(\theta_1 = \theta^{-1}, \sigma_a^2 = \theta^2\sigma^2)$. Apparently, then, for given values of γ_0 and γ_1, there are a *pair* of possible processes

$$w_t = (1 - \theta B)a_t$$

and

$$w_t = (1 - \theta^{-1}B)\alpha_t \tag{6.4.4}$$

with $\sigma_\alpha^2 = \sigma^2\theta^2$. If $-1 < \theta < 1$, then (6.4.4) is not an invertible representation. However, this model may be written

$$w_t = [(1 - \theta^{-1}B)(-\theta F)](-\theta^{-1}B\alpha_t)$$

Thus after setting $e_t = -\alpha_{t-1}/\theta$ the model becomes

$$w_t = (1 - \theta F)e_t \tag{6.4.5}$$

where e_t has the same variance as a_t. Thus (6.4.5) is simply the "backward" process, which is dual to the forward process

$$w_t = (1 - \theta B)a_t \tag{6.4.6}$$

Just as the shock a_t in (6.4.6) is expressible as a convergent sum of current and *previous* values of w,

$$a_t = w_t + \theta w_{t-1} + \theta^2 w_{t-2} + \cdots$$

so the shock e_t in (6.4.5) is expressible as a convergent sum of current and future values of w,

$$e_t = w_t + \theta w_{t+1} + \theta^2 w_{t+2} + \cdots$$

Thus the root θ^{-1} *does* produce an invertible process, but only if a representation of the shock e_t in terms of future values of w is permissible. The invertibility regions shown in Table 6.1 delimit acceptable values of the parameters, *given* that we express the shock in terms of *previous* history.

6.4.3 Use of the Backward Process to Determine Starting Values

Suppose that a time series $w_1, w_2, \ldots, w_n$ is available from a process

$$\phi(B)w_t = \theta(B)a_t \tag{6.4.7}$$

In Chapter 7, problems arise where we need to estimate the values w_0, w_{-1}, w_{-2}, and so on, of the series that occurred *before* the first observation was made. This happens because "starting values" are needed for certain basic recursive calculations used for estimating the parameters in the model. Now, suppose that we require to estimate w_{-l}, given $w_1, \ldots, w_n$. The discussion of Section 6.4.1 shows that the probability structure of $w_1, \ldots, w_n$ is equally explained by the forward model (6.4.7), or by the backward model

$$\phi(F)w_t = \theta(F)e_t \tag{6.4.8}$$

The value w_{-l} thus bears exactly the same probability relationship to the sequence $w_1, w_2, \ldots, w_n$, as does the value w_{n+l+1} to the sequence w_n, $w_{n-1}, w_{n-2}, \ldots, w_1$. Thus, to estimate a value $l + 1$ periods before observations started, we can first consider what would be the optimal estimate or forecast $l + 1$ periods after the series ended, and then apply this procedure to the *reversed* series. In other words, we "forecast" the reversed series. We call this "back forecasting."

APPENDIX A6.1 EXPECTED BEHAVIOR OF THE ESTIMATED AUTOCORRELATION FUNCTION FOR A NONSTATIONARY PROCESS

Suppose that a series of N observations $z_1, z_2, \ldots, z_N$ is generated by a nonstationary (0, 1, 1) process

$$\nabla z_t = (1 - \theta B)a_t$$

and the estimated autocorrelations r_k are computed, where

$$r_k = \frac{c_k}{c_0} = \frac{\sum_{t=1}^{N-k} (z_t - \bar{z})(z_{t+k} - \bar{z})}{\sum_{t=1}^{N} (z_t - \bar{z})^2}$$

Some idea of the behavior of these estimated autocorrelations may be obtained by deriving expected values for the numerator and denominator of this expression and considering the ratio. We will write, following [199]

$$\mathscr{E}[r_k] = \frac{E[c_k]}{E[c_0]}$$

$$= \frac{\Sigma_{t=1}^{N-k} E[(z_t - \bar{z})(z_{t+k} - \bar{z})]}{\Sigma_{t=1}^{N} E[(z_t - \bar{z})^2]}$$

After straightforward but tedious algebra we find that

$$\mathscr{E}[r_k] = \frac{(N-k)[(1-\theta)^2(N^2 - 1 + 2k^2 - 4kN) - 6\theta]}{N(N-1)[(N+1)(1-\theta)^2 + 6\theta]} \tag{A6.1.1}$$

For θ close to zero, $\mathscr{E}[r_k]$ will be close to unity, but for large values of θ it can be considerably smaller than unity, even for small values of k. Figure A6.1 illustrates this fact by showing values of $\mathscr{E}[r_k]$ for $\theta = 0.8$ with $N = 100$ and $N = 200$. Although, as anticipated for a nonstationary process, the ratios $\mathscr{E}[r_k]$ of expected values fail to damp out quickly, it will be seen that they do not approach the value 1 even for small lags.

Similar effects may be demonstrated whenever the parameters approach values where cancellation on both sides of the model would produce a stationary process. For instance, in the example above we can write the model as

$$(1 - B)z_t = [(1 - B) + \delta B]a_t$$

where $\delta = 0.2$. As δ tends to zero, the behavior of the process would be expected to come closer and closer to that of the white noise process $z_t = a_t$, for which the autocorrelation function is zero for lags $k > 0$.

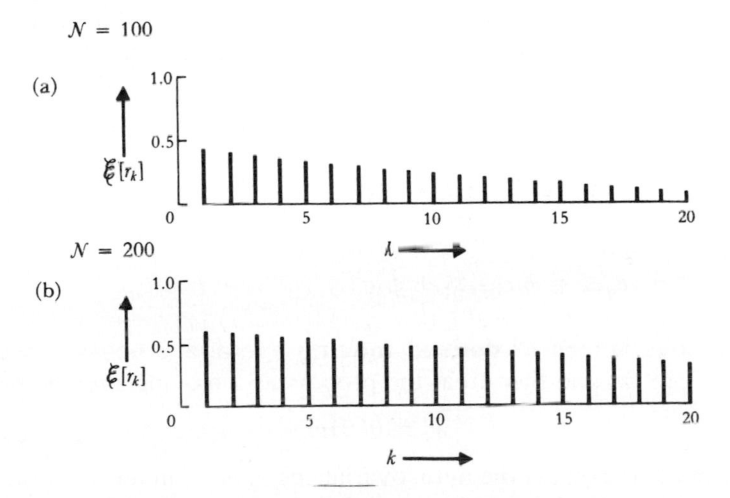

Figure A6.1 $\mathscr{E}[r_k] = E[c_k]/E[c_0]$ for series generated by $\nabla z_t = (1 - 0.8B)a_t$.

APPENDIX A6.2 GENERAL METHOD FOR OBTAINING INITIAL ESTIMATES OF THE PARAMETERS OF A MIXED AUTOREGRESSIVE–MOVING AVERAGE PROCESS

In Section 6.3 we have shown how to derive initial estimates for the parameters of simple ARMA models. In particular, Charts B, C, and D in Part Five have been prepared to enable the initial estimates to be read off quickly for the AR(2), MA(2), and ARMA(1, 1) processes. In this appendix we give a general method for obtaining initial estimates for a general ARMA(p, q) process $\phi(B)w_t = \theta(B)a_t$.

In the general case, calculation of the initial estimates of an ARMA(p, q) process is based on the first $p + q + 1$ autocovariances c_j [$j = 0, 1, \ldots, (p + q)$] of $w_t = \nabla^d z_t$, and proceeds in three stages.

1. The autoregressive parameters $\phi_1, \phi_2, \ldots, \phi_p$ are estimated from the autocovariances $c_{q-p+1}, \ldots, c_{q+1}, c_{q+2}, \ldots, c_{q+p}$.
2. Using the estimates $\hat{\boldsymbol{\phi}}$ obtained in (1), the first $q + 1$ autocovariances c'_j ($j = 0, 1, \ldots, q$) of the derived series

$$w'_t = w_t - \hat{\phi}_1 w_{t-1} - \cdots - \hat{\phi}_p w_{t-p}$$

   are calculated.
3. Finally, the autocovariances $c'_0, c'_1, \ldots, c'_q$ are used in an iterative calculation to compute initial estimates of the moving average parameters $\theta_1, \theta_2, \ldots, \theta_q$ and of the residual variance σ_a^2.

Initial estimates of autoregressive parameters. Making use of the result (3.4.3), initial estimates of the autoregressive parameters may be obtained by solving the p linear equations

$$\begin{aligned}
c_{q+1} &= \hat{\phi}_1 c_q + \hat{\phi}_2 c_{q-1} + \cdots + \hat{\phi}_p c_{q-p+1} \\
c_{q+2} &= \hat{\phi}_1 c_{q+1} + \hat{\phi}_2 c_q + \cdots + \hat{\phi}_p c_{q-p+2} \\
&\vdots \\
c_{q+p} &= \hat{\phi}_1 c_{q+p-1} + \hat{\phi}_2 c_{q+p-2} + \cdots + \hat{\phi}_p c_q
\end{aligned} \tag{A6.2.1}$$

Autocovariances of derived moving average process. We now write $w'_t = \phi(B)w_t$ and then treat the process as a moving average process

$$w'_t = \theta(B)a_t \tag{A6.2.2}$$

First, we need to express the autocovariances c'_j of w'_t in terms of the autocovariances c_j of w_t. It may be shown that

$$c'_j = \sum_{i=0}^{p} \phi_i^2 c_j + \sum_{i=1}^{p} (\phi_0\phi_i + \phi_1\phi_{i+1} + \cdots + \phi_{p-i}\phi_p)(c_{j+i} + c_{j-i}) \tag{A6.2.3}$$

for $j = 0, 1, \ldots, q$, where $\phi_0 = -1$.

Initial estimates of the moving average parameters. Using the autocovariance estimates c'_j, we can obtain initial estimates of the moving average parameters in the derived process (A6.2.2) by one of two iterative processes.

1. *Linearly convergent process.* Using the expressions

$$\gamma_0 = (1 + \theta_1^2 + \cdots + \theta_q^2)\sigma_a^2$$

$$\gamma_k = (-\theta_k + \theta_1\theta_{k+1} + \cdots + \theta_{q-k}\theta_q)\sigma_a^2 \qquad k \geq 1$$

for the autocovariance function of a MA(q) process, given in Section 3.3.2, we can compute estimates of the parameters $\sigma_a^2, \theta_q, \theta_{q-1}, \ldots, \theta_1$ in this precise order, using the iteration

$$\sigma_a^2 = \frac{c'_0}{1 + \theta_1^2 + \cdots + \theta_q^2}$$

$$\theta_j = -\left(\frac{c'_j}{\sigma_a^2} - \theta_1\theta_{j+1} - \theta_2\theta_{j+2} - \cdots - \theta_{q-j}\theta_q\right) \tag{A6.2.4}$$

with the convention that $\theta_0 = 0$. The parameters $\theta_1, \theta_2, \ldots, \theta_q$ are set equal to zero to start the iteration and the values of the θ_j and σ_a^2, to be used in any subsequent calculation, are the most up-to-date values available. For example, in the case $q = 2$, the equations (A6.2.4) are

$$\sigma_a^2 = \frac{c'_0}{1 + \theta_1^2 + \theta_2^2}$$

$$\theta_2 = -\frac{c'_2}{\sigma_a^2}$$

$$\theta_1 = -\left(\frac{c'_1}{\sigma_a^2} - \theta_1\theta_2\right)$$

2. *Quadratically convergent process.* A Newton–Raphson algorithm, which has superior convergence properties to method 1, has been given by Wilson [202]. We denote $\boldsymbol{\tau}' = (\tau_0, \tau_1, \ldots, \tau_q)$, where

$$\tau_0^2 = \sigma_a^2 \qquad \theta_j = \frac{-\tau_j}{\tau_0} \qquad j = 1, 2, \ldots, q \tag{A6.2.5}$$

Then, if $\boldsymbol{\tau}^i$ is the estimate of $\boldsymbol{\tau}$ obtained at the ith iteration, the new values at the $(i + 1)$ iteration are obtained from

$$\boldsymbol{\tau}^{i+1} = \boldsymbol{\tau}^i - (\mathbf{T}^i)^{-1}\mathbf{f}_i \tag{A6.2.6}$$

where $\mathbf{f}' = (f_0, f_1, \ldots, f_q)$, $f_j = \sum_{i=0}^{q-j} \tau_i\tau_{i+j} - c_j'$, and

$$\mathbf{T} = \begin{bmatrix} \tau_0 & \tau_1 & \cdots & \tau_{q-2} & \tau_{q-1} & \tau_q \\ \tau_1 & \tau_2 & \cdots & \tau_{q-1} & \tau_q & 0 \\ \tau_2 & \tau_3 & \cdots & \tau_q & 0 & 0 \\ \vdots & \vdots & & \vdots & \vdots & \vdots \\ \tau_q & 0 & \cdots & 0 & 0 & 0 \end{bmatrix} + \begin{bmatrix} \tau_0 & \tau_1 & \tau_2 & \cdots & \tau_q \\ 0 & \tau_0 & \tau_1 & \cdots & \tau_{q-1} \\ 0 & 0 & \tau_0 & \cdots & \tau_{q-2} \\ \vdots & \vdots & \vdots & & \vdots \\ 0 & 0 & 0 & \cdots & \tau_0 \end{bmatrix}$$

Knowing the values of the τ's at each iteration, the values of the parameters may be obtained from (A6.2.5).

Example. Consider the estimation of ϕ and in θ in the ARMA model

$$(1 - \phi B)\tilde{z}_t = (1 - \theta B)a_t$$

using values $c_0 = 1.25$, $c_1 = 0.50$, and $c_2 = 0.40$, which correspond to a process with $\phi = 0.8$, $\theta = 0.5$, and $\sigma_a^2 = 1.00$. The estimate of ϕ is obtained by substituting $p = q = 1$ in (A6.2.1), giving

$$c_2 = \phi c_1$$

so that $\hat{\phi} = 0.8$. Hence, using (A6.2.3), the first two covariances of the derived series

$$w_t' = \tilde{z}_t - \phi\tilde{z}_{t-1}$$

are

$$c_0' = (1 + \phi^2)c_0 - 2\phi c_1 = 1.25$$

$$c_1' = (1 + \phi^2)c_1 - \phi(c_2 + c_0) = -0.50$$

Substituting these values in (A6.2.4), the iterative process of method 1 is based on

$$\sigma_a^2 = \frac{1.25}{1 + \theta^2}$$

$$\theta = \frac{0.5}{\sigma_a^2}$$

TABLE A6.1 Convergence of Preliminary Estimates of σ_a^2 and θ_1 for a MA(1) Process

	Method 1		Method 2	
Iteration	σ_a^2	θ	σ_a^2	θ
0	—	0.000	1.250	0.000
1	1.250	0.400	2.250	0.667
2	1.077	0.464	1.210	0.545
3	1.029	0.486	1.012	0.503
4	1.011	0.494	1.000	0.500
5	1.004	0.498	—	—
6	1.002	0.499	—	—
7	1.001	0.500	—	—
8	1.000	0.500	—	—

Similarly, substituting in (A6.2.6), the iterative process of method 2 is based on

$$\begin{bmatrix} \tau_0^{i+1} \\ \tau_1^{i+1} \end{bmatrix} = \begin{bmatrix} \tau_0^i \\ \tau_1^i \end{bmatrix} - (\mathbf{T}^i)^{-1} \begin{bmatrix} f_0^i \\ f_1^i \end{bmatrix}$$

where

$$\begin{bmatrix} f_0^i \\ f_1^i \end{bmatrix} = \begin{bmatrix} (\tau_0^i)^2 + (\tau_1^i)^2 - c_0' \\ \tau_0^i \tau_1^i - c_1' \end{bmatrix}$$

and

$$\mathbf{T}^i = \begin{bmatrix} \tau_0^i & \tau_1^i \\ \tau_1^i & 0 \end{bmatrix} + \begin{bmatrix} \tau_0^i & \tau_1^i \\ 0 & \tau_0^i \end{bmatrix} = \begin{bmatrix} 2\tau_0^i & 2\tau_1^i \\ \tau_1^i & \tau_0^i \end{bmatrix}$$

The variance σ_a^2 and θ may then be calculated from $\sigma_a^2 = \tau_0^2$, $\theta = -\tau_1/\tau_0$. Table A6.1 shows how the iteration converged for methods 1 and 2.

7

MODEL ESTIMATION

The identification process having led to a tentative formulation for the model, we then need to obtain efficient estimates of the parameters. After the parameters have been estimated, the fitted model will be subjected to diagnostic checks and tests of goodness of fit. As pointed out by R. A. Fisher, for tests of goodness of fit to be relevant, it is necessary that efficient use of the data should have been made in the fitting process. If this is not so, inadequacy of fit may simply arise because of the inefficient fitting and not because the form of the model is inadequate. This chapter contains a general account of likelihood and Bayesian methods for estimation of the parameters in the stochastic model. Throughout the chapter, bold type is used to denote vectors and matrices. Thus $\mathbf{X} = \{x_{ij}\}$ is a matrix with x_{ij} an element in the ith row and jth column and $\mathbf{X}'$ is the transpose of the matrix.

7.1 STUDY OF THE LIKELIHOOD AND SUM OF SQUARES FUNCTIONS

7.1.1 Likelihood Function

Suppose that we have a sample of N observations $\mathbf{z}$ with which we associate an N-dimensional random variable, whose known probability distribution $p(\mathbf{z}|\boldsymbol{\xi})$ depends on some unknown parameters $\boldsymbol{\xi}$. We use the vector $\boldsymbol{\xi}$ to denote a general set of parameters and, in particular, it could refer to the $p + q + 1$ parameters $(\boldsymbol{\phi}, \boldsymbol{\theta}, \sigma)$ of the ARIMA model.

Before the data are available, $p(\mathbf{z}|\boldsymbol{\xi})$ will associate a density with each different outcome $\mathbf{z}$ of the experiment, for fixed $\boldsymbol{\xi}$. After the data have come to hand, we are led to contemplate the various values of $\boldsymbol{\xi}$ that might have given rise to the fixed set of observations $\mathbf{z}$ actually obtained. The appropriate function for this purpose is the *likelihood function* $L(\boldsymbol{\xi}|\mathbf{z})$, which is of the same form as $p(\mathbf{z}|\boldsymbol{\xi})$, but in which $\mathbf{z}$ is now fixed but $\boldsymbol{\xi}$ is variable. It is only the relative value of $L(\boldsymbol{\xi}|\mathbf{z})$ which is of interest, so that the likelihood function is usually regarded as containing an *arbitrary multiplicative constant.*

It is often convenient to work with the log-likelihood function $\ln[L(\boldsymbol{\xi}|\mathbf{z})] = l(\boldsymbol{\xi}|\mathbf{z})$, which contains an *arbitrary additive constant.* One reason that the likelihood function is of fundamental importance in estimation theory is because of the "likelihood principle," urged on somewhat different grounds by Fisher [91], Barnard [22], and Birnbaum [32]. This principle says that (given that the assumed model is correct) all that the *data* have to tell us about the parameters is contained in the likelihood function, all other aspects of the data being irrelevant. From a Bayesian point of view, the likelihood function is equally important, since it is the component in the posterior distribution of the parameters which comes from the data.

For a complete understanding of the estimation situation, it is necessary to make a thorough analytical and graphical study of the likelihood function, or in the Bayesian framework, the posterior distribution of the parameters, which in the situations we consider, is dominated by the likelihood. In many examples, for moderate and large samples, the log-likelihood function will be unimodal and can be approximated adequately over a sufficiently extensive region near the maximum by a quadratic function. In such cases the log-likelihood function can be described by its maximum and its second derivatives at the maximum. The values of the parameters that maximize the likelihood function, or equivalently the log-likelihood function, are called *maximum likelihood* (*ML*) *estimates.* The second derivatives of the log-likelihood function provide measures of "spread" of the likelihood function and can be used to calculate approximate standard errors for the estimates.

The limiting properties of maximum likelihood estimates are usually established for independent observations [164]. But as was shown by Whittle [196], they may be extended to cover stationary time series.

In what follows we shall assume that the reader is familiar with certain basic ideas in estimation theory. Appendices A7.1 and A7.2 summarize some important results in Normal distribution theory and linear least squares which are needed for this chapter. Some of the important earlier work on estimation of the parameters of time series models will be found in references [24], [26], [87], [98], [100], [127], [134], [135], [142], [162], [163], and [197].

7.1.2 Conditional Likelihood for an ARIMA Process

Let us suppose that the $N = n + d$ original observations $\mathbf{z}$ form a time series which we denote by $z_{-d+1}, \ldots, z_0, z_1, z_2, \ldots, z_n$. We assume that this series is generated by an ARIMA model of order (p, d, q). From these observations we can generate a series $\mathbf{w}$ of $n = N - d$ differences $w_1, w_2, \ldots, w_n$, where $w_t = \nabla^d z_t$. Thus the general problem of fitting the parameters $\boldsymbol{\phi}$ and $\boldsymbol{\theta}$ of the ARIMA model (6.1.1) is equivalent to fitting to the w's, the stationary, invertible* ARMA(p, q) model, which may be written

$$a_t = \tilde{w}_t - \phi_1\tilde{w}_{t-1} - \phi_2\tilde{w}_{t-2} - \cdots - \phi_p\tilde{w}_{t-p} + \theta_1 a_{t-1} + \theta_2 a_{t-2} + \cdots + \theta_p a_{t-q} \tag{7.1.1}$$

where $w_t = \nabla^d z_t$ and $\tilde{w}_t = w_t - \mu$ with $E[w_t] = \mu$.

For $d > 0$ it would often be appropriate to assume that $\mu = 0$ (see the discussion in Sections 4.1.3, 6.2.3, 6.3.5, and 6.3.8). When this is not appropriate, we suppose that $\bar{w} = \sum_{t=1}^{n} w_t/n$ is substituted for μ. For the sample sizes normally considered in time series analysis, this approximation will be adequate. However, if desired, μ may be included as an additional parameter to be estimated. The procedures we describe may then be used for simultaneous estimation of μ, along with the other parameters.

The w's cannot be substituted immediately in (7.1.1) to calculate the a's, because of the difficulty of starting up the difference equation. However, suppose that the p values $\mathbf{w}_*$ of the w's and the q values $\mathbf{a}_*$ of the a's prior to the commencement of the w series were given. Then the values of $a_1, a_2, \ldots, a_n$, conditional on this choice, could be calculated in turn from (7.1.1).

Thus, for any given choice of parameters $(\boldsymbol{\phi}, \boldsymbol{\theta})$ and of the starting values $(\mathbf{w}_*, \mathbf{a}_*)$, we could calculate successively a set of values $a_t(\boldsymbol{\phi}, \boldsymbol{\theta}|\mathbf{w}_*, \mathbf{a}_*, \mathbf{w})$, $t = 1, 2, \ldots, n$. Now, assuming that the a's are normally distributed, their probability density is

$$p(a_1, a_2, \ldots, a_n) \propto \sigma_a^{-n} \exp\left[-\left(\sum_{t=1}^{n} \frac{a_t^2}{2\sigma_a^2}\right)\right]$$

Given a particular set of data $\mathbf{w}$, the log-likelihood associated with the parameter values $(\boldsymbol{\phi}, \boldsymbol{\theta}, \sigma_a)$, *conditional* on the choice of $(\mathbf{w}_*, \mathbf{a}_*)$, would then be

$$l_*(\boldsymbol{\phi}, \boldsymbol{\theta}, \sigma_a) = -n \ln(\sigma_a) - \frac{S_*(\boldsymbol{\phi}, \boldsymbol{\theta})}{2\sigma_a^2} \tag{7.1.2}$$

where, following the previous discussion, no additive constant term need be included, and

*Special care is needed to ensure that estimates lie in the invertible region. See Appendix A7.6.

$$S_*(\boldsymbol{\phi}, \boldsymbol{\theta}) = \sum_{t=1}^{n} a_t^2(\boldsymbol{\phi}, \boldsymbol{\theta}|\mathbf{w}_*, \mathbf{a}_*, \mathbf{w}) \tag{7.1.3}$$

In the equation above, a subscript asterisk is used on the likelihood and sum of squares functions to emphasize that they are conditional on the choice of the starting values. We notice that the conditional log-likelihood l_* involves the data only through the conditional *sum of squares function*. It follows that contours of l_* for any fixed value of σ_a in the space of $(\boldsymbol{\phi}, \boldsymbol{\theta}, \sigma_a)$ are contours of S_*, that these maximum likelihood estimates are the same as the least squares estimates, and that in general we can, on the Normal assumption, study the behavior of the conditional likelihood by studying the conditional sum of squares function. In particular for any fixed σ_a, l_* is a linear function of S_*. The parameter values obtained by minimizing the conditional sum of squares function $S_*(\boldsymbol{\phi}, \boldsymbol{\theta})$ will be called *conditional least squares estimates*.

7.1.3 Choice of Starting Values for Conditional Calculation

We shall shortly discuss the calculation of the unconditional likelihood, which, strictly, is what we need for parameter estimation. However, for some purposes, when n is moderate or large, a sufficient approximation to the unconditional likelihood is obtained by using the conditional likelihood with suitable values substituted for the elements of $\mathbf{w}_*$ and $\mathbf{a}_*$ in (7.1.3). One procedure is to set the elements of $\mathbf{w}_*$ and of $\mathbf{a}_*$ equal to their unconditional expectations. The unconditional expectations of the elements of $\mathbf{a}_*$ are zero, and if the model contains no deterministic part, and in particular if $\mu = 0$, the unconditional expectations of the elements of $\mathbf{w}_*$ will also be zero.* However, this approximation can be poor if some of the roots of $\phi(B) = 0$ are close to the boundary of the unit circle, so that the process approaches nonstationarity. In this case, the initial data value w_1 could deviate considerably from its unconditional expectation, and the introduction of starting values of this sort could introduce a large transient, which would be slow to die out. In fitting a model of order (p, d, q), a more reliable approximation procedure, and one we shall employ sometimes, is to use (7.1.1) to calculate the a's *from* a_{p+1} *onward*, setting previous a's equal to zero. Thus actually occurring values are used for the w's throughout.

Using this method, we can sum the squares of only $n - p = N - p - d$ values of a_t, but the slight loss of information will be unimportant for long series. In cases where there are no autoregressive terms the two procedures are equivalent. For seasonal series, discussed in Chapter 9, the conditional

*If the assumption $E[w_t] = \mu \neq 0$ is appropriate, we can substitute $\bar{w}$ for each of the elements of $\mathbf{w}_*$.

approximation is not very satisfactory and the unconditional calculation becomes even more necessary. Next we illustrate the recursive calculation of the conditional sum of squares S_* with a simple example.

Calculation of the conditional sum of squares for a (0, 1, 1) process. Series B has been identified tentatively in Table 6.4 as an IMA(0, 1, 1) process:

$$\nabla z_t = (1 - \theta B)a_t \qquad -1 < \theta < 1 \tag{7.1.4}$$

that is,

$$a_t = w_t + \theta a_{t-1}$$

where $w_t = \nabla z_t$ and $E[w_t] = 0$. It will be recalled that in Chapter 6 a preliminary moment estimate (Table 6.5) was obtained which suggested that for these data, θ was close to zero.

The calculation of the first few a's is set out in Table 7.1 for the particular parameter value $\theta = 0.5$. The a's are calculated recursively from $a_t = w_t + 0.5a_{t-1}$, to two-decimal accuracy. In accordance with the discussion above, to start up the process, a_0 is set equal to zero. This value is shown in italic type. Proceeding in this way we find that

$$S_*(0.5) = \sum_{t=1}^{368} a_t^2(\theta = 0.5|a_0 = 0) = 27{,}694$$

Using values from $\theta = -0.5$ to $\theta = +0.5$ in steps of 0.1, the values for the conditional sum of squares $S_*(\theta)$ (given that $a_0 = 0$) are shown in Table 7.2.

7.1.4 Unconditional Likelihood; Sum of Squares Function; Least Squares Estimates

It is shown in Appendix A7.3 that corresponding to the $N = n + d$ observations assumed to be generated by an ARIMA model, the unconditional log-likelihood is given by

$$l(\boldsymbol{\phi}, \boldsymbol{\theta}, \sigma_a) = f(\boldsymbol{\phi}, \boldsymbol{\theta}) - n \ln(\sigma_a) - \frac{S(\boldsymbol{\phi}, \boldsymbol{\theta})}{2\sigma_a^2} \tag{7.1.5}$$

where $f(\boldsymbol{\phi}, \boldsymbol{\theta})$ is a function of $\boldsymbol{\phi}$ and $\boldsymbol{\theta}$. The *unconditional sum of squares function* is given by

$$S(\boldsymbol{\phi}, \boldsymbol{\theta}) = \sum_{t=1}^{n} [a_t|\mathbf{w}, \boldsymbol{\phi}, \boldsymbol{\theta}]^2 + [\mathbf{e}_*]'\boldsymbol{\Omega}^{-1}[\mathbf{e}_*] \tag{7.1.6}$$

where $[a_t|\mathbf{w}, \boldsymbol{\phi}, \boldsymbol{\theta}] = E[a_t|\mathbf{w}, \boldsymbol{\phi}, \boldsymbol{\theta}]$ denotes the expectation of a_t conditional on $\mathbf{w}$, $\boldsymbol{\phi}$, and $\boldsymbol{\theta}$. When the meaning is clear from the context, we shall further

TABLE 7.1 Recursive Calculation of a's for First 10 Values of Series B, Using $\theta = 0.5$

t	z_t	$w_t = \nabla z_t$	$a_t = w_t + 0.5a_{t-1}$
0	460		0
1	457	−3	−3.00
2	452	−5	−6.50
3	459	7	3.75
4	462	3	4.88
5	459	−3	−0.56
6	463	4	3.72
7	479	16	17.86
8	493	14	22.93
9	490	−3	8.46

TABLE 7.2 Sum of Squares Functions for Model $\nabla z_t = (1 - \theta B)a_t$ Fitted to Series B

θ	$\lambda = (1 - \theta)$	$S_*(\theta)$	$S(\theta)$	θ	$\lambda = (1 - \theta)$	$S_*(\theta)$	$S(\theta)$
−0.5	1.5	23,929	23,928	0.1	0.9	19,896	19,896
−0.4	1.4	21,595	21,595	0.2	0.8	20,851	20,851
−0.3	1.3	20,222	20,222	0.3	0.7	22,315	22,314
−0.2	1.2	19,483	19,483	0.4	0.6	24,471	24,468
−0.1	1.1	19,220	19,220	0.5	0.5	27,694	27,691
0.0	1.0	19,363	19,363				

abbreviate this conditional expectation to $[a_t]$. In (7.1.6) we also have $\mathbf{e}_* = (w_{1-p}, \ldots, w_0, a_{1-q}, \ldots, a_0)'$ to denote the vector of the $p + q$ initial values of the w_t and a_t processes needed prior to time $t = 1$, $\mathbf{\Omega}\sigma_a^2 = \text{cov}(\mathbf{e}_*)$ is the covariance matrix of $\mathbf{e}_*$, and $[\mathbf{e}_*] = ([w_{1-p}], \ldots, [w_0], [a_{1-q}], \ldots, [a_0])'$ denotes the vector of conditional expectations ("back-forecasts") of the initial values given $\mathbf{w}$, $\boldsymbol{\phi}$, and $\boldsymbol{\theta}$. An alternative way to represent the sum of squares is as $S(\boldsymbol{\phi}, \boldsymbol{\theta}) = \sum_{t=-\infty}^{n} [a_t]^2$, which in comparison with (7.1.6) indicates that $\sum_{t=-\infty}^{0} [a_t]^2 = [\mathbf{e}_*]'\mathbf{\Omega}^{-1}[\mathbf{e}_*]$.

Usually, $f(\boldsymbol{\phi}, \boldsymbol{\theta})$ is of importance only for small n. For moderate and large values of n, (7.1.5) is dominated by $S(\boldsymbol{\phi}, \boldsymbol{\theta})/2\sigma_a^2$, and thus the contours of the unconditional sum of squares function in the space of the parameters $(\boldsymbol{\phi}, \boldsymbol{\theta})$ are very nearly contours of likelihood and of log-likelihood. It follows, in particular, that the parameter estimates obtained by minimizing the sum of squares (7.1.6), which we call (*unconditional* or *exact*) *least squares estimates*, will usually provide very close approximations to the maximum likelihood estimates. From a Bayesian viewpoint, on assumptions discussed in Section 7.4, for all AR(p) and MA(q), essentially the posterior density is a function only of $S(\boldsymbol{\phi}, \boldsymbol{\theta})$. Hence very nearly the least squares estimates are those with maximum posterior density. In the remainder of this section and in Section 7.1.5, our main emphasis will be on the calculation, study, and use of the unconditional sum of squares function $S(\boldsymbol{\phi}, \boldsymbol{\theta})$, as defined in (7.1.6), and on calculating least squares estimates.

In the calculation of the unconditional sum of squares, the $[a]$'s are computed recursively by taking conditional expectations in (7.1.1). A preliminary back-calculation provides the values $[w_{-j}]$ and $[a_{-j}], j = 0, 1, 2, \ldots$ (i.e., the back-forecasts) needed to start off the forward recursion.

Calculation of the unconditional sum of squares for a moving average process. For illustration, we reconsider the IBM stock price example, again using the first 10* values of the series given in Table 7.1. With $w_t = \nabla z_t$, we have seen in Section 6.2 that the model identified is of order (0, 1, 1). Hence, according to (7.1.6), the only back-forecast that is needed for $S(\theta)$ is $[a_0]$. We begin by describing an *approximate*, but nevertheless accurate method to obtain $[a_0]$ which makes use of the interesting feature of the backward model for w_t and which is instructive in its own right. Recall from Section 6.4.3 that the model for w_t may be written in either the forward or backward forms

$$w_t = (1 - \theta B)a_t \qquad w_t = (1 - \theta F)e_t$$

and where again $\mu = E[w_t]$ is assumed equal to zero. Hence we can write

$$[e_t] = [w_t] + \theta[e_{t+1}] \tag{7.1.7}$$

$$[a_t] = [w_t] + \theta[a_{t-1}] \tag{7.1.8}$$

where $[w_t] = w_t$ for $t = 1, 2, \ldots, n$ and is the back-forecast of w_t for $t \leq 0$. These are the two basic equations we need in the computations. A convenient format for the calculations is shown in Table 7.3. We begin by entering in the table what we know:

1. the data values $z_0, z_1, \ldots, z_9$, from which we can calculate the first differences $w_1, w_2, \ldots, w_9$;
2. the values $[e_0], [e_{-1}], \ldots$, which are zero, since $e_0, e_{-1}, \ldots$ are distributed independently of $\mathbf{w}$;
3. the values $[a_{-1}], [a_{-2}], \ldots$, which are zero, because for any MA(q) process, $a_{-q}, a_{-q-1}, \ldots$ are distributed independently of $\mathbf{w}$. However, note that in general $[a_0], [a_{-1}], \ldots, [a_{-q+1}]$ will be nonzero and are obtained by back-forecasting. Thus, in the present example, $[a_0]$ is so obtained.

*In practice, of course, useful parameter estimates could not be obtained from as few as 10 observations. We utilize this data subset merely to illustrate the calculations.

TABLE 7.3 Calculation of the [a]'s from the First 10 Values of Series B, Using $\theta = 0.5$

t	z_t	$[a_t]$	$0.5[a_{t-1}]$	$[w_t]$	$0.5[e_{t+1}]$	$[e_t]$	Second cycle $[e_t]$	u_t
−1	[458.4]	0	0	0	0	0	0	
0	460	1.6	0	1.6	−1.6	0	0	−2.1
1	457	−2.2	0.8	−3.0	−0.1	−3.1	−3.1	−4.1
2	452	−6.1	−1.1	−5.0	4.8	−0.2	−0.2	−2.3
3	459	3.9	−3.0	7.0	2.6	9.6	9.6	8.5
4	462	5.0	2.0	3.0	2.3	5.3	5.2	9.5
5	459	−0.5	2.5	−3.0	7.6	4.6	4.4	9.2
6	463	3.7	−0.2	4.0	11.1	15.1	14.9	19.4
7	479	17.9	1.9	16.0	6.2	22.2	21.7	31.4
8	493	22.9	9.0	14.0	−1.5	12.5	11.4	27.5
9	490	8.5	11.5	−3.0	*0*	−3.0	−5.1	8.5
							−4.2	

Beginning at the end of the series, (7.1.7) is now used to compute the $[e_t]$'s for $t = 9, 8, 7, \ldots, 1$. We start this backward process by making the same approximation as was described previously for calculation of the conditional sum of squares. In the present instance, this amounts to setting $[e_{10}] = 0$. In general, the effect of this approximation will be to introduce a transient into the system which, because $\phi(B)$ and $\theta(B)$ are stationary operators, will for series of moderate length, almost certainly be negligible by the time the beginning of the series is reached and thus will not affect the calculation of the a's. As we see later, if desired, the adequacy of this approximation can be checked in any given case by performing a second iterative cycle.

Thus to start the recursion in Table 7.3, in the row corresponding to $t = 9$, we enter a zero (shown in italic type) in the sixth column for the unknown value $0.5[e_{10}]$. Then, using (7.1.7), we obtain

$$[e_9] = [w_9] + 0.5[e_{10}]$$

$$= w_9 + 0 = -3$$

so $0.5[e_9] = -1.5$ can be entered in the line $t = 8$, which enables us to compute $[e_8]$, and so on. Finally, we obtain

$$[e_0] = [w_0] + \theta[e_1]$$

that is,

$$0 = [w_0] - 1.6$$

which gives $[w_0] = 1.6$, and thereafter $[w_{-h}] = 0$, $h = 1, 2, 3, \ldots$.

Using (7.1.8) with $t = 0$, we obtain

$$[a_0] = [w_0] + \theta[a_{-1}] = 1.6 + (0.5)(0) = 1.6$$

and we can then continue the forward calculations of the remaining $[a]$'s, leading to $S(0.5) = \sum_{t=0}^{9} [a_t|0.5, \mathbf{w}]^2 = 1016.406$. Comparison of the values of a_t given in Tables 7.1 and 7.3 shows that in this particular example, the transient introduced by the change in the starting value has little effect for $t > 5$.

In comparison to the approximate method described above, the exact method to compute $[a_0]$ and hence to compute $S(\theta)$ is, in fact, presented for the general case in Appendix A7.3. For the MA(1) model for w_t, this method involves first computing the values $a_t(a_0 = 0)$, which we abbreviate as a_t^0, by the conditional method discussed in Section 7.1.3, as

$$a_t^0 = w_t + \theta a_{t-1}^0 \qquad t = 1, 2, \ldots, n$$

using $a_0^0 = 0$ as the initial value. Then a backward recursion is performed to obtain $u_t = a_t^0 + \theta u_{t+1}$, beginning from $t = n$, down to $t = 0$, with $u_{n+1} = 0$ as the starting value. Finally, then, the exact back-forecast $[a_0]$ is given by $[a_0] = -u_0(1 - \theta^2)/(1 - \theta^{2(n+1)})$, and using this starting value the $[a_t]$ are computed from the forward recursion

$$[a_t] = w_t + \theta[a_{t-1}] \qquad t = 1, 2, \ldots, n$$

as in (7.1.8) and the exact sum of squares is available as $S(\theta) = \sum_{t=0}^{n} [a_t]^2$. For comparison, in the present illustration, using the conditional values a_t^0 from Table 7.1, we obtain the values of u_t by the backward recursion for $t = 9, 8, \ldots, 0$, and these are displayed in the final column of Table 7.3. Hence we obtain the exact back-forecast value of a_0 as $[a_0] = -u_0(1 - \theta^2)/(1 - \theta^{2(n+1)}) = 1.549$, which is very similar to the approximate value of 1.545 obtained by the backward model approach, and this small difference has essentially no effect on the calculation of the remaining values $[a_t]$.

In practice, a second iterative cycle of the approximate backward model calculations could be used but would almost never be needed. However, we illustrate the procedure for the impractically short series of nine observations for the MA(1) model. For this, we use the value for a_9 of $[a_9] = 8.47$, computed in the first iteration, in the start of a new iteration to calculate the forward forecast of w_{10} using (7.1.8) as $[w_{10}] = -\theta[a_9] = -4.23$. This value may then be substituted in the backward equations (7.1.7) to obtain first $[e_{10}] = [w_{10}] = -4.23$ (since $[e_{11}] = 0$ for the backward MA(1) model), and eventually to obtain a new back-forecast $[w_0]$, hence to a new $[a_0] = [w_0]$, and so to new values $[a_t]$. Results for the second cycle of the backward recursion for the $[e_t]$'s are given in the next-to-last column of Table 7.3. The second cycle of calculations eventually leads to $[a_0] = 1.549$, which is very close to the value obtained in the first cycle of the backward model calculations and is essentially identical to the exact value. Thus a further calculation of the $[a]$'s would yield the same result as before.

Further, as follows by the reversibility properties of the process, in general $S(\boldsymbol{\phi}, \boldsymbol{\theta}|\mathbf{w})$ can equally well be computed from the sum of squares of the $[e_t]$'s. Using this fact, we find from Table 7.3 that

$$S(0.5) = \sum_{t=1}^{10} [e_t]^2 = 1016.406$$

from the second cycle results for the MA(1) example, and this agrees with the value found from the $[a_t]$'s.

Proceeding in the foregoing manner with the entire series, using the exact method, we find for the unconditional sum of squares

$$S(0.5) = \sum_{t=0}^{368} [a_t|0.5, \mathbf{w}]^2 = 27{,}691$$

which for this particular example is very close to the conditional value $S_*(0.5) = 27{,}694$. The unconditional sum of squares $S(\theta)$, for values of θ between -0.5 and $+0.5$, are given in Table 7.2 and are very close to the conditional values $S_*(\theta)$ for this particular example.

7.1.5 General Procedure for Calculating the Unconditional Sum of Squares

In the example above, w_t was a first-order moving average process, with zero mean. It followed that all forecasts for lead times greater than 1 were zero and consequently that only one preliminary value (the back-forecast $[w_0] = 1.6$) was required to start the recursive calculations using the backward model approximate approach, and only one back-forecast $[a_0]$ in the exact approach. For a qth-order moving average process, q nonzero preliminary values $[w_0], [w_{-1}], \ldots, [w_{1-q}]$ would be needed, or equivalently, the q values $[a_0], [a_{-1}], \ldots, [a_{1-q}]$ in the exact approach, with $S(\boldsymbol{\theta}) = \sum_{t=1-q}^{n} [a_t]^2$. Special procedures, which we discuss in Section 7.3.1, are available for estimating parameters in autoregressive models. However, we show in Appendix A7.3 that the procedure described in this section can supply the unconditional sum of squares for any ARIMA model.

Specifically, suppose that the w_t's are assumed to be generated by the stationary forward model

$$\phi(B)\tilde{w}_t = \theta(B)a_t \tag{7.1.9}$$

where $\nabla^d z_t = w_t$ and $\tilde{w}_t = w_t - \mu$. Then they could equally well have been generated by the backward model

$$\phi(F)\tilde{w}_t = \theta(F)e_t \tag{7.1.10}$$

As before, in the approximate method that utilizes the backward model, we could first employ (7.1.10) to supply back-forecasts $[\tilde{w}_{-j}|\mathbf{w}, \boldsymbol{\phi}, \boldsymbol{\theta}]$. Theorctically, the presence of the autoregressive operator ensures a series of such estimates that is infinite in extent. However, because of the stationary character of this operator, in practice, the estimates $[\tilde{w}_t]$ at and beyond some point $t = -Q$, with Q of moderate size, become essentially equal to zero.

Thus, to a sufficient approximation we can write

$$\tilde{w}_t = \phi^{-1}(B)\theta(B)a_t = \sum_{j=0}^{\infty} \psi_j a_{t-j} \simeq \sum_{j=0}^{Q} \psi_j a_{t-j}$$

This means that the original mixed process could be replaced by a moving average process of order Q and the procedure for moving averages outlined in Section 7.1.4 may be used.

Thus, in general, the dual set of equations for generating the conditional expectations $[a_t|\boldsymbol{\phi}, \boldsymbol{\theta}, \mathbf{w}]$ is obtained by taking conditional expectations in (7.1.10) and (7.1.9), that is,

$$\phi(F)[\tilde{w}_t] = \theta(F)[e_t] \tag{7.1.11}$$

is first used to generate the backward forecasts and then

$$\phi(B)[\tilde{w}_t] = \theta(B)[a_t] \tag{7.1.12}$$

is used to generate the $[a_t]$'s. If we find that the forecasts are negligible in magnitude beyond some lead time Q, the recursive calculation goes forward with

$$\begin{aligned} [e_{-j}|\boldsymbol{\phi}, \boldsymbol{\theta}, \mathbf{w}] &= 0 \qquad j = 0, 1, 2, \ldots \\ [a_{-j}|\boldsymbol{\phi}, \boldsymbol{\theta}, \mathbf{w}] &= 0 \qquad j > Q - 1 \end{aligned} \tag{7.1.13}$$

and the sum of squares is approximated by $S(\boldsymbol{\phi}, \boldsymbol{\theta}) = \Sigma_{t=1-Q}^{n} [a_t]^2$. As mentioned earlier, a second iterative cycle in this approximation method could be used, if desired.

However, as shown in Appendix A7.3, for the general model (7.1.9) the exact method can be used to obtain the sum of squares as

$$S(\boldsymbol{\phi}, \boldsymbol{\theta}) = \sum_{t=1-q}^{n} [a_t]^2 + ([\mathbf{w}_*] - \mathbf{C}'[\mathbf{a}_*])'\mathbf{K}^{-1}([\mathbf{w}_*] - \mathbf{C}'[\mathbf{a}_*]) \tag{7.1.14}$$

Here the vectors $[\mathbf{w}_*]' = ([w_{1-p}], \ldots, [w_0])$ and $[\mathbf{a}_*]' = ([a_{1-q}], \ldots, [a_0])$ are the exact back-forecasted values obtained as in (A7.3.12). They are given by $[\mathbf{e}_*] = ([\mathbf{w}_*]', [\mathbf{a}_*]')' = \mathbf{D}^{-1}\mathbf{F}'\mathbf{u}$, where the values u_t, $t = 1, \ldots, n$, of the vector $\mathbf{u}$ are obtained through the backward recursion $u_t = a_t^0 + \theta_1 u_{t+1} + \cdots + \theta_q u_{t+q}$ with zero initial values $u_{n+1} = \cdots = u_{n+q} = 0$, and the a_t^0 are the conditional values of the a_t computed from (7.1.12) using zero initial values, $a_{1-q}^0 = \cdots = a_0^0 = 0$ and $w_{1-p}^0 = \cdots = w_0^0 = 0$. The exact $[a_t]$'s are then calculated through the similar recursion

$$[a_t] = [w_t] - \phi_1[w_{t-1}] - \cdots - \phi_p[w_{t-p}] + \theta_1[a_{t-1}] + \cdots + \theta_q[a_{t-q}] \tag{7.1.15}$$

for $t = 1, 2, \ldots, n$ using the exact back-forecasts as starting values, with $[w_t] = w_t$ for $1 \le t \le n$. The matrices **C**, **K**, **D**, and **F** necessary for the computation in (7.1.14) are defined explicitly in Appendix A7.3.

Calculation of the unconditional sum of squares for a mixed autoregressive–moving average process. For illustration, consider the following $n = 12$ successive values of $w_t = \nabla^d z_t$:

t	1	2	3	4	5	6	7	8	9	10	11	12
w_t	1.1	4.3	3.0	−0.7	1.6	3.2	0.3	−1.9	−0.3	−0.3	0.8	2.0

Suppose that we wish to compute the unconditional sum of squares $S(\phi, \theta)$ associated with the ARIMA(1, d, 1) process

$$(1 - \phi B)w_t = (1 - \theta B)a_t$$

with $\nabla^d z_t = w_t$, it being assumed that the w's have mean zero.

Of course, estimates based on 12 observations would be almost valueless, but nevertheless, this short series serves to explain the nature of the calculation. We illustrate with the parameter values $\phi = 0.3$, $\theta = 0.7$. Then (7.1.15) may be written

$$[a_t] = [w_t] - 0.3[w_{t-1}] + 0.7[a_{t-1}] \tag{7.1.16}$$

where $[w_t] = w_t$ $(t = 1, 2, \ldots, n)$.

The layout of the necessary calculations is shown in Table 7.4. The data are first entered in the center column and then the conditional initial

TABLE 7.4 Calculation of $[a]$'s and of $S(0.3, 0.7)$ from 12 Values of a Series Assumed to be Generated by the Process $(1 - 0.3B)w_t = (1 - 0.7B)a_t$

t	a_t^0	$0.7a_{t-1}^0$	$-0.3w_{t-1}^0$	$[w_t]$	u_t	$0.7u_{t+1}$	$0.7[a_{t-1}]$	$[a_t]$
0	0			−2.46				−2.84
1	1.10	0	0	1.1	9.53	8.43	−1.99	−0.15
2	4.74	0.77	−0.33	4.3	12.05	7.31	−0.10	3.87
3	5.03	3.32	−1.29	3.0	10.44	5.41	2.71	4.42
4	1.92	3.52	−0.90	−0.7	7.73	5.81	3.09	1.49
5	3.15	1.34	0.21	1.6	8.30	5.15	1.04	2.85
6	4.93	2.21	−0.48	3.2	7.36	2.43	2.00	4.72
7	2.79	3.45	−0.96	0.3	3.47	0.68	3.30	2.64
8	−0.04	1.95	−0.09	−1.9	0.98	1.01	1.85	−0.14
9	0.24	−0.03	0.57	−0.3	1.45	1.21	−0.10	0.17
10	−0.04	0.17	0.09	−0.3	1.72	1.76	0.12	−0.09
11	0.86	−0.03	0.09	0.8	2.52	1.65	−0.06	0.83
12	2.36	0.60	−0.24	2.0	2.36	0	0.58	2.34

values $w_0^0 = 0$, $a_0^0 = 0$ are entered. The forward equation (7.1.16) is now started off in precisely the manner described in Section 7.1.3 for the conditional calculation to obtain the a_t^0 as $a_t^0 = w_t - \phi w_{t-1}^0 + \theta a_{t-1}^0$, with $w_t^0 = w_t$ for $1 \le t \le n$. Thus we begin by substituting 0 in line $t = 1$ for the values of $-0.3w_{t-1}^0$ and $0.7a_{t-1}^0$, resulting in $a_1^0 = w_1$. The recursive calculation now goes ahead using (7.1.16) to obtain the a_t^0. Next, values $u_t = a_t^0 + 0.7u_{t+1}$ are obtained through backward recursion beginning with $u_{13} = 0$, so that the value 0 is first entered in line $t = 12$ for the value of $0.7u_{t+1}$. Then the exact back-forecasts $[\mathbf{e}_*]' = ([w_0], [a_0])$ are obtained by solving the equations $\mathbf{D}[\mathbf{e}_*] = \mathbf{F}'\mathbf{u} = [\phi, -\theta]'u_1$, which in this case are

$$\begin{bmatrix} 5.864 & -6.099 \\ -6.099 & 7.648 \end{bmatrix} \begin{bmatrix} [w_0] \\ [a_0] \end{bmatrix} = \begin{bmatrix} 0.3 \\ -0.7 \end{bmatrix} u_1$$

leading to $[w_0] = -0.258u_1 = -2.462$ and $[a_0] = -0.297u_1 = -2.836$. The final forward recursion is now begun using (7.1.16) with these back-forecasted starting values, and the $[a_t]$ are computed. In Table 7.4, the calculations are shown to two decimals.

The unconditional sum of squares $S(\phi, \theta)$ is obtained as indicated in (A7.3.14) of Appendix A7.3 [see, also, (A7.3.19) for the specific case of the ARMA(1, 1) model] and above. Thus, since $\mathbf{C} = E[a_0 w_0]/\sigma_a^2 = 1$ in (7.1.14),

$$S(0.3, 0.7) = \sum_{t=0}^{12} [a_t]^2 + \frac{([w_0] - [a_0])^2}{\mathbf{K}} = 89.158$$

with $\mathbf{K} = \sigma_a^{-2}\gamma_0 - 1 = (\phi - \theta)^2/(1 - \phi^2)$.

We saw that in the fitting of a process of order (0, 1, 1) to the IBM series B, the *conditional* sums of squares provided a very close approximation to the unconditional value. That this is not always so can be seen from the present example.

We mentioned in Section 7.1.3 two conditional sums of squares that might be used as approximations to the unconditional values. These were obtained (1) by starting off the recursion at the first available observation, setting all unknown a's to zero and all the w's equal to their unconditional expectations; (2) by starting off the recursion of the pth observation using only observed values for the w's and zeros for unknown a's. Two such conditional sums of squares and the unconditional sum of squares (3) obtained from the $[a_t]$'s are compared below.

1. Unknown w's and a's set equal to zero:

$$\sum_{t=1}^{12} (a_t | 0.3, 0.7, w_0 = 0, a_0 = 0, \mathbf{w})^2 = 101.044$$

2. Unknown a's set equal to zero:

$$\sum_{t=2}^{12} (a_t | 0.3, 0.7, w_1 = 1.1, a_1 = 0, \mathbf{w})^2 = 82.441$$

3. Unconditional calculation:

$$\sum_{t=1}^{12} [a_t | 0.3, 0.7, \mathbf{w}]^2 + \frac{([w_0] - [a_0])^2}{\mathbf{K}} = 89.158$$

The sum of squares in (1) is a very poor approximation to (3), although the discrepancy, which is over 10% in this series of 12 values, would be diluted if the series were longer. This is because the transient introduced by the choice of starting value has essentially died out after 12 values, as can be seen from the results in Table 7.4. On the other hand, when allowance is made for the fact that the conditional sum of squares in (2) is based on only 11 rather than 12 squared quantities, it does seem to represent a more satisfactory approximation than (1). For reasons already given, if a conditional approximation is to be used, it should normally be of the form (2) rather than (1). However, as is discussed further in Chapter 9, for seasonal series, the conditional approximation becomes less satisfactory and the unconditional sum of squares should ordinarily be computed.

Of course, ultimately, the interest centers on the parameter values that *minimize* the sum of squares functions (or *maximize* the likelihood function), and these least squares (or maximum likelihood) estimates could be close even if the approximate and exact sums of squares functions differ. Simulation studies have been performed by Dent and Min [80] and Ansley and Newbold [14] to empirically investigate and compare the performance of the conditional least squares, unconditional least squares, and maximum likelihood estimators for ARMA models. Generally, the conditional and unconditional least squares estimators serve as satisfactory approximations to the maximum likelihood estimator for large sample sizes. However, simulation evidence suggests a preference for the maximum likelihood estimator for small or moderate sample sizes, especially if the moving average operator has a root close to the boundary of the invertibility region. Some additional information on the relative performance of the different estimators was provided by Hillmer and Tiao [110] and Osborn [152], who examined the *expected values* of the conditional sum of squares, the unconditional sum of squares, and the log-likelihood for a MA(1) model, as functions of the unknown parameter θ, for different sample sizes n. These studies provide some idea as to how the corresponding estimators will behave for various sample sizes, and the results are consistent with those obtained from simulation studies.

7.1.6 Graphical Study of the Sum of Squares Function

The sum of squares function $S(\theta)$ for the IBM data given in Table 7.2 is plotted in Figure 7.1. The overall minimum sum of squares is at about $\theta = -0.09$ ($\lambda = 1.09$), which is the least squares estimate and (on the assumption of Normality) a close approximation to the *maximum likelihood estimate* of the parameter θ.

The graphical study of the sum of squares functions is readily extended to two parameters by evaluating the sum of squares over a suitable grid of parameter values and plotting contours. As we have discussed in Section 7.1.4, on the assumption of Normality, the contours are very nearly likelihood contours. For most practical purposes, rough contours drawn in by eye on the computer output grid are adequate. However, devices for automatic plotting from computer output are common. Using these, the plotting of contours from grid values is readily arranged. Figure 7.2 shows a grid of $S(\lambda_0, \lambda_1)$ values for series B fitted with the IMA(0, 2, 2) process

$$\begin{aligned}\nabla^2 z_t &= (1 - \theta_1 B - \theta_2 B^2)a_t \\ &= [1 - (2 - \lambda_0 - \lambda_1)B - (\lambda_0 - 1)B^2]a_t\end{aligned} \tag{7.1.17}$$

or in the form

$$\nabla^2 z_t = (\lambda_0 \nabla + \lambda_1)a_{t-1} + \nabla^2 a_t$$

The minimum sum of squares in Figure 7.2 is at about $\hat{\lambda}_0 = 1.09$, $\hat{\lambda}_1 = 0.0$.

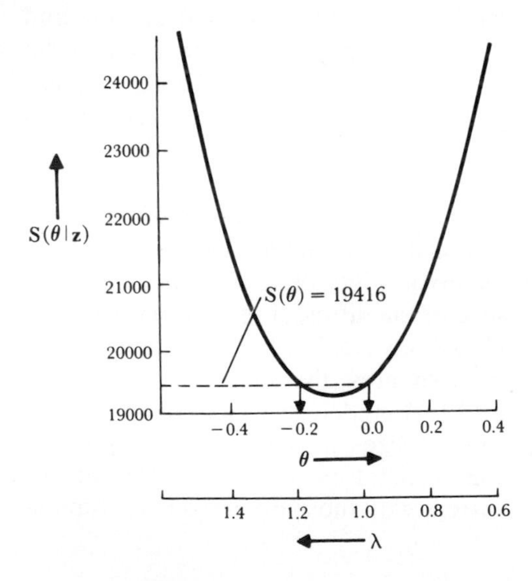

Figure 7.1 Plot of $S(\theta)$ for series B.

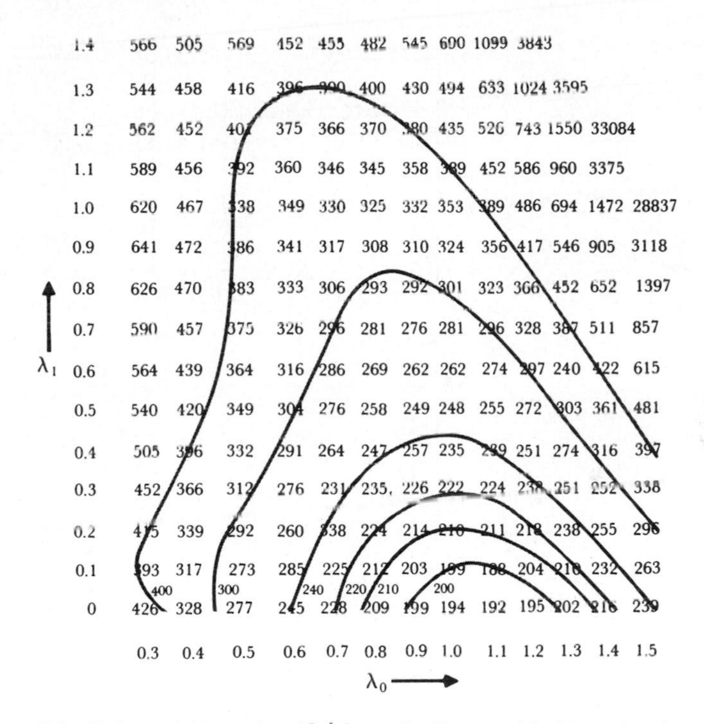

Figure 7.2 Values of $S(\lambda_0, \lambda_1) \times 10^{-2}$ for series B on a grid of (λ_0, λ_1) values and approximate contours.

The plot thus confirms that the preferred model in this case is an IMA(0, 1, 1) process. The device illustrated here, of fitting a model somewhat more elaborate than that expected to be needed, can provide a useful confirmation of the original identification. The elaboration of the model should be made, of course, in the direction "feared" to be necessary.

Three parameters. When we wish to study the joint estimation situation for three parameters, two-dimensional contour diagrams for a number of values of the third parameter can be drawn. For illustration, part of such a series of diagrams is shown in Figure 7.3 for series A, C, and D. In each case the "elaborated" model

$$\begin{aligned}\nabla^2 z_t &= (1 - \theta_1 B - \theta_2 B^2 - \theta_3 B^3) a_t \\ &= [1 - (2 - \lambda_{-1} - \lambda_0 - \lambda_1)B - (\lambda_0 + 2\lambda_{-1} - 1)B^2 + \lambda_{-1}B^3] a_t\end{aligned}$$

or

$$\nabla^2 z_t = (\lambda_{-1}\nabla^2 + \lambda_0 \nabla + \lambda_1) a_{t-1} + \nabla^2 a_t$$

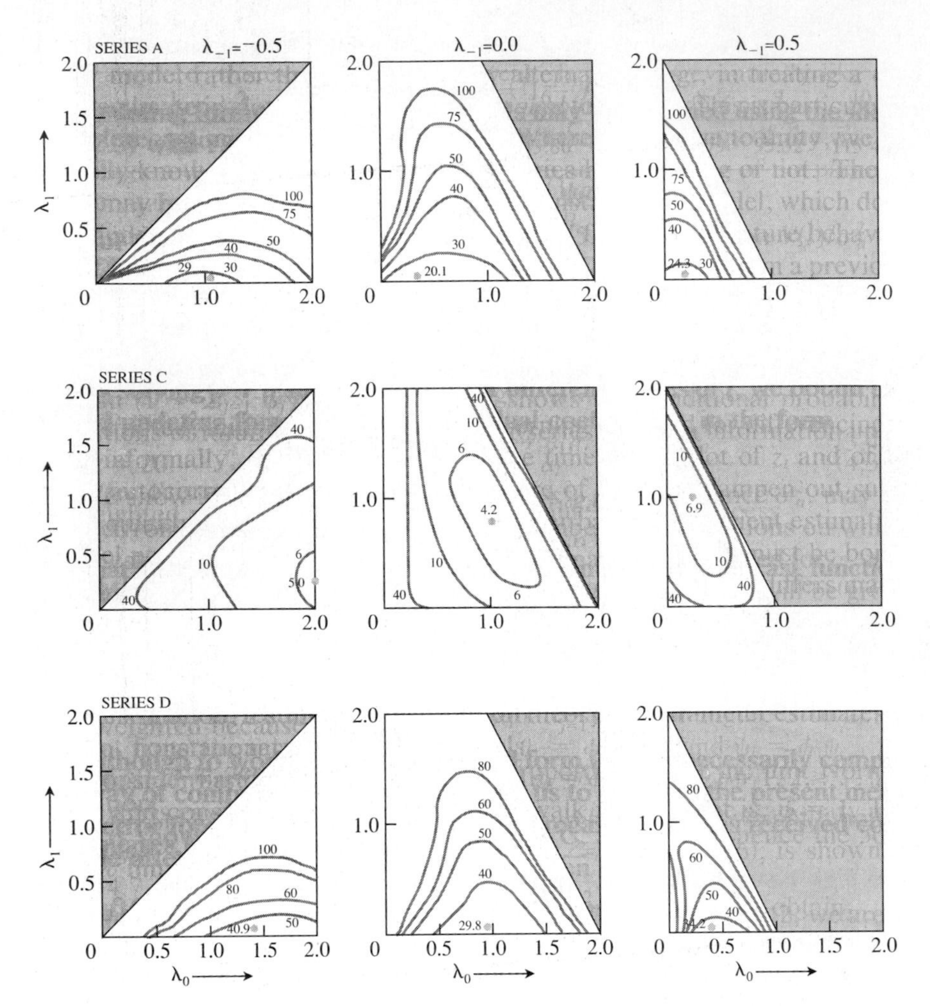

Figure 7.3 Sums of squares contours for series A, C, and D (shaded lines indicate boundaries of the invertibility regions).

has been fitted, leading to the conclusion that the best-fitting models of this type* are as shown in Table 7.5.

The inclusion of additional parameters (particularly λ_{-1}) in this fitting process is not strictly necessary, but we have included them to illustrate the effect of overfitting and to show how closely our identification seems to be confirmed for these series.

*We show later in Sections 7.2.5 that slightly better fits are obtained in some cases with closely related models containing "stationary" autoregressive terms.

TABLE 7.5 IMA Models Fitted to Series A, C, and D

Series	$\hat{\lambda}_{-1}$	$\hat{\lambda}_0$	$\hat{\lambda}_1$	Fitted Series
A	0	0.3	0	$\nabla z_t = 0.3a_{t-1} + \nabla a_t$
C	0	1.1	0.8	$\nabla^2 z_t = 1.1\nabla a_{t-1} + 0.8a_{t-1} + \nabla^2 a_t$
D	0	0.9	0	$\nabla z_t = 0.9a_{t-1} + \nabla a_t$

7.1.7 Description of "Well-Behaved" Estimation Situations; Confidence Regions

The likelihood function is not, of course, plotted merely to indicate maximum likelihood values. It is the entire course of this function that contains the totality of information coming from the data. In some fields of study, situations can occur where the likelihood function has two or more peaks (see, e.g., the example in [39]) and also where the likelihood function contains sharp ridges and spikes. All of these situations have logical interpretations. In each case, the likelihood function is trying to tell us something that we need to know. Thus the existence of two peaks of approximately equal heights implies that there are two sets of values of the parameters that might explain the data. The existence of obliquely oriented ridges means that a value of one parameter, considerably different from its maximum likelihood value, could explain the data if accompanied by a value of the other which deviated appropriately. Characteristics of this kind determine what may be called the *estimation situation*. To understand the estimation situation, we must examine the likelihood function both analytically and graphically.

Need for care in interpreting the likelihood function. Care is needed in interpreting the likelihood function. For example, results discussed later, which assume that the log-likelihood is approximately quadratic near its maximum, will clearly not apply to the three-parameter estimation situations depicted in Figure 7.3. However, these examples are exceptional because here we are deliberately *overfitting* the model. If the simpler model is justified, we should *expect* to find the likelihood function contours truncated near its maximum by a boundary in the higher-dimensional parameter space. However, quadratic approximations *could* be used if the simpler *identified* model rather than the overparameterized model was fitted.

Special care is needed when the maximum of the likelihood function may be on or near a boundary. Consider the situation shown in Figure 7.4 and suppose we know a priori that a parameter $\beta > \beta_0$. The maximum likelihood within the permissible range of β is at B, where $\beta = \beta_0$, not at A or at C. It will be noticed that the first derivative of the likelihood is in this case

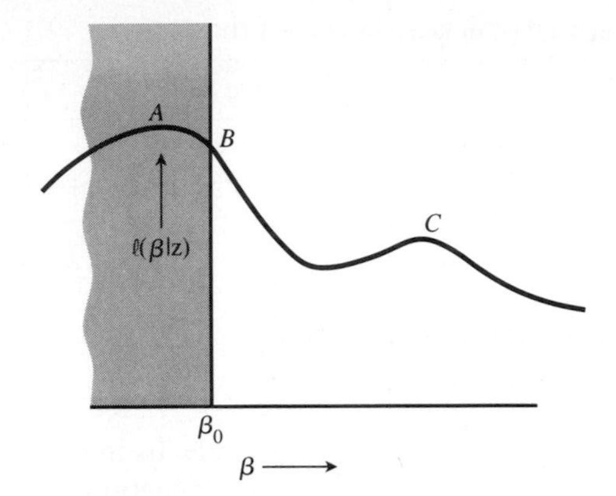

Figure 7.4 Hypothetical likelihood function with a constraint $\beta > \beta_0$.

nonzero at the maximum likelihood value and that the quadratic approximation is certainly not an adequate representation of the likelihood.

The treatment afforded the likelihood method has, in the past, sometimes left much to be desired, and ineptness by the practitioner has on occasion been mistaken for deficiency in the method. The treatment has often consisted of:

1. Differentiating the log-likelihood and setting first derivatives to zero to obtain the maximum likelihood (ML) estimates
2. Deriving approximate variances and covariances of these estimates from the second derivatives of the log-likelihood or from the expected values of the second derivatives

Mechanical application of the above can, of course, produce nonsensical answers. This is so, first, because of the elementary fact that setting derivatives to zero does not necessarily produce maxima, and second, because the information that the likelihood function contains is fully expressed by the ML estimates and by the second derivatives of the log-likelihood only if the quadratic approximation is adequate over the region of interest. To know whether this is so for a new estimation problem, a careful analytical and graphical investigation is usually required.

When a class of estimation problems (such as those arising from the estimation of parameters in ARMA models) is being investigated initially, it is important to plot the likelihood function rather extensively. After the behavior of a particular class of models is well understood, and knowledge of the situation indicates that it is safe to do so, we may take certain short cuts which we now consider. These results are described in greater detail in Appendices A7.3 and A7.4. We begin by considering expressions for the variances and covariances of maximum likelihood estimates, appropriate when the log-likelihood is approximately quadratic and the sample size is moderately large.

In what follows it is convenient to define a vector $\boldsymbol{\beta}$ whose $k = p + q$ elements are the autoregressive and moving average parameters $\boldsymbol{\phi}$ and $\boldsymbol{\theta}$. Thus the complete set of $p + q + 1 = k + 1$ parameters of the ARMA process may be written as $\boldsymbol{\phi}, \boldsymbol{\theta}, \sigma_a$; or as $\boldsymbol{\beta}, \sigma_a$; or simply as $\boldsymbol{\xi}$.

Variances and covariances of ML estimates. For the appropriately parameterized ARMA model, it will often happen that over the relevant* region of the parameter space, the log-likelihood will be approximately quadratic in the elements of $\boldsymbol{\beta}$ (i.e., of $\boldsymbol{\phi}$ and $\boldsymbol{\theta}$), so that

$$l(\boldsymbol{\xi}) = l(\boldsymbol{\beta}, \sigma_a) \simeq l(\hat{\boldsymbol{\beta}}, \sigma_a) + \frac{1}{2}\sum_{i=1}^{k}\sum_{j=1}^{k} l_{ij}(\beta_i - \hat{\beta}_i)(\beta_j - \hat{\beta}_j) \tag{7.1.18}$$

where, to the approximation considered, the derivatives

$$l_{ij} = \frac{\partial^2 l(\boldsymbol{\beta}, \sigma_a)}{\partial\beta_i\,\partial\beta_j} \tag{7.1.19}$$

are constant. For large n, the influence of the term $f(\boldsymbol{\phi}, \boldsymbol{\theta})$, or equivalently, $f(\boldsymbol{\beta})$ in (7.1.5) can be ignored in most cases. Hence $l(\boldsymbol{\beta}, \sigma_a)$ will be essentially quadratic in $\boldsymbol{\beta}$ if $S(\boldsymbol{\beta})$ is. Alternatively, $l(\boldsymbol{\beta}, \sigma_a)$ will be essentially quadratic in $\boldsymbol{\beta}$ if the conditional expectations $[a_t|\boldsymbol{\beta}, \mathbf{w}]$ in (7.1.6) are approximately locally linear in the element of $\boldsymbol{\beta}$.

For moderate and large samples, when the local quadratic approximation (7.1.18) is adequate, useful approximations to the variances and covariances of the estimates and approximate confidence regions may be obtained.

Information matrix for the parameters $\boldsymbol{\beta}$. The $(k \times k)$ matrix $-E[l_{ij}] = \mathbf{I}(\boldsymbol{\beta})$ is referred to [91], [196] as the *information matrix* for the parameters $\boldsymbol{\beta}$, where the expectation is taken over the distribution of $\mathbf{w}$. For a given value of σ_a, the *variance–covariance* matrix $\mathbf{V}(\hat{\boldsymbol{\beta}})$ for the ML estimates $\hat{\boldsymbol{\beta}}$ is, for large samples, given by the inverse of this information matrix, that is,

$$\mathbf{V}(\hat{\boldsymbol{\beta}}) \simeq \{-E[l_{ij}]\}^{-1} \tag{7.1.20}$$

For example, if $k = 2$, the large-sample variance–covariance matrix is

$$\mathbf{V}(\hat{\boldsymbol{\beta}}) = \begin{bmatrix} V(\hat{\beta}_1) & \text{cov}[\hat{\beta}_1, \hat{\beta}_2] \\ \text{cov}[\hat{\beta}_1, \hat{\beta}_2] & V(\hat{\beta}_2) \end{bmatrix} \simeq - \begin{bmatrix} E[l_{11}] & E[l_{12}] \\ E[l_{12}] & E[l_{22}] \end{bmatrix}^{-1}$$

In addition, the ML estimates $\hat{\boldsymbol{\beta}}$ obtained from a stationary ARMA process were shown to be asymptotically distributed as *multivariate Normal* with mean vector $\boldsymbol{\beta}$ and covariance matrix $\mathbf{I}^{-1}(\boldsymbol{\beta})$ (e.g., [142], [196], [215], [221]).

* Say over a 95% confidence region.

Now, using (7.1.5), we have

$$l_{ij} \simeq \frac{-S_{ij}}{2\sigma_a^2} \tag{7.1.21}$$

where

$$S_{ij} = \frac{\partial^2 S(\boldsymbol{\beta}|\mathbf{w})}{\partial \beta_i \, \partial \beta_j}$$

Furthermore, if for large samples, we approximate the expected values of l_{ij} or of S_{ij} by the values actually observed, then, using (7.1.20), we obtain

$$\mathbf{V}(\hat{\boldsymbol{\beta}}) \simeq \{-E[l_{ij}]\}^{-1} \simeq 2\sigma_a^2\{E[S_{ij}]\}^{-1} \simeq 2\sigma_a^2\{S_{ij}\}^{-1} \tag{7.1.22}$$

Thus, for $k = 2$,

$$\mathbf{V}(\hat{\boldsymbol{\beta}}) \simeq 2\sigma_a^2 \begin{bmatrix} \dfrac{\partial^2 S(\boldsymbol{\beta})}{\partial \beta_1^2} & \dfrac{\partial^2 S(\boldsymbol{\beta})}{\partial \beta_1 \, \partial \beta_2} \\ \dfrac{\partial^2 S(\boldsymbol{\beta})}{\partial \beta_1 \, \partial \beta_2} & \dfrac{\partial^2 S(\boldsymbol{\beta})}{\partial \beta_2^2} \end{bmatrix}^{-1}$$

If $S(\boldsymbol{\beta})$ were exactly quadratic in $\boldsymbol{\beta}$ over the relevant region of the parameter space, then, all the derivatives S_{ij} would be constant over this region. In practice, the S_{ij} will vary somewhat, and we shall usually suppose these derivatives to be determined at or near the point $\hat{\boldsymbol{\beta}}$. Now it is shown in the Appendices A7.3 and A7.4 that an estimate* of σ_a^2 is provided by

$$\hat{\sigma}_a^2 = \frac{S(\hat{\boldsymbol{\beta}})}{n} \tag{7.1.23}$$

and that for large samples, $\hat{\sigma}_a^2$ and $\hat{\boldsymbol{\beta}}$ are uncorrelated. Finally, the elements of (7.1.22) may be estimated from

$$\text{cov}[\hat{\beta}_i, \hat{\beta}_j] \simeq 2\hat{\sigma}_a^2 S^{ij} \tag{7.1.24}$$

where the matrix $\{S^{ij}\}$ is given by

$$\{S^{ij}\} = \{S_{ij}\}^{-1}$$

and the expression (7.1.24) is understood to define the variance $V(\hat{\beta}_i)$ when $j = i$.

Approximate confidence regions for the parameters. In particular, these results allow us to obtain the approximate variances of our estimates. By taking the square root of these variances, we obtain approximate standard deviations, which are usually called the *standard errors* of the

*Arguments can be advanced for using the divisor $n - k = n - p - q$ rather than n in (7.1.23), but for moderate sample sizes, this modification makes little difference.

estimates. The standard error of an estimate $\hat{\beta}_i$ is denoted by $SE[\hat{\beta}_i]$. When we have to consider several parameters simultaneously, we need some means of judging the precision of the estimates *jointly*. One means of doing this is to determine a *confidence region*. It may be shown (see, e.g., [201]) that a $1 - \varepsilon$ confidence region has the property that if repeated samples of size n are imagined to be drawn from the same population and a confidence region constructed from each such sample, a proportion $1 - \varepsilon$ of these regions will include the true parameter point.

If, for given σ_a, $l(\boldsymbol{\beta}, \sigma_a)$ is approximately quadratic in $\boldsymbol{\beta}$ in the neighborhood of $\hat{\boldsymbol{\beta}}$, then using (7.1.20) (see also Appendix A7.1), an approximate $1 - \varepsilon$ confidence region will be defined by

$$-\sum_{i,j} E[l_{ij}](\beta_i - \hat{\beta}_i)(\beta_j - \hat{\beta}_j) < \chi^2_\varepsilon(k) \tag{7.1.25}$$

where $\chi^2_\varepsilon(k)$ is the significance point exceeded by a proportion ε of the χ^2 distribution, having k degrees of freedom.

Alternatively, using the approximation (7.1.22) and substituting the estimate of (7.1.23) for σ_a^2, the approximate confidence region is given by*

$$\sum_{i,j} S_{ij}(\beta_i - \hat{\beta}_i)(\beta_j - \hat{\beta}_j) < 2\hat{\sigma}_a^2\chi^2_\varepsilon(k) \tag{7.1.26}$$

However, for a quadratic $S(\boldsymbol{\beta})$ surface

$$S(\boldsymbol{\beta}) - S(\hat{\boldsymbol{\beta}}) = \frac{1}{2}\sum_{i,j} S_{ij}(\beta_i - \hat{\beta}_i)(\beta_j - \hat{\beta}_j) \tag{7.1.27}$$

Thus, using (7.1.23) and (7.1.26), we finally obtain the result that the approximate $1 - \varepsilon$ confidence region is bounded by the contour on the sum of squares surface, for which

$$S(\boldsymbol{\beta}) = S(\hat{\boldsymbol{\beta}})\left[1 + \frac{\chi^2_\varepsilon(k)}{n}\right] \tag{7.1.28}$$

Examples of the calculation of approximate confidence intervals and regions.

1. *Series B.* For series B, values of $S(\lambda)$ and of its differences are shown in Table 7.6. The second difference of $S(\lambda)$ is not constant and thus $S(\lambda)$ is not strictly quadratic. However, in the range from $\lambda = 0.85$ to $\lambda = 1.35$, $\nabla^2(S)$ does not change greatly, so that (7.1.28) can be expected to provide a reasonably close approximation. With a minimum value $S(\hat{\lambda}) =$

*A somewhat closer approximation based on the F distribution, which takes account of the sampling distribution of $\hat{\sigma}_a^2$, may be employed. For moderate sample sizes this refinement makes little practical difference.

TABLE 7.6 $S(\lambda)$ and Its First and Second Differences for Various Values of λ for Series B

$\lambda = 1 - \theta$	$S(\lambda)$	$\nabla(S)$	$\Delta^2(S)$
1.5	23,928	2,333	960
1.4	21,595	1,373	634
1.3	20,222	739	476
1.2	19,483	263	406
1.1	19,220	−143	390
1.0	19,363	−533	422
0.9	19,896	−955	508
0.8	20,851	−1,463	691
0.7	22,314	−2,154	1069
0.6	24,468	−3,223	
0.5	27,691		

19,216, the critical value $S(\lambda)$, defining an approximate 95% confidence interval, is then given by

$$S(\lambda) = 19{,}216\left(1 + \frac{3.84}{368}\right) = 19{,}416$$

Reading off the values of λ corresponding to $S(\lambda) = 19{,}416$ in Figure 7.1, we obtain an approximate confidence interval $0.98 < \lambda < 1.19$.

Alternatively, we can employ (7.1.26). Using the second difference at $\lambda = 1.1$, given in Table 7.6, to approximate the derivative, we obtain

$$S_{11} = \frac{\partial^2 S}{\partial \lambda^2} \simeq \frac{390}{(0.1)^2}$$

Also, using (7.1.23), $\hat{\sigma}_a^2 = 19{,}216/368 = 52.2$. Thus the 95% confidence interval, defined by (7.1.26), is

$$\frac{390}{(0.1)^2}(\lambda - 1.09)^2 < 2 \times 52.2 \times 3.84$$

that is,

$$|\lambda - 1.09| < 0.10$$

Thus the interval is $0.99 < \lambda < 1.19$, which agrees closely with the previous calculation.

In this example, where there is only a single parameter λ, the use of (7.1.25) and (7.1.26) is equivalent to using an interval

$$\hat{\lambda} \pm u_{\varepsilon/2}\hat{\sigma}(\hat{\lambda})$$

where $u_{\varepsilon/2}$ is the deviate, which excludes a proportion $\varepsilon/2$ in the upper tail of the Normal distribution. An approximate standard error $\hat{\sigma}(\hat{\lambda}) = \sqrt{2\hat{\sigma}_a^2 S_{11}^{-1}}$ is obtained from (7.1.24). In the present example,

$$V(\hat{\lambda}) = 2\hat{\sigma}_a^2 S_{11}^{-1} = \frac{2 \times 52.2 \times 0.1^2}{390} = 0.00268$$

and the approximate standard error of $\hat{\lambda}$ is

$$\hat{\sigma}(\hat{\lambda}) = \sqrt{\text{var}[\hat{\lambda}]} = 0.052$$

Thus the approximate confidence interval is

$$\hat{\lambda} \pm 1.96\hat{\sigma}(\hat{\lambda}) = 1.09 \pm 0.10$$

as before.

Finally, we show later in Section 7.2 that it is possible to evaluate (7.1.20) analytically, for large samples from a MA(1) process, yielding

$$V(\hat{\lambda}) \simeq \frac{\lambda(2 - \lambda)}{n}$$

For the present example, substituting $\hat{\lambda} = 1.09$ for λ, we find that

$$V(\hat{\lambda}) \simeq 0.00269$$

which agrees closely with the previous estimate and so yields the same standard error of 0.052 and the same confidence interval.

2. *Series C.* In the identification of series C, one model that was entertained was a (0, 2, 2) process. To illustrate the application of (7.1.28) for more than one parameter, Figure 7.5 shows an approximate 95% confidence region (shaded) for λ_0 and λ_1 of series C. For this example $S(\hat{\lambda}) = 4.20$, $n = 224$, and $\chi^2_{0.05}(2) = 5.99$, so that the approximate 95% confidence region is bounded by the contour for which

$$S(\lambda_0, \lambda_1) = 4.20\left(1 + \frac{5.99}{224}\right) = 4.31$$

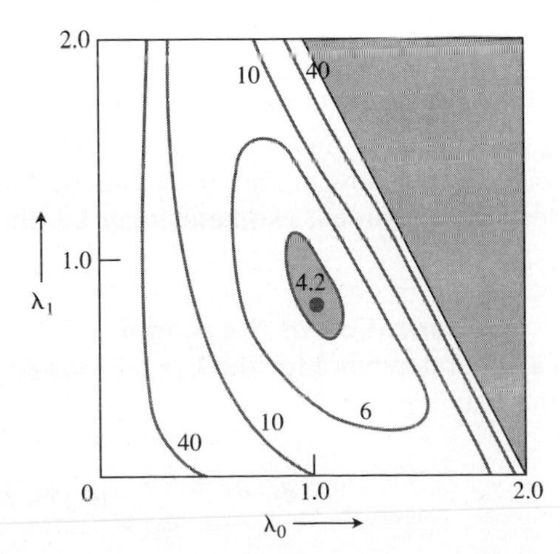

Figure 7.5 Sum of squares contours with shaded approximate 95% confidence region for series C, assuming a model of order (0, 2, 2).

7.2 NONLINEAR ESTIMATION

7.2.1 General Method of Approach

The plotting of the sum of squares function is of particular importance in the study of new estimation problems, because it ensures that any peculiarities in the estimation situation are shown up. When we are satisfied that anomalies are unlikely, other methods may be used.

We have seen that for most situations, the maximum likelihood estimates are closely approximated by the least squares estimates, which make

$$S(\boldsymbol{\phi}, \boldsymbol{\theta}) = \sum_{t=1}^{n} [a_t]^2 + [\mathbf{e}_*]'\boldsymbol{\Omega}^{-1}[\mathbf{e}_*]$$

a minimum, and in practice, this function could be approximated by a finite sum $\sum_{t=1-Q}^{n} [a_t]^2$ using the backward model approach.

In general, considerable simplification occurs in the minimization with respect to $\boldsymbol{\beta}$, of a sum of squares $\sum_{t=1}^{n} [f_t(\boldsymbol{\beta})]^2$, if each $f_t(\boldsymbol{\beta})$ $(t = 1, 2, \ldots, n)$ is a *linear* function of the parameters $\boldsymbol{\beta}$. We now show that the linearity status of $[a_t]$ is somewhat different in relation to the autoregressive parameters $\boldsymbol{\phi}$ and to the moving average parameters $\boldsymbol{\theta}$.

For the purely autoregressive process, $[a_t] = \phi(B)[\tilde{w}_t]$ and

$$\frac{\partial[a_t]}{\partial \phi_i} = -[\tilde{w}_{t-i}] + \phi(B)\frac{\partial[\tilde{w}_t]}{\partial \phi_i}$$

Now for $u > 0$, $[\tilde{w}_u] = \tilde{w}_u$ and $\partial[\tilde{w}_u]/\partial\phi_i = 0$, while for $u \leqslant 0$, $[\tilde{w}_u]$ and $\partial[\tilde{w}_u]/\partial\phi_i$ are both functions of $\boldsymbol{\phi}$. Thus, except for the effect of "starting values," $[a_t]$ is linear in the ϕ's. By contrast, for the pure moving average process,

$$[a_t] = \theta^{-1}(B)[\tilde{w}_t] \qquad \frac{\partial[a_t]}{\partial \theta_j} = \theta^{-2}(B)[\tilde{w}_{t-j}] + \theta^{-1}(B)\frac{\partial[\tilde{w}_t]}{\partial \theta_j}$$

so that the $[a_t]$'s are always nonlinear functions of the parameters.

We shall see in Section 7.3 that special simplifications occur in obtaining least squares and maximum likelihood estimates for the autoregressive process. We show in the present section how, by the iterative application of linear least squares, estimates may be obtained for any ARMA model.

Linearization of the model. In what follows, we continue to use $\boldsymbol{\beta}$ as a general symbol for the $k = p + q$ parameters $(\boldsymbol{\phi}, \boldsymbol{\theta})$. We need, then, to minimize

$$S(\boldsymbol{\phi}, \boldsymbol{\theta}) \simeq \sum_{t=1-Q}^{n} [a_t|\tilde{\mathbf{w}}, \boldsymbol{\beta}]^2 = \sum_{t=1-Q}^{n} [a_t]^2$$

Expanding $[a_t]$ in a Taylor series about its value corresponding to some guessed set of parameter values $\boldsymbol{\beta}_0' = (\beta_{1,0}, \beta_{2,0}, \ldots, \beta_{k,0})$, we have approximately

$$[a_t] = [a_{t,0}] - \sum_{i=1}^{k} (\beta_i - \beta_{i,0})x_{t,i} \tag{7.2.1}$$

where

$$[a_{t,0}] = [a_t|\mathbf{w}, \boldsymbol{\beta}_0]$$

and

$$x_{t,i} = -\left.\frac{\partial[a_t]}{\partial\beta_i}\right|_{\boldsymbol{\beta}=\boldsymbol{\beta}_0}$$

Now, if $\mathbf{X}$ is the $(n + Q) \times k$ matrix $\{x_{t,i}\}$, the $n + Q$ equations (7.2.1) may be expressed as

$$[\mathbf{a}_0] = \mathbf{X}(\boldsymbol{\beta} - \boldsymbol{\beta}_0) + [\mathbf{a}]$$

where $[\mathbf{a}_0]$ and $[\mathbf{a}]$ are column vectors with $n + Q$ elements.

The adjustments $\boldsymbol{\beta} - \boldsymbol{\beta}_0$, which minimize $S(\boldsymbol{\beta}) = S(\boldsymbol{\phi}, \boldsymbol{\theta}) = [\mathbf{a}]'[\mathbf{a}]$, may now be obtained by linear least squares, that is by "regressing" the $[a_0]$'s onto the x's. Because the $[a_t]$'s will not be exactly linear in the parameters $\boldsymbol{\beta}$, a single adjustment will not immediately produce least squares values. Instead, the adjusted values are substituted as new guesses and the process is repeated until convergence occurs. Convergence is faster if reasonably good guesses, such as may be obtained at the identification stage, are used initially. If sufficiently bad initial guesses are used, the process may not converge at all.

7.2.2 Numerical Estimates of the Derivatives

The derivatives $x_{t,i}$ may be obtained directly, as we illustrate later. However, for machine computation, a general nonlinear least squares routine has been found very satisfactory, in which the derivatives are obtained numerically. This is done by perturbing the parameters "one at a time." Thus, for a given model, the values $[a_t|\mathbf{w}, \beta_{1,0}, \beta_{2,0}, \ldots, \beta_{k,0}]$ for $t = 1 - Q, \ldots, n$ are calculated recursively, using whatever preliminary "backforecasts" may be needed. The calculation is then repeated for $[a_t|\mathbf{w}, \beta_{1,0} + \delta_1, \beta_{2,0}, \ldots, \beta_{k,0}]$, then for $[a_t|\mathbf{w}, \beta_{1,0}, \beta_{2,0} + \delta_2, \ldots, \beta_{k,0}]$, and so on. The negative of the required derivative is then given to sufficient accuracy using

$$x_{t,i} = \frac{[a_t|\mathbf{w}, \beta_{1,0}, \ldots, \beta_{i,0}, \ldots, \beta_{k,0}] - [a_t|\mathbf{w}, \beta_{1,0}, \ldots, \beta_{i,0} + \delta_i, \ldots, \beta_{k,0}]}{\delta_i} \tag{7.2.2}$$

The numerical method for obtaining derivatives described above has the advantage of universal applicability and requires us to program the calculation of the $[a_t]$'s only, not their derivatives. General nonlinear estimation routines, which essentially require only input instructions on how to compute the $[a_t]$'s, are now generally available [33]. In some versions it is necessary to choose the δ's in advance. In others, the program itself carries through a preliminary iteration to find suitable δ's. Some programs include special features to avoid overshoot and to speed up convergence [143].

Provided that the least squares solution is not on or near a constraining boundary, the value of $\mathbf{X} = \mathbf{X}_{\hat{\beta}}$ from the final iteration may be used to compute approximate variances, covariances, and confidence intervals. Thus $(\mathbf{X}_\beta'\mathbf{X}_\beta)^{-1}\sigma_a^2$ will approximate the variance–covariance matrix of the $\hat{\beta}$'s, and σ_a^2 will be estimated by $S(\hat{\boldsymbol{\beta}})/n$.

Application to a (0, 1, 1) process. As a simple illustration, consider the fitting of series A to a (0, 1, 1) process

$$w_t = \nabla z_t = (1 - \theta B)a_t$$

with $\mu = E[w_t] = 0$. The beginning of a calculation is shown in Table 7.7 for the guessed value $\theta_0 = 0.5$. The back-forecasted values for $[a_0]$ were actually obtained by setting $[e_7] = 0$ and using the back recursion $[e_t] = [w_t] + \theta[e_{t+1}]$. Greater accuracy would be achieved by beginning the back recursion further on in the series. Values of $[a_t]$ obtained by successive use of $[a_t] = \theta[a_{t-1}] + [w_t]$, for $\theta = 0.50$ and for $\theta = 0.51$, are shown in the fourth and fifth columns, together with values x_t, for the negative of the derivative, obtained using (7.2.2). To obtain a first adjustment for θ, we compute

$$\theta - \theta_0 = \frac{\sum_{t=0}^{n} [a_{t,0}]x_t}{\sum_{t=0}^{n} x_t^2}$$

TABLE 7.7 Illustration of Numerical Calculation of Derivatives for Data from Series A

t	z_t	w_t	$[a_{t,0}] = [a_t \mid 0.50]$	$[a_t \mid 0.51]$	$x_t = 100([a_t \mid 0.50] - [a_t \mid 0.51])$
0	17.0		0.2453	0.2496	−0.43
1	16.6	−0.40	−0.2773	−0.2727	−0.46
2	16.3	−0.30	−0.4387	−0.4391	0.04
3	16.1	−0.20	−0.4193	−0.4239	0.46
4	17.1	1.00	0.7903	0.7838	0.65
5	16.9	−0.20	0.1952	0.1997	−0.45
6	16.8	−0.10	−0.0024	0.0019	−0.43
7	17.4	0.60	0.5988	0.6010	−0.22
8	17.1	−0.30	−0.0006	0.0065	−0.71
9	17.0	−0.10	−0.1003	−0.0967	−0.36
10	16.7	−0.30	−0.3502	−0.3493	−0.09

In this example, using the entire series of 197 observations, convergence was obtained after four iterations. The course of the calculation was as follows:

Iteration	0	1	2	3	4	5
θ	0.50	0.63	0.68	0.69	0.70	0.70

In general, values of $\boldsymbol{\phi}$ and $\boldsymbol{\theta}$ that minimize $S(\boldsymbol{\phi}, \boldsymbol{\theta})$ can normally be found by this method to any degree of accuracy required. The method is especially attractive because, by its use, we do not need to program derivatives, and other than in the calculation of the $[a_t]$'s, no special provision need be made for end effects.

We now show that it is also possible to obtain derivatives directly, but additional recursive calculations are needed.

7.2.3 Direct Evaluation of the Derivatives

To illustrate the method, it will be sufficient to consider an ARMA(1, 1) process, which can be written in either of the forms

$$e_t = w_t - \phi w_{t+1} + \theta e_{t+1}$$

$$a_t = w_t - \phi w_{t-1} + \theta a_{t-1}$$

We have seen in Section 7.1.4 how the two versions of the model may be used in alternation, one providing initial values with which to start off a recursion with the other. We assume that a first computation has already been made yielding values of $[e_t]$, of $[a_t]$, and of $[w_0], [w_{-1}], \ldots, [w_{1-Q}]$, as in Section 7.1.5, and that $[w_{-Q}], [w_{-Q-1}], \ldots$ and hence $[a_{-Q}], [a_{-Q-1}], \ldots$ are negligible. We now show that a similar dual calculation may be used in calculating derivatives.

Using the notation $a_t^{(\phi)}$ to denote the partial derivative $\partial[a_t]/\partial\phi$, we obtain

$$e_t^{(\phi)} = w_t^{(\phi)} - \phi w_{t+1}^{(\phi)} + \theta e_{t+1}^{(\phi)} - [w_{t+1}] \tag{7.2.3}$$

$$a_t^{(\phi)} = w_t^{(\phi)} - \phi w_{t-1}^{(\phi)} + \theta a_{t-1}^{(\phi)} - [w_{t-1}] \tag{7.2.4}$$

$$e_t^{(\theta)} = w_t^{(\theta)} - \phi w_{t+1}^{(\theta)} + \theta e_{t+1}^{(\theta)} + [e_{t+1}] \tag{7.2.5}$$

$$a_t^{(\theta)} = w_t^{(\theta)} - \phi w_{t-1}^{(\theta)} + \theta a_{t-1}^{(\theta)} + [a_{t-1}] \tag{7.2.6}$$

Now

$$\left.\begin{aligned} [w_t] &= w_t \\ w_t^{(\phi)} &= w_t^{(\theta)} = 0 \end{aligned}\right\} \qquad t = 1, 2, \ldots, n \tag{7.2.7}$$

and

$$[e_{-j}] = 0 \qquad j = 0, 1, \ldots, n \tag{7.2.8}$$

Consider equations (7.2.3) and (7.2.4). By setting $e_{n+1}^{(\phi)} = 0$ in (7.2.3), we can begin a back recursion, which using (7.2.7) and (7.2.8) eventually allows us to compute $w_{-j}^{(\phi)}$ for $j = 0, 1, \ldots, Q-1$. Since $a_{-Q}^{(\phi)}, a_{-Q-1}^{(\phi)}, \ldots$ can be taken to be zero, we can now use (7.2.4) to compute recursively the required derivatives $a_t^{(\phi)}$. In a similar way, (7.2.5) and (7.2.6) can be used to calculate the derivatives $a_t^{(\theta)}$.

To illustrate, consider again the calculation of $x_t = -\partial[a_t]/\partial\theta$ for the first part of series A, performed wholly numerically in Table 7.7. Table 7.8 shows the corresponding calculations using

$$e_t^{(\theta)} = w_t^{(\theta)} + \theta e_{t+1}^{(\theta)} + [e_{t+1}]$$

$$-x_t = a_t^{(\theta)} = w_t^{(\theta)} + \theta a_{t-1}^{(\theta)} + [a_{t-1}]$$

The values of $[a_t]$ and $[e_t]$, which have already been computed, are first entered, and in the illustration the calculation of $e_t^{(\theta)}$ is begun by setting $e_6^{(\theta)} = 0$. It will be seen that the values for x_t agree very closely with those set out in Table 7.7, obtained by the purely numerical procedure.

7.2.4 General Least Squares Algorithm for the Conditional Model

An approximation we have sometimes used with long series is to set starting values for the a's, and hence for the x's, to their unconditional expectations of zero and then to proceed directly with the forward recursions. Thus for the previous example we could employ the equations

$$a_t = \theta a_{t-1} + w_t \qquad x_t = -a_t^{(\theta)} = \theta x_{t-1} - a_{t-1}$$

The effect is to introduce a transient into both the a_t and the x_t series, the latter being slower to die out since the x's depend on the a's. As one illustration, the values for the a's and the x's, obtained when this method was used

TABLE 7.8 Illustration of Recursive Calculation of Derivatives for Data from Series A

t	$[a_{t-1}]$	$\theta a_{t-1}^{(\theta)}$	$x_t = -a_t^{(\theta)}$	$e_t^{(\theta)}$	$\theta e_{t+1}^{(\theta)}$	$[e_{t+1}]$
0			−0.43	$(-w_0^{(\theta)} = -0.43)$	0.06	−0.49
1	0.25	0.22	−0.47	0.12	0.30	−0.18
2	−0.28	0.24	0.04	0.60	0.36	0.24
3	−0.44	−0.02	0.46	0.73	−0.15	0.88
4	−0.42	−0.23	0.65	−0.30	−0.05	−0.25
5	0.79	−0.33	−0.46	−0.10	*0*	−0.10
6	0.20	0.23	−0.43	*0*		
7	0.00	0.22	−0.22			
8	0.60	0.11	−0.71			
9	0.00	0.35	−0.35			
10	−0.10	−0.18	−0.08			

in the present example, were calculated. It was found that although not agreeing initially, the a_t's were in two-decimal agreement from $t = 4$ onward, and the x_t's from $t = 8$ onward. In some instances, where there is an abundance of data (say, 200 or more observations), the effect of the approximation can be nullified at the expense of some loss of information, by discarding, say, the first 10 calculated values.

If we adopt the approximation, an interesting general algorithm for this conditional model results. The general model may be written

$$a_t = \theta^{-1}(B)\phi(B)\tilde{w}_t$$

where $w_t = \nabla^d z_t$, $\tilde{w}_t = w_t - \mu$ and

$$\theta(B) = 1 - \theta_1 B - \cdots - \theta_i B^i - \cdots - \theta_q B^q$$

$$\phi(B) = 1 - \phi_1 B - \cdots - \phi_j B^j - \cdots - \phi_p B^p$$

If the first guesses for the parameters $\boldsymbol{\beta} = (\boldsymbol{\phi}, \boldsymbol{\theta})$ are $\boldsymbol{\beta}_0 = (\boldsymbol{\phi}_0, \boldsymbol{\theta}_0)$, then

$$a_{t,0} = \theta_0^{-1}(B)\phi_0(B)\tilde{w}_t$$

and

$$-\left.\frac{\partial a_t}{\partial \phi_j}\right|_{\beta_0} = u_{t,j} = u_{t-j} \qquad -\left.\frac{\partial a_t}{\partial \theta_i}\right|_{\beta_0} = v_{t,i} = v_{t-i}$$

where

$$u_t = \theta_0^{-1}(B)\tilde{w}_t \qquad = \phi_0^{-1}(B)a_{t,0} \tag{7.2.9}$$

$$v_t = -\theta_0^{-2}(B)\phi_0(B)\tilde{w}_t = -\theta_0^{-1}(B)a_{t,0} \tag{7.2.10}$$

The a's, u's, and v's may be calculated recursively, with starting values for a's, u's, and v's set equal to zero, as follows:

$$a_{t,0} = \tilde{w}_t - \phi_{1,0}\tilde{w}_{t-1} - \cdots - \phi_{p,0}\tilde{w}_{t-p} + \theta_{1,0}a_{t-1,0} + \cdots + \theta_{q,0}a_{t-q,0} \tag{7.2.11}$$

$$u_t = \theta_{1,0}u_{t-1} + \cdots + \theta_{q,0}u_{t-q} + \tilde{w}_t \tag{7.2.12}$$

$$= \phi_{1,0}u_{t-1} + \cdots + \phi_{p,0}u_{t-p} + a_{t,0} \tag{7.2.13}$$

$$v_t = \theta_{1,0}v_{t-1} + \cdots + \theta_{q,0}v_{t-q} - a_{t,0} \tag{7.2.14}$$

Corresponding to (7.2.1), the approximate linear regression equation becomes

$$a_{t,0} = \sum_{j=1}^{p}(\phi_j - \phi_{j,0})u_{t-j} + \sum_{i=1}^{q}(\theta_i - \theta_{i,0})v_{t-i} + a_t \tag{7.2.15}$$

The adjustments are then the regression coefficients of $a_{t,0}$ on the u_{t-j} and the v_{t-i}. By adding the adjustments to the first guesses $(\boldsymbol{\phi}_0, \boldsymbol{\theta}_0)$, a set of "second guesses" are formed and these now take the place of $(\boldsymbol{\phi}_0, \boldsymbol{\theta}_0)$ in a second iteration, in which new values of $a_{t,0}$, u_t, and v_t are computed, until convergence eventually occurs.

Alternative form for the algorithm. The approximate linear expansion (7.2.15) can be written in the form

$$a_{t,0} = \sum_{j=1}^{p} (\phi_j - \phi_{j,0})B^j\phi_0^{-1}(B)a_{t,0} - \sum_{i=1}^{q} (\theta_i - \theta_{i,0})B^i\theta_0^{-1}(B)a_{t,0} + a_t$$

$$= - [\phi(B) - \phi_0(B)]\phi_0^{-1}(B)a_{t,0} + [\theta(B) - \theta_0(B)]\theta_0^{-1}(B)a_{t,0} + a_t$$

that is,

$$a_{t,0} = -\phi(B)[\phi_0^{-1}(B)a_{t,0}] + \theta(B)[\theta_0^{-1}(B)a_{t,0}] + a_t \qquad (7.2.16)$$

which presents the algorithm in an interesting form.

Application to an IMA(0, 2, 2) process. To illustrate the calculation with the conditional approximation, consider the estimation of least squares values $\hat{\theta}_1$, $\hat{\theta}_2$ for series C using the model of order (0, 2, 2):

$$w_t = (1 - \theta_1 B - \theta_2 B^2)a_t$$

with

$$w_t = \nabla^2 z_t$$

$$a_{t,0} = w_t + \theta_{1,0}a_{t-1,0} + \theta_{2,0}a_{t-2,0}$$

$$v_t = -a_{t,0} + \theta_{1,0}v_{t-1} + \theta_{2,0}v_{t-2}$$

The calculations for initial values $\theta_{1,0} = 0.1$ and $\theta_{2,0} = 0.1$ are set out in Table 7.9.

The first adjustments to $\theta_{1,0}$ and $\theta_{2,0}$ are then found by "regressing" $a_{t,0}$ on v_{t-1} and v_{t-2}, and the process is repeated until convergence occurs. The iteration proceeded as shown in Table 7.10, using starting values $\theta_{1,0} = 0.1$, $\theta_{2,0} = 0.1$.

TABLE 7.9 Nonlinear Estimation of θ_1 and θ_2 for an IMA(0, 2, 2) Process

t	z_t	∇z_t	$\nabla^2 z_t = w_t$	$a_{t,0}$	v_{t-1}	v_{t-2}
1	26.6			*0*	*0*	*0*
2	27.0	0.4		*0*	*0*	*0*
3	27.1	0.1	−0.3	−0.300	*0*	*0*
4	27.2	0.1	0.0	−0.030	0.300	*0*
5	27.3	0.1	0.0	−0.033	0.060	0.300
6	26.9	−0.4	−0.5	−0.533	0.069	0.060
7	26.4	−0.5	−0.1	−0.156	0.546	0.069
8	26.0	−0.4	0.1	0.039	0.218	0.546
9	25.8	−0.2	0.2	0.189	0.038	0.218

TABLE 7.10 Convergence of Iterates of θ_1 and θ_2

Iteration	θ_1	θ_2
0	0.1000	0.1000
1	0.1247	0.1055
2	0.1266	0.1126
3	0.1286	0.1141
4	0.1290	0.1149
5	0.1292	0.1151
6	0.1293	0.1152
7	0.1293	0.1153
8	0.1293	0.1153

7.2.5 Summary of Models Fitted to Series A to F

In Table 7.11 we summarize the models fitted by the iterative least squares procedure of Sections 7.2.1 and 7.2.2 to series A to F. The models fitted were identified in Chapter 6 and have been summarized in Table 6.4. We see from Table 7.11 that in the case of series A, C, and D, two possible

TABLE 7.11 Summary of Models Fitted to Series A to F[a]

Series	Number of Observations	Fitted Models	Residual Variance[b]
A	197	$z_t - \underset{(\pm 0.04)}{0.92} z_{t-1} = 1.45 + a_t - \underset{(\pm 0.08)}{0.58} a_{t-1}$	0.097
		$\nabla z_t = a_t - \underset{(\pm 0.05)}{0.70} a_{t-1}$	0.101
B	369	$\nabla z_t = a_t + \underset{(\pm 0.05)}{0.09} a_{t-1}$	52.2
C	226	$\nabla z_t - \underset{(\pm 0.04)}{0.82} \nabla z_{t-1} = a_t$	0.018
		$\nabla^2 z_t = a_t - \underset{(\pm 0.07)}{0.13} a_{t-1} - \underset{(\pm 0.07)}{0.12} a_{t-2}$	0.019
D	310	$z_t - \underset{(\pm 0.03)}{0.87} z_{t-1} = 1.17 + a_t$	0.090
		$\nabla z_t = a_t - \underset{(\pm 0.06)}{0.06} a_{t-1}$	0.096
E	100	$z_t = 14.35 + \underset{(\pm 0.07)}{1.42} z_{t-1} - \underset{(\pm 0.07)}{0.73} z_{t-2} + a_t$	228
		$z_t = 11.31 + \underset{(\pm 0.10)}{1.57} z_{t-1} - \underset{(\pm 0.15)}{1.02} z_{t-2} + \underset{(\pm 0.10)}{0.21} z_{t-3} + a_t$	218
F	70	$z_t = 58.87 - \underset{(\pm 0.12)}{0.34} z_{t-1} + \underset{(\pm 0.12)}{0.19} z_{t-2} + a_t$	113

[a] The values (±) under each estimate denote the standard errors of those estimates.

[b] Obtained from $S(\hat{\phi}, \hat{\theta})/n$.

models were identified and subsequently fitted. For series A and D the alternative models involve the use of a stationary autoregressive operator $(1 - \phi B)$ instead of a nonstationary operator $(1 - B)$. Examination of Table 7.11 shows that in both cases the autoregressive model results in a slightly smaller residual variance, although as has been pointed out, the models are very similar. Even though a slightly better fit is possible with a stationary model, the IMA(0, 1, 1) model might be preferable in these cases on the grounds that unlike the stationary model, it does not assume that the series has a fixed mean. This is especially important in predicting future values of the series. For if the level does change, a model with $d > 0$ will continue to track it, whereas a model for which $d = 0$ will be tied to a mean level which may have become out of date.

The limits under the coefficients in Table 7.11 represent the standard errors of the estimates obtained from the covariance matrix $(\mathbf{X}'_{\hat{\beta}}\mathbf{X}_{\hat{\beta}})^{-1}\hat{\sigma}_a^2$, as described in Section 7.2.1. Note that the estimate $\hat{\phi}_3$ in the AR(3) process, fitted to the sunspot series E, is 2.1 times its standard error, indicating that a marginally better fit is obtained by the third-order autoregressive process, as compared with the second-order autoregressive process. This is in agreement with a conclusion reached by Moran [146].

7.2.6 Large-Sample Information Matrices and Covariance Estimates

Denote by $\mathbf{X} = [\mathbf{U} : \mathbf{V}]$, the $n \times (p + q)$ matrix of the time lagged u's and v's defined in (7.2.13) and (7.2.14), when the elements of $\boldsymbol{\beta}_0$ are the *true* values of the parameters, for a sample size n sufficiently large for end effects to be ignored. Then the information matrix for $(\boldsymbol{\phi}, \boldsymbol{\theta})$ for the mixed ARMA model is

$$\mathbf{I}(\boldsymbol{\phi}, \boldsymbol{\theta}) = E\begin{bmatrix} \mathbf{U}'\mathbf{U} & \mathbf{U}'\mathbf{V} \\ \mathbf{V}'\mathbf{U} & \mathbf{V}'\mathbf{V} \end{bmatrix} \boldsymbol{\sigma}_a^{-2} = E(\mathbf{X}'\mathbf{X})\sigma_a^{-2} \tag{7.2.17}$$

that is,

$$= n\sigma_a^{-2}\left[\begin{array}{cccc|cccc} \gamma_{uu}(0) & \gamma_{uu}(1) & \cdots & \gamma_{uu}(p-1) & \gamma_{uv}(0) & \gamma_{uv}(-1) & \cdots & \gamma_{uv}(1-q) \\ \gamma_{uu}(1) & \gamma_{uu}(0) & \cdots & \gamma_{uu}(p-2) & \gamma_{uv}(1) & \gamma_{uv}(0) & \cdots & \gamma_{uv}(2-q) \\ & \vdots & & \vdots & & \vdots & & \vdots \\ \gamma_{uu}(p-1) & \gamma_{uu}(p-2) & \cdots & \gamma_{uu}(0) & \gamma_{uv}(p-1) & \gamma_{uv}(p-2) & \cdots & \gamma_{uv}(p-q) \\ \hline \gamma_{uv}(0) & \gamma_{uv}(1) & \cdots & \gamma_{uv}(p-1) & \gamma_{vv}(0) & \gamma_{vv}(1) & \cdots & \gamma_{vv}(q-1) \\ \gamma_{uv}(-1) & \gamma_{uv}(0) & \cdots & \gamma_{uv}(p-2) & \gamma_{vv}(1) & \gamma_{vv}(0) & \cdots & \gamma_{vv}(q-2) \\ & \vdots & & \vdots & & \vdots & & \vdots \\ \gamma_{uv}(1-q) & \gamma_{uv}(2-q) & \cdots & \gamma_{uv}(p-q) & \gamma_{vv}(q-1) & \gamma_{vv}(q-2) & \cdots & \gamma_{vv}(0) \end{array}\right] \tag{7.2.18}$$

where $\gamma_{uu}(k)$ and $\gamma_{vv}(k)$ are the autocovariances for the u's and the v's and $\gamma_{uv}(k)$ are the cross covariances defined by

$$\gamma_{uv}(k) = \gamma_{vu}(-k) = E[u_t v_{t+k}] = E[v_t u_{t-k}]$$

The large-sample covariance matrix for the maximum likelihood estimates may be obtained using

$$\mathbf{V}(\hat{\boldsymbol{\phi}}, \hat{\boldsymbol{\theta}}) \simeq \mathbf{I}^{-1}(\boldsymbol{\phi}, \boldsymbol{\theta})$$

Estimates of $\mathbf{I}(\boldsymbol{\phi}, \boldsymbol{\theta})$ and hence of $\mathbf{V}(\hat{\boldsymbol{\phi}}, \hat{\boldsymbol{\theta}})$ may be obtained by evaluating the u's and v's with $\boldsymbol{\beta}_0 = \hat{\boldsymbol{\beta}}$ and omitting the expectation sign in (7.2.17), leading to $\hat{\mathbf{V}}(\hat{\boldsymbol{\phi}}, \hat{\boldsymbol{\theta}}) = (\mathbf{X}'_{\hat{\beta}}\mathbf{X}_{\hat{\beta}})^{-1}\hat{\sigma}_a^2$, or substituting standard sample estimates of the autocovariances and cross covariances in (7.2.18). Theoretical large-sample results can be obtained by noticing that with the elements of $\boldsymbol{\beta}_0$ equal to the true values of the parameters, equation (7.2.13) and (7.2.14) imply that the derived series u_t and v_t follow *autoregressive* processes defined by

$$\phi(B)u_t = a_t \qquad \theta(B)v_t = -a_t$$

It follows that the autocovariances which appear in (7.2.18) are those for pure autoregressive processes, and the cross covariances are the negative of those between two such processes generated by the same a's.

We illustrate the use of this result with a few examples.

Covariance matrix of parameter estimates for AR(p) and MA(q) processes. Let $\boldsymbol{\Gamma}_p(\boldsymbol{\phi})$ be the $p \times p$ autocovariance matrix of p successive observations from an AR(p) process with parameters $\boldsymbol{\phi}' = (\phi_1, \phi_2, \ldots, \phi_p)$. Then using (7.2.18), the $p \times p$ covariance matrix of the estimates $\hat{\boldsymbol{\phi}}$ is given by

$$\mathbf{V}(\hat{\boldsymbol{\phi}}) \simeq n^{-1}\sigma_a^2\boldsymbol{\Gamma}_p^{-1}(\boldsymbol{\phi}) \tag{7.2.19}$$

Let $\boldsymbol{\Gamma}_q(\boldsymbol{\theta})$ be the $q \times q$ autocovariance matrix of q successive observations from an AR(q) process with parameters $\boldsymbol{\theta}' = (\theta_1, \theta_2, \ldots, \theta_q)$. Then, using (7.2.18), the $q \times q$ covariance matrix of the estimates $\hat{\boldsymbol{\theta}}$ in an MA(q) model is

$$\mathbf{V}(\hat{\boldsymbol{\theta}}) \simeq n^{-1}\sigma_a^2\boldsymbol{\Gamma}_q^{-1}(\boldsymbol{\theta}) \tag{7.2.20}$$

It is occasionally useful to parameterize an ARMA process in terms of the zeros of $\phi(B)$ and $\theta(B)$. In this case a particularly simple form is obtained for the covariance matrix of the parameter estimates.

*** Covariances for the zeros of an ARMA process.** Consider the ARMA(p, q) process parameterized in terms of its zeros (assumed to be real), so that

* The line in the margin indicates that material may be omitted at first reading.

$$\prod_{i=1}^{p}(1 - G_iB)\tilde{w}_t = \prod_{j=1}^{q}(1 - H_jB)a_t$$

or

$$a_t = \prod_{i=1}^{p}(1 - G_iB)\prod_{j=1}^{q}(1 - H_jB)^{-1}\tilde{w}_t$$

The derivatives of the a's are then such that

$$u_{t,i} = -\frac{\partial a_t}{\partial G_i} = (1 - G_iB)^{-1}a_{t-1}$$

$$v_{t,j} = -\frac{\partial a_t}{\partial H_j} = -(1 - H_jB)^{-1}a_{t-1}$$

Hence, using (7.2.18), for large samples, the information matrix for the roots is such that

$n^{-1}\mathbf{I}(\mathbf{G}, \mathbf{H}) =$

$$\left[\begin{array}{cccc|ccc} (1-G_1^2)^{-1} & (1-G_1G_2)^{-1} & \cdots & (1-G_1G_p)^{-1} & -(1-G_1H_1)^{-1} & \cdots & -(1-G_1H_q)^{-1} \\ \vdots & \vdots & & \vdots & \vdots & & \vdots \\ (1-G_1G_p)^{-1} & (1-G_2G_p)^{-1} & \cdots & (1-G_p^2)^{-1} & -(1-G_pH_1)^{-1} & \cdots & -(1-G_pH_q)^{-1} \\ \hline -(1-G_1H_1)^{-1} & -(1-G_2H_1)^{-1} & \cdots & -(1-G_pH_1)^{-1} & (1-H_1^2)^{-1} & \cdots & (1-H_1H_q)^{-1} \\ \vdots & \vdots & & \vdots & \vdots & & \vdots \\ -(1-G_1H_q)^{-1} & -(1-G_2H_q)^{-1} & \cdots & -(1-G_pH_q)^{-1} & (1-H_1H_q)^{-1} & \cdots & (1-H_q^2)^{-1} \end{array}\right] \quad (7.2.21)$$

AR(2). In particular, therefore, for a second-order autoregressive process

$$(1 - G_1B)(1 - G_2B)\tilde{w}_t = a_t$$

$$\mathbf{V}(\hat{G}_1, \hat{G}_2) \simeq n^{-1}\begin{bmatrix} (1 - G_1^2)^{-1} & (1 - G_1G_2)^{-1} \\ (1 - G_1G_2)^{-1} & (1 - G_2^2)^{-1} \end{bmatrix}^{-1}$$

$$= \frac{1}{n}\frac{1 - G_1G_2}{(G_1 - G_2)^2}\begin{bmatrix} (1 - G_1^2)(1 - G_1G_2) & -(1 - G_1^2)(1 - G_2^2) \\ -(1 - G_1^2)(1 - G_2^2) & (1 - G_2^2)(1 - G_1G_2) \end{bmatrix} \quad (7.2.22)$$

Exactly parallel results will be obtained for a second-order moving average process.

ARMA(1, 1). Similarly, for the ARMA(1, 1) process,

$$(1 - \phi B)\tilde{w}_t = (1 - \theta B)a_t$$

on setting $\phi = G_1$ and $\theta = H_1$ in (7.2.21), we obtain

$$\begin{aligned} \mathbf{V}(\hat{\phi}, \hat{\theta}) &= n^{-1} \begin{bmatrix} (1-\phi^2)^{-1} & -(1-\phi\theta)^{-1} \\ (1-\phi\theta)^{-1} & (1-\theta^2)^{-1} \end{bmatrix}^{-1} \\ &= \frac{1}{n} \frac{1-\phi\theta}{(\phi-\theta)^2} \begin{bmatrix} (1-\phi^2)(1-\phi\theta) & (1-\phi^2)(1-\theta^2) \\ (1-\phi^2)(1-\theta^2) & (1-\theta^2)(1-\phi\theta) \end{bmatrix} \end{aligned} \tag{7.2.23}$$

The results for these two processes illustrate a duality property between the information matrices for the autoregressive model and the general ARMA model of order (p, q). Namely, suppose that the information matrix for parameters $(\mathbf{G}, \mathbf{H})$ associated with the ARMA(p, q) model

$$\prod_{i=1}^{p} (1 - G_i B)\tilde{w}_t = \prod_{j=1}^{q} (1 - H_j B) a_t$$

is denoted as $\mathbf{I}\{\mathbf{G}, \mathbf{H} | (p, q)\}$, and suppose, correspondingly, that the information matrix for the parameters $(\mathbf{G}, \mathbf{H})$ in the pure AR$(p + q)$ model

$$\prod_{i=1}^{p} (1 - G_i B) \prod_{j=1}^{q} (1 - H_j B)\tilde{w}_t = a_t$$

is denoted as

$$\mathbf{I}\{\mathbf{G}, \mathbf{H} | (p + q, 0)\} = \left[\begin{array}{c|c} \mathbf{I}_{GG} & \mathbf{I}_{GH} \\ \hline \mathbf{I}'_{GH} & \mathbf{I}_{HH} \end{array}\right]$$

where the matrix is partitioned after the pth row and column. Then, for moderate and large samples, we can see directly from (7.2.21) that

$$\mathbf{I}\{\mathbf{G}, \mathbf{H} | (p, q)\} \simeq \mathbf{I}\{\mathbf{G}, -\mathbf{H} | (p + q, 0)\} = \left[\begin{array}{c|c} \mathbf{I}_{GG} & -\mathbf{I}_{GH} \\ \hline -\mathbf{I}'_{GH} & \mathbf{I}_{HH} \end{array}\right] \tag{7.2.24}$$

Hence, since for moderate and large samples, the inverse of the information matrix provides a close approximation to the covariance matrix $\mathbf{V}(\hat{\mathbf{G}}, \hat{\mathbf{H}})$ of the parameter estimates, we have, correspondingly,

$$\mathbf{V}\{\hat{\mathbf{G}}, \hat{\mathbf{H}} | (p, q)\} \simeq \mathbf{V}\{\hat{\mathbf{G}}, -\hat{\mathbf{H}} | (p + q, 0)\} \tag{7.2.25}$$

7.3 SOME ESTIMATION RESULTS FOR SPECIFIC MODELS

In Appendices A7.3 and A7.4 some estimation results for special cases are derived. These, and results obtained earlier in this chapter, are summarized here for reference.

7.3.1 Autoregressive Processes

It is possible to obtain estimates of the parameters of a purely autoregressive process by solving certain *linear* equations. We show in Appendix A.7.4:

1. How exact least squares estimates may be obtained by solving a linear system of equations (see also Section 7.4.3).
2. How, by slight modification of the coefficients in these equations, a close approximation to the exact maximum likelihood equations may be obtained.
3. How conditional least squares estimates, as defined in Section 7.1.3, may be obtained by solving a system of linear equations of the form of the standard linear regression model normal equations.
4. How estimates that are approximations to the least squares estimates and to the maximum likelihood estimates may be obtained using the estimated autocorrelations as coefficients in the linear Yule–Walker equations.

The estimates obtained in (1) are, of course, identical with those given by direct minimization of $S(\boldsymbol{\phi})$, as described in general terms in Section 7.2. The estimates (4) are the well-known approximations due to Yule and Walker. They are useful as first estimates for use at the identification stage, but can differ appreciably from the estimates (1), (2), or (3), in some situations. To illustrate, we now compare the estimates (4), obtained from the Yule–Walker equations, with the least squares estimates (1), which have been summarized in Table 7.11.

Yule–Walker estimates. The Yule–Walker estimates (6.3.6) are

$$\hat{\boldsymbol{\phi}} = \mathbf{R}^{-1}\mathbf{r}$$

where

$$\mathbf{R} = \begin{bmatrix} 1 & r_1 & \cdots & r_{p-1} \\ r_1 & 1 & \cdots & r_{p-2} \\ \vdots & \vdots & & \vdots \\ r_{p-1} & r_{p-2} & \cdots & 1 \end{bmatrix} \qquad \mathbf{r} = \begin{bmatrix} r_1 \\ r_2 \\ \vdots \\ r_p \end{bmatrix} \tag{7.3.1}$$

In particular, the estimates for first- and second-order autoregressive processes are

$$\begin{aligned} &\text{AR(1):} \quad \hat{\phi}_1 = r_1 \\ &\text{AR(2):} \quad \hat{\phi}_1 = \frac{r_1(1 - r_2)}{1 - r_1^2} \qquad \hat{\phi}_2 = \frac{r_2 - r_1^2}{1 - r_1^2} \end{aligned} \tag{7.3.2}$$

it is shown in Appendix A7.4 that an approximation to $S(\hat{\boldsymbol{\phi}})$ is provided by

$$S(\hat{\boldsymbol{\phi}}) = \sum_{t=1}^{n} \tilde{w}_t^2(1 - \mathbf{r}'\hat{\boldsymbol{\phi}}) \tag{7.3.3}$$

whence

$$\hat{\sigma}_a^2 = \frac{S(\hat{\boldsymbol{\phi}})}{n} = c_0(1 - \mathbf{r}'\hat{\boldsymbol{\phi}}) \tag{7.3.4}$$

where c_0 is the sample variance of the w's. A parallel expression relates σ_a^2 and γ_0, the theoretical variance of the w's [see (3.2.8)], namely,

$$\sigma_a^2 = \gamma_0(1 - \boldsymbol{\rho}'\boldsymbol{\phi})$$

where the elements of $\boldsymbol{\rho}$ and of $\boldsymbol{\phi}$ are the theoretical values. Thus, from (7.2.19), the covariance matrix for the estimates $\hat{\boldsymbol{\phi}}$ is

$$\mathbf{V}(\hat{\boldsymbol{\phi}}) \simeq n^{-1}\sigma_a^2\boldsymbol{\Gamma}^{-1} = n^{-1}(1 - \boldsymbol{\rho}'\boldsymbol{\phi})\mathbf{P}^{-1} \tag{7.3.5}$$

where $\boldsymbol{\Gamma}$ and $\mathbf{P}$ are the autocovariance and autocorrelation matrices of p successive values of the AR(p) process, as defined in (2.1.7).

In particular, for first- and second-order autoregressive processes, we find that

$$\text{AR(1)}: \ V(\hat{\phi}) \simeq n^{-1}(1 - \phi^2) \tag{7.3.6}$$

$$\text{AR(2)}: \ \mathbf{V}(\hat{\phi}_1, \hat{\phi}_2) \simeq n^{-1}\begin{bmatrix} 1 - \phi_2^2 & -\phi_1(1 + \phi_2) \\ -\phi_1(1 + \phi_2) & 1 - \phi_2^2 \end{bmatrix} \tag{7.3.7}$$

Estimates of the variances and covariances are obtained by substituting estimates for the parameters in (7.3.5). Thus

$$\hat{\mathbf{V}}(\hat{\boldsymbol{\phi}}) = n^{-1}(1 - \mathbf{r}'\hat{\boldsymbol{\phi}})\mathbf{R}^{-1} \tag{7.3.8}$$

Examples. For series C, D, E, and F, which have been identified as possibly autoregressive (or in the case of series C, autoregressive in the first difference), the lag 1 and lag 2 autocorrelations are shown below.

Series	Tentative Identification	Degree of Differencing	Relevant Estimated Autocorrelations	n
C	(1, 1, 0)	1	$r_1 = 0.805$	225
D	(1, 0, 0)	0	$r_1 = 0.861$	310
E	(2, 0, 0)	0	$r_1 = 0.806$, $r_2 = 0.428$	100
F	(2, 0, 0)	0	$r_1 = -0.390$, $r_2 = 0.304$	70

Using (7.3.2), we obtain the Yule–Walker estimates shown in Table 7.12, together with their standard errors calculated from (7.3.6) and (7.3.7).

Using (7.3.7), the correlation coefficient between the estimates of the

TABLE 7.12 Yule–Walker Estimates for Series C to F

Series	Estimates	Standard Errors	Correlation Coefficients between Estimates
C	$\hat{\phi}_1 = 0.81$	± 0.04	
D	$\hat{\phi}_1 = 0.86$	± 0.03	
E	$\hat{\phi}_1 = 1.32$	± 0.08	-0.81
	$\hat{\phi}_2 = -0.63$	± 0.08	
F	$\hat{\phi}_1 = -0.32$	± 0.12	0.39
	$\hat{\phi}_2 = 0.18$	± 0.12	

two parameters of a second-order autoregressive process is given by

$$\rho(\hat{\phi}_1, \hat{\phi}_2) \simeq \frac{\operatorname{cov}[\hat{\phi}_1, \hat{\phi}_2]}{\sqrt{V(\hat{\phi}_1)\, V(\hat{\phi}_2)}} = \frac{-\phi_1}{1 - \phi_2} = -\rho_1 \simeq -r_1$$

We notice that in the case of series E, there is a high negative correlation between the estimates. This indicates that the confidence region for $\hat{\phi}_1$ and $\hat{\phi}_2$ will be attenuated along a NW-to-SE diagonal. This implies that the estimates are rather unstable and explains the relatively larger discrepancy between the least squares estimates of Table 7.11 and the Yule–Walker estimates of Table 7.12 for this particular series.

7.3.2 Moving Average Processes

Maximum likelihood estimates $\hat{\boldsymbol{\theta}}$ for moving average processes may, in simple cases, be obtained graphically, as illustrated in Section 7.1.6, or more generally, by the iterative calculation described in Section 7.2.1. From (7.2.20) it follows that for moderate and large samples, the covariance matrix for the estimates of the parameters of a qth-order moving average process is of the same form as the corresponding matrix for an autoregressive process of the same order.

Thus for first- and second-order moving average processes, we find, corresponding to (7.3.6) and (7.3.7),

$$\text{MA(1):}\quad V(\hat{\theta}) \simeq n^{-1}(1 - \theta^2) \tag{7.3.9}$$

$$\text{MA(2):}\quad \mathbf{V}(\hat{\theta}_1, \hat{\theta}_2) \simeq n^{-1}\begin{bmatrix} 1 - \theta_2^2 & -\theta_1(1 + \theta_2) \\ -\theta_1(1 + \theta_2) & 1 - \theta_2^2 \end{bmatrix} \tag{7.3.10}$$

7.3.3 Mixed Processes

Maximum likelihood estimates $(\hat{\boldsymbol{\phi}}, \hat{\boldsymbol{\theta}})$ for mixed processes, as for moving average processes, may be obtained graphically in simple cases, and more generally, by iterative calculation. For moderate and large samples, the covariance matrix may be obtained by evaluating and inverting the information matrix (7.2.18). In the important special case of the ARMA(1,1) process

$$(1 - \phi B)\tilde{w}_t = (1 - \theta B)a_t$$

we obtain, as in (7.2.23),

$$\mathbf{V}(\hat{\phi}, \hat{\theta}) \simeq n^{-1} \frac{1 - \phi\theta}{(\phi - \theta)^2} \begin{bmatrix} (1 - \phi^2)(1 - \phi\theta) & (1 - \phi^2)(1 - \theta^2) \\ (1 - \phi^2)(1 - \theta^2) & (1 - \theta^2)(1 - \phi\theta) \end{bmatrix} \tag{7.3.11}$$

It will be noted that when $\phi = \theta$, the variances of $\hat{\phi}$ and $\hat{\theta}$ are infinite. This is to be expected, for in this case the factor $(1 - \phi B) = (1 - \theta B)$ cancels on both sides of the model, which becomes

$$\tilde{w}_t = a_t$$

This is a particular case of *parameter redundancy*, which we discuss more fully in Section 7.3.5.

7.3.4 Separation of Linear and Nonlinear Components in Estimation

It is occasionally of interest to make an analysis in which the estimation of the parameters of the mixed model is separated into its basic linear and nonlinear parts. Consider the general mixed model, which we write as

$$a_t = \phi(B)\theta^{-1}(B)\tilde{w}_t$$

or

$$a_t = \phi(B)(\varepsilon_t|\boldsymbol{\theta}) \tag{7.3.12}$$

where

$$(\varepsilon_t|\boldsymbol{\theta}) = \theta^{-1}(B)\tilde{w}_t$$

that is,

$$\tilde{w}_t = \theta(B)(\varepsilon_t|\boldsymbol{\theta}) \tag{7.3.13}$$

For any given set of θ*'s*, the ε's may be calculated recursively from (7.3.13), which may be written

$$\varepsilon_t = \tilde{w}_t + \theta_1\varepsilon_{t-1} + \theta_2\varepsilon_{t-2} + \cdots + \theta_q\varepsilon_{t-q}$$

The recursion may be started by setting unknown ε's equal to zero. Having calculated the ε's, the conditional estimates $\hat{\boldsymbol{\phi}}_{\boldsymbol{\theta}}$ may readily be obtained. These are the estimated autoregressive parameters in the linear model (7.3.12), which may be written

$$a_t = \varepsilon_t - \phi_1\varepsilon_{t-1} - \phi_2\varepsilon_{t-2} - \cdots - \phi_p\varepsilon_{t-p} \tag{7.3.14}$$

As we have explained in Section 7.3.1, the least squares estimates of the autoregressive parameters may be found by direct solution of a simple set of linear equations and approximated by the use of the Yule–Walker equations. In simple cases we can examine the behavior of $S(\hat{\boldsymbol{\phi}}_{\boldsymbol{\theta}}, \boldsymbol{\theta})$ as well as obtain the absolute minimum sum of squares, by computing it on a grid of $\boldsymbol{\theta}$ values and plotting contours.

Example using series C. According to one tentative identification for series C, given in Table 6.4, it was possibly generated by a model of order (1, 1, 0)

$$(1 - \phi B)w_t = a_t$$

with $w_t = \nabla z_t$ and $E[w_t] = 0$. It was decided to examine the estimation situation for these data in relation to the somewhat more elaborate model

$$(1 - \phi B)w_t = (1 - \theta_1 B - \theta_2 B^2)a_t$$

Following the argument given above, the process may be thought of as resulting from a combination of the nonlinear model

$$\varepsilon_t = w_t + \theta_1\varepsilon_{t-1} + \theta_2\varepsilon_{t-2}$$

and the linear model

$$a_t = \varepsilon_t - \phi\varepsilon_{t-1}$$

For each choice of the nonlinear parameters $\boldsymbol{\theta} = (\theta_1, \theta_2)$ within the invertibility region, a set of ε's was calculated recursively. Using the Yule–Walker approximation, an estimate $\hat{\phi}_{\boldsymbol{\theta}} = r_1(\varepsilon)$ could now be obtained together with

$$S(\hat{\phi}_{\boldsymbol{\theta}}, \boldsymbol{\theta}) \simeq \sum_{t=1}^{n} \varepsilon_t^2[1 - r_1^2(\varepsilon)]$$

This sum of squares was plotted for a grid of values of θ_1 and θ_2 and its contours are shown in Figure 7.6. We see that a minimum close to $\theta_1 = \theta_2 = 0$ is indicated, at which point $r_1(\varepsilon) = 0.805$. Thus within the whole class of models of order (1, 1, 2), the simple (1, 1, 0) model,

$$(1 - 0.8B)\nabla z_t = a_t$$

is confirmed as providing an adequate representation.

7.3.5 Parameter Redundancy

The model

$$\phi(B)\tilde{w}_t = \theta(B)a_t$$

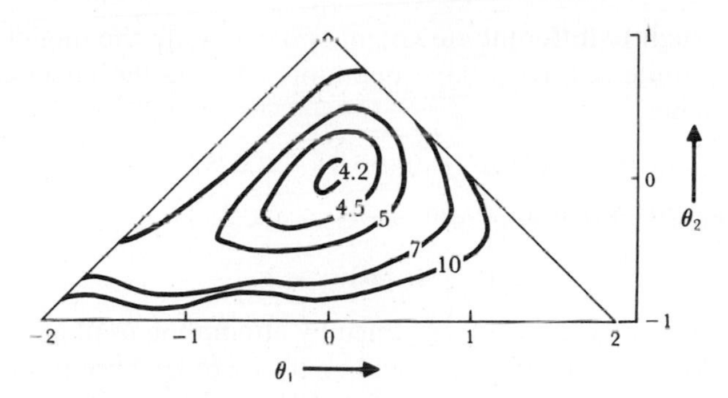

Figure 7.6 Contours of $S(\hat{\phi}_\theta, \boldsymbol{\theta})$ plotted over the admissible parameter space for the θ's.

is identical with the model

$$(1 - \alpha B)\phi(B)\tilde{w}_t = (1 - \alpha B)\theta(B)a_t$$

in which both autoregressive and moving average operators are multiplied by the same factor, $1 - \alpha B$. Serious difficulties in the estimation procedure will arise if a model is fitted that contains a redundant factor. Therefore, some care is needed in avoiding the situation where redundant or near-redundant factors occur. It is to be noted that the existence of redundancy is not necessarily obvious. For example, one can see the common factor in the ARMA(2, 1) model

$$(1 - 1.3B + 0.4B^2)\tilde{w}_t = (1 - 0.5B)a_t$$

only after factoring the left-hand side to obtain

$$(1 - 0.5B)(1 - 0.8B)\tilde{w}_t = (1 - 0.5B)a_t$$

that is,

$$(1 - 0.8B)\tilde{w}_t = a_t$$

In practice, it is not just exact cancellation that causes difficulties, but also near-cancellation. For example, suppose that the true model was

$$(1 - 0.4B)(1 - 0.8B)\tilde{w}_t = (1 - 0.5B)a_t \tag{7.3.15}$$

If an attempt was made to fit this model as ARMA(2, 1), extreme instability in the parameter estimates could be expected, because of near-cancellation of the factors $(1 - 0.4B)$ and $(1 - 0.5B)$, on the left and right sides. In this situation, combinations of parameter values yielding similar $[a]$'s and so similar likelihoods can be found, and a change of parameter value on the left can be nearly compensated by a suitable change on the right. The sum of squares contour surfaces in the three-dimensional parameter space will thus approach obliquely oriented cylinders, and a line of "near least squares" solutions rather than a clearly defined point minimum will be found.

From a slightly different viewpoint, we can write the model (7.3.15) in terms of an infinite autoregressive operator. Making the necessary expansion, we find that

$$(1 - 0.700B - 0.030B^2 - 0.015B^3 - 0.008B^4 - \cdots)\tilde{w}_t = a_t$$

Thus, very nearly, the model is

$$(1 - 0.7B)\tilde{w}_t = a_t \qquad (7.3.16)$$

The instability of the estimates, obtained by attempting to fit an ARMA(2, 1) model, would occur because we would be trying to fit three parameters in a situation that could almost be represented by one.

Preliminary identification as a means of avoiding parameter redundancy. A principal reason for going through the identification procedure prior to fitting the model is to avoid difficulties arising from parameter redundancy, or to be more positive, to achieve *parsimony* in parameterization. Thus in the example just considered, for a time series of only a few hundred observations, the estimated autocorrelation function from data generated by (7.3.15) would be indistinguishable from that from data generated by the simple autoregressive process (7.3.16).

Thus we should be led to the fitting of the AR(1) process, which would normally be entirely adequate. Only with time series of several thousand observations would the need for a more elaborate model be detected, and in these same circumstances, enough information would be available to obtain reasonably good estimates of the additional parameters.

Redundancy in the ARMA(1, 1) process. The simplest process where the possibility occurs of direct cancellation of factors is the ARMA(1, 1) process,

$$(1 - \phi B)\tilde{w}_t = (1 - \theta B)a_t$$

In particular, if $\phi = \theta$, then whatever common value they have,

$$\tilde{w}_t = a_t$$

and the model implies that $\tilde{w}$ is generated by a white noise process. The data then cannot supply information about the common parameter, and using (7.3.11), $\hat{\phi}$ and $\hat{\theta}$ have infinite variances. Furthermore, whatever the values of ϕ and θ, $S(\phi, \theta)$ must be constant on the line $\phi = \theta$ as it is for example in Figure 7.7, which shows a sum of squares plot for the data of series A. For these data the least squares values $\hat{\phi} = 0.92$ and $\hat{\theta} = 0.58$ correspond to a point that is not particularly close to the line $\phi = \theta$, and no difficulties occur in the estimation of these parameters.

In practice, if the identification technique we have recommended is adopted, these difficulties will be avoided. An ARMA(1,1) process in which

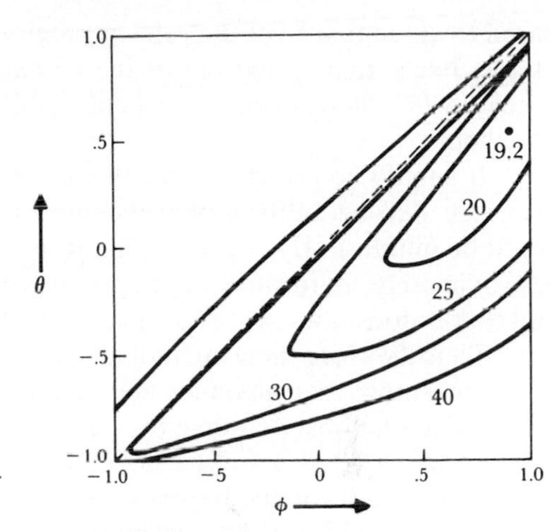

Figure 7.7 Sums of squares plot for series A.

ϕ is very nearly equal to θ will normally be identified as white noise, or if the difference is appreciable, as an AR(1) or MA(1) process with a single small coefficient.

In summary:

1. We should avoid mixed processes containing near common factors and we should be alert to the difficulties that can result.
2. We will automatically avoid such processes if we use identification and estimation procedures intelligently.

7.4 ESTIMATION USING BAYES' THEOREM

7.4.1 Bayes' Theorem

In this section we again employ the symbol $\boldsymbol{\xi}$ to represent a general vector of parameters. Bayes' theorem tells us that if $p(\boldsymbol{\xi})$ is the probability distribution for $\boldsymbol{\xi}$ prior to the collection of the data, then $p(\boldsymbol{\xi}|\mathbf{z})$, the distribution of $\boldsymbol{\xi}$ posterior to the data $\mathbf{z}$, is obtained by combining the prior distribution $p(\boldsymbol{\xi})$ and the likelihood $L(\boldsymbol{\xi}|\mathbf{z})$ in the following way:

$$p(\boldsymbol{\xi}|\mathbf{z}) = \frac{p(\boldsymbol{\xi})L(\boldsymbol{\xi}|\mathbf{z})}{\int p(\boldsymbol{\xi})L(\boldsymbol{\xi}|\mathbf{z})\, d\boldsymbol{\xi}} \tag{7.4.1}$$

The denominator merely ensures that $p(\boldsymbol{\xi}|\mathbf{z})$ integrates to 1. The important part of the expression is the numerator, from which we see that the posterior distribution is proportional to the prior distribution multiplied by the likelihood. Savage [173] has shown that prior and posterior probabilities can be interpreted as subjective probabilities. In particular, often before the data are available we have very little knowledge about $\boldsymbol{\xi}$ and we would be prepared to agree that over the relevant region, it would have appeared a priori just as likely that it had one value as another. In this case $p(\boldsymbol{\xi})$ could be taken as *locally* uniform, and hence $p(\boldsymbol{\xi}|\mathbf{z})$ would be proportional to the likelihood.

It should be noted that for this argument to hold, it is not necessary a priori for $\boldsymbol{\xi}$ to be uniform over its entire range (which for some parameters could be infinite). By requiring that it be *locally uniform* we mean that it be approximately uniform in the region in which the likelihood is appreciable and that it does not take an overwhelmingly large value outside that region.

Thus if ξ were the weight of a chair, we could certainly say a priori that it weighed more than an ounce and less than a ton. It is also likely that when we obtained an observation z by weighing the chair on a weighing machine, which had an error standard deviation σ, we could honestly say that we would have been equally happy with a priori values in the range $z \pm 3\sigma$. The exception would be if the weighing machine said that an apparently heavy chair weighed, say, 10 ounces. In this case the likelihood and the prior would be incompatible and we should not, of course, use Bayes' theorem to combine them, but would check the weighing machine and if this turned out to be accurate, inspect the chair more closely.

There is, of course, some arbitrariness in this idea. Suppose that we assumed the prior distribution of ξ to be locally uniform. This then implies that the distribution of any linear function of ξ is also locally uniform. However, the prior distribution of some nonlinear transformation $\alpha = \alpha(\xi)$ (such as $\alpha = \log \xi$) could *not* be exactly locally uniform. This arbitrariness will usually have very little effect if we are able to obtain fairly precise estimates of ξ. We will then be considering ξ only over a small range, and over such a range the transformation from ξ to, say, $\log \xi$ would often be very nearly linear.

Jeffreys [117] has argued that it is best to choose the metric $\alpha(\xi)$ so that Fisher's measure of information $I_\alpha = -E[\partial^2 l/\partial\alpha^2]$ is independent of the value of α, and hence of ξ. This is equivalent to choosing $\alpha(\xi)$ so that the limiting variance of its maximum likelihood estimate is independent of ξ and is achieved by choosing the prior distribution of ξ to be proportional to $\sqrt{I_\xi}$.

Jeffreys justified this choice of prior on the basis of its invariance to the parameterization employed. Specifically, with this choice, the posterior distributions for $\alpha(\xi)$ and for ξ, where $\alpha(\xi)$ and ξ are connected by a 1:1 transformation, are such that $p(\xi|\mathbf{z}) = p(\alpha|\mathbf{z})\, d\alpha/d\xi$. The same result may be obtained [58] by the following argument. If for large samples, the expected likelihood function for $\alpha(\xi)$ approaches a Normal curve, then the mean and variance of the curve summarize the information to be expected from the data. Suppose, now, that a transformation $\alpha(\xi)$ can be found in which the approximating Normal curve has nearly constant variance whatever the true values of the parameter. Then, in this parameterization, the *only* information in prospect from the data is conveyed by the *location* of the expected likelihood function. To say that we know essentially nothing a priori relative to this prospective observational information is to say that we regard different *locations* of α as equally likely a priori. Equivalently, we say that α should be taken as locally uniform.

The generalization of Jeffrey's rule to deal with several parameters is that the joint prior distribution of parameters $\boldsymbol{\xi}$ be taken proportional to

$$|I_{\xi}|^{1/2} = \left| - E\left[\frac{\partial^2 l}{\partial \xi_i \, \partial \xi_j}\right]\right|^{1/2} \tag{7.4.2}$$

It has been urged [119] that the likelihood itself is best considered and plotted in that metric α for which I_α is independent of α. If this is done, it will be noted that the likelihood function and the posterior density function with uniform prior are proportional.

7.4.2 Bayesian Estimation of Parameters

We now consider the estimation of the parameters in an ARIMA model from a Bayesian point of view. It is shown in Appendix A7.3 that the exact likelihood of a time series $\mathbf{z}$ of length $N = n + d$ from an ARIMA(p, d, q) process is of the form

$$L(\boldsymbol{\phi}, \boldsymbol{\theta}|\mathbf{z}) = \sigma_a^{-n} f(\boldsymbol{\phi}, \boldsymbol{\theta}) \exp\left[-\frac{S(\boldsymbol{\phi}, \boldsymbol{\theta})}{2\sigma_a^2}\right] \tag{7.4.3}$$

where

$$S(\boldsymbol{\phi}, \boldsymbol{\theta}) = \sum_{t=1}^{n} [a_t|\mathbf{w}, \boldsymbol{\phi}, \boldsymbol{\theta}]^2 + [\mathbf{e}_*]'\boldsymbol{\Omega}^{-1}[\mathbf{e}_*] \tag{7.4.4}$$

If we have no prior information about σ_a, $\boldsymbol{\phi}$, or $\boldsymbol{\theta}$, and since information about σ_a would supply no information about $\boldsymbol{\phi}$ and $\boldsymbol{\theta}$, it is sensible, following Jeffrey's, to employ a prior distribution for $\boldsymbol{\phi}$, $\boldsymbol{\theta}$, and σ_a of the form

$$p(\boldsymbol{\phi}, \boldsymbol{\theta}, \sigma_a) \propto |\mathbf{I}(\boldsymbol{\phi}, \boldsymbol{\theta})|^{1/2}\sigma_a^{-1}$$

It follows that the posterior distribution is

$$p(\boldsymbol{\phi}, \boldsymbol{\theta}, \sigma_a|\mathbf{z}) \propto \sigma_a^{-(n+1)}|\mathbf{I}(\boldsymbol{\phi}, \boldsymbol{\theta})|^{1/2} f(\boldsymbol{\phi}, \boldsymbol{\theta}) \exp\left[-\frac{S(\boldsymbol{\phi}, \boldsymbol{\theta})}{2\sigma_a^2}\right] \tag{7.4.5}$$

If we now integrate (7.4.5) from zero to infinity with respect to σ_a, we obtain the exact joint posterior distribution of the parameters $\boldsymbol{\phi}$ and $\boldsymbol{\theta}$ as

$$p(\boldsymbol{\phi}, \boldsymbol{\theta}|\mathbf{z}) \propto |\mathbf{I}(\boldsymbol{\phi}, \boldsymbol{\theta})|^{1/2} f(\boldsymbol{\phi}, \boldsymbol{\theta})\{S(\boldsymbol{\phi}, \boldsymbol{\theta})\}^{-n/2} \tag{7.4.6}$$

7.4.3 Autoregressive Processes

If z_t follows a process of order $(p, d, 0)$, then $w_t = \nabla^d z_t$ follows a pure autoregressive process of order p. It is shown in Appendix A7.4 that for such a process, the factors $|\mathbf{I}(\boldsymbol{\phi})|^{1/2}$ and $f(\boldsymbol{\phi})$, which in any case are dominated by the term in $S(\boldsymbol{\phi})$, essentially cancel. This yields the remarkably simple result that given the assumptions, the parameters $\boldsymbol{\phi}$ of the AR(p) process in w have the posterior distribution

$$p(\boldsymbol{\phi}|\mathbf{z}) \propto \{S(\boldsymbol{\phi})\}^{-n/2} \tag{7.4.7}$$

By this argument, then, the sum of squares contours which are approximate likelihood contours are, when nothing is known a priori, also contours of posterior probability.

Joint distribution of the autoregressive parameters. It is shown in Appendix A7.4 that for the pure AR process, the least squares estimates of the ϕ's which minimize $S(\boldsymbol{\phi}) = \boldsymbol{\phi}_u'\mathbf{D}\boldsymbol{\phi}_u$ are given by

$$\hat{\boldsymbol{\phi}} = \mathbf{D}_p^{-1}\mathbf{d} \tag{7.4.8}$$

where

$$\boldsymbol{\phi}_u = \begin{bmatrix} 1 \\ \hline \boldsymbol{\phi} \end{bmatrix} \qquad \mathbf{d} = \begin{bmatrix} D_{12} \\ D_{13} \\ \vdots \\ D_{1,p+1} \end{bmatrix} \qquad \mathbf{D}_p = \begin{bmatrix} D_{22} & D_{23} & \cdots & D_{2,p+1} \\ D_{23} & D_{33} & \cdots & D_{3,p+1} \\ \vdots & \vdots & \cdots & \vdots \\ D_{2,p+1} & D_{3,p+1} & \cdots & D_{p+1,p+1} \end{bmatrix}$$

$$\mathbf{D} = \left[\begin{array}{c|c} D_{11} & -\mathbf{d}' \\ \hline -\mathbf{d} & \mathbf{D}_p \end{array}\right] \tag{7.4.9}$$

and

$$D_{ij} = D_{ji} = \tilde{w}_i\tilde{w}_j + \tilde{w}_{i+1}\tilde{w}_{j+1} + \cdots + \tilde{w}_{n+1-j}\tilde{w}_{n+1-i} \tag{7.4.10}$$

It follows that

$$S(\boldsymbol{\phi}) = \nu s_a^2 + (\boldsymbol{\phi} - \hat{\boldsymbol{\phi}})'\mathbf{D}_p(\boldsymbol{\phi} - \hat{\boldsymbol{\phi}}) \tag{7.4.11}$$

where

$$s_a^2 = \frac{S(\hat{\boldsymbol{\phi}})}{\nu} \qquad \nu = n - p \tag{7.4.12}$$

and

$$S(\hat{\boldsymbol{\phi}}) = \hat{\boldsymbol{\phi}}_u'\mathbf{D}\hat{\boldsymbol{\phi}}_u = D_{11} - \hat{\boldsymbol{\phi}}'\mathbf{D}_p\hat{\boldsymbol{\phi}} = D_{11} - \mathbf{d}'\mathbf{D}_p^{-1}\mathbf{d} \tag{7.4.13}$$

Thus we can write

$$p(\boldsymbol{\phi}|\mathbf{z}) \propto \left[1 + \frac{(\boldsymbol{\phi} - \hat{\boldsymbol{\phi}})'\mathbf{D}_p(\boldsymbol{\phi} - \hat{\boldsymbol{\phi}})}{\nu s_a^2}\right]^{-n/2} \tag{7.4.14}$$

Equivalently,

$$p(\boldsymbol{\phi}|\mathbf{z}) \propto \left[1 + \frac{\frac{1}{2}\sum_i \sum_j S_{ij}(\phi_i - \hat{\phi}_i)(\phi_j - \hat{\phi}_j)}{\nu s_a^2}\right]^{-n/2} \tag{7.4.15}$$

where

$$S_{ij} = \frac{\partial^2 S(\boldsymbol{\phi})}{\partial \phi_i \, \partial \phi_j} = 2D_{i+1,j+1}$$

It follows that, a posteriori, the parameters of an autoregressive process have a multiple t-distribution (A7.1.13), with $\nu = n - p$ degrees of freedom.

In particular, for the special case $p = 1$, $(\phi - \hat{\phi})/s_{\hat{\phi}}$ is distributed *exactly* in a Student t-distribution with $n - 1$ degrees of freedom where, using the general results given above, $\hat{\phi}$ and $s_{\hat{\phi}}$ are given by

$$\hat{\phi} = \frac{D_{12}}{D_{22}} \qquad s_{\hat{\phi}} = \left[\frac{1}{(n-1)} \frac{D_{11}}{D_{22}} \left(1 - \frac{D_{12}^2}{D_{11}D_{22}}\right)\right]^{1/2} \tag{7.4.16}$$

The quantity $s_{\hat{\phi}}$, for large samples, tends to $[(1 - \phi^2)/n]^{1/2}$ and in the sampling theory framework is identical with the large-sample "standard error" for $\hat{\phi}$. However, when using this and similar expressions within the Bayesian framework, it is to be remembered that it is the parameters (ϕ in this case) that are random variables. Quantities such as $\hat{\phi}$ and $s_{\hat{\phi}}$, which are functions of data that have already occurred, are regarded as fixed.

Normal approximation. For samples of size $n > 50$, in which we are usually interested, the Normal approximation to the t-distribution is adequate. Thus, very nearly, $\boldsymbol{\phi}$ has a joint p-variate Normal distribution $N\{\hat{\boldsymbol{\phi}}, \mathbf{D}_p^{-1}s_a^2\}$ having mean $\hat{\boldsymbol{\phi}}$ and variance–covariance matrix $\mathbf{D}_p^{-1}s_a^2$.

Bayesian regions of highest probability density. In summarizing what the posterior distribution has to tell us about the probability of various $\boldsymbol{\phi}$ values, it is useful to indicate a region of *highest probability density*, called for short [61] an HPD region. A Bayesian $1 - \varepsilon$ HPD region has the properties:

1. Any parameter point inside the region has higher probability density than any point outside.
2. The total posterior probability mass within the region is $1 - \varepsilon$.

Since $\boldsymbol{\phi}$ has a multiple t-distribution, it follows, using the result (A7.1.4) that,

$$\Pr\{(\boldsymbol{\phi} - \hat{\boldsymbol{\phi}})'\mathbf{D}_p(\boldsymbol{\phi} - \hat{\boldsymbol{\phi}}) < ps_a^2 F_\varepsilon(p, \nu)\} = 1 - \varepsilon \tag{7.4.17}$$

defines the *exact* $1 - \varepsilon$ HPD region for $\boldsymbol{\phi}$. Now, for $\nu = n - p > 100$,

$$pF_\varepsilon(p, \nu) \simeq \chi_\varepsilon^2(p)$$

Also,

$$(\boldsymbol{\phi} - \hat{\boldsymbol{\phi}})'\mathbf{D}_p(\boldsymbol{\phi} - \hat{\boldsymbol{\phi}}) = \tfrac{1}{2} \sum_{i,j} S_{ij}(\phi_i - \hat{\phi}_i)(\phi_j - \hat{\phi}_j)$$

Thus, approximately, the HPD region defined in (7.4.17) is such that

$$\sum_{i,j} S_{ij}(\phi_i - \hat{\phi}_i)(\phi_j - \hat{\phi}_j) < 2s_a^2 \chi_\varepsilon^2(p) \tag{7.4.18}$$

which if we set $\hat{\sigma}_a^2 = s_a^2$ is identical with the confidence region defined by (7.1.26).

Although these approximate regions are identical, it will be remembered that their interpretation is different. From a sampling theory viewpoint, we say that if a confidence region is computed according to (7.1.26), then for each of a set of repeated samples, a proportion $1 - \varepsilon$ of these regions will include the true parameter point. From the Bayesian viewpoint, we are concerned only with the single sample $\mathbf{z}$ which has actually been observed. Assuming the relevance of the noninformative prior distribution that we have taken, the HPD region includes that proportion $1 - \varepsilon$ of the resulting probability distribution of $\boldsymbol{\phi}$, given $\mathbf{z}$, which has the highest density. In other words, the probability that the value of $\boldsymbol{\phi}$, which gave rise to the data $\mathbf{z}$, lies in the HPD region is $1 - \varepsilon$.

Using (7.4.11), (7.4.12), and (7.4.18), for large samples the approximate Bayesian HPD region is bounded by a contour for which

$$S(\boldsymbol{\phi}) = S(\hat{\boldsymbol{\phi}}) \left[1 + \frac{\chi_\varepsilon^2(p)}{n}\right] \tag{7.4.19}$$

which corresponds exactly with the confidence region defined by (7.1.28).

7.4.4 Moving Average Processes

If z follows an integrated moving average process of order $(0, d, q)$, then $w = \nabla^d z$ follows a pure moving average process of order q. Because of the duality in estimation results and in the information matrices, in particular, between the autoregressive model and the moving average model, it follows that in the moving average case the factors $|\mathbf{I}(\boldsymbol{\theta})|^{1/2}$ and $f(\boldsymbol{\theta})$ in (7.4.6), which in any case are dominated by $S(\boldsymbol{\theta})$, also cancel for large samples. Thus, corresponding to (7.4.7), we find that the parameters $\boldsymbol{\theta}$ of the MA(q) process in w have the posterior distribution

$$p(\boldsymbol{\theta}|\mathbf{z}) \propto [S(\boldsymbol{\theta})]^{-n/2} \tag{7.4.20}$$

Again the sum of squares contours are, for moderate samples, essentially *exact* contours of posterior density. However, because $[a_t]$ is not a linear function of the θ's, $S(\boldsymbol{\theta})$ will not be exactly quadratic in $\boldsymbol{\theta}$, though for large samples it will often be nearly so within the relevant ranges. In that case, we have approximately

$$S(\boldsymbol{\theta}) = \nu s_a^2 + \tfrac{1}{2} \sum_{i,j} S_{ij}(\theta_i - \hat{\theta}_i)(\theta_j - \hat{\theta}_j)$$

where $\nu s_a^2 = S(\hat{\boldsymbol{\theta}})$ and $\nu = n - q$. It follows, after substituting for $S(\boldsymbol{\theta})$ in (7.4.20) and using the exponential approximation, that:

1. For large samples, $\boldsymbol{\theta}$ is *approximately* distributed in a multi-Normal distribution $N\{\hat{\boldsymbol{\theta}}, 2\{S_{ij}\}^{-1}s_a^2\}$.
2. An approximate HPD region is defined by (7.4.18) or (7.4.19), with q replacing p and $\boldsymbol{\theta}$ replacing $\boldsymbol{\phi}$.

Example: posterior distribution of $\lambda = 1 - \theta$ for an IMA(0, 1, 1) process. To illustrate, Table 7.13 shows the calculation of the approximate posterior density distribution $p(\lambda|\mathbf{z})$ from the data of series B. The

TABLE 7.13 Calculation of Approximate Posterior Density $p(\lambda|\mathbf{z})$ for Series B

λ	$10^{57} \times [S(\lambda)]^{-184.5}$	$p(\lambda\|\mathbf{z})$
1.300	4	0.001
1.275	33	0.006
1.250	212	0.036
1.225	1,007	0.171
1.200	3,597	0.609
1.175	9,956	1.685
1.150	21,159	3.582
1.125	34,762	5.884
1.100	44,473	7.528
1.075	44,988	7.615
1.050	35,835	6.066
1.025	22,563	3.819
1.000	11,277	1.908
0.975	4,540	0.769
0.950	1,457	0.247
0.925	372	0.063
0.900	76	0.012
0.875	12	0.002
0.850	2	0.000
	236,325	40.000

second column of the table shows the calculated ordinates $[S(\lambda)]^{-n/2} \times 10^{57}$, at intervals in λ of length $h = 0.025$. Their sum Σ is 236,325. By dividing the ordinates by $h\Sigma$, we produce a posterior density function such that the area under the curve is, to a sufficient approximation, equal to 1.* The distribution is plotted in Figure 7.8. It is seen to be approximately Normal with its mode at $\hat{\lambda} = 1.09$ and having a standard deviation of about 0.05. A 95% Bayesian HPD interval covers essentially the same range, $0.98 < \lambda < 1.19$, as did the confidence interval.

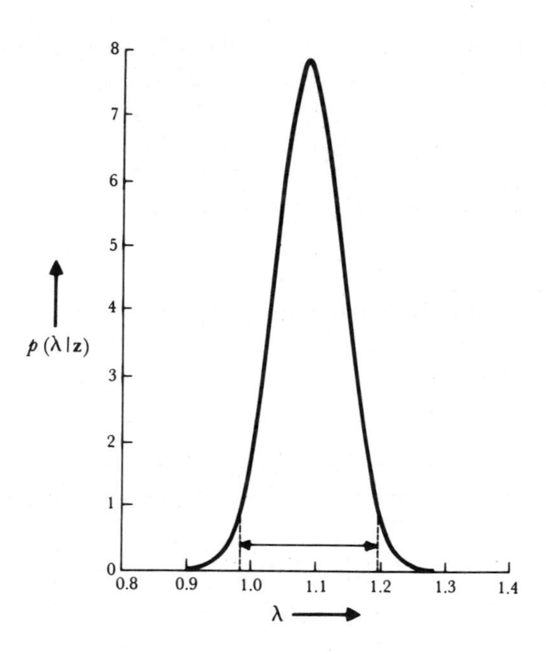

Figure 7.8 Posterior density $p(\lambda|\mathbf{z})$ for series B.

7.4.5 Mixed Processes

If z follows an ARIMA process of order (p, d, q), then $w = \nabla^d z$ follows an ARMA(p, q) process $\phi(B)\tilde{w}_t = \theta(B)a_t$. It can be shown that for such a process the factors $|\mathbf{I}(\boldsymbol{\phi}, \boldsymbol{\theta})|^{1/2}$ and $f(\boldsymbol{\phi}, \boldsymbol{\theta})$ in (7.4.5) do not exactly cancel. Instead, we can show, based on (7.2.24), that

$$|\mathbf{I}(\boldsymbol{\phi}, \boldsymbol{\theta})|^{1/2} f(\boldsymbol{\phi}, \boldsymbol{\theta}) = J(\boldsymbol{\phi}^* \mid \boldsymbol{\phi}, \boldsymbol{\theta}) \tag{7.4.21}$$

* The approximate numerical integration to achieve a "standardized" curve of unit area, which could be carried out more exactly if desired, makes it possible to read off the actual probability densities. For almost all practical purposes this refinement is not needed. It is sufficient to compute and plot on some convenient scale $[S(\lambda)]^{-n/2}$, which is *proportional* to the actual density.

In (7.4.21) the ϕ^*'s are the $p + q$ parameters obtained by multiplying the autoregressive and moving average operators

$$(1 - \phi_1^* B - \phi_2^* B^2 - \cdots - \phi_{p+q}^* B^{p+q}) = (1 - \phi_1 B - \cdots - \phi_p B^p) \times (1 - \theta_1 B - \cdots - \theta_q B^q)$$

and J is the Jacobian of the transformation from $\boldsymbol{\phi}^*$ to $(\boldsymbol{\phi}, \boldsymbol{\theta})$, that is,

$$p(\boldsymbol{\phi}, \boldsymbol{\theta} | \mathbf{z}) \propto J(\boldsymbol{\phi}^* | \boldsymbol{\phi}, \boldsymbol{\theta})[S(\boldsymbol{\phi}, \boldsymbol{\theta})]^{-n/2} \tag{7.4.22}$$

In particular, for the ARMA(1, 1) process, $\phi_1^* = \phi + \theta$, $\phi_2^* = -\phi\theta$, $J = |\phi - \theta|$, and

$$p(\phi, \theta | \mathbf{z}) \propto |\phi - \theta| [S(\phi, \theta)]^{-n/2} \tag{7.4.23}$$

In this case we see that the Jacobian will dominate in a region close to the line $\phi = \theta$ and will produce zero density on the line. This is sensible because the sum of squares $S(\phi, \theta)$ will take the finite value $\Sigma_{t=1}^n \tilde{w}_t^2$ for any $\phi = \theta$ and corresponds to our entertaining the possibility that $\tilde{w}_t$ is white noise. However, in our derivation we have not constrained the range of the parameters. The possibility that $\phi = \theta$ is thus associated with unlimited ranges for the (equal) parameters. The effect of limiting the parameter space by, for example, introducing the requirements for stationarity and invertibility $(-1 < \phi < 1, -1 < \theta < 1)$ would be to produce a small positive value for the density, but this refinement seems scarcely worthwhile.

The Bayesian analysis reinforces the point made in Section 7.3.5 that estimation difficulties will be encountered with the mixed model, and in particular with iterative solutions, when there is near-redundancy in the parameters. We have already seen that the use of preliminary identification will usually ensure that these situations are avoided.

7.5 LIKELIHOOD FUNCTION BASED ON THE STATE SPACE MODEL

Recall that in Section 5.5 it was indicated that the state space model formulation of the ARMA process and the associated Kalman filtering procedure could be useful for exact finite sample forecasting and also for exact likelihood function analysis. This approach has been described by Jones [123], Gardner, Harvey, and Phillips [94], and others. We now indicate how the Kalman filtering in the state space model form can be used as a convenient method for evaluation of the exact likelihood function, recursively, given n observations $w_1, w_2, \ldots, w_n$ from an ARMA(p, q) model.

The state space model form, as in Section 5.5, for the ARMA(p, q) model $\phi(B)w_t = \theta(B)a_t$, is given by $\boldsymbol{Y}_t = \boldsymbol{\Phi}\boldsymbol{Y}_{t-1} + \boldsymbol{\Psi}a_t$, and $w_t = \mathbf{H}\boldsymbol{Y}_t$, where $\boldsymbol{Y}_t' = (w_t, \hat{w}_t(1), \ldots, \hat{w}_t(r-1))$, $r = \max(p, q+1)$, $\mathbf{H} = (1, 0, \ldots, 0)$,

$$\mathbf{\Phi} = \begin{bmatrix} 0 & 1 & 0 & \cdots & 0 \\ 0 & 0 & 1 & \cdots & 0 \\ \cdot & \cdot & \cdot & \cdots & \cdot \\ 0 & 0 & \cdots & \cdots & 1 \\ \phi_r & \phi_{r-1} & \cdots & \cdots & \phi_1 \end{bmatrix}$$

and $\mathbf{\Psi}' = (1, \psi_1, \ldots, \psi_{r-1})$. The Kalman filter equations (5.5.5) through (5.5.8) provide one-step-ahead forecasts $\hat{\boldsymbol{Y}}_{t|t-1} = E[\boldsymbol{Y}_t | w_{t-1}, \ldots, w_1]$ of the state vector $\boldsymbol{Y}_t$ as well as the error covariance matrix $\mathbf{V}_{t|t-1} = E[(\boldsymbol{Y}_t - \hat{\boldsymbol{Y}}_{t|t-1})(\boldsymbol{Y}_t - \hat{\boldsymbol{Y}}_{t|t-1})']$. Specifically, for the state space form of the ARMA(p, q) model, these recursive equations are

$$\hat{\boldsymbol{Y}}_{t|t} = \hat{\boldsymbol{Y}}_{t|t-1} + \mathbf{K}_t(w_t - \hat{w}_{t|t-1}) \quad \text{with} \quad \mathbf{K}_t = \mathbf{V}_{t|t-1}\mathbf{H}'[\mathbf{H}\mathbf{V}_{t|t-1}\mathbf{H}']^{-1} \tag{7.5.1}$$

where $\hat{w}_{t|t-1} = \mathbf{H}\hat{\boldsymbol{Y}}_{t|t-1}$, and

$$\hat{\boldsymbol{Y}}_{t|t-1} = \mathbf{\Phi}\hat{\boldsymbol{Y}}_{t-1|t-1} \qquad \mathbf{V}_{t|t-1} = \mathbf{\Phi}\mathbf{V}_{t-1|t-1}\mathbf{\Phi}' + \sigma_a^2\mathbf{\Psi}\mathbf{\Psi}' \tag{7.5.2}$$

with

$$\mathbf{V}_{t|t} = [\mathbf{I} - \mathbf{K}_t\mathbf{H}]\mathbf{V}_{t|t-1} \tag{7.5.3}$$

for $t = 1, 2, \ldots, n$. In particular, then, the first component of the forecast vector is $\hat{w}_{t|t-1} = \mathbf{H}\hat{\boldsymbol{Y}}_{t|t-1} = E[w_t | w_{t-1}, \ldots, w_1]$, and the element $\sigma_a^2 v_t = \mathbf{H}\mathbf{V}_{t|t-1}\mathbf{H}' = E[(w_t - \hat{w}_{t|t-1})^2]$ is the forecast error variance.

To obtain the exact likelihood function of the vector of n observations $\mathbf{w}' = (w_1, w_2, \ldots, w_n)$ using the results above, we note that the joint distribution of w can be factored as

$$p(\mathbf{w}|\boldsymbol{\phi}, \boldsymbol{\theta}, \sigma_a) = \prod_{t=1}^{n} p(w_t | w_{t-1}, \ldots, w_1, \boldsymbol{\phi}, \boldsymbol{\theta}, \sigma_a) \tag{7.5.4}$$

where $p(w_t | w_{t-1}, \ldots, w_1, \boldsymbol{\phi}, \boldsymbol{\theta}, \sigma_a)$ denotes the conditional distribution of w_t given $w_{t-1}, \ldots, w_1$. Under Normality, this conditional distribution is Normal with (conditional) mean $\hat{w}_{t|t-1} = E[w_t | w_{t-1}, \ldots, w_1]$ and (conditional) variance $\sigma_a^2 v_t = E[(w_t - \hat{w}_{t|t-1})^2]$. Hence the joint distribution of $\mathbf{w}$ can be conveniently expressed as

$$p(\mathbf{w}|\boldsymbol{\phi}, \boldsymbol{\theta}, \sigma_a) = \prod_{t=1}^{n} (2\pi\sigma_a^2 v_t)^{-1/2} \exp\left[-\frac{1}{2\sigma_a^2}\sum_{t=1}^{n}\frac{(w_t - \hat{w}_{t|t-1})^2}{v_t}\right] \tag{7.5.5}$$

where the quantities $\hat{w}_{t|t-1}$ and $\sigma_a^2 v_t$ are easily determined recursively from the Kalman filter procedure. The initial values needed to start the Kalman filter recursions are given by $\hat{\boldsymbol{Y}}_{0|0} = 0$, an r-dimensional vector of zeros, and $\mathbf{V}_{0|0} = \text{cov}[\boldsymbol{Y}_0]$. The elements of $\mathbf{V}_{0|0}$ can readily be determined as a function of the autocovariances γ_k and the weights ψ_k of the ARMA(p, q) process

w_t, making use of the relation $w_{t+j} = \hat{w}_t(j) + \sum_{k=0}^{j-1} \psi_k a_{t+j-k}$ from Chapter 5. See [123] for further details. For example, in the case of an ARMA(1, 1) model for w_t, we have $Y_t' = (w_t, \hat{w}_t(1))$, so

$$\mathbf{V}_{0|0} = \text{cov}[Y_0] = \begin{bmatrix} \gamma_0 & \gamma_1 \\ \gamma_1 & \gamma_0 - \sigma_a^2 \end{bmatrix} = \sigma_a^2 \begin{bmatrix} \sigma_a^{-2}\gamma_0 & \sigma_a^{-2}\gamma_1 \\ \sigma_a^{-2}\gamma_1 & \sigma_a^{-2}\gamma_0 - 1 \end{bmatrix}$$

It also is generally the case that the one-step-ahead forecasts $\hat{w}_{t|t-1}$ and the corresponding error variances $\sigma_a^2 v_t$ rather quickly approach their steady-state forms, in which case the Kalman filter calculations at some stage (beyond time t_0, say) could be switched to the simpler form $\hat{w}_{t|t-1} = \sum_{i=1}^{p} \phi_i w_{t-i} - \sum_{i=1}^{q} \theta_i a_{t-i|t-i-1}$, and $\sigma_a^2 v_t = \text{var}[a_{t|t-1}] = \sigma_a^2$, for $t > t_0$, where $a_{t|t-1} = w_t - \hat{w}_{t|t-1}$. For example, refer to [94] for further details.

Innovations method. The likelihood function expressed in the form of (7.5.5) is generally referred to as the *innovations form*, and the quantities $a_{t|t-1} = w_t - \hat{w}_{t|t-1}$, $t = 1, \ldots, n$, are the (finite-sample) *innovations*. Calculation of the likelihood function in this form, based on the state space model representation of the ARMA process and associated Kalman filtering algorithms, has been proposed by many authors including Gardner, Harvey, and Phillips [94], Harvey and Phillips [107], and Jones [123]. The innovations form of the likelihood can also be obtained without directly using the state space model representation through the use of an "innovations algorithm" (e.g., see Ansley [11] and Brockwell and Davis [63]). This method essentially involves a Cholesky decomposition of an $n \times n$ band covariance matrix of the derived MA(q) process $w_t' = w_t - \phi_1 w_{t-1} - \cdots - \phi_p w_{t-p} = a_t - \theta_1 a_{t-1} - \cdots - \theta_q a_{t-q}$. More specifically, using the notation of Appendix A7.3 we write the ARMA model relations as $\mathbf{L}_\phi \mathbf{w} = \mathbf{L}_\theta \mathbf{a} + \mathbf{F}\mathbf{e}_*$. Then the covariance matrix of the vector of derived variables $\mathbf{L}_\phi \mathbf{w}$ is

$$\text{cov}[\mathbf{L}_\phi \mathbf{w}] = \text{cov}[\mathbf{L}_\theta \mathbf{a} + \mathbf{F}\mathbf{e}_*] = \sigma_a^2(\mathbf{L}_\theta \mathbf{L}_\theta' + \mathbf{F}\boldsymbol{\Omega}\mathbf{F}')$$

which is a band matrix. The innovations algorithm obtains the (square-root-free) Cholesky decomposition of the band matrix $\mathbf{L}_\theta \mathbf{L}_\theta' + \mathbf{F}\boldsymbol{\Omega}\mathbf{F}'$ as $\mathbf{G}\mathbf{D}\mathbf{G}'$, where $\mathbf{G}$ is a lower triangular band matrix with ones on the diagonal and $\mathbf{D}$ is a diagonal matrix with positive diagonal elements v_t, $t = 1, \ldots, n$. Hence, the quadratic form in the likelihood function is

$$\mathbf{w}'[\text{cov}(\mathbf{w})]^{-1}\mathbf{w} = \frac{1}{\sigma_a^2} \mathbf{w}'(\mathbf{L}_\phi^{-1}\mathbf{G}\mathbf{D}\mathbf{G}'\mathbf{L}_\phi'^{-1})^{-1}\mathbf{w}$$

$$= \frac{1}{\sigma_a^2} \mathbf{e}'\mathbf{D}^{-1}\mathbf{e} = \frac{1}{\sigma_a^2} \sum_{t=1}^{n} \frac{a_{t|t-1}^2}{v_t}$$

where $\mathbf{e} = \mathbf{G}^{-1}\mathbf{L}_\phi \mathbf{w} = (a_{1|0}, a_{2|1}, \ldots, a_{n|n-1})'$ is the vector of innovations, which are computed recursively from $\mathbf{G}\mathbf{e} = \mathbf{L}_\phi \mathbf{w}$.

The "innovations" state space approach to exact likelihood function determination has also been shown to be quite useful in dealing with estimation problems for ARMA models when some values z_t of the series are not observed, that is, there are missing values among the sequence of z_t values, for example, Jones [123], Harvey and Pierse [108], and Wincek and Reinsel [205]. The exact likelihood function calculated using the Kalman filtering approach via the recursive equations (7.5.1) to (7.5.3) can be maximized by using numerical optimization algorithms. These typically require some form of first partial derivatives of the log-likelihood with respect to the unknown parameters, and it is more satisfactory numerically to obtain analytical derivatives. From the form of the likelihood in (7.5.5), it is seen that this involves obtaining partial derivatives of the one-step predictions $\hat{w}_{t|t-1}$ and of the error variances $\sigma_a^2 v_t$ for each $t = 1, \ldots, n$. Wincek and Reinsel [205] give details on how the exact derivatives of $a_{t|t-1} = w_t - \hat{w}_{t|t-1}$ and $\sigma_a^2 v_t =$ var$[a_{t|t-1}]$ with respect to the model parameters $\boldsymbol{\phi}$, $\boldsymbol{\theta}$, and σ_a^2 can be obtained recursively through differentiation of equations of the Kalman filter procedure [the updating and prediction equations (7.5.1) to (7.5.3)]. This in turn leads to an explicit form of iterative calculations for the maximum likelihood estimation associated with the likelihood (7.5.5), similar to the nonlinear least squares procedures detailed in Section 7.2.

For illustration of the state space approach to the likelihood calculations, we refer to the example from Section 7.1.5 involving $n = 12$ values from an ARMA(1, 1) model, and indicate the results of the Kalman filter calculations with the parameter values $\phi = 0.3$, $\theta = 0.7$, as before. First, since for an ARMA(1, 1) model, $\sigma_a^{-2}\gamma_0 = (1 + \theta^2 - 2\phi\theta)/(1 - \phi^2)$, $\sigma_a^{-2}\gamma_1 = (1 - \phi\theta)(\phi - \theta)/(1 - \phi^2)$, $\sigma_a^{-2}\gamma_0 - 1 = (\phi - \theta)^2/(1 - \phi^2)$, we then have

$$\mathbf{V}_{0|0} = \sigma_a^2 \begin{bmatrix} 1.176 & -0.347 \\ -0.347 & 0.176 \end{bmatrix}$$

Also, in this case the Kalman gain matrix $\mathbf{K}_t$ in (7.5.1) will be of the form $\mathbf{K}_t = (1, k_{2t})'$, where k_{2t} is equal to the ratio of the element in the (2, 1) position to that in the (1, 1) position of the matrix $\mathbf{V}_{t|t-1}$. Following through the recursive calculations indicated in relation to (7.5.1) to (7.5.3), we find the resulting values of $a_{t|t-1} = w_t - \hat{w}_{t|t-1}$, $v_t = \sigma_a^{-2}$ var$[a_{t|t-1}]$, and k_{2t} as shown in Table 7.14. By comparison with the values of $[a_t]$ in Table 7.4, we find, in particular, that the $a_{t|t-1}$ and the v_t have essentially reached their steady-state forms from about $t = 7$ onward, since beyond this time the $a_{t|t-1}$ satisfy $a_{t|t-1} = w_t - 0.3w_{t-1} + 0.7a_{t-1|t-2}$ with $v_t = 1.00$ and the values $a_{t|t-1}$ agree with the values of $[a_t]$ in Table 7.4. Equivalently, the Kalman gain $\mathbf{K}_t$ approaches the steady-state value of $\boldsymbol{\Psi} = (1, \psi_1)'$, with $\psi_1 = \phi - \theta = -0.4$, from about $t = 7$ onward, which implies that the equations in (7.5.1) reduce to the steady-state form such that $\hat{w}_{t|t-1} = 0.3w_{t-1} - 0.7a_{t-1|t-2}$. In addition,

TABLE 7.14 Recursive Calculations from 12 Values Assumed to be Generated by the Model $(1 - 0.3B)w_t = (1 - 0.7B)a_t$ Using the Kalman Filter

t	$a_{t\|t-1}$	v_t	k_{2t}	t	$a_{t\|t-1}$	v_t	k_{2t}
1	1.10	1.18	−0.30	7	2.65	1.00	−0.40
2	4.62	1.07	−0.35	8	−0.14	1.00	−0.40
3	4.73	1.03	−0.38	9	0.17	1.00	−0.40
4	1.60	1.02	−0.39	10	−0.09	1.00	−0.40
5	2.91	1.01	−0.39	11	0.83	1.00	−0.40
6	4.74	1.00	−0.40	12	2.34	1.00	−0.40

we find that the Kalman filter approach leads to the exact sum of squares value of $S(0.3, 0.7) = \Sigma_{t=1}^{12} a_{t|t-1}^2/v_t = 89.158$, the same as obtained in Section 7.1.5 using different methods, as well as the determinant value $\Pi_{t=1}^{12} v_t = |\mathbf{M}_n^{(1,1)}|^{-1} = 1.345$, which is the same as the value obtained through the relation

$$|\mathbf{M}_n^{(1,1)}|^{-1} = 1 + \frac{1 - \theta^{2n}}{1 - \theta^2}\frac{(\phi - \theta)^2}{1 - \phi^2}$$

derived in Appendix A7.3 using the alternative approach.

APPENDIX A7.1 REVIEW OF NORMAL DISTRIBUTION THEORY

A7.1.1 Partitioning of a Positive-Definite Quadratic Form

Consider the positive-definite quadratic form $Q_p = \mathbf{x}'\Sigma^{-1}\mathbf{x}$. Suppose that the $p \times 1$ vector $\mathbf{x}$ is partitioned after the p_1th element, so that $\mathbf{x}' = (\mathbf{x}_1' \vdots \mathbf{x}_2') = (x_1, x_2, \ldots, x_{p_1} \vdots x_{p_1+1}, \ldots, x_p)$, and suppose that the $p \times p$ matrix Σ is also partitioned after the p_1th row and column, so that

$$\Sigma = \begin{bmatrix} \Sigma_{11} & \Sigma_{12} \\ \Sigma_{12}' & \Sigma_{22} \end{bmatrix}$$

Then, since

$$\mathbf{x}'\Sigma^{-1}\mathbf{x} = (\mathbf{x}_1' \vdots \mathbf{x}_2')$$

$$\times \begin{bmatrix} \mathbf{I} & -\Sigma_{11}^{-1}\Sigma_{12} \\ \mathbf{0} & \mathbf{I} \end{bmatrix} \begin{bmatrix} \Sigma_{11}^{-1} & 0 \\ \mathbf{0} & (\Sigma_{22} - \Sigma_{12}'\Sigma_{11}^{-1}\Sigma_{12})^{-1} \end{bmatrix} \begin{bmatrix} \mathbf{I} & \mathbf{0} \\ -\Sigma_{12}'\Sigma_{11}^{-1} & \mathbf{I} \end{bmatrix} \begin{pmatrix} \mathbf{x}_1 \\ \mathbf{x}_2 \end{pmatrix}$$

$Q_p = \mathbf{x}'\boldsymbol{\Sigma}^{-1}\mathbf{x}$ can always be written as a sum of two quadratic forms Q_{p_1} and Q_{p_2}, containing p_1 and p_2 elements, respectively, where

$$\begin{aligned} Q_p &= Q_{p_1} + Q_{p_2} \\ Q_{p_1} &= \mathbf{x}_1'\boldsymbol{\Sigma}_{11}^{-1}\mathbf{x}_1 \\ Q_{p_2} &= (\mathbf{x}_2 - \boldsymbol{\Sigma}_{12}'\boldsymbol{\Sigma}_{11}^{-1}\mathbf{x}_1)'(\boldsymbol{\Sigma}_{22} - \boldsymbol{\Sigma}_{12}'\boldsymbol{\Sigma}_{11}^{-1}\boldsymbol{\Sigma}_{12})^{-1}(\mathbf{x}_2 - \boldsymbol{\Sigma}_{12}'\boldsymbol{\Sigma}_{11}^{-1}\mathbf{x}_1) \end{aligned} \tag{A7.1.1}$$

We may also write for the determinant of $\boldsymbol{\Sigma}$

$$|\boldsymbol{\Sigma}| = |\boldsymbol{\Sigma}_{11}||\boldsymbol{\Sigma}_{22} - \boldsymbol{\Sigma}_{12}'\boldsymbol{\Sigma}_{11}^{-1}\boldsymbol{\Sigma}_{12}| \tag{A7.1.2}$$

A7.1.2 Two Useful Integrals

Let $\mathbf{z}'\mathbf{C}\mathbf{z}$ be a positive-definite quadratic form in $\mathbf{z}$, which has q elements, so that $\mathbf{z}' = (z_1, z_2, \ldots, z_q)$, where $-\infty < z_i < \infty$, $i = 1, 2, \ldots, q$, and let a, b, and m be positive real numbers. Then it may be shown that

$$\int_R \left(a + \frac{\mathbf{z}'\mathbf{C}\mathbf{z}}{b}\right)^{-(m+q)/2} d\mathbf{z} = \frac{(b\pi)^{q/2}\Gamma(m/2)}{a^{m/2}|\mathbf{C}|^{1/2}\Gamma[(m+q)/2]} \tag{A7.1.3}$$

where the q-fold integral extends over the entire $\mathbf{z}$ space R, and

$$\frac{\int_{\mathbf{z}'\mathbf{C}\mathbf{z} > qF_0} (1 + \mathbf{z}'\mathbf{C}\mathbf{z}/m)^{-(m+q)/2}\, d\mathbf{z}}{\int_R (1 + \mathbf{z}'\mathbf{C}\mathbf{z}/m)^{-(m+q)/2}\, d\mathbf{z}} = \int_{F_0}^{\infty} p(F|q, m)\, dF \tag{A7.1.4}$$

where the function $p(F|q, m)$ is called the F distribution with q and m degrees of freedom and is defined by

$$p(F|q, m) = \frac{(q/m)^{q/2}\Gamma[(m+q)/2]}{\Gamma(q/2)\Gamma(m/2)} F^{(q-2)/2}\left(1 + \frac{q}{m}F\right)^{-(m+q)/2}, \quad F > 0 \tag{A7.1.5}$$

If m tends to infinity, then

$$\left(1 + \frac{\mathbf{z}'\mathbf{C}\mathbf{z}}{m}\right)^{-(m+q)/2} \quad \text{tends to} \quad e^{-(\mathbf{z}'\mathbf{C}\mathbf{z})/2}$$

and writing $qF = \chi^2$, we obtain from (A7.1.4) that

$$\frac{\int_{\mathbf{z}'\mathbf{C}\mathbf{z} > \chi_0^2} e^{-(\mathbf{z}'\mathbf{C}\mathbf{z})/2}\, d\mathbf{z}}{\int_R e^{-(\mathbf{z}'\mathbf{C}\mathbf{z})/2}\, d\mathbf{z}} = \int_{\chi_0^2}^{\infty} p(\chi^2|q)\, d\chi^2 \tag{A7.1.6}$$

where the function $p(\chi^2|q)$ is called the χ^2 distribution with q degrees of freedom, and is defined by

$$p(\chi^2|q) = \frac{1}{2^{q/2}\Gamma(q/2)} (\chi^2)^{(q-2)/2} e^{-\chi^2/2}, \quad \chi^2 > 0 \tag{A7.1.7}$$

Here and elsewhere $p(x)$ is used as a *general* notation to denote a probability density function for a random variable x.

A7.1.3 Normal Distribution

The random variable x is said to be normally distributed with mean μ and standard deviation σ, or to have a distribution $N(\mu, \sigma^2)$, if its probability density is

$$p(x) = (2\pi)^{-1/2}(\sigma^2)^{-1/2}e^{-(x-\mu)^2/2\sigma^2} \tag{A7.1.8}$$

Thus the unit Normal deviate $u = (x - \mu)/\sigma$ has a distribution $N(0, 1)$. Table E in Part Five shows ordinates $p(u = u_\varepsilon)$ and values u_ε such that $\Pr\{u > u_\varepsilon\} = \varepsilon$ for chosen values of ε.

Multi-Normal distribution. The vector of random variables $\mathbf{x}' = (x_1, x_2, \ldots, x_p)$ is said to have a joint p-variate Normal distribution $N\{\boldsymbol{\mu}, \Sigma\}$ if its probability density function is

$$p(\mathbf{x}) = (2\pi)^{-p/2}|\Sigma|^{-1/2}e^{-(\mathbf{x}-\boldsymbol{\mu})'\Sigma^{-1}(\mathbf{x}-\boldsymbol{\mu})/2} \tag{A7.1.9}$$

The probability density contours are ellipsoids defined by $(\mathbf{x} - \boldsymbol{\mu})'\Sigma^{-1}(\mathbf{x} - \boldsymbol{\mu}) = \text{const}$. For illustration, the elliptical contours for a bivariate Normal distribution are shown in Figure A7.1.

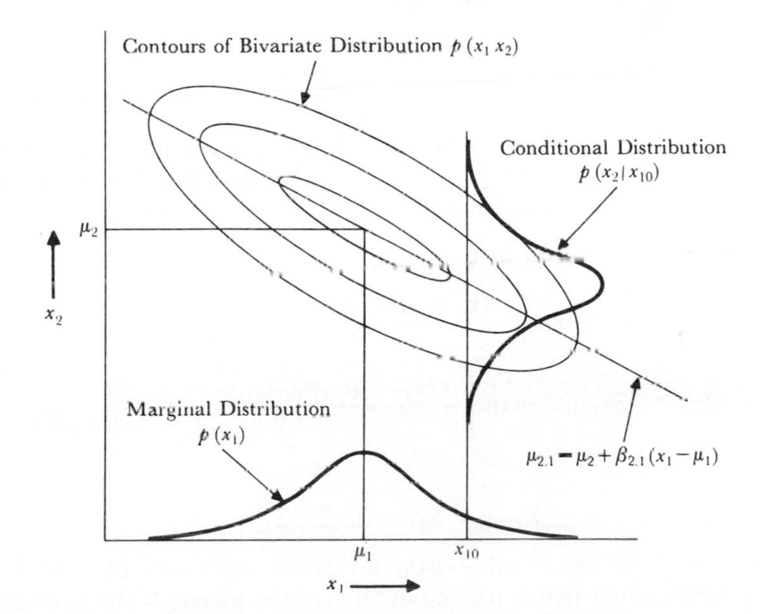

Figure A7.1 Contours of a bivariate Normal distribution showing the marginal distribution $p(x_1)$ and the conditional distribution $p(x_2|x_{10})$ at $x_1 = x_{10}$.

At the point $\mathbf{x} = \boldsymbol{\mu}$, the multivariate distribution has its maximum density

$$\max p(\mathbf{x}) = p(\boldsymbol{\mu}) = (2\pi)^{-p/2}|\boldsymbol{\Sigma}|^{-1/2}$$

The χ^2 distribution as the probability mass outside a density contour of the multivariate Normal. For the p-variate Normal distribution, (A7.1.9), the probability mass outside the density contour defined by

$$(\mathbf{x} - \boldsymbol{\mu})'\boldsymbol{\Sigma}^{-1}(\mathbf{x} - \boldsymbol{\mu}) = \chi_0^2$$

is given by the χ^2 integral with p degrees of freedom

$$\int_{\chi_0^2}^{\infty} p(\chi^2|p)\, d\chi^2$$

where the χ^2 density function is defined as in (A7.1.7). Table F in Part Five shows values of $\chi_\varepsilon^2(p)$, such that $\Pr\{\chi^2 > \chi_\varepsilon^2(p)\} = \varepsilon$ for chosen values of ε.

Marginal and conditional distributions for the multivariate Normal distribution. Suppose that the vector of $p = p_1 + p_2$ random variables is partitioned after the first p_1 elements, so that

$$\mathbf{x}' = (\mathbf{x}_1' : \mathbf{x}_2') = (x_1, x_2, \ldots, x_{p_1} : x_{p_1+1}, \ldots, x_{p_1+p_2})$$

and that the variance–covariance matrix is

$$\boldsymbol{\Sigma} = \begin{bmatrix} \boldsymbol{\Sigma}_{11} & \boldsymbol{\Sigma}_{12} \\ \boldsymbol{\Sigma}_{12}' & \boldsymbol{\Sigma}_{22} \end{bmatrix}$$

Then using (A7.1.1) and (A7.1.2), we can write the multivariate Normal distribution for the $p = p_1 + p_2$ variates as the *marginal* distribution of $\mathbf{x}_1$ multiplied by the *conditional* distribution of $\mathbf{x}_2$ given $\mathbf{x}_1$, that is,

$$\begin{aligned} p(\mathbf{x}) &= p(\mathbf{x}_1, \mathbf{x}_2) = p(\mathbf{x}_1)p(\mathbf{x}_2|\mathbf{x}_1) \\ &= (2\pi)^{-p_1/2}|\boldsymbol{\Sigma}_{11}|^{-1/2}\exp\left[-\frac{(\mathbf{x}_1 - \boldsymbol{\mu}_1)'\boldsymbol{\Sigma}_{11}^{-1}(\mathbf{x}_1 - \boldsymbol{\mu}_1)}{2}\right] \\ &\quad \times (2\pi)^{-p_2/2}|\boldsymbol{\Sigma}_{22.11}|^{-1/2}\exp\left[-\frac{(\mathbf{x}_2 - \boldsymbol{\mu}_{2.1})'\boldsymbol{\Sigma}_{22.11}^{-1}(\mathbf{x}_2 - \boldsymbol{\mu}_{2.1})}{2}\right] \end{aligned} \tag{A7.1.10}$$

where

$$\boldsymbol{\Sigma}_{22.11} = \boldsymbol{\Sigma}_{22} - \boldsymbol{\Sigma}_{12}'\boldsymbol{\Sigma}_{11}^{-1}\boldsymbol{\Sigma}_{12} \tag{A7.1.11}$$

and $\boldsymbol{\mu}_{2.1} = \boldsymbol{\mu}_2 + \boldsymbol{\beta}_{2.1}(\mathbf{x}_1 - \boldsymbol{\mu}_1) = E(\mathbf{x}_2|\mathbf{x}_1)$ define regression hyperplanes in $(p_1 + p_2)$-dimensional space, tracing the loci of the means of the p_2 elements of $\mathbf{x}_2$ as the p_1 elements of $\mathbf{x}_1$ vary. The $p_2 \times p_1$ matrix of regression coefficients is given by $\boldsymbol{\beta}_{2.1} = \boldsymbol{\Sigma}_{12}'\boldsymbol{\Sigma}_{11}^{-1}$.

Both marginal and conditional distributions for the multivariate Normal and therefore multivariate Normal distributions. It is seen that for the *multivariate Normal distribution*, the conditional distribution $p(\mathbf{x}_2|\mathbf{x}_1)$ is, *except for location* (i.e., mean value), identical whatever the value of $\mathbf{x}_1$.

Univariate marginals. In particular, the marginal density for a single element x_i $(i = 1, 2, \ldots, p)$ is $N(\mu_i, \sigma_i^2)$, a univariate Normal with mean equal to the ith element of $\boldsymbol{\mu}$ and variance equal to the ith diagonal element of Σ.

Bivariate Normal. For illustration, the marginal and conditional distributions for a bivariate Normal are shown in Figure A7.1. In this case, the marginal distribution of x_1 is $N(\mu_1, \sigma_1^2)$, while the conditional distribution of x_2 given x_1 is

$$N\left\{\mu_2 + \rho\frac{\sigma_2}{\sigma_1}(x_1 - \mu_1), \sigma_2^2(1 - \rho^2)\right\}$$

where $\rho = (\sigma_1/\sigma_2)\beta_{2.1}$ is the correlation coefficient between x_1 and x_2.

A7.1.4 Student's *t*-Distribution

The random variable x is said to be distributed in a scaled t distribution $t(\mu, s^2, \nu)$, with mean μ and scale parameter s and with ν degrees of freedom if

$$p(x) = (2\pi)^{-1/2}(s^2)^{-1/2}\left(\frac{\nu}{2}\right)^{-1/2}\Gamma\left(\frac{\nu+1}{2}\right)\Gamma^{-1}\left(\frac{\nu}{2}\right)\left[1 + \frac{(x-\mu)^2}{\nu s^2}\right]^{-(\nu+1)/2} \tag{A7.1.12}$$

Thus the standardized t deviate $t = (x - \mu)/s$ has distribution $t(0, 1, \nu)$. Table G in Part Five shows values t_ε such that $\Pr\{t > t_\varepsilon\} = \varepsilon$ for chosen values of ε.

Approach to Normal distribution. For large ν, the product

$$\left(\frac{\nu}{2}\right)^{-1/2}\Gamma\left(\frac{\nu+1}{2}\right)\Gamma^{-1}\left(\frac{\nu}{2}\right)$$

tends to unity, while the right-hand bracket in (A7.1.12) tends to $e^{-(1/2s^2)(x-\mu)^2}$. Thus if for large ν we write $s^2 = \sigma^2$, the t distribution tends to the Normal distribution (A7.1.8).

Multiple *t* distribution. Let $\boldsymbol{\mu}'$ be a $p \times 1$ vector $(\mu_1, \mu_2, \ldots, \mu_p)$ and S a $p \times p$ positive-definite matrix. Then the vector random variable $\mathbf{x}$ is said [73], [86] to have a scaled t distribution $t(\boldsymbol{\mu}, \mathbf{S}, \nu)$ with means $\boldsymbol{\mu}$, scaling matrix $\mathbf{S}$, and ν degrees of freedom if

$$p(\mathbf{x}) = (2\pi)^{-p/2}|\mathbf{S}|^{-1/2}\left(\frac{\nu}{2}\right)^{-p/2}\Gamma\left(\frac{\nu+p}{2}\right)\Gamma^{-1}\left(\frac{\nu}{2}\right) \times \left[1 + \frac{(\mathbf{x}-\boldsymbol{\mu})'\mathbf{S}^{-1}(\mathbf{x}-\boldsymbol{\mu})}{\nu}\right]^{-(\nu+p)/2} \tag{A7.1.13}$$

The probability contours of the multiple t distribution are ellipsoids defined by $(\mathbf{x}-\boldsymbol{\mu})'\mathbf{S}^{-1}(\mathbf{x}-\boldsymbol{\mu}) = \text{const.}$

Approach to the multi-Normal form. For large ν, the product

$$\left(\frac{\nu}{2}\right)^{-p/2}\Gamma\left(\frac{\nu+p}{2}\right)\Gamma^{-1}\left(\frac{\nu}{2}\right)$$

tends to unity; also, the right-hand bracket in (A7.1.13) tends to $e^{-(\mathbf{x}-\boldsymbol{\mu})'\mathbf{S}^{-1}(\mathbf{x}-\boldsymbol{\mu})/2}$. Thus if for large ν we write $\mathbf{S} = \boldsymbol{\Sigma}$, the multiple t tends to the multivariate Normal distribution (A7.1.9).

The F distribution as the probability mass outside a density contour of the multiple t distribution. Using (A7.1.4), the probability mass outside the density contour of the p-variate t distribution $t(\boldsymbol{\mu}, \mathbf{S}, \nu)$, defined by

$$(\mathbf{x}-\boldsymbol{\mu})'\mathbf{S}^{-1}(\mathbf{x}-\boldsymbol{\mu}) = pF_0$$

is given by the F integral with p and ν degrees of freedom

$$\int_{F_0}^{\infty} p(F|p, \nu)\, dF$$

where the density function for F is defined by (A7.1.5). For large ν, $\Pr\{F > F_\varepsilon(p, \nu)\} = \Pr\{\chi^2 > \chi^2_\varepsilon(p)\}$ with $pF = \chi^2$. Hence, as would be expected, the mass outside a density contour of the multiple t is, for large ν, equal to the mass outside the corresponding density contour of the multivariate Normal distribution, which the multiple t distribution approaches.

Marginal t distribution. If the $p = (p_1 + p_2)$-dimensional vector $\mathbf{x}$, distributed as in equation (A7.1.13), is partitioned after the p_1th element so that $\mathbf{x}' = (\mathbf{x}_1' \vdots \mathbf{x}_2')$, then with $\mathbf{S}$ similarly partitioned, we obtain

$$\mathbf{S} = \left[\begin{array}{c|c} \mathbf{S}_{11} & \mathbf{S}_{12} \\ \hline \mathbf{S}_{12}' & \mathbf{S}_{22} \end{array}\right]$$

Writing

$$\mathbf{S}_{22.11} = \mathbf{S}_{22} - \mathbf{S}_{12}'\mathbf{S}_{11}^{-1}\mathbf{S}_{12}$$

$$\boldsymbol{\mu}_{2.1} = \boldsymbol{\mu}_2 + \boldsymbol{\beta}_{2.1}(\mathbf{x}_1 - \boldsymbol{\mu}_1) \qquad \boldsymbol{\beta}_{2.1} = \mathbf{S}_{12}'\mathbf{S}_{11}^{-1}$$

then

$$p(\mathbf{x}_1, \mathbf{x}_2) = (2\pi)^{-(p_1+p_2)/2}|\mathbf{S}_{11}|^{-1/2}|\mathbf{S}_{22.11}|^{-1/2}\left(\frac{\nu}{2}\right)^{-(p_1+p_2)/2}$$

$$\times \Gamma\left(\frac{\nu + p_1 + p_2}{2}\right)\Gamma^{-1}\left(\frac{\nu}{2}\right)$$

$$\times \left[1 + \frac{(\mathbf{x}_1 - \boldsymbol{\mu}_1)'\mathbf{S}_{11}^{-1}(\mathbf{x}_1 - \boldsymbol{\mu}_1)}{\nu} + \frac{(\mathbf{x}_2 - \boldsymbol{\mu}_{2.1})'\mathbf{S}_{22.11}^{-1}(\mathbf{x}_2 - \boldsymbol{\mu}_{2.1})}{\nu}\right]^{-(\nu+p_1+p_2)/2}$$

Now, using the preliminary result (A7.1.3), with

$$a = 1 + \frac{(\mathbf{x}_1 - \boldsymbol{\mu}_1)'\mathbf{S}_{11}^{-1}(\mathbf{x}_1 - \boldsymbol{\mu}_1)}{\nu}$$

$b = \nu$, $m = \nu + p_1$, $q = p_2$, $\mathbf{C} = \mathbf{S}_{22.11}^{-1}$, we obtain

$$p(\mathbf{x}_1) = (2\pi)^{-p_1/2}|\mathbf{S}_{11}|^{-1/2}\left(\frac{\nu}{2}\right)^{-p_1/2}\Gamma\left(\frac{\nu + p_1}{2}\right)\Gamma^{-1}\left(\frac{\nu}{2}\right)$$
$$\times \left[1 + \frac{(\mathbf{x}_1 - \boldsymbol{\mu}_1)'\mathbf{S}_{11}^{-1}(\mathbf{x}_1 - \boldsymbol{\mu}_1)}{\nu}\right]^{-(\nu+p_1)/2} \quad \text{(A7.1.14)}$$

Thus, if a p-dimensional vector $\mathbf{x}$ has the multiple t distribution of equation (A7.1.13), the marginal distribution of any p_1 variables $\mathbf{x}_1$ is the p_1-variate t distribution $t(\boldsymbol{\mu}_1, \mathbf{S}_{11}, \nu)$.

Univariate marginals. In particular, the marginal distribution for a single element $x_i (i = 1, 2, \ldots, p)$ is $t(\mu_i, s_{ii}, \nu)$, a univariate t with mean equal to the ith element of $\boldsymbol{\mu}$ and scaling factor equal to the positive square root of the ith diagonal element of $\mathbf{S}$, and having ν degrees of freedom.

Conditional *t* distributions. The conditional distribution $p(\mathbf{x}_2|\mathbf{x}_1)$ can be obtained from the ratio of the joint distribution $p(\mathbf{x}_1, \mathbf{x}_2)$ and the marginal distribution $p(\mathbf{x}_1)$; thus

$$p(\mathbf{x}_2|\mathbf{x}_{10}) = \frac{p(\mathbf{x}_{10}, \mathbf{x}_2)}{p(\mathbf{x}_{10})}$$

Making the division, we have

$$p(\mathbf{x}_2|\mathbf{x}_{10}) = \text{const}\left\{1 + \frac{(\mathbf{x}_2 - \boldsymbol{\mu}_{2.1})'[c(\mathbf{x}_{10})\mathbf{S}_{22.11}^{-1}](\mathbf{x}_2 - \boldsymbol{\mu}_{2.1})}{\nu + p_1}\right\}^{-(\nu+p_1+p_2)/2} \tag{A7.1.15}$$

where $c(\mathbf{x}_{10})$ (which is a fixed constant for *given* $\mathbf{x}_{10}$) is given by

$$c(\mathbf{x}_{10}) = \frac{\nu + p_1}{\nu + (\mathbf{x}_{10} - \boldsymbol{\mu}_1)'\mathbf{S}_{11}^{-1}(\mathbf{x}_{10} - \boldsymbol{\mu}_1)} \tag{A7.1.16}$$

Thus the distribution of $\mathbf{x}_2$, for given $\mathbf{x}_1 = \mathbf{x}_{10}$, is the multiple t distribution

$$t\{\boldsymbol{\mu}_2 + \boldsymbol{\beta}_{2.1}(\mathbf{x}_{10} - \boldsymbol{\mu}_1), c^{-1}(\mathbf{x}_{10})\mathbf{S}_{22.11}, \nu + p_1\}$$

where $\boldsymbol{\beta}_{2.1} = \mathbf{S}_{12}'\mathbf{S}_{11}^{-1}$.

Bivariate *t* distribution. As before, some insight into the general multivariate situation can be gained by studying the bivariate distribution $t(\boldsymbol{\mu}, \mathbf{S}, \nu)$. A diagrammatic representation, parallel to Figure A7.1, is at first sight similar to the bivariate Normal. However, the marginal distributions are univariate t distributions having ν degrees of freedom, while the conditional distributions are t distributions with $\nu + 1$ degrees of freedom. Furthermore, the scale factor for the conditional distribution $p(x_2|x_{10})$, for example, would depend on x_{10}. This is to be contrasted with the conditional distribution for the Normal case, which has the same variance irrespective of x_{10}.

APPENDIX A7.2 REVIEW OF LINEAR LEAST SQUARES THEORY

A7.2.1 Normal Equations

The model is assumed to be

$$w_i = \beta_1 x_{i1} + \beta_2 x_{i2} + \cdots + \beta_k x_{ik} + e_i \tag{A7.2.1}$$

where the $w_i (i = 1, 2, \ldots, n)$ are observations obtained from an experiment in which the independent variables $x_{i1}, x_{i2}, \ldots, x_{ik}$ take on known *fixed* values, the β_i are unknown parameters to be estimated from the data, and the e_i are uncorrelated errors having zero means and the same variance σ^2.

The relations (A7.2.1) may be assembled into a matrix relation

$$\begin{bmatrix} w_1 \\ w_2 \\ \vdots \\ w_n \end{bmatrix} = \begin{bmatrix} x_{11} & x_{12} & \cdots & x_{1k} \\ x_{21} & x_{22} & \cdots & x_{2k} \\ \vdots & \vdots & & \vdots \\ x_{n1} & x_{n2} & \cdots & x_{nk} \end{bmatrix} \begin{bmatrix} \beta_1 \\ \beta_2 \\ \vdots \\ \beta_k \end{bmatrix} + \begin{bmatrix} e_1 \\ e_2 \\ \vdots \\ e_n \end{bmatrix}$$

or

$$\mathbf{w} = \mathbf{X}\boldsymbol{\beta} + \mathbf{e} \tag{A7.2.2}$$

where $\mathbf{X}$ is assumed to be of full rank k. Gauss's theorem of least squares may be stated [23] in the following form: The estimates $\hat{\boldsymbol{\beta}}' = (\hat{\beta}_1, \hat{\beta}_2, \ldots, \hat{\beta}_k)$ of the parameters $\boldsymbol{\beta}$, which are linear in the observations and unbiased for $\boldsymbol{\beta}$ and which minimize the mean square error among all such estimates of any linear function $\lambda_1\beta_1 + \lambda_2\beta_2 + \cdots + \lambda_k\beta_k$ of the parameters, are obtained by minimizing the sum of squares

$$S(\boldsymbol{\beta}) = \mathbf{e}'\mathbf{e} = (\mathbf{w} - \mathbf{X}\boldsymbol{\beta})'(\mathbf{w} - \mathbf{X}\boldsymbol{\beta}) \tag{A7.2.3}$$

To establish the minimum of $S(\boldsymbol{\beta})$, we note that the vector $\mathbf{w} - \mathbf{X}\boldsymbol{\beta}$ may be decomposed into two vectors $\mathbf{w} - \mathbf{X}\hat{\boldsymbol{\beta}}$ and $\mathbf{X}(\boldsymbol{\beta} - \hat{\boldsymbol{\beta}})$ according to

$$\mathbf{w} - \mathbf{X}\boldsymbol{\beta} = \mathbf{w} - \mathbf{X}\hat{\boldsymbol{\beta}} - \mathbf{X}(\boldsymbol{\beta} - \hat{\boldsymbol{\beta}}) \tag{A7.2.4}$$

Hence, provided that we choose $\hat{\boldsymbol{\beta}}$ so that

$$(\mathbf{X}'\mathbf{X})\hat{\boldsymbol{\beta}} = \mathbf{X}'\mathbf{w} \tag{A7.2.5}$$

it follows that

$$S(\boldsymbol{\beta}) = S(\hat{\boldsymbol{\beta}}) + (\boldsymbol{\beta} - \hat{\boldsymbol{\beta}})'\mathbf{X}'\mathbf{X}(\boldsymbol{\beta} - \hat{\boldsymbol{\beta}}) \tag{A7.2.6}$$

and the vectors $\mathbf{w} - \mathbf{X}\hat{\boldsymbol{\beta}}$ and $\mathbf{X}(\boldsymbol{\beta} - \hat{\boldsymbol{\beta}})$ are orthogonal. Since the second term on the right of (A7.2.6) is a positive-definite quadratic form, it follows that the minimum is attained when $\boldsymbol{\beta} = \hat{\boldsymbol{\beta}}$, where $\hat{\boldsymbol{\beta}} = (\mathbf{X}'\mathbf{X})^{-1}\mathbf{X}'\mathbf{w}$ is given by the *normal equations* (A7.2.5).

A7.2.2 Estimation of Residual Variance

Using (A7.2.3) and (A7.2.5), the sum of squares at the minimum is

$$S(\hat{\boldsymbol{\beta}}) = \mathbf{w}'\mathbf{w} - \hat{\boldsymbol{\beta}}'\mathbf{X}'\mathbf{X}\hat{\boldsymbol{\beta}} \tag{A7.2.7}$$

Furthermore, if we define

$$s^2 = \frac{S(\hat{\boldsymbol{\beta}})}{n - k} \tag{A7.2.8}$$

it may be shown [158] that $E[s^2] = \sigma^2$, and hence s^2 provides an unbiased estimate of σ^2.

A7.2.3 Covariance Matrix of Estimates

This is defined by

$$\begin{aligned}\mathbf{V}(\hat{\boldsymbol{\beta}}) &= \text{cov}[\hat{\boldsymbol{\beta}}, \hat{\boldsymbol{\beta}}'] \\ &= \text{cov}[(\mathbf{X}'\mathbf{X})^{-1}\mathbf{X}'\mathbf{w}, \mathbf{w}'\mathbf{x}(\mathbf{X}'\mathbf{X})^{-1}] \\ &= (\mathbf{X}'\mathbf{X})^{-1}\sigma^2\end{aligned} \tag{A7.2.9}$$

since $\text{cov}[\mathbf{w}, \mathbf{w}'] = \mathbf{I}\sigma^2$.

A7.2.4 Confidence Regions

Assuming Normality [158], the quadratic forms $S(\hat{\boldsymbol{\beta}})$ and $(\boldsymbol{\beta} - \hat{\boldsymbol{\beta}})'\mathbf{X}'\mathbf{X}(\boldsymbol{\beta} - \hat{\boldsymbol{\beta}})$ in (A7.2.6) are independently distributed as σ^2 times chi-squared random variables with $n - k$ and k degrees of freedom, respectively. Hence

$$\frac{(\boldsymbol{\beta} - \hat{\boldsymbol{\beta}})'\mathbf{X}'\mathbf{X}(\boldsymbol{\beta} - \hat{\boldsymbol{\beta}})}{S(\hat{\boldsymbol{\beta}})} \frac{n - k}{k}$$

is distributed as $F(k, n - k)$. Using (A7.2.8), it follows that

$$(\boldsymbol{\beta} - \hat{\boldsymbol{\beta}})'(\mathbf{X}'\mathbf{X})(\boldsymbol{\beta} - \hat{\boldsymbol{\beta}}) \leqslant ks^2F_\varepsilon(k, n - k) \tag{A7.2.10}$$

defines a $1 - \varepsilon$ confidence region for $\boldsymbol{\beta}$.

A7.2.5 Correlated Errors

Suppose that the errors $\mathbf{e}$ in (A7.2.2) have a *known* covariance matrix $\mathbf{V}$, where $\mathbf{V}^{-1} = \mathbf{P}\mathbf{P}'/\sigma^2$. Then (A7.2.2) may be written

$$\mathbf{P}'\mathbf{w} = \mathbf{P}'\mathbf{X}\boldsymbol{\beta} + \mathbf{P}'\mathbf{e}$$

or

$$\mathbf{w}^* = \mathbf{X}^*\boldsymbol{\beta} + \mathbf{e}^* \tag{A7.2.11}$$

The covariance matrix of $\mathbf{e}^* = \mathbf{P}'\mathbf{e}$ is

$$\text{cov}[\mathbf{P}'\mathbf{e}, \mathbf{e}'\mathbf{P}] = \mathbf{P}'\mathbf{V}\mathbf{P} = \mathbf{I}\sigma^2$$

Hence we may apply ordinary least squares theory with $\mathbf{V} = \mathbf{I}\sigma^2$ to the transformed model (A7.2.11), in which $\mathbf{w}$ is replaced by $\mathbf{w}^* = \mathbf{P}'\mathbf{w}$ and $\mathbf{X}$ by $\mathbf{X}^* = \mathbf{P}'\mathbf{X}$.

APPENDIX A7.3 EXACT LIKELIHOOD FUNCTION FOR MOVING AVERAGE AND MIXED PROCESSES

To obtain the required likelihood function for an MA(q) model, we have to derive the probability density function for a series $\mathbf{w}' = (w_1, w_2, \ldots, w_n)$ assumed to be generated by an invertible moving average model of order q:

$$\tilde{w}_t = a_t - \theta_1 a_{t-1} - \theta_2 a_{t-2} - \cdots - \theta_q a_{t-q} \qquad \text{(A7.3.1)}$$

where $\tilde{w}_t = w_t - \mu$, $\mu = E[w_t]$. Under the assumption that the a's and hence the $\tilde{w}$'s are normally distributed, the joint density may be written

$$p(\mathbf{w}|\boldsymbol{\theta}, \sigma_a, \mu) = (2\pi\sigma_a^2)^{-n/2}|\mathbf{M}_n^{(0,q)}|^{1/2} \exp\left[\frac{-\tilde{\mathbf{w}}'\mathbf{M}_n^{(0,q)}\tilde{\mathbf{w}}}{2\sigma_a^2}\right] \qquad \text{(A7.3.2)}$$

with $(\mathbf{M}_n^{(p,q)})^{-1}\sigma_a^2$ denoting the $n \times n$ covariance matrix of the w's for an ARMA(p, q) process. We now consider a convenient way of evaluating $\tilde{\mathbf{w}}'\mathbf{M}_n^{(0,q)}\tilde{\mathbf{w}}$, and for simplicity, we suppose that $\mu = 0$, so that $w_t = \tilde{w}_t$.

Using the model (A7.3.1), we can write down the n equations

$$w_t = a_t - \theta_1 a_{t-1} - \theta_2 a_{t-2} - \cdots - \theta_q a_{t-q} \qquad (t = 1, 2, \ldots, n)$$

These n equations can be conveniently expressed in matrix form in terms of the n dimensional vectors $\mathbf{w}' = (w_1, w_2, \ldots, w_n)$ and $\mathbf{a}' = (a_1, a_2, \ldots, a_n)$, and the q-dimensional vector of preliminary values $\mathbf{a}'_* = (a_{1-q}, a_{2-q}, \ldots, a_0)$ as

$$\mathbf{w} = \mathbf{L}_\theta \mathbf{a} + \mathbf{F}\mathbf{a}_*$$

where $\mathbf{L}_\theta$ is an $n \times n$ lower triangular matrix with 1's on the leading diagonal, $-\theta_1$ on the first subdiagonal, $-\theta_2$ on the second subdiagonal, and so on, with $\theta_i = 0$ for $i > q$. Further, $\mathbf{F}$ is an $n \times q$ matrix with the form $\mathbf{F} = (\mathbf{B}'_q, \mathbf{0}')'$, where $\mathbf{B}_q$ is $q \times q$ equal to

$$\mathbf{B}_q = -\begin{bmatrix} \theta_q & \theta_{q-1} & \cdots & \theta_1 \\ 0 & \theta_q & \cdots & \theta_2 \\ \cdots & \cdots & \cdots & \cdots \\ \cdots & \cdots & \cdots & \cdots \\ 0 & 0 & \cdots & \theta_q \end{bmatrix}$$

Now the joint distribution of the $n + q$ values which are the elements of $(\mathbf{a}', \mathbf{a}'_*)$ is

$$p(\mathbf{a}, \mathbf{a}_*|\sigma_a) = (2\pi\sigma_a^2)^{-(n+q)/2} \exp\left[-\frac{1}{2\sigma_a^2}(\mathbf{a}'\mathbf{a} + \mathbf{a}'_*\mathbf{a}_*)\right]$$

Noting that the transformation from $(\mathbf{a}, \mathbf{a}_*)$ to $(\mathbf{w}, \mathbf{a}_*)$ has unit Jacobian and $\mathbf{a} = \mathbf{L}_\theta^{-1}(\mathbf{w} - \mathbf{F}\mathbf{a}_*)$, the joint distribution of $\mathbf{w} = \mathbf{L}_\theta \mathbf{a} + \mathbf{F}\mathbf{a}_*$ and $\mathbf{a}_*$ is

$$p(\mathbf{w}, \mathbf{a}_* | \boldsymbol{\theta}, \sigma_a) = (2\pi\sigma_a^2)^{-(n+q)/2} \exp\left[-\frac{1}{2\sigma_a^2} S(\boldsymbol{\theta}, \mathbf{a}_*)\right]$$

where

$$S(\boldsymbol{\theta}, \mathbf{a}_*) = (\mathbf{w} - \mathbf{F}\mathbf{a}_*)'\mathbf{L}_\theta'^{-1}\mathbf{L}_\theta^{-1}(\mathbf{w} - \mathbf{F}\mathbf{a}_*) + \mathbf{a}_*'\mathbf{a}_* \qquad \text{(A7.3.3)}$$

Now let $\hat{\mathbf{a}}_*$ be the vector of values that minimize $S(\boldsymbol{\theta}, \mathbf{a}_*)$, which from generalized least squares theory can be shown to equal $\hat{\mathbf{a}}_* = \mathbf{D}^{-1}\mathbf{F}'\mathbf{L}_\theta'^{-1}\mathbf{L}_\theta^{-1}\mathbf{w}$, where $\mathbf{D} = \mathbf{I}_q + \mathbf{F}'\mathbf{L}_\theta'^{-1}\mathbf{L}_\theta^{-1}\mathbf{F}$. Then, using the result (A7.2.6), we have

$$S(\boldsymbol{\theta}, \mathbf{a}_*) = S(\boldsymbol{\theta}) + (\mathbf{a}_* - \hat{\mathbf{a}}_*)'\mathbf{D}(\mathbf{a}_* - \hat{\mathbf{a}}_*)$$

where

$$S(\boldsymbol{\theta}) = S(\boldsymbol{\theta}, \hat{\mathbf{a}}_*) = (\mathbf{w} - \mathbf{F}\hat{\mathbf{a}}_*)'\mathbf{L}_\theta'^{-1}\mathbf{L}_\theta^{-1}(\mathbf{w} - \mathbf{F}\hat{\mathbf{a}}_*) + \hat{\mathbf{a}}_*'\hat{\mathbf{a}}_* \qquad \text{(A7.3.4)}$$

is a function of the observations $\mathbf{w}$ but not of the preliminary values $\mathbf{a}_*$. Thus

$$p(\mathbf{w}, \mathbf{a}_* | \boldsymbol{\theta}, \sigma_a)$$

$$= (2\pi\sigma_a^2)^{-(n+q)/2} \exp\left\{-\frac{1}{2\sigma_a^2}\left[S(\boldsymbol{\theta}) + (\mathbf{a}_* - \hat{\mathbf{a}}_*)'\mathbf{D}(\mathbf{a}_* - \hat{\mathbf{a}}_*)\right]\right\}$$

However, since the joint distribution of $\mathbf{w}$ and $\mathbf{a}_*$ can be factored as

$$p(\mathbf{w}, \mathbf{a}_* | \boldsymbol{\theta}, \sigma_a) = p(\mathbf{w} | \boldsymbol{\theta}, \sigma_a) p(\mathbf{a}_* | \mathbf{w}, \boldsymbol{\theta}, \sigma_a)$$

it follows, similar to (A7.1.10), that

$$p(\mathbf{a}_* | \mathbf{w}, \boldsymbol{\theta}, \sigma_a) = (2\pi\sigma_a^2)^{-q/2}|\mathbf{D}|^{1/2} \exp\left[-\frac{1}{2\sigma_a^2}(\mathbf{a}_* - \hat{\mathbf{a}}_*)'\mathbf{D}(\mathbf{a}_* - \hat{\mathbf{a}}_*)\right] \qquad \text{(A7.3.5)}$$

$$p(\mathbf{w} | \boldsymbol{\theta}, \sigma_a) = (2\pi\sigma_a^2)^{-n/2}|\mathbf{D}|^{-1/2} \exp\left[-\frac{1}{2\sigma_a^2} S(\boldsymbol{\theta})\right] \qquad \text{(A7.3.6)}$$

We can now deduce the following:

1. From (A7.3.5), we see that $\hat{\mathbf{a}}_*$ is the conditional expectation of $\mathbf{a}_*$ given $\mathbf{w}$ and $\boldsymbol{\theta}$. Thus, using the notation introduced in Section 7.1.4, we obtain

$$\hat{\mathbf{a}}_* = [\mathbf{a}_* | \mathbf{w}, \boldsymbol{\theta}] = [\mathbf{a}_*]$$

whence $[\mathbf{a}] = \mathbf{L}_\theta^{-1}(\mathbf{w} - \mathbf{F}[\mathbf{a}_*])$ is the conditional expectation of $\mathbf{a}$ given $\mathbf{w}$ and $\boldsymbol{\theta}$, and using (A7.3.4),

$$S(\boldsymbol{\theta}) = [\mathbf{a}]'[\mathbf{a}] + [\mathbf{a}_*]'[\mathbf{a}_*] = \sum_{t=1-q}^{n} [a_t]^2 \qquad \text{(A7.3.7)}$$

To compute $S(\boldsymbol{\theta})$ the quantities $[a_t] = [a_t|\mathbf{w}, \boldsymbol{\theta}]$ may be obtained by using the estimates $[\mathbf{a}_*]' = ([a_{1-q}], [a_{2-q}], \ldots, [a_0])$ obtained as above by back-forecasting for preliminary values, and computing the elements $[a_1], [a_2], \ldots, [a_n]$ of $[\mathbf{a}]$ recursively from the relation $\mathbf{L}_\theta[\mathbf{a}] = \mathbf{w} - \mathbf{F}[\mathbf{a}_*]$ as

$$[a_t] = w_t + \theta_1[a_{t-1}] + \cdots + \theta_q[a_{t-q}] \qquad (t = 1, 2, \ldots, n)$$

Note that if the expression for $\hat{\mathbf{a}}_*$ is utilized in (A7.3.4), after rearranging we obtain

$$S(\boldsymbol{\theta}) = \mathbf{w}'\mathbf{L}_\theta'^{-1}(\mathbf{I}_n - \mathbf{L}_\theta^{-1}\mathbf{F}\mathbf{D}^{-1}\mathbf{F}'\mathbf{L}_\theta'^{-1})\mathbf{L}_\theta^{-1}\mathbf{w} = \mathbf{a}^{0\prime}\mathbf{a}^0 - \hat{\mathbf{a}}_*'\mathbf{D}\hat{\mathbf{a}}_*$$

where $\mathbf{a}^0 = \mathbf{L}_\theta^{-1}\mathbf{w}$ denotes the vector whose elements a_t^0 can be calculated recursively from $a_t^0 = w_t + \theta_1 a_{t-1}^0 + \cdots + \theta_q a_{t-q}^0$, $t = 1, 2, \ldots, n$, by setting the initial values $\mathbf{a}_*$ equal to zero. Hence the first term above, $S_*(\boldsymbol{\theta}) = \mathbf{a}^{0\prime}\mathbf{a}^0 = \Sigma_{t=1}^n (a_t^0)^2$, is the conditional sum of squares function, given $\mathbf{a}_* = 0$, as discussed in Section 7.1.2.

2. In addition, we find that

$$\mathbf{M}_n^{(0,q)} = \mathbf{L}_\theta'^{-1}(\mathbf{I}_n - \mathbf{L}_\theta^{-1}\mathbf{F}\mathbf{D}^{-1}\mathbf{F}'\mathbf{L}_\theta'^{-1})\mathbf{L}_\theta^{-1}$$

and $S(\boldsymbol{\theta}) = \mathbf{w}'\mathbf{M}_n^{(0,q)}\mathbf{w}$. Also, by comparing (A7.3.6) and (A7.3.2), we have

$$|\mathbf{D}|^{-1} = |\mathbf{M}_n^{(0,q)}|$$

3. The back-forecasts $\hat{\mathbf{a}}_* = [\mathbf{a}_*]$ can be calculated most conveniently from $\hat{\mathbf{a}}_* = \mathbf{D}^{-1}\mathbf{F}'\mathbf{u}$ (i.e., by solving $\mathbf{D}\hat{\mathbf{a}}_* = \mathbf{F}'\mathbf{u}$), where $\mathbf{u} = \mathbf{L}_\theta'^{-1}\mathbf{L}_\theta^{-1}\mathbf{w} = \mathbf{L}_\theta'^{-1}\mathbf{a}^0 = (u_1, u_2, \ldots, u_n)'$. Note that the elements u_t of $\mathbf{u}$ are calculated through a backward recursion as

$$u_t = a_t^0 + \theta_1 u_{t+1} + \cdots + \theta_q u_{t+q}$$

from $t = n$ down to $t = 1$, using zero starting values $u_{n+1} = \cdots = u_{n+q} = 0$, where the a_t^0 denote the estimates of the a_t conditional on the zero starting values $\mathbf{a}_* = 0$. Also, the vector $\mathbf{h} = \mathbf{F}'\mathbf{u}$ consists of the elements $h_j = -\Sigma_{i=1}^j \theta_{q-j+i}u_i$, $j = 1, \ldots, q$.

4. Finally, using (A7.3.6) and (A7.3.7), the unconditional likelihood is given exactly by

$$L(\boldsymbol{\theta}, \sigma_a|\mathbf{w}) = (\sigma_a^2)^{-n/2}|\mathbf{D}|^{-1/2} \exp\left\{-\frac{1}{2\sigma_a^2}\sum_{t=1-q}^{n}[a_t]^2\right\} \qquad \text{(A7.3.8)}$$

For example, in the MA(1) model with $q = 1$, we have $\mathbf{F}' = -(\theta, 0, \ldots, 0)$, an n-dimensional vector, and $\mathbf{L}_\theta$ is such that $\mathbf{L}_\theta^{-1}$ has first column equal to $(1, \theta, \theta^2, \ldots, \theta^{n-1})'$, so that

$$\mathbf{D} = 1 + \mathbf{F}'\mathbf{L}_\theta'^{-1}\mathbf{L}_\theta^{-1}\mathbf{F} = 1 + \theta^2 + \theta^4 + \cdots + \theta^{2n} = \frac{1 - \theta^{2(n+1)}}{1 - \theta^2}$$

In addition, the conditional values a_t^0 are computed recursively as $a_t^0 = w_t + \theta a_{t-1}^0$, $t = 1, 2, \ldots, n$, using the zero initial value $a_0^0 = 0$, and the values of the vector $\mathbf{u} = \mathbf{L}_\theta'^{-1}\mathbf{a}^0$ are computed in the backward recursion as $u_t = a_t^0 + \theta u_{t+1}$, from $t = n$ to $t = 1$, with $u_{n+1} = 0$. Then

$$\hat{\mathbf{a}}_* = [a_0] = -\mathbf{D}^{-1}\theta u_1 = -\mathbf{D}^{-1}u_0 = -\frac{u_0(1 - \theta^2)}{1 - \theta^{2(n+1)}}$$

where $u_0 = a_0^0 + \theta u_1 = \theta u_1$, and we have the exact likelihood

$$L(\theta, \sigma_a|\mathbf{w}) = (\sigma_a^2)^{-n/2}\frac{(1 - \theta^2)^{1/2}}{(1 - \theta^{2(n+1)})^{1/2}}\exp\left\{-\frac{1}{2\sigma_a^2}\sum_{t=0}^{n}[a_t]^2\right\} \tag{A7.3.9}$$

Extension to the autoregressive and mixed processes. The method outlined above may be readily extended to provide the unconditional likelihood for the general mixed model

$$\phi(B)\tilde{w}_t = \theta(B)a_t \tag{A7.3.10}$$

which, with $w_t = \nabla^d z_t$, defines the general ARIMA process. Details of the derivation have been presented by Newbold [149] and Ljung and Box [139], while an alternative approach to obtain the exact likelihood which uses the Cholesky decomposition of a band covariance matrix was given by Ansley [11]. First, assuming a zero mean for the process, the relations for the ARMA model may be written in matrix form, similar to before, as

$$\mathbf{L}_\phi\mathbf{w} = \mathbf{L}_\theta\mathbf{a} + \mathbf{F}\mathbf{e}_*$$

where $\mathbf{L}_\phi$ is an $n \times n$ matrix of the same form as $\mathbf{L}_\theta$ but with ϕ_i's in place of θ_i's, $\mathbf{e}_*' = (\mathbf{w}_*', \mathbf{a}_*') = (w_{1-p}, \ldots, w_0, a_{1-q}, \ldots, a_0)$ is the $(p + q)$-dimensional vector of initial values, and

$$\mathbf{F} = \begin{bmatrix} \mathbf{A}_p & \mathbf{B}_q \\ \mathbf{0} & \mathbf{0} \end{bmatrix}$$

with

$$\mathbf{A}_p = \begin{bmatrix} \phi_p & \phi_{p-1} & \cdots & \phi_1 \\ 0 & \phi_p & \cdots & \phi_2 \\ \cdots & \cdots & \cdots & \cdots \\ \cdots & \cdots & \cdots & \cdots \\ 0 & 0 & \cdots & \phi_p \end{bmatrix} \quad \text{and} \quad \mathbf{B}_q = -\begin{bmatrix} \theta_q & \theta_{q-1} & \cdots & \theta_1 \\ 0 & \theta_q & \cdots & \theta_2 \\ \cdots & \cdots & \cdots & \cdots \\ \cdots & \cdots & \cdots & \cdots \\ 0 & 0 & \cdots & \theta_q \end{bmatrix}$$

Let $\boldsymbol{\Omega}\sigma_a^2 = E[\mathbf{e}_*\mathbf{e}_*']$ denote the covariance matrix of $\mathbf{e}_*$. This matrix has the form

$$\boldsymbol{\Omega}\sigma_a^2 = \begin{bmatrix} \sigma_a^{-2}\boldsymbol{\Gamma}_p & \mathbf{C}' \\ \mathbf{C} & \mathbf{I}_q \end{bmatrix} \sigma_a^2$$

where $\boldsymbol{\Gamma}_p = E[\mathbf{w}_*\mathbf{w}_*']$ is a $p \times p$ matrix with (i, j)th element γ_{i-j}, and $\sigma_a^2\,\mathbf{C} = E[\mathbf{a}_*\mathbf{w}_*']$ has elements defined by $E[a_{i-q}w_{j-p}] = \sigma_a^2\psi_{j-i-p+q}$ for $j - i - p + q \geq 0$ and 0 otherwise. The ψ_k are the coefficients in the infinite MA operator $\psi(B) = \phi^{-1}(B)\theta(B) = \sum_{k=0}^{\infty} \psi_k B^k$, $\psi_0 = 1$, and are easily determined recursively through equations in Section 3.4. The autocovariances γ_k in $\boldsymbol{\Gamma}_p$ can directly be determined in terms of the coefficients ϕ_i and θ_i, and σ_a^2, through use of the first $(p + 1)$ equations (3.4.2) (see e.g., Ljung and Box [139]).

Similar to the result in (A7.3.3), since $\mathbf{a} = \mathbf{L}_\theta^{-1}(\mathbf{L}_\phi\mathbf{w} - \mathbf{F}\mathbf{e}_*)$ and $\mathbf{e}_*$ are independent, the joint distribution of $\mathbf{w}$ and $\mathbf{e}_*$ is

$$p(\mathbf{w}, \mathbf{e}_* | \boldsymbol{\phi}, \boldsymbol{\theta}, \sigma_a) = (2\pi\sigma_a^2)^{-(n+p+q)/2}|\boldsymbol{\Omega}|^{-1/2} \exp\left[-\frac{1}{2\sigma_a^2} S(\boldsymbol{\phi}, \boldsymbol{\theta}, \mathbf{e}_*)\right]$$

where

$$S(\boldsymbol{\phi}, \boldsymbol{\theta}, \mathbf{e}_*) = (\mathbf{L}_\phi\mathbf{w} - \mathbf{F}\mathbf{e}_*)'\mathbf{L}_\theta'^{-1}\mathbf{L}_\theta^{-1}(\mathbf{L}_\phi\mathbf{w} - \mathbf{F}\mathbf{e}_*) + \mathbf{e}_*'\boldsymbol{\Omega}^{-1}\mathbf{e}_*$$

Again, by generalized least squares theory, we can show that

$$S(\boldsymbol{\phi}, \boldsymbol{\theta}, \mathbf{e}_*) = S(\boldsymbol{\phi}, \boldsymbol{\theta}) + (\mathbf{e}_* - \hat{\mathbf{e}}_*)'\mathbf{D}(\mathbf{e}_* - \hat{\mathbf{e}}_*)$$

where

$$S(\boldsymbol{\phi}, \boldsymbol{\theta}) = S(\boldsymbol{\phi}, \boldsymbol{\theta}, \hat{\mathbf{e}}_*) = \hat{\mathbf{a}}'\hat{\mathbf{a}} + \hat{\mathbf{e}}_*'\boldsymbol{\Omega}^{-1}\hat{\mathbf{e}}_* \qquad \text{(A7.3.11)}$$

is the unconditional sum of squares function,

$$\hat{\mathbf{e}}_* = E[\mathbf{e}_* | \mathbf{w}, \boldsymbol{\phi}, \boldsymbol{\theta}] = [\mathbf{e}_*] = \mathbf{D}^{-1}\mathbf{F}'\mathbf{L}_\theta'^{-1}\mathbf{L}_\theta^{-1}\mathbf{L}_\phi\mathbf{w} \qquad \text{(A7.3.12)}$$

represent the conditional expectation of the preliminary values $\mathbf{e}_*$, with $\mathbf{D} = \boldsymbol{\Omega}^{-1} + \mathbf{F}'\mathbf{L}_\theta'^{-1}\mathbf{L}_\theta^{-1}\mathbf{F}$, and $\hat{\mathbf{a}} = [\mathbf{a}] = \mathbf{L}_\theta^{-1}(\mathbf{L}_\phi\mathbf{w} - \mathbf{F}\hat{\mathbf{e}}_*)$. By factorization of the joint distribution of $\mathbf{w}$ and $\mathbf{e}_*$ we can obtain

$$p(\mathbf{w} | \boldsymbol{\phi}, \boldsymbol{\theta}, \sigma_a) = (2\pi\sigma_a^2)^{-n/2}|\boldsymbol{\Omega}|^{-1/2}|\mathbf{D}|^{-1/2} \exp\left[-\frac{1}{2\sigma_a^2} S(\boldsymbol{\phi}, \boldsymbol{\theta})\right] \qquad \text{(A7.3.13)}$$

as the unconditional likelihood. It follows immediately from (A7.3.13) that the maximum likelihood estimate for σ_a^2 is given by $\hat{\sigma}_a^2 = S(\hat{\boldsymbol{\phi}}, \hat{\boldsymbol{\theta}})/n$, where $\hat{\boldsymbol{\phi}}$ and $\hat{\boldsymbol{\theta}}$ denote maximum likelihood estimates.

Again, we note that $S(\boldsymbol{\phi}, \boldsymbol{\theta}) = \sum_{t=1}^{n} [a_t]^2 + \hat{\mathbf{e}}_*'\boldsymbol{\Omega}^{-1}\hat{\mathbf{e}}_*$, and the elements $[a_1], [a_2], \ldots, [a_n]$ of $\hat{\mathbf{a}} = [\mathbf{a}]$ are computed recursively from the relation $\mathbf{L}_\theta[\mathbf{a}] = \mathbf{L}_\phi\mathbf{w} - \mathbf{F}[\mathbf{e}_*]$ as

$$[a_t] = w_t - \phi_1[w_{t-1}] - \cdots - \phi_p[w_{t-p}] + \theta_1[a_{t-1}] + \cdots + \theta_q[a_{t-q}]$$

for $t = 1, 2, \ldots, n$, using the back-forecasted values $[\mathbf{e}_*]$ for the preliminary values, with $[w_t] = w_t$ for $t = 1, 2, \ldots, n$. In addition, the back-forecasts $\hat{\mathbf{e}}_* = [\mathbf{e}_*]$ can be calculated from $\hat{\mathbf{e}}_* = \mathbf{D}^{-1}\mathbf{F}'\mathbf{u}$, where $\mathbf{u} = \mathbf{L}_\theta'^{-1}\mathbf{L}_\theta^{-1}\mathbf{L}_\phi\mathbf{w} = \mathbf{L}_\theta'^{-1}\mathbf{a}^0$, and the elements u_t of $\mathbf{u}$ are calculated through the backward recursion as

$$u_t = a_t^0 + \theta_1 u_{t+1} + \cdots + \theta_q u_{t+q}$$

with starting values $u_{n+1} = \cdots = u_{n+q} = 0$, and the a_t^0 are the elements of $\mathbf{a}^0 = \mathbf{L}_\theta^{-1}\mathbf{L}_\phi\mathbf{w}$ and denote the estimates of the a_t conditional on zero starting values $\mathbf{e}_* = 0$. Also, the vector $\mathbf{h} = \mathbf{F}'\mathbf{u}$ consists of the $p + q$ elements

$$h_j = \begin{cases} \sum_{i=1}^{j} \phi_{p-j+i} u_i & j = 1, \ldots, p \\ -\sum_{i=1}^{j-p} \theta_{q-j+p+i} u_i & j = p+1, \ldots, p+q \end{cases}$$

Finally, using (A7.1.1) and (A7.1.2), in $S(\boldsymbol{\phi}, \boldsymbol{\theta})$ we may write $\hat{\mathbf{e}}_*'\boldsymbol{\Omega}^{-1}\hat{\mathbf{e}}_* = \hat{\mathbf{a}}_*'\hat{\mathbf{a}} + (\hat{\mathbf{w}}_* - \mathbf{C}'\hat{\mathbf{a}}_*)'\mathbf{K}^{-1}(\hat{\mathbf{w}}_* - \mathbf{C}'\hat{\mathbf{a}}_*)$, so that we have

$$S(\boldsymbol{\phi}, \boldsymbol{\theta}) = \sum_{t=1-q}^{n} [a_t]^2 + (\hat{\mathbf{w}}_* - \mathbf{C}'\hat{\mathbf{a}}_*)'\mathbf{K}^{-1}(\hat{\mathbf{w}}_* - \mathbf{C}'\hat{\mathbf{a}}_*) \tag{A7.3.14}$$

where $\mathbf{K} = \sigma_a^{-2}\boldsymbol{\Gamma}_p - \mathbf{C}'\mathbf{C}$, as well as $|\boldsymbol{\Omega}| = |\mathbf{K}|$.

Therefore, in general, the likelihood associated with a series $\mathbf{z}$ of $n + d$ values generated by any ARIMA process is given by

$$L(\boldsymbol{\phi}, \boldsymbol{\theta}, \sigma_a|\mathbf{z}) = (2\pi\sigma_a^2)^{-n/2}|\mathbf{M}_n^{(p,q)}|^{1/2} \exp\left[-\frac{S(\boldsymbol{\phi}, \boldsymbol{\theta})}{2\sigma_a^2}\right] \tag{A7.3.15}$$

where

$$S(\boldsymbol{\phi}, \boldsymbol{\theta}) = \sum_{t=1}^{n} [a_t]^2 + \hat{\mathbf{e}}_*'\boldsymbol{\Omega}^{-1}\hat{\mathbf{e}}_*$$

and $|\mathbf{M}_n^{(p,q)}| = |\boldsymbol{\Omega}|^{-1}|\mathbf{D}|^{-1} = |\mathbf{K}|^{-1}|\mathbf{D}|^{-1}$. Also, by consideration of the mixed ARMA model as an infinite moving average $\tilde{w}_t = (1 + \psi_1 B + \psi_2 B^2 + \cdots)a_t$, and referring to results for the pure MA model, it follows that in the unconditional sum of squares function for the mixed model we have the relation that $\hat{\mathbf{e}}_*'\boldsymbol{\Omega}^{-1}\hat{\mathbf{e}}_* = \sum_{t=-\infty}^{0} [a_t]^2$. Hence we also have the representation $S(\boldsymbol{\phi}, \boldsymbol{\theta}) = \sum_{t=-\infty}^{n} [a_t]^2$, and in practice the values $[a_t]$ may be computed recursively with the summation proceeding from some point $t = 1 - Q$, beyond which the $[a_t]$'s are negligible.

Autoregressive processes as a special case. In the special case of a pure AR model of order p, $w_t - \sum_{i=1}^{p} \phi_i w_{t-i} = a_t$, the results above simplify somewhat. We then have $\mathbf{e}_* = \mathbf{w}_*$, $\boldsymbol{\Omega} = \sigma_a^{-2}\boldsymbol{\Gamma}_p$, $\mathbf{L}_\theta = \mathbf{I}_n$, $\mathbf{D} = \sigma_a^2\boldsymbol{\Gamma}_p^{-1} + \mathbf{F}'\mathbf{F} = \sigma_a^2\boldsymbol{\Gamma}_p^{-1} + \mathbf{A}_p'\mathbf{A}_p$, and $\hat{\mathbf{w}}_* = \mathbf{D}^{-1}\mathbf{F}'\mathbf{L}_\phi\mathbf{w} = \mathbf{D}^{-1}\mathbf{A}_p'\mathbf{L}_{11}\mathbf{w}_p$, where $\mathbf{w}_p' = (w_1,$

$w_2, \ldots, w_p)$ and $\mathbf{L}_{11}$ is the $p \times p$ upper left submatrix of $\mathbf{L}_\phi$. It can then be shown that the back-forecasts $\hat{w}_t$ are determined from the relations $\hat{w}_t = \phi_1 \hat{w}_{t+1} + \cdots + \phi_p \hat{w}_{t+p}$, $t = 0, -1, \ldots, 1 - p$, with $\hat{w}_t = w_t$ for $1 \le t \le n$, and hence these are the same as values obtained from the use of the backward model approach, as discussed in Section 7.1.4, for the special case of the AR model. Thus we obtain the exact sum of squares as $S(\boldsymbol{\phi}) = \Sigma_{t=1}^n [a_t]^2 + \sigma_a^2 \hat{\mathbf{w}}_*' \boldsymbol{\Gamma}_p^{-1} \hat{\mathbf{w}}_*$.

To illustrate, consider the first-order autoregressive process in w_t,

$$w_t - \phi w_{t-1} = a_t \tag{A7.3.16}$$

where w_t might be the dth difference $\nabla^d z_t$ of the actual observations and a series $\mathbf{z}$ of length $n + d$ observations is available. To compute the likelihood (A7.3.15), we require

$$\begin{aligned} S(\phi) &= \sum_{t=1}^{n} [a_t]^2 + (1 - \phi^2)\hat{w}_0^2 \\ &= \sum_{t=2}^{n} (w_t - \phi w_{t-1})^2 + (w_1 - \phi \hat{w}_0)^2 + (1 - \phi^2)\hat{w}_0^2 \end{aligned}$$

since $\boldsymbol{\Gamma}_1 = \gamma_0 = \sigma_a^2(1 - \phi^2)^{-1}$. Now because $\mathbf{D} = \sigma_a^2 \boldsymbol{\Gamma}_1^{-1} + \mathbf{A}_1' \mathbf{A}_1 = \sigma_a^2 \gamma_0^{-1} + \phi^2 = 1$, and hence $\hat{w}_0 = \phi w_1$, substituting this into the last two terms of $S(\phi)$ above, it reduces to

$$S(\phi) = \sum_{t=2}^{n} (w_t - \phi w_{t-1})^2 + (1 - \phi^2) w_1^2 \tag{A7.3.17}$$

a result that may be obtained more directly by methods discussed in Appendix A7.4.

ARMA(1, 1) model as an example. As an example for the mixed model, consider the ARMA(1, 1) model

$$w_t - \phi w_{t-1} = a_t - \theta a_{t-1} \tag{A7.3.18}$$

Then we have $\mathbf{e}_*' = (w_0, u_0)$, $\mathbf{A}_1 = \phi$, $\mathbf{B}_1 = -\theta$, and

$$\sigma_a^2 \boldsymbol{\Omega} = \sigma_a^2 \begin{bmatrix} \sigma_a^{-2}\gamma_0 & 1 \\ 1 & 1 \end{bmatrix}$$

with $\sigma_a^{-2}\gamma_0 = (1 + \theta^2 - 2\phi\theta)/(1 - \phi^2)$. Thus we have

$$\mathbf{D} = \boldsymbol{\Omega}^{-1} + \mathbf{F}'\mathbf{L}_\theta'^{-1}\mathbf{L}_\theta^{-1}\mathbf{F} = \frac{1}{\sigma_a^{-2}\gamma_0 - 1} \begin{bmatrix} 1 & -1 \\ -1 & \sigma_a^{-2}\gamma_0 \end{bmatrix} + \frac{1 - \theta^{2n}}{1 - \theta^2} \begin{bmatrix} \phi^2 & -\phi\theta \\ -\phi\theta & \theta^2 \end{bmatrix}$$

and the back-forecasts are obtained as $\hat{\mathbf{e}}_* = \mathbf{D}^{-1}\mathbf{h}$, where $\mathbf{h}' = (h_1, h_2) = (\phi, -\theta)u_1$, the u_t are obtained from the backward recursion $u_t = a_t^0 + \theta u_{t+1}$,

$u_{n+1} = 0$, and $a_t^0 = w_t - \phi w_{t-1}^0 + \theta a_{t-1}^0$, $t = 1, 2, \ldots, n$, are obtained using the zero initial values $w_0^0 = a_0^0 = 0$, with $w_t^0 = w_t$ for $1 \leq t \leq n$. Thus the exact sum of squares is obtained as

$$S(\phi, \theta) = \sum_{t=0}^{n} [a_t]^2 + \frac{(\hat{w}_0 - \hat{a}_0)^2}{\sigma_a^{-2}\gamma_0 - 1} \tag{A7.3.19}$$

with $[a_t] = w_t - \phi[w_{t-1}] + \theta[a_{t-1}]$, $t = 1, 2, \ldots, n$, and $\sigma_a^{-2}\gamma_0 - 1 = \mathbf{K} = (\phi - \theta)^2/(1 - \phi^2)$. In addition, we have $|\mathbf{M}_n^{(1,1)}| = \{|\mathbf{K}|\,|\mathbf{D}|\}^{-1}$, with

$$|\mathbf{K}|\,|\mathbf{D}| = 1 + \frac{(1 - \theta^{2n})}{(1 - \theta^2)}\frac{(\phi - \theta)^2}{(1 - \phi^2)}$$

APPENDIX A7.4 EXACT LIKELIHOOD FUNCTION FOR AN AUTOREGRESSIVE PROCESS

We now suppose that a given series $\mathbf{w}' = (w_1, w_2, \ldots, w_n)$ is generated by the pth-order stationary autoregressive model

$$w_t - \phi_1 w_{t-1} - \phi_2 w_{t-2} - \cdots - \phi_p w_{t-p} = a_t$$

where, temporarily, the w's are assumed to have mean $\mu = 0$, but as before, the argument can be extended to the case where $\mu \neq 0$. Assuming Normality for the a's and hence for the w's, the joint probability density function of the w's is

$$p(\mathbf{w}|\boldsymbol{\phi}, \sigma_a) = (2\pi\sigma_a^2)^{-n/2}|\mathbf{M}_n^{(p,0)}|^{1/2} \exp\left[-\frac{\mathbf{w}'\mathbf{M}_n^{(p,0)}\mathbf{w}}{2\sigma_a^2}\right] \tag{A7.4.1}$$

and because of the reversible character of the general process, the $n \times n$ matrix $\mathbf{M}_n^{(p,0)}$ is symmetric about *both* of its principal diagonals. We shall say that such a matrix is *doubly* symmetric. Now

$$p(\mathbf{w}|\boldsymbol{\phi}, \sigma_a) = p(w_{p+1}, w_{p+2}, \ldots, w_n|\mathbf{w}_p, \boldsymbol{\phi}, \sigma_a)p(\mathbf{w}_p|\boldsymbol{\phi}, \sigma_a)$$

where $\mathbf{w}_p' = (w_1, w_2, \ldots, w_p)$. The first factor on the right may be obtained by making use of the distribution

$$p(a_{p+1}, \ldots, a_n) = (2\pi\sigma_a^2)^{-(n-p)/2} \exp\left[-\frac{1}{2\sigma_a^2}\sum_{t=p+1}^{n} a_t^2\right] \tag{A7.4.2}$$

For fixed $\mathbf{w}_p$, $(a_{p+1}, \ldots, a_n)$ and $(w_{p+1}, \ldots, w_n)$ are related by the transformation

$$\begin{aligned} a_{p+1} &= w_{p+1} - \phi_1 w_p - \cdots - \phi_p w_1 \\ &\vdots \\ a_n &= w_n - \phi_1 w_{n-1} - \cdots - \phi_p w_{n-p} \end{aligned}$$

which has unit Jacobian. Thus we obtain

$$p(w_{p+1}, \ldots, w_n | \mathbf{w}_p, \boldsymbol{\phi}, \sigma_a)$$

$$= (2\pi\sigma_a^2)^{-(n-p)/2} \exp\left[-\frac{1}{2\sigma_a^2} \sum_{t=p+1}^{n} (w_t - \phi_1 w_{t-1} - \cdots - \phi_p w_{t-p})^2\right]$$

Also,

$$p(\mathbf{w}_p | \boldsymbol{\phi}, \sigma_a) = (2\pi\sigma_a^2)^{-p/2} |\mathbf{M}_p^{(p,0)}|^{1/2} \exp\left[-\frac{1}{2\sigma_a^2} \mathbf{w}_p' \mathbf{M}_p^{(p,0)} \mathbf{w}_p\right]$$

Thus

$$p(\mathbf{w} | \boldsymbol{\phi}, \sigma_a) = (2\pi\sigma_a^2)^{-n/2} |\mathbf{M}_p^{(p,0)}|^{1/2} \exp\left[\frac{-S(\boldsymbol{\phi})}{2\sigma_a^2}\right] \tag{A7.4.3}$$

where

$$S(\boldsymbol{\phi}) = \sum_{i=1}^{p} \sum_{j=1}^{p} m_{ij}^{(p)} w_i w_j + \sum_{t=p+1}^{n} (w_t - \phi_1 w_{t-1} - \cdots - \phi_p w_{t-p})^2 \tag{A7.4.4}$$

Also,

$$\mathbf{M}_p^{(p,0)} = \{m_{ij}^{(p)}\} = \{\gamma_{|i-j|}\}^{-1} \sigma_a^2 = \begin{bmatrix} \gamma_0 & \gamma_1 & \cdots & \gamma_{p-1} \\ \gamma_1 & \gamma_0 & \cdots & \gamma_{p-2} \\ \vdots & \vdots & & \vdots \\ \gamma_{p-1} & \gamma_{p-2} & \cdots & \gamma_0 \end{bmatrix}^{-1} \sigma_a^2 \tag{A7.4.5}$$

where $\gamma_0, \gamma_1, \ldots, \gamma_{p-1}$ are the theoretical autocovariances of the process, and $|\mathbf{M}_p^{(p,0)}| = |\mathbf{M}_n^{(p,0)}|$.

Now let $n = p + 1$, so that

$$\mathbf{w}_{p+1}' \mathbf{M}_{p+1}^{(p,0)} \mathbf{w}_{p+1} = \sum_{i=1}^{p} \sum_{j=1}^{p} m_{ij}^{(p)} w_i w_j$$

$$+ (w_{p+1} - \phi_1 w_p - \phi_2 w_{p-1} - \cdots - \phi_p w_1)^2$$

Then

$$\mathbf{M}_{p+1}^{(p)} = \left[\begin{array}{ccc|c} & & & 0 \\ & \mathbf{M}_p^{(p)} & & 0 \\ & & & \vdots \\ \hline 0 & 0 & \cdots & 0 \end{array}\right] + \left[\begin{array}{ccc|c} \phi_p^2 & \phi_p\phi_{p-1} & \cdots & -\phi_p \\ \phi_p\phi_{p-1} & \phi_{p-1}^2 & \cdots & -\phi_{p-1} \\ \vdots & \vdots & & \vdots \\ & & & -\phi_1 \\ \hline -\phi_p & -\phi_{p-1} & \cdots & 1 \end{array}\right]$$

and the elements of $\mathbf{M}_p^{(p)} = \mathbf{M}_p^{(p,0)}$ can now be deduced from the consideration that both $\mathbf{M}_p^{(p)}$ and $\mathbf{M}_{p+1}^{(p)}$ are doubly summetric. Thus, for example,

$$\mathbf{M}_2^{(1)} = \begin{bmatrix} m_{11}^{(1)} + \phi_1^2 & -\phi_1 \\ -\phi_1 & 1 \end{bmatrix} = \begin{bmatrix} 1 & -\phi_1 \\ -\phi_1 & m_{11}^{(1)} + \phi_1^2 \end{bmatrix}$$

and after equating elements in the two matrices, we have

$$\mathbf{M}_1^{(1)} = m_{11}^{(1)} = 1 - \phi_1^2$$

Proceeding in this way, we find for processes of orders 1 and 2:

$$\mathbf{M}_1^{(1)} = 1 - \phi_1^2 \qquad |\mathbf{M}_1^{(1)}| = 1 - \phi_1^2$$

$$\mathbf{M}_2^{(2)} = \begin{bmatrix} 1 - \phi_2^2 & -\phi_1(1 + \phi_2) \\ -\phi_1(1 + \phi_2) & 1 - \phi_2^2 \end{bmatrix}$$

$$|\mathbf{M}_2^{(2)}| = (1 + \phi_2)^2[(1 - \phi_2)^2 - \phi_1^2]$$

For example, when $p = 1$,

$$p(\mathbf{w}|\phi, \sigma_a)$$

$$= (2\pi\sigma_a^2)^{-n/2}(1 - \phi^2)^{1/2} \exp\left\{-\frac{1}{2\sigma_a^2}\left[(1 - \phi^2)w_1^2 + \sum_{t=2}^{n}(w_t - \phi w_{t-1})^2\right]\right\}$$

which checks with the result obtained in (A7.3.17). The process of generation above must lead to matrices $\mathbf{M}_p^{(p)}$, whose elements are *quadratic* in the ϕ's.

Thus it is clear from (A7.4.4) that not only is $S(\boldsymbol{\phi}) = \mathbf{w}'\mathbf{M}_n^{(p)}\mathbf{w}$ a quadratic form in the w's, but it is also quadratic in the parameters $\boldsymbol{\phi}$. Writing $\boldsymbol{\phi}_u' = (1, \phi_1, \phi_2, \ldots, \phi_p)$, it is clearly true that for some $(p + 1) \times (p + 1)$ matrix $\mathbf{D}$ whose elements are quadratic functions of the w's,

$$\mathbf{w}'\mathbf{M}_n^{(p)}\mathbf{w} = \boldsymbol{\phi}_u'\mathbf{D}\boldsymbol{\phi}_u$$

Now write

$$\mathbf{D} = \begin{bmatrix} D_{11} & -D_{12} & -D_{13} & \cdots & -D_{1,p+1} \\ -D_{12} & D_{22} & D_{23} & \cdots & D_{2,p+1} \\ \vdots & \vdots & \vdots & & \vdots \\ -D_{1,p+1} & D_{2,p+1} & D_{3,p+1} & \cdots & D_{p+1,p+1} \end{bmatrix} \tag{A7.4.6}$$

Inspection of (A7.4.4) shows that the elements D_{ij} are "symmetric" sums of squares and lagged products, defined by

$$D_{ij} = D_{ji} = w_i w_j + w_{i+1}w_{j+1} + \cdots + w_{n+1-j}w_{n+1-i} \tag{A7.4.7}$$

where the sum D_{ij} contains $n - (i - 1) - (j - 1)$ terms.

Finally, we can write the *exact* probability density, and hence the exact likelihood, as

$$p(\mathbf{w}|\boldsymbol{\phi}, \sigma_a) = L(\boldsymbol{\phi}, \sigma_a|\mathbf{w}) = (2\pi\sigma_a^2)^{-n/2}|\mathbf{M}_p^{(p)}|^{1/2} \exp\left[\frac{-S(\boldsymbol{\phi})}{2\sigma_a^2}\right] \tag{A7.4.8}$$

where

$$S(\boldsymbol{\phi}) = \mathbf{w}_p'\mathbf{M}_p^{(p)}\mathbf{w}_p + \sum_{t=p+1}^{n} (w_t - \phi_1 w_{t-1} - \cdots - \phi_p w_{t-p})^2 = \boldsymbol{\phi}_u'\mathbf{D}\boldsymbol{\phi}_u \tag{A7.4.9}$$

and the log-likelihood is

$$l(\boldsymbol{\phi}, \sigma_a|\mathbf{w}) = -\frac{n}{2}\ln(\sigma_a^2) + \frac{1}{2}\ln|\mathbf{M}_p^{(p)}| - \frac{S(\boldsymbol{\phi})}{2\sigma_a^2} \tag{A7.4.10}$$

Maximum likelihood estimates. Differentiating with respect to σ_a and each of the ϕ's in (A7.4.10), we obtain

$$\frac{\partial l}{\partial \sigma_a} = -\frac{n}{\sigma_a} + \frac{S(\boldsymbol{\phi})}{\sigma_a^3} \tag{A7.4.11}$$

$$\frac{\partial l}{\partial \phi_j} = M_j + \sigma_a^{-2}(D_{1,j+1} - \phi_1 D_{2,j+1} - \cdots - \phi_p D_{p+1,j+1}) \qquad j = 1, 2, \ldots, p \tag{A7.4.12}$$

where

$$M_j = \frac{\partial(\frac{1}{2}\ln|\mathbf{M}_p^{(p)}|)}{\partial \phi_j}$$

Hence maximum likelihood estimates may be obtained by equating these expressions to zero and solving the resultant equations.

We have at once from (A7.4.11)

$$\hat{\sigma}_a^2 = \frac{S(\hat{\boldsymbol{\phi}})}{n} \tag{A7.4.13}$$

Estimates of $\boldsymbol{\phi}$. A difficulty occurs in dealing with the equations (A7.4.12) since, in general, the quantities M_j ($j = 1, 2, \ldots, p$) are complicated functions of the ϕ's. We consider briefly four alternative approximations.

1. *Least squares estimates.* Whereas the expected value of $S(\boldsymbol{\phi})$ is proportional to n, the value of $|\mathbf{M}_p^{(p)}|$ is independent of n and for moderate or large samples, (A7.4.8) is dominated by the term in $S(\boldsymbol{\phi})$ and the term in $|\mathbf{M}_p^{(p)}|$ is, by comparison, small.

If we ignore the influence of this term, then

$$l(\boldsymbol{\phi}, \sigma_a|\mathbf{w}) \simeq -\frac{n}{2}\ln(\sigma_a^2) - \frac{S(\boldsymbol{\phi})}{2\sigma_a^2} \tag{A7.4.14}$$

and the estimates $\hat{\boldsymbol{\phi}}$ of $\boldsymbol{\phi}$ obtained by maximization of (A7.4.14) are the least squares estimates obtained by minimizing $S(\boldsymbol{\phi})$. Now, from (A7.4.9), $S(\boldsymbol{\phi}) = \boldsymbol{\phi}_u'\mathbf{D}\boldsymbol{\phi}_u$, where $\mathbf{D}$ is a $(p + 1) \times (p + 1)$ matrix of symmetric sums of squares and products, defined in (A7.4.7). Thus, on differentiating, the minimizing values are

$$\begin{aligned}
D_{12} &= \hat{\phi}_1 D_{22} + \hat{\phi}_2 D_{23} + \cdots + \hat{\phi}_p D_{2,p+1} \\
D_{13} &= \hat{\phi}_1 D_{23} + \hat{\phi}_2 D_{33} + \cdots + \hat{\phi}_p D_{3,p+1} \\
&\vdots \\
D_{1,p+1} &= \hat{\phi}_1 D_{2,p+1} + \hat{\phi}_2 D_{3,p+1} + \cdots + \hat{\phi}_p D_{p+1,p+1}
\end{aligned} \tag{A7.4.15}$$

which, in an obvious matrix notation, can be written

$$\mathbf{d} = \mathbf{D}_p \hat{\boldsymbol{\phi}}$$

so that

$$\hat{\boldsymbol{\phi}} = \mathbf{D}_p^{-1}\mathbf{d}$$

These least squares estimates also maximize the posterior density (7.4.15).

2. *Approximate maximum likelihood estimates.* We now recall an earlier result (3.2.3) which may be written

$$\gamma_j - \phi_1\gamma_{j-1} - \phi_2\gamma_{j-2} - \cdots - \phi_p\gamma_{j-p} = 0 \qquad j > 0 \tag{A7.4.16}$$

Also, on taking expectations in (A7.4.12) and using the fact that $E[\partial l/\partial\phi_j] = 0$, we obtain

$$\begin{aligned}
M_j\sigma_a^2 + (n - j)\gamma_j - (n - j - 1)\phi_1\gamma_{j-1} - (n - j - 2)\phi_2\gamma_{j-2} \\
- \cdots - (n - j - p)\phi_p\gamma_{j-p} = 0
\end{aligned} \tag{A7.4.17}$$

After multiplying (A7.4.16) by n and subtracting the result from (A7.4.17), we obtain

$$M_j\sigma_a^2 = j\gamma_j - (j + 1)\phi_1\gamma_{j-1} - \cdots - (j + p)\phi_p\gamma_{j-p}$$

Therefore, on using $D_{i+1,j+1}/(n - j - i)$ as an estimate of $\gamma_{|j-i|}$, a natural estimate of $M_j\sigma_a^2$ is

$$j\frac{D_{1,j+1}}{n - j} - (j + 1)\phi_1\frac{D_{2,j+1}}{n - j - 1} - \cdots - (j + p)\phi_p\frac{D_{p+1,j+1}}{n - j - p}$$

Substituting this estimate in (A7.4.12) yields

$$\frac{\partial l}{\partial\phi_j} \simeq n\sigma_a^{-2}\left(\frac{D_{1,j+1}}{n - j} - \phi_1\frac{D_{2,j+1}}{n - j - 1} - \cdots - \phi_p\frac{D_{p+1,j+1}}{n - j - p}\right) \tag{A7.4.18}$$

$$j = 1, 2, \ldots, p$$

leading to a set of linear equations of the form (A7.4.15), but now with

$$D_{ij}^* = \frac{nD_{ij}}{n - (i - 1) - (j - 1)}$$

replacing D_{ij}.

3. *Conditional least squares estimates.* For moderate and relatively large n, we might also consider the conditional sum of squares function, obtained by adopting the procedure in Section 7.1.3. This yields the sum of squares given in the exponent of the expression in (A7.4.2),

$$S_*(\boldsymbol{\phi}) = \sum_{t=p+1}^{n} (w_t - \phi_1 w_{t-1} - \cdots - \phi_p w_{t-p})^2$$

and is thus the sum of squares associated with the conditional distribution of $w_{p+1}, \ldots, w_n$, given $\mathbf{w}_p' = (w_1, w_2, \ldots, w_p)$. Conditional least squares estimates are obtained by minimizing $S_*(\boldsymbol{\phi})$, which is a standard linear least squares regression problem associated with the linear model $w_t = \phi_1 w_{t-1} + \phi_2 w_{t-2} + \cdots + \phi_p w_{t-p} + a_t$, $t = p + 1, \ldots, n$. This results in the familiar least squares estimates $\hat{\boldsymbol{\phi}} = \tilde{\mathbf{D}}_p^{-1}\tilde{\mathbf{d}}$, as in (A7.2.5), where $\tilde{\mathbf{D}}_p$ has (i, j)th element $\tilde{D}_{ij} = \Sigma_{t=p+1}^{n} w_{t-i}w_{t-j}$ and $\tilde{\mathbf{d}}$ has ith element $\tilde{d}_i = \Sigma_{t=p+1}^{n} w_{t-i}w_t$.

4. *Yule–Walker estimates.* Finally, if n is moderate or large, as an approximation, we may replace the symmetric sums of squares and products in (A7.4.15) by n times the appropriate autocovariance estimate. For example, D_{ij}, where $|i - j| = k$, would be replaced by $nc_k = \Sigma_{t=1}^{n-k} \tilde{w}_t \tilde{w}_{t+k}$. On dividing by nc_0 throughout in the resultant equations, we obtain the following relations expressed in terms of the estimated autocorrelations $r_k = c_k/c_0$:

$$\begin{aligned}
r_1 &= \hat{\phi}_1 \quad + \hat{\phi}_2 r_1 \quad + \cdots + \hat{\phi}_p r_{p-1} \\
r_2 &= \hat{\phi}_1 r_1 + \hat{\phi}_2 \quad\;\; + \cdots + \hat{\phi}_p r_{p-2} \\
&\;\;\vdots \\
r_p &= \hat{\phi}_1 r_{p-1} + \hat{\phi}_2 r_{p-2} + \cdots + \hat{\phi}_p
\end{aligned}$$

These are the well-known Yule–Walker equations.

In the matrix notation (7.3.1) they can be written $\mathbf{r} = \mathbf{R}\hat{\boldsymbol{\phi}}$, so that

$$\hat{\boldsymbol{\phi}} = \mathbf{R}^{-1}\mathbf{r} \tag{A7.4.19}$$

which corresponds to the equations (3.2.7), with $\mathbf{r}$ substituted for $\boldsymbol{\rho}_p$ and $\mathbf{R}$ for $\mathbf{P}_p$.

To illustrate the differences among the four estimates, take the case $p = 1$. Then $M_1\sigma_a^2 = -\gamma_1$ and, corresponding to (A7.4.12), the exact maximum likelihood estimate of ϕ is the solution of

$$-\gamma_1 + \sum_{t=2}^{n} w_t w_{t-1} - \phi \sum_{t=2}^{n-1} w_t^2 = 0$$

Note that $\gamma_1 = \sigma_a^2\phi/(1 - \phi^2)$ and on substituting the maximum likelihood solution $\hat{\sigma}_a^2 = S(\phi)/n$ with $S(\phi) = D_{11} - 2\phi D_{12} + \phi^2 D_{22}$, as obtained from (A7.4.11), for σ_a^2 in this expression for γ_1 in the likelihood equation above, we arrive at a *cubic* equation in ϕ whose solution yields the maximum likelihood estimate of ϕ.

Approximation (1) corresponds to ignoring the term γ_1 altogether, yielding

$$\hat{\phi} = \frac{\sum_{t=2}^{n} w_t w_{t-1}}{\sum_{t=2}^{n-1} w_t^2} = \frac{D_{12}}{D_{22}}$$

(2) corresponds to substituting the estimate $\sum_{t=2}^{n} w_t w_{t-1}/(n - 1)$ for γ_1, yielding

$$\hat{\phi} = \frac{\sum_{t=2}^{n} w_t w_{t-1}/(n-1)}{\sum_{t=2}^{n-1} w_t^2/(n-2)} = \frac{n-2}{n-1}\frac{D_{12}}{D_{22}}$$

(3) corresponds to the standard linear model least squares estimate obtained by regression of w_t on w_{t-1} for $t = 2, 3, \ldots, n$, so that

$$\hat{\phi} = \frac{\sum_{t=2}^{n} w_t w_{t-1}}{\sum_{t=2}^{n} w_{t-1}^2} = \frac{D_{12}}{D_{22} + w_1^2}$$

(4) replaces numerator and denominator by standard autocovariance estimates (2.1.10), yielding

$$\hat{\phi} = \frac{\sum_{t=2}^{n} w_t w_{t-1}}{\sum_{t=1}^{n} w_t^2} = \frac{c_1}{c_0} = r_1 = \frac{D_{12}}{D_{11}}$$

Usually, as in this example, for moderate and large samples, the differences between the estimates given by the various approximations will be small. We normally use the least squares estimates given by (1). These estimates can of course be computed directly from (A7.4.15). However, assuming that a computer package is available, it is scarcely worthwhile to treat autoregressive processes separately, and we have found it simplest, even when the fitted process is autoregressive, to employ the general iterative algorithm described in Section 7.2.1, which computes least squares estimates for any ARMA process.

Estimate of σ_a^2. Using approximation (4) with (A7.4.9) and (A7.4.13),

$$\hat{\sigma}_a^2 = \frac{S(\hat{\boldsymbol{\phi}})}{n} = c_0(1 \vdots \hat{\boldsymbol{\phi}}')\left[\begin{array}{c|c} 1 & -\mathbf{r}' \\ \hline -\mathbf{r} & \mathbf{R} \end{array}\right]\left(\begin{array}{c} 1 \\ \hline \hat{\boldsymbol{\phi}} \end{array}\right)$$

On multiplying out the right-hand side and recalling that $\mathbf{r} - \mathbf{R}\hat{\boldsymbol{\phi}} = 0$, we find that

$$\hat{\sigma}_a^2 = c_0(1 - \mathbf{r}'\hat{\boldsymbol{\phi}}) = c_0(1 - \mathbf{r}'\mathbf{R}^{-1}\mathbf{r}) = c_0(1 - \hat{\boldsymbol{\phi}}'\mathbf{R}\hat{\boldsymbol{\phi}}) \qquad \text{(A7.4.20a)}$$

It is readily shown that σ_a^2 can be similarly written in terms of the *theoretical* correlations

$$\sigma_a^2 = \gamma_0(1 - \boldsymbol{\rho}'\boldsymbol{\phi}) = \gamma_0(1 - \boldsymbol{\rho}'\mathbf{P}_p^{-1}\boldsymbol{\rho}) = \gamma_0(1 - \boldsymbol{\phi}'\mathbf{P}_p\boldsymbol{\phi}) \tag{A7.4.20b}$$

agreeing with the result (3.2.8).

Parallel expressions for $\hat{\sigma}_a^2$ may be obtained for approximations (1), (2), and (3).

Information matrix. Differentiating for a second time in (A7.4.11) and (A7.4.18), we obtain

$$-\frac{\partial^2 l}{\partial \sigma_a^2} = -\frac{n}{\sigma_a^2} + \frac{3S(\boldsymbol{\phi})}{\sigma_a^4} \tag{A7.4.21a}$$

$$\frac{\partial^2 l}{\partial \sigma_a \, \partial \phi_j} \simeq -2\sigma_a^{-1}\frac{\partial l}{\partial \phi_j} \tag{A7.4.21b}$$

$$-\frac{\partial^2 l}{\partial \phi_i \, \partial \phi_j} \simeq \frac{n}{\sigma_a^2}\frac{D_{i+1,j+1}}{n-i-j} \tag{A7.4.21c}$$

Now, since

$$E\left[\frac{\partial l}{\partial \phi_j}\right] = 0$$

it follows that for moderate or large samples,

$$E\left[-\frac{\partial^2 l}{\partial \sigma_a \, \partial \phi_j}\right] \simeq 0$$

and

$$|\mathbf{I}(\boldsymbol{\phi}, \sigma_a)| \simeq |\mathbf{I}(\boldsymbol{\phi})| I(\sigma_a)$$

where

$$I(\sigma_a) = E\left[-\frac{\partial^2 l}{\partial \sigma_a^2}\right] = \frac{2n}{\sigma_a^2}$$

Now, using (A7.4.21c), we have

$$\mathbf{I}(\boldsymbol{\phi}) = -E\left[\frac{\partial^2 l}{\partial \phi_i \, \partial \phi_j}\right] \simeq \frac{n}{\sigma_a^2}\boldsymbol{\Gamma}_p = \frac{n\gamma_0}{\sigma_a^2}\mathbf{P}_p = n(\mathbf{M}_p^{(p)})^{-1} \tag{A7.4.22}$$

Hence

$$\left|\mathbf{I}(\boldsymbol{\phi}, \sigma_a)\right| \sim \frac{2n^{p+1}}{\sigma_a^2}\left|\mathbf{M}_p^{(p,0)}\right|^{-1}$$

Variances and covariances of estimates of autoregressive parameters. Now, in circumstances fully discussed in [196], the inverse of the information matrix supplies the asymptotic variance–covariance matrix of the maximum likelihood (ML) estimates. Moreover if the log-likelihood is approximately quadratic and the maximum is not close to a boundary, even if the sample size is only moderate, the elements of this matrix will normally provide adequate approximations to the variances and covariances of the estimates.

Thus, using (A7.4.22) and (A7.4.20b) gives

$$\begin{aligned}\mathbf{V}(\hat{\boldsymbol{\phi}}) &= \mathbf{I}^{-1}(\hat{\boldsymbol{\phi}}) \simeq n^{-1}\mathbf{M}_p^{(p)} = n^{-1}\sigma_a^2\boldsymbol{\Gamma}_p^{-1} \\ &= n^{-1}(1 - \boldsymbol{\rho}'\mathbf{P}_p^{-1}\boldsymbol{\rho})\mathbf{P}_p^{-1} \\ &= n^{-1}(1 - \boldsymbol{\phi}'\mathbf{P}_p\boldsymbol{\phi})\mathbf{P}_p^{-1} = n^{-1}(1 - \boldsymbol{\rho}'\boldsymbol{\phi})\mathbf{P}_p^{-1}\end{aligned} \tag{A7.4.23}$$

In particular, for autoregressive processes of first and second order,

$$V(\hat{\phi}) \simeq n^{-1}(1 - \phi^2)$$

$$\mathbf{V}(\hat{\phi}_1, \hat{\phi}_2) \simeq n^{-1}\begin{bmatrix} 1 - \phi_2^2 & -\phi_1(1 + \phi_2) \\ -\phi_1(1 + \phi_2) & 1 - \phi_2^2 \end{bmatrix} \tag{A7.4.24}$$

Estimates of the variances and covariances may be obtained by substituting estimates for the parameters in (A7.4.24). For example, we may substitute r's for ρ's and $\hat{\boldsymbol{\phi}}$ for $\boldsymbol{\phi}$ in (A7.4.23) to obtain

$$\hat{\mathbf{V}}(\hat{\boldsymbol{\phi}}) = n^{-1}(1 - \mathbf{r}'\hat{\boldsymbol{\phi}})\mathbf{R}^{-1} \tag{A7.4.25}$$

APPENDIX A7.5 EXAMPLES OF THE EFFECT OF PARAMETER ESTIMATION ERRORS ON PROBABILITY LIMITS FOR FORECASTS

The variances and probability limits for the forecasts given in Section 5.2.4 are based on the assumption that the parameters $(\boldsymbol{\phi}, \boldsymbol{\theta})$ in the ARIMA model are known exactly. In practice, it is necessary to replace these by their estimates $(\hat{\boldsymbol{\phi}}, \hat{\boldsymbol{\theta}})$. To gain some insight into the effect of estimation errors on the variance of the forecast errors, we consider the special cases of the nonstationary IMA(0, 1, 1) and the stationary first-order autoregressive processes. It is shown that for these processes and for parameter estimates based on series of moderate length, the effect of such estimation errors is small.

IMA(0, 1, 1) processes. Writing the model $\nabla z_t = a_t - \theta a_{t-1}$ for $t + l$, $t + l - 1, \ldots, t + 1$, and summing, we obtain

$$z_{t+l} - z_t = a_{t+l} + (1 - \theta)(a_{t+l-1} + \cdots + a_{t+1}) - \theta a_t$$

Denote by $\hat{z}_t(l|\theta)$ the lead l forecast when the parameter θ is known exactly. On taking conditional expectations at time t, for $l = 1, 2, \ldots$, we obtain

$$\hat{z}_t(1|\theta) = z_t - \theta a_t$$

$$\hat{z}_t(l|\theta) = \hat{z}_t(1|\theta) \qquad l \geq 2$$

Hence the lead l forecast error is

$$\begin{aligned} e_t(l|\theta) &= z_{t+l} - \hat{z}_t(l|\theta) \\ &= a_{t+l} + (1 - \theta)(a_{t+l-1} + \cdots + a_{t+1}) \end{aligned}$$

and the variance of the forecast error at lead time l is

$$V(l) = E_t[e_t^2(l|\theta)] = \sigma_a^2[1 + (l - 1)\lambda^2] \tag{A7.5.1}$$

where $\lambda = 1 - \theta$.

However, if θ is replaced by its estimate $\hat{\theta}$, obtained from a time series consisting of n values of $w_t = \nabla z_t$, then

$$\hat{z}_t(1|\hat{\theta}) = z_t - \hat{\theta}\hat{a}_t$$

$$\hat{z}_t(l|\hat{\theta}) = \hat{z}_t(1|\hat{\theta}) \qquad l \geq 2$$

where $\hat{a}_t = z_t - \hat{z}_{t-1}(1|\hat{\theta})$. Hence the lead l forecast error using $\hat{\theta}$ is

$$\begin{aligned} e_t(l|\hat{\theta}) &= z_{t+l} - \hat{z}_t(l|\hat{\theta}) \\ &= z_{t+l} - z_t + \hat{\theta}\hat{a}_t \\ &= e_t(l|\theta) - (\theta a_t - \hat{\theta}\hat{a}_t) \end{aligned} \tag{A7.5.2}$$

Since $\nabla z_t = (1 - \theta B)a_t = (1 - \hat{\theta}B)\hat{a}_t$, it follows that

$$\hat{a}_t = \left(\frac{1 - \theta B}{1 - \hat{\theta}B}\right) a_t$$

and on eliminating $\hat{a}_t$ from (A7.5.2), we obtain

$$e_t(l|\hat{\theta}) = e_t(l|\theta) - \frac{\theta - \hat{\theta}}{1 - \hat{\theta}B} a_t$$

Now

$$\begin{aligned} \frac{\theta - \hat{\theta}}{1 - \hat{\theta}B} a_t &= \frac{\theta - \hat{\theta}}{1 - \theta B}\left[1 + \frac{(\theta - \hat{\theta})B}{1 - \theta B}\right]^{-1} a_t \\ &\simeq \frac{\theta - \hat{\theta}}{1 - \theta B}\left[1 - \frac{(\theta - \hat{\theta})B}{1 - \theta B}\right] a_t \\ &= (\theta - \hat{\theta})(a_t + \theta a_{t-1} + \theta^2 a_{t-2} + \cdots) \\ &\quad - (\theta - \hat{\theta})^2(a_{t-1} + 2\theta a_{t-2} + 3\theta^2 a_{t-3} + \cdots) \end{aligned} \tag{A7.5.3}$$

On the assumption that the forecast and the estimate $\hat{\theta}$ are based on essentially nonoverlapping data, $\hat{\theta}$ and $a_t, a_{t-1}, \ldots$ are independent. Also, $\hat{\theta}$ will be approximately normally distributed about θ with variance $(1 - \theta^2)/$

n, for moderate-sized samples. On these assumptions the variance of the expression in (A7.5.3) may be shown to be

$$\frac{\sigma_a^2}{n}\left(1 + \frac{3}{n}\frac{1 + \theta^2}{1 - \theta^2}\right)$$

Thus provided that $|\theta|$ is not close to unity,

$$\text{var}[e_t(l|\hat{\theta})] \simeq \sigma_a^2[1 + (l - 1)\lambda^2] + \frac{\sigma_a^2}{n} \tag{A7.5.4}$$

Clearly, the proportional change in the variance will be greatest for $l = 1$, when the exact forecast error reduces to σ_a^2. In this case, for parameter estimates based on a series of moderate length, the probability limits will be increased by a factor $(n + 1)/n$.

First-order autoregressive processes. Writing the model $\tilde{z}_t = \phi\tilde{z}_{t-1} + a_t$ at time $t + l$ and taking conditional expectations at time t, the lead l forecast, given the true value of the parameter, is

$$\hat{\tilde{z}}_t(l|\phi) = \phi\hat{\tilde{z}}_t(l - 1|\phi) = \phi^l\tilde{z}_t$$

Similarly,

$$\hat{\tilde{z}}_t(l|\hat{\phi}) = \hat{\phi}\hat{\tilde{z}}_t(l - 1|\hat{\phi}) = \hat{\phi}^l\tilde{z}_t$$

and hence

$$e_t(l|\hat{\phi}) = e_t(l|\phi) + (\phi^l - \hat{\phi}^l)\tilde{z}_t$$

It follows that

$$E_t[e_t^2(l|\hat{\phi})] = E_t[e_t^2(l|\phi)] + \tilde{z}_t^2E_t[(\phi^l - \hat{\phi}^l)^2]$$

so that *on the average*

$$\text{var}[e_t(l|\hat{\phi})] \simeq \sigma_a^2\frac{(1 - \phi^{2l})}{(1 - \phi^2)} + \sigma_a^2\frac{E[(\phi^l - \hat{\phi}^l)^2]}{1 - \phi^2} \tag{A7.5.5}$$

using (5.4.16). When $l = 1$,

$$\text{var}[e_t(1|\hat{\phi})] \simeq \sigma_a^2 + \frac{\sigma_a^2}{1 - \phi^2}\frac{1 - \phi^2}{n}$$

$$= \sigma_a^2\left(1 + \frac{1}{n}\right) \tag{A7.5.6}$$

For $l > 1$, we have

$$\phi^l - \hat{\phi}^l = \phi^l - \{\phi - (\phi - \hat{\phi})\}^l = \phi^l - \phi^l\left(1 - \frac{\phi - \hat{\phi}}{\phi}\right)^l \simeq l\phi^{l-1}(\phi - \hat{\phi})$$

Thus, on the average,

$$\text{var}[e_t(l|\hat{\phi})] \simeq \text{var}[e_t(l|\phi)] + \frac{l^2\phi^{2(l-1)}}{n}\sigma_a^2$$

and the discrepancy is again of order n^{-1}.

Related approximation results for the effect of parameter estimation errors on forecast variances have been given by Yamamoto [210] for the general AR(p) model. In particular, the approximation for one-step-ahead forecasts,

$$\text{var}[e_t(1|\hat{\boldsymbol{\phi}})] \simeq \sigma_a^2\left(1 + \frac{p}{n}\right)$$

is readily obtained for the AR model of order p.

APPENDIX A7.6 SPECIAL NOTE ON ESTIMATION OF MOVING AVERAGE PARAMETERS

If the least squares iteration is allowed to stray outside the invertibility region, parameter values can readily be found that apparently provide sums of squares smaller than the true minimum. However, these do not provide appropriate estimates and are quite meaningless. To illustrate, suppose that a series has been generated by the first-order moving average model $w_t = (1 - \theta B)a_t$ with $-1 < \theta < 1$. Then the series could equally well have been generated by the corresponding backward process $w_t = (1 - \theta F)e_t$ with $\sigma_e^2 = \sigma_a^2$. Now the latter process can also be written as $w_t = (1 - \theta^{-1}B)\alpha_t$ where now θ^{-1} is *outside* the invertibility region. However, in this representation $\sigma_\alpha^2 = \sigma_a^2\theta^2$ and is itself a function of θ. Therefore, a valid estimate of θ^{-1} will not be provided by minimizing $\Sigma\, \alpha_t^2 = \theta^2 \,\Sigma\, a_t^2$. Indeed, this has its minimum at $\theta^{-1} = \infty$.

The difficulty may be avoided:

1. By using as starting values rough preliminary estimates within the invertibility region obtained at the identification stage
2. By checking that all moving average estimates, obtained after convergence has apparently occurred, lie within the invertibility region

It is also possible to write least squares programs such that estimates are constrained to lie within the invertibility region.

8

MODEL DIAGNOSTIC CHECKING

The model having been identified and the parameters estimated, *diagnostic checks* are then applied to the fitted model. One useful method of checking a model is to *overfit*, that is, to estimate the parameters in a model somewhat more general than that which we believe to be true. This method assumes that we can guess the direction in which the model is likely to be inadequate. Therefore, it is necessary to supplement this approach by less specific checks applied to the residuals from the fitted model. These allow the data themselves to suggest modifications to the model. We shall describe two such checks that employ (1) the autocorrelation function of the residuals and (2) the cumulative periodogram of the residuals.

8.1 CHECKING THE STOCHASTIC MODEL

8.1.1 General Philosophy

Suppose that using a particular time series, the model has been identified and the parameters estimated using the methods described in Chapters 6 and 7. The question remains of deciding whether this model is adequate. If there should be evidence of serious inadequacy, we shall need to know how the model should be modified in the next iterative cycle. What we are doing is described only partially by the words "testing goodness of fit." We need to discover *in what way* a model is inadequate, so as to suggest appropriate modification. To illustrate, by reference to familiar procedures outside time series analysis, the scrutiny of residuals for the analysis of variance, described by Anscombe and Tukey [9], [10], and the criticism of factorial experiments, leading to Normal plotting and other methods due to Daniel [76], would be called *diagnostic checks*.

No model form ever represents the truth absolutely. It follows that given sufficient data, statistical tests can discredit models which could nevertheless be entirely adequate for the purpose at hand. Alternatively, tests can fail to indicate serious departures from assumptions because these tests are insensitive to the types of discrepancies that occur. The best policy is to devise the most sensitive statistical procedures possible but be prepared, for sufficient reason, to employ models that exhibit slight lack of fit. Know the facts as clearly as they can be known—then use judgment.

Clearly, diagnostic checks must be such that they *place the model in jeopardy*. That is, they must be sensitive to discrepancies which are likely to happen. No system of diagnostic checks can ever be comprehensive, since it is always possible that characteristics in the data of an unexpected kind could be overlooked. However, if diagnostic checks, which have been thoughtfully devised, are applied to a model fitted to a reasonably large body of data and fail to show serious discrepancies, then we shall rightly feel more comfortable about using that model.

8.1.2 Overfitting

One technique that can be used for diagnostic checking is *overfitting*. Having identified what is believed to be a correct model, we actually fit a more elaborate one. This puts the identified model in jeopardy, because the more elaborate model contains additional parameters covering feared directions of discrepancy. Careful thought should be given to the question of how the model should be augmented. In particular, in accordance with the discussion on model redundancy in Section 7.3.5, it would be foolish to add factors *simultaneously* to both sides of the ARMA model. If the analysis fails to show that the additions are needed, we shall not, of course, have proved that our model is correct. A model is only capable of being "proved" in the biblical sense of being put to the test. As was recommended by Saint Paul in his first epistle to the Thessalonians, what we can do is to "Prove all things; hold fast to that which is good."

Example of overfitting. As an example, we consider again some IBM stock price data. For this analysis, data were employed that are listed as series B′ in the "Collection of Time Series" in Part Five. This series consists of IBM stock prices for the period* June 29, 1959–June 30, 1960.

* The IBM stock data previously considered, referred to as series B, covers a different period, May 17, 1961–November 2, 1962.

The (0, 1, 1) model

$$\nabla z_t = (1 - \theta B)a_t$$

with $\hat{\lambda}_0 = 1 - \hat{\theta} = 0.90$, was identified and fitted to the 255 available observations.

The (0, 1, 1) model can equally well be expressed in the form

$$\nabla z_t = \lambda_0 a_{t-1} + \nabla a_t$$

The extended model that was considered in the overfitting procedure was the (0, 3, 3) process

$$\nabla^3 z_t = (1 - \theta_1 B - \theta_2 B^2 - \theta_3 B^3)a_t$$

or using (4.3.21), in the form

$$\nabla^3 z_t = (\lambda_0 \nabla^2 + \lambda_1 \nabla + \lambda_2)a_{t-1} + \nabla^3 a_t$$

The immediate motivation for extending the model in this particular way, was to test a suggestion made by Brown [64] that the series should be forecasted by an adaptive *quadratic* forecast function. Now, it was shown in Chapter 5 that an IMA(0, q, q) process has for its optimal forecasting function an adaptive polynomial of degree $q - 1$. Thus for the extended (0, 3, 3) model above, the optimal lead l forecast function is the quadratic polynomial in l,

$$\hat{z}_t(l) = b_0^{(t)} + b_1^{(t)} l + b_2^{(t)} l^2$$

where the coefficients $b_0^{(t)}$, $b_1^{(t)}$, and $b_2^{(t)}$ are adjusted as each new piece of data becomes available.

However, the model we have identified is an IMA(0, 1, 1) process, yielding a forecast function

$$\hat{z}_t(l) = b_0^{(t)} \tag{8.1.1}$$

This is "a polynomial in l" of degree zero. Hence the model implies that the forecast $\hat{z}_t(l)$ is independent of l; that is, the forecast at any particular time t is the same for one step ahead, two steps ahead, and so on. In other words, the series contains information only on the future *level* of the series, and nothing about slope or curvature. At first sight this is somewhat surprising because, using hindsight, quite definite linear and curvilinear trends appear to be present in the series. Therefore, it is worthwhile to check whether nonzero values of λ_1 and λ_2, which would produce predictable trends, actually occur. Sum of squares grids for $S(\lambda_1, \lambda_2 | \lambda_0)$ are shown for $\lambda_0 = 0.7, 0.9$, and 1.1, in Figure 8.1, from which it can be seen that the minimum is close to $\hat{\lambda}_0 = 0.9$, $\hat{\lambda}_1 = 0$, and $\hat{\lambda}_2 = 0$. It is also clear that values of $\lambda_1 > 0$ and $\lambda_2 > 0$ lead to higher sums of squares and therefore departures from the identified IMA(0, 1, 1) model in these directions are counterindicated. This implies, in particular, that a quadratic forecast function would give worse instead of

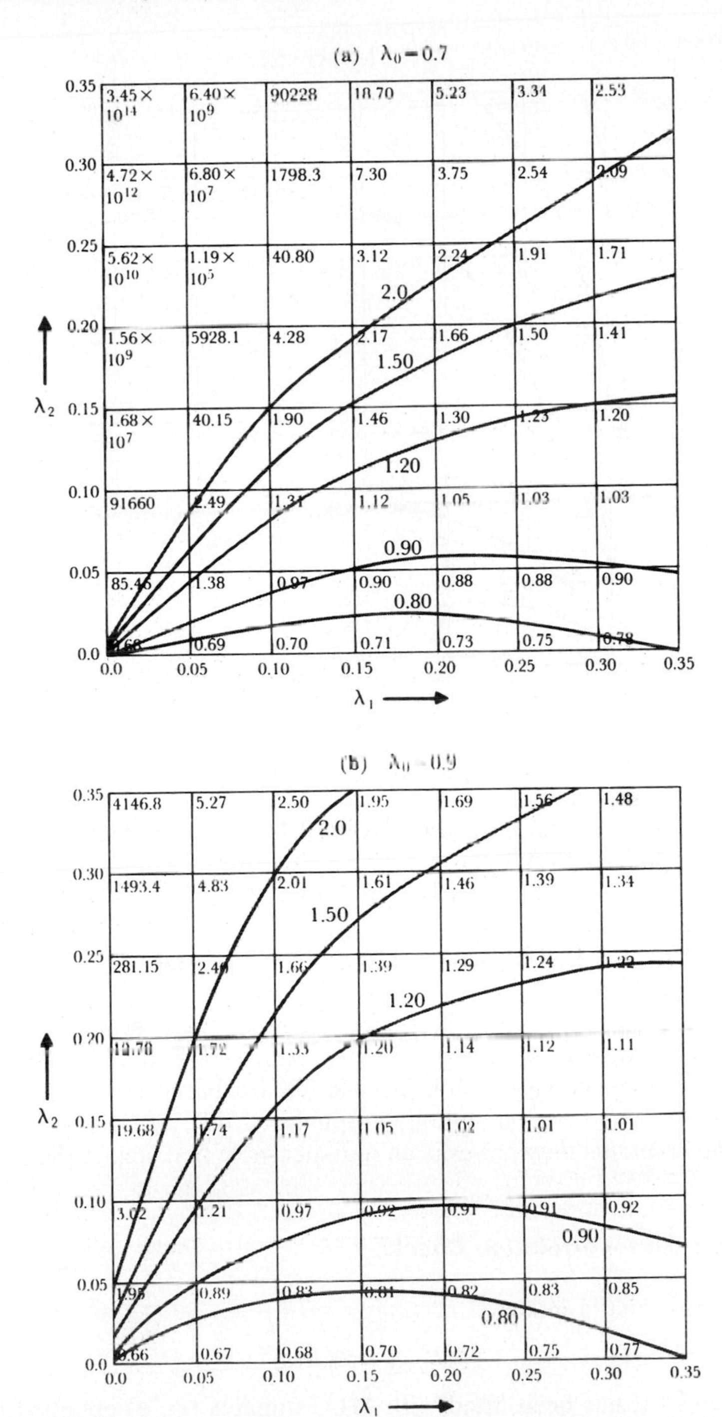

Figure 8.1 Sum of squares grids and contours for series B′ for extended model of order (0, 3, 3).

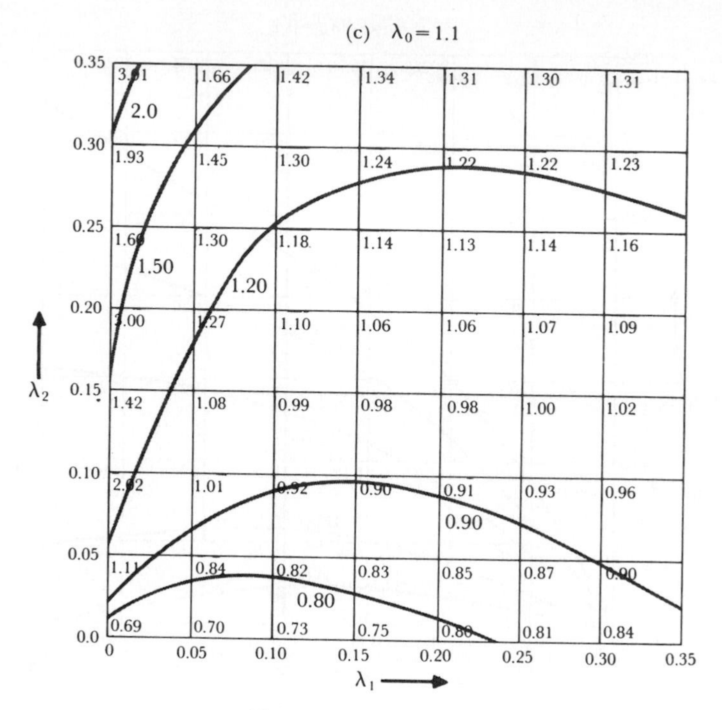

Figure 8.1 (*cont.*)

better forecasts than those obtained from (8.1.1), as was indeed shown to be the case in Section A5.3.3.

8.2 DIAGNOSTIC CHECKS APPLIED TO RESIDUALS

The method of overfitting, by extending the model in a particular direction, assumes that we know what kind of discrepancies are to be feared. Procedures less dependent upon such knowledge are based on the analysis of *residuals*. It cannot be too strongly emphasized that *visual inspection of a plot of the residuals themselves* is an indispensable first step in the checking process.

8.2.1 Autocorrelation Check

Suppose that a model

$$\phi(B)\tilde{w}_t = \theta(B)a_t$$

with $w_t = \nabla^d z_t$, has been fitted with ML estimates $(\hat{\boldsymbol{\phi}}, \hat{\boldsymbol{\theta}})$ obtained for the parameters. Then we shall refer to the quantities

$$\hat{a}_t = \hat{\theta}^{-1}(B)\hat{\phi}(B)\tilde{w}_t \tag{8.2.1}$$

as the *residuals*. The residuals are computed recursively from $\hat{\theta}(B)\hat{a}_t = \hat{\phi}(B)\tilde{w}_t$ as

$$\hat{a}_t = \tilde{w}_t - \sum_{j=1}^{p} \hat{\phi}_j \tilde{w}_{t-j} + \sum_{j=1}^{q} \hat{\theta}_j \hat{a}_{t-j} \qquad t = 1, 2, \ldots, n$$

using either zero initial values (conditional method) or back-forecasted initial values (exact method) for the initial $\hat{a}$'s and $\tilde{w}$'s. Now it is possible to show that if the model is adequate,

$$\hat{a}_t = a_t + O\left(\frac{1}{\sqrt{n}}\right)$$

As the series length increases, the $\hat{a}_t$'s become close to the white noise a_t's. Therefore, one might expect that study of the $\hat{a}_t$'s could indicate the existence and nature of model inadequacy. In particular, recognizable patterns in the estimated autocorrelation function of the $\hat{a}_t$'s could point to appropriate modifications in the model. This point is discussed further in Section 8.3.

Now suppose the form of the model were correct and that we *knew* the true parameter values $\boldsymbol{\phi}$ and $\boldsymbol{\theta}$. Then, using (2.1.13) and a result of Anderson [6], the estimated autocorrelations $r_k(a)$, of the a's, would be uncorrelated and distributed approximately normally about zero with variance n^{-1}, and hence with a standard error of $n^{-1/2}$. We could use these facts to assess approximately the statistical significance of apparent departures of these autocorrelations from zero.

Now, in practice, we do not know the *true* parameter values. We have only the estimates $(\hat{\boldsymbol{\phi}}, \hat{\boldsymbol{\theta}})$, from which, using (8.2.1), we can calculate not the a's but the $\hat{a}$'s. The autocorrelations $r_k(\hat{a})$ of the $\hat{a}$'s can yield valuable evidence concerning lack of fit and the possible nature of model inadequacy. However, it was pointed out by Durbin [88] that it might be dangerous to assess the statistical significance of apparent discrepancies of these autocorrelations $r_k(\hat{a})$ from their theoretical zero values on the basis of a standard error $n^{-1/2}$, appropriate to the $r_k(a)$'s. He was able to show, for example, that for the AR(1) process with parameter ϕ, the variance of $r_1(\hat{a})$ is $\phi^2 n^{-1}$, which can be very substantially *less* than n^{-1}. The large-sample variances and covariances for all the autocorrelations of the $\hat{a}$'s from any ARMA process were subsequently derived by Box and Pierce [56]. They showed that while in all cases, a reduction in variance can occur for low lags and that at these low lags the $r_k(\hat{a})$'s can be highly correlated, these effects usually disappear rather quickly at high lags. Thus the use of $n^{-1/2}$ as the standard error for $r_k(\hat{a})$ would underestimate the statistical significance of apparent departures from zero of the autocorrelations at low lags but could usually be employed for moderate or high lags.

For illustration, the large-sample one- and two-standard-error limits of the $r_k(\hat{a})$'s, for two first-order autoregressive processes and two second-order autoregressive processes, are shown in Figure 8.2. These also supply the corresponding approximate standard errors for moving average processes with the same parameters (as indicated in Figure 8.2).

It may be concluded that except at moderately high lags, $n^{-1/2}$ must be regarded as supplying an upper bound for the standard errors of the $r_k(\hat{a})$'s rather than the standard errors themselves. If for low lags we use the standard error $n^{-1/2}$ for the $r_k(\hat{a})$'s we may seriously *underestimate* the significance of apparent discrepancies.

8.2.2 Portmanteau Lack-of-Fit Test

Rather than consider the $r_k(\hat{a})$'s individually, an indication is often needed of whether, say, the first 20 autocorrelations of the $\hat{a}$'s *taken as a whole,* indicate inadequacy of the model. Suppose that we have the first K autocorrelations* $r_k(\hat{a})$ ($k = 1, 2, \ldots, K$) from any ARIMA(p, d, q) process; then it is possible to show [56] that if the fitted model is appropriate,

$$Q = n \sum_{k=1}^{K} r_k^2(\hat{a}) \tag{8.2.2}$$

is approximately distributed as χ^2 ($K - p - q$), where $n = N - d$ is the number of w's used to fit the model. On the other hand, if the model is inappropriate, the average values of Q will be inflated. Therefore, an approximate, general, or "portmanteau" test of the hypothesis of model adequacy, designed to take account of the difficulties discussed above, may be made by referring an observed value of Q to a table of the percentage points of χ^2 (such as Table F in Part Five).

However, it has been argued by Ljung and Box [138] that the chi-squared distribution does not provide a sufficiently accurate approximation to the distribution of the statistic Q under the null hypothesis, with values of Q tending to be somewhat smaller than expected under the chi-squared distribution. Empirical evidence to support this was presented by Davies, Triggs, and Newbold [78]. Consequently, Ljung and Box proposed a modified form of the statistic (modified Ljung–Box–Pierce statistic),

$$\tilde{Q} = n(n + 2) \sum_{k=1}^{K} (n - k)^{-1} r_k^2(\hat{a}) \tag{8.2.3}$$

* It is assumed here that k is taken sufficiently large so that the weights ψ_j in the model, written in the form

$$\tilde{w}_t = \phi^{-1}(B)\theta(B)a_t = \psi(B)a_t$$

will be negligibly small after $j = K$.

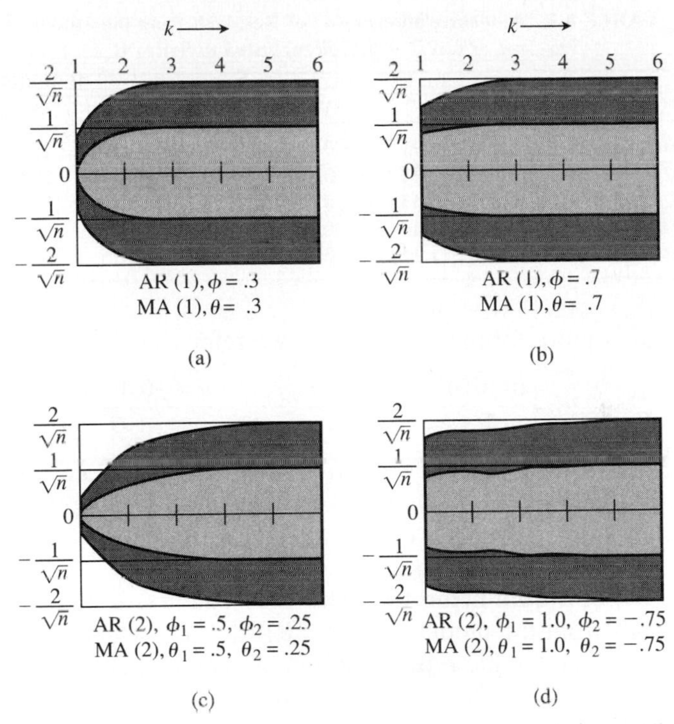

Figure 8.2 Standard error limits for residual autocorrelations $r_k(\hat{a})$.

such that the modified statistic has, approximately, the mean $E[\tilde{Q}] \approx K - p - q$ of the $\chi^2(K - p - q)$ distribution. The motivation for (8.2.3) is that a more accurate value for the variance of $r_k(a)$ from a white noise series is $(n - k)/n^2$ rather than $1/n$. This modified form of the portmanteau test statistic has been recommended for use as having a null distribution that is much closer to the $\chi^2(K - p - q)$ distribution for typical sample sizes n, although Davies et al. [78] have pointed out that the variance of $\tilde{Q}$ may exceed that of the corresponding chi-squared distribution.

The (0, 2, 2) model fitted to series C. To illustrate the portmanteau criterion (8.2.3), Table 8.1 shows the first 25 autocorrelations $r_k(\hat{a})$ of the residuals from the IMA(0, 2, 2) process $\nabla^2 z_t = (1 - 0.13B - 0.12B^2)a_t$, which was one of the models fitted to series C in Chapter 7. Since there are $n = 224$ w's, the approximate upper bound for the standard error of a single autocorrelation is $1/\sqrt{224} \approx 0.07$. Compared with this standard error bound, the values $r_3(\hat{a}) = -0.125$, $r_9(\hat{a}) = -0.130$, $r_{11}(\hat{a}) = -0.129$, $r_{17}(\hat{a}) = 0.153$, $r_{22}(\hat{a}) = 0.132$, and $r_{25}(\hat{a}) = -0.127$, are all rather large. Of course, occasional large deviations occur even in random series, but taking these results as a whole, there must certainly be a suspicion of lack of fit of the model.

TABLE 8.1 Autocorrelations $r_k(\hat{a})$ of Residuals from the Model $\nabla^2 z_t = (1 - 0.13B - 0.12B^2)a_t$ Fitted to Series C

k	$r_k(\hat{a})$	k	$r_k(\hat{a})$	k	$r_k(\hat{a})$	k	$r_k(\hat{a})$	k	$r_k(\hat{a})$
1	0.020	6	−0.033	11	−0.129	16	−0.050	21	0.007
2	0.032	7	0.022	12	0.063	17	0.153	22	0.132
3	−0.125	8	−0.056	13	−0.084	18	−0.092	23	0.012
4	−0.078	9	−0.130	14	0.022	19	−0.005	24	−0.012
5	−0.011	10	0.093	15	−0.006	20	−0.015	25	−0.127

To make a more formal assessment, we refer

$$\tilde{Q} = (224)(226)\left[\frac{(0.020)^2}{223} + \frac{(0.032)^2}{222} + \cdots + \frac{(-0.127)^2}{199}\right] = 36.2$$

to a χ^2 table with 23 degrees of freedom. The 10% and 5% points for χ^2, with 23 degrees of freedom, are 32.0 and 35.2, respectively. Therefore, there is some doubt as to the adequacy of this model.

The (1, 1, 0) model fitted to series C. The first 25 autocorrelations of the residuals from the model $(1 - 0.82B)\nabla z_t = a_t$, which we decided in Chapter 7 gave a preferable representation of series C, are shown in Table 8.2. For this model, $\tilde{Q} = (225)(227)\Sigma_{k=1}^{25} r_k^2(\hat{a})/(225 - k) = 31.3$. Comparison with the χ^2 table for 24 degrees of freedom shows that there is no ground here for questioning this model.

Table 8.3 summarizes the values of the criterion $\tilde{Q}$, based on $K = 25$ residual autocorrelations for the models fitted to series A to F in Table 7.11. However, especially for series of shorter length n, such as series E and F, a somewhat smaller value of K in (8.2.3) is recommended for use in practice, since the asymptotic theory involved in the distribution of the statistic $\tilde{Q}$ relies on K growing (but only slowly, such that $K/n \to 0$) as the series length n increases (see related discussion at the end of Section 8.2.4).

Inspection of Table 8.3 shows that only two suspiciously large values of $\tilde{Q}$ occur. One is the value 36.2 obtained after fitting a (0, 2, 2) model to series C, which we have discussed already. The other is the value $\tilde{Q} = 38.8$

TABLE 8.2 Autocorrelations $r_k(\hat{a})$ of Residuals from the Model $(1 - 0.82B)\nabla z_t$ Fitted to Series C

k	$r_k(\hat{a})$	k	$r_k(\hat{a})$	k	$r_k(\hat{a})$	k	$r_k(\hat{a})$	k	$r_k(\hat{a})$
1	−0.007	6	0.019	11	−0.098	16	−0.039	21	0.001
2	−0.002	7	0.073	12	0.074	17	0.165	22	0.129
3	−0.061	8	−0.030	13	−0.054	18	−0.083	23	0.014
4	−0.014	9	−0.097	14	0.034	19	−0.004	24	−0.017
5	0.047	10	0.133	15	0.002	20	−0.009	25	−0.129

TABLE 8.3 Summary of Results of Portmanteau Test Applied to Residuals of Various Models Fitted to Series A to F

Series	$n = N - d$	Fitted Model	$\tilde{Q}$	Degrees of Freedom
A	197	$z_t - 0.92z_{t-1} = 1.45 + a_t - 0.58a_{t-1}$	28.4	23
	196	$\nabla z_t = a_t - 0.70a_{t-1}$	31.9	24
B	368	$\nabla z_t = a_t + 0.09a_{t-1}$	38.8	24
C	225	$\nabla z_t - 0.82\nabla z_{t-1} = a_t$	31.3	24
	224	$\nabla^2 z_t = a_t - 0.13a_{t-1} - 0.12a_{t-2}$	36.2	23
D	310	$z_t - 0.87z_{t-1} = 1.17 + a_t$	11.5	24
	309	$\nabla z_t = a_t - 0.06a_{t-1}$	18.8	24
E	100	$z_t - 1.42_{t-1} + 0.73z_{t-2} = 14.35 + a_t$	26.8	23
	100	$z_t - 1.57z_{t-1} + 1.02z_{t-2} - 0.21z_{t-3} = 11.31 + a_t$	20.0	22
F	70	$z_t + 0.34z_{t-1} - 0.19z_{t-2} = 58.87 + a_t$	14.7	23

obtained after fitting a (0, 1, 1) model to series B. This suggests some model inadequacy, since the 5% and 2.5% points for χ^2 with 24 degrees of freedom are 36.4 and 39.3, respectively. We consider the possible nature of this inadequacy in the next section.

8.2.3 Model Inadequacy Arising from Changes in Parameter Values

One interesting form of model inadequacy, which may be imagined, occurs when the *form* of model remains the same but the parameters change over a prolonged period of time. Evidence exists that might explain in these terms the possible inadequency of the (0, 1, 1) model fitted to the IBM data.

Table 8.4 shows the results obtained by fitting (0, 1, 1) processes separately to the first and second halves of series B as well as to the complete series. Denoting the estimates obtained from the two halves by $\hat{\lambda}^{(1)}$ and $\hat{\lambda}^{(2)}$, we find that the standard error of $\hat{\lambda}^{(1)} - \hat{\lambda}^{(2)}$ is $\sqrt{(0.070)^2 + (0.074)^2} = 0.102$.

TABLE 8.4 Comparison of IMA(0, 1, 1) Models Fitted to First and Second Halves of Series B

	n	$\hat{\theta}$	$\hat{\lambda} = 1 - \hat{\theta}$	$\hat{\sigma}(\hat{\lambda}) = \left[\frac{\hat{\lambda}(2 - \hat{\lambda})}{n}\right]^{1/2}$	Residual Variance $\hat{\sigma}_a^2$	$\tilde{Q}$	Degrees of Freedom
First half	184	−0.29	1.29	±0.070	26.3	24.6	24
Second half	183	−0.03	1.03	±0.074	77.3	37.1	24
Complete	368	−0.09	1.09	±0.052	52.2	38.8	24

Since the difference $\hat{\lambda}^{(1)} - \hat{\lambda}^{(2)} = 0.26$ is 2.6 times its standard error, it is likely that a real change in λ has occurred. Inspection of the $\tilde{Q}$ values suggests that the (0, 1, 1) model, with parameters appropriately modified for different time periods, might explain the series more exactly. The estimation results for the residual variances $\hat{\sigma}_a^2$ also strongly indicate that a real change in variability has occurred between the two halves of the series. Upon closer examination of the data, these changes in the characteristics of the series appear to have occurred starting around time $t = 236$.

8.2.4 Score Tests for Model Checking

As an alternative to the direct use of overfitting in model checking, the Lagrange multiplier or score test procedure, which is also closely related to the portmanteau test procedure, may be employed. The general score test procedure was presented by Silvey [179], and its use for ARIMA models has been discussed by Godfrey [96] and Poskitt and Tremayne [159]. A computational advantage of the score test procedure is that it requires maximum likelihood estimation of parameters only under the null model under test, but it yields tests asymptotically equivalent to the corresponding likelihood ratio tests obtained by directly overfitting the model. Furthermore, the test statistic is easily computed in the form of the sample size n times a coefficient of determination from a particular "auxiliary" regression.

Hence, we assume that an ARMA(p, q) model has been fit by ML estimation to the observations $\tilde{w}_t$, and we want to assess the adequacy of the model by testing this null model against the alternative of an ARMA($p + r$, q) model or of an ARMA(p, $q + r$) model. That is, for the ARMA($p + r$, q) alternative we test $H_0 : \phi_{p+1} = \cdots = \phi_{p+r} = 0$, while for the ARMA($p$, $q + r$) alternative we test $H_0 : \theta_{q+1} = \cdots = \theta_{q+r} = 0$. The score test procedure is based on the first partial derivatives, or scores, of the log-likelihood function with respect to the model parameters of the alternative model, but evaluated at the ML estimates obtained under the null model. The log-likelihood function is essentially given by $l = -(n/2)\ln(\sigma_a^2) - (1/2\sigma_a^2)\sum_{t=1}^n a_t^2$. So the partial derivatives of l with respect to the parameters $(\boldsymbol{\phi}, \boldsymbol{\theta})$ are

$$\frac{\partial l}{\partial \phi_j} = -\frac{1}{\sigma_a^2}\sum_{t=1}^n \frac{\partial a_t}{\partial \phi_j} a_t$$

$$\frac{\partial l}{\partial \theta_j} = -\frac{1}{\sigma_a^2}\sum_{t=1}^n \frac{\partial a_t}{\partial \theta_j} a_t$$

As in (7.2.9) and (7.2.10), we have

$$-\frac{\partial a_t}{\partial \phi_j} = u_{t-j} \qquad -\frac{\partial a_t}{\partial \theta_j} = v_{t-j}$$

where $u_t = \theta^{-1}(B)\tilde{w}_t = \phi^{-1}(B)a_t$ and $v_t = -\theta^{-1}(B)a_t$. Given residuals $\hat{a}_t$, obtained from ML fitting of the null model, as

$$\hat{a}_t = \tilde{w}_t - \sum_{j=1}^{p} \hat{\phi}_j \tilde{w}_{t-j} + \sum_{j=1}^{q} \hat{\theta}_j \hat{a}_{t-j} \qquad t = 1, 2, \ldots, n$$

the u_t's and v_t's evaluated under the ML estimates of the null model can be calculated recursively, starting with initial values set equal to zero, for example, as

$$u_t = \tilde{w}_t + \hat{\theta}_1 u_{t-1} + \cdots + \hat{\theta}_q u_{t-q}$$

$$v_t = -\hat{a}_t + \hat{\theta}_1 v_{t-1} + \cdots + \hat{\theta}_q v_{t-q}$$

The score vector of first partial derivatives with respect to all the model parameters $\boldsymbol{\beta}$ can be expressed as

$$\frac{\partial l}{\partial \boldsymbol{\beta}} = \frac{1}{\sigma_a^2} \mathbf{X}'\mathbf{a} \tag{8.2.4}$$

where $\mathbf{a} = (a_1, \ldots, a_n)'$ and $\mathbf{X}$ denotes the $n \times (p + q + r)$ matrix whose tth row consists of $(u_{t-1}, \ldots, u_{t-p-r}, v_{t-1}, \ldots, v_{t-q})$ in the case of the ARMA$(p + r, q)$ alternative and $(u_{t-1}, \ldots, u_{t-p}, v_{t-1}, \ldots, v_{t-q-r})$ in the case of the ARMA$(p, q + r)$ alternative model. Then, similar to (7.2.17), since the large-sample information matrix for $\boldsymbol{\beta}$ can be consistently estimated by $\hat{\sigma}_a^{-2}\mathbf{X}'\mathbf{X}$, where $\hat{\sigma}_a^2 = n^{-1}\sum_{t=1}^{n} \hat{a}_t^2 = n^{-1}\hat{\mathbf{a}}'\hat{\mathbf{a}}$, it follows that the score test statistic for testing that the additional r parameters are equal to zero is

$$\Lambda = \frac{\hat{\mathbf{a}}'\mathbf{X}(\mathbf{X}'\mathbf{X})^{-1}\mathbf{X}'\hat{\mathbf{a}}}{\hat{\sigma}_a^2} \tag{8.2.5}$$

Godfrey [96] noted that the computation of the test statistic in (8.2.5) can be given the interpretation as being equal to n times the coefficient of determination in an auxiliary regression equation. That is, if the alternative model is ARMA$(p + r, q)$ we consider the auxiliary regression equation

$$\hat{a}_t = \alpha_1 u_{t-1} + \cdots + \alpha_{p+r} u_{t-p-r} + \beta_1 v_{t-1} + \cdots + \beta_q v_{t-q} + \varepsilon_t$$

while if the alternative model is ARMA$(p, q + r)$, we consider the regression equation

$$\hat{a}_t = \alpha_1 u_{t-1} + \cdots + \alpha_p u_{t-p} + \beta_1 v_{t-1} + \cdots + \beta_{q+r} v_{t-q-r} + \varepsilon_t$$

Let $\hat{\varepsilon}_t$ denote the residuals from the ordinary least squares estimation of this regression equation. Then from (8.2.5) it is seen that Λ can be expressed, essentially, as

$$\Lambda = \frac{n\left(\sum_{t=1}^{n} \hat{a}_t^2 - \sum_{t=1}^{n} \hat{\varepsilon}_t^2\right)}{\sum_{t=1}^{n} \hat{a}_t^2} = n\left(1 - \frac{\sum_{t=1}^{n} \hat{\varepsilon}_t^2}{\sum_{t=1}^{n} \hat{a}_t^2}\right)$$

which is n times the coefficient of determination of the regression of the $\hat{a}_t$'s on the u_{t-j}'s and the v_{t-j}'s. Under the null hypothesis that the fitted ARMA(p, q) model is correct, the statistic Λ has an asymptotic χ^2 distribution with r degrees of freedom, and the null model is rejected as inadequate for large values of Λ. As argued by Godfrey [96] and others, rejection of the null model by the score test procedure should not be taken as evidence to adopt the specific alternative model involved, but simply as evidence against the adequacy of the fitted model. Correspondingly, it is felt that a score test procedure of the general form above will have reasonable power to detect inadequacies in the fitted model, even when the correct model is not that which is specified under the alternative. In particular, for example, Poskitt and Tremayne [159] established that the score test against an ARMA($p + r$, q) model alternative is asymptotically identical to a test against an ARMA(p, $q + r$) alternative. Hence the score test procedure may not be sensitive to the particular model specified under the alternative, but its performance will, of course, depend on the choice of the number r of additional parameters specified.

We also note an alternative form for the score statistic Λ. By the ML estimation procedure, it follows that the first partial derivatives, $\partial l/\partial \phi_j$, $j = 1, \ldots, p$, and $\partial l/\partial \theta_j$, $j = 1, \ldots, q$, will be identically equal to zero when evaluated at the ML estimates. Hence the score vector, $\partial l/\partial \boldsymbol{\beta}$, will contain only r nonzero elements when evaluated at the ML estimates from the null model, these being the partial derivatives with respect to the additional r parameters of the alternative model. Thus the score statistic in (8.2.5) can also be viewed as a quadratic form in these r nonzero values, whose matrix in the quadratic form is a consistent estimate of the inverse of the covariance matrix of these r score values when evaluated at the ML estimates obtained under the null model. Since these r score values are asymptotically Normal with zero means under the null model, the validity of the asymptotic $\chi^2(r)$ distribution under the null hypothesis is easily seen. It has also been noted by Newbold [150] that the score test statistic procedure against the alternative of r additional parameters is closely related to an appropriate test statistic based on the first r residual autocorrelations $r_k(\hat{a})$ from the fitted model. The test statistic is essentially a quadratic form in these first r residual autocorrelations, but of a more complex form than the portmanteau statistic in (8.2.2). As a direct illustration, suppose that the fitted or null model is a pure AR(p) model, and the alternative in the score test procedure is an ARMA(p, r) model. Then it follows from above that the variables v_{t-j} are identical to $-\hat{a}_{t-j}$, since $\theta(B) \equiv 1$ under the null model. Hence the nonzero elements of the score vector in (8.2.4) are equal to $-n$ times the first r residual autocorrelations, $r_1(\hat{a}), \ldots, r_r(\hat{a})$ from the fitted model, and the score test is thus directly seen to be a quadratic form in these first r residual autocorrelations. This connection, among other factors, led Ljung [137] to recommend the use of the modified portmanteau test statistic in (8.2.3) with relatively small values of K (e.g., $K = 5$), for model-checking purposes.

8.2.5 Cumulative Periodogram Check

In some situations, particularly in the fitting of seasonal time series, which are discussed in Chapter 9, it may be feared that we have not adequately taken into account the *periodic* characteristics of the series. Therefore, we are on the lookout for periodicities in the residuals. The autocorrelation function will not be a sensitive indicator of such departures from randomness, because periodic effects will typically dilute themselves among several autocorrelations. The periodogram, on the other hand, is specifically designed for the detection of periodic patterns in a background of white noise.

The periodogram of a time series a_t, $t = 1, 2, \ldots, n$, as defined in Section 2.2.1 is

$$I(f_i) = \frac{2}{n}\left[\left(\sum_{t=1}^{n} a_t \cos(2\pi f_i t)\right)^2 + \left(\sum_{t=1}^{n} a_t \sin(2\pi f_i t)\right)^2\right] \qquad (8.2.6)$$

where $f_i = i/n$ is the frequency. Thus it is a device for correlating the a_t's with sine and cosine waves of different frequencies. A pattern with given frequency f_i in the residuals is reinforced when correlated with a sine or cosine wave at that same frequency, and so produces a large value of $I(f_i)$.

Cumulative periodogram. It has been shown by Bartlett [26] (see also [122]) that the *cumulative periodogram* provides an effective means for the detection of periodic nonrandomness.

The power spectrum $p(f)$ for white noise has a constant value $2\sigma_a^2$ over the frequency domain 0 to 0.5 cycle. Consequently, the cumulative spectrum for white noise

$$P(f) = \int_0^f p(g)\, dg \qquad (8.2.7)$$

plotted against f is a straight line running from (0, 0) to $(0.5, \sigma_a^2)$, that is, $P(f)/\sigma_a^2$ is a straight line running from (0, 0) to (0.5, 1).

As mentioned in Section 2.2.3, $I(f)$ provides an estimate of the power spectrum at frequency f. In fact, for white noise, $E[I(f)] = 2\sigma_a^2$, and hence the estimate is unbiased. It follows that $(1/n)\,\Sigma_{i=1}^{j} I(f_i)$ provides an unbiased estimate of the integrated spectrum $P(f_j)$, and

$$C(f_j) = \frac{\Sigma_{i=1}^{j} I(f_i)}{ns^2} \qquad (8.2.8)$$

an estimate of $P(f_j)/\sigma_a^2$, where s^2 is an estimate of σ_a^2. We shall refer to $C(f_j)$ as the *normalized cumulative periodogram*.

Now if the model were adequate and the parameters known *exactly*, the a's could be computed from the data and would yield a white noise series. For a white noise series, the plot of $C(f_j)$ against f_j would be scattered about a straight line joining the points (0, 0) and (0.5, 1). On the other hand, model inadequacies would produce nonrandom a's, whose cumulative periodogram could show systematic deviations from this line. In particular, periodicities in the a's would tend to produce a series of neighboring values of $I(f_j)$ which were large. These large ordinates would reinforce each other in $C(f_j)$ and form a bump on the expected straight line.

In practice, we do not know the exact values of the parameters, but only their estimated values. We do not have the a's, but only the estimated residual $\hat{a}$'s. However, for large samples, the periodogram for the $\hat{a}$'s will have similar properties to that for the a's. Thus careful inspection of the periodogram of the $\hat{a}$'s can provide a useful additional diagnostic check, particularly for indicating periodicities inadequately taken account of.

Example: series C. We have seen that series C is well fitted by the model of order (1, 1, 0)

$$(1 - 0.82B)\nabla z_t = a_t$$

and somewhat less well by the IMA(0, 2, 2) model

$$\nabla^2 z_t = (1 - 0.13B - 0.12B^2)a_t$$

which is almost equivalent to it. We illustrate the cumulative periodogram test by showing what happens when we analyze the residual a's after fitting to the series the inadequate IMA(0, 1, 1) model

$$\nabla z_t = (1 - \theta B)a_t$$

When the model is thus restricted, the least squares estimate of θ is found to be -0.65. The normalized cumulative periodogram plot of the residuals from this model is shown in Figure 8.3(a). Study of the figure shows immediately that there are marked departures from linearity in the cumulative periodogram. These departures are very pronounced at low frequencies and are what might be expected if, for example, there were insufficient differencing. Figure 8.3(b) shows the corresponding plot for the best-fitting IMA(0, 2, 2) process. The points of the cumulative periodogram now cluster more closely about the expected line, although, as we have seen in Table 8.3, other evidence points to the inadequacy of this model.

It is wise to indicate on the diagram the period as well as the frequency. This makes for easy identification of the bumps that occur when residuals contain periodicities. For example, in monthly sales data, bumps

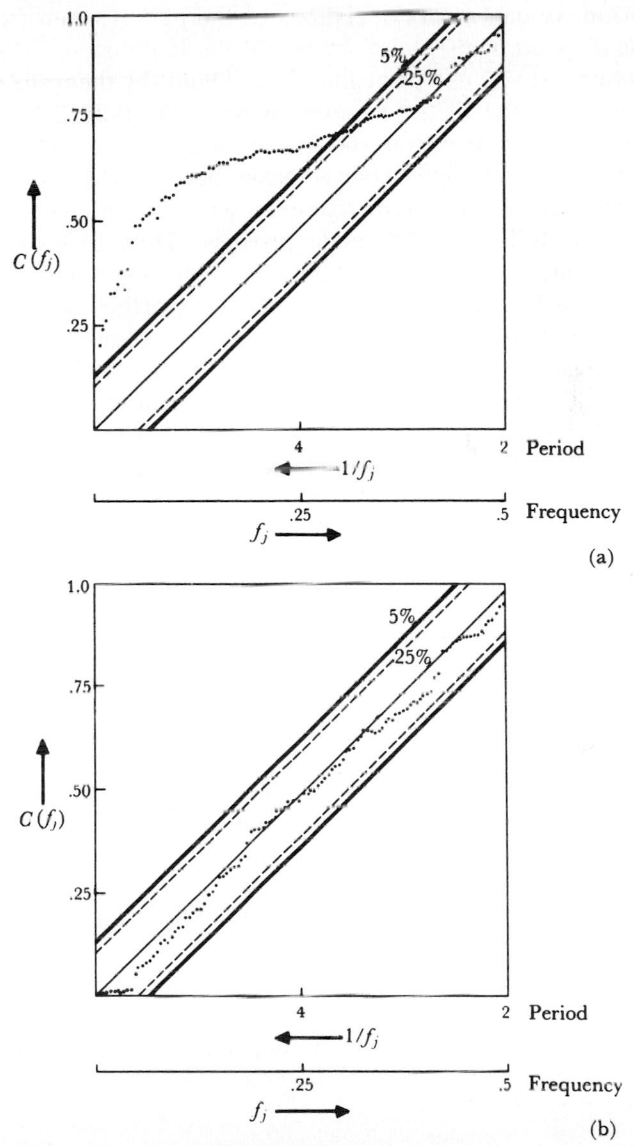

Figure 8.3 Series C: cumulative periodograms of residuals from best-fitting processes (a) of order (0, 1, 1) and (b) of order (0, 2, 2).

near periods 12, 24, 36, and so on, might indicate that seasonal effects were inadequately accounted for.

The probability relationship between the cumulative periodogram and the integrated spectrum is precisely the same as that between the empirical cumulative frequency function and the cumulative distribution function.

For this reason we can assess deviations of the periodogram from that expected if the $\hat{a}$'s were white noise, by use of the Kolmogorov–Smirnov test [99]. Using this test, we can place limit lines about the theoretical line. The limit lines are such that if the $\hat{a}_t$ series were white noise, the cumulative periodogram would deviate from the straight line sufficiently to cross these limits only with the stated probability. Now, because the $\hat{a}$'s are fitted values and not the true a's, we know that even when the model is correct they will not precisely follow a white noise process. Thus, as a test for model inadequacy, application of the Kolmogorov–Smirnov limits will indicate only approximate probabilities. However, it is worthwhile to show these limits on the cumulative periodogram to provide a rough guide as to what deviations to regard with scepticism and what to take more note of.

The limit lines are such that for a truly random series, they would be crossed a proportion ε of the time. They are drawn at distances $\pm K_\varepsilon/\sqrt{q}$ above and below the theoretical line, where $q = (n - 2)/2$ for n even and $(n - 1)/2$ for n odd. Approximate values for K_ε are given in Table 8.5.

TABLE 8.5 Coefficients for Calculating Approximate Probability Limits for Cumulative Periodogram Test

ε	0.01	0.05	0.10	0.25
K_ε	1.63	1.36	1.22	1.02

For series C, $q = (224 - 2)/2 = 111$, and the 5% limit lines inserted on Figure 8.3 deviate from the theoretical line by amounts $\pm 1.36/\sqrt{111} = \pm 0.13$. Similarly, the 25% limit lines deviate by $\pm 1.02/\sqrt{111} = \pm 0.10$.

Conclusions. Each of the checking procedures mentioned above has essential advantages and disadvantages. Checks based on the study of the estimated autocorrelation function and the cumulative periodogram, although they can point out *unsuspected* peculiarities of the series, may not be particularly sensitive. Tests for specific departures by overfitting are more sensitive but may fail to warn of trouble other than that specifically anticipated.

8.3 USE OF RESIDUALS TO MODIFY THE MODEL

8.3.1 Nature of the Correlations in the Residuals When an Incorrect Model Is Used

When the autocorrelation function of the residuals from some fitted model has indicated model inadequacy, it becomes necessary to consider in what way the model ought to be modified. In Section 8.3.2 we show how the autocorrelations of the residuals can be used to suggest such modifications. By way of introduction, we consider the effect of fitting an incorrect model on the autocorrelation function of the residuals.

Suppose that the correct model is

$$\phi(B)\tilde{w}_t = \theta(B)a_t$$

but that an incorrect model

$$\phi_0(B)\tilde{w}_t = \theta_0(B)b_t$$

is used. Then the residuals b_t, in the incorrect model, will be correlated and since

$$b_t = \theta_0^{-1}(B)\theta(B)\phi_0(B)\phi^{-1}(B)a_t \tag{8.3.1}$$

the autocovariance generating function of the b's will be

$$\sigma_a^2[\theta_0^{-1}(B)\theta_0^{-1}(F)\theta(B)\theta(F)\phi_0(B)\phi_0(F)\phi^{-1}(B)\phi^{-1}(F)] \tag{8.3.2}$$

For example, suppose in an IMA(0, 1, 1) process that instead of the correct value θ, we use some other value θ_0. Then the residuals b_t would follow the mixed process of order (1, 0, 1),

$$(1 - \theta_0 B)b_t = (1 - \theta B)a_t$$

and using (3.4.8), we have

$$\rho_1 = \frac{(1 - \theta\theta_0)(\theta_0 - \theta)}{1 + \theta^2 - 2\theta\theta_0}$$

$$\rho_j = \rho_1\theta_0^{j-1} \qquad j = 2, 3, \ldots$$

For example, suppose that in the IMA(0, 1, 1) process,

$$\nabla z_t = (1 - \theta B)a_t$$

we took $\theta_0 = 0.8$ when the correct value was $\theta = 0$. Then

$$\theta_0 = 0.8 \qquad \theta = 0.0$$

$$\rho_1 = 0.8 \qquad \rho_j = 0.8^j$$

Thus the b's would be highly correlated and would follow the autoregressive process

$$(1 - 0.8B)b_t = a_t$$

8.3.2 Use of Residuals to Modify the Model

Suppose that the residuals b_t from the model

$$\phi_0(B)\nabla^{d_0}z_t = \theta_0(B)b_t \tag{8.3.3}$$

appear to be nonrandom. Using the autocorrelation function of b_t, the methods of Chapter 6 may now be applied to identify a model

$$\bar{\phi}(B)\nabla^{\bar{d}} b_t = \bar{\theta}(B) a_t \tag{8.3.4}$$

for the b_t series. On eliminating b_t between (8.3.3) and (8.3.4), we arrive at a new model

$$\phi_0(B)\bar{\phi}(B)\nabla^{d_0}\nabla^{\bar{d}} z_t = \theta_0(B)\bar{\theta}(B) a_t \tag{8.3.5}$$

which can now be fitted and diagnostically checked.

For example, suppose that a series had been wrongly identified as an IMA(0, 1, 1) process and fitted to give the model

$$\nabla z_t = (1 + 0.6B) b_t \tag{8.3.6}$$

Suppose also that a model

$$\nabla b_t = (1 - 0.8B) a_t \tag{8.3.7}$$

was identified for this residual series. Then on eliminating b_t between (8.3.6) and (8.3.7), we would obtain

$$\nabla^2 z_t = (1 - 0.2B - 0.48B^2) a_t$$

which would suggest that an IMA(0, 2, 2) process should now be entertained.

9

SEASONAL MODELS

In Chapters 3 to 8 we have considered the properties of a class of linear stochastic models, which are of value in representing stationary and nonstationary time series, and we have seen how these models may be used for forecasting. We then considered the practical problems of identification, fitting, and diagnostic checking which arise when relating these models to actual data. In the present chapter we apply these methods to analyzing and forecasting seasonal series and also provide an opportunity to show how the ideas of the previous chapters fit together.

9.1 PARSIMONIOUS MODELS FOR SEASONAL TIME SERIES

Figure 9.1 shows the totals of international airline passengers for 1952, 1953, and 1954. It is part of a longer series (12 years of data) quoted by Brown [64] and listed as series G in the "Collection of Time Series" in Part Five. The series shows a marked seasonal pattern since travel is at its highest in the late summer months, while a secondary peak occurs in the spring. Many other series, particularly sales data, show similar seasonal characteristics.

In general, we say that a series exhibits periodic behavior with period s, when similarities in the series occur after s basic time intervals. In the example above, the basic time interval is 1 month and the period is $s = 12$ months. However, examples occur when s can take on other values. For example, $s = 4$ for quarterly data showing seasonal effects within years. It

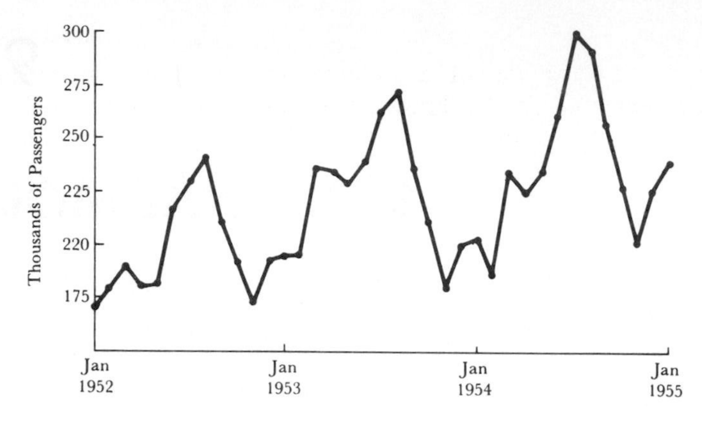

Figure 9.1 Totals of international airline passengers in thousands (part of series G).

sometimes happens that there is more than one periodicity. Thus, because bills tend to be paid monthly, we would expect weekly business done by a bank to show a periodicity of about 4 within months, while monthly business shows a periodicity of 12.

9.1.1 Fitting versus Forecasting

One of the deficiencies in the analysis of time series in the past has been the confusion between *fitting* a series and *forecasting* it. For example, suppose that a time series has shown a tendency to increase over a particular period and also to follow a seasonal pattern. A common method of analysis is to decompose the series arbitrarily into three components: a "trend," a "seasonal component," and a "random component." The trend might be fitted by a polynomial and the seasonal component by a Fourier series. A forecast was then made by projecting these fitted functions.

Such methods can give extremely misleading results. For example, we have already seen that the behavior of IBM stock prices (series B) is closely approximated by the random walk model $\nabla z_t = a_t$, that is,

$$z_t = z_0 + \sum_{j=0}^{t-1} a_{t-j} \tag{9.1.1}$$

which implies that $\hat{z}_t(l) = z_t$. In other words, the best forecast of future values of the stock is very nearly today's price. Now, it is true that short lengths of series B do look as if they might be fitted by quadratic curves. This simply reflects the fact that a sum of random deviates can sometimes have this appearance. However, there is no basis for the use of a quadratic forecast function, which produces very poor forecasts. Of course, genuine systematic effects that can be explained physically should be taken into account by the inclusion of a suitable deterministic component in the model. For example, if it is known that heat is being steadily added to a system, it would be sensible to explain the resulting increase in temperature by means of a suitable deterministic function of time, in addition to the stochastic component.

9.1.2 Seasonal Models Involving Adaptive Sines and Cosines

The general linear model

$$\tilde{z}_t = \sum_{j=1}^{\infty} \pi_j \tilde{z}_{t-j} + a_t = \sum_{j=1}^{\infty} \psi_j a_{t-j} + a_t \tag{9.1.2}$$

with suitable values for the coefficients π_j and ψ_j, is entirely adequate to describe many seasonal time series. The problem is to choose a suitable system of *parsimonious parameterization* for such models. As we have said before, this is not a mathematical problem but a question of finding out how the world tends to behave. To do this one can only proceed by trying out ideas on actual time series and developing those concepts that seem fruitful.

We have seen that for nonseasonal series, it is usually possible to obtain a useful and parsimonious representation in the form

$$\varphi(B)\tilde{z}_t = \theta(B)a_t \tag{9.1.3}$$

Moreover, the generalized autoregressive operator $\varphi(B)$ determines the eventual forecast function, which is the solution of the difference equation

$$\varphi(B)\hat{z}_t(l) = 0$$

where B is understood to operate on l. In representing seasonal behavior, we shall want the forecast function to trace out a periodic pattern. Our first thought might be that $\varphi(B)$ should produce a forecast function consisting of a mixture of sines and cosines, and possibly mixed with polynomial terms, to allow for changes in the level of the series and changes in the seasonal pattern. Such a forecast function could arise perfectly naturally within the structure of the general model (9.1.3). For example, with monthly data, a forecast function that is a sine wave with a 12-month period, adaptive in phase and amplitude, will satisfy the difference equation

$$(1 - \sqrt{3}B + B^2)\hat{z}_t(l) = 0$$

where B is understood to operate on l. However, it is not true that periodic behavior is necessarily represented *economically* by mixtures of sines and cosines. Many sine–cosine components would, for example, be needed to represent sales data affected by Christmas, Easter, and other seasonal buying. To take an extreme case, sales of fireworks in Britain are largely confined to the weeks immediately prior to November 5, when the abortive attempt of Guy Fawkes to blow up the Houses of Parliament is celebrated. An attempt to represent the "single spike" of fireworks sales data directly by sines and cosines might be unprofitable. It is clear that a more careful consideration of the problem is needed.

Now, in our previous analysis, we have not necessarily estimated *all* the components of $\varphi(B)$. Where differencing d times was needed to induce stationarity, we have written $\varphi(B) = \phi(B)(1 - B)^d$, which is equivalent to setting d roots of the equation $\varphi(B) = 0$ equal to unity. When such a representation proved adequate, we could proceed with the simpler analysis of $w_t = \nabla^d z_t$. Thus we have used $\nabla = 1 - B$ as a simplifying operator. In other problems, different types of simplifying operators might be appropriate. For example, the consumption of fuel oil for heat is highly dependent on ambient temperature which, because the earth rotates around the sun, is known to follow approximately a sine wave with period 12 months. In analyzing sales of fuel oil, it might then be sensible to introduce $1 - \sqrt{3}\,B + B^2$ as a simplifying operator, constituting one of the contributing components of the generalized autoregressive operator $\varphi(B)$. If such a representation proved useful, we could then proceed with the simpler analysis of $w_t = (1 - \sqrt{3}\,B + B^2)z_t$. This operator, it may be noted, is of the homogeneous nonstationary variety, having zeros $e^{\pm(i2\pi/12)}$ on the unit circle.

9.1.3 General Multiplicative Seasonal Model

Simplifying operator 1 − B^s. The fundamental fact about seasonal time series with period s is that observations which are s intervals apart are similar. Therefore, one might expect that the operation $B^s z_t = z_{t-s}$ would play a particularly important role in the analysis of seasonal series, and furthermore, since nonstationarity is to be expected in the series $z_t, z_{t-s}, z_{t-2s}, \ldots$, the simplifying operation $\nabla_s z_t = (1 - B^s)z_t = z_t - z_{t-s}$ might be useful. This stable nonstationary operator $1 - B^s$ has s zeros $e^{i(2\pi k/s)}$ $(k = 0, 1, \ldots, s - 1)$ evenly spaced on the unit circle. Furthermore, the eventual forecast function satisfies $(1 - B^s)\hat{z}_t(l) = 0$ and so may (but need not) be represented by a full complement of sines and cosines,

$$\hat{z}_t(l) = b_0^{(t)} + \sum_{j=1}^{[s/2]} \left[b_{1j}^{(t)} \cos\left(\frac{2\pi jl}{s}\right) + b_{2j}^{(t)} \sin\left(\frac{2\pi jl}{s}\right) \right]$$

where the b's are adaptive coefficients, and where $[s/2] = \frac{1}{2}s$ if s is even and $[s/2] = \frac{1}{2}(s - 1)$ if s is odd.

Multiplicative model. When we have a series exhibiting seasonal behavior with known periodicity s, it is of value to set down the data in the form of a table containing s columns, such as Table 9.1, which shows the

TABLE 9.1 Natural Logarithms of Monthly Passenger Totals (Measured in Thousands) in International Air Travel (Series G)

	Jan.	Feb.	Mar.	Apr.	May	June	July	Aug.	Sept.	Oct.	Nov.	Dec.
1949	4.718	4.771	4.883	4.860	4.796	4.905	4.997	4.997	4.913	4.779	4.644	4.771
1950	4.745	4.836	4.949	4.905	4.828	5.004	5.136	5.136	5.063	4.890	4.736	4.942
1951	4.977	5.011	5.182	5.094	5.147	5.182	5.293	5.293	5.215	5.088	4.984	5.112
1952	5.142	5.193	5.263	5.199	5.209	5.384	5.438	5.489	5.342	5.252	5.147	5.268
1953	5.278	5.278	5.464	5.460	5.434	5.493	5.576	5.606	5.468	5.352	5.193	5.303
1954	5.318	5.236	5.460	5.425	5.455	5.576	5.710	5.680	5.557	5.434	5.313	5.434
1955	5.489	5.451	5.587	5.595	5.598	5.753	5.897	5.849	5.743	5.613	5.468	5.628
1956	5.649	5.624	5.759	5.746	5.762	5.924	6.023	6.004	5.872	5.724	5.602	5.724
1957	5.753	5.707	5.875	5.852	5.872	6.045	6.142	6.146	6.001	5.849	5.720	5.817
1958	5.829	5.762	5.892	5.852	5.894	6.075	6.196	6.225	6.001	5.883	5.737	5.820
1959	5.886	5.835	6.006	5.981	6.040	6.157	6.306	6.326	6.138	6.009	5.892	6.004
1960	6.033	5.969	6.038	6.133	6.157	6.282	6.433	6.407	6.230	6.133	5.966	6.068

logarithms of the airline data. For seasonal data special care is needed in selecting an appropriate transformation. In this example (see Section 9.3.5) data analysis supports the use of the logarithm.

The arrangement of Table 9.1 emphasizes the fact that in periodic data, there are not one but two time intervals of importance. For this example, these intervals correspond to months and to years. Specifically, we expect relationships to occur (a) between observations for successive months in a particular year and (b) between the observations for the same month in successive years. The situation is somewhat like that in a two-way analysis of variance model, where similarities can be expected between observations in the same column and between observations in the same row.

Referring to the airline data of Table 9.1, the seasonal effect implies that an observation for a particular month, say April, is related to the observations for previous Aprils. Suppose that the tth observation z_t is for the month of April. We might be able to link this observation z_t to observations in previous Aprils by a model of the form

$$\Phi(B^s)\nabla_s^D z_t = \Theta(B^s)\alpha_t \tag{9.1.4}$$

where $s = 12$, $\nabla_s = 1 - B^s$ and $\Phi(B^s)$, $\Theta(B^s)$ are polynomials in B^s of degrees P and Q, respectively, and satisfying stationarity and invertibility conditions. Similarly, a model

$$\Phi(B^s)\nabla_s^D z_{t-1} = \Theta(B^s)\alpha_{t-1} \tag{9.1.5}$$

might be used to link the current behavior for March with previous March observations, and so on, for each of the 12 months. Moreover, it would usually be reasonable to assume that the parameters $\boldsymbol{\Phi}$ and $\boldsymbol{\Theta}$ contained in these monthly models would be approximately the same for each month.

Now the error components $\alpha_t, \alpha_{t-1}, \ldots,$ in these models would not in general be uncorrelated. For example, the total of airline passengers in April 1960, while related to previous April totals, would also be related to totals in March 1960, February 1960, January 1960, and so on. Thus we would expect that α_t in (9.1.4) would be related to α_{t-1} in (9.1.5) and to α_{t-2}, and so on. Therefore, to take care of such relationships, we introduce a second model

$$\phi(B)\nabla^d\alpha_t = \theta(B)a_t \tag{9.1.6}$$

where now a_t is a white noise process and $\phi(B)$ and $\theta(B)$ are polynomials in B of degrees p and q, respectively, and satisfying stationarity and invertibility conditions, and $\nabla = \nabla_1 = 1 - B$.

Substituting (9.1.6) in (9.1.4), we finally obtain a general *multiplicative* model

$$\phi_p(B)\Phi_P(B^s)\nabla^d\nabla_s^D z_t = \theta_q(B)\Theta_Q(B^s)a_t \tag{9.1.7}$$

where for this particular example, $s = 12$. Also, in (9.1.7) the subscripts p,

P, q, Q have been added to remind the reader of the orders of the various operators. The resulting multiplicative process will be said to be *of order* $(p, d, q) \times (P, D, Q)_s$. A similar argument can be used to obtain models with three or more periodic components to take care of multiple seasonalities.

9.2 REPRESENTATION OF THE AIRLINE DATA BY A MULTIPLICATIVE (0, 1, 1) × (0, 1, 1)$_{12}$ MODEL

In the remainder of this chapter we consider the basic forms for seasonal models of the kind just introduced and their potential for forecasting. We also consider the problems of identification, estimation, and diagnostic checking that arise in relating such models to data. No new principles are needed to do this, merely an application of the procedures and ideas we have already discussed in detail in Chapters 6 to 8. We proceed in Section 9.2 by discussing a particular example in considerable detail. In Section 9.3 we discuss those aspects of the general seasonal model that call for special mention.

The detailed illustration in this section will consist of relating a model of order $(0, 1, 1) \times (0, 1, 1)_{12}$ to the airline data of series G. In Section 9.2.1 we consider the model itself; in Section 9.2.2, its forecasting; in Section 9.2.3, its identification; in Section 9.2.4, its fitting; and finally in Section 9.2.5, its diagnostic checking.

9.2.1 Multiplicative (0, 1, 1) × (0, 1, 1)$_{12}$ Model

We have already seen that a simple and widely applicable stochastic model for the analysis of nonstationary time series, which contains no seasonal component, is the IMA(0, 1, 1) process. Suppose, following the argument of Section 9.1.3, that we employed such a model,

$$\nabla_{12} z_t = (1 - \Theta B^{12})\alpha_t$$

for linking z's 1 year apart. Suppose further that we employed a similar model

$$\nabla \alpha_t = (1 - \theta B) a_t$$

for linking α's 1 month apart, where in general θ and Θ will have different values. Then, on combining these expressions, we would obtain the seasonal multiplicative model

$$\nabla \nabla_{12} z_t = (1 - \theta B)(1 - \Theta B^{12}) a_t \qquad (9.2.1)$$

of order $(0, 1, 1) \times (0, 1, 1)_{12}$. The model written explicitly is

$$z_t - z_{t-1} - z_{t-12} + z_{t-13} = a_t - \theta a_{t-1} - \Theta a_{t-12} + \theta\Theta a_{t-13} \qquad (9.2.2)$$

The invertibility region for this model, required by the condition that the roots of $(1 - \theta B)(1 - \Theta B^{12}) = 0$ lie outside the unit circle, is defined by the inequalities

$$-1 < \theta < 1 \qquad -1 < \Theta < 1$$

Note that the moving average operator $(1 - \theta B)(1 - \Theta B^{12}) = 1 - \theta B - \Theta B^{12} + \theta\Theta B^{13}$, on the right of (9.2.1), is of order $q + sQ = 1 + 12(1) = 13$.

We shall show in Sections 9.2.3, 9.2.4, and 9.2.5 that the logged airline data is well fitted by a model of the form (9.2.1), where to a sufficient approximation, $\hat{\theta} = 0.4$, $\hat{\Theta} = 0.6$, and $\hat{\sigma}_a^2 = 1.34 \times 10^{-3}$. It is convenient as a preliminary to consider how, using this model and with these parameter values inserted, future values of the series may be forecast.

9.2.2 Forecasting

In Chapter 4 we saw that there are three basically different ways of considering the general model, each giving rise in Chapter 5 to a different way of viewing the forecast. We consider now these three approaches for the forecasting of the seasonal model (9.2.1).

Difference equation approach. Forecasts are best *computed* directly from the difference equation itself. Thus since

$$z_{t+l} = z_{t+l-1} + z_{t+l-12} - z_{t+l-13} + a_{t+l} - \theta a_{t+l-1} - \Theta a_{t+l-12} + \theta\Theta a_{t+l-13} \tag{9.2.3}$$

after setting $\theta = 0.4$, $\Theta = 0.6$, the minimum mean square error forecast at lead time l and origin t is given immediately by

$$\hat{z}_t(l) = [z_{t+l-1} + z_{t+l-12} - z_{t+l-13} + a_{t+l} - 0.4a_{t+l-1} - 0.6a_{t+l-12} + 0.24a_{t+l-13}] \tag{9.2.4}$$

As in Chapter 5, we refer to

$$[z_{t+l}] = E[z_{t+l}|\theta, \Theta, z_t, z_{t-1}, \ldots]$$

as the conditional expectation of z_{t+l} taken at origin t. In this expression the parameters are supposed exactly known, and knowledge of the series z_t, $z_{t-1}, \ldots$ is supposed to extend into the remote past.

Practical application depends upon the following facts:

1. Invertible models fitted to actual data usually yield forecasts that depend appreciably only on recent values of the series.
2. The forecasts are insensitive to small changes in parameter values such as are introduced by estimation errors.

Now

$$[z_{t+j}] = \begin{cases} z_{t+j} & j \leqslant 0 \\ \hat{z}_t(j) & j > 0 \end{cases} \tag{9.2.5}$$

$$[a_{t+j}] = \begin{cases} a_{t+j} & j \leqslant 0 \\ 0 & j > 0 \end{cases} \tag{9.2.6}$$

Thus, to obtain the forecasts, as in Chapter 5, we simply replace unknown z's by forecasts and unknown a's by zeros. The known a's are, of course, the one-step-ahead forecast errors already computed, that is, $a_t = z_t - \hat{z}_{t-1}(1)$.

For example, to obtain the three-months-ahead forecast, we have

$$z_{t+3} = z_{t+2} + z_{t-9} - z_{t-10} + a_{t+3} - 0.4a_{t+2} - 0.6a_{t-9} + 0.24a_{t-10}$$

Taking conditional expectations at the origin t,

$$\hat{z}_t(3) = \hat{z}_t(2) + z_{t-9} - z_{t-10} - 0.6a_{t-9} + 0.24a_{t-10}$$

that is,

$$\hat{z}_t(3) = \hat{z}_t(2) + z_{t-9} - z_{t-10} - 0.6[z_{t-9} - \hat{z}_{t-10}(1)] + 0.24[z_{t-10} - \hat{z}_{t-11}(1)]$$

Hence

$$\hat{z}_t(3) = \hat{z}_t(2) + 0.4z_{t-9} - 0.76z_{t-10} + 0.6\hat{z}_{t-10}(1) - 0.24\hat{z}_{t-11}(1) \tag{9.2.7}$$

which expresses the forecast in terms of previous z's and previous forecasts of z's. Although separate expressions for each lead time may readily be written down, computation of the forecasts is best carried out by using the single expression (9.2.4) directly, the elements of its right-hand side being defined by (9.2.5) and (9.2.6).

Figure 9.2 shows the forecasts for lead times up to 36 months, all made at the arbitrarily selected origin, July 1957. We see that the simple model, containing only two parameters, faithfully reproduces the seasonal pattern and supplies excellent forecasts. It is to be remembered, of course, that like all predictions obtained from the general linear stochastic model, the forecast function is adaptive. When changes occur in the seasonal pattern, these will be appropriately projected into the forecast. It will be noticed that when the 1-month-ahead forecast is too high, there is a tendency for all future forecasts from the point to be high. This is to be expected because, as has been noted in Appendix A5.1, forecast errors from the same origin, but for different lead times, are highly correlated. Of course, a forecast for a long lead time, such as 36 months, may necessarily contain a fairly large error.

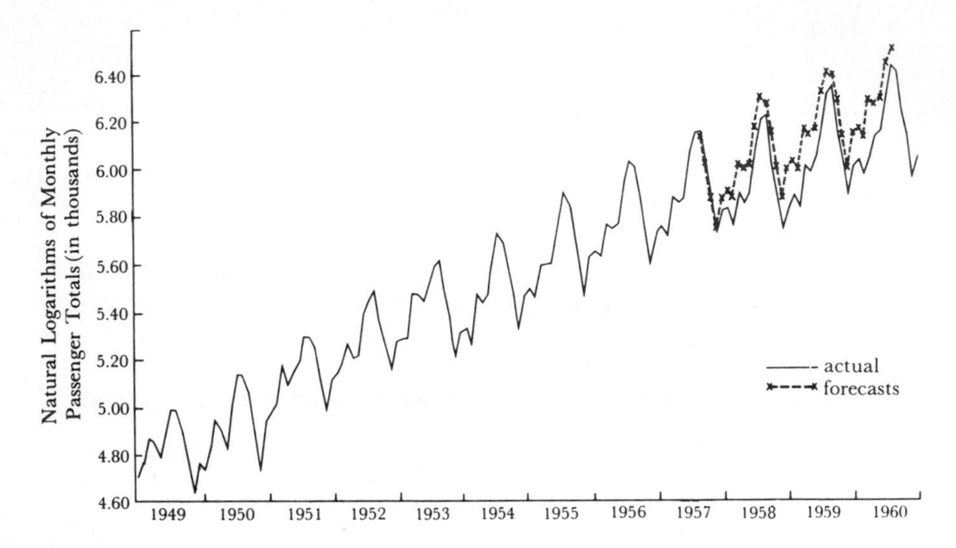

Figure 9.2 Series G, with forecasts for 1, 2, 3, . . . , 36 months ahead, all made from an arbitrarily selected origin, July 1957.

However, in practice, an initially remote forecast will be updated continually, and as the lead shortens, greater accuracy will be possible.

The preceding forecasting procedure is robust to moderate changes in the values of the parameters. Thus if we used $\theta = 0.5$ and $\Theta = 0.5$, instead of the values $\theta = 0.4$ and $\Theta = 0.6$, the forecasts would not be greatly affected. This is true even for forecasts made several steps ahead (e.g., 12 months). The approximate effect on the one-step-ahead forecasts of modifying the values of the parameters can be seen by studying the sum of squares surface. Thus we know that the approximate confidence region for the k parameters $\boldsymbol{\beta}$ is bounded, in general, by the contour $S(\boldsymbol{\beta}) = S(\hat{\boldsymbol{\beta}})[1 + \chi_\varepsilon^2(k)/n]$, which includes the true parameter point with probability $1 - \varepsilon$. Therefore, we know that, had the *true* parameter values been employed, with this same probability the mean square of the one-step-ahead forecast errors could not have been increased by a factor greater than $1 + \chi_\varepsilon^2(k)/n$.

Forecast function, its updating, and the forecast error variance. As we have said in Chapter 5, in practice, the difference equation procedure is by far the simplest and most convenient way for actually *computing* forecasts and updating them. However, the difference equation itself does not reveal very much about the *nature* of the forecasts so computed and about their updating. It is to throw light on these aspects (and not to provide alternative computational procedures) that we now consider the forecasts from other points of view.

Forecast Function. Using (5.1.12) yields

$$z_{t+l} = \hat{z}_t(l) + e_t(l) \tag{9.2.8}$$

where

$$e_t(l) = a_{t+l} + \psi_1 a_{t+l-1} + \cdots + \psi_{l-1} a_{t+1}$$

Now, the moving average operator on the right of (9.2.1) is of order 13. Hence, from (5.3.2) and for $l > 13$, the forecasts satisfy the difference equation

$$(1 - B)(1 - B^{12})\hat{z}_t(l) = 0 \qquad l > 13 \tag{9.2.9}$$

where in this equation B operates on l.

We now write $l = (r, m) = 12r + m$, $r = 0, 1, 2, \ldots$ and $m = 1, 2, \ldots, 12$, to represent a lead time of r years and m months, so that, for example, $l = 15 = (1, 3)$. Then the forecast function, which is the solution of (9.2.9), with starting conditions given by the first 13 forecasts, is of the form

$$\hat{z}_t(l) = \hat{z}_t(r, m) = b_{0,m}^{(t)} + r b_1^{(t)} \qquad l > 0 \tag{9.2.10}$$

This forecast function contains 13 adjustable coefficients $b_{0,1}^{(t)}, b_{0,2}^{(t)}, \ldots, b_{0,12}^{(t)}, b_1^{(t)}$. These represent 12 monthly contributions and one yearly contribution and are determined by the first 13 forecasts. The nature of this function is more clearly understood from Figure 9.3, which shows a forecast function of this kind, but with the period $s = 5$, so that there are six adjustable coefficients $b_{0,1}^{(t)}, b_{0,2}^{(t)}, \ldots, b_{0,5}^{(t)}, b_1^{(t)}$.

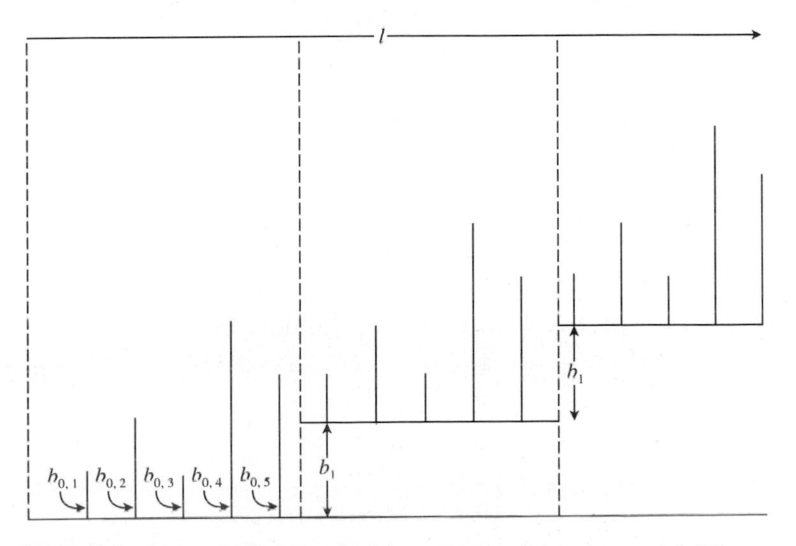

Figure 9.3 Seasonal forecast function generated by the model $\nabla\nabla_s z_t = (1 - \theta B)(1 - \Theta B^s)a_t$, with $s = 5$.

Equivalently, since $\hat{z}_t(l)$ satisfies (9.2.9) and the roots of $(1 - B)(1 - B^{12}) = 0$ are $1, 1, -1, e^{\pm(i2\pi j/12)}, j = 1, \ldots, 5$, on the unit circle, the forecast function, as in (5.3.3), can be represented as

$$\hat{z}_t(l) = \sum_{j=1}^{5} \left[b_{1j}^{(t)} \cos\left(\frac{2\pi jl}{12}\right) + b_{2j}^{(t)} \sin\left(\frac{2\pi jl}{12}\right)\right] + b_{16}^{(t)}(-1)^l + b_0^{(t)} + b_1^{*(t)} l$$

This shows that $\hat{z}_t(l)$ consists of a mixture of sinusoids at the seasonal frequencies $2\pi j/12$, $j = 1, \ldots, 6$, plus a linear trend with slope $b_1^{*(t)}$. The coefficients $b_{1j}^{(t)}$, $b_{2j}^{(t)}$, $b_0^{(t)}$, and $b_1^{*(t)}$ in the expression above are all adaptive with regard to the forecast origin t, being determined by the first 13 forecasts. In comparison to (9.2.10), it is clear, for example, that $b_1^{(t)} = 12 b_1^{*(t)}$ and represents the *annual* rate of change in the forecasts $\hat{z}_t(l)$, whereas $b_1^{*(t)}$ is the *monthly* rate of change.

The ψ weights. To determine updating formulas and to obtain the variance of the forecast error $e_t(l)$, we need the ψ weights in the form $z_t = \sum_{j=0}^{\infty} \psi_j a_{t-j}$ of the model. We can write the moving average operator in (9.2.1) in the form

$$(1 - \theta B)(1 - \Theta B^{12}) = (\nabla + \lambda B)(\nabla_{12} + \Lambda B^{12})$$

where $\lambda = 1 - \theta$, $\Lambda = 1 - \Theta$, $\nabla_{12} = 1 - B^{12}$. Hence the model (9.2.1) may be written

$$\nabla\nabla_{12} z_t = (\nabla + \lambda B)(\nabla_{12} + \Lambda B^{12}) a_t$$

By equating coefficients in $\nabla\nabla_{12}\psi(B) = (\nabla + \lambda B)(\nabla_{12} + \Lambda B^{12})$, it can be seen that the ψ weights satisfy $\psi_0 = 1$, $\psi_1 - \psi_0 = \lambda - 1$, $\psi_{12} - \psi_{11} - \psi_0 = \Lambda - 1$, $\psi_{13} - \psi_{12} - \psi_1 + \psi_0 = (1 - \lambda)(1 - \Lambda)$, and $\psi_j - \psi_{j-1} - \psi_{j-12} + \psi_{j-13} = 0$ otherwise. Thus the ψ weights for this process are

$$\psi_1 = \psi_2 = \cdots = \psi_{11} = \lambda \qquad \psi_{12} = \lambda + \Lambda$$

$$\psi_{13} = \psi_{14} = \cdots = \psi_{23} = \lambda(1 + \Lambda) \qquad \psi_{24} = \lambda(1 + \Lambda) + \Lambda$$

$$\psi_{25} = \psi_{26} = \cdots = \psi_{35} = \lambda(1 + 2\Lambda) \qquad \psi_{36} = \lambda(1 + 2\Lambda) + \Lambda$$

and so on. Writing ψ_j as $\psi_{r,m} = \psi_{12r+m}$, where $r = 0, 1, 2, \ldots$ and $m = 1, 2, \ldots, 12$, refer, respectively, to years and months, we obtain

$$\psi_{r,m} = \lambda(1 + r\Lambda) + \delta\Lambda \tag{9.2.11}$$

where

$$\delta = \begin{cases} 1, & \text{when } m = 12 \\ 0, & \text{when } m \neq 12 \end{cases}$$

Updating. The general updating formula (5.2.5) is

$$\hat{z}_{t+1}(l) = \hat{z}_t(l + 1) + \psi_l a_{t+1}$$

Thus, if $m \neq s = 12$,

$$b_{0,m}^{(t+1)} + rb_1^{(t+1)} = b_{0,m+1}^{(t)} + rb_1^{(t)} + (\lambda + r\lambda\Lambda)a_{t+1}$$

and on equating coefficients of r, the updating formulas are

$$\begin{aligned} b_{0,m}^{(t+1)} &= b_{0,m+1}^{(t)} + \lambda a_{t+1} \\ b_1^{(t+1)} &= b_1^{(t)} + \lambda\Lambda a_{t+1} \end{aligned} \tag{9.2.12}$$

Alternatively, if $m = s = 12$,

$$b_{0,12}^{(t+1)} + rb_1^{(t+1)} = b_{0,1}^{(t)} + (r + 1)b_1^{(t)} + (\lambda + \Lambda + r\lambda\Lambda)a_{t+1}$$

and in this case,

$$\begin{aligned} b_{0,12}^{(t+1)} &= b_{0,1}^{(t)} + b_1^{(t)} + (\lambda + \Lambda)a_{t+1} \\ b_1^{(t+1)} &= b_1^{(t)} + \lambda\Lambda a_{t+1} \end{aligned} \tag{9.2.13}$$

In studying these relations it should be remembered that $b_{0,m}^{(t+1)}$ will be the updated version of $b_{0,m+1}^{(t)}$. Thus if the origin t was January of a particular year, $b_{0,2}^{(t)}$ would be the coefficient for March. After a month had elapsed we should move the forecast origin to February and the updated version for the March coefficient would now be $b_{0,1}^{(t+1)}$.

Forecast Error Variance. Knowledge of the ψ weights enables us to calculate the variance of the forecast errors at any lead time l, using the result (5.1.16), namely

$$V(l) = (1 + \psi_1^2 + \cdots + \psi_{l-1}^2)\sigma_a^2 \tag{9.2.14}$$

Setting $\lambda = 0.6$, $\Lambda = 0.4$, $\sigma_a^2 = 1.34 \times 10^{-3}$ in (9.2.11) and (9.2.14), the estimated standard deviations $\hat{\sigma}(l)$ of the log forecast errors of the airline data for lead times 1 to 36 are shown in Table 9.2.

Forecasts as a weighted average of previous observations. If we write the model in the form

$$z_t = \sum_{j=1}^{\infty} \pi_j z_{t-j} + a_t$$

the one-step-ahead forecast is

$$\hat{z}_t(1) = \sum_{j=1}^{\infty} \pi_j z_{t+1-j}$$

The π weights may be obtained by equating coefficients in

$$(1 - B)(1 - B^{12}) = (1 - \theta B)(1 - \Theta B^{12})(1 - \pi_1 B - \pi_2 B^2 - \cdots)$$

TABLE 9.2 Estimated Standard Deviations of Forecast Errors for Logarithms of Airline Series at Various Lead Times

Forecast Lead Times	$\hat{\sigma}(l) \times 10^{-2}$	Forecast Lead Times	$\hat{\sigma}(l) \times 10^{-2}$	Forecast Lead Times	$\hat{\sigma}(l) \times 10^{-2}$
1	3.7	13	9.0	25	14.4
2	4.3	14	9.5	26	15.0
3	4.8	15	10.0	27	15.5
4	5.3	16	10.5	28	16.0
5	5.8	17	10.9	29	16.4
6	6.2	18	11.4	30	17.0
7	6.6	19	11.7	31	17.4
8	6.9	20	12.1	32	17.8
9	7.2	21	12.6	33	18.3
10	7.6	22	13.0	34	18.7
11	8.0	23	13.3	35	19.2
12	8.2	24	13.6	36	19.6

Thus

$$
\begin{aligned}
\pi_j &= \theta^{j-1}(1-\theta) \qquad j = 1, 2, \ldots, 11 \\
\pi_{12} &= \theta^{11}(1-\theta) + (1-\Theta) \\
\pi_{13} &= \theta^{12}(1-\theta) - (1-\theta)(1-\Theta) \\
(1 - \theta B &- \Theta B^{12} + \theta\Theta B^{13})\pi_j = 0 \qquad j \geqslant 14
\end{aligned}
\tag{9.2.15}
$$

These weights are plotted in Figure 9.4 for the parameter values $\theta = 0.4$ and $\Theta = 0.6$.

The reason that the weight function takes the particular form shown in the figure may be understood as follows: The process (9.2.1) may be written

$$
a_{t+1} = \left(1 - \frac{\lambda B}{1 - \theta B}\right)\left(1 - \frac{\Lambda B^{12}}{1 - \Theta B^{12}}\right) z_{t+1} \tag{9.2.16}
$$

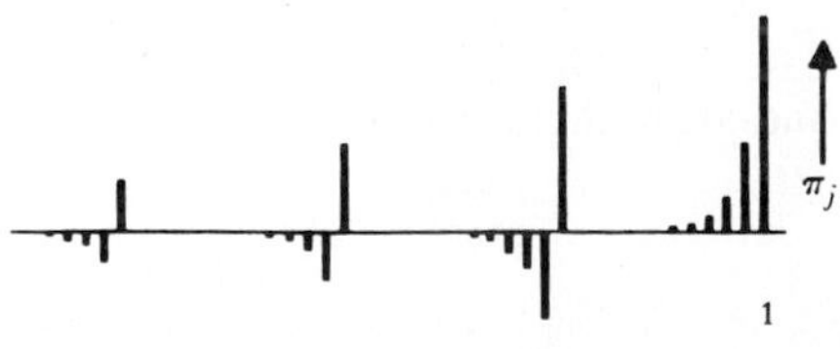

Figure 9.4 π weights for $(0, 1, 1) \times (0, 1, 1)_{12}$ process fitted to series G ($\theta = 0.4$, $\Theta = 0.6$).

We now use the notation $\text{EWMA}_\lambda(z_t)$ to mean an exponentially weighted moving average, with parameter λ, of values $z_t, z_{t-1}, z_{t-2}, \ldots$, so that

$$\text{EWMA}_\lambda\,(z_t) = \frac{\lambda}{1-\theta B} z_t = \lambda z_t + \lambda\theta z_{t-1} + \lambda\theta^2 z_{t-2} + \cdots$$

Similarly, we use $\text{EWMA}_\Lambda(z_t)$ to mean an exponentially weighted moving average, with parameter Λ, of values $z_t, z_{t-12}, z_{t-24}, \ldots$, so that

$$\text{EWMA}_\Lambda(z_t) = \frac{\Lambda}{1-\Theta B^{12}} z_t = \Lambda z_t + \Lambda\Theta z_{t-12} + \Lambda\Theta^2 z_{t-24} + \cdots$$

Substituting $\hat{z}_t(1) = z_{t+1} - a_{t+1}$ in (9.2.16), we obtain

$$\hat{z}_t(1) = \text{EWMA}_\lambda(z_t) + \text{EWMA}_\Lambda(z_{t-11} - \text{EWMA}_\lambda z_{t-12}) \qquad (9.2.17)$$

Thus the forecast is an EWMA taken over previous months, modified by a second EWMA of discrepancies found between similar monthly EWMAs and actual performance in previous years. As a particular case, if $\theta = 0$ ($\lambda = 1$), (9.2.17) would reduce to

$$\begin{aligned}\hat{z}_t(1) &= z_t + \text{EWMA}_\Lambda(z_{t-11} - z_{t-12})\\ &= z_t + \Lambda[(z_{t-11} - z_{t-12}) + \Theta\,(z_{t-23} - z_{t-24}) + \cdots]\end{aligned}$$

which shows that first differences are forecast as the seasonal EWMA of first differences for similar months from previous years.

For example, suppose that we were attempting to predict December sales for a department store. These sales would include a heavy component from Christmas buying. The first term on the right of (9.2.17) would be an EWMA taken over previous months up to November. However, we know this will be an underestimate, so we correct it by taking a second EWMA over previous years of the *discrepancies* between actual December sales and the corresponding monthly EWMAs taken over previous months in those years.

The forecasts for lead times $l > 1$ can be generated from the π weights by substituting forecasts of shorter lead time for unknown values. Thus

$$\hat{z}_t(2) = \pi_1\hat{z}_t(1) + \pi_2 z_t + \pi_3 z_{t-1} + \cdots$$

Alternatively, explicit values for the weights applied directly to $z_t, z_{t-1}, z_{t-2}, \ldots$ may be computed, for example, from (5.3.9) or from (A5.2.3).

9.2.3 Identification

The identification of the nonseasonal IMA(0, 1, 1) process depends upon the fact that, after taking first differences, the autocorrelations for all lags beyond the first are zero. For the multiplicative $(0, 1, 1) \times (0, 1, 1)_{12}$ process (9.2.1), the only nonzero autocorrelations of $\nabla\nabla_{12}z_t$ are those at lags

1, 11, 12, and 13. In fact, from (9.2.2) the model is viewed as $w_t = a_t - \theta a_{t-1} - \Theta a_{t-12} + \theta\Theta a_{t-13}$, an MA model of order 13 for $w_t = \nabla\nabla_{12} z_t$. So the autocovariances of w_t are obtained directly from results in Section 3.3, and hence they are

$$
\begin{aligned}
\gamma_0 &= [1 + \theta^2 + \Theta^2 + (\theta\Theta)^2]\sigma_a^2 = (1 + \theta^2)(1 + \Theta^2)\sigma_a^2 \\
\gamma_1 &= [-\theta - \Theta(\theta\Theta)]\sigma_a^2 = -\theta(1 + \Theta^2)\sigma_a^2 \\
\gamma_{11} &= \theta\Theta\sigma_a^2 \\
\gamma_{12} &= [-\Theta - \theta(\theta\Theta)]\sigma_a^2 = -\Theta(1 + \theta^2)\sigma_a^2 \\
\gamma_{13} &= \theta\Theta\sigma_a^2
\end{aligned}
\qquad (9.2.18)
$$

In particular, these expressions imply that $\rho_1 = -\theta/(1 + \theta^2)$ and $\rho_{12} = -\Theta/(1 + \Theta^2)$, so that the value ρ_1 is unaffected by the presence of the seasonal MA factor $(1 - \Theta B^{12})$ in the model (9.2.1), while the value of ρ_{12} is unaffected by the nonseasonal or regular MA factor $(1 - \theta B)$ in (9.2.1).

Table 9.3 shows the estimated autocorrelations of the logged airline data for (a) the original logged series, z_t, (b) the logged series differenced with respect to months only, ∇z_t, (c) the logged series differenced with respect to years only, $\nabla_{12} z_t$, and (d) the logged series differenced with respect to months and years, $\nabla\nabla_{12} z_t$. The autocorrelations for z are large and fail to die out at higher lags. While simple differencing reduces the correlations in general, a very heavy periodic component remains. This is evidenced particularly by very large correlations at lags 12, 24, 36, and 48. Simple differencing with respect to period 12 results in correlations which are first persistently positive and then persistently negative. By contrast, the differencing $\nabla\nabla_{12}$ markedly reduces correlation throughout.

The autocorrelations of $\nabla\nabla_{12} z_t$ are seen to exhibit spikes at lags 1 and 12, compatible with the features in (9.2.18) corresponding to the model (9.2.1). However, the autocorrelations for $\nabla_{12} z_t$ might be viewed as dying out at a (slow) exponential rate, for low lags, after lag 1, so there is also the possibility that $\nabla_{12} z_t$ may follow a model with nonseasonal operator of the form of an ARMA(1, 1), with the value of ϕ relatively close to 1, rather than the nonstationary IMA(1, 1) model as in (9.2.1). The distinction between these two model possibilities may not be substantial, in practice, as has been discussed in Chapter 6, and the latter model possibility will not be explored further here. The choice between the nonstationary and stationary AR(1) factor could, in fact, be tested by methods similar to those described in Section 6.3.6.

On the assumption that the model is of the form (9.2.1), the variances for the estimated higher lag autocorrelations are approximated by Bartlett's formula (2.1.13), which in this case becomes

$$\mathrm{var}[r_k] \simeq \frac{1 + 2(\rho_1^2 + \rho_{11}^2 + \rho_{12}^2 + \rho_{13}^2)}{n} \qquad k > 13 \qquad (9.2.19)$$

TABLE 9.3 Estimated Autocorrelations of Various Differences of the Logged Airline Data

	Lag	Autocorrelation											
(a) z	1–12	0.95	0.90	0.85	0.81	0.78	0.76	0.74	0.73	0.73	0.74	0.76	0.76
	13–24	0.72	0.66	0.62	0.58	0.54	0.52	0.50	0.49	0.50	0.50	0.52	0.52
	25–36	0.48	0.44	0.40	0.36	0.34	0.31	0.30	0.29	0.30	0.30	0.31	0.32
	37–48	0.29	0.24	0.21	0.17	0.15	0.12	0.11	0.10	0.10	0.11	0.12	0.13
(b) ∇z	1–12	0.20	−0.12	−0.15	−0.32	−0.08	0.03	−0.11	−0.34	−0.12	−0.11	0.21	0.84
	13–24	0.22	−0.14	−0.12	−0.28	−0.05	0.01	−0.11	−0.34	−0.11	−0.08	0.20	0.74
	25–36	0.20	−0.12	−0.10	−0.21	−0.06	0.02	−0.12	−0.29	−0.13	−0.04	0.15	0.66
	37–48	0.19	−0.13	−0.06	−0.16	−0.06	0.01	−0.11	−0.28	−0.11	−0.03	0.12	0.59
(c) $\nabla_{12} z$	1–12	0.71	0.62	0.48	0.44	0.39	0.32	0.24	0.19	0.15	−0.01	−0.12	−0.24
	13–24	−0.14	−0.14	−0.10	−0.15	−0.10	−0.11	−0.14	−0.16	−0.11	−0.08	0.00	−0.05
	25–36	−0.10	−0.09	−0.13	−0.15	−0.19	−0.20	−0.19	−0.15	−0.22	−0.23	−0.27	−0.22
	37–48	−0.18	−0.16	−0.14	−0.10	−0.05	0.02	0.04	0.10	0.15	0.22	0.29	0.30
(d) $\nabla\nabla_{12} z$	1–12	−0.34	0.11	−0.20	0.02	0.06	0.03	−0.06	0.00	0.18	−0.08	0.06	−0.39
	13–24	0.15	−0.06	0.15	−0.14	0.07	0.02	−0.01	−0.12	0.04	−0.09	0.22	−0.02
	25–36	−0.10	0.05	−0.03	0.05	−0.02	−0.05	−0.05	0.20	−0.12	0.08	−0.15	−0.01
	37–48	0.05	0.03	−0.02	−0.03	−0.07	0.10	−0.09	0.03	−0.04	−0.04	0.11	−0.05

Substituting estimated correlations for the ρ's and setting $n = 144 - 13 = 131$ in (9.2.19), where $n = 131$ is the number of differences $\nabla\nabla_{12}z_t$, we obtain a standard error $\hat{\sigma}(r) \simeq 0.11$.

In Table 9.4 observed frequencies of the 35 autocorrelations r_k, $k > 13$, are compared with those for a Normal distribution having mean zero and standard deviation 0.11. This rough check suggests that the model is worthy of further investigation.

TABLE 9.4 Comparison of Observed and Expected Frequencies for Autocorrelations of $\nabla\nabla_{12}z_t$ at Lags Greater Than 13

	Expected from Normal Distribution Mean Zero and Std. Dev. 0.11	Observed[a]
$0 < \lvert r \rvert < 0.11$	23.9	27.5
$0.11 < \lvert r \rvert < 0.22$	9.5	7.0
$0.22 < \lvert r \rvert$	1.6	0.5
	35.0	35.0

[a] Observations on the cell boundary are allocated half to each adjacent cell.

Preliminary estimates. As with the nonseasonal model, by equating observed correlations to their expected values, approximate values can be obtained for the parameters θ and Θ. On substituting the sample estimates $r_1 = -0.34$ and $r_{12} = -0.39$ in the expressions

$$\rho_1 = \frac{-\theta}{1 + \theta^2} \qquad \rho_{12} = \frac{-\Theta}{1 + \Theta^2}$$

we obtain rough estimates $\hat{\theta} \simeq 0.39$ and $\hat{\Theta} \simeq 0.48$. A table summarizing the behavior of the autocorrelation function for some specimen seasonal models, useful in identification and in obtaining preliminary estimates of the parameters, is given in Appendix A9.1.

9.2.4 Estimation

Contours of the sum of squares function $S(\theta, \Theta)$ for the airline data fitted to the model (9.2.1) are shown in Figure 9.5, together with the appropriate 95% confidence region. The least squares estimates are seen to be very nearly $\hat{\theta} = 0.4$ and $\hat{\Theta} = 0.6$. The grid of values for $S(\theta, \Theta)$ was computed using the technique described in Chapter 7. It was shown there that given n observations $\mathbf{w}$ from a linear process defined by

$$\phi(B)w_t = \theta(B)a_t$$

the quadratic form $\mathbf{w}'\mathbf{M}_n\mathbf{w}$, which appears in the exponent of the likelihood, can always be expressed in terms of a sum of squares of the conditional

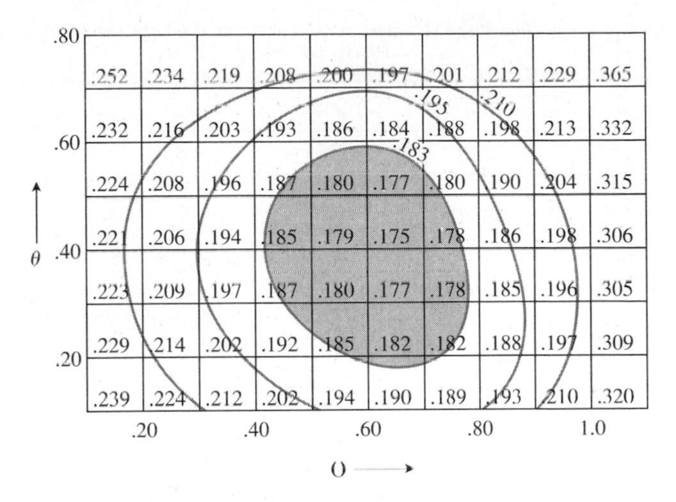

Figure 9.5 Series G fitted by the model $\nabla\nabla_{12}z_t = (1 - \theta B)(1 - \Theta B^{12})a_t$: contours of $S(\theta, \Theta)$ with shaded 95% confidence region.

expectation of a's and a quadratic function of the conditional expectation of the $p + q$ initial values $\mathbf{e}_* = (w_{1-p}, \ldots, w_0, a_{1-q}, \ldots, a_0)'$, that is,

$$\mathbf{w}'\mathbf{M}_n\mathbf{w} = S(\boldsymbol{\phi}, \boldsymbol{\theta}) = \sum_{t=-\infty}^{n} [a_t]^2 = \sum_{t=1}^{n} [a_t]^2 + [\mathbf{e}_*]'\boldsymbol{\Omega}^{-1}[\mathbf{e}_*]$$

where $[a_t] = [a_t|\mathbf{w}, \boldsymbol{\phi}, \boldsymbol{\theta}]$ and $[\mathbf{e}_*] = [\mathbf{e}_*|\mathbf{w}, \boldsymbol{\phi}, \boldsymbol{\theta}]$. Furthermore, it was shown that $S(\boldsymbol{\phi}, \boldsymbol{\theta})$ plays a central role in the estimation of the parameters $\boldsymbol{\phi}$ and $\boldsymbol{\theta}$ from both a sampling theory and a likelihood or Bayesian point of view.

The computation for seasonal models follows precisely the same course as described in Section 7.1.5 for nonseasonal models. We illustrate by considering the computation of $S(\theta, \Theta)$ for the airline data in relation to the model

$$\nabla\nabla_{12}z_t = w_t = (1 - \theta B)(1 - \Theta B^{12})a_t$$

A format for the computation of the $[a]$'s is shown in Table 9.5. If there are N observations of z, then in general, with a difference operator $\nabla^d\nabla_s^D$, we can compute $n = N - d - sD$ values of w. Therefore, it is convenient to use a numbering system so that the first observation in the z series has a subscript $1 - d - sD$. The first observation in the w series then has a subscript 1 and the last has subscript n. There are $N = 144$ observations in the airline series. Accordingly, in Table 9.5 these are designated as $z_{-12}, z_{-11}, \ldots, z_{131}$. The w's, obtained by differencing, then form the series $w_1, w_2, \ldots, w_n$, where $n = 131$. These values are set out in the center of the table.

TABLE 9.5 Airline Data: Computation Table for the $[a]$'s, and Hence for $S(\theta, \Theta)$

z_t	t	$[a_t]$	$[w_t]$	$[e_t]$	a_t^0	u_t
		0	0	0		
z_{-12}	-12	$[a_{-12}]$	$[w_{-12}]$	0	0	
z_{-11}	-11	$[a_{-11}]$	$[w_{-11}]$	0	0	
⋮	⋮	⋮	⋮	⋮	⋮	
z_0	0	$[a_0]$	$[w_0]$	0	0	
z_1	1	$[a_1]$	w_1	$[e_1]$	a_1^0	u_1
z_2	2	$[a_2]$	w_2	$[e_2]$	a_2^0	u_2
⋮	⋮	⋮	⋮	⋮	⋮	⋮
z_{131}	131	$[a_{131}]$	w_{131}	$[e_{131}]$	a_{131}^0	u_{131}

As noted in Section 7.1.5, an approximate method to obtain the $[a_t]$ using the corresponding backward model can be used for computational convenience. The fundamental formulas, on which backward and forward recursions are based in this method, may be obtained, as before, by taking conditional expectations in the backward and forward forms of the model. In this instance they yield

$$[e_t] = [w_t] + \theta[e_{t+1}] + \Theta[e_{t+12}] - \theta\Theta[e_{t+13}] \tag{9.2.20}$$

$$[a_t] = [w_t] + \theta[a_{t-1}] + \Theta[a_{t-12}] - \theta\Theta[a_{t-13}] \tag{9.2.21}$$

In general, for seasonal models, we might have a stationary autoregressive operator $\phi(B)\Phi(B^s)$ of degree $(p + sP)$. If we wished the back computation of Section 7.1.5 to begin as far back in the series as possible, the recursion would begin with the calculation of an approximate value for $[e_{n-p-sP}]$, obtained by setting unknown $[e]$'s equal to zero. In the present example, $p = P = 0$ and hence, using (9.2.20), we can begin with

$$[e_{131}] = w_{131} + (\theta \times 0) + (\Theta \times 0) - (\theta\Theta \times 0)$$

$$[e_{130}] = w_{130} + (\theta \times [e_{131}]) + (\Theta \times 0) - (\theta\Theta \times 0)$$

and so on, until $[e_1]$ is obtained. Recalling that $[e_{-j}] = 0$ when $j \geqslant 0$, we can now use (9.2.20) to compute the back-forecasts $[w_0]$, $[w_{-1}]$, . . . , $[w_{-12}]$. Furthermore, values of $[w_{-j}]$ for $j > 12$ are all zero, and since each $[a_t]$ is a function of previously occurring $[w]$'s, it follows (and otherwise is obvious directly from the form of the model) that $[a_{-j}] = 0, j > 12$. Thus (9.2.21) may now be used directly to compute the $[a]$'s, and hence to calculate $S(\theta, \Theta) = \sum_{t=-12}^{131} [a_t]^2$. In almost all cases of interest, the transients introduced by the approximation at the beginning of the back recursion will have negligible effect on the calculation of the preliminary $[w]$'s, so that $S(\theta, \Theta)$ computed in this way will be virtually exact. However, it is possible, as indicated in Section 7.1.5, to continue the "up and down" iteration. The next iteration would involve recomputing the $[e]$'s, starting off the iteration using forecasts $[w_{n+1}], [w_{n+2}], \ldots, [w_{n+13}]$ obtained from the $[a]$'s already calculated.

Alternatively, the exact method to compute the $[a_t]$ conveniently, as discussed in Section 7.1.5, can be employed. For the present model, this involves first computing the conditional estimates of the a_t, using zero initial values $a^0_{-12} = a^0_{-11} = \cdots = a^0_0 = 0$, through recursive calculations similar to (9.2.21), as

$$a_t^0 = w_t + \theta a_{t-1}^0 + \Theta a_{t-12}^0 - \theta\Theta a_{t-13}^0 \qquad t = 1, \ldots, n$$

Then a backward recursion is used to obtain values of u_t from the a_t^0 as

$$u_t = a_t^0 + \theta u_{t+1} + \Theta u_{t+12} - \theta\Theta u_{t+13} \qquad t = n, \ldots, 1$$

using zero initial values $u_{n+1} = \cdots = u_{n+13} = 0$. Finally, the exact back-forecasts for the vector of initial values $\mathbf{a}_*' = (a_{-12}, \ldots, a_0)$ is obtained by solving the equations $\mathbf{D}[\mathbf{a}_*] = \mathbf{F}'\mathbf{u}$, as described in general in (A7.3.12) of Appendix A7.3. We note that if the vector $\mathbf{F}'\mathbf{u}$ is denoted as $\mathbf{h} = \mathbf{F}'\mathbf{u} = (h_{-12}, h_{-11}, \ldots, h_0)'$, the values h_{-j} are computed as $h_{-j} = -(\theta u_{-j+1} + \Theta u_{-j+12} - \theta\Theta u_{-j+13})$, where the convention $u_{-j} = 0, j \geqslant 0$, must be used. Once the back-forecasted values are obtained, the remaining $[a_t]$ values for $t = 1, 2, \ldots, n$ are obtained recursively from (9.2.21) exactly as in the previous method, and hence the exact sum of squares $S(\theta, \Theta) = \sum_{t=-12}^{131} [a_t]^2$ is obtained. The format for computation of the $[a_t]$ by this method is also shown in the last two columns of Table 9.5.

Iterative calculation of least squares estimates $\hat{\theta}, \hat{\Theta}$. As discussed in Section 7.2, while it is essential to plot sums of squares surfaces in a new situation, or whenever difficulties arise, an iterative linearization technique may be used in straightforward situations to supply the least squares estimates and their approximate standard errors. The procedure has been set out in Section 7.2.1, and no new difficulties arise in estimating the parameters of seasonal models.

For the present example, we can write approximately

$$a_{t,0} = (\theta - \theta_0)x_{t,1} + (\Theta - \Theta_0)x_{t,2} + a_t$$

where

$$x_{t,1} = -\left.\frac{\partial a_t}{\partial \theta}\right|_{\theta_0,\,\Theta_0} \qquad x_{t,2} = -\left.\frac{\partial a_t}{\partial \Theta}\right|_{\theta_0,\,\Theta_0}$$

and where θ_0 and Θ_0 are guessed values and $a_{t,0} = [a_t | \theta_0, \Theta_0]$. As explained and illustrated in Section 7.2.2, the derivatives are most easily computed numerically. Proceeding in this way and using as starting values the preliminary estimates $\hat{\theta} = 0.39$, $\hat{\Theta} = 0.48$ obtained in Section 9.2.3 from the estimated autocorrelations, the iteration proceeded as in Table 9.6. Alternatively, proceeding as in Section 7.2.3, the derivatives could be obtained to any degree of required accuracy by recursive calculation.

TABLE 9.6 Iterative Estimation of θ and Θ for the Logged Airline Data

Iteration	θ	Θ
Starting values	0.390	0.480
1	0.404	0.640
2	0.395	0.612
3	0.396	0.614
4	0.396	0.614

Thus values of the parameters correct to two decimals, which is the most that would be needed in practice, are available in three iterations. The estimated variance of the residuals is $\hat{\sigma}_a^2 = 1.34 \times 10^{-3}$. From the inverse of the matrix of sums of squares and products of the x's on the last iteration, the standard errors of the estimates may now be calculated. The least squares estimates followed by their standard errors are then

$$\hat{\theta} = 0.40 \pm 0.08$$

$$\hat{\Theta} = 0.61 \pm 0.07$$

agreeing closely with the values obtained from the sum of squares plot.

Large-sample variances and covariances for the estimates. As in Section 7.2.6, large-sample formulas for the variances and covariances of the parameter estimates may be obtained. In this case, from the model equation $w_t = a_t - \theta a_{t-1} - \Theta a_{t-12} + \theta\Theta a_{t-13}$, the derivatives $x_{t,1} = -\partial a_t/\partial\theta$ are seen to satisfy

$$x_{t,1} - \theta x_{t-1,1} - \Theta x_{t-12,1} + \theta\Theta x_{t-13,1} + a_{t-1} - \Theta a_{t-13} = 0$$

hence $(1 - \theta B)(1 - \Theta B^{12})x_{t,1} = -(1 - \Theta B^{12})a_{t-1}$, or simply $(1 - \theta B)x_{t,1} = -a_{t-1}$. Thus, using a similar derivation for $x_{t,2} = -\partial a_t/\partial\Theta$, we obtain that

$$x_{t,1} \simeq -(1 - \theta B)^{-1}a_{t-1} = -\sum_{j=0}^{\infty} \theta^j B^j a_{t-1}$$

$$x_{t,2} \simeq -(1 - \Theta B^{12})^{-1}a_{t-12} = -\sum_{i=0}^{\infty} \Theta^i B^{12i} a_{t-12}$$

Therefore, for large samples, the information matrix is

$$\mathbf{I}(\theta, \Theta) = n \begin{bmatrix} (1 - \theta^2)^{-1} & \theta^{11}(1 - \theta^{12}\Theta)^{-1} \\ \theta^{11}(1 - \theta^{12}\Theta)^{-1} & (1 - \Theta^2)^{-1} \end{bmatrix}$$

Provided that $|\theta|$ is not close to unity, the off-diagonal term is negligible and approximate values for the variances and covariances of $\hat{\theta}$ and $\hat{\Theta}$ are

$$\begin{gathered} V(\hat{\theta}) \simeq n^{-1}(1 - \theta^2) \qquad V(\hat{\Theta}) \simeq n^{-1}(1 - \Theta^2) \\ \text{cov}[\hat{\theta}, \hat{\Theta}] \simeq 0 \end{gathered} \tag{9.2.22}$$

In the present example, substituting the values $\hat{\theta} = 0.40$, $\hat{\Theta} = 0.61$, and $n = 131$, we obtain

$$V(\hat{\theta}) \simeq 0.0064 \qquad V(\hat{\Theta}) \simeq 0.0048$$

and

$$\sigma(\hat{\theta}) \simeq 0.08 \qquad \sigma(\hat{\Theta}) \simeq 0.07$$

which, to this accuracy, are identical with the values obtained directly from the iteration. It is also interesting to note that the parameter estimates $\hat{\theta}$ and $\hat{\Theta}$, associated with months and with years, respectively, are virtually uncorrelated.

9.2.5 Diagnostic Checking

Before proceeding further, we check the adequacy of fit of the model by examining the residuals from the fitted process.

Autocorrelation check. The estimated autocorrelations of the residuals $\hat{a}_t = \nabla\nabla_{12}z_t + 0.40\hat{a}_{t-1} + 0.61\hat{a}_{t-12} - 0.24\hat{a}_{t-13}$ are shown in Table 9.7. A number of individual correlations appear rather large compared with the upper bound 0.09 of their standard error, and the value $r_{23} = 0.22$, which is about 2.5 times this upper bound, is particularly discrepant. However, among 48 random deviates one would expect some large deviations.

An overall check is provided by the quantity $\tilde{Q} = n(n+2)\sum_{k=1}^{24} r_k^2(\hat{a})/(n-k)$, which (see Section 8.2.2) is approximately distributed as χ^2 with 22 degrees of freedom, since two parameters have been fitted. The observed value of $\tilde{Q}$ is $131 \times 0.1950 = 25.5$, and on the hypothesis of adequacy of the model, deviations greater than this would be expected in about 27% of cases. The check does not provide any evidence of inadequacy in the model.

Periodogram check. The cumulative periodogram (see Section 8.2.5) for the residuals is shown in Figure 9.6. The Kolmogorov–Smirnov

TABLE 9.7 Estimated Autocorrelations of the Residuals from Fitting the Model $\nabla\nabla_{12}z_t = (1 - 0.40B)(1 - 0.61B^{12})a_t$ to Logged Airline Data (Series G)

Lag k	Autocorrelation $r_k(\hat{a})$												Standard Error (Upper Bound)
1–12	0.02	0.02	−0.13	−0.14	0.05	0.06	−0.07	−0.04	0.10	−0.08	0.02	−0.01	0.09
13–24	0.03	0.04	0.05	−0.16	0.03	0.00	−0.11	−0.10	−0.03	−0.03	0.22	0.03	0.09
25–36	−0.02	0.06	−0.04	−0.06	−0.05	−0.08	−0.05	0.12	−0.13	0.00	−0.06	−0.02	0.09
37–48	0.11	0.07	−0.02	−0.05	−0.10	−0.02	−0.04	0.00	−0.08	0.03	0.04	0.06	0.09

$$(n + 2) \sum_{k=1}^{24} r_k^2(\hat{a})/(n - k) = 0.1950$$

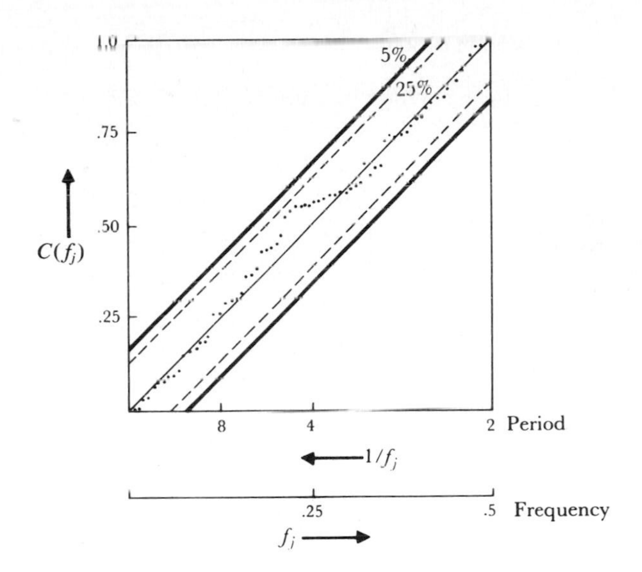

Figure 9.6 Cumulative periodogram check on residuals from model $\nabla\nabla_{12}z_t = (1 - 0.40B)(1 - 0.61B^{12})a_t$ fitted to series G.

5% and 25% probability limits, which as we have seen in Section 8.2.5 supply a very rough guide to the significance of apparent deviations, fail in this instance to indicate any significant departure from the assumed model.

9.3 SOME ASPECTS OF MORE GENERAL SEASONAL MODELS

9.3.1 Multiplicative and Nonmultiplicative Models

In previous sections we discussed methods of dealing with seasonal time series, and in particular, we examined an example of a multiplicative model. We have seen how this can provide a useful representation with remarkably few parameters. It now remains to study other seasonal models of this kind, and insofar as new considerations arise, the associated processes of identification, estimation, diagnostic checking, and forecasting.

Suppose, in general, that we have a seasonal effect associated with period s. Then the general class of multiplicative models may be typified in the manner shown in Figure 9.7. In the multiplicative model it is supposed that the "between periods" development of the series is represented by some model

$$\Phi_P(B^s)\nabla_s^D z_{r,m} = \Theta_Q(B^s)\alpha_{r,m}$$

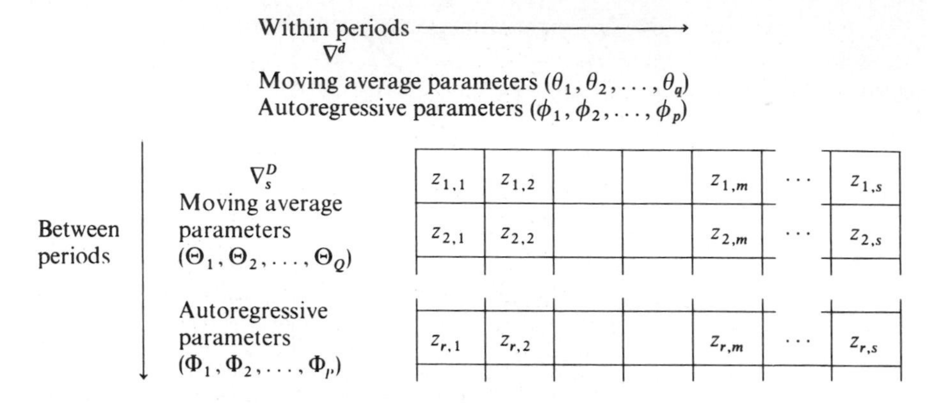

Figure 9.7 Two-way table for multiplicative seasonal model.

while "within periods" the α's are related by

$$\phi_p(B)\nabla^d \alpha_{r,m} = \theta_q(B) a_{r,m}$$

Obviously, we could change the order in which we considered the two types of models and in either case obtain the general multiplicative model

$$\phi_p(B)\Phi_P(B^s)\nabla^d \nabla_s^D z_{r,m} = \theta_q(B)\Theta_Q(B^s) a_{r,m} \tag{9.3.1}$$

where $a_{r,m}$ is a white noise process with zero mean. In practice, the usefulness of models such as (9.3.1) depends on how far it is possible to parameterize actual time series parsimoniously in these terms. In fact, this has been possible for a variety of seasonal time series coming from widely different sources [18].

It is not possible to obtain a completely adequate fit with multiplicative models for all series. One modification that is sometimes useful allows the mixed moving average operator to be nonmultiplicative. By this is meant that we replace the operator $\theta_q(B)\Theta_Q(B^s)$ on the right-hand side of (9.3.1) by a more general moving average operator $\theta_{q^*}^*(B)$. Alternatively, or in addition, it may be necessary to replace the autoregressive operator $\phi_p(B)\Phi_P(B^s)$ on the left by a more general autoregressive operator $\phi_{p^*}^*(B)$. Some specimens of nonmultiplicative models are given in Appendix A9.1. These are numbered 4, 4a, 5, and 5a.

In those cases where a nonmultiplicative model is found necessary, experience suggests that the best-fitting multiplicative model can provide a good starting point from which to construct a better nonmultiplicative model. The situation is reminiscent of the problems encountered in analyzing two-way analysis of variance tables, where additivity of row and column constants may or may not be an adequate assumption, but may provide a good point of departure.

Our general strategy for relating multiplicative or nonmultiplicative models to data is that which we have already discussed and illustrated in some detail in Section 9.2. Using the autocorrelation function for guidance:

1. The series is differenced with respect to ∇ and ∇_s, so as to produce stationarity.
2. By inspection of the autocorrelation function of the suitably differenced series, a tentative model is selected.
3. From the values of appropriate autocorrelations of the differenced series, preliminary estimates of the parameters are obtained. These can be used as starting values in the search for the least squares estimates.
4. After fitting, the diagnostic checking process applied to the residuals either may lead to the acceptance of the tentative model or, alternatively, may suggest ways in which it can be improved, leading to refitting and repetition of the diagnostic checks.

As a few practical guidelines for model specification, we note that for seasonal series the order of seasonal differencing D needed would almost never be greater than 1, and especially for monthly series with $s = 12$, the orders P and Q of the seasonal AR and MA operators $\Phi(B^s)$ and $\Theta(B^s)$ would rarely need to be greater than 1. This is particularly so when the series length of available data is not sufficient to warrant such a complicated form of model with $P > 1$ or $Q > 1$.

9.3.2 Identification

A useful aid in model identification is the list in Appendix A9.1, giving the covariance structure of $w_t = \nabla^d \nabla_s^D z_t$ for a number of simple seasonal models. This list makes no claim to be comprehensive. However, it is believed that it does include some of the frequently encountered models, and the reader should have no difficulty in discovering the characteristics of others that seem representationally useful. It should be emphasized that rather simple models (such as models 1 and 2 in Appendix A9.1) have provided adequate representations for many seasonal series.

Since the multiplicative seasonal ARMA models for the differences $w_t = \nabla \nabla_s z_t$ may be viewed merely as special forms of ARMA models with orders $p + sP$ and $q + sQ$, their autocovariances can be derived from the principles of Chapter 3, as was done in Section 9.2.3 for the MA model $w_t = a_t - \theta a_{t-1} - \Theta a_{t-12} + \theta\Theta a_{t-13}$. To illustrate the derivation of autocovariances for models such as those given in Appendix A9.1, consider the model $(1 - \phi B)w_t = (1 - \Theta B^s)a_t$, which is a special form of ARMA model with AR order 1 and MA order s. First, since the ψ weights for this model for w_t

satisfy $\psi_j - \phi\psi_{j-1} = 0, j = 1, \ldots, s-1$, we have $\psi_j = \phi^j, j = 1, \ldots, s-1$, as well as $\psi_s = \phi^s - \Theta$ and $\psi_j = \phi\psi_{j-1}, j > s$. From Section 3.4 we know that the autocovariances for w_t will satisfy

$$\gamma_0 = \phi\gamma_1 + \sigma_a^2(1 - \Theta\psi_s)$$

$$\gamma_j = \phi\gamma_{j-1} - \sigma_a^2\Theta\psi_{s-j} \qquad j = 1, \ldots, s \tag{9.3.2}$$

$$\gamma_j = \phi\gamma_{j-1} \qquad j > s$$

Solving the first two equations for γ_0 and γ_1, we obtain

$$\gamma_0 = \sigma_a^2 \frac{1 - \Theta(\phi^s - \Theta) - \phi^s\Theta}{1 - \phi^2} = \sigma_a^2 \frac{1 + \Theta^2 - 2\phi^s\Theta}{1 - \phi^2}$$

$$\gamma_1 = \sigma_a^2 \frac{\phi[1 - \Theta(\phi^s - \Theta)] - \phi^{s-1}\Theta}{1 - \phi^2} = \sigma_a^2 \frac{\phi(1 + \Theta^2 - \phi^s\Theta) - \phi^{s-1}\Theta}{1 - \phi^2}$$

with $\gamma_j = \phi\gamma_{j-1} - \sigma_a^2\Theta\phi^{s-j} = \phi^j\gamma_0 - \sigma_a^2\Theta\phi^{s-j}(1 + \phi^{2j})/(1 - \phi^2), j = 1, \ldots,$ s and $\gamma_j = \phi\gamma_{j-1} = \phi^{j-s}\gamma_s, j > s$. Hence, in particular, for monthly data with $s = 12$ and $|\phi|$ not too close to 1, the autocorrelation function ρ_j for this process will behave, for low lags, similarly to that of a regular AR(1) process, $\rho_j \simeq \phi^j$ for small j, while the value of ρ_{12} will be close to $-\Theta/(1 + \Theta^2)$.

A fact of considerable utility in deriving the autocovariances of a multiplicative process is that for such a process, the autocovariance generating function (3.1.11) is the product of the generating functions of the components. Thus in (9.3.1) if the component models for $\nabla^d z_t$ and $\nabla_s^D \alpha_t$,

$$\phi_p(B)\nabla^d z_t = \theta_q(B)\alpha_t \qquad \Phi_P(B^s)\nabla_s^D \alpha_t = \Theta_Q(B^s)a_t$$

have autocovariance generating function $\gamma(B)$ and $\Gamma(B^s)$, the autocovariance generating function for $w_t = \nabla^d\nabla_s^D z_t$ in (9.3.1) is

$$\gamma(B)\Gamma(B^s)$$

Another point to be remembered is that it may be useful to parametrize more general models in terms of their departures from related multiplicative forms in a manner now illustrated.

The three-parameter nonmultiplicative operator

$$1 - \theta_1 B - \theta_{12}B^{12} - \theta_{13}B^{13} \tag{9.3.3}$$

employed in models 4 and 5 may be written

$$(1 - \theta_1 B)(1 - \theta_{12}B^{12}) - \kappa B^{13}$$

where

$$\kappa = \theta_1\theta_{12} - (-\theta_{13})$$

An estimate of κ that was large compared with its standard error would indicate the need for a nonmultiplicative model in which the value of θ_{13} is not tied to the values of θ_1 and θ_{12}. On the other hand, if κ is small, then on writing $\theta_1 = \theta$, $\theta_{12} = \Theta$, the model approximates the multiplicative $(0, 1, 1) \times (0, 1, 1)_{12}$ model.

9.3.3 Estimation

No new problems arise in the estimation of the parameters of general seasonal models. The unconditional sum of squares is computed quite generally by the methods set out fully in Section 7.1.5 and illustrated further in Section 9.2.4. As always, contour plotting can illuminate difficult situations. In well-behaved situations, iterative least squares with numerical determination of derivatives yield rapid convergence to the least squares estimates, together with approximate variances and covariances of the estimates. Recursive procedures can be derived in each case which allow direct calculation of derivatives, if desired.

Large-sample variances and covariances of the estimates. The large-sample information matrix $\mathbf{I}(\boldsymbol{\phi}, \boldsymbol{\theta}, \boldsymbol{\Phi}, \boldsymbol{\Theta})$ is given by evaluating $E[\mathbf{X}'\mathbf{X}]$, where, as in Section 7.2.6, $\mathbf{X}$ is the $n \times (p + q + P + Q)$ matrix of derivatives with reversed signs. Thus for the general multiplicative model

$$a_t = \theta^{-1}(B)\Theta^{-1}(B^s)\phi(B)\Phi(B^s)w_t$$

where

$$w_t = \nabla^d \nabla_s^D z_t$$

the required derivatives are

$$\frac{\partial a_t}{\partial \theta_i} = \theta^{-1}(B)B^i a_t \qquad \frac{\partial a_t}{\partial \Theta_i} = \Theta^{-1}(B^s)B^{si} a_t$$

$$\frac{\partial a_t}{\partial \phi_j} = -\phi^{-1}(B)B^j a_t \qquad \frac{\partial a_t}{\partial \Phi_j} = -\Phi^{-1}(B^s)B^{sj} a_t$$

Approximate variances and covariances of the estimates are obtained as before, by inverting the matrix $\mathbf{I}(\boldsymbol{\phi}, \boldsymbol{\theta}, \boldsymbol{\Phi}, \boldsymbol{\Theta})$.

9.3.4 Eventual Forecast Functions for Various Seasonal Models

We now consider the characteristics of the eventual forecast functions for a number of seasonal models. For a seasonal model with single periodicity s, the eventual forecast function at origin t for lead time l is the solution of the difference equation

$$\phi(B)\Phi(B^s)\nabla^d \nabla_s^D \hat{z}_t(l) = 0$$

Table 9.8 shows this solution for various choices of the difference equation; also shown is the number of initial values on which the behavior of the forecast function depends.

In Figure 9.8 the behavior of each forecast function is illustrated for $s = 4$. It will be convenient to regard the lead time $l = rs + m$ as referring to a forecast r years and m quarters ahead. In the diagram, an appropriate number of initial values (required to start the forecast off and indicated by bold dots) has been set arbitrarily and the course of the forecast function traced to the end of the fourth period. When the difference equation involves an autoregressive parameter, its value has been set equal to 0.5.

The constants $b_{0,m}$, b_1, and so on, appearing in the solutions in Table 9.8, should strictly be indicated by $b_{0,m}^{(t)}$, $b_1^{(t)}$, and so on, since each one depends on the origin t of the forecast, and these constants are adaptively modified each time the origin changes. The superscript t has been omitted temporarily to simplify notation. The operator labeled (1) is stationary, containing a fixed mean μ. It is autoregressive in the seasonal pattern, which decays with each period, approaching closer and closer to the mean.

Operator (2) is nonstationary in the seasonal component. The forecasts for a particular quarter are linked from year to year by a polynomial of degree 0. Thus the basic forecast of the seasonal component is exactly reproduced in forecasts of future years.

TABLE 9.8 Eventual Forecast Functions for Various Autoregressive Operators

Autoregressive Operator	Eventual Forecast Function $\hat{z}(r, m)$[a]	Number of Initial Values on Which Forecast Function Depends
(1) $1 - \Phi B^s$	$\mu + (b_{0,m} - \mu)\Phi^r$	s
(2) $1 - B^s$	$b_{0,m}$	s
(3) $(1 - B)(1 - \Phi B^s)$	$b_0 + (b_{0,m} - b_0)\Phi^r + b_1 \left\{\frac{1 - \Phi^r}{1 - \Phi}\right\}$	$s + 1$
(4) $(1 - B)(1 - B^s)$	$b_{0,m} + b_1 r$	$s + 1$
(5) $(1 - \phi B)(1 - B^s)$	$b_{0,m} + b_1 \phi^{m-1} \left\{\frac{1 - \phi^{sr}}{1 - \phi^s}\right\}$	$s + 1$
(6) $(1 - B)(1 - B^s)^2$	$b_{0,m} + b_{1,m} r + \frac{1}{2} b_2 r(r - 1)$	$2s + 1$
(7) $(1 - B)^2(1 - B^s)$	$b_{0,m} + [b_1 + (m - 1)b_2]r + \frac{1}{2} b_2 sr(r - 1)$	$s + 2$

[a] Coefficients b are all adaptive and depend upon forecast origin t.

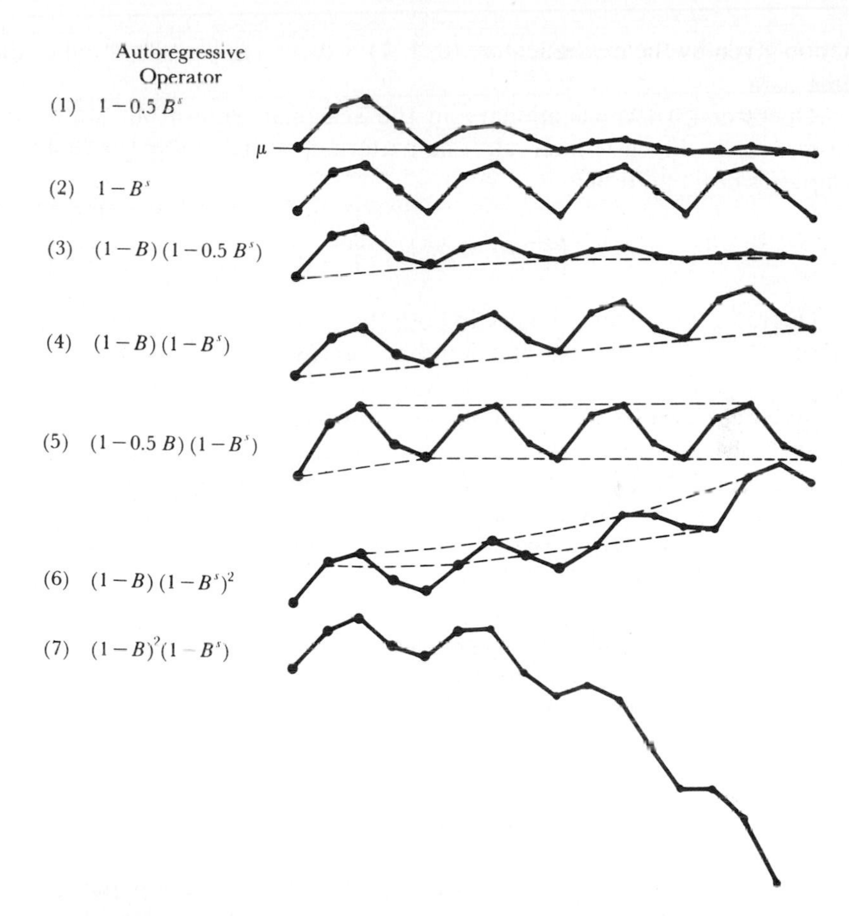

Figure 9.8 Behavior of the seasonal forecast function for various choices of the general seasonal autoregressive operator.

Operator (3) is nonstationary with respect to the basic interval but stationary in the seasonal component. Figure 9.8 (3) shows the general level of the forecast approaching asymptotically the new level

$$b_0 + \frac{b_1}{1 - \Phi}$$

where, at the same time, the superimposed predictable component of the stationary seasonal effect dies out exponentially.

Operator (4) is the limiting case of the operator (3) as Φ approaches unity. The operator is nonstationary with respect to both the basic interval and the periodic component. The basic initial forecast pattern is reproduced, as is the incremental yearly increase. This is the type of forecast function given by the multiplicative $(0, 1, 1) \times (0, 1, 1)_{12}$ process fitted to the airline data.

Operator (5) is nonstationary in the seasonal pattern but stationary with respect to the basic interval. The pattern approaches exponentially an asymptotic basic pattern

$$\hat{z}_t(\infty, m) = b_{0,m} + \frac{b_1 \phi^{m-1}}{1 - \phi^s}$$

Operator (6) is nonstationary in both the basic interval and the seasonal component. An overall quadratic trend occurs over years, and a particular kind of modification occurs in the seasonal pattern. Individual quarters not only have their own level $b_{0,m}$ but also their own rate of change of level $b_{1,m}$. Therefore, when this kind of forecast function is appropriate, we can have a situation where, for example, as the lead time is increased, the difference in summer over spring sales can be forecast to increase from one year to the next, while at the same time the difference in autumn over summer sales can be forecast to decrease.

Operator (7) is again nonstationary in both basic interval and in the seasonal component, and there is again a quadratic tendency over years with the incremental changes in the forecasts from one quarter to the next changing linearly. However, in this case they are restricted to have a common *rate* of change.

9.3.5 Choice of Transformation

It is particularly true for seasonal models that the weighted averages of previous data values, which comprise the forecasts, may extend far back into the series. Care is therefore needed in choosing a transformation in terms of which a parsimonious linear model will closely apply over a sufficient stretch of the series. Simple graphical analysis can often suggest such a transformation. Thus an appropriate transformation may be suggested by determining in what metric the amplitude of the seasonal component is roughly independent of the level of the series. To illustrate how a data-based transformation may be chosen more exactly, denote the *untransformed* airline data by x, and let us assume that some power transformation [$z = x^\lambda$ for $\lambda \neq 0$, $z = \ln(x)$ for $\lambda = 0$] may be needed to make the model (9.2.1) appropriate. Then, as suggested in Section 4.1.3, the approach of Box and Cox [38] may be followed, and the maximum likelihood value obtained by fitting the model to $x^{(\lambda)} = (x^\lambda - 1)/\lambda \dot{x}^{\lambda-1}$ for various values of λ, and choosing the value of λ that results in the smallest residual sum of squares S_λ. In this expression $\dot{x}$ is the geometric mean of the series, and it is easily shown that $x^{(0)} = \dot{x}\ln(x)$.

For the airline data we find

λ	S_λ	λ	S_λ	λ	S_λ
−0.4	13,825.5	−0.1	11,627.2	0.2	11,784.3
−0.3	12,794.6	0	11,458.1	0.3	12,180.0
−0.2	12,046.0	0.1	11,554,3	0.4	12,633.2

The maximum likelihood value is thus close to $\lambda = 0$, confirming for this particular example the appropriateness of the logarithmic transformation.

9.4 STRUCTURAL COMPONENT MODELS AND DETERMINISTIC SEASONAL COMPONENTS

As mentioned at the beginning of Section 9.1.1, a traditional method to represent a seasonal time series has been to decompose the series into trend, seasonal, and noise components, as $z_t = T_t + S_t + N_t$, where the trend T_t and seasonal S_t are represented as deterministic functions of time using polynomial and sinusoidal functions, respectively. More recently, such "structural component" models, but with the trend and seasonal components following stochastic rather than deterministic models, have become increasingly popular for modeling, forecasting, and seasonal adjustment of time series (e.g., Harvey and Todd [109], Gersch and Kitagawa [95], Kitagawa and Gersch [128], and Hillmer and Tiao [111]). For example, for monthly data, in such a model approach the trend might be assumed to follow the model $(1 - B)T_t = (1 - \theta_T B)a_t$ or $(1 - B)^2 T_t = (1 - \theta_{T1}B - \theta_{T2}B^2)a_t$, while the seasonal might follow $(1 - B^{12})S_t = b_t$, where a_t and b_t are independent white noise processes. An appeal of this modeling approach, especially for seasonal adjustment issues, is that Kalman filtering and smoothing methods based on state space formulations of the model, as discussed in Section 5.5, can be employed.

However, it should be noted that such structural models have an equivalent ARIMA model representation. To illustrate, consider $z_t = T_t + S_t + N_t$, where it is assumed that

$$(1 - B)T_t = (1 - \theta_T B)a_t \qquad (1 - B^{12})S_t = (1 - \Theta_S B^{12})b_t$$

and $N_t = c_t$ is white noise. Then we have

$$\begin{aligned}(1 - B)&(1 - B^{12})z_t \\ &= (1 - B^{12})(1 - \theta_T B)a_t + (1 - B)(1 - \Theta_S B^{12})b_t + (1 - B)(1 - B^{12})c_t\end{aligned}$$

and according to the developments in Appendix A4.3, the right-hand-side expression above can be represented as the MA model $(1 - \theta_1 B - \theta_{12} B^{12} -$

$\theta_{13}B^{13})\varepsilon_t$, where ε_t is white noise, since the right-hand side will have nonzero autocovariances only at the lags 0, 1, 11, 12, and 13. Under additional structure, the MA operator could have the multiplicative form, but in general we see that the foregoing structural model, $z_t = T_t + S_t + N_t$, has an equivalent ARIMA model representation as

$$(1 - B)(1 - B^{12})z_t = (1 - \theta_1 B - \theta_{12}B^{12} - \theta_{13}B^{13})\varepsilon_t$$

9.4.1 Deterministic Seasonal and Trend Components and Common Factors

Now in some cases, particularly for series arising in physical sciences, it could be approximately true that a seasonal or trend component would be deterministic. For example, suppose that the seasonal component could in fact be represented as

$$S_t = \beta_0 + \sum_{j=1}^{6} \left[\beta_{1j} \cos\left(\frac{2\pi jt}{12}\right) + \beta_{2j} \sin\left(\frac{2\pi jt}{12}\right)\right]$$

where the β coefficients are deterministic (fixed). We note that this can be viewed as a special case of the previous examples, since S_t satisfies $(1 - B^{12})S_t = 0$. Now, ignoring the trend component for the present, in a model such as $z_t = S_t + N_t$, where $(1 - B^{12})S_t = 0$ and $N_t = (1 - \theta_N B)c_t$, say, we find that z_t follows the seasonal ARIMA model

$$(1 - B^{12})z_t = (1 - \theta_N B)(1 - B^{12})c_t$$

However, we now notice the presence of a common factor of $1 - B^{12}$ in both the generalized AR operator and the MA operator of this model; equivalently, we might say that $\Theta = 1$ for the seasonal MA operator $\Theta(B^{12}) = (1 - \Theta B^{12})$. This is caused by and, in fact, is indicative of the presence of the deterministic (seasonal) component S_t in the original form of the model.

In general, the presence of deterministic seasonal or trend components in the structure of a time series z_t is characterized by common factors of $(1 - B^s)$ or $(1 - B)$ in the generalized AR operator and the MA operator of the model. We can state the result more formally as follows. Suppose that z_t follows the model $\varphi(B)z_t = \theta_0 + \theta(B)a_t$, and the operators $\varphi(B)$ and $\theta(B)$ contain a common factor $G(B)$, so that $\varphi(B) = G(B)\varphi_1(B)$ and $\theta(B) = G(B)\theta_1(B)$. Hence the model is

$$G(B)\varphi_1(B)z_t = \theta_0 + G(B)\theta_1(B)a_t \tag{9.4.1}$$

Let $G(B) = 1 - g_1 B - \cdots - g_r B^r$ and suppose that this polynomial has roots $G_1^{-1}, \ldots, G_r^{-1}$ (assume distinct). Then the common factor $G(B)$ can be canceled from both sides of the model above, but a term of the form $\sum_{i=1}^{r} c_i G_i^t$ is added to the model. Thus the model (9.4.1) can be expressed in equivalent form as

$$\varphi_1(B)z_t = c_{0t} + \sum_{i=1}^{r} c_i G_i^t + \theta_1(B)a_t \tag{9.4.2}$$

where the c_i are constants, and c_{0t} is a term that satisfies $G(B)c_{0t} = \theta_0$. Modifications of the result for the case where some of the roots G_i^{-1} are repeated are straightforward.

Thus it is seen that an equivalent representation for the model above is $\varphi_1(B)z_t = x_t + \theta_1(B)a_t$, where x_t is a deterministic function of t that satisfies $G(B)x_t = \theta_0$. Note that roots in $G(B)$ corresponding to "stationary factors," such that $|G_i| < 1$, will make a negligible contribution to the component x_t, and hence these terms may be ignored. Thus only those factors whose roots correspond to nonstationary "differencing" and other "simplifying" operators, such as $(1 - B)$ and $(1 - B^s)$, with roots such that $|G_i| = 1$, need to be included in the deterministic component x_t in the statement of the equivalent form of the model. These common factors will, of course, give rise to deterministic functions in x_t that are of the form of polynomials, sine and cosine functions, and products of these, depending on the roots of the common factor $G(B)$.

Examples. For a few simple examples, the model $(1 - B)z_t = \theta_0 + (1 - B)\theta_1(B)a_t$ has an equivalent form $z_t = c_1 + \theta_0 t + \theta_1(B)a_t$, which occurs upon cancellation of the common factor $(1 - B)$, while the model $(1 - \sqrt{3}B + B^2)z_t = \theta_0 + (1 - \sqrt{3}B + B^2)\theta_1(B)a_t$ has an equivalent model form as $z_t = c_0 + c_1 \cos(2\pi t/12) + c_2 \sin(2\pi t/12) + \theta_1(B)a_t$, where $(1 - \sqrt{3} + 1)c_0 = \theta_0$.

In practice, possible common factors with roots on the unit circle and the associated cancellation will only be approximate, and it would then be useful to estimate the tentative model with deterministic components that is implied by the cancellation directly by a combination of regression model and ARIMA time series methods. One additional consequence of the presence of deterministic factors for forecasting is that at least some of the coefficients $b_j^{(t)}$ in the general forecast function $\hat{z}_t(l)$ for z_{t+l} in (5.3.3) will not be adaptive but will be deterministic (fixed) constants. Results such as those discussed above concerning the relationship between common factors with roots on the unit circle in the generalized AR and the MA operators of ARIMA models and the presence of deterministic polynomial and sinusoidal components have been discussed by Abraham and Box [2], Harvey [106], and Bell [27].

9.4.2 Models with Regression Terms and Time Series Error Terms

The previous discussion motivates consideration of models for time series which include regression terms, such as deterministic sine and cosine functions to represent seasonal behavior, and a time series error or "noise" term which may be autocorrelated. We consider the case where the error term N_t can be assumed to be a stationary ARMA process; otherwise, we may need to consider differencing of the original series. So we suppose that the series w_t follows the regression model

$$w_t = \beta_1 x_{t1} + \beta_2 x_{t2} + \cdots + \beta_k x_{tk} + N_t \qquad t = 1, \ldots, n \tag{9.4.3}$$

where $x_{t1}, \ldots, x_{tk}$ are values of k explanatory or predictor variables, and the errors N_t are assumed to be generated by an ARMA(p, q) model, $\phi(B)N_t = \theta(B)a_t$. Recall that results for the traditional linear regression model were reviewed briefly in Appendix A7.2. In that notation, the model (9.4.3) may be written as $\mathbf{w} = \mathbf{X}\boldsymbol{\beta} + \boldsymbol{N}$, and $\mathbf{V} = \text{cov}[\boldsymbol{N}]$. In the traditional model, we have $\mathbf{V} = \sigma^2\mathbf{I}$ and the least squares estimator of $\boldsymbol{\beta}$, $\hat{\boldsymbol{\beta}} = (\mathbf{X}'\mathbf{X})^{-1}\mathbf{X}'\mathbf{w}$ has the usual properties such as $\text{cov}[\hat{\boldsymbol{\beta}}] = \sigma^2(\mathbf{X}'\mathbf{X})^{-1}$.

However, in the case of autocorrelated errors, this property no longer holds, and instead we find that the least squares estimator has covariance matrix given by $\text{cov}[\hat{\boldsymbol{\beta}}] = (\mathbf{X}'\mathbf{X})^{-1}\mathbf{X}'\mathbf{V}\mathbf{X}(\mathbf{X}'\mathbf{X})^{-1}$. Hence the usual sampling properties and inference procedures, such as usual standard error formulas for the estimates $\hat{\beta}_i$, t-statistics, and confidence intervals, will no longer be valid since they are all based on an incorrect assumption concerning the properties of the errors N_t in (9.4.3), leading to an incorrect form for the covariance matrix of the least squares estimator $\hat{\boldsymbol{\beta}}$.

Example. As a simple example to illustrate the situation, consider the model $w_t = \beta_1 x_t + N_t$, where N_t follows the MA(1) model $N_t = (1 - \theta B)a_t$. Then we know that N_t has only nonzero autocovariances of $\gamma_0 = (1 + \theta^2)\sigma_a^2$ and $\gamma_1 = -\theta\sigma_a^2$. The least squares estimator of β_1 is $\hat{\beta}_1 = \Sigma_{t=1}^n x_t w_t / \Sigma_{t=1}^n x_t^2$, and its actual variance is

$$\text{var}[\hat{\beta}_1] = \frac{\mathbf{X}'\mathbf{V}\mathbf{X}}{(\Sigma_{t=1}^n x_t^2)^2} = \frac{\gamma_0 \Sigma_{t=1}^n x_t^2 + 2\gamma_1 \Sigma_{t=1}^{n-1} x_t x_{t+1}}{(\Sigma_{t=1}^n x_t^2)^2}$$

This expression can be simplified to

$$\text{var}[\hat{\beta}_1] = \frac{\gamma_0}{\Sigma_{t=1}^n x_t^2}[1 + 2\rho_1 r_1(x)]$$

where $r_1(x) = \Sigma_{t=1}^{n-1} x_t x_{t+1} / \Sigma_{t=1}^n x_t^2$ can be interpreted as a lag 1 sample autocorrelation of the x_t, uncorrected for the mean. By contrast, if the errors N_t in the model were presumed to be uncorrelated, we would take $\rho_1 = 0$ and would incorrectly be led to believe that the variance of $\hat{\beta}_1$ was $\gamma_0/\Sigma_{t=1}^n x_t^2$. Hence use of this form for the variance could lead to incorrect inferences concerning β_1. We also see in this simple example that the effect of autocorrelation ρ_1 on the variance of the least squares estimator depends on the nature of autocorrelation in the explanatory variable series x_t through the factor $\rho_1 r_1(x)$. Hence when the series x_t and the noise N_t have lag 1 autocorrelation of the same sign, the variance of the least squares estimator will be inflated over what it would be with uncorrelated errors, but when they have autocorrelation of opposite sign, the reverse is true.

When fitting a regression model to time series data, one should always consider the possibility of autocorrelation in the error term. Often, a reasonable approach to identify an appropriate model for the error N_t is first to obtain the least squares estimate $\hat{\boldsymbol{\beta}}$, then obtain the corresponding regression model residuals

$$\hat{N}_t = w_t - \hat{\beta}_1 x_{t1} - \hat{\beta}_2 x_{t2} - \cdots - \hat{\beta}_k x_{tk} \tag{9.4.4}$$

This residual series can be examined by the usual time series modeling methods, such as examination of its sample autocorrelation and partial autocorrelation functions, to identify an appropriate ARMA model for N_t. This preliminary identification technique would typically be adequate to specify a tentative model for the error term N_t, especially when the explanatory variables x_{ti} are of the form of deterministic functions such as sine and cosine functions, or polynomial terms, since in such cases it is known (e.g., Anderson [8, Sec. 10.2]) that the least squares estimator for $\boldsymbol{\beta}$ is an asymptotically efficient estimator relative to the best linear estimator. In addition, it is known that the sample autocorrelations and partial autocorrelations calculated from the residuals of the preliminary least squares fit are asymptotically equivalent to those obtained from the actual noise series N_t (e.g., Anderson [8, Sec. 10.3] and Fuller [93, Sec. 9.3]). In general, the best linear estimator of $\boldsymbol{\beta}$, known as the generalized least squares estimator, is given by $\hat{\boldsymbol{\beta}}_G = (\mathbf{X}'\mathbf{V}^{-1}\mathbf{X})^{-1}\mathbf{X}'\mathbf{V}^{-1}\mathbf{w}$, and it has covariance matrix given by $\text{cov}[\hat{\boldsymbol{\beta}}_G] = (\mathbf{X}'\mathbf{V}^{-1}\mathbf{X})^{-1}$. This estimator also corresponds to the maximum likelihood estimator under the assumption of Normality of the errors when the covariance matrix $\mathbf{V}$ is known.

In practice, based on the particular ARMA model specified for the noise term N_t, one can determine the form of the covariance matrix $\mathbf{V}$ of $\mathbf{N}$ and can find a matrix $\mathbf{P}'$, which can be chosen to be lower triangular, such that $\mathbf{P}'\mathbf{V}\mathbf{P} = \sigma_a^2\mathbf{I}$, that is, $\mathbf{V}^{-1} = \mathbf{P}\mathbf{P}'/\sigma_a^2$. Then, as in Section A7.2.5, the generalized least squares estimator can in fact be obtained from transformed variables $\mathbf{w}^* = \mathbf{P}'\mathbf{w}$ and $\mathbf{X}^* = \mathbf{P}'\mathbf{X}$ as $\hat{\boldsymbol{\beta}}_G = (\mathbf{X}^{*\prime}\mathbf{X}^*)^{-1}\mathbf{X}^{*\prime}\mathbf{w}^*$, with $\text{cov}[\hat{\boldsymbol{\beta}}_G] = \sigma_a^2(\mathbf{X}^{*\prime}\mathbf{X}^*)^{-1}$. In practice, since the parameters ϕ_i and θ_i of the time series model for N_t are not known, one must iterate between the computation of $\hat{\boldsymbol{\beta}}_G$ using the current estimates of the time series model parameters to form the transformation matrix $\hat{\mathbf{P}}'$, and maximum likelihood or exact least squares procedures, as discussed in Chapter 7, to obtain time series model parameter estimates based on the estimated noise series $\hat{N}_t$ constructed from the current estimate of $\boldsymbol{\beta}$. The computational procedure used to determine the exact sum of squares function for the specified ARMA model will also essentially determine the nature of the transformation matrix $\mathbf{P}'$ that is needed in the calculation of the generalized least squares estimator.

Example. We take the simple example of an AR(1) model, $(1 - \phi B)N_t = a_t$, for the noise N_t, for illustration. Then the $n \times n$ matrix $\mathbf{P}'$ is such that its (1, 1) element is equal to $(1 - \phi^2)^{1/2}$, its remaining diagonal elements are equal to 1, its elements on the diagonal just below the main diagonal are equal to $-\phi$, and all remaining elements are equal to zero. Hence under this transformation, the transformed variables are $w_1^* = (1 - \phi^2)^{1/2}w_1$, and $w_t^* = w_t - \phi w_{t-1}$, $t = 2, 3, \ldots, n$, and similarly for the transformed explanatory variables x_{ti}^*. In effect, with AR(1) errors, the model (9.4.3) has been transformed by applying the operator $(1 - \phi B)$ through the equation, to obtain

$$w_t - \phi w_{t-1} = \beta_1(x_{t1} - \phi x_{t-1,1}) + \beta_2(x_{t2} - \phi x_{t-1,2}) + \cdots + \beta_k(x_{tk} - \phi x_{t-1,k}) + a_t$$

or $w_t^* = \beta_1 x_{t1}^* + \beta_2 x_{t2}^* + \cdots + \beta_k x_{tk}^* + a_t$, which has the form of the usual regression model with uncorrelated errors a_t, so that ordinary least squares regression procedures apply to the transformed model.

Generalization of the transformation procedure for higher-order AR models is straightforward, so that, apart from special treatment for the initial p observations, the transformed variables are $w_t^* = \phi(B)w_t = w_t - \phi_1 w_{t-1} - \cdots - \phi_p w_{t-p}$ and $x_{ti}^* = \phi(B)x_{ti} = x_{ti} - \phi_1 x_{t-1,i} - \cdots - \phi_p x_{t-p,i}$, $i = 1, \ldots, k$. The exact form of the transformation in the case of mixed ARMA models will be more complicated (an approximate form is $w_t^* \simeq \theta^{-1}(B)\phi(B)w_t$ and so on) but, as noted, it is determined through the same procedure as is used to construct the exact sum of squares function for the ARMA model.

Forecasting for regression models with time series errors is straightforward when future values $x_{t+l,i}$ of the explanatory variables are assumed to be known, as would be the case for deterministic functions such as sine and cosine functions, for example. Then, based on forecast origin t, the lead l forecast of

$$w_{t+l} = \beta_1 x_{t+l,1} + \cdots + \beta_k x_{t+l,k} + N_{t+l}$$

based on past values through time t, is

$$\hat{w}_t(l) = \beta_1 x_{t+l,1} + \beta_2 x_{t+l,2} + \cdots + \beta_k x_{t+l,k} + \hat{N}_t(l) \tag{9.4.5}$$

where $\hat{N}_t(l)$ is the usual l-step-ahead forecast of N_{t+l} from the ARMA(p, q) model, $\phi(B)N_t = \theta(B)a_t$, based on the past values of the noise series N_t. The forecast error is

$$e_t(l) = w_{t+l} - \hat{w}_t(l) = N_{t+l} - \hat{N}_t(l) = \sum_{i=0}^{l-1} \psi_i a_{t+l-i} \tag{9.4.6}$$

with $V(l) = \text{var}[e_t(l)] = \sigma_a^2 \sum_{i=0}^{l-1} \psi_i^2$, just the forecast error and its variance from the ARMA model for the noise series N_t, where the ψ_i are the coefficients in $\psi(B) = \phi^{-1}(B)\theta(B)$ for the noise model.

Example. For an example, in the model $w_t = \beta_0 + \beta_1 \cos(2\pi t/12) + \beta_2 \sin(2\pi t/12) + N_t$, where $(1 - \phi B)N_t = a_t$, the forecasts are

$$\hat{w}_t(l) = \beta_0 + \beta_1 \cos\left[\frac{2\pi(t+l)}{12}\right] + \beta_2 \sin\left[\frac{2\pi(t+l)}{12}\right] + \hat{N}_t(l)$$

with $\hat{N}_t(l) = \phi^l N_t$. Note that these forecasts are similar in functional form to those that would be obtained in an ARMA(1, 3) model (with zero constant term) for the series $(1 - B)(1 - \sqrt{3}B + B^2)w_t$, except that the β coefficients in the forecast function for the regression model case are deterministic, not adaptive, as was noted at the end of Section 9.4.1.

In practice, estimates of $\boldsymbol{\beta}$, as well as of the time series model parameters, would be used to obtain the estimated noise series $\hat{N}_t$ from which forecasts of future values would be made. The effect of these parameter estimation errors on the variance of the corresponding forecast error was investigated by Baillie [20] for the case of regression models with autoregressive errors, generalizing a study by Yamamoto [210] on parameter estimation error effects on forecasting for pure AR models.

More detailed discussions of procedures for regression models with time series errors are given by Harvey and Phillips [107] and by Wincek and Reinsel [205], who include the possibility of missing data. A state space approach with associated Kalman filtering calculations, as discussed in Section 7.5, can be employed for the regression model with time series errors, and this corresponds to one particular choice for the transformation matrix $\mathbf{P}'$ in the discussion above. A specific application of the use of regression models with time series errors to model calendar effects in seasonal time series was given by Bell and Hillmer [28], and Reinsel and Tiao [167] used regression models with time series errors to model atmospheric ozone data for estimation of trends.

For an illustrative example from [167], consider the time series z_t of monthly averages of atmospheric total column ozone measured at the station Aspendale, Australia, for the period 1958–1984. This series is highly seasonal, so in terms of ARIMA modeling, the seasonal differences $w_t = (1 - B^{12})z_t$ were considered. Based on the sample ACF and PACF characteristics of w_t, the following model was specified and estimated,

$$(1 - 0.48B - 0.22B^2)(1 - B^{12})z_t = (1 - 0.99B^{12})a_t$$

and the model was found to be adequate. Hence we see that this model contains a near common seasonal difference factor $(1 - B^{12})$, and consequently, it is equivalent to the model which contains a deterministic seasonal component,

$$z_t = \beta_0 + \sum_{j=1}^{6} \left[\beta_{1j} \cos\left(\frac{2\pi jt}{12}\right) + \beta_{2j} \sin\left(\frac{2\pi jt}{12}\right)\right] + N_t$$

where N_t follows the AR(2) model, $(1 - 0.48B - 0.22B^2)N_t = a_t$. This form of model was, in fact, estimated in [167] by using regression model-time series estimation methods similar to those discussed above.

APPENDIX A9.1. AUTOCOVARIANCES FOR SOME SEASONAL MODELS

See Table A9.1.

TABLE A9.1 Autocovariances for Some Seasonal Models

Model	(Autocovariance of w_t)/σ_a^2	Special Characteristics
(1) $w_t = (1 - \theta B)(1 - \Theta B^s)a_t$ $w_t = a_t - \theta a_{t-1} - \Theta a_{t-s} + \theta\Theta a_{t-s-1}$ $s \geq 3$	$\gamma_0 = (1 + \theta^2)(1 + \Theta^2)$ $\gamma_1 = -\theta(1 + \Theta^2)$ $\gamma_{s-1} = \theta\Theta$ $\gamma_s = -\Theta(1 + \theta^2)$ $\gamma_{s+1} = \gamma_{s-1}$ All other autocovariances are zero.	(a) $\gamma_{s-1} = \gamma_{s+1}$ (b) $\rho_{s-1} = \rho_{s+1} = \rho_1\rho_s$
(2) $(1 - \Phi B^s)w_t = (1 - \theta B)(1 - \Theta B^s)a_t$ $w_t - \Phi w_{t-s} = a_t - \theta a_{t-1} - \Theta a_{t-s} + \theta\Theta a_{t-s-1}$ $s \geq 3$	$\gamma_0 = (1 + \theta^2)\left[1 + \frac{(\Theta - \Phi)^2}{1 - \Phi^2}\right]$ $\gamma_1 = -\theta\left[1 + \frac{(\Theta - \Phi)^2}{1 - \Phi^2}\right]$ $\gamma_{s-1} = \theta\left[\Theta - \Phi - \frac{\Phi(\Theta - \Phi)^2}{1 - \Phi^2}\right]$ $\gamma_s = -(1 + \theta^2)\left[\Theta - \Phi - \frac{\Phi(\Theta - \Phi)^2}{1 - \Phi^2}\right]$ $\gamma_{s+1} = \gamma_{s-1}$ $\gamma_j = \Phi\gamma_{j-s} \quad j \geq s + 2$ For $s \geq 4$, $\gamma_2, \gamma_3, \ldots, \gamma_{s-2}$ are all zero.	(a) $\gamma_{s-1} = \gamma_{s+1}$ (b) $\gamma_j = \Phi\gamma_{j-s} \quad j \geq s + 2$
(3) $w_t = (1 - \theta_1 B - \theta_2 B^2)$ $\times (1 - \Theta_1 B^s - \Theta_2 B^{2s})a_t$ $w_t = a_t - \theta_1 a_{t-1} - \theta_2 a_{t-2} - \Theta_1 a_{t-s}$ $+ \theta_1\Theta_1 a_{t-s-1} + \theta_2\Theta_1 a_{t-s-2}$ $-\Theta_2 a_{t-2s} + \theta_1\Theta_2 a_{t-2s-1}$ $+ \theta_2\Theta_2 a_{t-2s-2}$ $s \geq 5$	$\gamma_0 = (1 + \theta_1^2 + \theta_2^2)(1 + \Theta_1^2 + \Theta_2^2)$ $\gamma_1 = -\theta_1(1 - \theta_2)(1 + \Theta_1^2 + \Theta_2^2)$ $\gamma_2 = -\theta_2(1 + \Theta_1^2 + \Theta_2^2)$ $\gamma_{s-2} = \theta_2\Theta_1(1 - \Theta_2)$ $\gamma_{s-1} = \theta_1\Theta_1(1 - \theta_2)(1 - \Theta_2)$ $\gamma_s = -\Theta_1(1 + \theta_1^2 + \theta_2^2)(1 - \Theta_2)$ $\gamma_{s+1} = \gamma_{s-1}$ $\gamma_{s+2} = \gamma_{s-2}$ $\gamma_{2s-2} = \theta_2\Theta_2$ $\gamma_{2s-1} = \theta_1\Theta_2(1 - \theta_2)$ $\gamma_{2s} = -\Theta_2(1 + \theta_1^2 + \theta_2^2)$ $\gamma_{2s+1} = \gamma_{2s-1}$ $\gamma_{2s+2} = \gamma_{2s-2}$ All other autocovariances are zero.	(a) $\gamma_{s-2} = \gamma_{s+2}$ (b) $\gamma_{s-1} = \gamma_{s+1}$ (c) $\gamma_{2s-2} = \gamma_{2s+2}$ (d) $\gamma_{2s-1} = \gamma_{2s+1}$

TABLE A9.1 (*cont.*)

Model	(Autocovariance of w_t)/σ_a^2	Special Characteristics
(3a) *Special case of model 3* $w_t = (1 - \theta_1 B - \theta_2 B^2)(1 - \Theta B^s)a_t$ $w_t = a_t - \theta_1 a_{t-1} - \theta_2 a_{t-2} - \Theta a_{t-s}$ $+ \theta_1 \Theta a_{t-s-1} + \theta_2 \Theta a_{t-s-2}$ $s \geqslant 5$	$\gamma_0 = (1 + \theta_1^2 + \theta_2^2)(1 + \Theta^2)$ $\gamma_1 = -\theta_1(1 - \theta_2)(1 + \Theta^2)$ $\gamma_2 = -\theta_2(1 + \Theta^2)$ $\gamma_{s-2} = \theta_2 \Theta$ $\gamma_{s-1} = \theta_1 \Theta(1 - \theta_2)$ $\gamma_s = -\Theta(1 + \theta_1^2 + \theta_2^2)$ $\gamma_{s+1} = \gamma_{s-1}$ $\gamma_{s+2} = \gamma_{s-2}$ All other autocovariances are zero.	(a) $\gamma_{s-2} = \gamma_{s+2}$ (b) $\gamma_{s-1} = \gamma_{s+1}$
(3b) *Special case of model 3* $w_t = (1 - \theta B)(1 - \Theta_1 B^s - \Theta_2 B^{2s})a_t$ $w_t = a_t - \theta a_{t-1} - \Theta_1 a_{t-s} + \theta\Theta_1 a_{t-s-1}$ $- \Theta_2 a_{t-2s} + \theta\Theta_2 a_{t-2s-1}$ $s \geqslant 3$	$\gamma_0 = (1 + \theta^2)(1 + \Theta_1^2 + \Theta_2^2)$ $\gamma_1 = -\theta(1 + \Theta_1^2 + \Theta_2^2)$ $\gamma_{s-1} = \theta\Theta_1(1 - \Theta_2)$ $\gamma_s = -\Theta_1(1 + \theta^2)(1 - \Theta_2)$ $\gamma_{s+1} = \gamma_{s-1}$ $\gamma_{2s-1} = \theta\Theta_2$ $\gamma_{2s} = -\Theta_2(1 + \theta^2)$ $\gamma_{2s+1} = \gamma_{2s-1}$ All other autocovariances are zero.	(a) $\gamma_{s-1} = \gamma_{s+1}$ (b) $\gamma_{2s-1} = \gamma_{2s+1}$
(4) $w_t = (1 - \theta_1 B - \theta_s B^s - \theta_{s+1} B^{s+1})a_t$ $w_t = a_t - \theta_1 a_{t-1} - \theta_s a_{t-s}$ $- \theta_{s+1} a_{t-s-1}$ $s \geqslant 3$	$\gamma_0 = 1 + \theta_1^2 + \theta_s^2 + \theta_{s+1}^2$ $\gamma_1 = -\theta_1 + \theta_s \theta_{s+1}$ $\gamma_{s-1} = \theta_1 \theta_s$ $\gamma_s = \theta_1 \theta_{s+1} - \theta_s$ $\gamma_{s+1} = -\theta_{s+1}$ All other autocovariances are zero.	(a) In general, $\gamma_{s-1} \neq \gamma_{s+1}$ $\gamma_1 \gamma_s \neq \gamma_{s+1}$
(4a) *Special case of model 4* $w_t = (1 - \theta_1 B - \theta_s B^s)a_t$ $w_t = a_t - \theta_1 a_{t-1} - \theta_s a_{t-s}$ $s \geqslant 3$	$\gamma_0 = 1 + \theta_1^2 + \theta_s^2$ $\gamma_1 = -\theta_1$ $\gamma_{s-1} = \theta_1 \theta_s$ $\gamma_s = -\theta_s$ All other autocovariances are zero.	(a) Unlike model 4, $\gamma_{s+1} = 0$

(5) $(1 - \Phi B^s)w_t = (1 - \theta_1 B - \theta_s B^s - \theta_{s+1}B^{s+1})a_t$ $w_t - \Phi w_{t-s} = a_t - \theta_1 a_{t-1} - \theta_s a_{t-s} - \theta_{s+1}a_{t-s-1}$ $s \geqslant 3$	$\gamma_0 = 1 + \theta_1^2 + \dfrac{(\theta_s - \Phi)^2}{1 - \Phi^2} + \dfrac{(\theta_{s+1} + \theta_1\Phi)^2}{1 - \Phi^2}$ $\gamma_1 = -\theta_1 + \dfrac{(\theta_s - \Phi)(\theta_{s+1} + \theta_1\Phi)}{1 - \Phi^2}$ $\gamma_{s-1} = (\theta_s - \Phi)\left[\theta_1 + \Phi\dfrac{\theta_{s+1} + \Phi\theta_1}{1 - \Phi^2}\right]$ $\gamma_s = -(\theta_s - \Phi)\left[1 - \Phi\dfrac{\theta_s - \Phi}{1 - \Phi^2}\right] - (\theta_{s+1} + \theta_1\Phi)\left[\theta_1 + \Phi\dfrac{\theta_{s+1} + \theta_1\Phi}{1 - \Phi^2}\right]$ $\gamma_{s+1} = -(\theta_{s+1} + \theta_1\Phi)\left[1 - \Phi\dfrac{\theta_s - \Phi}{1 - \Phi^2}\right]$ $\gamma_j = \Phi\gamma_{j-s} \quad j \geqslant s + 2$ For $s \geqslant 4$, $\gamma_2, \ldots, \gamma_{s-2}$ are all zero.	(a) $\gamma_{s-1} \neq \gamma_{s+1}$ (b) $\gamma_j = \Phi\gamma_{j-s} \quad j \geqslant s + 2$
(5a) *Special case of model 5* $(1 - \Phi B^s)w_t = (1 - \theta_1 B - \theta_s B^s)a_t$ $w_t - \Phi w_{t-s} = a_t - \theta_1 a_{t-1} - \theta_s a_{t-s}$ $s \geqslant 3$	$\gamma_0 = 1 + \dfrac{\theta_1^2 + (\theta_s - \Phi)^2}{1 - \Phi^2}$ $\gamma_1 = -\theta_1\left[1 - \Phi\dfrac{\theta_s - \Phi}{1 - \Phi^2}\right]$ $\gamma_{s-1} = \dfrac{\theta_1(\theta_s - \Phi)}{1 - \Phi^2}$ $\gamma_s = \dfrac{\Phi\theta_1^2 - (\theta_s - \Phi)(1 - \Phi\theta_s)}{1 - \Phi^2}$ $\gamma_j = \Phi\gamma_{j-s} \quad j \geqslant s - 1$ For $s \geqslant 4$, $\gamma_2, \ldots, \gamma_{s-2}$ are all zero.	(a) Unlike model 5, $\gamma_{s-1} = \Phi\gamma_1$

Part Three

TRANSFER FUNCTION MODEL BUILDING

Suppose that X measures the level of an *input* to a system. For example, X might be the concentration of some constituent in the feed to a chemical process. Suppose that the level of X influences the level of a system *output* Y. For example, Y might be the yield of product from the chemical process. It will usually be the case that because of the inertia of the system, a change in X from one level to another will have no immediate effect on the output but, instead, will produce delayed response with Y eventually coming to equilibrium at a new level. We refer to such a change as a *dynamic* response. A model that describes this dynamic response is called a *transfer function model.* We shall suppose that observations of input and output are made at equispaced intervals of time. The associated transfer function model will then be called a *discrete* transfer function model.

Models of this kind can describe not only the behavior of industrial processes but also that of economic and business systems. Transfer function model building is important because it is only when the dynamic characteristics of a system are understood that intelligent direction, manipulation, and control of the system is possible.

Even under carefully controlled conditions, influences other than X will affect Y. We refer to the combined effect on Y of such influences as the *disturbance* or the *noise*. A model such as can be related to real data must take account not only of the dynamic relationship associating X and Y but also of the noise infecting the system. Such joint models are obtained by combining a deterministic transfer function model with a stochastic noise model.

In Chapter 10 following we introduce a class of linear transfer function models capable of representing many of the dynamic relationships commonly met in practice. In Chapter 11 we show how, taking account of corrupting noise, they may be related to data. This relating of model to data is accomplished by processes of *identification, estimation,* and *diagnostic checking*, which closely parallel those already described. In Chapter 12 we describe how simple pulse and step indicator variables can be used as inputs in transfer function models to represent and assess the effects of unusual *intervention* events on the behavior of a time series Y.

10

TRANSFER FUNCTION MODELS

In this chapter we introduce a class of discrete linear transfer function models. These models can be used to represent commonly occurring dynamic situations and are parsimonious in their use of parameters.

10.1 LINEAR TRANSFER FUNCTION MODELS

We suppose that pairs of observations (X_t, Y_t) are available at equispaced intervals of time of an input X and an output Y from some dynamic system, as illustrated in Figure 10.1. In some situations, both X and Y are essentially continuous but are observed only at discrete times. It then makes sense to consider not only what the data has to tell us about the model representing transfer from one discrete series to another, but also what the discrete model might be able to tell us about the corresponding continuous model. In other examples the discrete series are all that exist, and there is no background continuous process. Where we relate continuous and discrete systems, we shall use the basic sampling interval as the unit of time. That is, periods of time will be measured by the number of sampling intervals they occupy. Also, a discrete observation X_t will be deemed to have occurred "at time t."

When we consider the value of a continuous variable, say Y at time t, we denote it by $Y(t)$. If t happens to be a time at which a discrete variable Y is observed, its value is denoted by Y_t. When we wish to emphasize the dependence of a discrete output Y, not only on time but also on the level of the input X, we write $Y_t(X)$.

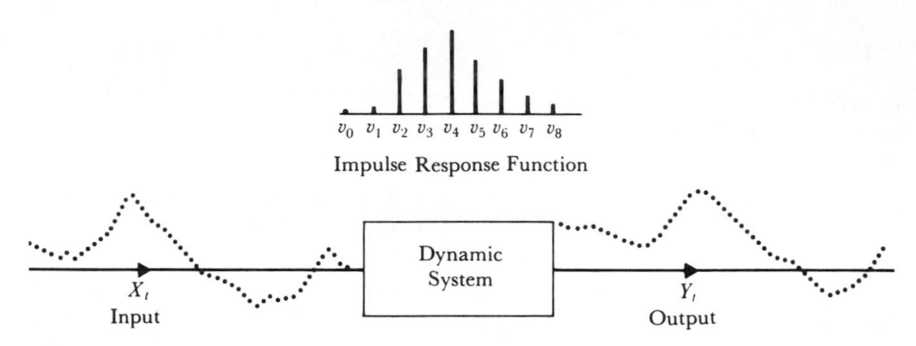

Figure 10.1 Input to, and output from, a dynamic system.

10.1.1 Discrete Transfer Function

With suitable inputs and outputs, which are left to the imagination of the reader, the dynamic system of Figure 10.1 might represent an industrial process, the economy of a country, or the behavior of a particular corporation or government department.

From time to time, we refer to the *steady-state* level of the output obtained when the input is held at some fixed value. By this we mean the value $Y_\infty(X)$ at which the discrete output from a stable system *eventually* comes to equilibrium when the input is held at the fixed level X. Very often, over the range of interest, the relationship between $Y_\infty(X)$ and X will be approximately linear. Hence, if we use Y and X to denote *deviations* from convenient origins situated on the line, we can write the steady-state relationship as

$$Y_\infty = gX \tag{10.1.1}$$

where g is called the *steady-state gain*, and it is understood that Y_∞ is a function of X.

Now, suppose the level of the input is being varied and that X_t and Y_t represent *deviations* at time t from equilibrium. Then it frequently happens that to an adequate approximation, the inertia of the system can be represented by a *linear filter* of the form

$$\begin{aligned} Y_t &= v_0 X_t + v_1 X_{t-1} + v_2 X_{t-2} + \cdots \\ &= (v_0 + v_1 B + v_2 B^2 + \cdots) X_t \\ &= v(B) X_t \end{aligned} \tag{10.1.2}$$

in which the output deviation at some time t is represented as a linear aggregate of input deviations at times t, $t - 1$, ... The operator $v(B)$ is called the *transfer function* of the filter.

Impulse response function. The weights $v_0, v_1, v_2, \ldots$ in (10.1.2) are called the *impulse response function* of the system. This is because the v_j may be regarded as the output or *response* at times $j \geq 0$ to a unit *pulse* input at time 0, that is, an input X_t such that $X_t = 1$ if $t = 0$, $X_t = 0$ otherwise. The impulse response function is shown in Figure 10.1 in the form of a bar chart. When there is no immediate response, one or more of the initial v's, say $v_0, v_1, \ldots, v_{b-1}$, will be equal to zero.

According to (10.1.2), the output deviation can be regarded as a linear aggregate of a series of superimposed impulse response functions scaled by the deviations X_t. This is illustrated in Figure 10.2, which shows a hypothetical impulse response function and the transfer it induces from the input to the output. In the situation illustrated, the input and output are initially in equilibrium. The deviations that occur in the input at times $t = 1$, $t = 2$, and $t = 3$ produce impulse response patterns of deviations in the output, which add together to produce the overall output response.

Relation between the incremental changes. Denote by

$$y_t = Y_t - Y_{t-1} = \nabla Y_t$$

and by

$$x_t = X_t - X_{t-1} = \nabla X_t$$

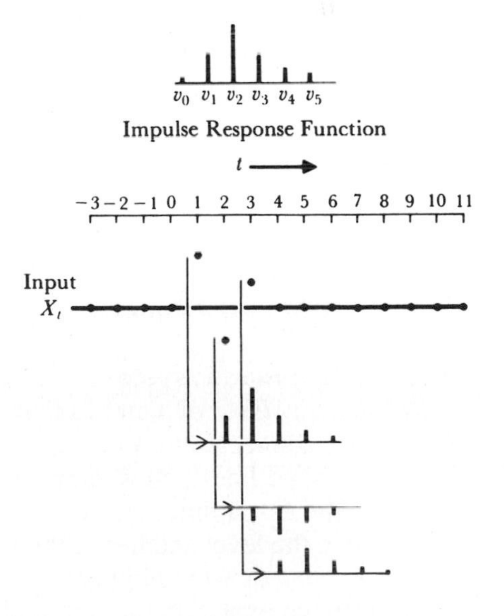

Figure 10.2 Linear transfer from input X to output Y.

the *incremental changes* in Y and X. We often wish to relate such changes. On differencing (10.1.2), we obtain

$$y_t = v(B)x_t$$

Thus we see that the incremental changes y_t and x_t satisfy the same transfer function model as do Y_t and X_t.

Stability. If the infinite series $v_0 + v_1B + v_2B^2 + \cdots$ converges for $|B| \leq 1$, or equivalently, if the v_j are absolutely summable, so that $\Sigma_{j=0}^{\infty} |v_j| < \infty$, then the system is said to be *stable*. We shall be concerned here only with stable systems and consequently, impose this condition on the models we study. The stability condition implies that a finite incremental change in the input results in a finite incremental change in the output.

Now, suppose that X is held indefinitely at the value +1. Then, according to (10.1.1), Y will adjust and maintain itself at the value g. On substituting in (10.1.2) the values $Y_t = g$, $1 = X_t = X_{t-1} = X_{t-2} = \cdots$, we obtain

$$\sum_{j=0}^{\infty} v_j = g \tag{10.1.3}$$

Thus for a stable system the sum of the impulse response weights converges and is equal to the steady-state gain of the system.

Parsimony. It would often be unsatisfactory to parameterize the system in terms of the v's of (10.1.2). To be thus prodigal in the use of parameters could, at the estimation stage, lead to inaccurate and unstable estimation of the transfer function. Furthermore, it is usually inappropriate to estimate the weights v_j directly because for many real situations the v's would be functionally related, as we now see.

10.1.2 Continuous Dynamic Models Represented by Differential Equations

First-order dynamic system. Consider Figure 10.3. Suppose that at time t, $X(t)$ is the *volume* of water in tank A and $Y_1(t)$ the volume of water in tank B, which is connected to A by a pipe. For the time being we ignore tank C, shown by dashed lines. Now suppose that water can be forced in or out of A through pipe P and that mechanical devices are available which make it possible to force the level and hence the volume X in A to follow any desired pattern *irrespective* of what happens in B.

Now if the volume X in the first tank is held at some *fixed* level, water will flow from one tank to the other until the levels are equal. If we now reset the volume X to some other value, again a flow between the tanks will

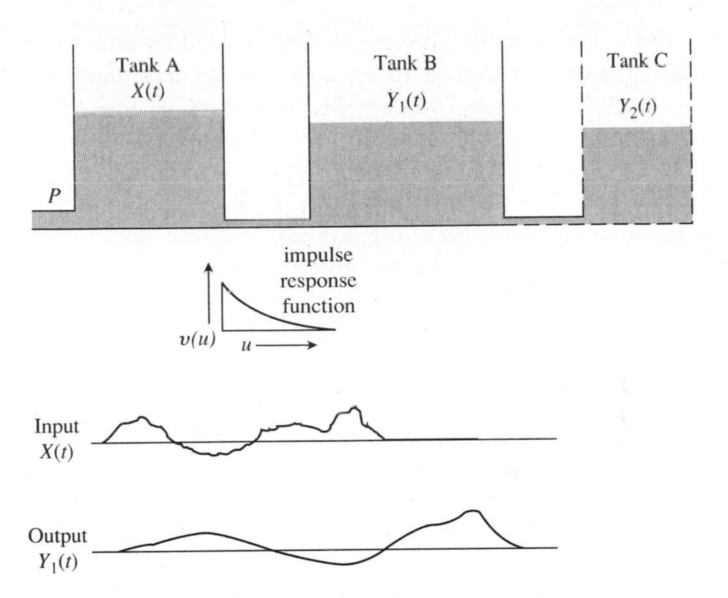

Figure 10.3 Representation of a simple dynamic system.

occur until equilibrium is reached. The volume in B at equilibrium as a function of the fixed volume in A yields the steady-state relationship

$$Y_{1\infty} = g_1 X \tag{10.1.4}$$

In this case the steady-state gain g_1 physically represents the ratio of the cross-sectional areas of the two tanks. If the levels are not in equilibrium at some time t, it is to be noted that the difference in the water level between the tanks is proportional to $g_1 X(t) - Y_1(t)$.

Suppose now that by forcing liquid in and out of pipe P, the volume $X(t)$ is made to follow a pattern like that labeled "Input $X(t)$" in Figure 10.3. Then the volume $Y_1(t)$ in B will correspondingly change in some pattern such as that labeled on the figure as "Output $Y_1(t)$." In general, the function $X(t)$ that is responsible for driving the system is called the *forcing function*.

To relate output and input, we note that to a close approximation, the rate of flow through the pipe will be proportional to the difference in head. That is,

$$\frac{dY_1}{dt} = \frac{1}{T_1}[g_1 X(t) - Y_1(t)] \tag{10.1.5}$$

where T_1 is a constant. The differential equation (10.1.5) may be rewritten in the form

$$(1 + T_1 D) Y_1(t) = g_1 X(t) \tag{10.1.6}$$

where $D = d/dt$. The dynamic system so represented by a first-order differential equation is often referred to as a first-order dynamic system. The constant T_1 is called the *time constant* of the system. The same first-order model can approximately represent the behavior of many simple systems. For example, $Y_1(t)$ might be the outlet temperature of water from a water heater, and $X(t)$ the flow rate of water into the heater.

It is possible to show (see, e.g., [122]) that the solution of a linear differential equation such as (10.1.6) can be written in the form

$$Y_1(t) = \int_0^\infty v(u)X(t-u)du \tag{10.1.7}$$

where in general $v(u)$ is the (continuous) impulse response function. We see that $Y_1(t)$ is generated from $X(t)$ as a continuously weighted aggregate, just as Y_t is generated from X_t as a discretely weighted aggregate in (10.1.2). Furthermore, we see that the role of weight function played by $v(u)$ in the continuous case is precisely parallel to that played by v_j in the discrete situation. For the particular first-order system defined by (10.1.6),

$$v(u) = g_1 T_1^{-1} e^{-u/T_1}$$

Thus the impulse response in this case undergoes simple exponential decay, as indicated in Figure 10.3.

In the continuous case, determination of the output for a completely arbitrary forcing function, such as shown in Figure 10.3, is normally accomplished by simulation on an analog computer, or by using numerical procedures on a digital machine. Solutions are available analytically only for special forcing functions. Suppose, for example, that with the hydraulic system empty, $X(t)$ was suddenly raised to a level $X(t) = 1$ and maintained at that value. Then we shall refer to the forcing function, which was at a steady level of zero and changed instantaneously to a steady level of unity, as a (unit) *step function*. The response of the system to such a function, called the *step response* to the system, is derived by solving the differential equation (10.1.6) with a unit step input, to obtain

$$Y_1(t) = g_1(1 - e^{-t/T_1}) \tag{10.1.8}$$

Thus the level in tank B rises exponentially in the manner shown in Figure 10.4. Now, when $t = T_1$, $Y_1(t) = g_1(1 - e^{-1}) = 0.632g_1$. Thus the time constant T_1 is the time required after the initiation of a step input for the first-order system (10.1.6) to reach 63.2% of its final equilibrium level.

Sometimes there is an initial period of pure *delay* or *dead time* before the response to a given input change begins to take effect. For example, if there were a long length of pipe between A and B in Figure 10.3, a sudden change in level in A could not begin to take effect until liquid had flowed down the pipe. Suppose that the delay thus introduced occupies τ units of time. Then the response of the delayed system would be represented by a

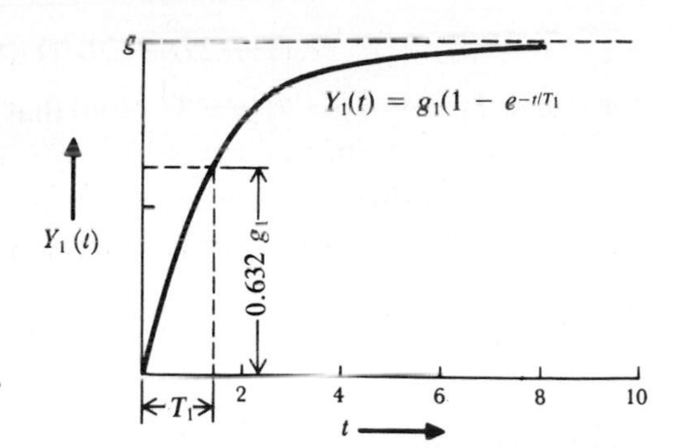

Figure 10.4 Response of a first-order system to a unit step change.

differential equation like (10.1.6), but with $t - \tau$ replacing t on the right-hand side, so that

$$(1 + T_1D)Y_1(t) = g_1X(t - \tau) \tag{10.1.9}$$

The corresponding impulse and step response functions for this system would be of precisely the same shape as for the undelayed system, but the functions would be translated along the horizontal axis a distance τ.

Second-order dynamic system. Consider Figure 10.3 once more. Imagine a three-tank system in which a pipe leads from tank B to a third tank C, the volume of liquid in which is denoted by $Y_2(t)$. Let T_2 be the time constant for the additional system and g_2 its steady-state gain. Then $Y_2(t)$ and $Y_1(t)$ are related by the differential equation

$$(1 + T_2D)Y_2(t) = g_2Y_1(t)$$

After substitution in (10.1.6), we obtain a *second-order* differential equation linking the output from the third tank and the input to the first:

$$[1 + (T_1 + T_2)D + T_1T_2D^2]Y_2(t) = gX(t) \tag{10.1.10}$$

where $g = g_1g_2$. For such a system the impulse response function is a mixture of exponentials

$$v(u) = \frac{g(e^{-u/T_1} - e^{-u/T_2})}{T_1 - T_2} \tag{10.1.11}$$

and the response to a unit step is given by

$$Y_2(t) = g\left(1 - \frac{T_1e^{-t/T_1} - T_2e^{-t/T_2}}{T_1 - T_2}\right) \tag{10.1.12}$$

The continuous curve R in Figure 10.5 shows the response to a unit step for the system

$$(1 + 3D + 2D^2)Y_2(t) = 5X(t)$$

for which $T_1 = 1$, $T_2 = 2$, $g = 5$. Note that unlike the first-order system, the second-order system has a step response that has zero slope initially.

A more general second-order system is defined by

$$(1 + \Xi_1 D + \Xi_2 D^2)Y(t) = gX(t) \tag{10.1.13}$$

where

$$\Xi_1 = T_1 + T_2 \qquad \Xi_2 = T_1 T_2 \tag{10.1.14}$$

and the constants T_1 and T_2 may be complex. If we write

$$T_1 = \frac{1}{\zeta} e^{i\lambda} \qquad T_2 = \frac{1}{\zeta} e^{-i\lambda} \tag{10.1.15}$$

then (10.1.13) becomes

$$\left(1 + \frac{2 \cos \lambda}{\zeta} D + \frac{1}{\zeta^2}\right) Y(t) = gX(t) \tag{10.1.16}$$

The impulse response function (10.1.11) then reduces to

$$v(u) = g \frac{\zeta e^{-\zeta u \cos\lambda} \sin (\zeta u \sin \lambda)}{\sin \lambda} \tag{10.1.17}$$

and the response (10.1.12) to a unit step, to

$$Y(t) = g \left[1 - \frac{e^{-\zeta t \cos\lambda} \sin (\zeta t \sin \lambda + \lambda)}{\sin \lambda}\right] \tag{10.1.18}$$

The continuous curve C in Figure 10.5 shows the response to a unit step for the system

$$(1 + \sqrt{2}D + 2D^2)Y(t) = 5X(t)$$

for which $\lambda = \pi/3$ and $\zeta = \sqrt{2}/2$. It will be noticed that the response overshoots the value $g = 5$ and then comes to equilibrium as a damped sine wave. This behavior is typical of underdamped systems, as they are called. In general, a second-order system is said to be *overdamped, critically damped,* or *underdamped*, depending on whether the constants T_1 and T_2 are real, real and equal, or complex. The overdamped system has a step response which is a mixture of exponentials, given by (10.1.12), and will always remain below the asymptote $Y(\infty) = g$. As with the first-order system, the response can be made subject to a period of dead time by replacing t on the right-hand side of (10.1.13) by $t - \tau$. Many quite complicated dynamic systems can be closely approximated by such second-order systems with delay.

More elaborate linear dynamic systems can be represented by allowing not only the level of the forcing function $X(t)$ but also its rate of change

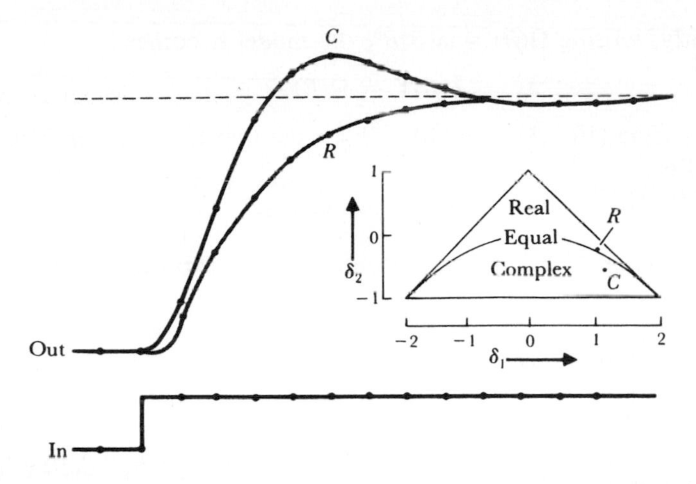

Figure 10.5 Step responses of coincident, discrete and continuous second-order systems having characteristic equations with real roots (curve R) and complex roots (curve C).

dX/dt and higher derivatives to influence the behavior of the system. Thus a general model for representing (continuous) dynamic systems is the linear differential equation

$$(1 + \Xi_1 D + \cdots + \Xi_R D^R)Y(t) = g(1 + H_1 D + \cdots + H_S D^S)X(t - \tau) \tag{10.1.19}$$

10.2 DISCRETE DYNAMIC MODELS REPRESENTED BY DIFFERENCE EQUATIONS

10.2.1 General Form of the Difference Equation

Corresponding to the continuous representation (10.1.19), discrete dynamic systems are often parsimoniously represented by the general linear *difference* equation

$$(1 + \xi_1 \nabla + \cdots + \xi_r \nabla^r)Y_t = g(1 + \eta_1 \nabla + \cdots + \eta_s \nabla^s)X_{t-b} \tag{10.2.1}$$

which we refer to as a transfer function model of order (r, s). The difference equation (10.2.1) may also be written in terms of the backward shift operator $B = 1 - \nabla$ as

$$(1 - \delta_1 B - \cdots - \delta_r B^r)Y_t = (\omega_0 - \omega_1 B - \cdots - \omega_s B^s)X_{t-b} \tag{10.2.2}$$

or as

$$\delta(B)Y_t = \omega(B)X_{t-b}$$

Equivalently, writing $\Omega(B) = \omega(B)B^b$, the model becomes

$$\delta(B)Y_t = \Omega(B)X_t \tag{10.2.3}$$

Comparing (10.2.3) with (10.1.2) we see that the transfer function for this model is

$$v(B) = \delta^{-1}(B)\Omega(B) \tag{10.2.4}$$

Thus the transfer function is represented by the ratio of two polynomials in B.

Dynamics of ARIMA stochastic models. The ARIMA model

$$\varphi(B)z_t = \theta(B)a_t$$

used for the representation of a time series $\{z_t\}$ relates z_t and a_t by the linear filtering operation

$$z_t = \varphi^{-1}(B)\theta(B)a_t$$

where a_t is white noise. Thus the ARIMA model postulates that a time series can be usefully represented as an output from a dynamic system to which the input is white noise and for which the transfer function can be parsimoniously expressed as the ratio of two polynomials in B.

Stability of the discrete models. The requirement of stability for the discrete transfer function models exactly parallels that of stationarity for the ARMA stochastic models. In general, for stability we require that the roots of the characteristic equation

$$\delta(B) = 0$$

with B regarded as a variable, lie outside the unit circle. In particular, this implies that for the first-order model, the parameter δ_1 satisfies

$$-1 < \delta_1 < 1$$

and for the second-order model (see, e.g., Figure 10.5), the parameters δ_1, δ_2 satisfy

$$\delta_2 + \delta_1 < 1$$

$$\delta_2 - \delta_1 < 1$$

$$-1 < \delta_2 < 1$$

On writing (10.2.2) in full as

$$Y_t = \delta_1 Y_{t-1} + \cdots + \delta_r Y_{t-r} + \omega_0 X_{t-b} - \omega_1 X_{t-b-1} - \cdots - \omega_s X_{t-b-s}$$

we see that if X_t is held indefinitely at a value $+1$, Y_t will eventually reach the value

$$g = \frac{\omega_0 - \omega_1 - \cdots - \omega_s}{1 - \delta_1 - \cdots - \delta_r} \tag{10.2.5}$$

which expresses the steady-state gain in terms of the parameters of the model.

10.2.2 Nature of the Transfer Function

If we employ a transfer function model defined by the difference equation (10.2.2), then substituting

$$Y_t = v(B)X_t \tag{10.2.6}$$

in (10.2.2), we obtain the identity

$$(1 - \delta_1 B - \delta_2 B^2 - \cdots - \delta_r B^r)(v_0 + v_1 B + v_2 B^2 + \cdots) = (\omega_0 - \omega_1 B - \cdots - \omega_s B^s)B^b \tag{10.2.7}$$

On equating coefficients of B, we find

$$v_j = \begin{cases} 0 & j < b \\ \delta_1 v_{j-1} + \delta_2 v_{j-2} + \cdots + \delta_r v_{j-r} + \omega_0 & j = b \\ \delta_1 v_{j-1} + \delta_2 v_{j-2} + \cdots + \delta_r v_{j-r} - \omega_{j-b} & j = b+1, b+2, \ldots, b+s \\ \delta_1 v_{j-1} + \delta_2 v_{j-2} + \cdots + \delta_r v_{j-r} & j > b+s \end{cases} \tag{10.2.8}$$

The weights $v_{b+s}, v_{b+s-1}, \ldots, v_{b+s-r+1}$ supply r starting values for the difference equation

$$\delta(B)v_j = 0 \qquad j > b + s$$

The solution

$$v_j = f(\boldsymbol{\delta}, \boldsymbol{\omega}, j)$$

of this difference equation applies to all values v_j for which $j \geqslant b + s - r + 1$.

Thus, in general, the impulse response weights v_j consist of:

1. b zero values $v_0, v_1, \ldots, v_{b-1}$.
2. A further $s - r + 1$ values $v_b, v_{b+1}, \ldots, v_{b+s-r}$ following no fixed pattern (no such values occur if $s < r$).
3. Values v_j with $j \geqslant b + s - r + 1$ following the pattern dictated by the rth-order difference equation which has r starting values v_{b+s}, $v_{b+s-1}, \ldots, v_{b+s-r+1}$. Starting values v_j for $j < b$ will, of course, be zero.

Step response. We now write $V(B)$ for the generating function of the step response weights V_j, which represent the *response* at times $j \geq 0$ to a unit *step* input at time 0, $X_t = 1$ if $t \geq 0$, $X_t = 0$ if $t < 0$, so that $V_j = \sum_{i=0}^{j} v_i$ for $j \geq 0$. Thus

$$V(B) = V_0 + V_1 B + V_2 B^2 + \cdots = v_0 + (v_0 + v_1)B + (v_0 + v_1 + v_2)B^2 + \cdots \tag{10.2.9}$$

and

$$v(B) = (1 - B)V(B) \tag{10.2.10}$$

Substitution of (10.2.10) in (10.2.7) yields the identity

$$(1 - \delta_1^* B - \delta_2^* B^2 - \cdots - \delta_{r+1}^* B^{r+1})(V_0 + V_1 B + V_2 B^2 + \cdots) = (\omega_0 - \omega_1 B - \cdots - \omega_s B^s)B^b \tag{10.2.11}$$

with

$$(1 - \delta_1^* B - \delta_2^* B^2 - \cdots - \delta_{r+1}^* B^{r+1}) = (1 - B)(1 - \delta_1 B - \cdots - \delta_r B^r) \tag{10.2.12}$$

The identity (10.2.11) for the step response weights V_j exactly parallels the identity (10.2.7) for the impulse response weights, except that the left-hand operator $\delta^*(B)$ is of order $r + 1$ instead of r.

Using the results of (10.2.8), it follows that the step response function is defined by:

1. b zero values $V_0, V_1, \ldots, V_{b-1}$.
2. A further $s - r$ values $V_b, V_{b+1}, \ldots, V_{b+s-r-1}$ following no fixed pattern (no such values occur if $s < r + 1$).
3. Values V_j, with $j \geqslant b + s - r$, which follow the pattern dictated by the $(r + 1)$th-order difference equation $\delta^*(B)V_j = 0$, which has $r + 1$ starting values $V_{b+s}, V_{b+s-1}, \ldots, V_{b+s-r}$. Starting values V_j for $j < b$ will, of course, be zero.

10.2.3 First- and Second-Order Discrete Transfer Function Models

Details of transfer function models for all combinations of $r = 0, 1, 2$ and $s = 0, 1, 2$ are shown in Table 10.1. Specific examples of the models, with bar charts showing step response and impulse response, are given in Figure 10.6. The equations at the end of Table 10.1 allow the parameters $\boldsymbol{\xi}$, g, $\boldsymbol{\eta}$ of the ∇ form of the model to be expressed in terms of the parameters $\boldsymbol{\delta}$, $\boldsymbol{\omega}$ of the B form. These equations are given for the most general of the models considered, namely that for which $r = 2$ and $s = 2$. All the other models are special cases of this one, and the corresponding equations for these are obtained by setting appropriate parameters to zero. For example, if $r = 1$ and $s = 1$, $\xi_2 = \eta_2 = \delta_2 = \omega_2 = 0$, then

$$\delta_1 = \frac{\xi_1}{1 + \xi_1} \qquad \omega_0 = \frac{g(1 + \eta_1)}{1 + \xi_1} \qquad \omega_1 = \frac{g\eta_1}{1 + \xi_1}$$

In Table 10.2 the starting values for the difference equations satisfied by the impulse and step responses, respectively, are indicated by circles on the bar charts.

Discussion of the tabled models. The models, whose properties are summarized in Table 10.1 and Figure 10.6, will repay careful study since they are useful in representing many commonly met dynamic systems. In all the models the operator B^b on the right ensures that the first nonzero term in the impulse response function is v_b. In the examples in Figure 10.6 the value of g is supposed equal to 1, and b is supposed equal to 3.

Models with r = 0. With r and s both equal to zero, the impulse response consists of a single value $v_b = \omega_0 = g$. The output is proportional to the input, but is displaced by b time intervals. More generally, if we have an operator of order s on the right, the instantaneous input will be delayed b intervals and will be spread over $s + 1$ values in proportion to $v_b = \omega_0$, $v_{b+1} = -\omega_1, \ldots, v_{b+s} = -\omega_s$. The step response is obtained by summing the impulse response and eventually satisfies the difference equation $(1 - B)V_j = 0$ with starting value $V_{b+s} = g = \omega_0 - \omega_1 - \cdots - \omega_s$.

Models with r = 1. With $s = 0$, the impulse response tails off exponentially (geometrically) from the initial starting value $v_b = \omega_0 = g/(1 + \xi_1) = g(1 - \delta_1)$. The step response increases exponentially until it attains the value $g = 1$. If the exponential step response is extrapolated backwards as indicated by the dashed line, it cuts the time axis at time $b - 1$. This corresponds to the fact that $V_{b-1} = 0$ as well as $V_b = v_b$ are starting values for the appropriate difference equation $(1 - \delta B)(1 - B)V_j = 0$.

With $s = 1$, there is an initial value $v_b = \omega_0 = g(1 + \eta_1)/(1 + \xi_1)$ of the impulse response which does not follow a pattern. The exponential pattern induced by the difference equation $v_j = \delta_1 v_{j-1}$ associated with the left-hand operator begins with the starting value $v_{b+1} = (\delta_1\omega_0 - \omega_1) = g(\xi_1 - \eta_1)/(1 + \xi_1)^2$. The step response function follows an exponential curve, determined by the difference equation $(1 - \delta B)(1 - B)V_j = 0$, which approaches g asymptotically from the starting value $V_b = v_b$ and $V_{b+1} = v_b + v_{b+1}$. An exponential curve projected by the dashed line backwards through the points will, in general, cut the time axis at some intermediate point in the time interval. We show in Section 10.3 that certain discrete models, which approximate continuous first-order systems having *fractional* periods of delay, may in fact be represented by a first-order difference equation with an operator of order $s = 1$ on the right.

TABLE 10.1 Impulse Response Functions for Transfer Function Models of the Form $\delta_r(B)Y_t = \omega_s(B)B^bX_t$

rsb	∇ Form	*B* Form	Impulse Response v_j	
00*b*	$Y_t = gX_{t-b}$	$Y_t = \omega_0 B^b X_t$	0 ω_0 0	$j < b$ $j = b$ $j > b$
01*b*	$Y_t = g(1 + \eta_1\nabla)X_{t-b}$	$Y_t = (\omega_0 - \omega_1 B)B^b X_t$	0 ω_0 $-\omega_1$ 0	$j < b$ $j = b$ $j = b + 1$ $j > b + 1$
02*b*	$Y_t = g(1 + \eta_1\nabla + \eta_2\nabla^2)X_{t-b}$	$Y_t = (\omega_0 - \omega_1 B - \omega_2 B^2)B^b X_t$	0 ω_0 $-\omega_1$ $-\omega_2$ 0	$j < b$ $j = b$ $j = b + 1$ $j = b + 2$ $j > b + 2$
10*b*	$(1 + \xi_1\nabla)Y_t = gX_{t-b}$	$(1 - \delta_1 B)Y_t = \omega_0 B^b X_t$	0 ω_0 $\delta_1 v_{j-1}$	$j < b$ $j = b$ $j > b$
11*b*	$(1 + \xi_1\nabla)Y_t = g(1 + \eta_1\nabla)X_{t-b}$	$(1 - \delta_1 B)Y_t = (\omega_0 - \omega_1 B)B^b X_t$	0 ω_0 $\delta_1\omega_0 - \omega_1$ $\delta_1 v_{j-1}$	$j < b$ $j = b$ $j = b + 1$ $j > b + 1$
12*b*	$(1 + \xi_1\nabla)Y_t = g(1 + \eta_1\nabla + \eta_2\nabla^2)X_{t-b}$	$(1 - \delta_1 B)Y_t = (\omega_0 - \omega_1 B - \omega_2 B^2)B^b X_t$	0 ω_0 $\delta_1\omega_0 - \omega_1$ $\delta_1^2\omega_0 - \delta_1\omega_1 - \omega_2$ $\delta_1 v_{j-1}$	$j < b$ $j = b$ $j = b + 1$ $j = b + 2$ $j > b + 2$

			v_j	
20b	$(1 + \xi_1\nabla + \xi_2\nabla^2)Y_t = gX_{t-b}$	$(1 - \delta_1 B - \delta_2 B^2)Y_t = \omega_0 B^b X_t$	0	$j < b$
			ω_0	$j = b$
			$\delta_1 v_{j-1} + \delta_2 v_{j-2}$	$j > b$
21b	$(1 + \xi_1\nabla + \xi_2\nabla^2)Y_t = g(1 + \eta_1\nabla)X_{t-b}$	$(1 - \delta_1 B - \delta_2 B^2)Y_t = (\omega_0 - \omega_1 B)B^b X_t$	0	$j < b$
			ω_0	$j = b$
			$\delta_1\omega_0 - \omega_1$	$j = b + 1$
			$\delta_1 v_{j-1} + \delta_2 v_{j-2}$	$j > b + 1$
22b	$(1 + \xi_1\nabla + \xi_2\nabla^2)Y_t = g(1 + \eta_1\nabla + \eta_2\nabla^2)X_{t-b}$	$(1 - \delta_1 B - \delta_2 B^2)Y_t = (\omega_0 - \omega_1 B - \omega_2 B^2)B^b X_t$	0	$j < b$
			ω_0	$j = b$
			$\delta_1\omega_0 - \omega_1$	$j = b + 1$
			$(\delta_1^2 + \delta_2)\omega_0 - \delta_1\omega_1 - \omega_2$	$j = b + 2$
			$\delta_1 v_{j-1} + \delta_2 v_{j-2}$	$j > b + 2$

$$\xi_1 = \frac{\delta_1 + 2\delta_2}{1 - \delta_1 - \delta_2}, \quad \xi_2 = \frac{-\delta_2}{1 - \delta_1 - \delta_2} \qquad \delta_1 = \frac{\xi_1 + 2\xi_2}{1 + \xi_1 + \xi_2}, \quad \delta_2 = \frac{-\xi_2}{1 + \xi_1 + \xi_2}$$

$$g = \frac{\omega_0 - \omega_1 - \omega_2}{1 - \delta_1 - \delta_2} \qquad \omega_0 = \frac{g(1 + \eta_1 + \eta_2)}{1 + \xi_1 + \xi_2}$$

$$\eta_1 = \frac{\omega_1 + 2\omega_2}{\omega_0 - \omega_1 - \omega_2} \qquad \omega_1 = \frac{g(\eta_1 + 2\eta_2)}{1 + \xi_1 + \xi_2}$$

$$\eta_2 = \frac{-\omega_2}{\omega_0 - \omega_1 - \omega_2} \qquad \omega_2 = \frac{-g\eta_2}{1 + \xi_1 + \xi_2}$$

$$1 - \delta_1 - \delta_2 = (1 + \xi_1 + \xi_2)^{-1}$$

With $s = 2$, there are two values v_b and v_{b+1} for the impulse response that do not follow a pattern, followed by exponential fall off beginning with v_{b+2}. Correspondingly, there is a single preliminary value V_b in the step response that does not coincide with the exponential curve projected by the dashed line. This curve is, as before, determined by the difference equation $(1 - \delta B)(1 - B)V_j = 0$ but with starting values V_{b+1} and V_{b+2}.

Models with r = 2. The flexibility of the model with $s = 0$ is limited because the first starting value of the impulse response is fixed to be zero. More useful models are obtained for $s = 1$ and $s = 2$. The use of these models in approximating continuous second-order systems is discussed in Section 10.3 and in Appendix A10.1.

The behavior of the dynamic weights v_j, which eventually satisfy

$$v_j - \delta_1 v_{j-1} - \delta_2 v_{j-2} = 0 \qquad j > b + s \tag{10.2.13}$$

depends on the nature of the roots S_1^{-1} and S_2^{-1}, of the *characteristic equation*

$$1 - \delta_1 B - \delta_2 B^2 = (1 - S_1 B)(1 - S_2 B) = 0$$

This dependence is shown in Table 10.2. As in the continuous case, the model may be overdamped, critically damped, or underdamped, depending on the nature of the roots of the characteristic equation.

TABLE 10.2 Dependence of Nature of Second-Order System on the Roots of $1 - \delta_1 B - \delta_2 B^2 = 0$

Roots (S_1^{-1}, S_2^{-1})	Condition	Damping
Real	$\delta_1^2 + 4\delta_2 > 0$	Overdamped
Real and equal	$\delta_1^2 + 4\delta_2 = 0$	Critically damped
Complex	$\delta_1^2 + 4\delta_2 < 0$	Underdamped

When the roots are complex, the solution of (10.2.13) will follow a damped sine wave, as in the examples of second-order systems in Figure 10.6. When the roots are real, the solution will be the sum of two exponentials. As in the continuous case considered in Section 10.1.2, the system can then be thought of as equivalent to two discrete first-order systems arranged in series and having parameters S_1 and S_2.

The weights V_j for the step response eventually satisfy a difference equation

$$(V_j - g) - \delta_1(V_{j-1} - g) - \delta_2(V_{j-2} - g) = 0$$

which is of the same form as (10.2.13). Thus the behavior of the step response V_j about its asymptotic value g parallels the behavior of the impulse

Figure 10.6 Examples of Impulse and Step Response Functions with Gain $g = 1$.

r, s, b	∇ Form	B Form	Impulse Response v_j	Step Response $V_j = \sum_{i=0}^{j} v_i$
003	$Y_t = X_{t-3}$	$Y_t = B^3 X_t$	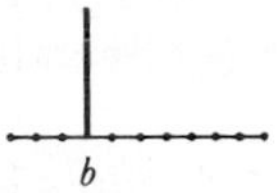 b	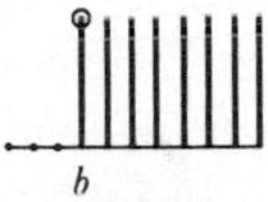 b
013	$Y_t = (1 - .5\nabla) X_{t-3}$	$Y_t = (.5 + .5B) B^3 X_t$	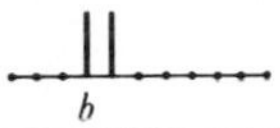 b	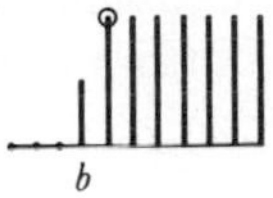 b
023	$Y_t = (1 - \nabla + .25\nabla^2) X_{t-3}$	$Y_t = (.25 + .50B + .25B^2) B^3 X_t$	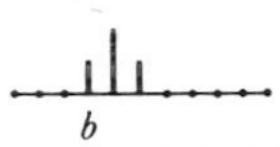 b	 b
103	$(1 + \nabla) Y_t = X_{t-3}$	$(1 - .5B) Y_t = .5B^3 X_t$	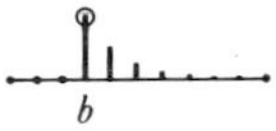 b	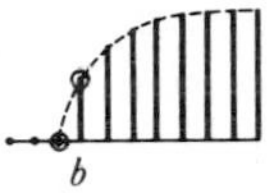 b
113	$(1 + \nabla) Y_t = (1 - .5\nabla) X_{t-3}$	$(1 - .5B) Y_t = (.25 + .25B) B^3 X_t$	 b	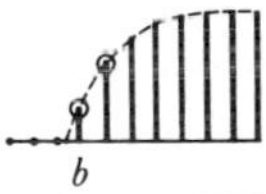 b
123	$(1 + \nabla) Y_t = (1 - \nabla + .25\nabla^2) X_{t-3}$	$(1 - .5B) Y_t = (.125 + .25B + .125B^2) B^3 X_t$	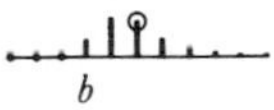 b	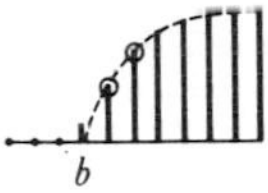 b
203	$(1 - .25\nabla + .5\nabla^2) Y_t = X_{t-3}$	$(1 - .6B + .4B^2) Y_t = .8B^3 X_t$	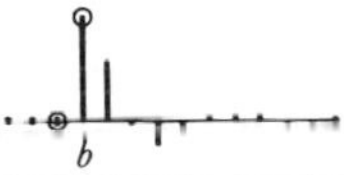 b	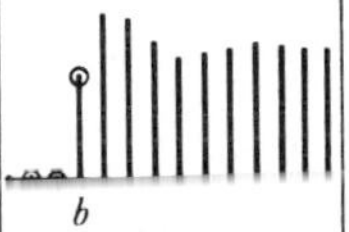 b
213	$(1 - .25\nabla + .5\nabla^2) Y_t = (1 - .5\nabla) X_{t-3}$	$(1 - .6B + .4B^2) Y_t = (.4 + .4B) B^3 X$	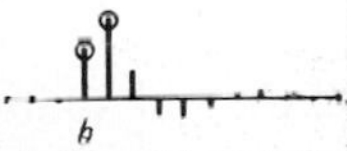 b	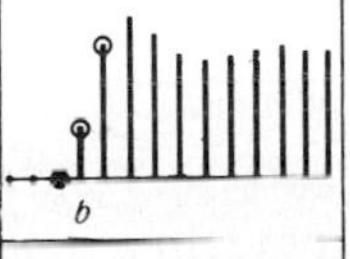 b
223	$(1 - .25\nabla + .5\nabla^2) Y_t = (1 - \nabla + .25\nabla^2) X_{t-3}$	$(1 - .6B + .4B^2) Y_t = (.2 + .4B + .2B^2) B^3 X_t$	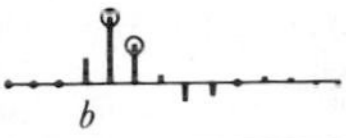 b	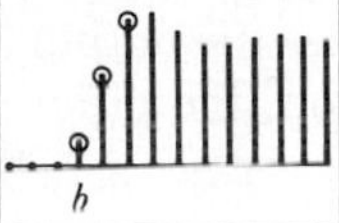 b

response about the time axis. In the situation where there are complex roots the step response "overshoots" the value g and then oscillates about this value until it reaches equilibrium. When the roots are real and positive, the step response, which is the sum of two exponential terms, approaches its asymptote g without crossing it. However, if there are negative real roots, the step response may overshoot and oscillate as it settles down to its equilibrium value.

In Figure 10.5 the dots indicate two discrete step responses, labeled R and C, respectively, in relation to a discrete step input indicated by dots at the bottom of the figure. The difference equation models* corresponding to R and C are:

$$R: \quad (1 - 0.97B + 0.22B^2)Y_t = 5(0.15 + 0.09B)X_{t-1}$$

$$C: \quad (1 - 1.15B + 0.49B^2)Y_t = 5(0.19 + 0.15B)X_{t-1}$$

Also shown in Figure 10.5 is a diagram of the stability region with the parameter points (δ_1, δ_2) marked for each of the two models. Note that the system described by model R, which has real positive roots, has no overshoot while that for model C, which has complex roots, does have overshoot.

10.2.4 Recursive Computation of Output for Any Input

It would be extremely tedious if it were necessary to use the impulse response form (10.1.2) of the model to compute the output for a given input. Fortunately, this is not necessary. Instead, we may employ the difference equation model directly. In this way it is a simple matter to compute the output recursively for any input whatsoever. For example, consider the model with $r = 1$, $s = 0$, $b = 1$ and with $\xi = 1$ and $g = 5$. Thus

$$(1 + \nabla)Y_t = 5X_{t-1}$$

or equivalently,

$$(1 - 0.5B)Y_t = 2.5X_{t-1} \tag{10.2.14}$$

Table 10.3 shows the calculation of Y_t when the input X_t is (a) a unit pulse input; (b) a unit step input; (c) a "general" input. In all cases it is assumed that the output has the initial value $Y_0 = 0$. To perform the recursive calculation, the difference equation is written out with Y_t on the left. Thus

$$Y_t = 0.5Y_{t-1} + 2.5X_{t-1}$$

*The parameters in these models were in fact selected, in a manner to be discussed in Section 10.3.2, so that at the discrete points, the step responses exactly matched those of the continuous systems introduced in Section 10.1.2.

TABLE 10.3 Calculation of Output from Discrete First-Order System for Impulse, Step, and General Input

	(a) Impulse Input		(b) Step Input		(c) General Input	
t	Input X_t	Output Y_t	Input X_t	Output Y_t	Input X_t	Output Y_t
0	0	0	0	0	0	0
1	1	0	1	0	1.5	0
2	0	2.50	1	2.50	0.5	3.75
3	0	1.25	1	3.75	2.0	3.12
4	0	0.62	1	4.38	1.0	6.56
5	0	0.31	1	4.69	−2.5	5.78
6	0	0.16	1	4.84	0.5	−3.36

and, for example, in the case of the "general" input

$$Y_1 = 0.5 \times 0 + 2.5 \times 0 = 0$$

$$Y_2 = 0.5 \times 0 + 2.5 \times 1.5 = 3.75$$

and so on. These inputs and outputs are plotted in Figure 10.7(a), (b), and (c).

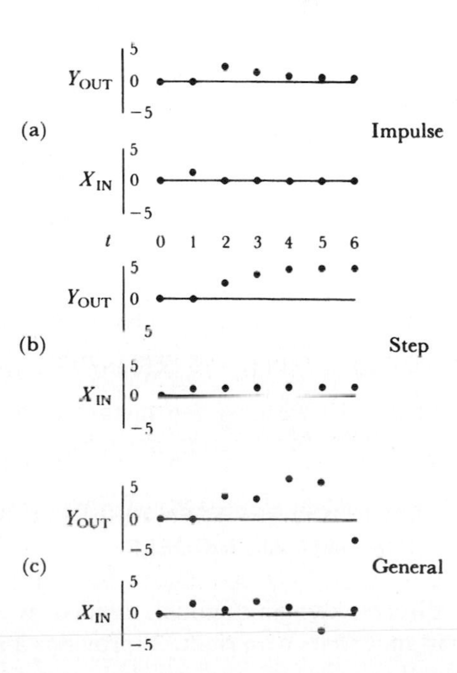

Figure 10.7 Response of a first-order system to an impulse, a step, and a "general" input.

In general, we see that having written the transfer function model in the form

$$Y_t = \delta_1 Y_{t-1} + \cdots + \delta_r Y_{t-r} + \omega_0 X_{t-b} - \omega_1 X_{t-b-1} - \cdots - \omega_s X_{t-b-s}$$

it is an easy matter to compute the discrete output for any discrete input. To start off the recursion we need to know certain initial values. This need is not, of course, a shortcoming of the method of calculation but comes about because with a transfer function model, the initial values of Y *will* depend on values of X that occurred before observation was begun. In practice, when the necessary initial values are not known, we can substitute mean values for unknown Y's and X's (zeros if these quantities are considered as deviations from their means). The early calculated values will then depend upon this choice of the starting values. However, for a stable system, the effect of this choice will be negligible after a period sufficient for the impulse response to become negligible. If this period is p_0 time intervals, an alternative procedure is to compute $Y_{p_0}, Y_{p_0+1}, \ldots$ directly from the impulse response until enough values are available to set the recursion going.

10.2.5 Transfer Function Models with Added Noise

In practice, the output Y could not be expected to follow exactly the pattern determined by the transfer function model, even if that model were entirely adequate. Disturbances of various kinds other than X normally corrupt the system. A disturbance might originate at any point in the system, but it is often convenient to consider it in terms of its net effect on the output Y, as indicated in Figure 1.5. If we assume that the disturbance, or noise N_t, is independent of the level of X and is additive with respect to the influence of X, we can write

$$Y_t = \delta^{-1}(B)\omega(B)X_{t-b} + N_t \qquad (10.2.15)$$

If the noise model can be represented by an ARIMA(p, d, q) process

$$N_t = \varphi^{-1}(B)\theta(B)a_t$$

where a_t is white noise, the model (10.2.15) can be written finally as

$$Y_t = \delta^{-1}(B)\omega(B)X_{t-b} + \varphi^{-1}(B)\theta(B)a_t \qquad (10.2.16)$$

In Chapter 11 we describe methods for identifying, fitting, and checking combined transfer function-noise models of the form (10.2.16).

10.3 RELATION BETWEEN DISCRETE AND CONTINUOUS MODELS

The discrete dynamic model, defined by a linear difference equation, is of importance in its own right. It provides a sensible class of transfer functions and needs no other justification. In many examples no question will arise of attempting to relate the discrete model to a supposed underlying continuous model because no underlying continuous series properly exists. However, in some cases, for example where instantaneous observations are taken periodically on a chemical reactor, the discrete record can be used to tell us something about the continuous system. In particular, control engineers are used to thinking in terms of the time constants and dead times of continuous systems and may best understand the results of the discrete analysis when so expressed.

As before, we denote a continuous output and input at time t by $Y(t)$ and $X(t)$, respectively. Suppose that the output and input are related by the linear filtering operation

$$Y(t) = \int_0^\infty v(u)X(t - u)du$$

Suppose now that only discrete observations (X_t, Y_t), (X_{t-1}, Y_{t-1}), . . . of output and input are available at equispaced intervals of time t, $t - 1$, . . . and that the discrete output and input are related by the discrete linear filter

$$Y_t = \sum_{j=0}^{\infty} v_j X_{t-j}$$

Then, for certain special cases, and with appropriate assumptions, useful relationships may be established between the discrete and continuous models.

10.3.1 Response to a Pulsed Input

A special case, which is of importance in the design of the discrete control schemes discussed in Part Four, arises when the opportunity for adjustment of the process occurs immediately after observation of the output, so that the input variable is allowed to remain at the same level between observations. The typical appearance of the resulting square wave, or *pulsed input* as we shall call it, is shown in Figure 10.8. We denote the fixed level at which the input is held during the period $t - 1 < \tau < t$ by X_{t-1+}.

Consider a continuous linear system that has b whole periods of delay plus a fractional period c of further delay. Thus, in terms of previous notation, $b + c = \tau$. Then we can represent the output from the system as

$$Y(t) = \int_0^\infty v(u)X(t - u)\, du$$

where the impulse response function $v(u)$ is zero for $u < b + c$. Now for a pulsed input, as shown in Figure 10.9, the output at time t will be given

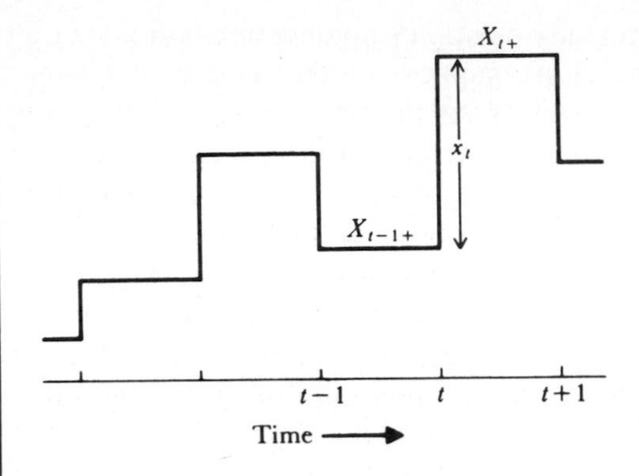

Figure 10.8 Example of a pulsed input.

exactly by

$$Y(t) = \left[\int_{b+c}^{b+1} v(u)\,du\right]X_{t-b-1+} + \left[\int_{b+1}^{b+2} v(u)\,du\right]X_{t-b-2+} + \cdots$$

Thus

$$Y(t) = Y_t = v_b X_{t-b-1+} + v_{b+1}X_{t-b-2+} + \cdots$$

Therefore, for a *pulsed input*, there exists a discrete linear filter which is such that at times $t, t-1, t-2, \ldots$, the continuous output $Y(t)$ *exactly* equals the discrete output.

Given a pulsed input, consider the output Y_t from a discrete model

$$\xi(\nabla)Y_t = \eta(\nabla)X_{t-b-1+} \tag{10.3.1}$$

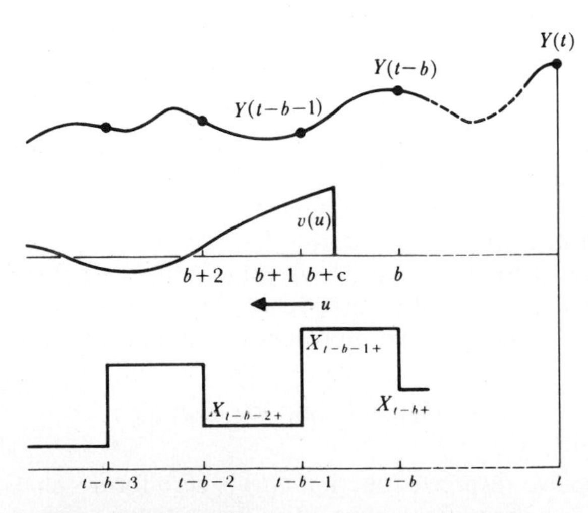

Figure 10.9 Transfer to output from a pulsed input.

of order (r, r) in relation to the continuous output from the Rth-order model

$$(1 + \Xi_1 D + \Xi_2 D^2 + \cdots + \Xi_R D^R)Y(t) = X(t - b - c) \quad (10.3.2)$$

subject to the same input. It is shown in Appendix A10.1 that for suitably chosen values of the parameters (Ξ, c), the outputs will coincide exactly if $R = r$. Furthermore, if $c = 0$, the output from the continuous model (10.3.2) will be identical at the discrete times with that of a discrete model (10.3.1) of order $(r, r - 1)$. We refer to the related continuous and discrete models as *discretely coincident* systems. If, then, a discrete model of the form (10.3.1) of order (r, r) has been obtained, then on the assumption *that the continuous model would be represented by the* rth-*order differential equation (10.3.2)*, the parameters, and in particular the time constants for the discretely coincident continuous system, may be written explicitly in terms of the parameters of the discrete model.

The parameter relationships for a delayed second-order system have been derived in Appendix A10.1. From these the corresponding relationships for simpler systems may be obtained by setting appropriate constants equal to zero, as we shall now discuss.

10.3.2 Relationships for First- and Second-Order Coincident Systems

Undelayed first-order system

***B* Form.** The continuous system satisfying

$$(1 + TD)Y(t) = gX(t) \quad (10.3.3)$$

is, for a pulsed input, discretely coincident with the discrete system satisfying

$$(1 - \delta B)Y_t = \omega_0 X_{t-1+} \quad (10.3.4)$$

where

$$\delta = e^{-1/T} \qquad T = (-\ln \delta)^{-1} \qquad \omega_0 = g(1 - \delta) \quad (10.3.5)$$

∇ *Form*. Alternatively, the difference equation may be written

$$(1 + \xi\nabla)Y_t = gX_{t-1+} \quad (10.3.6)$$

where

$$\xi = \frac{\delta}{1 - \delta} \quad (10.3.7)$$

To illustrate, we reconsider the example of Section 10.2.4 for the "general" input. The output for this case is calculated in Table 10.3(c) and plotted in Figure 10.7(c). Suppose that, in fact, we had a continuous system

$$(1 + 1.44D)Y(t) = 5X(t)$$

Then this would be discretely coincident with the discrete model (10.2.14) actually considered, namely,

$$(1 - 0.5B)Y_t = 2.5X_{t-1+}$$

If the input and output were continuous and the input were pulsed, the actual course of the response would be that shown by the continuous lines in Figure 10.10. The output would in fact follow a series of exponential curves. Each dashed line shows the further course that the response would take if no further change in the input were made. The curves correspond exactly at the discrete sample points with the discrete output already calculated in Table 10.3(c) and plotted in Figure 10.7(c).

Delayed first-order system

***B* Form.** The continuous system satisfying

$$(1 + TD)Y(t) = gX(t - b - c) \tag{10.3.8}$$

is, for a pulsed input, discretely coincident with the discrete system satisfying

$$(1 - \delta B)Y_t = (\omega_0 - \omega_1 B)X_{t-b-1+} \tag{10.3.9}$$

where

$$\delta = e^{-1/T} \qquad \omega_0 = g(1 - \delta^{1-c}) \qquad \omega_1 = g(\delta - \delta^{1-c}) \tag{10.3.10}$$

∇ *Form.* Alternatively, the difference equation may be written

$$(1 + \xi\nabla)Y_t = g(1 + \eta\nabla)X_{t-b-1+} \tag{10.3.11}$$

where

$$\xi = \frac{\delta}{1 - \delta} \qquad -\eta = \frac{\delta(\delta^{-c} - 1)}{1 - \delta} \tag{10.3.12}$$

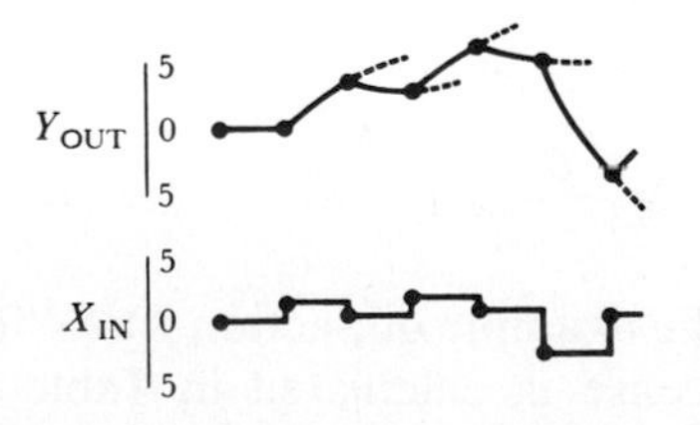

Figure 10.10 Continuous response of the system $(1 + 1.44D)Y(t) = 5X(t)$ to a pulsed input.

Now

$$(1 + \eta\nabla)X_{t-b-1+} = (1 + \eta)X_{t-b-1+} - \eta X_{t-b-2+} \tag{10.3.13}$$

can be regarded as an interpolation at an increment $(-\eta)$ between X_{t-b-1+} and X_{t-b-2+}. Table 10.4 allows the corresponding parameters $(\xi, -\eta)$ and (T, c) of the discrete and continuous models to be determined for a range of alternatives.

TABLE 10.4 Values of $-\eta$ for Various Values of T and c for a First-Order System with Delay; Corresponding Values of ξ and δ

			$-\eta$ for				
δ	ξ	T	$c = 0.9$	$c = 0.7$	$c = 0.5$	$c = 0.3$	$c = 0.1$
0.9	9.00	9.49	0.90	0.69	0.49	0.29	0.10
0.8	4.00	4.48	0.89	0.68	0.47	0.28	0.09
0.7	2.33	2.80	0.88	0.66	0.46	0.26	0.09
0.6	1.50	1.95	0.88	0.64	0.44	0.25	0.08
0.5	1.00	1.44	0.87	0.62	0.41	0.23	0.07
0.4	0.67	1.09	0.85	0.60	0.39	0.21	0.06
0.3	0.43	0.83	0.84	0.57	0.35	0.19	0.05
0.2	0.25	0.62	0.82	0.52	0.31	0.15	0.04
0.1	0.11	0.43	0.77	0.45	0.24	0.11	0.03

Undelayed second-order system

***B* Form.** The continuous system satisfying

$$(1 + T_1D)(1 + T_2D)Y(t) = gX(t) \tag{10.3.14}$$

is, for a pulsed input, discretely coincident with the system

$$(1 - \delta_1B - \delta_2B^2)Y_t = (\omega_0 - \omega_1B)X_{t-1+} \tag{10.3.15}$$

or equivalently, with the system

$$(1 - S_1B)(1 - S_2B)Y_t = (\omega_0 - \omega_1B)X_{t-1+} \tag{10.3.16}$$

where

$$S_1 = e^{-1/T_1} \qquad S_2 = e^{-1/T_2}$$
$$\omega_0 = g(T_1 - T_2)^{-1}[T_1(1 - S_1) - T_2(1 - S_2)] \tag{10.3.17}$$
$$\omega_1 = g(T_1 - T_2)^{-1}[T_1S_2(1 - S_1) - T_2S_1(1 - S_2)]$$

∇ *Form*. Alternatively, the difference equation may be written

$$(1 + \xi_1\nabla + \xi_2\nabla^2)Y_t = g(1 + \eta_1\nabla)X_{t-1+} \tag{10.3.18}$$

where

$$-\eta_1 = (1 - S_1)^{-1}(1 - S_2)^{-1}(T_1 - T_2)^{-1}[T_2S_1(1 - S_2) - T_1S_2(1 - S_1)] \tag{10.3.19}$$

may be regarded as the increment of an interpolation between X_{t-1+} and X_{t-2+}. Values for ξ_1 and ξ_2 in terms of the δ's can be obtained directly using the results given in Table 10.1.

As a specific example, Figure 10.5 shows the step response for two discrete systems we have considered before, together with the corresponding continuous responses from the discretely coincident systems.

The pairs of models are, for Curve C,

Continuous: $(1 + 1.41D + 2D^2)Y(t) = 5X(t)$

Discrete: $(1 - 1.15B + 0.49B^2)Y_t = 5(0.19 + 0.15B)X_{t-1+}$

and for Curve R,

Continuous: $(1 + 2D)(1 + D)Y(t) = 5X(t)$

Discrete: $(1 - 0.97B + 0.22B^2)Y_t = 5(0.15 + 0.09B)X_{t-1+}$

The continuous curves were drawn using (10.1.18) and (10.1.12), which give the continuous step responses for second-order systems having, respectively, complex and real roots.

The discrete representation of the response of a second-order continuous system with delay to a pulsed input is given in Appendix A10.1.

10.3.3 Approximating General Continuous Models by Discrete Models

Perhaps we should emphasize once more that the discrete transfer function models do not need to be justified in terms of, or related to, continuous systems. They are of importance in their own right in allowing a discrete output to be calculated from a discrete input. However, in some instances, such relationships are of interest.

For continuous systems the pulsed input arises of itself in control problems when the convenient way to operate is to make an observation on the output Y and then immediately to make any adjustment that may be needed on the input variable X. Thus the input variable stays at a fixed level between observations, and we have a pulsed input. The relationships established in the previous sections may then be applied immediately. In particular, these relationships indicate that with the notation we have used, the *undelayed* discrete system is represented by

$$\xi(\nabla)Y_t = \eta(\nabla)X_{t-1+}$$

in which the subscript $t - 1+$ on X is one step behind the subscript t on Y.

Use of discrete models when continuous records are available. Even though we have a continuous record of input and output, it may be convenient to determine the dynamic characteristics of the system by discrete methods, as we describe in Chapter 11. Thus if pairs of values are read off with a sufficiently short sampling interval, very little is lost by replacing the continuous record by the discrete one.

One way in which the discrete results may then be used to approximate the continuous transfer function is to treat the input as though it were pulsed, that is to treat the input record as if the discrete input observed at time j extended from just after $j - \frac{1}{2}$ to $j + \frac{1}{2}$, as in Figure 10.11. Thus $X(t) = X_j \; (j - \frac{1}{2} < t \leq j + \frac{1}{2})$. We can then relate the discrete result to that of the continuous record by using the pulsed input equations with X_t replacing X_{t+} and with $b + c - \frac{1}{2}$ replacing $b + c$, that is, with one half a time period subtracted from the delay. The continuous record will normally be read at a sufficiently small sampling interval so that sudden changes do not occur between the sampled points. In this case the approximation will be very close.

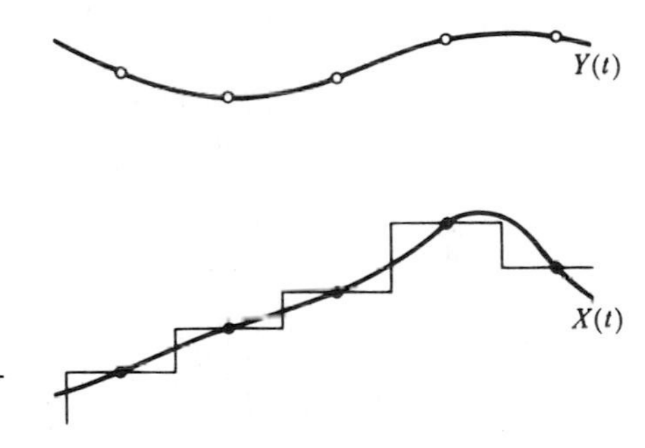

Figure 10.11 Replacement of continuous input by pulsed input.

APPENDIX A10.1 CONTINUOUS MODELS WITH PULSED INPUTS

We showed in Section 10.3.1 (see also Figure 10.9) that for a pulsed input, the output from any delayed continuous linear system

$$Y(t) = \int_0^{\infty} v(u)X(t - u)\, du$$

where $v(u) = 0$, $u < b + c$, is exactly given at the discrete times $t, t-1, t-2, \ldots$ by the discrete linear filter

$$Y_t = v(B)X_{t-1+}$$

where the weights $v_0, v_1, \ldots, v_{b-1}$ are zero and the weights $v_b, v_{b+1}, \ldots$ are given by

$$v_b = \int_{b+c}^{b+1} v(u)\, du \tag{A10.1.1}$$

$$v_{b+j} = \int_{b+j}^{b+j+1} v(u)\, du \qquad j \geqslant 1 \tag{A10.1.2}$$

Now suppose that the dynamics of the continuous system is represented by the Rth-order linear differential equation

$$\Xi(D)Y(t) = gX(t - b - c) \tag{A10.1.3}$$

which may be written in the form

$$\prod_{h=1}^{R} (1 + T_h D)Y(t) = gX(t - b - c)$$

where $T_1, T_2, \ldots, T_R$ may be real or complex. We now show that for a pulsed input, the output from this continuous system is discretely coincident with that from a discrete difference equation model of order (r, r), or of order $(r, r - 1)$ if $c = 0$. Now $v(u)$ is zero for $u < b + c$ and for $u \geqslant b + c$ is in general nonzero and satisfies the differential equation

$$\prod_{h=1}^{R} (1 + T_h D)v(u - b - c) = 0 \qquad u \geqslant b + c$$

Thus

$$v(u) = 0 \qquad u < b + c$$

$$v(u) = \alpha_1 e^{-(u-b-c)/T_1} + \alpha_2 e^{-(u-b-c)/T_2} + \cdots + \alpha_R e^{-(u-b-c)/T_R} \qquad u \geqslant b + c$$

whence, using (A10.1.1) and (A10.1.2),

$$v_b = \sum_{h=1}^{R} \alpha_h T_h [1 - e^{-(1-c)/T_h}] \tag{A10.1.4}$$

$$v_{b+j} = \sum_{h=1}^{R} \alpha_h T_h (1 - e^{-1/T_h}) e^{c/T_h}\, e^{-j/T_h} \qquad j \geqslant 1 \tag{A10.1.5}$$

It will be noted that in the particular case when $c = 0$, the weights v_{b+j} are given by (A10.1.2) for $j = 0$ as well as for $j > 0$.

Now consider the difference equation model of order (r, s),

$$\delta(B)Y_t = \omega(B)B^b X_{t-1+} \tag{A10.1.6}$$

If we write

$$\Omega(B) = \omega(B)B^b$$

the discrete transfer function $v(B)$ for this model satisfies

$$\delta(B)v(B) = \Omega(B) \tag{A10.1.7}$$

As we have observed in (10.2.8), by equating coefficients in (A10.1.7) we obtain b zero weights $v_0, v_1, \ldots, v_{b-1}$, and if $s \geqslant r$, a further $s - r + 1$ values $v_b, v_{b+1}, \ldots, v_{b+s-r}$ which do not follow a pattern. The weights v_j eventually satisfy

$$\delta(B)v_j = 0 \qquad j > b + s \tag{A10.1.8}$$

with $v_{b+s}, v_{b+s-1}, \ldots, v_{b+s-r+1}$ supplying the required r starting values. Now write

$$\delta(B) = \prod_{h=1}^{r} (1 - S_h B)$$

where $S_1^{-1}, S_2^{-1}, \ldots, S_r^{-1}$ are the roots of the equation $\delta(B) = 0$. Then the solution of (A10.1.8) is of the form

$$v_j = A_1(\boldsymbol{\omega})S_1^j + A_2(\boldsymbol{\omega})S_2^j + \cdots + A_r(\boldsymbol{\omega})S_r^j \qquad j > b + s - r \tag{A10.1.9}$$

where the coefficients $A_h(\boldsymbol{\omega})$ are suitably chosen so that the solutions of (A10.1.9) for $j = s - r + 1, s - r + 2, \ldots, s$ generate the starting values $v_{b+s-r+1}, \ldots, v_{b+s}$ and the notation $A_h(\boldsymbol{\omega})$ is used as a reminder that the A's are functions of $\omega_0, \omega_1, \ldots, \omega_s$. Thus if we set $s = r$, for given parameters $(\boldsymbol{\omega}, \boldsymbol{\delta})$ in (A10.1.6), and hence for given parameters $(\boldsymbol{\omega}, \mathbf{S})$, there will be a corresponding set of values $A_h(\boldsymbol{\omega})$ $(h = 1, 2, \ldots, r)$ that produce the appropriate r starting values $v_{b+1}, v_{b+2}, \ldots, v_{b+r}$. Furthermore, we know that $v_b = \omega_0$. Thus

$$v_b = \omega_0 \tag{A10.1.10}$$

$$v_{b+j} = \sum_{h=1}^{r} A_h(\boldsymbol{\omega})S_h^j \tag{A10.1.11}$$

and we can equate the values of the weights in (A10.1.4) and (A10.1.5), which come from the differential equation, to those in (A10.1.10) and (A10.1.11), which come from the difference equation. To do this we must set

$$R - r \qquad S_h = e^{1/T_h}$$

and the remaining $r + 1$ equations

$$\omega_0 = \sum_{h=1}^{r} \alpha_h T_h (1 - S_h^{1-c})$$

$$A_h(\boldsymbol{\omega}) = \alpha_h T_h (1 - S_h) S_h^{c}$$

determine $c, \alpha_1, \alpha_2, \ldots, \alpha_r$ in terms of the S's and ω's.

When $c = 0$, we set $s = r - 1$, and for given parameters $(\boldsymbol{\omega}, \mathbf{S})$ in the difference equation, there will then be a set of r values $A_h(\boldsymbol{\omega})$ which are functions of $\omega_0, \omega_1, \ldots, \omega_{r-1}$, which produce the r starting values $v_b, v_{b+1}, \ldots, v_{b+r-1}$ and which can be equated to the values given by (A10.1.5) for $j = 0, 1, \ldots, r-1$. To do this we set

$$R = r \qquad S_h = e^{-1/T_h}$$

and the remaining r equations

$$A_h(\boldsymbol{\omega}) = \alpha_h T_h(1 - S_h)$$

determine $\alpha_1, \alpha_2, \ldots, \alpha_r$ in terms of the S's and ω's.

It follows, in general, that for a pulsed input the output at times $t, t-1, \ldots$ from the continuous rth-order dynamic system defined by

$$\Xi(D)Y(t) = gX(t - b - c) \qquad \text{(A10.1.12)}$$

is identical to the output from a discrete model

$$\xi(\nabla)Y_t = g\eta(\nabla)X_{t-b-1+} \qquad \text{(A10.1.13)}$$

of order (r, r) with the parameters suitably chosen. Furthermore, if $c = 0$, the output from the continuous model (A10.1.12) is identical at the discrete times to that of a model (A10.1.13) of order $(r, r-1)$.

We now derive the discrete model corresponding to the second-order system with delay, from which the results given in Section 10.3.2 may be obtained as special cases.

Second-order system with delay. Suppose that the differential equation relating input and output for a continuous system is given by

$$(1 + T_1D)(1 + T_2D)Y(t) = gX(t - b - c) \qquad \text{(A10.1.14)}$$

Then the continuous impulse response function is

$$v(u) = g(T_1 - T_2)^{-1}(e^{-(u-b-c)/T_1} - e^{-(u-b-c)/T_2}) \qquad u > b + c \qquad \text{(A10.1.15)}$$

For a pulsed input, the output at discrete times $t, t-1, t-2, \ldots$ will be related to the input by the difference equation

$$(1 + \xi_1\nabla + \xi_2\nabla^2)Y_t = g(1 + \eta_1\nabla + \eta_2\nabla^2)X_{t-b-1+} \qquad \text{(A10.1.16)}$$

with suitably chosen values of the parameters. This difference equation can also be written

$$(1 - \delta_1B - \delta_2B^2)Y_t = (\omega_0 - \omega_1B - \omega_2B^2)X_{t-b-1+}$$

or

$$(1 - S_1B)(1 - S_2B)Y_t = (\omega_0 - \omega_1B - \omega_2B^2)X_{t-b-1+} \qquad \text{(A10.1.17)}$$

Using (A10.1.1) and (A10.1.2) and writing

$$S_1 = e^{-1/T_1} \qquad S_2 = e^{-1/T_2}$$

we obtain

$$v_b = \int_{b+c}^{b+1} v(u)\, du = g(T_1 - T_2)^{-1}[T_1(1 - S_1^{1-c}) - T_2(1 - S_2^{1-c})]$$

$$v_{b+j} = \int_{b+j}^{b+j+1} v(u)\, du = g(T_1 - T_2)^{-1}[T_1 S_1^{-c}(1 - S_1)S_1^j - T_2 S_2^{-c}(1 - S_2)S_2^j]$$

$$j \geqslant 1$$

Thus

$$\begin{aligned}(T_1 - T_2)v(B) = {} & gB^b T_1[1 - S_1^{1-c} + S_1^{-c}(1 - S_1)(1 - S_1 B)^{-1} S_1 B] \\ & -gB^b T_2[1 - S_2^{1-c} + S_2^{-c}(1 - S_2)(1 - S_2 B)^{-1} S_2 B]\end{aligned}$$

But from (A10.1.17),

$$v(B) = \frac{B^b(\omega_0 - \omega_1 B - \omega_2 B^2)}{(1 - S_1 B)(1 - S_2 B)}$$

whence we obtain

$$\omega_0 = g(T_1 - T_2)^{-1}[T_1(1 - S_1^{1-c}) - T_2(1 - S_2^{1-c})]$$

$$\omega_1 = g(T_1 - T_2)^{-1}[(S_1 + S_2)(T_1 - T_2) + T_2 S_2^{1-c}(1 + S_1) - T_1 S_1^{1-c}(1 + S_2)] \tag{A10.1.18}$$

$$\omega_2 = gS_1 S_2(T_1 - T_2)^{-1}[T_2(1 - S_2^{-c}) - T_1(1 - S_1^{-c})]$$

and

$$\delta_1 = S_1 + S_2 = e^{-1/T_1} + e^{-1/T_2} \qquad \delta_2 = -S_1 S_2 = -e^{-(1/T_1)-(1/T_2)} \tag{A10.1.19}$$

Complex roots. If T_1 and T_2 are complex, corresponding expressions are obtained by substituting

$$T_1 = \zeta^{-1} e^{i\lambda} \qquad T_2 = \zeta^{-1} e^{-i\lambda} \qquad (i^2 = -1)$$

yielding

$$\omega_0 = g\left\{1 - \frac{e^{-\zeta(1-c)\cos\lambda} \sin[\zeta(1 - c)\sin\lambda + \lambda]}{\sin\lambda}\right\}$$

$$\omega_2 = g\delta_2\left[1 - \frac{e^{\zeta c \cos\lambda} \sin(-\zeta c \sin\lambda + \lambda)}{\sin\lambda}\right] \tag{A10.1.20}$$

$$\omega_1 = \omega_0 - \omega_2 - (1 - \delta_1 - \delta_2)g$$

where

$$\delta_1 = 2e^{-\zeta\cos\lambda} \cos(\zeta \sin \lambda) \qquad \text{(A10.1.21)}$$
$$\delta_2 = -e^{-2\zeta\cos\lambda}$$

APPENDIX A10.2 NONLINEAR TRANSFER FUNCTIONS AND LINEARIZATION

The linearity (or additivity) of the transfer function models we have considered implies that the overall response to the sum of a number of individual inputs will be the sum of the individual responses to those inputs. Specifically, that if $Y_t^{(1)}$ is the response at time t to an input history $\{X_t^{(1)}\}$ and $Y_t^{(2)}$ is the response at time t to an input history $\{X_t^{(2)}\}$ the response at time t to an input history $\{X_t^{(1)} + X_t^{(2)}\}$ would be $Y_t^{(1)} + Y_t^{(2)}$, and similarly for continuous inputs and outputs. In particular, if the input level is multiplied by some constant, the output level is multiplied by this same constant. In practice, this assumption is probably never quite true, but it supplies a useful approximation for many practical situations.

Models for nonlinear systems may sometimes be obtained by allowing the parameters to depend upon the level of the input in some prescribed manner. For example, suppose that a system were being studied over a range where Y had a maximum η, and for any X the steady-state relation could be approximated by the quadratic expression

$$Y_\infty = \eta - \tfrac{1}{2}k(\mu - X)^2$$

where Y and X are, as before, deviations from a convenient origin. Then

$$g(X) = \frac{dY_\infty}{dX} = k(\mu - X)$$

and the dynamic behavior of the system might then be capable of representation by the first-order difference equation (10.3.4) but with variable gain proportional to $k(\mu - X)$. Thus

$$Y_t = \delta Y_{t-1} + k(\mu - X_{t-1+})(1 - \delta)X_{t-1+} \qquad \text{(A10.2.1)}$$

Dynamics of a simple chemical reactor. It sometimes happens that we can make a theoretical analysis of a physical problem which will yield the appropriate form for the transfer function. In particular this allows us to see very specifically what is involved in the linearized approximation.

As an example, suppose that a pure chemical A is continuously fed through a stirred tank reactor, and in the presence of a catalyst a certain proportion of it is changed to a product B, with no change of overall volume; hence the material continuously leaving the reactor consists of a mixture of B and unchanged A.

Suppose that initially the system is in equilibrium and that with quantities measured in suitable units:

1. μ is the rate at which A is fed to the reactor (and consequently is also the rate at which the mixture of A and B leaves the reactor).
2. η is the proportion of unchanged A at the outlet, so that $1 - \eta$ is the proportion of the product B at the outlet.
3. V is the volume of the reactor.
4. k is a constant determining the rate at which the product B is formed.

Suppose that the reaction is "first order" with respect to A which means that the rate at which B is formed and A is used up is proportional to the amount of A present. Then the rate of formation of B is $kV\eta$, but the rate at which B is leaving the outlet is $\mu(1 - \eta)$, and since the system is in equilibrium,

$$\mu(1 - \eta) = kV\eta \tag{A10.2.2}$$

Now, suppose that the equilibrium of the system is disturbed, the rate of feed to the reactor at time t being $\mu + X(t)$ and the corresponding concentration of A in the outlet being $\eta + Y(t)$. Now the rate of chemical formation of B, which now equals $kV(\eta + Y(t))$, will in general no longer exactly balance the rate at which B is flowing out of the system, which now equals $[\mu + X(t)][1 - \eta - Y(t)]$. The difference in these two quantities is the rate of increase in the amount of B within the reactor, which equals $-V[dY(t)/dt]$. Thus

$$-V\frac{dY(t)}{dt} = kV[\eta + Y(t)] - [\mu + X(t)][1 - \eta - Y(t)] \tag{A10.2.3}$$

Using (A10.2.2) and rearranging, (A10.2.3) may be written

$$(kV + \mu + VD)Y(t) = X(t)[1 - \eta - Y(t)]$$

or

$$(1 + TD)Y(t) = g\left(1 - \frac{Y(t)}{1 - \eta}\right)X(t) \tag{A10.2.4}$$

where

$$T = \frac{V}{kV + \mu} \qquad g = \frac{1 - \eta}{kV + \mu} \tag{A10.2.5}$$

Now (A10.2.4) is a nonlinear differential equation, since it contains a term $X(t)$ multiplied by $Y(t)$. However, in some practical circumstances, it could be adequately approximated by a linear differential equation, as we now show.

Processes operate under a wide range of conditions, but certainly a not unusual situation might be one where $100(1 - \eta)$, the percentage conversion of feed A to product B, was, say, 80%, and $100Y(t)$, the percentage fluctuation that was of practical interest, was 4%. In this case the factor $1 - Y(t)/(1 - \eta)$ would vary from 0.95 to 1.05 and, to a good approximation, could be replaced by unity. The nonlinear differential equation (A10.2.4) could then be replaced by the linear first-order differential equation

$$(1 + TD)Y(t) = gX(t)$$

where T and g are as defined in Section 10.1.2. If the system were observed at discrete intervals of time, this equation could be approximated by a linear difference equation.

Situations can obviously occur when nonlinearities are of importance. This is particularly true of optimization studies, where the range of variation for the variables may be large. A device that is sometimes useful when the linear assumption is not adequate is to represent the dynamics by a set of linear models applicable over different ranges of the input variables. However, for discrete systems it is often less clumsy to work directly with a nonlinear difference equation that can be "solved" recursively rather than analytically. For example, we might replace the nonlinear differential equation (A10.2.4) by the nonlinear difference equation

$$(1 + \xi_1\nabla)Y_t = g(1 + \eta_{12}Y_{t-1})X_{t-1}$$

11

IDENTIFICATION, FITTING, AND CHECKING OF TRANSFER FUNCTION MODELS

In Chapter 10 a parsimonious class of discrete linear transfer function models was introduced:

$$Y_t - \delta_1 Y_{t-1} - \cdots - \delta_r Y_{t-r} = \omega_0 X_{t-b} - \omega_1 X_{t-b-1} - \cdots - \omega_s X_{t-b-s}$$

or

$$Y_t = \delta^{-1}(B)\omega(B)X_{t-b}$$

In these models X_t and Y_t were deviations from equilibrium of the system input and output. In practice the system will be infected by disturbances, or noise, whose net effect is to corrupt the output predicted by the transfer function model by an amount N_t. The combined transfer function–noise model may then be written as

$$Y_t = \delta^{-1}(B)\omega(B)X_{t-b} + N_t$$

In this chapter, methods are described for identifying, fitting, and checking transfer function–noise models when simultaneous pairs of observations $(X_1, Y_1), (X_2, Y_2), \ldots, (X_N, Y_N)$ of the input and output are available at discrete equispaced times $1, 2, \ldots, N$.

Engineering methods for estimating transfer functions are usually based on the choice of special inputs to the system, for example, step and sine wave inputs [211] and "pulse" inputs [114]. These methods have been useful when the system is affected by small amounts of noise but are less satisfactory otherwise. In the presence of appreciable noise, it is necessary to use statistical methods for estimating the transfer function. Two previous approaches that have been tried for this problem are direct estimation of the impulse response in the time domain, and direct estimation of the gain and phase characteristics in the frequency domain, as described, for example, in [62], [115], and [122]. These methods are often unsatisfactory because they involve the estimation of too many parameters. For example, to determine the gain and phase characteristics, it is necessary to estimate two parameters at each frequency. The approach adopted in this chapter is to estimate the parameters in parsimonious difference equation models. Throughout most of the chapter we assume that the input X_t is itself a stochastic process. Models of the kind discussed are useful in representing and forecasting certain multiple time series.

11.1 CROSS CORRELATION FUNCTION

In the same way that the autocorrelation function was used to identify stochastic models, the data analysis tool employed for the identification of transfer function models is the *cross correlation function* between the input and output. In this section we describe the basic properties of the cross correlation function and in the next section show how it can be used to identify transfer function models.

11.1.1 Properties of the Cross Covariance and Cross Correlation Functions

Bivariate stochastic processes. We have seen in Chapter 2 that to analyze a statistical time series, it is useful to regard it as a realization of a hypothetical population of time series called a stochastic process. Now, suppose that we wish to describe an input time series X_t and the corresponding output time series Y_t from some physical system. For example, Figure 11.1 shows continuous data representing the (coded) input gas feed rate and corresponding output CO_2 concentration from a gas furnace. Then we can regard this pair of time series as realizations of a hypothetical population of pairs of time series, called a *bivariate stochastic process* (X_t, Y_t). We shall assume that the data are read off at equispaced times yielding a pair of discrete time series, generated by a discrete bivariate process, and that values of the time series at times $t_0 + h, t_0 + 2h, \ldots, t_0 + Nh$ are denoted by $(X_1, Y_1), (X_2, Y_2), \ldots, (X_N, Y_N)$.

In this chapter, extensive illustrative use is made of the gas furnace data read at intervals of 9 seconds (see Figure 11.1). The values (X_t, Y_t) so obtained are listed as series J in the "Collection of Time Series" in Part Five.

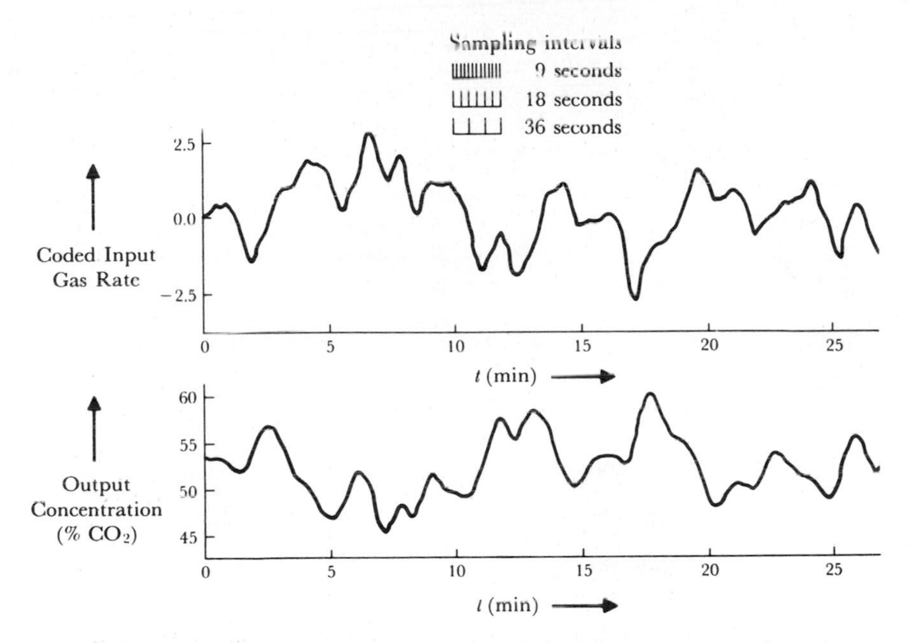

Figure 11.1 Input gas rate and output CO_2 concentration from a gas furnace.

Cross covariance and cross correlation functions. We have seen in Chapter 2 that a stationary Gaussian stochastic process can be described by its mean μ and autocovariance function γ_k, or equivalently, by its mean μ, variance σ^2, and autocorrelation function ρ_k. Moreover, since $\gamma_k = \gamma_{-k}$ and $\rho_k = \rho_{-k}$, the autocovariance and autocorrelation functions need only be plotted for nonnegative values of the lag $k = 0, 1, 2, \ldots$.

In general, a bivariate stochastic process (X_t, Y_t) need not be stationary. However, as in Chapter 4, we assume that the appropriately differenced process (x_t, y_t), where $x_t = \nabla^d X_t$, $y_t = \nabla^d Y_t$, is stationary. The stationarity assumption implies in particular that the constituent processes x_t and y_t have constant means μ_x and μ_y and constant variances σ_x^2 and σ_y^2. If, in addition, it is assumed that the bivariate process is Gaussian, or Normal, it is uniquely characterized by its means μ_x, μ_y and its covariance matrix. Figure 11.2 shows the different kinds of covariances that need to be considered.

The autocovariance coefficients of each constituent series at lag k are defined by the usual formula

$$\gamma_{xx}(k) = E[(x_t - \mu_x)(x_{t+k} - \mu_x)] = E[(x_t - \mu_x)(x_{t-k} - \mu_x)]$$

$$\gamma_{yy}(k) = E[(y_t - \mu_y)(y_{t+k} - \mu_y)] = E[(y_t - \mu_y)(y_{t-k} - \mu_y)]$$

where we now use the extended notation $\gamma_{xx}(k)$ and $\gamma_{yy}(k)$ for the autocovariances of the x and y series. The only other covariances that can appear

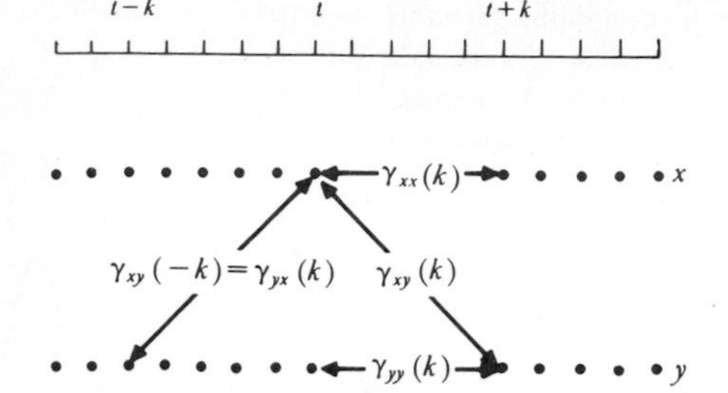

Figure 11.2 Autocovariances and cross covariances of a bivariate stochastic process.

in the covariance matrix are the *cross covariance* coefficients between x and y at lag $+k$:

$$\gamma_{xy}(k) = E[(x_t - \mu_x)(y_{t+k} - \mu_y)] \qquad k = 0, 1, 2, \ldots \tag{11.1.1}$$

and the cross covariance coefficients between y and x at lag $+k$:

$$\gamma_{yx}(k) = E[(y_t - \mu_y)(x_{t+k} - \mu_x)] \qquad k = 0, 1, 2, \ldots \tag{11.1.2}$$

Note that, in general, $\gamma_{xy}(k)$ will not be the same as $\gamma_{yx}(k)$. However, since

$$\gamma_{xy}(k) = E[(x_{t-k} - \mu_x)(y_t - \mu_y)] = E[(y_t - \mu_y)(x_{t-k} - \mu_x)] = \gamma_{yx}(-k)$$

we need only define one function $\gamma_{xy}(k)$ for $k = 0, \pm 1, \pm 2, \ldots$. The function $\gamma_{xy}(k)$, defined for $k = 0, \pm 1, \pm 2, \ldots$, is called the *cross covariance function* of the bivariate process. Similarly, the dimensionless quantity

$$\rho_{xy}(k) = \frac{\gamma_{xy}(k)}{\sigma_x \sigma_y} \qquad k = 0, \pm 1, \pm 2, \ldots \tag{11.1.3}$$

is called the *cross correlation* coefficient at lag k, and the function $\rho_{xy}(k)$, defined for $k = 0, \pm 1, \pm 2, \ldots$, the *cross correlation function* of the bivariate process.

Since $\rho_{xy}(k)$ is not in general equal to $\rho_{xy}(-k)$, the cross correlation function, in contrast to the autocorrelation function, is not symmetric about $k = 0$. In fact it will often happen that the cross correlation function is zero over some range $-\infty$ to i or i to $+\infty$. For example, consider the cross covariance function between a and z for the "delayed" first-order autoregressive process

$$(1 - \phi B)\tilde{z}_t = a_{t-b} \qquad -1 < \phi < 1, \quad b > 0$$

where a_t has zero mean. Then since

$$\tilde{z}_{t+k} = a_{t+k-b} + \phi a_{t+k-b-1} + \phi^2 a_{t+k-b-2} + \cdots$$

the cross covariance function between a and z is

$$\gamma_{az}(k) = E[a_t \tilde{z}_{t+k}] = \begin{cases} \phi^{k-b}\sigma_a^2 & k \geqslant b \\ 0 & k < b \end{cases}$$

Hence for the delayed autoregressive process, the cross correlation function is

$$\rho_{az}(k) = \begin{cases} \phi^{k-b}\dfrac{\sigma_a}{\sigma_z} = \phi^{k-b}(1-\phi^2)^{1/2} & k \geqslant b \\ 0 & k < b \end{cases}$$

Figure 11.3 shows this cross correlation function when $b = 2$ and $\phi = 0.6$.

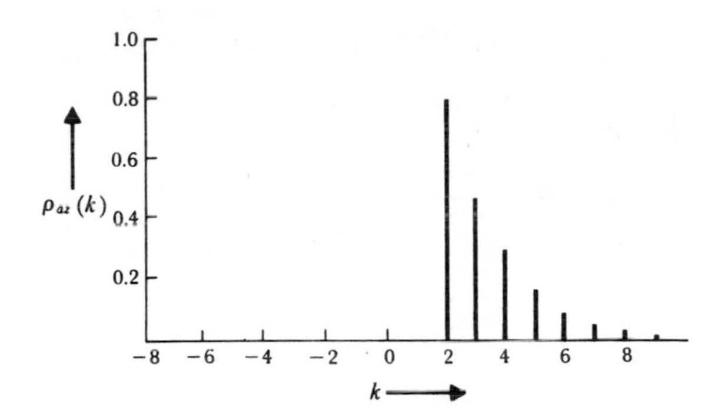

Figure 11.3 Cross correlation function between a and z for delayed autoregressive process $\tilde{z}_t - 0.6\tilde{z}_{t-1} = a_{t-2}$

11.1.2 Estimation of the Cross Covariance and Cross Correlation Functions

We assume that after differencing the original input and output time series d times, there are $n = N - d$ pairs of values $(x_1, y_1), (x_2, y_2), \ldots, (x_n, y_n)$ available for analysis. Then it is shown, for example in [122], that an estimate $c_{xy}(k)$ of the cross covariance coefficient at lag k is provided by

$$c_{xy}(k) = \begin{cases} \dfrac{1}{n}\displaystyle\sum_{t=1}^{n-k}(x_t - \bar{x})(y_{t+k} - \bar{y}) & k = 0, 1, 2, \ldots \\ \dfrac{1}{n}\displaystyle\sum_{t=1}^{n+k}(y_t - \bar{y})(x_{t-k} - \bar{x}) & k = 0, -1, -2, \ldots \end{cases} \tag{11.1.4}$$

where $\bar{x}$, $\bar{y}$ are the sample means of the x series and y series, respectively. Similarly, the estimate $r_{xy}(k)$ of the cross correlation coefficient $\rho_{xy}(k)$ at lag k may be obtained by substituting in (11.1.3) the estimates $c_{xy}(k)$ for $\gamma_{xy}(k)$, $s_x = \sqrt{c_{xx}(0)}$ for σ_x and $s_y = \sqrt{c_{yy}(0)}$ for σ_y, yielding

$$r_{xy}(k) = \frac{c_{xy}(k)}{s_x s_y} \qquad k = 0, \pm 1, \pm 2, \ldots \tag{11.1.5}$$

Example. In practice we would need at least 50 pairs of observations to obtain a useful estimate of the cross correlation function. However, to illustrate the formulas (11.1.4) and (11.1.5), we compute an estimate of the cross correlation function at lags +1 and −1 for the following series of five pairs of observations:

t	1	2	3	4	5
x_t	11	7	8	12	14
y_t	7	10	6	7	10

Now $\bar{x} = 10.4$, $\bar{y} = 8$, so that the deviations from the means are

t	1	2	3	4	5
$x_t - \bar{x}$	0.6	−3.4	−2.4	1.6	3.6
$y_t - \bar{y}$	−1.0	2.0	−2.0	−1.0	2.0

Hence

$$\sum_{t=1}^{4} (x_t - \bar{x})(y_{t+1} - \bar{y}) = (0.6)(2.0) + (-3.4)(-2.0) + (-2.4)(-1.0) + (1.6)(2.0) = 13.60$$

and $c_{xy}(1) = 13.60/5 = 2.720$. Using $s_x = 2.577$ and $s_y = 1.673$, we obtain

$$r_{xy}(1) = \frac{c_{xy}(1)}{s_x s_y} = \frac{2.720}{(2.577)(1.673)} = 0.63$$

Similarly, $\Sigma_{t=1}^{4} (y_t - \bar{y})(x_{t+1} - \bar{x}) = -8.20$. Hence $c_{xy}(-1) = -1.640$ and

$$r_{xy}(-1) = \frac{-1.640}{(2.577)(1.673)} = -0.38$$

Figure 11.4 shows the estimated cross correlation function $r_{XY}(k)$ between the input and output for the discrete gas furnace data obtained by reading the continuous data of Figure 11.1 at intervals of 9 seconds. Note that the cross correlation function is not symmetrical about zero and has a well-defined peak at $k = +5$, indicating that the output lags behind the input. The cross correlations are negative. This is to be expected since (see Figure 11.1) an *increase* in the coded input produces a *decrease* in the output.

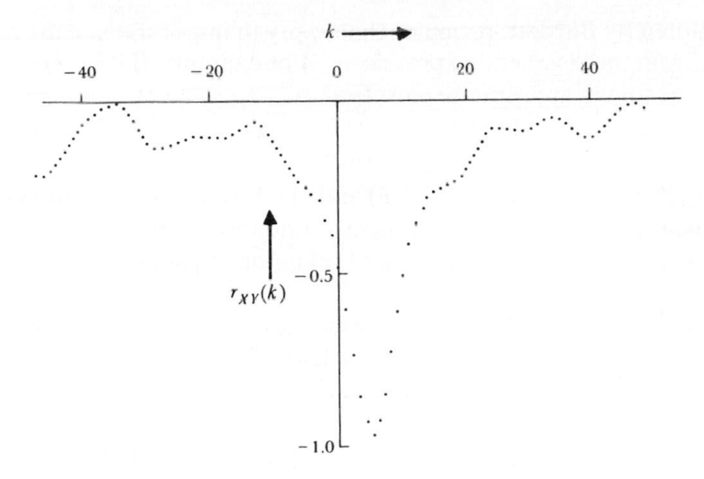

Figure 11.4 Cross correlation function between input and output for coded gas furnace data read at 9-second intervals.

11.1.3 Approximate Standard Errors of Cross Correlation Estimates

A crude check as to whether certain values of the cross correlation function $\rho_{xy}(k)$ could be effectively zero may be made by comparing the corresponding cross correlation estimates with their approximate standard errors obtained from a formula by Bartlett [26]. He shows that the covariance between two cross correlation estimates $r_{xy}(k)$ and $r_{xy}(k+l)$ is, on the Normal assumption, and $k \geqslant 0$, given by

$$\begin{aligned}
&\text{cov}[r_{xy}(k), r_{xy}(k+l)] \\
&\quad \simeq (n-k)^{-1} \sum_{v=-\infty}^{+\infty} \{\rho_{xx}(v)\rho_{yy}(v+l) + \rho_{xy}(-v)\rho_{xy}(v+2k+l) \\
&\quad + \rho_{xy}(k)\rho_{xy}(k+l)[\rho_{xy}^2(v) + \tfrac{1}{2}\rho_{xx}^2(v) + \tfrac{1}{2}\rho_{yy}^2(v)] \\
&\quad - \rho_{xy}(k)[\rho_{xx}(v)\rho_{xy}(v+k+l) + \rho_{xy}(-v)\rho_{yy}(v+k+l)] \\
&\quad - \rho_{xy}(k+l)[\rho_{xx}(v)\rho_{xy}(v+k) + \rho_{xy}(-v)\rho_{yy}(v+k)]\}
\end{aligned} \tag{11.1.6}$$

In particular, setting $l = 0$,

$$\begin{aligned}
\text{var}[r_{xy}(k)] \simeq (n-k)^{-1} &\sum_{v=-\infty}^{+\infty} \{\rho_{xx}(v)\rho_{yy}(v) + \rho_{xy}(k+v)\rho_{xy}(k-v) \\
&+ \rho_{xy}^2(k)[\rho_{xy}^2(v) + \tfrac{1}{2}\rho_{xx}^2(v) + \tfrac{1}{2}\rho_{yy}^2(v)] \\
&- 2\rho_{xy}(k)[\rho_{xx}(v)\rho_{xy}(v+k) + \rho_{xy}(-v)\rho_{yy}(v+k)]\}
\end{aligned} \tag{11.1.7}$$

As noted by Bartlett, formulas that apply to important special cases are derivable from these general expressions. For example, if it is supposed that $x_t \equiv y_t$, it becomes appropriate to set

$$\rho_{xx}(v) = \rho_{yy}(v) = \rho_{xy}(v) = \rho_{xy}(-v)$$

On making this substitution in (11.1.6) and (11.1.7) one obtains an expression for the covariance between two autocorrelation estimates and more particularly, the expression for the variance of an autocorrelation estimate given in (2.1.11).

It is often the case that two processes are appreciably cross correlated only over some rather narrow range of lags. Suppose it is postulated that $\rho_{xy}(v)$ is nonzero *only* over some range $Q_1 \leqslant v \leqslant Q_2$. Then:

1. If neither k, $k + l$ nor $k + \frac{1}{2}l$ are included in this range, all terms in (11.1.6) except the first are zero, and

$$\text{cov}[r_{xy}(k), r_{xy}(k + l)] \simeq (n - k)^{-1} \sum_{v=-\infty}^{\infty} \rho_{xx}(v)\rho_{yy}(v + l) \tag{11.1.8}$$

2. If k is not included in this range, then in a similar way (11.1.7) reduces to

$$\text{var}[r_{xy}(k)] \simeq (n - k)^{-1} \sum_{v=-\infty}^{\infty} \rho_{xx}(v)\rho_{yy}(v) \tag{11.1.9}$$

In particular, on the hypothesis that the two processes have *no cross correlation,* it follows that the simple formulas (11.1.8) and (11.1.9) apply for *all* lags k and $k + l$.

Another special case of some interest occurs when two processes are *not cross correlated and one is white noise.* Suppose that $y_t = a_t$ is generated by a white noise process but x_t is autocorrelated; then from (11.1.8),

$$\text{cov}[r_{xa}(k), r_{xa}(k + l)] \simeq (n - k)^{-1}\rho_{xx}(l) \tag{11.1.10}$$

$$\text{var}[r_{xa}(k)] \simeq (n - k)^{-1} \tag{11.1.11}$$

whence it follows that

$$\rho[r_{xa}(k), r_{xa}(k + l)] \simeq \rho_{xx}(l) \tag{11.1.12}$$

Thus in this case the cross correlations have the *same* autocorrelation function as the process generating x_t. Thus even though a_t and x_t are *not* cross correlated, the cross correlation function can be expected to vary about zero with standard deviation $(n - k)^{-1/2}$ *in a systematic pattern* typical of the behavior of the autocorrelation function $\rho_{xx}(l)$. Finally, if two processes are *both* white noise and are not cross correlated, the covariance between cross correlations will be zero.

11.2 IDENTIFICATION OF TRANSFER FUNCTION MODELS

We now show how to *identify* a combined transfer function–noise model

$$Y_t = \delta^{-1}(B)\omega(B)X_{t-b} + N_t$$

for a linear system corrupted by noise N_t at the output and assumed to be generated by an ARIMA process that is statistically independent* of the input X_t. Specifically, the objective at this stage is to obtain some idea of the orders r and s of the denominator and numerator operators in the transfer function model and to derive initial guesses for the parameters $\boldsymbol{\delta}$, $\boldsymbol{\omega}$, and the delay parameter b. In addition, we aim to make rough guesses of the parameters p, d, q of the ARIMA process describing the noise at the output and to obtain initial estimates of the parameters $\boldsymbol{\phi}$ and $\boldsymbol{\theta}$ in that model. The tentative transfer function and noise models so obtained can then be used as a starting point for more efficient estimation methods described in Section 11.3.

Outline of the identification procedure. Suppose that the transfer function model

$$Y_t = v(B)X_t + N_t \tag{11.2.1}$$

may be parsimoniously parametrized in the form

$$Y_t = \delta^{-1}(B)\omega(B)X_{t-b} + N_t \tag{11.2.2}$$

where $\delta(B) = 1 - \delta_1 B - \delta_2 B^2 - \cdots - \delta_r B^r$ and $\omega(B) = \omega_0 - \omega_1 B - \cdots - \omega_s B^s$. The identification procedure is as follows.

1. Derive rough estimates $\hat{v}_j$ of the impulse response weights v_j in (11.2.1).
2. Use the estimates $\hat{v}_j$ so obtained to make guesses of the orders r and s of the denominator and numerator operators in (11.2.2) and of the delay parameter b.
3. Substitute the estimates $\hat{v}_j$ in the equations (10.2.8) with values of r, s, and b obtained from (2) to obtain initial estimates of the parameters $\boldsymbol{\delta}$ and $\boldsymbol{\omega}$ in (11.2.2).

* When the input is at our choice, we can guarantee that it is independent of N_t by *generating* X_t according to some random process.

Knowing the $\hat{v}_j$, values of b, r, and s may be guessed using the following facts established in Section 10.2.2. For a model of the form of (11.2.2) the impulse response weights v_j consist of:

1. b zero values $v_0, v_1, \ldots, v_{b-1}$.
2. A further $s - r + 1$ values $v_b, v_{b+1}, \ldots, v_{b+s-r}$ following no fixed pattern (no such values occur if $s < r$).
3. Values v_j with $j \geqslant b + s - r + 1$ that follow the pattern dictated by an rth-order difference equation that has r starting values $v_{b+s}, \ldots, v_{b+s-r+1}$. Starting values v_j for $j < b$ will, of course, be zero.

Differencing of the input and output. The basic tool that is employed here in the identification process is the cross correlation function between input and output. When the processes are nonstationary it is assumed that stationarity can be induced by suitable differencing. Nonstationary behavior is suspected if the estimated auto- and cross-correlation functions of the (X_t, Y_t) series fail to damp out quickly. We assume that a degree of differencing* d necessary to induce stationarity has been achieved when the estimated auto- and cross correlations $r_{xx}(k)$, $r_{yy}(k)$, and $r_{xy}(k)$ of $x_t = \nabla^d X_t$ and $y_t = \nabla^d Y_t$ damp out quickly. In practice, d is usually 0, 1, or 2.

Identification of the impulse response function without prewhitening. Suppose that after differencing d times, the model (11.2.1) can be written in the form

$$y_t = v_0 x_t + v_1 x_{t-1} + v_2 x_{t-2} + \cdots + n_t \tag{11.2.3}$$

where $y_t = \nabla^d Y_t$, $x_t = \nabla^d X_t$, and $n_t = \nabla^d N_t$ are stationary processes with zero means. Then, on multiplying throughout in (11.2.3) by x_{t-k} for $k \geqslant 0$, we obtain

$$x_{t-k} y_t = v_0 x_{t-k} x_t + v_1 x_{t-k} x_{t-1} + \cdots + x_{t-k} n_t \tag{11.2.4}$$

If we make the further assumption that x_{t-k} is uncorrelated with n_t for all k, taking expectations in (11.2.4) yields the set of equations

$$\gamma_{xy}(k) = v_0 \gamma_{xx}(k) + v_1 \gamma_{xx}(k-1) + \cdots \qquad k = 0, 1, 2, \ldots \tag{11.2.5}$$

Suppose that the weights v_j are effectively zero beyond $k = K$. Then the first $K + 1$ of the equations (11.2.5) can be written

$$\boldsymbol{\gamma}_{xy} = \boldsymbol{\Gamma}_{xx} \mathbf{v} \tag{11.2.6}$$

* The procedures outlined can equally well be used when different degrees of differencing are employed for input and output.

where

$$\boldsymbol{\gamma}_{xy} = \begin{bmatrix} \gamma_{xy}(0) \\ \gamma_{xy}(1) \\ \vdots \\ \gamma_{xy}(K) \end{bmatrix} \qquad \mathbf{v} = \begin{bmatrix} v_0 \\ v_1 \\ \vdots \\ v_K \end{bmatrix}$$

$$\boldsymbol{\Gamma}_{xx} = \begin{bmatrix} \gamma_{xx}(0) & \gamma_{xx}(1) & \cdots & \gamma_{xx}(K) \\ \gamma_{xx}(1) & \gamma_{xx}(0) & \cdots & \gamma_{xx}(K-1) \\ \vdots & \vdots & \vdots & \vdots \\ \gamma_{xx}(K) & \gamma_{xx}(K-1) & \cdots & \gamma_{xx}(0) \end{bmatrix}$$

Substituting estimates $c_{xx}(k)$ of the autocovariance function of the input and estimates $c_{xy}(k)$ of the cross covariance function between the input and output, (11.2.6) provides $K + 1$ linear equations for the first $K + 1$ weights. However, these equations, which do not in general provide efficient estimates, are cumbersome to solve and in any case require knowledge of the point K beyond which v_j is effectively zero. The sample version of equations (11.2.6) represents essentially the least squares normal equations from linear regression of y_t on $x_t, x_{t-1}, \ldots, x_{t-K}$, in which it is assumed, implicitly, that the noise n_t in (11.2.3) is not autocorrelated. This is one source of the inefficiency in this identification method, which may be called the *regression method.* To improve the efficiency of this method, Liu and Hanssens [217] (see, also, Pankratz [218, Ch. 5]) suggest performing the generalized least squares estimation of the regression equation $y_t = v_0 x_t + v_1 x_{t-1} + \cdots + v_K x_{t-K} + n_t$ assuming the noise n_t follows some autocorrelated time series ARMA model. They also discuss generalization of this method of identification of impulse response functions to the case with multiple input processes $X_{1,t}, X_{2,t}, \ldots, X_{m,t}$ in the model, that is, $Y_t = v_1(B)X_{1,t} + \cdots + v_m(B)X_{m,t} + N_t$.

11.2.1 Identification of Transfer Function Models by Prewhitening the Input

Considerable simplification in the identification process would occur if the input to the system were white noise. Indeed, as is discussed in more detail in Section 11.6, when the choice of the input is at our disposal, there is much to recommend such an input. When the original input follows some other stochastic process, simplification is possible by *prewhitening*.

Suppose that the suitably differenced input process x_t is stationary and is capable of representation by some member of the general linear class of autoregressive–moving average models. Then, given a set of data, we can carry out our usual identification and estimation methods to obtain a model for the x_t process

$$\theta_x^{-1}(B)\phi_x(B)x_t = \alpha_t \tag{11.2.7}$$

which, to a close approximation, transforms the correlated input series x_t to the uncorrelated white noise series α_t. At the same time, we can obtain an estimate s_α^2 of σ_α^2 from the sum of squares of the $\hat{\alpha}$'s. If we now apply this *same* transformation to y_t to obtain

$$\beta_t = \theta_x^{-1}(B)\phi_x(B)y_t$$

then the model (11.2.3) may be written

$$\beta_t = v(B)\alpha_t + \varepsilon_t \tag{11.2.8}$$

where ε_t is the transformed noise series defined by

$$\varepsilon_t = \theta_x^{-1}(B)\phi_x(B)n_t \tag{11.2.9}$$

On multiplying (11.2.8) on both sides by α_{t-k} and taking expectations, we obtain

$$\gamma_{\alpha\beta}(k) = v_k\sigma_\alpha^2 \tag{11.2.10}$$

where $\gamma_{\alpha\beta}(k) = E[\alpha_{t-k}\beta_t]$ is the cross covariance at lag $+k$ between α and β. Thus

$$v_k = \frac{\gamma_{\alpha\beta}(k)}{\sigma_\alpha^2}$$

or in terms of the cross correlations,

$$v_k = \frac{\rho_{\alpha\beta}(k)\sigma_\beta}{\sigma_\alpha} \qquad k = 0, 1, 2, \ldots \tag{11.2.11}$$

Hence, after prewhitening the input, the cross correlation function between the prewhitened input and correspondingly transformed output is directly proportional to the impulse response function. We note that the effect of prewhitening is to convert the nonorthogonal set of equations (11.2.6) into the orthogonal set (11.2.10).

In practice, we do not know the theoretical cross correlation function $\rho_{\alpha\beta}(k)$, so we must substitute estimates in (11.2.11) to give

$$\hat{v}_k = \frac{r_{\alpha\beta}(k)s_\beta}{s_\alpha} \qquad k = 0, 1, 2, \ldots \tag{11.2.12}$$

The preliminary estimates $\hat{v}_k$ so obtained are again, in general, statistically inefficient but can provide a rough basis for selecting suitable operators $\delta(B)$ and $\omega(B)$ in the transfer function model. We now illustrate this identification and preliminary estimation procedure with an actual example.

11.2.2 Example of the Identification of a Transfer Function Model

In an investigation on adaptive optimization [136], a gas furnace was employed in which air and methane combined to form a mixture of gases containing CO_2 (carbon dioxide). The air feed was kept constant, but the methane feed rate could be varied in any desired manner and the resulting CO_2 concentration in the off-gases measured. The continuous data of Figure 11.1 were collected to provide information about the dynamics of the system over a region of interest where it was known that an approximately linear steady-state relationship applied. The continuous stochastic input series $X(t)$ shown in the top half of Figure 11.1 was generated by passing white noise through a linear filter. The process had mean zero and during the realization that was used for this experiment, varied from -2.5 to $+2.5$. It was desired that the actual methane gas feed rate should cover a range from 0.5 to 0.7 cubic foot per minute. To ensure this, the input gas feedrate was caused to follow the process

$$\text{methane gas input feed} = 0.60 - 0.04X(t)$$

For simplicity we shall work throughout with the "coded" input $X(t)$. The final transfer function expressed in terms of the actual feed rate is readily obtained by substitution. Series J in the "Collection of Time Series" in Part Five shows 296 successive pairs of observations (X_t, Y_t) read off from the continuous records at 9-second intervals. In this particular experiment the nature of the input disturbance was known because it was deliberately induced. However, we proceed as if it were not. The estimated auto- and cross correlation functions of X_t and Y_t damped out fairly quickly, confirming that no differencing was necessary. The usual identification and fitting procedure applied to the input X_t indicated that it is well described by a third-order autoregressive process

$$(1 - \phi_1 B - \phi_2 B^2 - \phi_3 B^3)X_t = \alpha_t$$

with $\hat{\phi}_1 = 1.97$, $\hat{\phi}_2 = -1.37$, $\hat{\phi}_3 = 0.34$, and $s_\alpha^2 = 0.0353$. Hence the transformations

$$\alpha_t = (1 - 1.97B + 1.37B^2 - 0.34B^3)X_t$$

$$\beta_t = (1 - 1.97B + 1.37B^2 - 0.34B^3)Y_t$$

were applied to the input and output series to yield the series α_t and β_t with $s_\alpha = 0.188$, $s_\beta = 0.358$. The estimated cross correlation function between α_t and β_t is shown in Table 11.1, together with the estimate (11.2.12) of the impulse response function,

$$\hat{v}_k = \frac{0.358}{0.188} r_{\alpha\beta}(k)$$

TABLE 11.1 Estimated Cross Correlation Function after Prewhitening and Approximate Impulse Response Function for Gas Furnace Data

k	$r_{\alpha\beta}(k)$	$\hat{\sigma}(r)$	$\hat{v}_k$	$r_{\beta\beta}(k)$	k	$r_{\alpha\beta}(k)$	$\hat{\sigma}(r)$	$\hat{v}_k$	$r_{\beta\beta}(k)$
0	−0.01	0.06	−0.02	1.00	6	−0.27	0.06	−0.52	0.12
1	0.05	0.06	0.10	0.22	7	−0.17	0.06	−0.32	0.04
2	−0.03	0.06	−0.06	0.36	8	−0.03	0.06	−0.06	0.09
3	−0.28	0.05	−0.53	0.13	9	0.03	0.06	0.06	0.00
4	−0.33	0.06	−0.63	0.08	10	−0.05	0.06	−0.10	0.10
5	−0.46	0.05	−0.88	0.01					

The approximate standard errors for the cross correlations $r_{\alpha\beta}(k)$ shown in Table 11.1 are the square roots of the variances obtained from (11.1.7):

1. With cross correlations up to lag + 2 and from lag + 8 onward assumed equal to zero
2. With autocorrelations $\rho_{\alpha\alpha}(k)$ assumed zero for $k > 0$
3. With autocorrelations $\rho_{\beta\beta}(k)$ assumed zero for $k > 4$
4. With estimated correlations $r_{\alpha\beta}(k)$ and $r_{\beta\beta}(k)$ from Table 11.1 replacing theoretical values

Also shown in Table 11.1 are the values of $r_{\beta\beta}(k)$ that are needed in (11.1.7). The estimated cross correlations together with one and two standard error limits centered on zero are plotted in Figure 11.5. For this example the standard errors differ very little from the approximate values $n^{-1/2} = 0.06$ appropriate to the hypothesis that the series are uncorrelated.

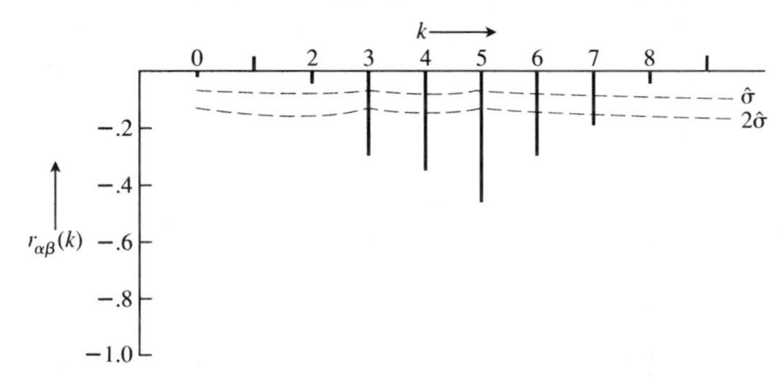

Figure 11.5 Estimated cross correlation function for coded gas furnace data after prewhitening.

The values $\hat{v}_0$, $\hat{v}_1$, and $\hat{v}_2$ are small compared with their standard errors, suggesting that $b = 3$ (that there are two whole periods of delay). Using the results of Section 11.2.1, the subsequent pattern of the $\hat{v}$'s might be accounted for by a model with (r, s, b) either equal to (1, 2, 3) or to (2, 2, 3).

The first model would imply that v_3 and v_4 were preliminary values following no fixed pattern and that v_5 provided the starting value for an exponential decay determined by the difference equation $v_j - \delta v_{j-1} = 0, j > 5$. The second model would imply that v_3 was a single preliminary value and that v_4 and v_5 provided the starting values for a pattern of double exponential decay determined by the difference equation $v_j - \delta_1 v_{j-1} - \delta_2 v_{j-2} = 0, j > 5$.

Thus the preliminary identification suggests a transfer function model

$$(1 - \delta_1 B - \delta_2 B^2)Y_t = (\omega_0 - \omega_1 B - \omega_2 B^2)X_{t-b} \qquad (11.2.13)$$

or some simplification of it, probably with $b = 3$.

Preliminary estimates. Assuming the model (11.2.13) with $b = 3$, the equations (10.2.8) for the impulse response function are

$$\begin{aligned} v_j &= 0 \qquad j < 3 \\ v_3 &= \omega_0 \\ v_4 &= \delta_1 v_3 - \omega_1 \\ v_5 &= \delta_1 v_4 + \delta_2 v_3 - \omega_2 \\ v_6 &= \delta_1 v_5 + \delta_2 v_4 \\ v_7 &= \delta_1 v_6 + \delta_2 v_5 \end{aligned} \qquad (11.2.14)$$

Substituting the estimates $\hat{v}_k$ from Table 11.1 in the last two of these equations, we obtain

$$\begin{aligned} 0.88\hat{\delta}_1 - 0.63\hat{\delta}_2 &= -0.52 \\ -0.52\hat{\delta}_1 - 0.88\hat{\delta}_2 &= -0.32 \end{aligned}$$

which give preliminary estimates $\hat{\delta}_1 = 0.57$, $\hat{\delta}_2 = 0.02$. If these values are now substituted in the second, third, and fourth of equations (11.2.14), we obtain

$$\hat{\omega}_0 = \hat{v}_3 = -0.53$$

$$\hat{\omega}_1 = \hat{\delta}_1\hat{v}_3 - \hat{v}_4 = (0.57)(-0.53) + 0.63 = 0.33$$

$$\hat{\omega}_2 = \hat{\delta}_1\hat{v}_4 + \hat{\delta}_2\hat{v}_3 - \hat{v}_5 = (0.57)(-0.63) + (0.02)(-0.53) + 0.88 = 0.51$$

Thus the preliminary identification suggests a tentative transfer function model

$$(1 - 0.57B - 0.02B^2)Y_t = -(0.53 + 0.33B + 0.51B^2)X_{t-3}$$

The estimates so obtained can be used as starting values for the more efficient iterative estimation methods which will be described in Section 11.3. Note that the estimate $\hat{\delta}_2$ is very small and suggests that this parameter may be omitted, but we shall retain it for the time being.

11.2.3 Identification of the Noise Model

Reverting to the general case, suppose that (where necessary, after suitable differencing) the model could be written

$$y_t = v(B)x_t + n_t$$

where

$$n_t = \nabla^d N_t$$

Given that a preliminary estimate $\hat{v}(B)$ of the transfer function has been obtained in the manner discussed in Section 11.2.2, an estimate of the noise series is provided by

$$\hat{n}_t = y_t - \hat{v}(B)x_t$$

that is,

$$\hat{n}_t = y_t - \hat{v}_0 x_t - \hat{v}_1 x_{t-1} - \hat{v}_2 x_{t-2} - \cdots$$

Alternatively, $\hat{v}(B)$ may be replaced by the tentative transfer function model $\hat{\delta}^{-1}(B)\hat{\omega}(B)B^b$ determined by preliminary identification. Thus

$$\hat{n}_t = y_t - \hat{\delta}^{-1}(B)\hat{\omega}(B)x_{t-b}$$

and $\hat{n}_t$ may be computed by first calculating $\hat{\mathcal{Y}}_t = \hat{\delta}^{-1}(B)\hat{\omega}(B)x_{t-b}$ recursively through $\hat{\delta}(B)\hat{\mathcal{Y}}_t = \hat{\omega}(B)x_{t-b}$ as

$$\hat{\mathcal{Y}}_t = \hat{\delta}_1 \hat{\mathcal{Y}}_{t-1} + \cdots + \hat{\delta}_r \hat{\mathcal{Y}}_{t-r} + \hat{\omega}_0 x_{t-b} - \hat{\omega}_1 x_{t-b-1} - \cdots - \hat{\omega}_s x_{t-b-s} \tag{11.2.15}$$

and then computing the noise series from $\hat{n}_t = y_t - \hat{\mathcal{Y}}_t$. In either case, study of the estimated autocorrelation function of $\hat{n}_t$ can lead to identification of the noise model.

It is also possible to identify the noise using the correlation functions for the input and output, after prewhitening, in the following way. Suppose that the input could be exactly prewhitened to give

$$\beta_t = v(B)\alpha_t + \varepsilon_t \tag{11.2.16}$$

where the known relationship

$$\varepsilon_t = \theta_x^{-1}(B)\phi_x(B)n_t \tag{11.2.17}$$

would link ε_t and n_t. If a stochastic model could be found for ε_t, then, using (11.2.17), a model could be deduced for n_t and hence for N_t. If we now write $v(B)\alpha_t = u_t$, so that $\beta_t = u_t + \varepsilon_t$, and provided that our independence assumption concerning x_t and n_t, and hence concerning u_t and ε_t, is justified, we can write

$$\gamma_{\beta\beta}(k) = \gamma_{uu}(k) + \gamma_{\varepsilon\varepsilon}(k) \tag{11.2.18}$$

Since α_t is white noise, $\gamma_{uu}(k)$ may be obtained using the result (3.1.8), which gives the autocorrelation function of a linear process. Thus

$$\gamma_{uu}(k) = \sigma_\alpha^2 \sum_{j=0}^{\infty} v_j v_{j+k} = \frac{1}{\sigma_\alpha^2} \sum_{j=0}^{\infty} \gamma_{\alpha\beta}(j)\gamma_{\alpha\beta}(j+k)$$

using (11.2.10). Hence, using (11.2.18), the autocovariances of ε_t may be obtained from $\gamma_{\varepsilon\varepsilon}(k) = \gamma_{\beta\beta}(k) - \gamma_{uu}(k)$, with autocorrelations

$$\rho_{\varepsilon\varepsilon}(k) = \frac{\gamma_{\varepsilon\varepsilon}(k)}{\gamma_{\varepsilon\varepsilon}(0)} = \frac{\rho_{\beta\beta}(k) - \gamma_{uu}(k)/\gamma_{\beta\beta}(0)}{1 - \gamma_{uu}(0)/\gamma_{\beta\beta}(0)}$$

$$= \frac{\rho_{\beta\beta}(k) - \sum_{j=0}^{\infty} \rho_{\alpha\beta}(j)\rho_{\alpha\beta}(j+k)}{1 - \sum_{j=0}^{\infty} \rho_{\alpha\beta}^2(j)}$$

Now, in practice, it is necessary to *estimate* the prewhitening transformation. Having made the approximate prewhitening transformation, rough values for $\rho_{\varepsilon\varepsilon}(k)$ could be obtained by substituting the estimates $r_{\alpha\beta}(j)$ of the cross correlation function between transformed input and output and $r_{\beta\beta}(j)$ of the autocorrelation function of the transformed output.

Application to the gas furnace example. Table 11.2 shows the first 12 values of the sample autocorrelations and partial autocorrelations of the noise series $\hat{N}_t = Y_t - \hat{\mathcal{Y}}_t$, where $\hat{\mathcal{Y}}_t = \hat{\delta}^{-1}(B)\hat{\omega}(B)X_{t-3}$ is computed as in (11.2.15) using the preliminary estimates for the transfer function model obtained previously. That is, the values are computed as

$$\hat{\mathcal{Y}}_t - 0.57\hat{\mathcal{Y}}_{t-1} = (0.53X_{t-3} + 0.33X_{t-4} + 0.51X_{t-5})$$

The partial autocorrelations of $\hat{N}_t$ indicate that a second-order autoregressive model might be an adequate representation, and the least squares estimates obtained from the $\hat{N}_t$ values for the AR(2) model yield

$$(1 - 1.54B + 0.64B^2)N_t = a_t \tag{11.2.19}$$

with $\hat{\sigma}_a^2 = 0.057$.

TABLE 11.2 Autocorrelation and Partial Autocorrelation Functions of the Noise in Gas Furnace Data

k	r_k	$\hat{\phi}_{kk}$	k	r_k	$\hat{\phi}_{kk}$
1	0.89	0.89	7	0.01	−0.02
2	0.71	−0.43	8	−0.03	0.01
3	0.51	−0.13	9	−0.05	−0.01
4	0.32	0.02	10	−0.04	0.08
5	0.17	0.04	11	−0.03	−0.06
6	0.07	−0.02	12	−0.03	−0.10

Thus the analysis of this section and Section 11.2.2 suggests the identification

$$Y_t = \frac{\omega_0 - \omega_1 B - \omega_2 B^2}{1 - \delta_1 B - \delta_2 B^2} X_{t-3} + \frac{1}{1 - \phi_1 B - \phi_2 B^2} a_t \qquad (11.2.20)$$

for the gas furnace model. Furthermore, the initial estimates $\hat{\omega}_0 = -0.53$, $\hat{\omega}_1 = 0.33$, $\hat{\omega}_2 = 0.51$, $\hat{\delta}_1 = 0.57$, $\hat{\delta}_2 = 0.02$, $\hat{\phi}_1 = 1.54$, $\hat{\phi}_2 = -0.64$ can be used as rough starting values for the nonlinear estimation procedures that we describe in Section 11.3.

11.2.4 Some General Considerations in Identifying Transfer Function Models

Some general remarks can now be made concerning the procedures for identifying transfer function and noise models that we have just described.

1. For many practical situations, when the effect of noise is appreciable, a delayed first- or second-order system such as that given by (11.2.13), or some simplification of it, would often provide as elaborate a model as could be justified for the data.
2. Efficient estimation is only possible assuming the model *form* to be known. The estimates $\hat{v}_k$ given by (11.2.12) are in general *necessarily* inefficient therefore. They are employed at the identification stage because they are easily computed and can indicate a form of model worthy to be fitted by more elaborate means.
3. Even if these were efficient estimates, the number of $\hat{v}$'s required to trace out the impulse response function fully would usually be considerably larger than the number of parameters in a transfer function model. In cases where the δ's and ω's in an adequate transfer function model could be estimated accurately, nevertheless, the estimates of the corresponding v's could have large variances and be highly correlated.
4. The variance of

$$r_{\alpha\beta}(k) = \hat{v}_k \frac{s_\alpha}{s_\beta}$$

is of order $1/n$. Thus we can expect that the estimates $r_{\alpha\beta}(k)$, and hence the $\hat{v}_k$, will be buried in noise unless σ_α is reasonably large compared with the residual noise, or unless n is large. Thus the identification procedure requires that the variation in the input X shall be reasonably large compared with the variation due to the noise and/or that a large volume of data is available. These requirements are satisfied by the gas furnace data for which, as we show in Section 11.3, the initial identification is remarkably good. When these requirements are not satisfied, the identification procedure may fail. Usually, this will mean that only very rough estimates are possible with the available data. However, some kind of rudimentary modeling may be possible by postulating a plausible but simple transfer function/noise model, fitting directly by the least squares procedures of the next section, and applying diagnostic checks leading to elaboration of the model when this proves necessary.

5. It should, perhaps, be emphasized that the prewhitened series α_t and β_t (their cross correlation function, $r_{\alpha\beta}(k)$, in particular) are used only for the purpose of identification of the form of the transfer function model. Once the model form is identified, the original series X_t and Y_t (not the prewhitened series) are used for efficient estimation of the parameters of the model, for forecasting, and so on.
6. An alternative method of identification of the transfer function–noise model relation has been proposed by Haugh and Box [216], and similar ideas are also discussed by Priestley [220, Ch. 9]. The general method, which might be referred to as "double-prewhitening," involves prewhitening *both* input and output series. That is, separate univariate ARIMA models are built for both the input and the output processes, X_t and Y_t, respectively, and then the cross correlation structure of the resulting (univariate white noise) residuals from these models are examined. From this examination, a transfer function model that relates the two residual series may be identified and hence, from knowledge of the filters (from the observed series to the white noise residuals) in the individual univariate ARIMA models for X_t and Y_t, a model in terms of the original series can be specified. However, while sometimes useful, in some cases this procedure can tend to become overly complicated in terms of the final form of the transfer function model specified, due to the presence of the univariate prewhitening factors.

Lack of uniqueness of the model. Suppose that a particular dynamic system is represented by the model

$$Y_t = \delta^{-1}(B)\omega(B)X_{t-b} + \varphi^{-1}(B)\theta(B)a_t \qquad (11.2.21)$$

Then it could equally well be represented by

$$L(B)Y_t = L(B)\delta^{-1}(B)\omega(B)X_{t-b} + L(B)\,\varphi^{-1}(B)\theta(B)a_t \qquad (11.2.22)$$

where $L(B)$ could be an arbitrary common factor, and hence would be redundant. Similar to the discussion in Section 7.3.5 on parameter redundancy for ARMA models, for uniqueness of model parameterization in (11.2.21) it is implied that the possibility of common factors in the operators $\delta(B)$ and $\omega(B)$, or in the $\varphi(B)$ and $\theta(B)$ operators, must be avoided. The chance that we may iterate toward a model of unnecessarily complicated form is reduced if we base our strategy on the following considerations:

1. Since rather simple transfer function models of first or second order, with or without delay, are often adequate, iterative model building should begin with a fairly simple model, looking for further simplification if this is possible, and reverting to more complicated models only as the need is demonstrated.
2. One should be always on the look out for the possibility of removing a factor common to two or more of the operators on Y_t, X_t, and a_t. In practice, we shall be dealing with estimated coefficients, which may be subject to rather large errors, so that only approximate factorization can be expected, and considerable imagination may need to be exerted to spot a possible factorization. The factored model may be refitted and checked to show whether the simplification can be justified.
3. When simplification by factorization is possible, but is overlooked, the least squares estimation procedure may become extremely unstable since the minimum will tend to lie on a line or surface in the parameter space rather than at a point. Conversely, instability in the solution can point to the possibility of simplification of the model. As emphasized previously, one reason for carrying out the identification procedure before fitting the model is to avoid redundancy or, conversely, to achieve *parsimony* in parameterization.

An alternative method of identifying transfer function models, which is capable of ready generalization to deal with multiple inputs, is given in Appendix A11.1.

Remark. If the operator $L(B)$ in (11.2.22) were set equal to $\varphi(B)\delta(B)$, we would obtain

$$\varphi(B)\delta(B)Y_t = \varphi(B)\omega(B)X_{t-b} + \delta(B)\theta(B)a_t \qquad (11.2.23)$$

which can be written as $\delta^*(B)Y_t = \omega^*(B)X_{t-b} + \theta^*(B)a_t$. Models of the general form of (11.2.23) have been referred to as ARMAX models in the econometrics literature (e.g., Hannan and Deistler [101], Hannan, Dunsmuir, and Deistler [102], and Reinsel [166]). As can be seen, there is a reasonable possibility of the occurrence of common factors among the operators in this form.

11.3 FITTING AND CHECKING TRANSFER FUNCTION MODELS

11.3.1 Conditional Sum of Squares Function

We now consider the problem of efficiently and simultaneously estimating the parameters b, $\boldsymbol{\delta}$, $\boldsymbol{\omega}$, $\boldsymbol{\phi}$, and $\boldsymbol{\theta}$ in the tentatively identified model

$$y_t = \delta^{-1}(B)\omega(B)x_{t-b} + n_t \tag{11.3.1}$$

where $y_t = \nabla^d Y_t$, $x_t = \nabla^d X_t$, $n_t = \nabla^d N_t$ are all stationary processes and

$$n_t = \phi^{-1}(B)\theta(B)a_t \tag{11.3.2}$$

It is assumed that $n = N - d$ pairs of values are available for the analysis and that Y_t and X_t (y_t and x_t if $d > 0$) denote deviations from expected values. These expected values may be estimated along with the other parameters, but for the lengths of time series normally worth analyzing it will usually be sufficient to use the sample means as estimates. When $d > 0$ it will frequently be true that expected values for y_t and x_t are zero.

If starting values $\mathbf{x}_0$, $\mathbf{y}_0$, and $\mathbf{a}_0$ prior to the commencement of the series were available, then given the data, for any choice of the parameters (b, $\boldsymbol{\delta}$, $\boldsymbol{\omega}$, $\boldsymbol{\phi}$, $\boldsymbol{\theta}$) and of the starting values ($\mathbf{x}_0$, $\mathbf{y}_0$, $\mathbf{a}_0$) we could calculate, successively, values of $a_t = a_t(b, \boldsymbol{\delta}, \boldsymbol{\omega}, \boldsymbol{\phi}, \boldsymbol{\theta}|\mathbf{x}_0, \mathbf{y}_0, \mathbf{a}_0)$ for $t = 1, 2, \ldots, n$. Under the Normal assumption for the a's, a close approximation to the maximum likelihood estimates of the parameters can be obtained by minimizing the *conditional sum of squares function*,

$$S_0(b, \boldsymbol{\delta}, \boldsymbol{\omega}, \boldsymbol{\phi}, \boldsymbol{\theta}) = \sum_{t=1}^{n} a_t^2(b, \boldsymbol{\delta}, \boldsymbol{\omega}, \boldsymbol{\phi}, \boldsymbol{\theta}|\mathbf{x}_0, \mathbf{y}_0, \mathbf{a}_0) \tag{11.3.3}$$

Three-stage procedure for calculating the *a*'s. Given appropriate starting values, the generation of the a's *for any particular choice of the parameter values* may be accomplished using the following three-stage procedure.

First, the output $\mathcal{Y}_t$ from the transfer function model may be computed from

$$\mathcal{Y}_t = \delta^{-1}(B)\omega(B)x_{t-b}$$

that is, from

$$\delta(B)\mathcal{Y}_t = \omega(B)x_{t-b}$$

or from

$$\mathcal{Y}_t - \delta_1\mathcal{Y}_{t-1} - \cdots - \delta_r\mathcal{Y}_{t-r} = \omega_0 x_{t-b} - \omega_1 x_{t-b-1} - \cdots - \omega_s x_{t-b-s} \tag{11.3.4}$$

Having calculated the $\mathcal{Y}_t$ series, then using (11.3.1), the noise series n_t can be obtained from

$$n_t = y_t - \mathcal{Y}_t \tag{11.3.5}$$

Finally, the a's can be obtained from (11.3.2) written in the form

$$a_t = \theta^{-1}(B)\phi(B)n_t$$

that is,

$$a_t = \theta_1 a_{t-1} + \cdots + \theta_q a_{t-q} + n_t - \phi_1 n_{t-1} - \cdots - \phi_p n_{t-p} \tag{11.3.6}$$

Starting values. As discussed in Section 7.1.3 for stochastic model estimation, the effect of transients can be minimized if the difference equations are started off from a value of t for which all previous x's and y's are known. Thus $\mathcal{y}_t$ in (11.3.4) is calculated from $t = u + 1$ onward, where u is the larger of r and $s + b$. This means that n_t will be available from n_{u+1} onward; hence, if unknown a's are set equal to their unconditional expected values of zero, the a's may be calculated from a_{u+p+1} onward. Thus the conditional sum of squares function is

$$S_0(b, \boldsymbol{\delta}, \boldsymbol{\omega}, \boldsymbol{\phi}, \boldsymbol{\theta}) = \sum_{t=u+p+1}^{n} a_t^2(b, \boldsymbol{\delta}, \boldsymbol{\omega}, \boldsymbol{\phi}, \boldsymbol{\theta} | \mathbf{x}_0, \mathbf{y}_0, \mathbf{a}_0) \tag{11.3.7}$$

Example using the gas furnace data. For these data the model (11.2.20), namely

$$Y_t = \frac{\omega_0 - \omega_1 B - \omega_2 B^2}{1 - \delta_1 B - \delta_2 B^2} X_{t-3} + \frac{1}{1 - \phi_1 B - \phi_2 B^2} a_t$$

has been identified. Equations (11.3.4), (11.3.5), and (11.3.6) then become

$$\mathcal{Y}_t = \delta_1 \mathcal{Y}_{t-1} + \delta_2 \mathcal{Y}_{t-2} + \omega_0 X_{t-3} - \omega_1 X_{t-4} - \omega_2 X_{t-5} \tag{11.3.8}$$

$$N_t = Y_t - \mathcal{Y}_t \tag{11.3.9}$$

$$a_t = N_t - \phi_1 N_{t-1} - \phi_2 N_{t-2} \tag{11.3.10}$$

Thus (11.3.8) can be used to generate $\mathcal{Y}_t$ from $t = 6$ onward and (11.3.10) to generate a_t from $t = 8$ onward. The slight loss of information that results will not be important for a sufficiently long length of series. For example, since $N = 296$ for the gas furnace data, the loss of seven values at the beginning of the series is of little practical consequence. For illustration, Table 11.3 shows the calculation of the first few values of a_t for the coded gas furnace data with

$$b = 3 \quad \delta_1 = 0.1 \quad \delta_2 = 0.1 \quad \omega_0 = 0.1 \quad \omega_1 = -0.1 \quad \omega_2 = -0.1$$
$$\phi_1 = 0.1 \quad \phi_2 = 0.1$$

The X_t and Y_t values in the second and fourth columns were obtained by subtracting the means $\overline{X} = -0.057$ and $\overline{Y} = 53.51$ from the values of the series given in the "Collection of Time Series" in Part Five.

In the above we have assumed that $b = 3$. To estimate b, the values of

TABLE 11.3 Calculation of First Few Values a_t for Gas Furnace Data When $b = 3$, $\delta_1 = 0.1$, $\delta_2 = 0.1$, $\omega_0 = 0.1$, $\omega_1 = -0.1$, $\omega_2 = -0.1$, $\phi_1 = 0.1$, $\phi_2 = 0.1$

t	X_t	$\mathcal{Y}_t$	Y_t	N_t	a_t
1	−0.052	—	0.29	—	—
2	0.057	—	0.09	—	—
3	0.235	—	−0.01	—	—
4	0.396	—	−0.01	—	—
5	0.430	—	−0.11	—	—
6	0.498	0.024	−0.41	−0.434	—
7	0.518	0.071	−0.81	−0.881	—
8	0.405	0.116	−1.11	−1.226	−1.094
9	0.184	0.151	−1.31	−1.461	−1.250
10	−0.123	0.171	−1.51	−1.681	−1.412

$\boldsymbol{\delta}$, $\boldsymbol{\omega}$, $\boldsymbol{\phi}$, and $\boldsymbol{\theta}$, which minimize the conditional sum of squares, can be calculated for each value of b in the likely range and the overall minimum with respect to b, $\boldsymbol{\delta}$, $\boldsymbol{\omega}$, $\boldsymbol{\phi}$, and $\boldsymbol{\theta}$ obtained.

11.3.2 Nonlinear Estimation

A nonlinear least squares algorithm, analogous to that given for fitting the stochastic model in Section 7.2.4, can be used to obtain the least squares estimates and their approximate standard errors. The algorithm will behave well when the sum of squares function is very roughly quadratic. However, the procedure can sometimes run into trouble, in particular if the parameters are very highly correlated (if, for example, the model approaches singularity due to near-common-factorization), or in some cases, if estimates are near a boundary of the permissible parameter space. In difficult cases the estimation situation may be clarified by plotting sums of squares contours for selected two-dimensional sections of the parameter space.

The algorithm may be derived as follows: At any stage of the iteration, and for some fixed value of the delay parameter b, let the best guesses available for the remaining parameters be denoted by

$$\boldsymbol{\beta}_0' = (\delta_{1,0}, \ldots, \delta_{r,0}; \omega_{0,0}, \ldots, \omega_{s,0}; \phi_{1,0}, \ldots, \phi_{p,0}; \theta_{1,0}, \ldots, \theta_{q,0})$$

Now let $a_{t,0}$ be that value computed from the model, as in Section 11.3.1, for the guessed parameter values $\boldsymbol{\beta}_0$ and denote the negative of the derivatives of a_t with respect to the parameters as follows:

$$d_{i,t}^{(\delta)} = -\left.\frac{\partial a_t}{\partial \delta_i}\right|_{\boldsymbol{\beta}_0} \qquad d_{j,t}^{(\omega)} = -\left.\frac{\partial a_t}{\partial \omega_j}\right|_{\boldsymbol{\beta}_0} \qquad d_{g,t}^{(\phi)} = -\left.\frac{\partial a_t}{\partial \phi_g}\right|_{\boldsymbol{\beta}_0} \qquad d_{h,t}^{(\theta)} = -\left.\frac{\partial a_t}{\partial \theta_h}\right|_{\boldsymbol{\beta}_0} \tag{11.3.11}$$

Then a Taylor series expansion of $a_t = a_t(\boldsymbol{\beta})$ about parameter values $\boldsymbol{\beta} = \boldsymbol{\beta}_0$ can be rearranged in the form

$$\begin{aligned} a_{t,0} \simeq & \sum_{i=1}^{r} (\delta_i - \delta_{i,0}) d_{i,t}^{(\delta)} + \sum_{j=0}^{s} (\omega_j - \omega_{j,0}) d_{j,t}^{(\omega)} \\ & + \sum_{g=1}^{p} (\phi_g - \phi_{g,0}) d_{g,t}^{(\phi)} + \sum_{h=1}^{q} (\theta_h - \theta_{h,0}) d_{h,t}^{(\theta)} + a_t \end{aligned} \tag{11.3.12}$$

We proceed as in Section 7.2 to obtain adjustments $\delta_i - \delta_{i,0}$, $\omega_j - \omega_{j,0}$, and so on, by fitting this linearized equation by standard linear least squares. By adding the adjustments to the first guesses $\boldsymbol{\beta}_0$, a set of second guesses can be formed and the process repeated until convergence is reached.

As with stochastic models (see Chapter 7 and especially Section 7.2.3) the derivatives may be computed recursively. However, it seems simplest to work with a standard nonlinear least squares computer program in which derivatives are determined numerically and an option is available of "constrained iteration" to prevent instability (see Chapter 7). It is then necessary only to program the computation of a_t itself.

The covariance matrix of the estimates may be obtained from the converged value of the matrix $(\boldsymbol{X}'_{\hat{\boldsymbol{\beta}}}\boldsymbol{X}_{\hat{\boldsymbol{\beta}}})^{-1}\hat{\sigma}_a^2$ as described in Section 7.2.2; in addition, the least squares estimates $\hat{\boldsymbol{\beta}}$ have been shown to have a multivariate Normal asymptotic distribution (e.g., Pierce [156], Reinsel [166]). If b, which is an integer, needs to be estimated, the iteration may be run to convergence for a series of values of b and that value of b giving the minimum sum of squares, selected. One special feature (see, e.g., Pierce [156]) of the covariance matrix of the least squares estimates $\hat{\boldsymbol{\beta}}$ is that it will be approximately a block diagonal matrix whose two blocks on the diagonal consist of the covariance matrices of the parameters $(\hat{\delta}_1, \ldots, \hat{\delta}_r, \hat{\omega}_0, \ldots, \hat{\omega}_s)'$ and $(\hat{\phi}_1, \ldots, \hat{\phi}_p, \hat{\theta}_1, \ldots, \hat{\theta}_q)'$, respectively. Thus the parameter estimates of the transfer function part of the model are approximately uncorrelated with the estimates of the noise part of the model, which results from the assumed independence between the input X_t and the white noise a_t in the model.

More exact sum of squares and exact likelihood function methods could also be employed in the estimation of the transfer function–noise models (11.3.1) and (11.3.2), as in the case of the ARMA models discussed in Chapter 7 (see, e.g., Newbold [148]). This is especially so for dealing with the noise part of the model, where for given values of b, $\boldsymbol{\delta}$, and $\boldsymbol{\omega}$, the noise series n_t can be constructed as in (11.3.5), and then the exact sum of squares and exact likelihood methods for ARMA(p,q) models can be applied directly to the n_t values for $t = u + 1, \ldots, n$. The state space model–Kalman filtering approach to the exact likelihood evaluation, as in Section 7.5, could also be used. However, for moderate and large n and nonseasonal data, there will generally be little difference between the conditional and exact methods.

11.3.3 Use of Residuals for Diagnostic Checking

Serious model inadequacy can usually be detected by examining:

1. The autocorrelation function $r_{\hat{a}\hat{a}}(k)$ of the residuals $\hat{a}_t = a_t(\hat{b}, \hat{\boldsymbol{\delta}}, \hat{\boldsymbol{\omega}}, \hat{\boldsymbol{\phi}}, \hat{\boldsymbol{\theta}})$ from the fitted model.
2. Certain cross correlation functions involving input and residuals: in particular, the cross correlation function $r_{\alpha\hat{a}}(k)$ between prewhitened input α_t and the residuals $\hat{a}_t$.

Suppose, if necessary after suitable differencing, that the model can be written

$$\begin{aligned} y_t &= \delta^{-1}(B)\omega(B)x_{t-b} + \phi^{-1}(B)\theta(B)a_t \\ &= v(B)x_t + \psi(B)a_t \end{aligned} \tag{11.3.13}$$

Now, suppose that we select an incorrect model leading to residuals a_{0t}, where

$$y_t = v_0(B)x_t + \psi_0(B)a_{0t}$$

Then

$$a_{0t} = \psi_0^{-1}(B)[v(B) - v_0(B)]x_t + \psi_0^{-1}(B)\psi(B)a_t \tag{11.3.14}$$

whence it is apparent in general that if a wrong model is selected, the a_{0t}'s will be autocorrelated and the a_{0t}'s will be cross correlated with the x_t's and hence with the α_t's, which generate the x_t's.

Now consider what happens in two special cases: (1) when the transfer function model is correct but the noise model is incorrect, and (2) when the transfer function model is incorrect.

Transfer function model correct, noise model incorrect. If $v_0(B) = v(B)$ but $\psi_0(B) \neq \psi(B)$, then (11.3.14) becomes

$$a_{0t} = \psi_0^{-1}(B)\psi(B)a_t \tag{11.3.15}$$

Therefore, the a_{0t}'s would *not* be cross correlated with x_t's or with α_t's. However, the a_{0t} process would be autocorrelated, and the form of the autocorrelation function could indicate appropriate modification of the noise structure.

Transfer function model incorrect. From (11.3.14) it is apparent that if the transfer function model were incorrect, not only would the a_{0t}'s be cross correlated with the x_t's (and α_t's), but *also the a_{0t}'s would be autocorrelated.* This would be true even if the noise model were correct, for then (11.3.14) would become

$$a_{0t} = \psi^{-1}(B)[v(B) - v_0(B)]x_t + a_t \tag{11.3.16}$$

Whether or not the noise model was correct, a cross correlation analysis could indicate the modifications needed in the transfer function model. This aspect is clarified by considering the model after prewhitening. If the output and the input are assumed to be transformed so that the input is white noise, then, as in (11.2.8), we may write the model as

$$\beta_t = v(B)\alpha_t + \varepsilon_t$$

where $\beta_t = \theta_x^{-1}(B)\phi_x(B)y_t$, $\varepsilon_t = \theta_x^{-1}(B)\phi_x(B)n_t$. Now, consider the quantities

$$\varepsilon_{0t} = \beta_t - v_0(B)\alpha_t$$

Since $\varepsilon_{0t} = [v(B) - v_0(B)]\alpha_t + \varepsilon_t$, arguing as in Section 11.2.1, the cross correlations between the ε_{0t}'s and the α_t's measure the discrepancy between the correct and incorrect impulse functions. Specifically, as in (11.2.11),

$$v_k - v_{0k} = \frac{\rho_{\alpha\varepsilon_0}(k)\sigma_{\varepsilon_0}}{\sigma_\alpha} \qquad k = 0, 1, 2, \ldots \tag{11.3.17}$$

11.3.4 Specific Checks Applied to the Residuals

In practice, we do not know the process parameters exactly but must apply our checks to residuals $\hat{a}_t$ computed after least squares fitting. Even if the functional form of the fitted model were adequate, the parameter estimates would differ somewhat from the true values and the distribution of the correlations of the residual $\hat{a}_t$'s would also differ to some extent from that of the autocorrelations of the a_t's. Therefore, some caution is necessary in using the results of the previous sections to suggest the behavior of residual correlations. The brief discussion which follows is based on a more detailed study given in [157].

Autocorrelation checks. Suppose that, a transfer function–noise model having been fitted by least squares and the residual $\hat{a}_t$'s calculated by substituting least squares estimates for the parameters, the estimated autocorrelation function $r_{\hat{a}\hat{a}}(k)$ of these residuals is computed. Then, as we have seen:

1. If the autocorrelation function $r_{\hat{a}\hat{a}}(k)$ shows marked correlation patterns, this suggests model inadequacy.
2. If the cross correlation checks do not indicate inadequacy of the transfer function model, the inadequacy is probably in the fitted noise model $n_t = \psi_0(B)a_t$.

In the latter case, identification of a subsidiary model

$$\hat{a}_{0t} = \mathrm{T}(B)a_t$$

to represent the correlation of the residuals from the primary model can, in accordance with (11.3.15), indicate roughly the form

$$n_t = \psi_0(B)\mathrm{T}(B)a_t$$

to take for the modified noise model. However, in making assessments of whether an apparent discrepancy of estimated autocorrelations from zero is, or is not, likely to point to a nonzero theoretical value, certain facts must be borne in mind analogous to those discussed in Section 8.2.1.

Suppose that after allowing for starting values, $m = n - u - p$ values of the $\hat{a}_t$'s are actually available for this computation. Then if the model was correct in functional form and the *true parameter values were substituted*, the residuals would be white noise and the estimated autocorrelations would be distributed mutually independently about zero with variance $1/m$. When estimates are substituted for the parameter values, the distributional properties of the correlations at low lag are affected. In particular, the variance of these estimated low lag correlations can be considerably less than $1/m$, and the values can be highly correlated. Thus, with k small, comparison of an estimated autocorrelation $r_{\hat{a}\hat{a}}(k)$ with a "standard error" $1/\sqrt{m}$ could greatly underestimate its significance. Also, ripples in the estimated autocorrelation function at low lags can arise simply because of the high induced correlation between these estimates. If the amplitude of such low lag ripples is small compared with $1/\sqrt{m}$, they could have arisen by chance alone and need not be indicative of some real pattern in the theoretical autocorrelations.

A helpful overall check, which takes account of these distributional effects produced by fitting, is as follows. Consider the first K estimated autocorrelations $r_{\hat{a}\hat{a}}(1), \ldots, r_{\hat{a}\hat{a}}(K)$ and let K be taken sufficiently large so that if the model is written as $y_t = v(B)x_t + \psi(B)a_t$, the weights ψ_j can be expected to be negligible for $j > K$. Then if the functional form of the model is adequate, the quantity

$$Q = m \sum_{k=1}^{K} r_{\hat{a}\hat{a}}^2(k) \tag{11.3.18}$$

is approximately distributed as χ^2 with $K - p - q$ degrees of freedom. Note that the degrees of freedom in χ^2 depends on the number of parameters in the noise model but not on the number of parameters in the transfer function model. By referring Q to a table of percentage points of χ^2, we can obtain an approximate test of the hypothesis of model adequacy. However, the modified statistic $\tilde{Q} = m(m + 2) \sum_{k=1}^{K} (m - k)^{-1} r_{\hat{a}\hat{a}}^2(k)$, as defined in (8.2.3) of Section 8.2.2 for the ARIMA model situation, would be recommended for use in practice instead of (11.3.18) because $\tilde{Q}$ provides a closer approximation to the chi-squared distribution than Q under the null hypothesis of model adequacy.

Cross correlation check. As we have seen in Section 11.3.3:

1. A pattern of markedly nonzero cross correlations $r_{x\hat{a}}(k)$ suggests inadequacy of the transfer function model.
2. A somewhat different cross correlation analysis can suggest the *type* of modification needed. Specifically, if the fitted transfer function is $\hat{v}_0(B)$ and we consider the cross correlations between quantities $\hat{\varepsilon}_{0t} = \beta_t - \hat{v}_0(\mathrm{B})\alpha_t$ and α_t, rough estimates of the discrepancies $v_k - v_{0k}$ are given by

$$\frac{r_{\alpha\hat{\varepsilon}_0}(k)\, s_{\hat{\varepsilon}_0}}{s_\alpha}$$

Suppose that the model were of the correct functional form and *true* parameter values had been substituted. The residuals would be white noise uncorrelated with the x's and, using (11.1.11), the variance of the $r_{xa}(k)$ for an effective length of series m would be approximately $1/m$. However, unlike the autocorrelations $r_{aa}(k)$, these cross correlations will not be approximately uncorrelated. In general, if the x's are autocorrelated, so are the cross correlations $r_{xa}(k)$. In fact, as has been seen in (11.1.12), on the assumption that the x's and the a's have no cross correlation, the correlation coefficient between $r_{xa}(k)$ and $r_{xa}(k + l)$ is

$$\rho[r_{xa}(k), r_{xa}(k + l)] \simeq \rho_{xx}(l) \tag{11.3.19}$$

That is, approximately, the cross correlations have *the same* autocorrelation function as does the original input series x_t. Thus, when the x_t's are autocorrelated, a perfectly adequate transfer function model will give rise to cross correlations $r_{x\hat{a}}(k)$ which although small in magnitude may show *pronounced patterns*. This effect is eliminated if the check is made by computing cross correlations $r_{\alpha\hat{a}}(k)$ with the *prewhitened* input α_t.

As with the autocorrelations, when estimates are substituted for parameter values, the distributional properties of the autocorrelations are affected. However, a rough overall test of the hypothesis of model adequacy, similar to the autocorrelation test, can be obtained based on the sizes of the cross correlations. To employ the check, the cross correlations $r_{\alpha\hat{a}}(k)$ for $k = 0, 1, 2, \ldots, K$ between the input α_t in *prewhitened* form and the residuals $\hat{a}_t$ are estimated, and K is chosen sufficiently large so that the weights v_j and ψ_j in (11.3.13) can be expected to be negligible for $j > K$. The effects resulting from the use of estimated parameters in calculating residuals are, as before, principally confined to correlations of low order whose variances are considerably less than m^{-1} and which may be highly correlated even when the input is white noise.

However, it is true (see [157]) that

$$S = m \sum_{k=0}^{K} r_{\alpha\hat{a}}^2(k) \tag{11.3.20}$$

is approximately distributed as χ^2 with $K + 1 - (r + s + 1)$ degrees of freedom, where $(r + s + 1)$ is the number of parameters fitted in the transfer function model. Note that the number of degrees of freedom is independent of the number of parameters fitted in the noise model. Based on studies of the behavior of the Q statistic discussed in Chapter 8, the modified statistic, $\tilde{S} = m(m + 2) \sum_{k=0}^{K} (m - k)^{-1} r_{\alpha\hat{a}}^2(k)$, might be suggested for use in place of S in (11.3.20) because it may more accurately approximate the χ^2 distribution under the null model, although detailed investigations of its performance have not been made (however, see empirical results in Poskitt and Tremayne [219]).

11.4 SOME EXAMPLES OF FITTING AND CHECKING TRANSFER FUNCTION MODELS

11.4.1 Fitting and Checking of the Gas Furnace Model

We now illustrate the approach described in Section 11.3 to the fitting of the model

$$Y_t = \frac{\omega_0 - \omega_1 B - \omega_2 B^2}{1 - \delta_1 B - \delta_2 B^2} X_{t-3} + \frac{1}{1 - \phi_1 B - \phi_2 B^2} a_t$$

which was identified for the gas furnace data in Sections 11.2.2 and 11.2.3.

Nonlinear estimation. Using the initial estimates $\hat{\omega}_0 = -0.53$, $\hat{\omega}_1 = 0.33$, $\hat{\omega}_2 = 0.51$, $\hat{\delta}_1 = 0.57$, $\hat{\delta}_2 = 0.02$, $\hat{\phi}_1 = 1.54$, $\hat{\phi}_2 = -0.64$ derived in Sections 11.2.2 and 11.2.3 with the conditional least squares algorithm described in Section 11.3.2, least squares values, to two decimals, were achieved in four iterations. To test whether the iteration would converge in unfavorable circumstances, Table 11.4 shows how the iteration proceeded with all starting values taken to be either +0.1 or −0.1. The fact that, even then, convergence was achieved in 10 iterations with as many as seven parameters in the model is encouraging.

The last line in Table 11.4 shows the rough preliminary estimates obtained at the identification stage in Sections 11.2.2 and 11.2.3. It is seen that for this example, they are in close agreement with the least squares esti-

TABLE 11.4 Convergence of Nonlinear Least Squares Fit of Gas Furnace Data

Iteration	ω_0	ω_1	ω_2	δ_1	δ_2	ϕ_1	ϕ_2	Sum of Squares
0	0.10	−0.10	−0.10	0.10	0.10	0.10	0.10	13,601
1	−0.46	0.63	0.60	0.14	0.27	1.33	−0.27	273.1
2	−0.52	0.45	0.31	0.40	0.52	1.37	−0.43	92.5
3	−0.63	0.60	0.01	0.12	0.73	1.70	−0.76	31.8
4	−0.54	0.50	0.29	0.24	0.42	1.70	−0.81	19.7
5	−0.50	0.31	0.51	0.63	0.09	1.56	−0.68	16.84
6	−0.53	0.38	0.53	0.54	0.01	1.54	−0.64	16.60
7	−0.53	0.37	0.51	0.56	0.01	1.53	−0.63	16.60
8	−0.53	0.37	0.51	0.56	0.01	1.53	−0.63	16.60
9	−0.53	0.37	0.51	0.57	0.01	1.53	−0.63	16.60
Preliminary estimates	−0.53	0.33	0.51	0.57	0.02	1.54	−0.64	

mates given on the previous line. Thus the final fitted transfer function model is

$$(1 - 0.57B - 0.01B^2)Y_t = -(0.53 + 0.37B + 0.51B^2)X_{t-3} \qquad (11.4.1)$$
$$(\pm 0.21)\ (\pm 0.14) \qquad (\pm 0.08)(\pm 0.15)\ (\pm 0.16)$$

and the fitted noise model is

$$(1 - 1.53B + 0.63B^2)N_t = a_t \qquad (11.4.2)$$
$$(\pm 0.05)\ (\pm 0.05)$$

with $\hat{\sigma}_a^2 = 0.0561$, where the limits in parentheses are the ±one standard error limits obtained from the nonlinear estimation procedure.

Diagnostic checking. Before accepting the model above as an adequate representation of the system, autocorrelation and cross correlation checks should be applied, as described in Section 11.3.4. The first 36 lags of the residual autocorrelations are tabulated in Table 11.5, together with the upper bound $1/\sqrt{m}$ for their standard errors ($m = 289$) assuming that the model is adequate. There seems to be no evidence of model inadequacy from the behavior of individual autocorrelations. This is confirmed by calculating the $\tilde{Q}$ criterion, which is

$$\tilde{Q} = (289)(291)\ \Sigma_{k=1}^{36}\ (289 - k)^{-1} r_{\hat{a}\hat{a}}^2(k) = 43.8$$

Comparison of $\tilde{Q}$ with the χ^2 table for $K - p - q = 36 - 2 - 0 = 34$ degrees of freedom provides no grounds for questioning model adequacy.

The first 36 lags of the cross correlation function $r_{x\hat{a}}(k)$ between the input and estimated residuals are given in Table 11.6(a), together with the

TABLE 11.5 Estimated Autocorrelation Function $r_{\hat{a}\hat{a}}(k)$ of Residuals from Fitted Gas Furnace Model

Lag k	$r_{\hat{a}\hat{a}}(k)$												Upper Bound to Standard Error
1–12	0.02	0.06	−0.07	−0.05	−0.05	0.12	0.03	0.03	−0.08	0.05	0.02	0.10	±0.06
13–24	−0.04	0.05	−0.09	−0.01	−0.08	0.00	−0.12	0.00	−0.01	0.08	0.02	−0.01	±0.06
25–36	0.04	−0.02	0.02	0.09	−0.12	0.06	−0.03	−0.06	0.11	0.02	0.03	0.06	±0.06

TABLE 11.6 Estimated Cross Correlation Functions for Gas Furnace Data

Lag k													Upper Bound to Standard Error
	(a) $r_{x\hat{a}}(k)$ between the input and the output residuals												
0–11	0.00	0.00	0.00	0.00	0.00	0.00	−0.01	−0.02	−0.03	−0.05	−0.06	−0.05	±0.06
12–23	−0.03	−0.03	−0.03	−0.07	−0.10	−0.12	−0.12	−0.10	−0.04	−0.01	−0.01	−0.02	±0.06
24–35	−0.03	−0.04	−0.04	−0.02	−0.01	0.02	0.04	0.05	0.06	0.07	0.07	0.06	±0.06
	(b) $r_{\alpha\hat{a}}(k)$ between the prewhitened input and the output residuals												
0–11	−0.06	0.03	−0.01	0.00	0.01	0.01	0.01	−0.04	0.02	0.07	−0.03	−0.02	±0.06
12–23	−0.03	−0.11	0.02	0.04	0.04	0.01	0.01	−0.15	−0.03	−0.07	−0.08	0.02	±0.06
24–35	−0.01	0.02	0.05	−0.07	0.00	0.04	−0.15	0.04	0.03	−0.02	0.00	−0.03	±0.06

upper bound $1/\sqrt{m}$ for their standard errors. It will be seen that although the cross correlations $r_{x\hat{a}}(k)$ are not especially large compared with the upper bound of their standard errors, they are themselves highly autocorrelated. This is to be expected because as indicated by (11.3.19), the cross correlations follow the same stochastic process as does the input x_t, and as we have already seen, for this example the input was highly autocorrelated.

The corresponding cross correlations between $\hat{a}_t$ and the prewhitened input α_t are given in Table 11.6(b). The $\tilde{S}$ criterion yields

$$\tilde{S} = (289)(291)\ \Sigma_{k=0}^{35}\ (289 - k)^{-1} r_{\alpha\hat{a}}^2(k) = 32.1$$

Comparison of $\tilde{S}$ with the χ^2 table for $K + 1 - (r + s + 1) = 36 - 5 = 31$ degrees of freedom again provides no evidence that the model is inadequate.

Step and impulse responses. The estimate $\hat{\delta}_2 = 0.01$ in (11.4.1) is very small compared with its standard error ± 0.14, and the parameter δ_2 can in fact be omitted from the model without affecting the estimates of the remaining parameters to the accuracy considered. The final form of the combined transfer function–noise model for the gas furnace data is

$$Y_t = \frac{-(0.53 + 0.37B + 0.51B^2)}{1 - 0.57B} X_{t-3} + \frac{1}{1 - 1.53B + 0.63B^2} a_t$$

The step and impulse response functions corresponding to the transfer function model

$$(1 - 0.57B)Y_t = -(0.53 + 0.37B + 0.51B^2)X_{t-3}$$

are given in Figure 11.6. Using (10.2.5), the steady-state gain of the coded data is

$$g = \frac{-(0.53 + 0.37 + 0.51)}{1 - 0.57} = -3.3$$

The results agree very closely with those obtained by cross spectral analysis [122].

Choice of sampling interval. When a choice is available, the sampling interval should be taken as fairly short compared with the time constants expected for the system. When in doubt the analysis can be repeated with several trial sampling intervals. In the choice of sampling interval it is the noise at the output that is important, and its variance should approach a minimum value as the interval is shortened. Thus in the gas furnace example that we have used for illustration, a pen recorder was used to provide a continuous record of input and output. The discrete data that we have actually analyzed were obtained by reading off values from this continuous record at points separated by 9-second intervals. This interval was chosen

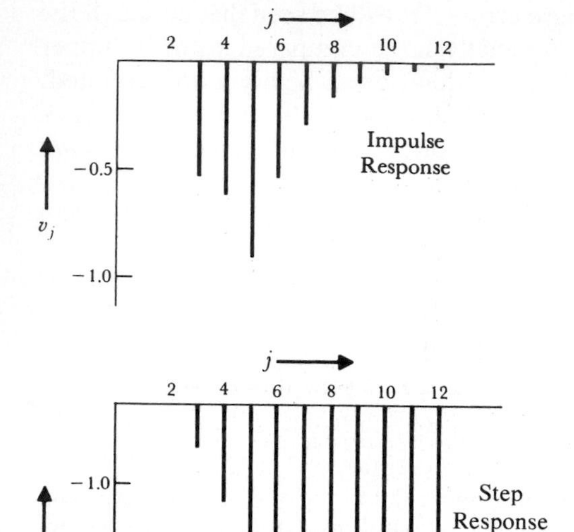

Figure 11.6 Impulse and step responses for transfer function model $(1 - 0.57B)Y_t = -(0.53 + 0.37B + 0.51B^2)X_{t-3}$ fitted to coded gas furnace data.

because inspection of the traces shown in Figure 11.1 suggested that it ought to be adequate to allow all the variation (apart from slight pen chatter) which occurred in input and output to be taken account of. The use of this kind of common sense is usually a reliable guide in choosing the interval. The estimated mean square error for the gas furnace data (obtained by dividing $\Sigma(Y - \hat{Y})^2$ by the appropriate number of degrees of freedom, with $\hat{Y}$ the fitted value) is shown for various time intervals in Table 11.7. These values are also plotted in Figure 11.7. Little change in mean square error occurs until the interval is almost 40 seconds, when a very rapid rise occurs. There is little difference in the mean square error, or indeed the plotted step response, for the 9-, 18-, and 27-second intervals, but a considerable change

TABLE 11.7 Mean Square Error at the Output for Various Choices of the Sampling Interval

	Interval (seconds)						
	9	18	27	36	45	54	72
Number of data points N	296	148	98	74	59	49	37
M.S. error	0.71	0.78	0.74	0.95	0.97	1.56	7.11

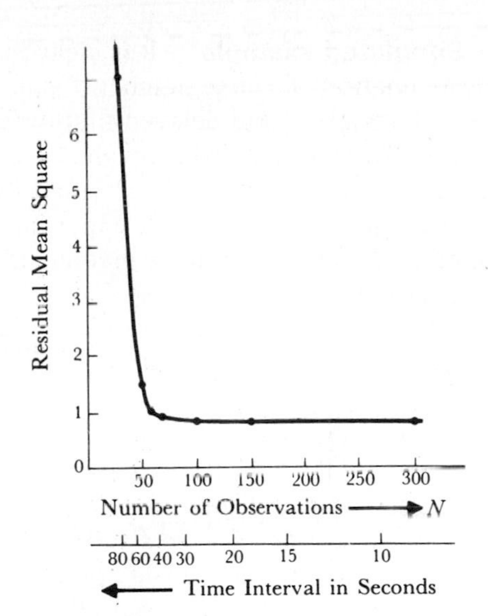

Figure 11.7 Mean square error at the output for various choices of sampling interval.

occurs when the 36-second interval is used. It will be seen that the 9-second interval we have used in this example is, in fact, conservative.

11.4.2 Simulated Example with Two Inputs

The fitting of models involving more than one input series involves no difficulty in principle, except for the increase in the number of parameters that has to be handled. For example, for two inputs we can write the model as

$$y_t = \delta_1^{-1}(B)\omega_1(B)x_{1,t-b_1} + \delta_2^{-1}(B)\omega_2(B)x_{2,t-b_2} + n_t$$

with

$$n_t = \phi^{-1}(B)\theta(B)a_t$$

where $y_t = \nabla^d Y_t$, $x_{1,t} = \nabla^d X_{1,t}$, $x_{2,t} = \nabla^d X_{2,t}$ and $n_t = \nabla^d N_t$ are stationary processes. To compute the a_t's, we first calculate for specified values of the parameters b_1, $\boldsymbol{\delta}_1$, $\boldsymbol{\omega}_1$,

$$\mathscr{Y}_{1,t} = \delta_1^{-1}(B)\omega_1(B)x_{1,t-b_1} \tag{11.4.3}$$

and for specified values of b_2, $\boldsymbol{\delta}_2$, $\boldsymbol{\omega}_2$,

$$\mathscr{Y}_{2,t} = \delta_2^{-1}(B)\omega_2(B)x_{2,t-b_2} \tag{11.4.4}$$

Then the noise n_t can be calculated from

$$n_t = y_t - \mathscr{Y}_{1,t} - \mathscr{Y}_{2,t} \tag{11.4.5}$$

and finally, a_t from

$$a_t = \theta^{-1}(B)\phi(B)n_t \tag{11.4.6}$$

Simulated example. It is clear that even simple situations can lead to the estimation of a large number of parameters. The example below, with two input variables and delayed first-order models, has eight unknown parameters. To determine whether the iterative nonlinear least squares procedure described in Section 11.3.2 could be used to obtain estimates of the parameters in such models, an experiment was performed using manufactured data, details of which are given in [51]. The data were generated from the model written in ∇ form as

$$Y_t = \beta + g_1 \frac{1 + \eta_1 \nabla}{1 + \xi_1 \nabla} X_{1,t-1} + g_2 \frac{1 + \eta_2 \nabla}{1 + \xi_2 \nabla} X_{2,t-1} + \frac{1}{1 - \phi_1 B} a_t \tag{11.4.7}$$

with $\beta = 60$, $g_1 = 13.0$, $\eta_1 = -0.6$, $\xi_1 = 4.0$, $g_2 = -5.5$, $\eta_2 = -0.6$, $\xi_2 = 4.0$, $\phi_1 = 0.5$, and $\sigma_a^2 = 9.0$. The input variables X_1 and X_2 were changed according to a randomized 2^2 factorial design replicated three times. Each input condition was supposed held fixed for 5 minutes and output observations taken every minute. The data are plotted in Figure 11.8 and appear as series K in the "Collection of Time Series" in Part Five.

The constrained iterative nonlinear least squares program, described in Chapter 7, was used to obtain the least squares estimates, so that it was only necessary to calculate the a's. Thus for specified values of the parameters $g_1, g_2, \xi_1, \xi_2, \eta_1, \eta_2$, the values $\mathcal{Y}_{1,t}$ and $\mathcal{Y}_{2,t}$ can be obtained from

$$(1 + \xi_1 \nabla)\mathcal{Y}_{1,t} = g_1(1 + \eta_1 \nabla)X_{1,t-1}$$

$$(1 + \xi_2 \nabla)\mathcal{Y}_{2,t} = g_2(1 + \eta_2 \nabla)X_{2,t-1}$$

and can be used to calculate

$$N_t = Y_t - \mathcal{Y}_{1,t} - \mathcal{Y}_{2,t}$$

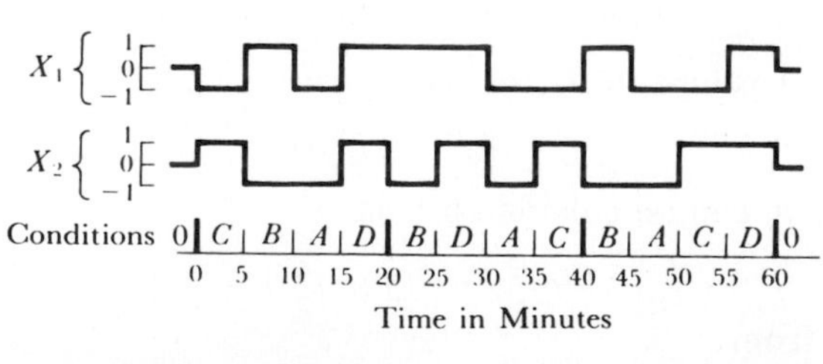

Figure 11.8 Data for simulated two-input example (series K).

TABLE 11.8 Convergence of Nonlinear Least Squares Procedure for Two Inputs Using Guessed Initial Estimates

Itera-tion	β	g_1	g_2	η_1	η_2	ξ_1	ξ_2	ϕ_1	Sum of Squares
0	59.19	10.00	−7.00	−0.50	−0.50	1.00	1.00	0.10	2,046.8
1	59.20	9.07	−6.37	−0.58	−0.56	1.33	1.31	0.24	1,085.4
2	59.24	8.38	−5.35	−0.70	−0.59	2.03	1.75	0.39	621.5
3	59.35	9.24	−3.98	−0.75	−0.55	3.45	1.95	0.36	503.5
4	59.41	11.90	−3.40	−0.75	−0.56	5.21	1.66	0.22	463.7
5	59.39	12.03	−3.52	−0.80	−0.57	4.99	1.76	0.21	461.8
6	59.39	12.08	−3.53	−0.79	−0.56	5.03	1.77	0.21	461.8
7	59.39	12.07	−3.53	−0.79	−0.56	5.03	1.77	0.21	461.8
8	59.39	12.07	−3.53	−0.79	−0.56	5.03	1.77	0.21	461.8

Finally, for specified values of ϕ_1, a_t can be calculated from

$$a_t = N_t - \phi_1 N_{t-1}$$

It was assumed that the process inputs had been maintained at their center conditions for some time before the start of the experiment, so that $\mathcal{Y}_{1,t}$, $\mathcal{Y}_{2,t}$, and hence N_t, may be computed from $t = 0$ onward and a_t from $t = 1$.

Two runs were made of the nonlinear least squares procedure using two different sets of initial values. In the first the parameters were chosen as representing what a person reasonably familiar with the process might guess for initial values. In the second, the starting value for β was chosen to be the mean $\overline{Y}$ of all observations and all other starting values were set equal to 0.1. Thus the second run represents a much more extreme situation than would normally arise in practice. Table 11.8 shows that with the first set of initial values, convergence occurs after five iterations, and Table 11.9 shows

TABLE 11.9 Convergence of Nonlinear Least Squares Procedure for Two Inputs Using Extreme Initial Values

Itera-tion	β	g_1	g_2	η_1	η_2	ξ_1	ξ_2	ϕ_1	Sum of Squares
0	59.19	0.10	0.10	0.10	0.10	0.10	0.10	0.10	2,496.4
1	59.19	0.24	−0.07	−1.57	0.48	1.77	−0.28	0.15	2,190.5
2	59.22	1.62	−0.29	−2.09	−2.24	−0.07	0.26	0.29	1,473.6
3	59.21	1.80	−0.77	−1.75	0.58	0.20	−0.10	0.56	1,016.8
4	59.21	3.01	−1.31	−1.15	−0.83	0.91	0.22	0.72	743.1
5	59.31	6.17	−2.82	−0.93	−0.65	3.03	1.20	0.67	611.4
6	59.61	15.83	−3.25	−0.70	−0.66	8.88	1.64	0.26	534.2
7	59.47	10.31	−3.48	−0.74	−0.56	3.52	1.63	0.23	501.9
8	59.41	11.89	−3.41	−0.74	−0.58	5.01	1.65	0.20	462.8
9	59.39	12.07	−3.52	−0.79	−0.57	5.04	1.76	0.21	461.8
10	59.39	12.07	−3.53	−0.79	−0.56	5.03	1.77	0.21	461.8
11	59.39	12.07	−3.53	−0.79	−0.56	5.03	1.77	0.21	461.8

that convergence with the second set occurs after nine iterations. These results suggest that in realistic circumstances, multiple inputs can be handled without serious estimation difficulties.

11.5 FORECASTING USING LEADING INDICATORS

Frequently, forecasts of a time series $Y_t, Y_{t-1}, \ldots$ may be considerably improved by using information coming from some associated series $X_t, X_{t-1}, \ldots$. This is particularly true if changes in Y tend to be *anticipated* by changes in X, in which case economists call X a "leading indicator" for Y.

To obtain an optimal forecast using information from both Y and X, we first build a transfer function–noise model connecting Y and X in the manner already outlined. Suppose, using previous notations, that an adequate model is

$$Y_t = \delta^{-1}(B)\omega(B)X_{t-b} + \varphi^{-1}(B)\theta(B)a_t \qquad b \geqslant 0 \tag{11.5.1}$$

In general, the noise component of this model, which is assumed statistically independent of the input X_t, is nonstationary with

$$\varphi(B) = \phi(B)\nabla^d$$

so that if

$$\nabla^d Y_t = y_t \quad \text{and} \quad \nabla^d X_t = x_t$$

$$y_t = \delta^{-1}(B)\omega(B)x_{t-b} + \phi^{-1}(B)\theta(B)a_t$$

Also, we shall assume that an adequate stochastic model for the leading series is

$$X_t = \varphi_x^{-1}(B)\theta_x(B)\alpha_t \tag{11.5.2}$$

so that with

$$\varphi_x(B) = \phi_x(B)\nabla^d$$

$$x_t = \phi_x^{-1}(B)\theta_x(B)\alpha_t$$

11.5.1 Minimum Mean Square Error Forecast

Now (11.5.1) may be written

$$Y_t = \nu(B)\alpha_t + \psi(B)a_t \tag{11.5.3}$$

with the a's and the α's statistically independent, and $\nu(B) = \delta^{-1}(B)\omega(B)B^b\varphi_x^{-1}(B)\theta_x(B)$. Arguing as in Section 5.1.1, suppose that the forecast $\hat{Y}_t(l)$ of Y_{t+l} made at origin t is of the form

$$\hat{Y}_t(l) = \sum_{j=0}^{\infty} \nu_{l+j}^0 \alpha_{t-j} + \sum_{j=0}^{\infty} \psi_{l+j}^0 a_{t-j}$$

Then

$$Y_{t+l} - \hat{Y}_t(l) = \sum_{i=0}^{l-1} (\nu_i \alpha_{t+l-i} + \psi_i a_{t+l-i})$$

$$+ \sum_{j=0}^{\infty} [(\nu_{l+j} - \nu_{l+j}^0)\alpha_{t-j} + (\psi_{l+j} - \psi_{l+j}^0)a_{t-j}]$$

and

$$E[Y_{t+l} - \hat{Y}_t(l)]^2 = (\nu_0^2 + \nu_1^2 + \cdots + \nu_{l-1}^2)\sigma_\alpha^2 + (1 + \psi_1^2 + \cdots + \psi_{l-1}^2)\sigma_a^2$$

$$+ \sum_{j=0}^{\infty} [(\nu_{l+j} - \nu_{l+j}^0)^2\sigma_\alpha^2 + (\psi_{l+j} - \psi_{l+j}^0)^2\sigma_a^2]$$

which is minimized only if $\nu_{l+j}^0 = \nu_{l+j}$ and $\psi_{l+j}^0 = \psi_{l+j}$. Thus the minimum mean square error forecast $\hat{Y}_t(l)$ of Y_{t+l} at origin t is given by the conditional expectation of Y_{t+l} at time t. Theoretically, this expectation is conditional on knowledge of the series from the infinite past up to the present origin t. As in Chapter 5, such results are of practical use because, usually, the forecasts depend appreciably only on *recent* past values of X and Y.

Computation of the forecast. Now (11.5.1) may be written

$$\varphi(B)\delta(B)Y_t = \varphi(B)\omega(B)X_{t-b} + \delta(B)\theta(B)a_t$$

which we shall write as

$$\delta^*(B)Y_t = \omega^*(B)X_{t-b} + \theta^*(B)a_t$$

Then, using square brackets to denote conditional expectations at time t, and writing $p^* = p + d$, we have for the lead l forecast

$$\begin{aligned} \hat{Y}_t(l) = [Y_{t+l}] = \delta_1^*[Y_{t+l-1}] + \cdots + \delta_{p^*+r}^*[Y_{t+l-p^*-r}] + \omega_0^*[X_{t+l-b}] \\ - \omega_{p^*+s}^*[X_{t+l-b-p^*-s}] + [a_{t+l}] - \theta_1^*[a_{t+l-1}] \\ - \cdots - \theta_{q+r}^*[a_{t+l-q-r}] \end{aligned} \qquad (11.5.4)$$

where

$$[Y_{t+j}] = \begin{cases} Y_{t+j} & j \leqslant 0 \\ \hat{Y}_t(j) & j > 0 \end{cases}$$

$$[X_{t+j}] = \begin{cases} X_{t+j} & j \leqslant 0 \\ \hat{X}_t(j) & j > 0 \end{cases} \qquad (11.5.5)$$

$$[a_{t+j}] = \begin{cases} a_{t+j} & j \leqslant 0 \\ 0 & j > 0 \end{cases}$$

and a_t is calculated from (11.5.1), which if $b \geqslant 1$, is equivalent to

$$a_t = Y_t - \hat{Y}_{t-1}(1)$$

Thus, by appropriate substitutions, the minimum mean square error forecast is readily computed directly using (11.5.4) and (11.5.5). The forecasts $\hat{X}_t(j)$ are obtained in the usual way (see Section 5.2) utilizing the model (11.5.2).

It is important to note that the conditional expectations in (11.5.4) and (11.5.5) are taken with respect to values in *both* series Y_t and X_t through time t, but because of the assumed independence between input X_t and noise N_t in (11.5.1), it follows in particular that we will have

$$\hat{X}_t(j) = E[X_{t+j}|X_t, X_{t-1}, \ldots, Y_t, Y_{t-1}, \ldots] = E[X_{t+j}|X_t, X_{t-1}, \ldots]$$

That is, given the past values of the input series X_t, the optimal forecasts of its future values depend only on the past X's and cannot be improved by the additional knowledge of the past Y's; hence the optimal values $\hat{X}_t(j)$ can be obtained directly from the univariate model (11.5.2).

Variance of the forecast. The ν weights and the ψ weights of (11.5.3) may be obtained explicitly by equating coefficients in

$$\delta(B)\varphi_x(B)\nu(B) = \omega(B)\theta_x(B)B^b$$

and in

$$\varphi(B)\psi(B) = \theta(B)$$

The variance of the lead l forecast error is then given by

$$V(l) = E[Y_{t+l} - \hat{Y}_t(l)]^2 = \sigma_\alpha^2 \sum_{j=b}^{l-1} \nu_j^2 + \sigma_a^2 \sum_{j=0}^{l-1} \psi_j^2 \qquad (11.5.6)$$

Forecasts as a weighted aggregate of previous observations. For any given example, it is instructive to consider precisely how the forecasts of future values of Y utilize the previous values of the X and Y series. We have seen in Section 5.3.3 how the forecasts may be written as linear aggregates of previous values of the series. Thus for forecasts of the leading indicator, we could write

$$\hat{X}_t(l) = \sum_{j=1}^{\infty} \pi_j^{(l)} X_{t-j+1} \qquad (11.5.7)$$

The weights $\pi_j^{(1)} = \pi_j$ arise when the model (11.5.2) is written in the form

$$\alpha_t = X_t - \pi_1 X_{t-1} - \pi_2 X_{t-2} - \cdots$$

and may thus be obtained by explicitly equating coefficients in

$$\varphi_x(B) = (1 - \pi_1 B - \pi_2 B^2 - \cdots)\theta_x(B)$$

Also, using (5.3.9),

$$\pi_j^{(l)} = \pi_{j+l-1} + \sum_{h=1}^{l-1} \pi_h \pi_j^{(l-h)} \tag{11.5.8}$$

In a similar way, we can write the transfer function model (11.5.1) in the form

$$a_t = Y_t - \sum_{j=1}^{\infty} P_j Y_{t-j} - \sum_{j=1}^{\infty} Q_j X_{t-j} \qquad b > 0 \tag{11.5.9}$$

[It should be noted that if the transfer function between the leading indicator X and the output Y is such that $\nu_j = 0$ for $j < b$, then $Q_1, Q_2, \ldots, Q_{b-1}$ in (11.5.9) will be zero.]

Now (11.5.9) may be written

$$a_t = \left(1 - \sum_{j=1}^{\infty} P_j B^j\right) Y_t - \sum_{j=1}^{\infty} Q_j B^j X_t$$

Comparison with (11.5.1) shows that the P and Q weights may be obtained by equating coefficients in the expressions

$$\theta(B)\left(1 - \sum_{j=1}^{\infty} P_j B^j\right) = \varphi(B)$$

$$\theta(B)\delta(B) \sum_{j=1}^{\infty} Q_j B^j = \varphi(B)\omega(B)B^b$$

On substituting $t + l$ for t in (11.5.9), and taking conditional expectations at origin t, we have the lead l forecast in the form

$$\hat{Y}_t(l) = \sum_{j=1}^{\infty} P_j [Y_{t+l-j}] + \sum_{j=1}^{\infty} Q_j [X_{t+l-j}] \tag{11.5.10}$$

Now the lead 1 forecast is

$$\hat{Y}_t(1) = \sum_{j=1}^{\infty} P_j Y_{t+1-j} + \sum_{j=1}^{\infty} Q_j X_{t+1-j}$$

Also, the quantities in square brackets in (11.5.10) are either known values of the X and Y series or forecasts which are linear functions of these known values.

Thus we can write the forecast in terms of the values of the series which have already occurred at time t in the form

$$\hat{Y}_t(l) = \sum_{j=1}^{\infty} P_j^{(l)} Y_{t+1-j} + \sum_{j=1}^{\infty} Q_j^{(l)} X_{t+1-j} \tag{11.5.11}$$

where the coefficients $P_j^{(l)}$, $Q_j^{(l)}$ may be computed recursively as follows:

$$P_j^{(1)} = P_j \qquad Q_j^{(1)} = Q_j$$

$$P_j^{(l)} = P_{j+l-1} + \sum_{h=1}^{l-1} P_h P_j^{(l-h)} \tag{11.5.12}$$

$$Q_j^{(l)} = Q_{j+l-1} + \sum_{h=1}^{l-1} \{P_h Q_j^{(l-h)} + Q_h \pi_j^{(l-h)}\}$$

11.5.2 Forecast of CO_2 Output from Gas Furnace

For illustration, consider the gas furnace data shown in Figure 11.1. For this example, the fitted model (see Section 11.4.1) was

$$Y_t = \frac{-(0.53 + 0.37B + 0.51B^2)}{1 - 0.57B} X_{t-3} + \frac{a_t}{1 - 1.53B + 0.63B^2}$$

and $(1 - 1.97B + 1.37B^2 - 0.34B^3)X_t = \alpha_t$. The forecast function, written in the form (11.5.4), is thus

$$\begin{aligned} \hat{Y}_t(l) = [Y_{t+l}] = {} & 2.1[Y_{t+l-1}] - 1.5021[Y_{t+l-2}] + 0.3591[Y_{t+l-3}] \\ & - 0.53[X_{t+l-3}] + 0.4409[X_{t+l-4}] - 0.2778[X_{t+l-5}] \\ & + 0.5472[X_{t+l-6}] - 0.3213[X_{t+l-7}] \\ & + [a_{t+l}] - 0.57[a_{t+l-1}] \end{aligned}$$

Figure 11.9 shows the forecasts for lead times $l = 1, 2, \ldots, 12$ made at origin $t = 206$. The π, P, and Q weights for the model arc given in Table 11.10.

TABLE 11.10 π, P, and Q **Weights for Gas Furnace Model**

j	π_j	P_j	Q_j	j	π_j	P_j	Q_j
1	1.97	1.53	0	7	0	0	−0.07
2	−1.37	−0.63	0	8	0	0	−0.04
3	0.34	0	−0.53	9	0	0	−0.02
4	0	0	0.14	10	0	0	−0.01
5	0	0	−0.20	11	0	0	−0.01
6	0	0	0.43				

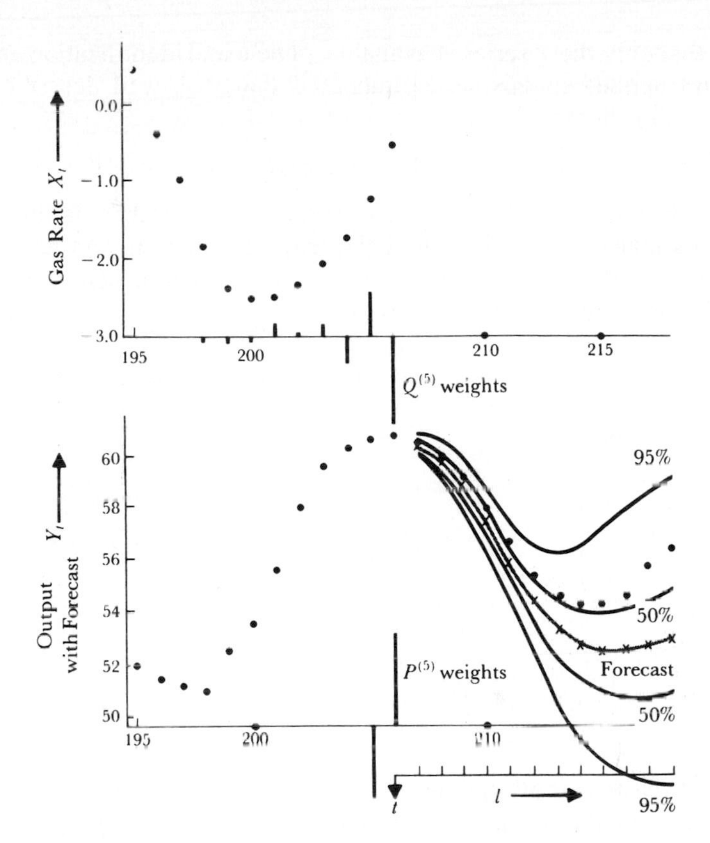

Figure 11.9 Forecast of CO_2 output from a gas furnace using input and output series.

Figure 11.9 shows the weights $P_j^{(5)}$ and $Q_j^{(5)}$ appropriate to the lead 5 forecast. The weights ν_i and ψ_i of (11.5.3) are listed in Table 11.11. Using estimates $\hat{\sigma}_\alpha^2 = 0.0353$ and $\hat{\sigma}_a^2 = 0.0561$, obtained in Sections 11.2.2 and 11.4.1, respectively, (11.5.6) may be employed to obtain variances of the forecast errors and the 50% and 95% probability limits shown in Figure 11.9.

To illustrate the advantages of using a leading indicator in forecasting,

TABLE 11.11 ν **and** ψ **Weights for Gas Furnace Model**

i	ν_i	ψ_i	i	ν_i	ψ_i
0	0	1	6	−5.33	0.89
1	0	1.53	7	−6.51	0.62
2	0	1.71	8	−6.89	0.39
3	−0.53	1.65	9	−6.57	0.20
4	−1.72	1.45	10	−5.77	0.06
5	−3.55	1.18	11	−4.73	−0.03

assume that only the Y series is available. The usual identification and fitting procedure applied to this series indicated that it is well described by an ARMA(4, 2) process,

$$(1 - 2.42B + 2.38B^2 - 1.16B^3 + 0.23B^4)Y_t = (1 - 0.31B + 0.47B^2)\varepsilon_t$$

with $\hat{\sigma}_\varepsilon^2 = 0.1081$. Table 11.12 shows estimated standard deviations of forecast errors made with and without the leading indicator. As might be expected, for short lead times use of the leading indicator can produce forecasts of considerably greater accuracy.

TABLE 11.12 Estimated Standard Deviations of Forecast Errors Made with and without the Leading Indicator

l	With leading indicator	Without leading indicator	l	With leading indicator	Without leading indicator
1	0.23	0.33	7	1.52	2.74
2	0.43	0.77	8	1.96	2.86
3	0.59	1.30	9	2.35	2.95
4	0.72	1.82	10	2.65	3.01
5	0.86	2.24	11	2.87	3.05
6	1.12	2.54	12	3.00	3.08

Univariate modeling check. To further confirm the univariate modeling results for the series Y_t, we can use results from Appendix A4.3 to obtain the nature of the univariate ARIMA model for Y_t that is implied by the transfer function–noise model between Y_t and X_t and the univariate AR(3) model for X_t. These models imply that

$$\begin{aligned}(1 - 0.57B)(1 - 1.53B + 0.63B^2)Y_t \\ = -(0.53 + 0.37B + 0.51B^2)(1 - 1.53B + 0.63B^2)X_{t-3} \\ + (1 - 0.57B)a_t \qquad (11.5.13)\end{aligned}$$

But since

$$\varphi_X(B) = 1 - 1.97B + 1.37B^2 - 0.34B^2 \simeq (1 - 1.46B + 0.60B^2)(1 - 0.52B)$$

in the AR(3) model for X_t, the right-hand side of (11.5.13) reduces approximately to $-(0.53 + 0.37B + 0.51B^2)(1 - 0.52B)^{-1}\alpha_{t-3} + (1 - 0.57B)a_t$, and hence we obtain

$$\begin{aligned}(1 - 0.52B)(1 - 0.57B)(1 - 1.53B + 0.63B^2)Y_t \\ = -(0.53 + 0.37B + 0.51B^2)\alpha_{t-3} + (1 - 0.52B)(1 - 0.57B)a_t\end{aligned}$$

The results of Appendix A4.3 imply that the right-hand side of this last equation has an MA(2) model representation as $(1 - \theta_1 B - \theta_2 B^2)\varepsilon_t$, with nonzero autocovariances $\gamma_0 = 0.1516$, $\gamma_1 = -0.0657$, and $\gamma_2 = 0.0262$. Hence the implied univariate model for Y_t would be ARMA(4, 2), with approximate AR operator equal to $(1 - 2.62B + 2.59B^2 - 1.14B^3 + 0.19B^4)$, and from methods of Appendix A6.2, the MA(2) operator would be $(1 - 0.44B + 0.21B^2)$, with $\sigma_\varepsilon^2 = 0.1220$. This model result is in good agreement with the univariate model actually identified and fitted to the series Y_t, which gives an additional check and provides further support to the transfer function–noise model that has been specified for the gas furnace data.

11.5.3 Forecast of Nonstationary Sales Data Using a Leading Indicator

As a second illustration, consider the data on sales Y_t in relation to a leading indicator X_t, plotted in Figure 11.10 and listed as series M in the "Collection of Time Series" in Part Five. The data are typical of that arising in business forecasting and are well fitted by the nonstationary model*

$$y_t = 0.035 + \frac{4.82x_{t-3}}{1 - 0.72B} + (1 - 0.54B)a_t$$

$$x_t = (1 - 0.32B)\alpha_t$$

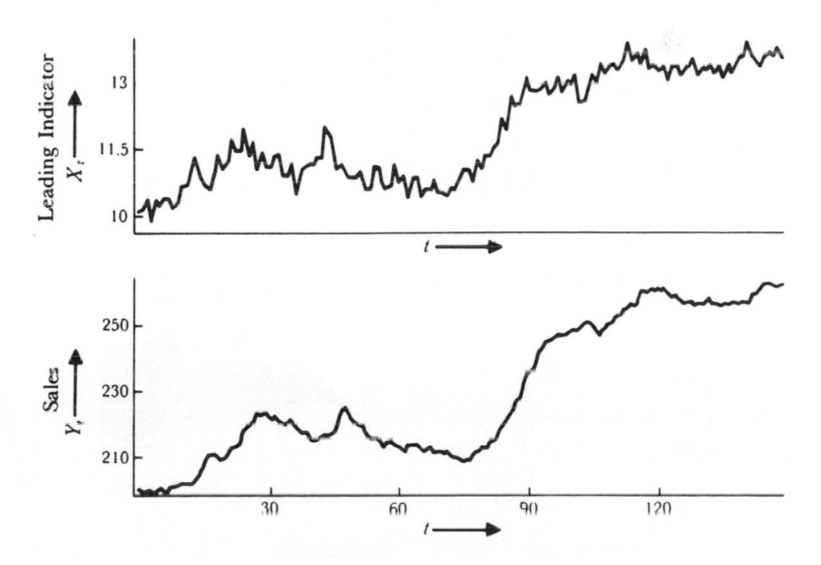

Figure 11.10 Sales data with leading indicator.

* Using data the latter part of which is listed as series M.

with y_t and x_t first differences of the series. The forecast function, in the form (11.5.4), is then

$$\hat{Y}_t(l) = [Y_{t+l}] = 1.72[Y_{t+l-1}] - 0.72[Y_{t+l-2}] + 0.0098 + 4.82[X_{t+l-3}]$$
$$- 4.82[X_{t+l-4}] + [a_{t+l}] - 1.26[a_{t+l-1}]$$
$$+ 0.3888[a_{t+l-2}]$$

Figure 11.11 shows the forecasts for lead times $l = 1, 2, \ldots, 12$ made at origin $t = 89$. The weights ν_j and ψ_j are given in Table 11.13.

Using the estimates $\hat{\sigma}^2_\alpha = 0.0676$ and $\hat{\sigma}^2_a = 0.0484$, obtained in fitting the above model, the variance of the forecast error may be found from (11.5.6). The 50% and 95% probability limits are shown in Figure 11.11. It will be seen that in this particular example, the use of the leading indicator allows very accurate forecasts to be obtained for lead times $l = 1$, 2, and 3. The π, P, and Q weights for this model are given in Table 11.14. The weights $P_j^{(5)}$ and $Q_j^{(5)}$ appropriate to the lead 5 forecasts are shown in Figure 11.11.

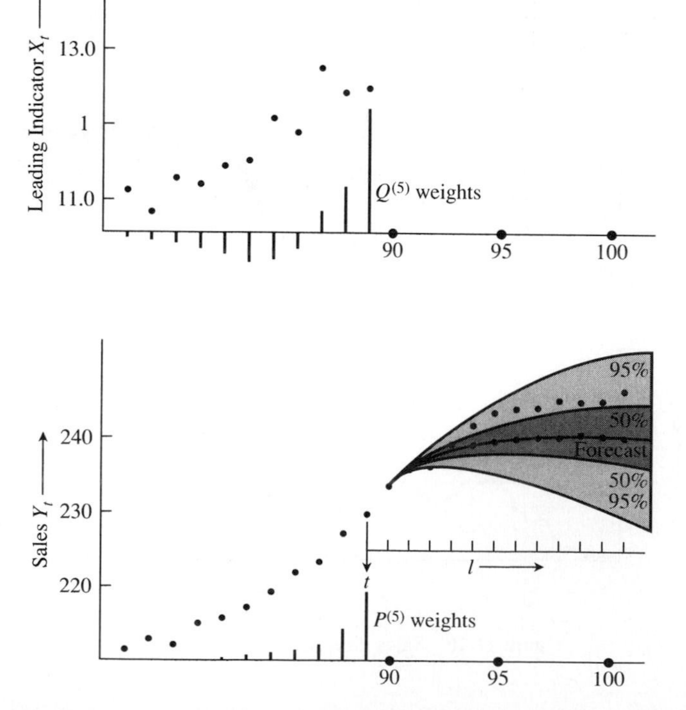

Figure 11.11 Forecast of sales at origin $t = 89$ with P and Q weights for lead 5 forecast.

TABLE 11.13 ν **and** ψ **Weights for Nonstationary Model**

j	ν_j	ψ_j	j	ν_j	ψ_j
0	0	1	6	9.14	0.46
1	0	0.46	7	9.86	0.46
2	0	0.46	8	10.37	0.46
3	4.82	0.46	9	10.75	0.46
4	6.75	0.46	10	11.02	0.46
5	8.14	0.46	11	11.21	0.46

TABLE 11.14 π, P, **and** Q **Weights for Nonstationary Model**

j	π_j	P_j	Q_j	j	π_j	P_j	Q_j
1	0.68	0.46	0	9	0.00	0.00	0.74
2	0.22	0.25	0	10	0.00	0.00	−0.59
3	0.07	0.13	4.82	11	0.00	0.00	−0.29
4	0.02	0.07	1.25	12	0.00	0.00	−0.13
5	0.01	0.04	−0.29	13	0.00	0.00	0.06
6	0.00	0.02	−0.86	14	0.00	0.00	−0.02
7	0.00	0.01	−0.97	15	0.00	0.00	0.00
8	0.00	0.01	−0.89				

11.6 SOME ASPECTS OF THE DESIGN OF EXPERIMENTS TO ESTIMATE TRANSFER FUNCTIONS

In some engineering applications the form of the input X_t can be deliberately chosen so as to obtain good estimates of the parameters in the transfer function–noise model

$$Y_t = \delta^{-1}(B)\omega(B)X_{t-b} + N_t$$

The estimation of the transfer function is equivalent to estimation of a dynamic "regression" model and the methods that can be used are very similar to those used in ordinary nondynamic regression. As might be expected, the same problems [34] face us.

As with static regression, it is very important to be clear on the objective of the investigation. In some situations we want to answer the question: If the input X is merely observed (but not interfered with) what can this tell us of the present and future behavior of the output Y under *normal* conditions of process operation? In other situations the appropriate question is: If the input X is *changed* in some specific way, what *change* will be induced in the present and future behavior of the output Y? The types of data we need to answer these two questions are different.

To answer the first question unambiguously, we must use data, obtained by observing, *but not interfering with,* the normal operation of the system. By contrast, the second question can only be answered unambiguously from data in which *deliberate* changes have been induced into the input of the system; that is, the data must be specially generated by a *designed experiment.*

Clearly, if X is to be used as a control variable, that is, a variable which may be used to manipulate the output, we need to answer the second question. To understand how we can design experiments to obtain valid estimates of the parameters of a cause-and-effect relationship, it is necessary to examine the assumptions of the analysis.

A critical assumption is that the X_t's are distributed independently of the N_t's. When this assumption is violated:

1. The estimates we obtain are, in general, not even consistent. Specifically, as the sample size is made large, the estimates converge not on the true values but on other values differing from the true values by an unknown amount.
2. The violation of this assumption is not detectable by examining the data. Therefore, the possibility that in any particular situation the independence assumption may not be true is a particularly disturbing one. The only way it is possible to guarantee its truth is by deliberately *designing* the experiment rather than using data that have simply "happened." Specifically, we must deliberately generate and feed into the process an input X_t, which we know to be uncorrelated with N_t because we have generated it by some external random process.

The input X_t can, of course, be autocorrelated; it is necessary only that it should not be cross-correlated with N_t. To satisfy this requirement, we could, for example, draw a set of random deviates α_t and use them to generate any desired input process $X_t = \psi_X(B)\alpha_t$.

Alternatively, we can choose a fixed "design," for example the factorial design used in Section 11.4.2, and randomize the order in which the runs are made. Appendix A11.2 contains a preliminary discussion of some elementary design problems, and it is sufficient to expose some of the difficulties in the practical selection of the "optimal" stochastic input. In particular, as is true in a wider context: (1) it is difficult to decide what is a sensible criterion for optimality, and (2) the choice of "optimal" input depends on the values of the unknown parameters that are to be optimally estimated. In general, a white noise input has distinct advantages in simplifying identification, and if nothing very definite were known about the system under study, would provide a sensible initial choice of input.

APPENDIX A11.1 USE OF CROSS SPECTRAL ANALYSIS FOR TRANSFER FUNCTION MODEL IDENTIFICATION

In this appendix we show that an alternative method for identifying transfer function models, which does not require prewhitening of the input, can be based on spectral analysis. Furthermore, it is easily generalized to multiple inputs.

A11.1.1 Identification of Single Input Transfer Function Models

Suppose that the transfer function $v(B)$ is *defined* so as to allow the possibility of nonzero impulse response weights v_j for j a negative integer, so that

$$v(B) = \sum_{k=-\infty}^{\infty} v_k B^k$$

Then if, corresponding to (11.2.3), the transfer function–noise model is

$$y_t = v(B)x_t + n_t$$

equations (11.2.5) become

$$\gamma_{xy}(k) = \sum_{j=-\infty}^{\infty} v_j \gamma_{xx}(k-j) \qquad k = 0, \pm 1, \pm 2, \ldots \tag{A11.1.1}$$

We now define a *cross covariance generating function*

$$\gamma^{xy}(B) = \sum_{k=-\infty}^{\infty} \gamma_{xy}(k) B^k \tag{A11.1.2}$$

which is analogous to the autocovariance generating function (3.1.10). On multiplying throughout in (A11.1.1) by B^k and summing, we obtain

$$\gamma^{xy}(B) = v(B)\gamma^{xx}(B) \tag{A11.1.3}$$

If we now substitute $B = e^{-i2\pi f}$ in (A11.1.2), we obtain the cross spectrum $p_{xy}(f)$ between input and output. Making the same substitution in (A11.1.3) yields

$$v(e^{-i2\pi f}) = \frac{p_{xy}(f)}{p_{xx}(f)} \qquad -\tfrac{1}{2} \leqslant f < \tfrac{1}{2} \tag{A11.1.4}$$

where

$$v(e^{-i2\pi f}) = G(f)e^{i2\pi\phi(f)} = \sum_{k=-\infty}^{\infty} v_k e^{-i2\pi f k} \tag{A11.1.5}$$

is called the *frequency response function* of the system and is the Fourier transform of the impulse response function. Since $v(e^{-i2\pi f})$ is complex, we write it as a product involving a *gain function* $G(f)$ and a *phase function* $\phi(f)$. (A11.1.4) shows that the frequency response function is the ratio of the cross spectrum to the input spectrum. Methods for estimating the frequency response function $v(e^{-i2\pi f})$ are described in [122]. Knowing $v(e^{-i2\pi f})$, the impulse response function v_k can then be obtained from

$$v_k = \int_{-1/2}^{1/2} v(e^{-i2\pi f})e^{i2\pi fk}\, df \qquad \text{(A11.1.6)}$$

Using a similar approach, the autocovariance generating function of the noise n_t is

$$\gamma^{nn}(B) = \gamma^{yy}(B) - \frac{\gamma^{xy}(B)\gamma^{xy}(F)}{\gamma^{xx}(B)} \qquad \text{(A11.1.7)}$$

On substituting $B = e^{-i2\pi f}$ in (A11.1.7), we obtain the expression

$$p_{nn}(f) = p_{yy}(f)[1 - \kappa_{xy}^2(f)] \qquad \text{(A11.1.8)}$$

for the spectrum of the noise, where

$$\kappa_{xy}^2(f) = \frac{|p_{xy}(f)|^2}{p_{xx}(f)p_{yy}(f)}$$

and $\kappa_{xy}(f)$, the *coherency spectrum,* behaves like a correlation coefficient at each frequency f. Knowing the noise spectrum, the noise autocovariance function $\gamma_{nn}(k)$ may then be obtained from

$$\gamma_{nn}(k) = 2\int_0^{1/2} p_{nn}(f)\cos(2\pi fk)\, df$$

By substituting estimates of the spectra such as are described in [122], estimates of the impulse response weights v_k and noise autocorrelation function are obtained. These can be used to identify the transfer function model and noise model as described in Sections 11.2.1 and 6.2.1.

A11.1.2 Identification of Multiple Input Transfer Function Models

We now generalize the model

$$\begin{aligned} Y_t &= v(B)X_{t-b} + N_t \\ &= \delta^{-1}(B)\omega(B)X_{t-b} + N_t \end{aligned}$$

to allow for several inputs $X_{1,t}, X_{2,t}, \ldots, X_{m,t}$. Thus

$$Y_t = v_1(B)X_{1,t} + \cdots + v_m(B)X_{m,t} + N_t \qquad \text{(A11.1.9)}$$

$$= \delta_1^{-1}(B)\omega_1(B)X_{1,t-b_1} + \cdots + \delta_m^{-1}(B)\omega_m(B)X_{m,t-b_m} + N_t \tag{A11.1.10}$$

where $v_j(B)$ is the generating function of the impulse response weights relating $X_{j,t}$ to the output. We assume, as before, that after differencing, (A11.1.9) may be written

$$y_t = v_1(B)x_{1,t} + \cdots + v_m(B)x_{m,t} + n_t$$

Multiplying throughout by $x_{1,t-k}, x_{2,t-k}, \ldots, x_{m,t-k}$ in turn, taking expectations and forming the generating functions, we obtain

$$\begin{aligned}
\gamma^{x_1y}(B) &= v_1(B)\gamma^{x_1x_1}(B) + v_2(B)\gamma^{x_1x_2}(B) + \cdots + v_m(B)\gamma^{x_1x_m}(B) \\
\gamma^{x_2y}(B) &= v_1(B)\gamma^{x_2x_1}(B) + v_2(B)\gamma^{x_2x_2}(B) + \cdots + v_m(B)\gamma^{x_2x_m}(B) \\
&\vdots \\
\gamma^{x_my}(B) &= v_1(B)\gamma^{x_mx_1}(B) + v_2(B)\gamma^{x_mx_2}(B) + \cdots + v_m(B)\gamma^{x_mx_m}(B)
\end{aligned} \tag{A11.1.11}$$

On substituting $B = e^{-i2\pi f}$, the spectral equations are obtained. For example, with $m = 2$,

$$p_{x_1y}(f) = H_1(f)p_{x_1x_1}(f) + H_2(f)p_{x_1x_2}(f)$$

$$p_{x_2y}(f) = H_1(f)p_{x_2x_1}(f) + H_2(f)p_{x_2x_2}(f)$$

and the frequency response functions $H_1(f) = v_1(e^{-i2\pi f})$, $H_2(f) = v_2(e^{-i2\pi f})$ can be calculated as described in [122]. The impulse response weights can then be obtained using the inverse transformation (A11.1.6).

APPENDIX A11.2 CHOICE OF INPUT TO PROVIDE OPTIMAL PARAMETER ESTIMATES

Suppose that the input to a dynamic system can be made to follow an imposed stochastic process which is at choice. For example, it might be an autoregressive process, a moving average process, or white noise. To illustrate the problems involved in the optimal selection of this stochastic process, it is sufficient to consider an elementary example.

A11.2.1 Design of Optimal Inputs for a Simple System

Suppose that a system is under study for which the transfer function–noise model is assumed to be

$$Y_t = \beta_1 Y_{t-1} + \beta_2 X_{t-1} + a_t \qquad |\beta_1| < 1 \tag{A11.2.1}$$

where a_t is white noise. It is supposed also that the input and output processes are stationary and that X_t and Y_t denote deviations of these processes from their respective means. For large samples, and associated with any fixed probability, the approximate area of the Bayesian HPD region for β_1 and β_2, and also of the corresponding confidence region, is proportional to $\Delta^{-1/2}$, where

$$\Delta = \begin{vmatrix} E[Y_t^2] & E[Y_tX_t] \\ E[Y_tX_t] & E[X_t^2] \end{vmatrix}$$

We shall proceed by attempting to find the design minimizing the area of the region and thus maximizing Δ. Now

$$E[Y_t^2] = \sigma_Y^2 = \sigma_X^2\beta_2^2\frac{1+2q}{1-\beta_1^2} + \frac{\sigma_a^2}{1-\beta_1^2} \qquad \text{(A11.2.2)}$$

$$E[Y_tX_t] = \sigma_X^2\frac{\beta_2}{\beta_1}q$$

$$E[X_t^2] = \sigma_X^2$$

where

$$q = \sum_{i=1}^{\infty}\beta_1^i\rho_i \qquad \sigma_X^2\rho_i = E[X_tX_{t-i}]$$

The value of the determinant may be written in terms of σ_X^2 as

$$\Delta = \frac{\sigma_X^2\sigma_a^2}{1-\beta_1^2} + \frac{\beta_2^2\sigma_X^4}{(1-\beta_1^2)^2} - \frac{\sigma_X^4\beta_2^2}{\beta_1^2}\left(q - \frac{\beta_1^2}{1-\beta_1^2}\right)^2 \qquad \text{(A11.2.3)}$$

Thus, as might be expected, the area of the region can be made small by making σ_X^2 large (i.e., by varying the input variable over a wide range). In practice, there may be limits to the amount of variation that can be allowed in X. Let us proceed by first supposing that σ_X^2 is to be held fixed at some specified value.

Solution with σ_X^2 fixed. With $(1-\beta_1^2) > 0$ and for any fixed σ_X^2, we see from (A11.2.3) that Δ is maximized by setting

$$q = \frac{\beta_1^2}{1-\beta_1^2}$$

that is,

$$\beta_1\rho_1 + \beta_1^2\rho_2 + \beta_1^3\rho_3 + \cdots = \beta_1^2 + \beta_1^4 + \beta_1^6 + \cdots$$

There are an infinity of ways in which, for given β_1, this equality could be achieved. One solution is

$$\rho_i = \beta_1^i$$

Thus one way to maximize Δ for fixed σ_X^2 would be to force the input to follow the autoregressive process

$$(1 - \beta_1 B)X_t = \varepsilon_t$$

where ε_t is a white noise process with variance $\sigma_\varepsilon^2 = \sigma_X^2(1 - \beta_1^2)$.

Solution with σ_Y^2 fixed. So far we have supposed that σ_Y^2 is unrestricted. In some cases we might wish to avoid too great a variation in the output rather than in the input. Suppose that σ_Y^2 is held equal to some fixed acceptable value but that σ_X^2 is unrestricted. Then the value of the determinant Δ can be written in terms of σ_Y^2 as

$$\Delta = \frac{\sigma_Y^4}{\beta_2^2}\left[\frac{\sigma_Y^2 - \sigma_a^2}{\sigma_Y^2} - \frac{\beta_1^2}{s^2}\left(\frac{q + s}{1 + 2q}\right)^2\right] \tag{A11.2.4}$$

where

$$s = \frac{\beta_1^2 r}{1 + \beta_1^2 r} \tag{A11.2.5}$$

and

$$r = \frac{\sigma_Y^2}{\sigma_Y^2 - \sigma_a^2} \tag{A11.2.6}$$

The maximum is achieved by setting

$$q = -s = -\beta_1^2 r/(1 + \beta_1^2 r) \tag{A11.2.7}$$

that is,

$$\beta_1\rho_1 + \beta_1^2\rho_2 + \beta_1^3\rho_3 + \cdots = -\beta_1^2 r + \beta_1^4 r^2 - \beta_1^6 r^3 + \cdots$$

There are again an infinity of ways of satisfying this equality. In particular we can make

$$\rho_i = (-\beta_1 r)^i \tag{A11.2.8}$$

by forcing the input to follow the autoregressive process

$$(1 + \beta_1 r B)X_t = \varepsilon_t \tag{A11.2.9}$$

where ε_t is a white noise process with variance $\sigma_\varepsilon^2 = \sigma_X^2(1 - \beta_1^2 r^2)$. Since r is essentially positive, the sign of the parameter $\beta_1 r$ of this autoregressive process is opposite to that obtained for the optimal input with σ_X^2 fixed.

Solution with $\sigma_Y^2 \times \sigma_X^2$ fixed. In practice it might happen that excessive variations in input and output were both to be avoided. If it were true that a given *percentage* decrease in the variance of X was equally as desirable as the same *percentage* decrease in the variance of Y, it would be sensible to maximize Δ subject to a fixed value of the product $\sigma_X^2 \times \sigma_Y^2$. The determinant is

$$\Delta = \sigma_X^2 \sigma_Y^2 - \frac{\sigma_X^4 \beta_2^2 q^2}{\beta_1^2} \qquad \text{(A11.2.10)}$$

which is maximized for fixed $\sigma_X^2 \sigma_Y^2$ only if $q = 0$. Once again there is an infinity of solutions. However, by using a white noise input, Δ is maximized *whatever the value of* β_1. For such an input, using (A11.2.2), σ_X^2 is the positive root of

$$\sigma_X^4 \beta_2^2 + \sigma_X^2 \sigma_a^2 - k(1 - \beta_1^2) = 0 \qquad \text{(A11.2.11)}$$

where $k = \sigma_X^2 \sigma_Y^2$.

A11.2.2 Numerical Example

Suppose that we were studying the first-order dynamic system (A11.2.1) with $\beta_1 = 0.50$, $\beta_2 = 1.00$, so that

$$Y_t = 0.50 Y_{t-1} + 1.00 X_{t-1} + a_t$$

where $\sigma_a^2 = 0.2$.

σ_X^2 fixed, σ_Y^2 unrestricted. Suppose at first that the design is chosen to maximize Δ with $\sigma_X^2 = 1.0$. Then one optimal choice for the input X_t will be the autoregressive process

$$(1 - 0.5B)X_t = \varepsilon_t$$

where the white noise process ε_t would have variance $\sigma_\varepsilon^2 = \sigma_X^2(1 - \beta_1^2) = 0.75$. Using (A11.2.2), the variance σ_Y^2 of the output would be 2.49, and the scheme will achieve a Bayesian region for β_1 and β_2 whose area is proportional to $\Delta^{-1/2} = 0.70$.

σ_Y^2 fixed, σ_X^2 unrestricted. The scheme above is optimal under the assumption that the input variance is $\sigma_X^2 = 1$ and the output variance is unrestricted. This output variance then turns out to be $\sigma_Y^2 = 2.49$. If, instead, the input variance were unrestricted, then with a *fixed* output variance of 2.49, we could, of course, do considerably better. In fact, using (A11.2.6), $r = 1.087$, so that from (A11.2.9), one optimal choice for the unrestricted input would be the autoregressive process

$$(1 + 0.54B)X_t = \varepsilon_t$$

where in this case ε_t is a white noise process with $\sigma_\varepsilon^2 = 2.05$. The variance σ_X^2 of the input would now be increased to 2.91, and $\Delta^{-1/2}$, which measures the area of the Bayesian region, would be reduced to $\Delta^{-1/2} = 0.42$.

Product $\sigma_Y^2 \times \sigma_X^2$ fixed. Finally, we consider a scheme that attempts to control both σ_Y^2 and σ_X^2 by minimizing Δ with $\sigma_Y^2 \times \sigma_X^2$ fixed. In the previous example in which σ_Y^2 was fixed, we found that $\Delta^{-1/2} = 0.42$ with $\sigma_X^2 = 2.91$ and $\sigma_Y^2 = 2.49$, so that the product $2.91 \times 2.49 = 7.25$. If our objective had been to minimize $\Delta^{-1/2}$ while keeping this product equal to 7.25, we could have made an optimal choice *without knowledge of* β_1 by choosing a white noise input $X_t = \varepsilon_t$. Using (A11.2.11), $\sigma_X^2 = \sigma_\varepsilon^2 = 2.29$, $\sigma_Y^2 = 3.16$ and in this case, as expected, $\Delta^{-1/2} = 0.37$ is slightly smaller than in the previous example.

It is worth considering this example in terms of spectral ideas. To optimize with σ_X^2 fixed we have used an autoregressive input with ϕ positive that has high power at low frequencies. Since the gain of the system is high at low frequencies, this achieves maximum transfer from X to Y and so induces large variations in Y. When σ_Y^2 is fixed, we have introduced an input that is an autoregressive process with ϕ negative. This has high power at high frequencies. Since there is minimum transfer from X to Y at high frequencies, the disturbance in X must now be made large at these frequencies. When the product $\sigma_X^2 \times \sigma_Y^2$ is fixed, the "compromise" input white noise is indicated and does not require knowledge of β_1. This final maximization is equivalent to minimizing the correlation between the estimates $\hat{\beta}_1$ and $\hat{\beta}_2$, and in fact the correlation between these estimates is zero when a white noise input is used.

Conclusions. This investigation shows:

1. The optimal choice of design rests heavily on how we define "optimal."
2. Both in the case where σ_X^2 is held fixed and in that where σ_Y^2 is held fixed, the optimal choices require specific stochastic processes whose parameters are functions of the *unknown* dynamic parameters. Thus we are in the familiar paradoxical situation where we can do a better job of data gathering only to the extent that we already know something about the answer we seek. A sequential approach, where we improve the design as we find out more about the parameters, is a possibility worth further investigation. In particular, a pilot investigation using a possibly nonoptimal input, say white noise, could be used to generate data from which preliminary estimates of the dynamic parameters could be obtained. These estimates could then be used to specify a further input using one of our previous criteria.
3. The use of white noise is shown, *for the simple case investigated,* to be optimal for a sensible criterion of optimality, and its use as an input requires no prior knowledge of the parameters.

12

INTERVENTION ANALYSIS MODELS AND OUTLIER DETECTION

'ime series are often affected by special events or circumstances such as ɔolicy changes, strikes, advertising promotions, environmental regulations, and similar events, which we shall refer to as *intervention* events. In this chapter we describe the method of intervention analysis which can account for the expected effects of these interventions. For this, the transfer function models of the previous chapters are used, but in the intervention analysis model the input series will be in the form of a simple pulse or step indicator function to indicate the presence or absence of the event. Initially, it will be assumed that the timing of the intervention event is known. Later, methods for the related problem of detection of outlying or unusual behavior in a time series at an unknown point of time is also discussed.

12.1 INTERVENTION ANALYSIS METHODS

12.1.1 Models for Intervention Analysis

In the setting of intervention analysis, it is assumed that an intervention event has occurred at a known point in time T of a time series. It is of interest to determine whether there is any evidence of a change or effect, of an expected kind, on the time series Y_t under study associated with the event. We consider the use of transfer function models to model the nature of and estimate the magnitude of the effects of the intervention, and hence to account for the possible unusual behavior in the time series related to the event. Based on the study by Box and Tiao [59], the type of model to be considered has the form

$$Y_t = \frac{\omega(B)B^b}{\delta(B)} \xi_t + N_t \tag{12.1.1}$$

where the term $\mathcal{Y}_t = \delta^{-1}(B)\omega(B)B^b\xi_t$ represents the effects of the intervention event in terms of the deterministic input series ξ_t, and N_t is the noise series which represents the background observed series Y_t without the intervention effects. It is assumed that N_t follows an ARIMA(p, d, q) model, $\varphi(B)N_t = \theta(B)a_t$, with $\varphi(B) = \phi(B)(1 - B)^d$. Multiplicative seasonal ARIMA models as presented in Chapter 9 can also be included for N_t, but special note of the seasonal models will not be made in this chapter.

There are two common types of deterministic input variables ξ_t that have been found useful to represent the impact of intervention events on a time series. Both of these are indicator variables taking only the values 0 and 1 to denote the nonoccurrence and occurrence of intervention. One type is a *step function* at time T, given by

$$S_t^{(T)} = \begin{cases} 0 & t < T \\ 1 & t \geq T \end{cases} \tag{12.1.2}$$

which would typically be used to represent the effects of an intervention that are expected to remain permanently after time T to some extent. The other type is a *pulse function* at T, given by

$$P_t^{(T)} = \begin{cases} 0 & t \neq T \\ 1 & t = T \end{cases} \tag{12.1.3}$$

which could represent the effects of an intervention that are temporary or transient and will die out after time T. These indicator input variables are used in many situations where the effects of the intervention cannot be represented as the response to a quantitative variable, because such a quantitative variable does not exist or it is impractical or impossible to obtain measurements on such a variable.

Because of the deterministic nature of the indicator input series ξ_t in (12.1.1), unlike the transfer function model situation of Chapter 11, identification of the structure of the intervention model operator $\upsilon(B) = \delta^{-1}(B)\omega(B)B^b$ cannot be based on the technique of prewhitening. Instead, it is desirable to postulate the form of the intervention model through consideration of the mechanisms that might cause the change or effect and the implied form of the change that would be expected. In addition, the identification may be aided by direct inspection of the data to suggest the form of effect due to the known event, and supplementary evidence may sometimes be available from examination of the residuals from a model fitted before the intervention term is introduced.

Several different response patterns $\mathcal{Y}_t = \delta^{-1}(B)\omega(B)B^b\xi_t$ are possible through different choices of the transfer function. Figure 12.1 shows the responses for various simple transfer functions with both step and pulse indicators as input. For example, the model $\mathcal{Y}_t = \omega B S_t^{(T)}$ in Figure 12.1(a) can be used to represent a permanent step change in level of unknown magnitude ω after time T, while the form

$$\mathcal{Y}_t = \frac{\omega B}{1 - \delta B} S_t^{(T)} \qquad 0 < \delta < 1 \tag{12.1.4}$$

in Figure 12.1(b), which implies that $\mathcal{Y}_t = \omega(1 - \delta^{t-T})/(1 - \delta)$, $t \geq T$, corresponds to a gradual change with rate δ that eventually approaches the long-run change in level equal to $\omega/(1 - \delta)$. Similarly, the model

$$\mathcal{Y}_t = \frac{\omega_1 B}{1 - \delta B} P_t^{(T)} \qquad 0 < \delta < 1 \tag{12.1.5}$$

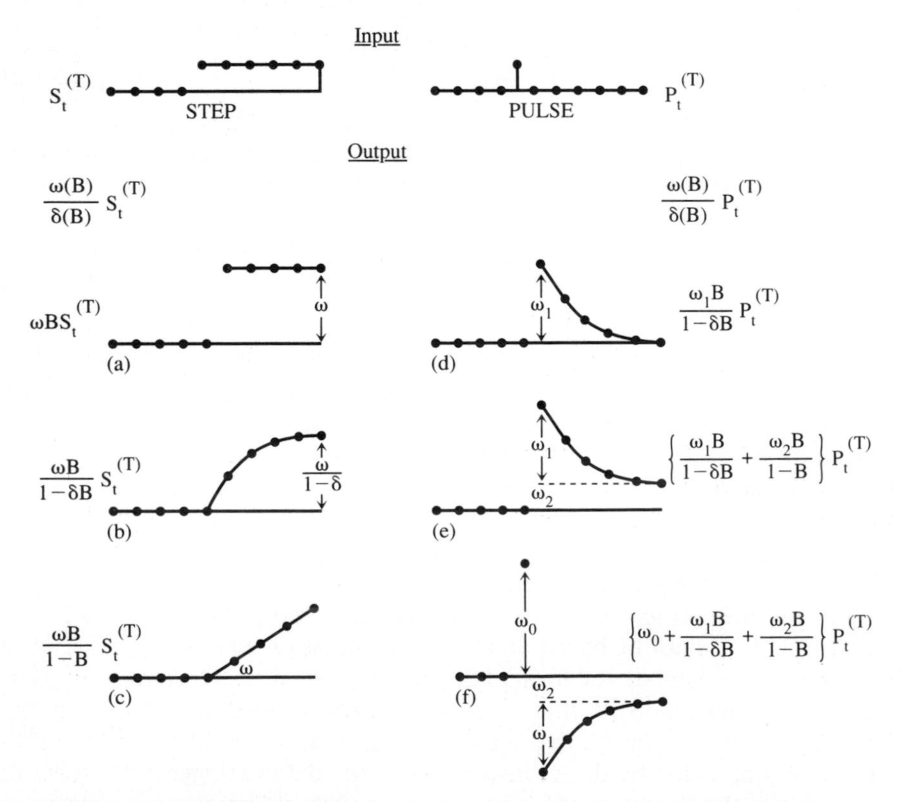

Figure 12.1 Responses to a step and a pulse input: (a), (b), (c) response to a step input for various simple transfer function models; (d), (e), (f) response to a pulse input for some models of interest.

in Figure 12.1(d), which implies that $\mathcal{Y}_t = \omega_1 \delta^{t-T-1}$, $t > T$, would represent a sudden "pulse" change after time T of unknown magnitude ω_1, followed by a gradual decay of rate δ back to the original preintervention level with no permanent effect. More complex forms of response can be obtained by various linear combinations of the simpler forms, such as in the case of Figure 12.1(f). It is also noted that since $(1 - B)S_t^{(T)} = P_t^{(T)}$, any of the transfer function models that involve $S_t^{(T)}$ could equally well be represented in terms of $P_t^{(T)}$.

The following additional points concerning the intervention models are worthy of note. The function $\mathcal{Y}_t$ represents the additional effect of the intervention event over the noise or "background" series N_t. Hence, when possible, the model $N_t = [\theta(B)/\varphi(B)]a_t$ for the noise is identified based on the usual procedures applied to the time series observations available before the date of the intervention, that is, Y_t, $t < T$. Also, it is assumed in model (12.1.1) that only the level of the series is affected by the intervention and, in particular, that the form and the parameters of the time series model for N_t are the same before and after the intervention. One should also recognize that there can be considerable differences in the accuracy with which the intervention model parameters can be estimated depending on whether the noise N_t is stationary or nonstationary, as well as on whether permanent or transitory effects are postulated.

In general, the parameter estimates and their standard errors for the intervention model

$$Y_t = \frac{\omega(B)B^b}{\delta(B)} \xi_t + \frac{\theta(B)}{\varphi(B)} a_t \qquad (12.1.6)$$

are obtained by the least squares method of estimation for transfer function–noise models as described in Section 11.3. Diagnostics checking of the residuals $\hat{a}_t$ from the model similar to methods previously discussed would also be employed to assess the adequacy of the fitted model.

12.1.2 Example of Intervention Analysis

Box and Tiao [59] considered the monthly time series consisting of the rate of change in the U.S. consumer price index (CPI) for the period July 1953 through December 1972. Beginning in September 1971, Phase I economic control went into effect for 3 months, and after that Phase II was in effect. It was of interest to investigate the possible effect of the Phase I and II controls on the rate of change in the CPI.

Inspection of the sample autocorrelation functions of the rate of change of the CPI and its first differences for the 218 monthly observations prior to Phase I suggested a noise model of the form

$$(1 - B)N_t = (1 - \theta B)a_t \qquad (12.1.7)$$

with maximum likelihood estimates $\hat{\theta} = 0.84$ and $\hat{\sigma}_a^2 = 0.0019$. Examination of the residuals and their autocorrelations reveals no obvious inadequacies in this model.

Then, to address the question of the possible effects of Phase I and II controls, it is assumed that Phase I and II are expected to produce changes in the level of the rate of change of the CPI, and that the form of the noise model remains the same. Based on these assumptions, the appropriate model to assess the impact of the controls is

$$Y_t = \omega_1 \xi_{1t} + \omega_2 \xi_{2t} + \frac{1 - \theta B}{1 - B} a_t \tag{12.1.8}$$

where

$$\xi_{1t} = \begin{cases} 1 & t = \text{September, October, or November 1971} \\ 0 & \text{otherwise} \end{cases}$$

$$\xi_{2t} = \begin{cases} 1 & t \geqslant \text{December 1971} \\ 0 & \text{otherwise} \end{cases}$$

The nonlinear least squares estimates of the parameters in model (12.1.8) were obtained, with standard errors in parentheses, as $\hat{\theta} = 0.85(0.05)$, $\hat{\omega}_1 = -0.0022(0.0010)$, $\hat{\omega}_2 = -0.0008(0.0009)$. Hence the analysis suggests that a drop in the rate of increase of the CPI is associated with Phase I, but the effect of Phase II is much less certain.

There have been many other examples of the use of intervention analysis, including study of the effects of regulations for engine design changes in new cars on oxidant pollution levels in the Los Angeles area [59], the effect of a change in policy in relation to debt collection on bad debt collections [121], the effectiveness of seat belt legislation on road deaths [31], and the impact of the Arab oil embargo on electricity consumption in the United States [145].

12.1.3 Nature of the MLE for a Simple Level Change Parameter Model

It is instructive to consider the nature of the maximum likelihood estimator of the intervention parameters, such as those in (12.1.8), for some relatively simple situations. We consider the simple model

$$Y_t = \omega \xi_t + N_t \tag{12.1.9}$$

where $N_t = \varphi^{-1}(B)\theta(B)a_t$. This model can be written, formally, as

$$\pi(B) Y_t = \omega \pi(B) \xi_t + a_t \tag{12.1.10}$$

where $\pi(B) = \theta^{-1}(B)\varphi(B) = 1 - \sum_{i=1}^{\infty} \pi_i B^i$. Letting $w_t = \pi(B) Y_t$ and $x_t = \pi(B)\xi_t$, we can write (12.1.10) in the form of a simple linear model $w_t = \omega x_t +$

a_t, $t = 1, 2, \ldots, n$. Hence the maximum likelihood estimator of ω is, approximately,

$$\hat{\omega} = \frac{\sum_{t=1}^{n} x_t w_t}{\sum_{t=1}^{n} x_t^2} \tag{12.1.11}$$

with $\text{var}(\hat{\omega}) = \sigma_a^2 / \sum_{t=1}^{n} x_t^2$.

Let us consider a specific situation of (12.1.9) with $\xi_t = BS_t^{(T)}$, a step function at T. Then $x_t = \pi(B)\, BS_t^{(T)} = 1 - \sum_{i=1}^{t-T-1} \pi_i$, $t > T + 1$, with $x_{T+1} = 1$ and $x_t = 0$ for $t \leq T$. For the discussion to follow, we suppose that n is large, and a relatively large number of observations are available before and after the intervention time T.

Now suppose that the noise N_t in (12.1.9) follows the IMA(0, 1, 1) model, $(1 - B)N_t = (1 - \theta B)a_t$, so that $\pi(B) = (1 - \theta B)^{-1}(1 - B)$ with $\pi_i = (1 - \theta)\theta^{i-1}$, $i \geq 1$. Then $x_t = (1 - \theta B)^{-1}(1 - B)BS_t^{(T)} = (1 - \theta B)^{-1}BP_t^{(T)} = \theta^{t-T-1}$, $t \geq T + 1$, and hence

$$\sum_{t=1}^{n} x_t^2 = \sum_{t=T+1}^{n} \theta^{2(t-T-1)} = \frac{1 - \theta^{2(n-T)}}{1 - \theta^2} \simeq \frac{1}{1 - \theta^2}$$

Also, $w_t = (1 - \theta B)^{-1}(1 - B)Y_t = Y_t - (1 - \theta) \sum_{i=1}^{\infty} \theta^{i-1} Y_{t-i} = Y_t - \overline{Y}_{t-1}$, where $\overline{Y}_{t-1}$ is an exponentially weighted moving average of values prior to time t. Following the results in [59], it can then be shown that

$$\sum_{t=1}^{n} x_t w_t = \sum_{t=T+1}^{n} \theta^{t-T-1} w_t \simeq (1 + \theta)^{-1}\left(\sum_{s=0}^{\infty} \theta^s Y_{T+1+s} - \sum_{s=0}^{\infty} \theta^s Y_{T-s}\right)$$

Therefore, in this situation, the maximum likelihood estimator of ω is

$$\hat{\omega} = \frac{\sum_{t=1}^{n} x_t w_t}{\sum_{t=1}^{n} x_t^2} \simeq (1 - \theta)\left(\sum_{s=0}^{\infty} \theta^s Y_{T+1+s} - \sum_{s=0}^{\infty} \theta^s Y_{T-s}\right) \tag{12.1.12}$$

with $\text{var}[\hat{\omega}] \simeq \sigma_a^2(1 - \theta^2)$. The estimator $\hat{\omega}$ can thus be interpreted as a contrast between two exponentially weighted moving averages, one consisting of the observations after the intervention and the other for the observations before the intervention.

Now, as a second case, suppose that the noise instead follows the ARI(1, 1, 0) model, so that $\pi(B) = (1 - \phi B)(1 - B)$. Then $x_t = (1 - \phi B)(1 - B)BS_t^{(T)} = (1 - \phi B)BP_t^{(T)} = 0$, $t > T + 2$, with $x_{T+1} = 1$ and $x_{T+2} = -\phi$, so that $\sum_{t=1}^{n} x_t^2 = 1 + \phi^2$. In addition, $w_t = (1 - \phi B)(1 - B)Y_t$ and it follows that

$$\begin{aligned}\sum_{t=1}^{n} x_t w_t &= (1 - \phi B)(1 - B)Y_{T+1} - \phi(1 - \phi B)(1 - B)Y_{T+2} \\ &= (1 - \phi B)(1 - \phi F)(1 - B)Y_{T+1} \\ &= [(1 + \phi + \phi^2)Y_{T+1} - \phi Y_{T+2}] - [(1 + \phi + \phi^2)Y_T - \phi Y_{T-1}]\end{aligned}$$

Thus for this case we have

$$\hat{\omega} = \frac{\sum_{t=1}^{n} x_t w_t}{\sum_{t=1}^{n} x_t^2}$$
$$= (1 + \phi^2)^{-1}\{[(1 + \phi + \phi^2)Y_{T+1} - \phi Y_{T+2}] - [(1 + \phi + \phi^2)Y_T - \phi Y_{T-1}]\} \quad (12.1.13)$$

with $\text{var}[\hat{\omega}] = \sigma_a^2/(1 + \phi^2)$. Again, the estimator $\hat{\omega}$ can be viewed as a contrast between two weighted averages of the same form, one of the postintervention observations Y_{T+1} and Y_{T+2} and the other of the preintervention observations Y_T and Y_{T-1}, but the weighted averages are only finite in extent because the noise model contains only an AR factor $(1 - \phi B)$ and no MA factor as in the previous case.

Finally, we consider a simpler situation of model (12.1.9), in which the noise is *stationary*, for example, an AR(1) model $(1 - \phi B)N_t = a_t$. In this situation we obtain $x_t = (1 - \phi B)BS_t^{(T)} = 1 - \phi$ for $t > T + 1$ with $x_{T+1} = 1$ and $w_t = (1 - \phi B)Y_t = Y_t - \phi Y_{t-1}$. Then it readily follows that

$$\hat{\omega} = \frac{\sum_{t=1}^{n} x_t w_t}{\sum_{t=1}^{n} x_t^2}$$
$$\simeq \frac{(1 - \phi)\sum_{t=T+1}^{n}(Y_t - \phi Y_{t-1})}{(n - T)(1 - \phi)^2} \simeq \overline{Y}_2 \quad (12.1.14)$$

where $\overline{Y}_2 = (n - T)^{-1}\sum_{t=T+1}^{n} Y_t$ denotes an unweighted average of all observations after the intervention, with $\text{var}[\hat{\omega}] = \sigma_a^2/[1 + (n - T - 1)(1 - \phi)^2] \simeq \sigma_a^2/[(n - T)(1 - \phi)^2]$. Notice that because of the stationarity of the noise, we have an *unweighted* average of postintervention observations and also that there is no adjustment for the preintervention observations because they are assumed to be *stationary* about a *known mean of zero*. Also note that in the stationary case, the variance of $\hat{\omega}$ decreases proportionally with $1/(n - T)$, whereas in the previous nonstationary noise situations, $\text{var}[\hat{\omega}]$ is essentially a constant not dependent on the sample size. This reflects the differing degrees of accuracy in the estimators of intervention model parameters, such as the level shift parameter ω, that can be expected in large samples between the nonstationary noise and the stationary noise model situations.

Specifically, in the model (12.1.9), with $\xi_t = BS_t^{(T)}$ equal to a step input, suppose that the noise process N_t is nonstationary ARIMA with $d = 1$, so that $\phi(B)(1 - B)N_t = \theta(B)a_t$. Then, by applying the differencing operator $(1 - B)$, the model

$$Y_t = \omega BS_t^{(T)} + N_t \quad (12.1.15)$$

can also be expressed as

$$y_t = \omega BP_t^{(T)} + n_t \quad (12.1.16)$$

where $y_t = (1 - B)Y_t$ and $n_t = (1 - B)N_t$, and hence n_t is a stationary ARMA(p, q) process. Therefore, the MLE of ω for the original model (12.1.15) with a (permanent) *step* input effect and *nonstationary* noise (d = 1) will have similar features to the MLE in the model (12.1.16) which has a (transitory) *pulse* input effect and *stationary* noise.

Of course, the model (12.1.9) can be generalized to allow for an unknown nonzero mean ω_0 before the intervention, $Y_t = \omega_0 + \omega\xi_t + N_t$, with $\xi_t = BS_t^{(T)}$, so that ω represents the *change in mean level* after the intervention. Then, for the stationary AR(1) noise model case, for example, similar to (12.1.14), it can be shown that the MLE of ω *is* $\hat{\omega} \simeq \overline{Y}_2 - \overline{Y}_1$, where $\overline{Y}_1 = T^{-1}\Sigma_{t=1}^{T} Y_t$ denotes the sample mean of all preintervention observations.

12.2 OUTLIER ANALYSIS FOR TIME SERIES

Time series observations may sometimes be affected by unusual events, disturbances, or errors that create spurious effects in the series and result in extraordinary patterns in the observations that are not in accord with most observations in the time series. Such unusual observations may be referred to as *outliers*. They may be the result of unusual external events such as strikes, sudden political or economic changes, sudden changes in a physical system, and so on, or simply due to recording or gross errors in measurement. The presence of such outliers in a time series can have substantial effects on the behavior of sample autocorrelations, partial autocorrelations, estimates of ARMA model parameters, forecasting, and can even affect the specification of the model. If the time of occurrence T of an intervention event that results in the outlying behavior is known, the unusual effects can often be accounted for by the use of intervention analysis techniques discussed in Section 12.1. However, since in practice the presence of outliers is often not known at the start of the analysis, additional procedures for detection of outliers and assessment of their possible impacts are important. In this section we discuss some useful models for representing outliers and corresponding methods, similar to the methods of intervention analysis, for detection of outliers. A few recent references that deal with the topics of outlier detection, influence of outliers, and robust methods of estimation are [67], [69], [144], and [190].

12.2.1 Models for Additive and Innovational Outliers

Following the work of Fox [92], we consider two simple intervention models to represent two different types of outliers that might occur in practice. These are the *additive outlier* (AO) and the *innovational outlier* (IO) models. Let z_t denote the underlying time series process which is free of the impact of outliers, and let Y_t denote the observed time series. We assume that z_t follows the ARIMA(p, d, q) model $\varphi(B)z_t = \theta(B)a_t$. Then an additive outlier (AO) at time T, or "observational outlier," is modeled as

$$Y_t = \omega P_t^{(T)} + z_t = \omega P_t^{(T)} + \frac{\theta(B)}{\varphi(B)} a_t \tag{12.2.1}$$

where $P_t^{(T)} = 1$ if $t = T$, $P_t^{(T)} = 0$ if $t \neq T$, denotes the pulse indicator at time T. An innovational outlier (IO) at time T, or "innovational shock," is modeled as

$$Y_t = \frac{\theta(B)}{\varphi(B)} (\omega P_t^{(T)} + a_t) = \omega \frac{\theta(B)}{\varphi(B)} P_t^{(T)} + z_t \tag{12.2.2}$$

Hence an AO affects the level of the observed time series only at time T, $Y_T = \omega + z_T$, by an unknown additive amount ω, while an IO represents an extraordinary random shock at time T, $a_T + \omega = a_T^*$, which affects all succeeding observations $Y_T, Y_{T+1}, \ldots$ through the dynamics of the system described by $\theta(B)/\varphi(B)$. More generally, an observed time series Y_t might be affected by outliers of different types at several points of time $T_1, T_2, \ldots, T_k$, and the multiple outlier model of the general form

$$Y_t = \sum_{j=1}^{k} \omega_j v_j(B) P_t^{(T_j)} + z_t \tag{12.2.3}$$

could be considered, where $v_j(B) = 1$ for an AO at time T_j and $v_j(B) = \theta(B)/\varphi(B)$ for an IO at time T_j. Problems of interest associated with these types of outlier models are to identify the timing and the type of outliers and to estimate the magnitude ω of the outlier effect, so that the analysis of the time series will adjust for these outlier effects.

12.2.2 Estimation of Outlier Effect for Known Timing of the Outlier

We first consider the estimation of the impact ω of an AO in (12.2.1) and that of an IO in (12.2.2), respectively, in the situation where the parameters of the time series model for the underlying process z_t are assumed known. To motivate iterative procedures that have been proposed for the general case, it will also be assumed that the timing T of the outlier is given.

Let $\pi(B) = \theta^{-1}(B)\varphi(B) = 1 - \sum_{i=1}^{\infty} \pi_i B^i$ and define $e_t = \pi(B) Y_t$ for $t = 1, 2, \ldots, n$, in terms of the observed series Y_t. Then we can write the outlier models, (12.2.2) and (12.2.1), respectively, as

$$\text{IO:} \quad e_t = \omega P_t^{(T)} + a_t \tag{12.2.4a}$$

$$\text{AO:} \quad e_t = \omega \pi(B) P_t^{(T)} + a_t = \omega x_{1t} + a_t \tag{12.2.4b}$$

where for the AO model, $x_{1t} = \pi(B)P_t^{(T)} = -\pi_i$ if $t = T + i \geq T$, $x_{1t} = 0$ if $t < T$, with $\pi_0 = -1$. Thus we see from (12.2.4) that the information about an IO is contained solely in the "residual" e_T at the particular time T, whereas that for an AO is spread over the stretch of residuals $e_T, e_{T+1}, e_{T+2}, \ldots$, with generally decreasing weights $1, -\pi_1, -\pi_2, \ldots$, because the π_i are absolutely summable due to the invertibility of the MA operator $\theta(B)$.

From least squares principles, the least squares estimator of the outlier impact ω in the IO model is simply the residual at time T:

$$\text{IO:} \quad \hat{\omega}_{I,T} = e_T \tag{12.2.5a}$$

with $\text{var}[\hat{\omega}_{I,T}] = \sigma_a^2$, while that in the AO model is the linear combination of $e_T, e_{T+1}, \ldots$:

$$\text{AO:} \quad \hat{\omega}_{A,T} = \frac{e_T - \sum_{i=1}^{n-T} \pi_i e_{T+i}}{\sum_{i=0}^{n-T} \pi_i^2} = \frac{\pi^*(F)e_T}{\tau^2} \tag{12.2.5b}$$

with $\text{var}[\hat{\omega}_{A,T}] = \sigma_a^2/\tau^2$, where $\tau^2 = \sum_{i=0}^{n-T} \pi_i^2$ and $\pi^*(F) = 1 - \pi_1 F - \pi_2 F^2 - \cdots - \pi_{n-T}F^{n-T}$. The notation in (12.2.5) reflects the fact that the estimates depend upon the time T. Note that in an underlying autoregressive model $\varphi(B)z_t = a_t$, since then $\pi^*(B) = \pi(B) = \varphi(B)$ for $T < n - p - d$, and $e_t = \varphi(B)Y_t$, in terms of the observations Y_t, the estimate $\hat{\omega}_{A,T}$ in (12.2.5b) can be written as

$$\hat{\omega}_{A,T} = \frac{\varphi(F)\varphi(B)Y_T}{\tau^2}$$

Since $\tau^2 \geq 1$, it is seen in general that $\text{var}[\hat{\omega}_{A,T}] \leq \text{var}[\hat{\omega}_{I,T}] = \sigma_a^2$, and in some cases $\text{var}[\hat{\omega}_{A,T}]$ can be much smaller than σ_a^2. For example, in an MA(1) model for z_t, the variance of $\hat{\omega}_{A,T}$ would be $\sigma_a^2(1 - \theta^2)/(1 - \theta^{2(n-T+1)}) \simeq \sigma_a^2(1 - \theta^2)$ when $n - T$ is large.

Significance tests for the presence of an outlier of type AO or IO at the given time T can be formulated as a test of $\omega = 0$ in either model (12.2.1) or (12.2.2), against $\omega \neq 0$. The likelihood ratio test criteria can be derived for both situations and essentially take the form of the standardized statistics

$$\lambda_{I,T} = \frac{\hat{\omega}_{I,T}}{\sigma_a} \quad \text{and} \quad \lambda_{A,T} = \frac{\tau\hat{\omega}_{A,T}}{\sigma_a} \tag{12.2.6}$$

respectively, for IO and AO types. Under the null hypothesis that $\omega = 0$, both statistics in (12.2.6) will have the standard Normal distribution.

12.2.3 Iterative Procedure for Outlier Detection

In practice, the time T of a possible outlier as well as the model parameters are unknown. To address the problem of detection of outliers at unknown times, iterative procedures that are relatively convenient computationally have been proposed by Chang, Tiao, and Chen [69] to identify and adjust for the effects of outliers.

At the first stage of this procedure, the ARIMA model is estimated for the observed time series Y_t in the usual way assuming that the series contains no outliers. The residuals $\hat{e}_t$ from the model are obtained as $\hat{e}_t = \hat{\theta}^{-1}(B)\hat{\varphi}(B)Y_t = \hat{\pi}(B)Y_t$, and $\hat{\sigma}_a^2 = n^{-1}\Sigma_{t=1}^n \hat{e}_t^2$ is obtained. Then the statistics as in (12.2.6), $\hat{\lambda}_{I,t} = \hat{\omega}_{I,t}/\hat{\sigma}_a$ and $\hat{\lambda}_{A,t} = \hat{\tau}\hat{\omega}_{A,t}/\hat{\sigma}_a$, are computed for each time $t = 1, 2, \ldots, n$, as well as $\hat{\lambda}_T = \max_t[\max(|\hat{\lambda}_{I,t}|, |\hat{\lambda}_{A,t}|)]$, where T denotes the time when this maximum occurs. The possibility of an outlier of type IO is identified at time T if $\hat{\lambda}_T = |\hat{\lambda}_{I,T}| > c$, where c is a prespecified constant with typical values for c of 3.0, 3.5, or 4.0. The effect of this IO can be eliminated from the residuals by defining $\tilde{e}_T = \hat{e}_T - \hat{\omega}_{I,T} = 0$ at T. If $\hat{\lambda}_T = |\hat{\lambda}_{A,T}| > c$, the possibility of an AO is identified at T, and its impact is estimated by $\hat{\omega}_{A,T}$ as in (12.2.5b). The effect of this AO can be removed from the residuals by defining $\tilde{e}_t = \hat{e}_t - \hat{\omega}_{A,T}\hat{\pi}(B)P_t^{(T)} = \hat{e}_t + \hat{\omega}_{A,T}\hat{\pi}_{t-T}$ for $t \geqslant T$. In either case, a new estimate $\tilde{\sigma}_a^2$ is computed from the modified residuals $\tilde{e}_t$.

If any outliers are identified, the modified residuals $\tilde{e}_t$ and modified estimate $\tilde{\sigma}_a^2$, but the same parameters $\hat{\pi}(B) = \hat{\theta}^{-1}(B)\hat{\varphi}(B)$ are used to compute new statistics $\hat{\lambda}_{I,t}$ and $\hat{\lambda}_{A,t}$. The preceding steps are then repeated until all outliers are identified. Suppose that this procedure identifies outliers at k time points $T_1, T_2, \ldots, T_k$. Then the overall outlier model as in (12.2.3),

$$Y_t = \sum_{j=1}^{k} \omega_j v_j(B) P_t^{(T_j)} + \frac{\theta(B)}{\varphi(B)} a_t \tag{12.2.7}$$

is estimated for the observed series Y_t, where $v_j(B) = 1$ for an AO and $v_j(B) = \theta(B)/\varphi(B)$ for an IO at time T_j. A revised set of residuals

$$\hat{e}_t = \hat{\theta}^{-1}(B)\hat{\varphi}(B)[Y_t - \Sigma_{j=1}^k \hat{\omega}_j \hat{v}_j(B) P_t^{(T_j)}]$$

and a new $\hat{\sigma}_a^2$ are obtained from this fitted model. The previous steps of the procedure can then be repeated with new residuals, until all outliers are identified and a final model of the general form of (12.2.7) is estimated. If desired, a modified time series of observations in which the effects of the outliers have been removed can be constructed as $\tilde{z}_t = Y_t - \Sigma_{j=1}^k \hat{\omega}_j \hat{v}_j(B) P_t^{(T_j)}$.

The procedure above can be implemented, with few modifications, to any existing software capable of estimation of ARIMA and transfer function–noise models. The technique can be a useful tool in the identification of potential time series outliers that if undetected could have a negative impact on the effectiveness of modeling and estimation. However, there should be some cautions concerning the systematic use of such "outlier adjustment" procedures, particularly in regard to the overall interpretation of results, the appropriateness of a general model specification for "outliers" such as (12.2.7) which treats the outliers as deterministic constants, and the possibilities for "overspecification" in the number of outliers. Whenever possible, it would always be highly desirable to search for the causes or sources of the outliers that may be identified by the foregoing procedure, so that the outlying behavior can be better understood and properly accounted for in the analysis.

12.2.4 Examples of Analysis of Outliers

We consider two numerical examples to illustrate the application of the outlier analysis procedures discussed in the previous sections. For computational convenience, conditional least squares estimation methods are used throughout in these examples.

Series D. The first example involves series D, which represents "uncontrolled" viscosity readings every hour from a chemical process. In Chapter 7 an AR(1) model $(1 - \phi B)z_t = \theta_0 + a_t$ has been suggested and fitted to this series. In the outlier detection procedure, the model is first estimated assuming that no outliers are present, and the results are given in Table 12.1(a). Then the AO and IO statistics as in (12.2.6) are computed for each time point t, using $\hat{\sigma}_a^2 = 0.08949$. Based on a critical value of $c = 3.5$, we are lead to identification of an IO of rather large magnitude at time $T = 217$. The effect of this IO is removed by modifying the residual at T, a new estimate $\tilde{\sigma}_a^2 = 0.08414$ is obtained, and new outlier statistics are computed using $\tilde{\sigma}_a$. At this stage, no outliers are identified. Then the time series parameters and the outlier parameter ω in model (12.2.2), that is, in the model

$$Y_t = \frac{1}{1 - \phi B}[\theta_0 + \omega P_t^{(T)} + a_t]$$

TABLE 12.1 Outlier Detection and Parameter Estimation Results for Series C and D Examples

	Parameter[a]						Outlier			
	$\hat{\theta}_0$	$\hat{\phi}$	$\hat{\omega}_1$	$\hat{\omega}_2$	$\hat{\omega}_3$	$\hat{\sigma}_a^2$	Time	$\hat{\omega}$	$\hat{\lambda}$	Type
(a) Series D										
Cycle 1	1.269	0.862				0.0895	217	−1.28	−4.29	IO
	(0.258)	(0.028)								
Final	1.181	0.872	−1.296			0.0841				
	(0.251)	(0.027)	(0.292)							
(b) Series C										
Cycle 1		0.813				0.0179	58	0.76	5.65	IO
		(0.038)					59	−0.51	−4.16	IO
							60	−0.44	−3.74	IO
Final		0.851	0.745	−0.551	−0.455	0.0132				
		(0.035)	(0.116)	(0.120)	(0.116)					

[a] Standard errors of parameter estimates are in parentheses.

are estimated simultaneously, and the estimates are given in Table 12.1(a). Repeating the outlier–detection procedure based on these new parameter estimates and corresponding residuals does not reveal any other outliers. Hence only one extreme IO is identified, and adjusting for this IO does not result in much change in the estimate $\hat{\phi}$ of the time series model parameter but gives about a 6% reduction in the estimate of σ_a^2. Several other potential outliers, at times $t = 29$, 113, 115, 171, 268, and 272, were also suggested during the outlier procedure as having values of the test statistics $\hat{\lambda}$ slightly greater than 3.0 in absolute value, but adjustment for such values did not affect the estimates of the model substantially.

Series C. The second example we consider is series C, the "uncontrolled" temperature readings every minute in a chemical process. The model previously identified and fitted to this series is the ARIMA(1, 1, 0), $(1 - \phi B)(1 - B)z_t = a_t$. The estimation results for this model obtained assuming there are no outliers are given in Table 12.1(b). Proceeding with the sequence of calculations of the outlier test statistics and using the critical value of $c = 3.5$, we first identify an IO at time 58. The residual at time 58 is modified, we obtain a new estimate $\tilde{\sigma}_a^2 = 0.01521$, and next an IO at time 59 is identified. This residual is modified, a new estimate $\tilde{\sigma}_a^2 = 0.01409$ is obtained, and then another IO at time 60 is indicated. After this, no further outliers are identified. These innovational outliers at times 58, 59, and 60 are rather apparent in Figure 12.2, which shows a time series plot of the first 100 residuals from the initial model fit before any adjustment for outliers.

Then the time series outlier model

$$(1 - B)Y_t = \frac{1}{1 - \phi B}[\omega_1 P_t^{(58)} + \omega_2 P_t^{(59)} + \omega_3 P_t^{(60)} + a_t]$$

is estimated for the series, and the results are presented in Table 12.1(b). No other outliers are detected when the outlier procedure is repeated with the new model parameter estimates. In this example we see that adjustment for the outliers has a little more effect on the estimate $\hat{\phi}$ of the time series parameter than in the previous case, and it reduces the estimate of σ_a^2 substantially by about 26%.

12.3 ESTIMATION FOR ARMA MODELS WITH MISSING VALUES

In many situations in practice, the values of a time series z_t will not be observed at equally spaced times because there will be "missing values" corresponding to some time points. In this section we discuss briefly the maximum likelihood estimation of parameters in an ARIMA(p, d, q) model for such situations, through consideration of the calculation of the exact

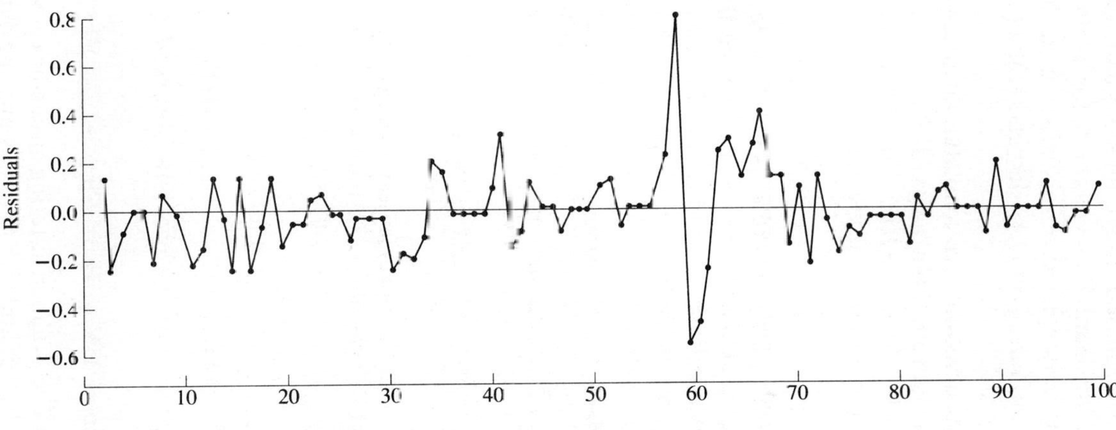

Figure 12.2 First 100 residuals from the ARIMA(1, 1, 0) model fitted to series C.

Gaussian likelihood function for the observed data. It is shown that in missing data situations, the likelihood function can readily and conveniently be constructed using the state space form of the model and associated Kalman filtering procedures, as discussed in Sections 5.5 and 7.5, but modified to accommodate the missing data. These methods for evaluation of the likelihood in cases of irregularly spaced observations have been examined by Jones [123], Harvey and Pierse [108], Ansley and Kohn [12], [13], and Wincek and Reinsel [205].

We suppose n observations are available at integer times $t_1 < t_2 < \cdots < t_n$, not equally spaced, from an ARIMA(p, d, q) process, which follows the model $\phi(B)(1 - B)^d z_t = \theta(B)a_t$. From Section 5.5, the process z_t has the state space model formulation given by

$$\boldsymbol{Y}_t = \boldsymbol{\Phi}\boldsymbol{Y}_{t-1} + \boldsymbol{\Psi}a_t \tag{12.3.1}$$

with $z_t = \mathbf{H}\boldsymbol{Y}_t = [1, 0, \ldots, 0]\boldsymbol{Y}_t$, where $\boldsymbol{Y}_t$ is the r-dimensional state vector and $r = \max(p + d, q + 1)$. Let $\Delta_i = t_i - t_{i-1}$ denote the time difference between successive observations $z_{t_{i-1}}$ and z_{t_i}, $i = 2, \ldots, n$. By successive substitutions, Δ_i times, on the right-hand side of (12.3.1), we obtain

$$\boldsymbol{Y}_{t_i} = \boldsymbol{\Phi}^{\Delta_i}\boldsymbol{Y}_{t_{i-1}} + \sum_{j=0}^{\Delta_i - 1} \boldsymbol{\Phi}^j\boldsymbol{\Psi}a_{t_i - j} \equiv \boldsymbol{\Phi}_i^*\boldsymbol{Y}_{t_{i-1}} + \mathbf{a}_{t_i}^* \tag{12.3.2}$$

where $\boldsymbol{\Phi}_i^* = \boldsymbol{\Phi}^{\Delta_i}$ and $\mathbf{a}_{t_i}^* = \Sigma_{j=0}^{\Delta_i - 1}\boldsymbol{\Phi}^j\boldsymbol{\Psi}a_{t_i - j}$, with

$$\text{cov}[\mathbf{a}_{t_i}^*] = \Sigma_i = \sigma_a^2 \sum_{j=0}^{\Delta_i - 1} \boldsymbol{\Phi}^j\boldsymbol{\Psi}\boldsymbol{\Psi}'\boldsymbol{\Phi}'^j$$

Thus (12.3.2) together with the observation equation $z_{t_i} = \mathbf{H}\boldsymbol{Y}_{t_i}$ constitute a state space model form for the *observed* time series data $z_{t_1}, z_{t_2}, \ldots, z_{t_n}$.

Therefore, the Kalman filter recursive equations as in (5.5.5)–(5.5.8) can be directly employed to obtain the state predictors $\hat{\boldsymbol{Y}}_{t_i|t_{i-1}}$ and their error covariance matrices $\mathbf{V}_{t_i|t_{i-1}}$. So we can obtain the predictors

$$\hat{z}_{t_i|t_{i-1}} = E[z_{t_i}|z_{t_{i-1}}, \ldots, z_{t_1}] = \mathbf{H}\hat{\boldsymbol{Y}}_{t_i|t_{i-1}} \tag{12.3.3}$$

for the observations z_{t_i} based on the previous *observed* data and their error variances

$$\sigma_a^2 v_i = \mathbf{H}\mathbf{V}_{t_i|t_{i-1}}\mathbf{H}' = E[(z_{t_i} - \hat{z}_{t_i|t_{i-1}})^2] \tag{12.3.4}$$

readily from the recursive Kalman filtering procedure. More specifically, the updating equations (5.5.5) and (5.5.6) in this missing data setting take the form

$$\hat{\boldsymbol{Y}}_{t_i|t_i} = \hat{\boldsymbol{Y}}_{t_i|t_{i-1}} + \mathbf{K}_i(z_{t_i} - \mathbf{H}\hat{\boldsymbol{Y}}_{t_i|t_{i-1}}) \tag{12.3.5}$$

with

$$\mathbf{K}_i = \mathbf{V}_{t_i|t_{i-1}}\mathbf{H}'[\mathbf{H}\mathbf{V}_{t_i|t_{i-1}}\mathbf{H}']^{-1} \tag{12.3.6}$$

while the prediction equations (5.5.7) are given by

$$\hat{\mathbf{Y}}_{t_i|t_{i-1}} = \mathbf{\Phi}_i^* \hat{\mathbf{Y}}_{t_{i-1}|t_{i-1}} = \mathbf{\Phi}^{\Delta_i} \hat{\mathbf{Y}}_{t_{i-1}|t_{i-1}}, \quad \mathbf{V}_{t_i|t_{i-1}} = \mathbf{\Phi}_i^* \mathbf{V}_{t_{i-1}|t_{i-1}} \mathbf{\Phi}_i^{*\prime} + \Sigma_i \tag{12.3.7}$$

with

$$\mathbf{V}_{t_i|t_i} = [\mathbf{I} - \mathbf{K}_i \mathbf{H}] \mathbf{V}_{t_i|t_{i-1}} \tag{12.3.8}$$

Notice that the calculation of the prediction equations (12.3.7) can be interpreted as computation of the successive one-step-ahead predictions

$$\hat{\mathbf{Y}}_{t_{i-1}+j|t_{i-1}} = \mathbf{\Phi} \hat{\mathbf{Y}}_{t_{i-1}+j-1|t_{i-1}}$$

$$\mathbf{V}_{t_{i-1}+j|t_{i-1}} = \mathbf{\Phi} \mathbf{V}_{t_{i-1}+j-1|t_{i-1}} \mathbf{\Phi}' + \sigma_a^2 \mathbf{\Psi}\mathbf{\Psi}'$$

for $j = 1, \ldots, \Delta_i$, without any updating since there are no observations available between the time points t_{i-1} and t_i to provide any additional information for updating.

The exact likelihood for the vector of observations $\mathbf{z}' = (z_{t_1}, z_{t_2}, \ldots, z_{t_n})$ is obtained directly from the quantities in (12.3.3) and (12.3.4) because the joint density of $\mathbf{z}$ can be expressed as the product of the conditional densities of the z_{t_i}, given $z_{t_{i-1}}, \ldots, z_{t_1}$, for $i = 2, \ldots, n$, which are Gaussian with (conditional) means and variances given by (12.3.3)–(12.3.4). Hence the joint density of the observations $\mathbf{z}$ can be expressed as

$$p(\mathbf{z}|\boldsymbol{\phi}, \boldsymbol{\theta}, \sigma_a) = \prod_{i=1}^{n} (2\pi\sigma_a^2 v_i)^{-1/2} \exp\left[-\frac{1}{2\sigma_a^2} \sum_{i=1}^{n} \frac{(z_{t_i} - \hat{z}_{t_i|t_{i-1}})^2}{v_i}\right] \tag{12.3.9}$$

In (12.3.9), the quantities $\hat{z}_{t_i|t_{i-1}}$ and $\sigma_a^2 v_i$ are directly determined from the recursive filtering calculations (12.3.5)–(12.3.8). In the case of a stationary ARMA(p, q) model, the initial conditions required to start the filtering procedure can be determined readily (see, e.g., Jones [123]). However, for the nonstationary ARIMA model situation, some additional assumptions need to be specified concerning the process and initial conditions, and appropriate methods for such cases have been examined by Ansley and Kohn [13].

As a simple example to illustrate the missing data methods, consider the stationary AR(1) model $(1 - \phi B)z_t = a_t$. Then (12.3.2) directly becomes (see, e.g., [168])

$$z_{t_i} = \phi^{\Delta_i} z_{t_{i-1}} + \sum_{j=0}^{\Delta_i - 1} \phi^j a_{t_i - j} \tag{12.3.10}$$

and it is readily determined that

$$\hat{z}_{t_i|t_{i-1}} = \phi^{\Delta_i} z_{t_{i-1}} \quad \text{and} \quad \sigma_i^2 = \sigma_a^2 v_i = \frac{\sigma_a^2(1 - \phi^{2\Delta_i})}{1 - \phi^2} \tag{12.3.11}$$

Hence the likelihood for the observed data in the first-order autoregressive model with missing values is as given in (12.3.9), with these expressions for $\hat{z}_{t_i|t_{i-1}}$ and $\sigma_a^2 v_i$.

Estimation of missing values of an ARMA process. A related problem of interest that often arises in the context of missing values for time series is that of estimation of the missing values. Earlier examinations of this problem of interpolation of missing values for ARIMA time series from a least squares viewpoint were made by Brubacher and Tunnicliffe Wilson [66], Damsleth [75], and Abraham [1]. Within the framework of the state space formulation, estimates of missing values and their corresponding error variances can be derived conveniently through the use of recursive smoothing methods associated with the Kalman filter, which are described in general terms in Anderson and Moore [5], for example. These methods have been considered more specifically for the ARIMA model missing data situation by Harvey and Pierse [108] and by Kohn and Ansley [129].

However, in the case of a pure autoregressive model, $\phi(B)z_t = a_t$, some rather simple and explicit interpolation results are available. For example, in an AR(p) process with a single missing value at time T surrounded by at least p consecutive observed values both before and after time T, it is well known (see, e.g., [66]) that the optimal interpolation of the missing value z_T is given by

$$\hat{z}_T = -d_0^{-1} \sum_{j=1}^{p} d_j(z_{T-j} + z_{T+j}) \tag{12.3.12}$$

where $d_j = \sum_{i=j}^{p} \phi_i \phi_{i-j}$, $\phi_0 = -1$, and $d_0 = 1 + \sum_{i=1}^{p} \phi_i^2$, with $E[(z_T - \hat{z}_T)^2] = \sigma_a^2 d_0^{-1} = \sigma_a^2(1 + \sum_{i=1}^{p} \phi_i^2)^{-1}$. Notice that the value in (12.3.12) can be expressed as $\hat{z}_T = z_T - [\phi(B)\phi(F)z_T/d_0]$, with interpolation error equal to

$$\hat{e}_T = z_T - \hat{z}_T = \frac{\phi(B)\phi(F)z_T}{d_0} \tag{12.3.13}$$

In the general ARMA model situation, Bruce and Martin [67], among others, have noted a close connection between the likelihood function construction in the case of missing values and the formulation of the consecutive data model likelihood with additive outliers (AO) specified for each time point that corresponds to a missing value. Hence, in effect, in such a time series additive outlier (AO) model for consecutive data, the estimate of the outlier effect parameter ω corresponds to the interpolation error in the missing data situation. For example, in the autoregressive model situation, compare the result in (12.3.13) with the result given following (12.2.5b) for the AO model. Furthermore, the sum of squares function in the likelihood (12.3.9) for the missing data situation is equal to the sum of squares obtained from a complete set of consecutive observations in which an AO has been assumed at each time point where a missing value occurs and for which the likelihood is evaluated at the maximum likelihood estimates for each of the corresponding AO outlier effect parameters ω, for given values of the time series model parameters $\boldsymbol{\phi}$ and $\boldsymbol{\theta}$. As an illustration, for the simple AR(1) model situation with a single isolated missing value at time T, from (12.3.11) the relevant term in the missing data sum of squares function is

$$\frac{(z_{T+1} - \phi^2 z_{T-1})^2}{1 + \phi^2} \equiv [(z_T - \hat{\omega}) - \phi z_{T-1}]^2 + [z_{T+1} - \phi(z_T - \hat{\omega})]^2 \tag{12.3.4}$$

$$= (\hat{z}_T - \phi z_{T-1})^2 + (z_{T+1} - \phi \hat{z}_T)^2$$

where

$$\hat{\omega} = z_T - \frac{\phi}{1 + \phi^2}(z_{T-1} + z_{T+1}) = z_T - \hat{z}_T$$

is the maximum likelihood estimate of the outlier effect ω in the AO model (12.2.1), and the latter expressions in (12.3.14) represent the sum of squares terms in the consecutive data situation but with an AO modeled at time T.

Part Four

DESIGN OF DISCRETE CONTROL SCHEMES

In earlier chapters we studied the modeling of discrete time series and dynamic systems. We saw how once adequate models have been obtained, they may be put to use to yield forecasts of time series and to characterize the transfer function of a dynamic system. However, the models and the methods for their manipulation are of much wider importance than even these applications indicated. The ideas we have outlined are of importance in the analysis of a wide class of stochastic–dynamic systems occurring for example, in economics, engineering, commerce, and in organizational studies.

It is obviously impossible to illustrate every application. Rather, it is hoped that the theory and examples of this book may help the reader to adapt the general methods to his own particular problems. In doing this, the dynamic and stochastic models we have discussed will often act as *building blocks* which can be linked together to represent the particular system under study. Also, techniques of identification, estimation, and diagnostic checking, similar to those we have illustrated, will be needed to establish the model. Finally, recursive calculations and the ideas considered under the general heading of forecasting will have wider application in working out the consequences of a model once it has been fitted.

We shall conclude this book by illustrating these possibilities in one further application—the design of feedback control schemes. In working through Chapter 13, it is the exercise of bringing together the previously discussed ideas in a fresh application, quite as much as the detailed results, which we hope will be of value.

13

ASPECTS OF PROCESS CONTROL

The term *process control* is used in different ways. Shewhart charts and other quality control charts are frequently employed in industries concerned with the manufacture of discrete "parts" in what is called *statistical process control* (SPC). By contrast, various forms of feedback and feedforward adjustment are used, particularly in the process and chemical industries, in what we shall call *engineering process control* (EPC). Because the adjustments made by engineering process control are usually computed and applied automatically, this type of control is sometimes called *automatic process control* (APC). However, the manner in which adjustments are applied is a matter of convenience, so we will not use that terminology here. The object of this chapter is to draw on the earlier discussions in this book to provide insight into the statistical aspects of these control methods and to appreciate better their relationships and objectives.

We first discuss *process monitoring* using, for example, Shewhart control charts and contrast this with techniques for *process adjustment*. In particular, a common adjustment problem is to maintain an output variable close to a target value by manipulation of an input variable, to obtain feedback control. We consider this problem first in a purely intuitive way and then relate this to some of the previously discussed stochastic and transfer function models to yield control schemes producing minimum mean square error (MMSE) at the output. This leads to a discussion of discrete schemes which are analogs of the proportional–integral (PI) schemes of engineering control and we show how simple charts may be devised for *manually adjusting* processes with PI control.

It turns out that minimum mean square error control often requires excessively large adjustments of the input variable. ''Optimal'' constrained schemes are, therefore, introduced which require much smaller adjustments at the expense of only minor increases in the output mean square error. These constrained schemes are generally not PI schemes, but in certain important cases, it turns out that appropriately chosen PI schemes can often closely approximate their behavior.

Particularly in industries concerned with the manufacture of parts, there may be a fixed cost associated with adjusting the process and, in some cases, a monitoring cost associated with obtaining an observation. We therefore discuss bounded adjustment schemes for feedback control which minimize overall cost in these circumstances. Finally, we consider a general procedure for monitoring control schemes for possible changes in parameter values using Cuscore charts. More general discussion is given in the appendices and references.

13.1 PROCESS MONITORING AND PROCESS ADJUSTMENT

Process control is no less than an attempt to cancel out the effect of a fundamental physical law—the second law of thermodynamics, which implies that if left to itself, the entropy or disorganization of any system can never decrease and will usually increase. Statistical process control (SPC) and engineering process control (EPC) are two complementary approaches to combat this law. SPC attempts to *remove* disturbances using *process monitoring*, while EPC attempts to *compensate* them using *process adjustment* (see also Box and Kramer [52]).

13.1.1 Process Monitoring

The SPC strategy for stabilization of a process is to standardize procedures and raw materials and to use hypothesis-generating devices (such as graphs, check sheets, Pareto charts, cause–effect diagrams, etc.) to track down and eliminate causes of trouble (see, e.g., Ishikawa [116]). Since searching for assignable causes is tedious and expensive, it usually makes sense to wait until ''statistically significant'' deviations from the stable model occur before instituting this search. This is achieved by the use of process *monitoring charts* such as Shewhart charts, Cusum charts, and Roberts' EWMA charts. The philosophy is ''don't fix it when it ain't broke''—don't needlessly tamper with the process (see, e.g., Deming [79]).

Figure 13.1 shows an example of process monitoring using a Shewhart control chart. Condoms were tested by taking a sample of 50 items every 2 hours from routine production, inflating them to a very high fixed pressure

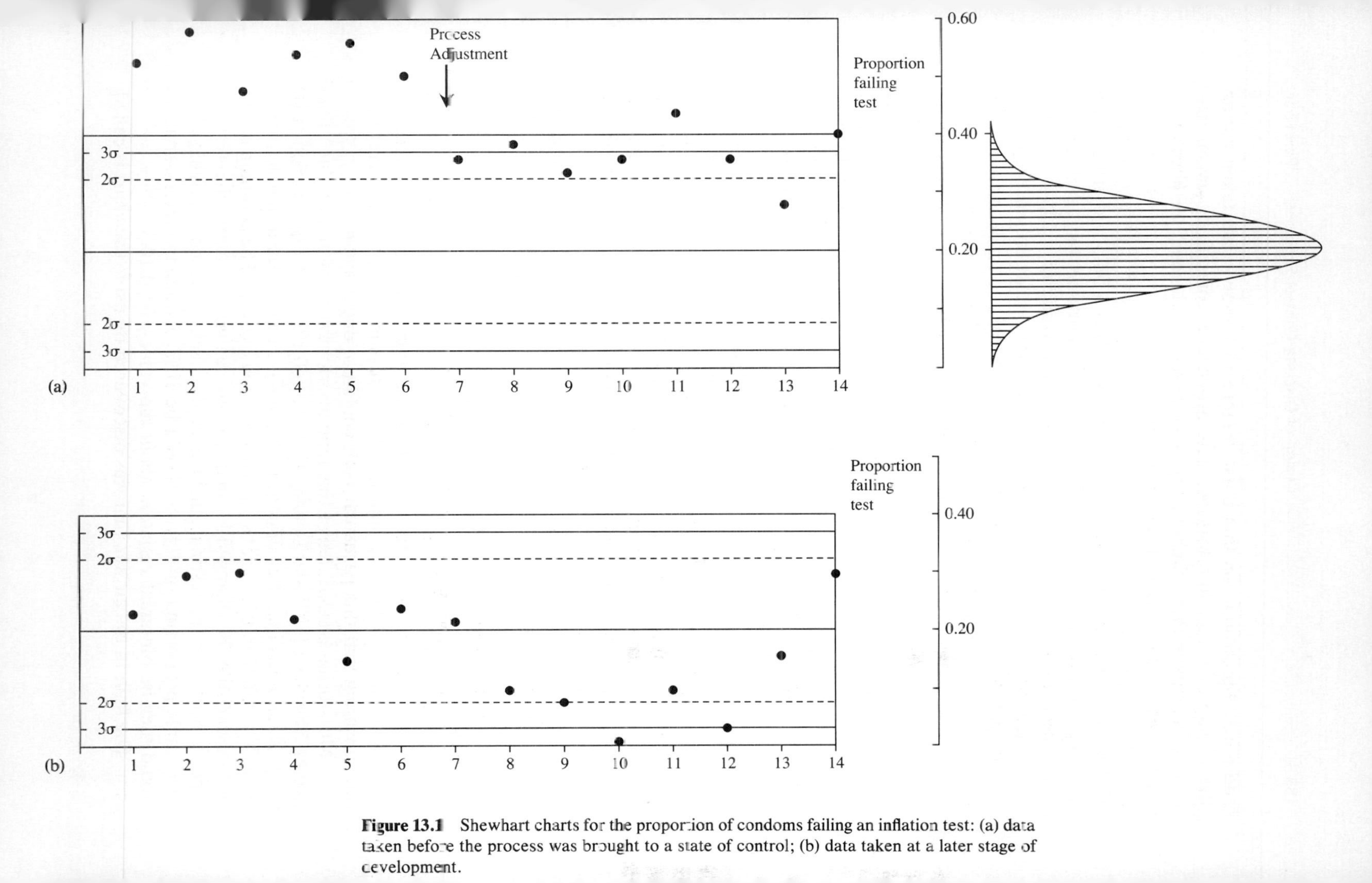

Figure 13.1 Shewhart charts for the proportion of condoms failing an inflation test: (a) data taken before the process was brought to a state of control; (b) data taken at a later stage of development.

and noting the proportion that burst. Figure 13.1 shows data taken during the startup of a machine making these articles. Studies from similar machines had shown that a high-quality product was produced if the proportion failing this very severe test was $p = 0.20$.

The *reference* distribution indicated by the bars on the right of Figure 13.1(a) characterizes desired process behavior. It is a binomial distribution showing the probabilities of getting various proportions failing in random samples of $n = 50$ when p stays constant at a value of 0.20. If the data behaved like a random sequence from this reference distribution, we should say the process appeared to be in a *state of control* and no action would be called for. By contrast, if the data did not have this appearance, showing outliers or suspicious patterns, we might have reason to suppose that something else was going on. In practice, the whole reference distribution would not usually be shown. Instead, upper and lower control limits and warning lines would be set. When, as in this case, a normal curve (shown as a continuous line) provides a close approximation to the reference distribution, these are usually set at $\pm 2\sigma$ and $\pm 3\sigma$ with $\sigma = \sqrt{p\,(1 - p)/n}$, the standard deviation of the sample proportion from a binomial distribution. In this example, with $p = 0.20$ and $n = 50$, this gives $\sigma = 0.057$. Figure 13.1(a) shows that during the startup phase the process was badly out of control, with the proportion of items failing the test initially as high as 50%. A process adjustment made after 12 hours of operation brought the number of defectives down to around 40%, but further changes were needed to get the process to a state of control at the desired level of $p = 0.20$. By a series of management actions, this was eventually achieved and Figure 13.1(b) shows the operation of the process at a later stage of development. Although for the most part, the system now appears to be in the desired state of control, notice that the tenth point on the chart fell *below the lower* $\pm 3\sigma$ line. Subsequent investigation showed that the testing procedure was responsible for this aberrant point. A fault in the air line had developed and the condoms tested at about this time were inadvertently submitted to a much reduced air pressure, resulting in a falsely low value of the proportion defective. Corrective action was taken and the system was modified so that the testing machine would not function unless the air pressure was at the correct setting, ensuring that this particular fault could not occur again.

Monitoring procedures of this kind are obviously of great value. Following Shewhart [177] and Deming [79], we refer to the natural variation in the process when in a state of control (binomial variation for a sample of $n = 50$ with $p = 0.20$ in this case) as due to *common causes*. The common cause system can only be changed by management action that alters the system. Thus a new type of testing machine might be introduced for which the acceptable proportion of defects should be 10%. Common cause variation would then be binomial variation about the value $p = 0.10$

The fault in the air line that was discovered by using the chart is called a *special** *cause*. By suitable "detective" work, it is often possible for the plant operators to track down and eliminate special causes. The objectives of process monitoring are thus (1) to establish and continually confirm that the desired common cause system remains in operation and (2) to look for deviations unlikely to be due to chance that can lead to the tracking and elimination of assignable causes of trouble.

13.1.2 Process Adjustment

Although we must always make a dedicated endeavor to remove causes of variation such as unsatisfactory testing methods, differences in raw materials, differences in operators, and so on, some processes cannot be fully brought to a satisfactory state of stability in this way. Despite our best efforts, there remains a tendency for the process to wander off target. This may be due to known but uncontrollable phenomena such as variations in ambient temperature, humidity, and feedstock quality, or to causes currently unknown. In such circumstances, some system of process *adjustment or regulation* may be necessary in which manipulation of some additional variable is used to *compensate* for deviations in the quality characteristic.

To fix ideas, we first introduce a simple feedback adjustment scheme relying on a purely empirical argument and leave theoretical justification until later. Consider the measurements shown in Figure 13.2 of the thickness of a very thin metallic film taken at equally spaced units of time. The quality characteristic was badly out of control but standard procedures failed to stabilize it (Box [36]). Suppose that the *disturbance* N_t is defined as the deviation of this quality characteristic from its target value T when *no adjustment is made*; that is, N_t is the underlying noise process. Suppose also that there is a manipulable variable—deposition rate X—which can be used conveniently to adjust the thickness and that a unit change in X will produce g units of change in thickness and will take full effect in one time interval. If at time t, X was set equal to X_t, then at time $t + 1$ the deviation from target, $\varepsilon_{t+1} = Y_{t+1} - T$, after adjustment would be

$$\varepsilon_{t+1} = gX_t + N_{t+1} \tag{13.1.1}$$

Now suppose that *at time t* you can, in some way or other, compute an estimate (forecast) $\hat{N}_t(1)$ *of* N_{t+1} and that this forecast has an error $e_t(1)$, so that

$$N_{t+1} = \hat{N}_t(1) + e_t(1) \tag{13.1.2}$$

*Also called an "assignable" cause. However, we are sometimes faced with a system that is demonstrably not in a state of control and yet no causative reason can be found. So we will stay with Deming in his less optimistic word "special."

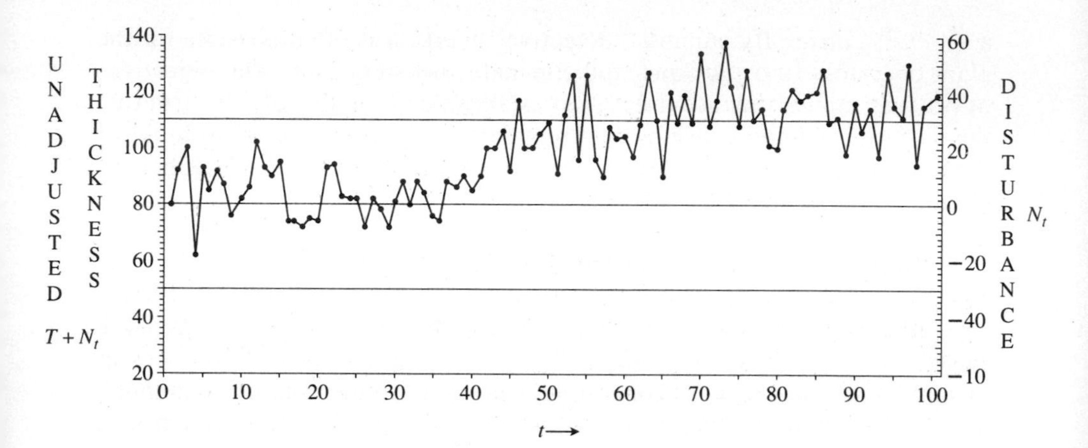

Figure 13.2 One hundred successive values of the thickness of a metallic film when no adjustment was applied.

Then using (13.1.1) and (13.1.2),

$$\varepsilon_{t+1} = gX_t + \hat{N}_t(1) + e_t(1) \tag{13.1.3}$$

If, in particular, X can be adjusted so that at time t,

$$X_t = -\frac{1}{g}\hat{N}_t(1) \tag{13.1.4}$$

then for the adjusted process

$$\varepsilon_{t+1} = e_t(1) \tag{13.1.5}$$

Thus the deviation from target ε_{t+1} for the *adjusted* process would now be the error $e_t(1)$ in *forecasting* N_{t+1}, instead of the deviation N_{t+1} measured when the process is not adjusted.

If we used measurements of one or more of the known disturbing *input* factors (e.g., ambient temperature) to calculate the estimate $\hat{N}_t(1)$ of N_{t+1}, we would have an example of *feedforward* control. If the estimate $\hat{N}_t(1)$ of N_{t+1} directly or indirectly used only present and past values of the *output* disturbance $N_t, N_{t-1}, N_{t-2}, \ldots$, equation (13.1.4) would define a system of *feedback* control. A system of mixed *feedback–feedforward* control would employ both kinds of data. For simplicity, we will here consider only the feedback case.

13.2 PROCESS ADJUSTMENT USING FEEDBACK CONTROL

Empirical introduction. It might often be reasonable to use for the estimate $\hat{N}_t(1)$ in (13.1.4) some kind of weighted average of past values N_t,

$N_{t-1}, N_{t-2}, \ldots$. In particular, an *exponentially* weighted moving average (EWMA) has intuitive appeal since recently occurring data are given most weight. Suppose, then, that $\hat{N}_t(1)$ is an EWMA,

$$\hat{N}_t(1) = \lambda(N_t + \theta N_{t-1} + \theta^2 N_{t-2} + \cdots) \qquad 0 \leq \theta \leq 1 \qquad (13.2.1)$$

where θ is the smoothing constant and $\lambda = 1 - \theta$.

We first consider the situation where, as has usually been the case in the process industries, adjustments are continually made as each observation comes to hand. Then using equation (13.1.4), the *adjustment* (*change* in deposition rate) made at time t would be given by

$$X_t - X_{t-1} = -\frac{1}{g}[\hat{N}_t(1) - \hat{N}_{t-1}(1)] \qquad (13.2.2)$$

Now with $e_{t-1}(1) = N_t - \hat{N}_{t-1}(1)$ the forecast error, the updating formula for an EWMA forecast can be written

$$\hat{N}_t(1) - \hat{N}_{t-1}(1) = \lambda e_{t-1}(1) \qquad (13.2.3)$$

Therefore, for any feedback scheme in which the compensatory variable X was set so as to cancel out an EWMA of the noise $\{N_t\}$, the required adjustment should be such that

$$X_t - X_{t-1} = -\frac{\lambda}{g} e_{t-1}(1) = -\frac{\lambda}{g} \varepsilon_t \qquad (13.2.4)$$

For the metal deposition process, $g = 1.2$, $\lambda = 0.2$, and $T = 80$, so that the adjustment equation is

$$X_t - X_{t-1} = -\tfrac{1}{6}\varepsilon_t \qquad (13.2.5)$$

13.2.1 Feedback Adjustment Chart

This kind of adjustment is very easily applied, as is shown in Figure 13.3. This shows a manual feedback adjustment chart (Box and Jenkins [48]) for the metallic thickness example given previously. To use it the operator records the latest value of thickness and reads off on the adjustment scale the appropriate amount by which he or she should now increase or decrease the deposition rate. For example, the first recorded thickness of 80 is on target, so no action is called for. The second value of 92 is 12 units above the target, so $\varepsilon_2 = 12$, corresponding on the left-hand scale to a deposition rate adjustment of $X_2 - X_1 = -2$. Thus the operator should now reduce the deposition rate by 2 units from its previous level.

Notice that the successive recorded thickness values shown on this chart are the readings that would actually occur *after adjustment*; the underlying disturbance is, of course, not seen on this chart. In this example, over the recorded period of observation the chart produces a more than fivefold

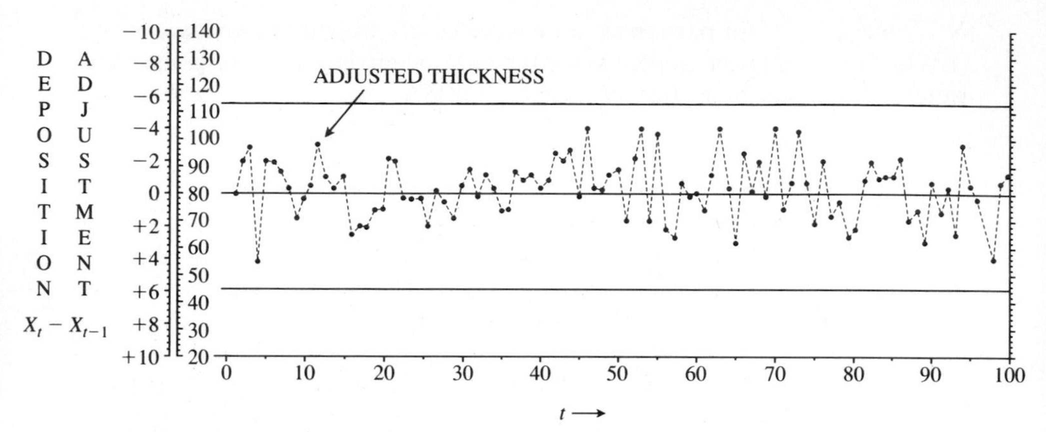

Figure 13.3 Manual adjustment chart for thickness which allows the operator to read off the appropriate change in deposition rate.

reduction in mean square error; the standard deviation of the adjusted thickness being now only about $\sigma_\varepsilon = 11$. Notice that:

1. The chart is no more difficult to use than a Shewhart chart.
2. While the "intuitive" adjustment would be $-(1/g)\ \varepsilon_t = -\frac{5}{6}\ \varepsilon_t$ (corresponding to what Deming calls "tinkering"), the adjustment given by equation (13.2.4) is $-(\lambda/g)\ \varepsilon_t = -\frac{1}{6}\ \varepsilon_t$. Thus it uses a discounted or "damped" *estimate* $\lambda\varepsilon_t$ of the deviation from target to determine the appropriate adjustment, where the discount factor λ is $1 - \theta$, with θ the smoothing constant of the EWMA estimate of the noise.
3. By summing equation (13.2.4) we see that the *total* adjustment at time t is

$$X_t = k_0 + k_I \sum_{i=1}^{t} \varepsilon_i \tag{13.2.6}$$

   with $k_0 = X_0$ and $k_I = -\lambda/g$. This adjustment procedure thus depends on the *cumulative sum* of the adjustment errors ε_i and the constant k_I determines how much the "intuitive" adjustment is discounted.
4. It follows from the previous argument that the adjustment is also equivalent to estimating at each time t the next value of the total unadjusted disturbance N_{t+1} by an *exponentially weighted average* of its past values and using this estimate to make an appropriate adjustment. This is illustrated for the metallic thickness example in Figure 13.4. Notice that in this preliminary discussion we have not explicitly assumed any particular time series model or claimed any particular optimal proper-

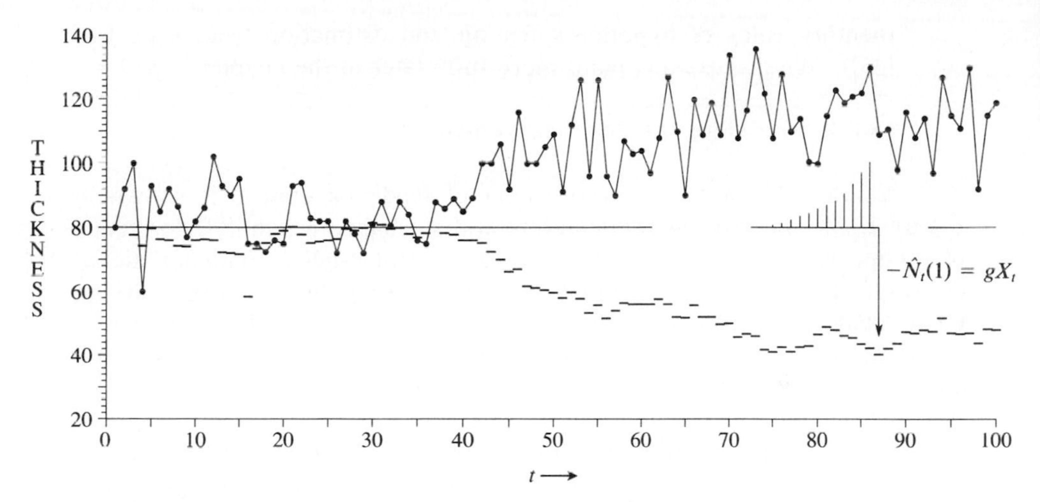

Figure 13.4 Dashes indicate the total adjustment $\hat{N}_t(1) = -gX_t$ achieved by the manual adjustment chart of Figure 13.3.

ties for the procedure. That the procedure can be discussed in such terms accounts, to some extent, for its remarkable robustness, which we discuss later.

In summary, then:

1. By *process monitoring* we mean the use of, for example, Shewhart charts and/or Cusum or Cuscore charts (Box and Ramírez [57]). These are devices for continually checking a model that represents the desired ideal stable state of the system: for example, normal, independent, identically distributed (NIID) variation about a fixed target T. The use of such charts can lead to the elimination of special causes pointed to by discrepant behavior. The judgment that behavior is sufficiently discrepant to merit attention is decided by a process analogous to *hypothesis testing*. Its properties are described in terms of probabilities (e.g., the probability of a point falling outside the 3σ limits of a Shewhart chart).
2. By *process adjustment* we mean the use of feedback and feedforward control or some combination of these to maintain the process as close as possible to some desired target value. Process adjustment employs a system of statistical *estimation* (forecasting) rather than of hypothesis testing, and its properties are described, for example, by output *mean square error*. Process monitoring and process adjustment are complementary rather than competitive corresponding to the comple-

mentary roles of hypothesis testing and estimation (see, e.g., Box [35]). We discuss this point more fully later in the chapter.

13.2.2 Modeling the Feedback Loop

A somewhat more general system of feedback control is shown in Figure 13.5. The process is affected by a disturbance which in the absence of compensatory action would cause the output quality characteristic to deviate from target by an amount N_t. Thus $\{N_t\}$ is a time series exemplifying what would happen at the output if no control were applied. In fact, a compensating variable X_t (deposition rate in our example) can be manipulated to cancel out this disturbance as far as is possible. Changes in X will pass through the process and be acted on by its dynamics to produce at time t an amount of compensation $\mathcal{Y}_t$ at the output (again measured as a deviation from target). To the extent that this compensation $\mathcal{Y}_t$ fails to cancel out the disturbance N_t, there will be an error, or deviation from target $\varepsilon_t = Y_t - T$, equal to $\varepsilon_t = N_t + \mathcal{Y}_t$. The controller is some means (automatic or manual) that brings into effect the control equation $X_t = f(\varepsilon_t, \varepsilon_{t-1}, \ldots)$, which adjusts the output depending on present and past errors.

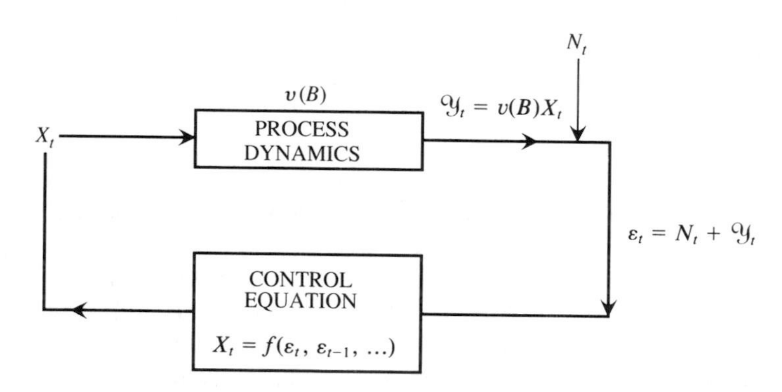

Figure 13.5 Feedback control loop.

A device that has been used in the process industries for many years is the *three-term controller*. Controllers of this kind are usually operated automatically and employ continuous rather than discrete measurement and adjustment. If ε_t is the error at the output at time t, control action could, in particular, be made proportional to ε itself, to its integral with respect to time, or to its derivative with respect to time. A three-term controller uses a linear combination of these modes of control action, so that if X_t indicates the level of the manipulated variable at time t, the control equation is of the form

$$X_t = k_0 + k_D \frac{dc_t}{dt} + k_P \varepsilon_t + k_I \int \varepsilon_t \, dt \tag{13.2.7}$$

where k_D, k_P, and k_I are constants.

Frequently, only one or two of these three modes of action are used. In particular, if only k_I is nonzero ($k_D = 0$, $k_P = 0$), we have *integral* control. If only k_I and k_P are nonzero ($k_D = 0$), we have *proportion–integral* (PI) control.

Notice that in the example we have just discussed, where the result of any adjustment fully takes effect at the output in one time interval, the dynamics of the process are represented by $\mathcal{Y}_t = gX_{t-1} = gBX_t$. The control equation $X_t = k_0 + k_I \Sigma_{i=1}^{t} \varepsilon_i$ in (13.2.6) is then the discrete analog of the control engineer's *integral* control.

In general, the discrete analog of (13.2.7) is

$$X_t = k_0 + k_D \nabla \varepsilon_t + k_P \varepsilon_t + k_I \sum_{i=1}^{t} \varepsilon_i$$

or in terms of the adjustment to be made,

$$\begin{aligned} x_t = X_t - X_{t-1} &= k_D \nabla^2 \varepsilon_t + k_P \nabla \varepsilon_t + k_I \varepsilon_t \\ &= c_1 \varepsilon_t + c_2 \varepsilon_{t-1} + c_3 \varepsilon_{t-2} \end{aligned}$$

where c_1, c_2, and c_3 are suitable constants. Not unexpectedly, control equations of this type are of considerable practical value.

13.2.3 Simple Models for Disturbances and Dynamics

So far we introduced a simple system of feedback control on purely empirical grounds. The *efficiency* of any such system will depend on the nature of the disturbance and the dynamics of the process. From a theoretical point of view, we can consider very general models for noise and dynamics and then proceed to find the control equation that "optimizes" the system in accordance with some criterion. However, the practical effectiveness of such models are usually determined by whether they, and the "optimization" criterion, make broad *scientific* sense and by their robustness to likely deviations from the ideal. We have already borne this in mind in first discussing control procedures from a purely commonsense point of view and shall continue to do so when choosing models for the disturbance and for process dynamics.

Characterizing appropriate disturbance models with a variogram. A tool that helps to characterize process disturbances is the standardized variogram, which measures the variance of the difference between observations m steps apart compared to the variance of the difference of observations one step apart:

$$G_m = \frac{\text{var}[N_{t+m} - N_t]}{\text{var}[N_{t+1} - N_t]} \equiv \frac{V_m}{V_1} \tag{13.2.8}$$

For a stationary process, G_m is a simple function of the autocorrelation function. In fact, then, $G_m = (1 - \rho_m)/(1 - \rho_1)$. However, the variogram can be used to characterize nonstationary as well as stationary behavior. Figure 13.6 shows realizations of 100 observations initially on target generated by (a) a white noise process, (b) a first-order autoregressive process, and (c, d, e, f) IMA (0, 1, 1) processes with $\lambda = 0.1, 0.2, 0.3, 0.4$, respectively. The corresponding standardized variograms for these time series models are also shown.

In some imaginary world we might, once and for all, set the controls of a machine and give a set of instructions to an ever-alert and never-forgetting

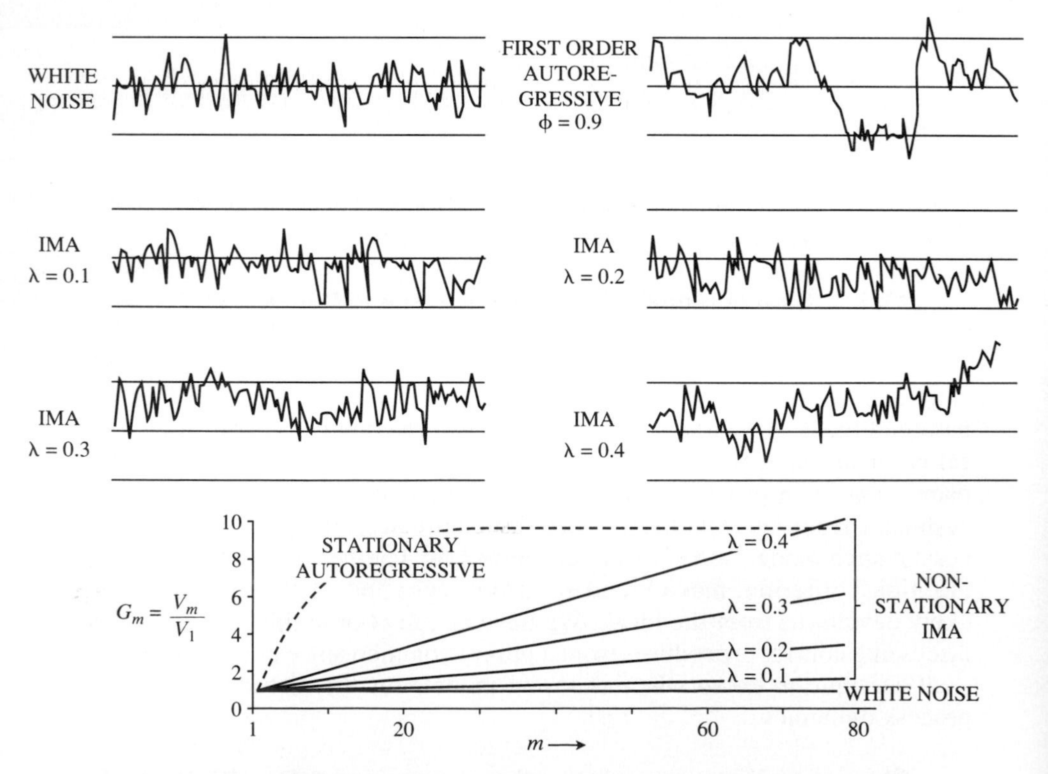

Figure 13.6 Realizations of white noise, autoregressive, and IMA (0, 1, 1) time series with theoretical variograms.

operator, and this would yield a perfectly stable process from that point on. In such a case the disturbance might be represented by a "white noise" series, and its corresponding standardized variogram G_m would be independent of m and equal to 1. But, in reality, left to themselves, machines involved in production are slowly losing adjustment and wearing out, and left to themselves, people tend, gradually, to forget instructions and miscommunicate. Thus for an *uncontrolled* disturbance, some kind of monotonically increasing variogram would be expected. We cannot obtain such a variogram from a linear *stationary model*, for although G_m can initially increase with m, it will always approach an asymptote. That this can happen quite quickly, even when successive observations are highly positively correlated, is illustrated by the variogram shown in the figure for the first-order stationary autoregressive time series model $N_t = 0.9N_{t-1} + a_t$. In this example, even though successive deviations N_t from the target value have autocorrelation 0.9, G_m is already within 5% of its asymptotic value after only 20 lags. This implies that, for example, when generated by such a model, observations 100 steps apart differ little more than those 20 steps apart.

A model that can approximate the behavior of an uncontrolled system which *continuously* increases its entropy may be arrived at by thinking of the disturbance as containing two parts, a transitory part b_t and a nontransitory part w_t:

$$N_t = b_t + w_t \tag{13.2.9}$$

The transitory part b_t is associated only with the tth observation and is supposed independent of observations taken at every other time. Typical sources contributing to b_t are measurement and sampling errors. We represent this transitory part by random drawings from a distribution having mean zero and variance σ_b^2.

"Sticky innovation" model. The evolving nontransitory part w_t represents innovations that enter the system from time to time and *get stuck* there. These "sticky" innovations can arise from a multitude of causes, such as wear, corrosion, and human miscommunication. Thus a car tire hits a sharp stone and *from that point onward* the tread is slightly damaged; a tiny crater caused by corrosion appears on the surface of a driving shaft and *remains there;* certain details in the standard procedure for taking blood pressure in a hospital are forgotten and from that point on *permanently omitted or changed.* It is these nontransitory or sticky innovations that constitute the unwanted "signal" we wish to cancel out. Every system is subject to such influences. They continuously drive the increase in entropy if nothing is done to combat them. Such a sticky innovation model was suggested by Barnard [21] and has a variogram that increases linearly with

m. A special case of this model, which may also be used to approximate it, is the IMA(0, 1, 1) model

$$N_t - N_{t-1} = a_t - \theta a_{t-1} \tag{13.2.10}$$

Since for this model the EWMA of equation (13.2.1) with smoothing parameter θ provides a minimum mean square error (MMSE) forecast with forecast error $e_{t-1}(1) = a_t$, the corresponding discrete "integral" controller of (13.2.6) with $k_I = -\lambda/g$ produces MMSE control with $\varepsilon_t = a_t$. As we discuss later more formally, this is then a special case of the general MMSE linear feedback control scheme.

Dynamics. In the discussion of the integral control scheme of equation (13.2.6), we supposed that any change made at the input of the system would have its full effect at the output in one time interval. The assumed dynamic equation for the response effect $\mathcal{Y}_t$ was, therefore,

$$\mathcal{Y}_t = gBX_{t+} \tag{13.2.11}$$

A somewhat more general assumption is that the system can be described by the first-order difference equation

$$(1 + \xi\nabla)\mathcal{Y}_t = gBX_{t+} \tag{13.2.12}$$

[see, e.g., (10.3.6)] or, equivalently,

$$(1 - \delta B)\mathcal{Y}_t = (1 - \delta)gBX_{t+} \qquad -1 < \delta < 1 \tag{13.2.13}$$

where $\xi = \delta/(1 - \delta)$ or, equivalently, $\delta = \xi/(1 + \xi)$. In that case at time $t + 1$ [cf. (13.1.1)], the deviation from target after adjustment is

$$\varepsilon_{t+1} = \mathcal{Y}_{t+1} + N_{t+1}$$

so that

$$\varepsilon_{t+1} = \frac{(1 - \delta)g}{1 - \delta B} X_{t+} + \hat{N}_t(1) + e_t(1)$$

where $\hat{N}_t(1)$ is some forecast of N_{t+1} made at time t with forecast error $e_t(1)$. Then if we use the adjustment equation

$$X_{t+1} - X_t = x_t = -\frac{1 - \delta B}{(1 - \delta)g}[\hat{N}_t(1) - \hat{N}_{t-1}(1)]$$

the deviation ε_{t+1} from the target is equal to the forecast error $e_t(1)$. Thus again we substitute the error in *forecasting* N_{t+1} for the deviation N_{t+1} itself. In particular, if $\hat{N}_t(1)$ is an EWMA forecast with smoothing parameter θ and if $\lambda = 1 - \theta$, then using (13.2.3),

$$x_t = (1 - B)X_t = -\frac{\lambda(1 - \delta B)}{g(1 - \delta)}\varepsilon_t = -\frac{\lambda(1 - \delta) + \delta\nabla}{g(1 - \delta)}\varepsilon_t \tag{13.2.14}$$

Finally, if N_t can be represented by an IMA (0, 1, 1) process with parameter θ, then $\varepsilon_t = a_t$ and this adjustment will yield MMSE control. After summing (13.2.14), we obtain

$$X_t = k_0 + k_P \varepsilon_t + k_I \sum_{i=1}^{t} \varepsilon_i \tag{13.2.15}$$

in which

$$k_P = -\frac{\lambda}{g}\xi \quad \text{and} \quad k_I = -\frac{\lambda}{g}$$

The control equation (13.2.15) yields the discrete analog of continuous proportional–integral (PI) control mentioned earlier and will hereafter be referred to as (discrete) PI control.

Notice that despite their interesting ramifications, the adjustment equations corresponding to discrete integral control and proportion–integral control are extremely simple and intuitive. For discrete integral control

$$x_t = c_1 \varepsilon_t \qquad (\text{with } c_1 = k_I)$$

and for proportional–integral control

$$x_t = c_1 \varepsilon_t + c_2 \varepsilon_{t-1} \qquad (\text{with } c_1 = k_I + k_P \text{ and } c_2 = -k_P)$$

They thus make the adjustment x_t depend linearly on the last error and the last two errors, respectively.

13.2.4 General Minimum Mean Square Error Feedback Control Schemes

Arguing as earlier, it is not difficult to derive theoretical minimum mean square error feedback control schemes for the more general stochastic and linear dynamic models discussed in Chapters 4 and 10. Suppose that the response to the series of adjustments in the manipulable input variable X_t is represented by the dynamic transfer function relation (10.2.3), written as

$$\mathcal{Y}_t = L_1^{-1}(B)L_2(B)B^{f+1}X_{t+}$$

where $L_1(B)$ and $L_2(B)$ are polynomials in B. This relation allows for f periods of pure dead time in the response. In addition, assume that the noise or process disturbances $\{N_t\}$ may be represented by the linear stochastic ARIMA process defined by

$$N_t = \varphi^{-1}(B)\theta(B)a_t = \left(1 + \sum_{i=1}^{\infty} \psi_i B^i\right)a_t$$

where a_t is a white noise process. Then the error at the output, $\varepsilon_{t+f+1} = Y_{t+f+1} - T$, at time $t + f + 1$ can be written

$$\varepsilon_{t+f+1} = \mathcal{Y}_{t+f+1} + N_{t+f+1} = L_1^{-1}(B)L_2(B)X_{t+} + N_{t+f+1}$$

Now we can write $N_{t+f+1} = \hat{N}_t(f + 1) + e_t(f + 1)$, where $\hat{N}_t(f + 1)$ is the *forecast* at time t of N_{t+f+1} and $e_t(f + 1)$ is the error of the forecast for $f + 1$ steps ahead. The noise N_{t+f+1} is not known at time t, but its minimum mean square error forecast $\hat{N}_t(f + 1)$ can be deduced from the error sequence $\varepsilon_t, \varepsilon_{t-1}, \varepsilon_{t-2}, \ldots$ which is observed. Thus it follows that the control equation $X_{t+} = -L_1(B)L_2^{-1}(B)N_t(f + 1)$ will produce at time $t + f + 1$ a level at the output that will cancel out the forecast of the noise $f + 1$ periods ahead, and the error at the output will then be $\varepsilon_{t+f+1} = e_t(f + 1)$, the error of the forecast. To express the control equation in terms of the error sequence ε_t's, we can write

$$\varepsilon_t = e_{t-f-1}(f + 1) = a_t + \psi_1 a_{t-1} + \cdots + \psi_f a_{t-f} = L_4(B)a_t$$

and

$$\hat{N}_t(f + 1) = \psi_{f+1}a_t + \psi_{f+2}a_{t-1} + \cdots = L_3(B)a_t$$

where the operators $L_3(B)$ and $L_4(B)$ are determined from knowledge of the model $N_t = \varphi^{-1}(B)\theta(B)a_t = \psi(B)a_t$ for the noise process. Hence we have

$$\hat{N}_t(f + 1) = L_3(B)L_4^{-1}(B)\varepsilon_t$$

Therefore, the minimum mean square error (MMSE) feedback control equation is then

$$X_{t+} = -\frac{L_1(B)L_3(B)}{L_2(B)L_4(B)}\varepsilon_t \tag{13.2.16}$$

Alternatively, as is usually convenient, we can define the control action in terms of the *adjustment* $x_t = X_{t+} - X_{t-1+}$ to be made at time t as

$$x_t = -\frac{L_1(B)L_3(B)(1 - B)}{L_2(B)L_4(B)}\varepsilon_t$$

Example: model with dead time. In particular, the more general dynamic model used above allows for "dead time"—that is, pure delay in response to adjustment. To illustrate the application of equation (13.2.16), consider a first-order system affected by between f and $f + 1$ unit intervals of pure delay so that

$$(1 - \delta B)\mathcal{Y}_t = g(1 - \delta)[(1 - \nu) + \nu B]B^f X_{t-1} \tag{13.2.17}$$

Combining this with the IMA(0, 1, 1) disturbance model of equation (13.2.10), we can use the general derivation above to obtain the minimum mean square control scheme. In terms of the general model, we have $L_2(B)/L_1(B) = g(1 - \delta)(1 - \nu\nabla)/(1 - \delta B)$, and the IMA noise model yields $\hat{N}_t(f + 1) - \hat{N}_{t-1}(f + 1) = \lambda a_t$, so that $L_3(B) = \lambda/(1 - B)$, and also

$$e_{t-f-1}(f + 1) = [1 + \lambda(B + B^2 + \cdots + B^f)]a_t \equiv L_4(B)a_t$$

Hence, for the adjustment x_t, we have the relation $L_2(B)L_4(B)x_t = -L_1(B)L_3(B)(1 - B)\varepsilon_t$ and we obtain the MMSE control equation as

$$(1 - \nu\nabla)[1 + \lambda(B + B^2 + \cdots + B^f)]x_t = -\frac{\lambda}{g(1 - \delta)}(1 - \delta B)\varepsilon_t$$

Thus this optimal control scheme is not PI but is of the form

$$x_t = c_1 x_{t-1} + c_2 x_{t-2} + \cdots + c_f x_{t-f-1} + c(\varepsilon_t - \delta\varepsilon_{t-1}) \qquad (13.2.18)$$

where $c = -\lambda/[g(1 - \delta)] = k_I + k_P$.

An interesting example by Fearn and Maris [90] describes a MMSE scheme of this kind applied to the control of gluten addition to bread making flour in a flour mill where the object was to maintain the protein content of the flour as close as possible to target value. A careful process study showed that to an adequate approximation for this process $\delta = 0$, $\nu = 0$, $f = 1$, and $\lambda = 0.25$ ($\theta = 0.75$). The adjustment equation was thus

$$x_t = -0.25\, x_{t-1} - \frac{0.25}{g}\varepsilon_t \qquad (13.2.19)$$

The scheme was tested extensively and the authors remarked that it worked well over a wide range of manufacturing conditions and was robust to moderate changes in the parameters.

The flour milling example does not yield a PI scheme. Notice, however, that the adjustment equation can be written $x_t = -(1 + \lambda B)^{-1}(\lambda/g)\varepsilon_t = -(1 - \lambda B + \lambda^2 B^2 - \cdots)(\lambda/g)\varepsilon_t$. For the rather small value $\lambda = 0.25$ if we truncate the expansion after the first term we obtain the PI scheme $x_t = c_1\varepsilon_t + c_2\varepsilon_{t-1}$ with $c_1 = -\lambda/g$ and $c_2 = \lambda^2/g$. In practice, the behavior of this PI scheme will be almost identical to that of (13.2.19). More generally, we shall find that PI schemes have an importance in addition to that conferred on them by their producing MMSE schemes for certain simple models. We therefore next consider how PI schemes can be put in effect using simple *feedback control charts*.

13.2.5 Manual Adjustment for Discrete Proportional-Integral Schemes

The equation for the adjustment $x_t = X_t - X_{t-1}$ for the discrete PI scheme (13.2.15) may also be written

$$x_t = -G(1 + P\nabla)\varepsilon_t \qquad (13.2.20)$$

where

$$-G = k_I \quad \text{and} \quad P = \frac{k_P}{k_I} \qquad (13.2.21)$$

or equivalently, $k_I = -G$ and $k_P = -PG$, and P is zero for pure integral control. In the special case where the stochastic and dynamic models are defined by (13.2.10) and (13.2.12), respectively, the PI control equation (13.2.15) yields MMSE when $G = \lambda/g$ and $P = \xi$.

Equation (13.2.20) shows how we can make a manual adjustment chart to put PI control into effect. We have already illustrated the use of such a chart for the metallic thickness example in Figure 13.3. For further illustration we adapt an example discussed by Box, Hunter, and Hunter [41]. In a dyeing process, the quality characteristic of interest was the color index. Deviations ε_t from the desired target value of $T = 9$ were compensated by changing the dye addition rate X. For this example, the disturbance in the color index was approximated by an IMA (0, 1, 1) model with $\lambda = 0.3$, and a change of 1 unit in the dye addition rate X eventually produced a change of 0.06 unit in the color index so that $g = 0.06$.

Suppose at first that ξ were zero so that the dynamic model was simply $Y_t = gBX_{t+}$, implying that a change in the input X was fully effective at the output in one time interval. Then

$$-G = k_I = -\frac{\lambda}{g} = -\frac{0.30}{0.06} = -5 \quad \text{and} \quad k_P = 0 \tag{13.2.22}$$

The MMSE integral feedback equation would be

$$X_t = k_0 - G\sum_{i=1}^{t} \varepsilon_i = k_0 - 5\sum_{i=1}^{t} \varepsilon_i \tag{13.2.23}$$

and at time t the corresponding *adjustment* would be

$$x_t = -G\varepsilon_t = -5\varepsilon_t \tag{13.2.24}$$

Appropriate action is read off the manual adjustment chart in Figure 13.7 with scales such that one unit deviation in the color index corresponds to $-G = -5$ units of adjustment of the dye addition rate. Action is taken after each observation by recording the value of the color index (indicated by a filled dot) and reading off on the left-hand scale the required adjustment to the dye addition rate. Thus in the diagram at time 1:30 P.M., the color index was 9.14 calling for a reduction of -0.7 in the dye addition rate.

Now consider the case where, due perhaps to incomplete mixing of the dye, the process was subject to inertia which was approximated by a first-order dynamic system as in (13.2.13) with $\delta = 0.2$ and consequently, $\xi = \delta/(1 - \delta) = 0.25$. Thus, as before, $G = 0.3/0.06 = 5$ and now $P = \xi = 0.25$. Thus the appropriate MMSE control equation (13.2.15) would call for proportional–integral action such that

$$X_t = k_0 - 1.25\varepsilon_t - 5\sum_{i=1}^{t} \varepsilon_i \tag{13.2.25}$$

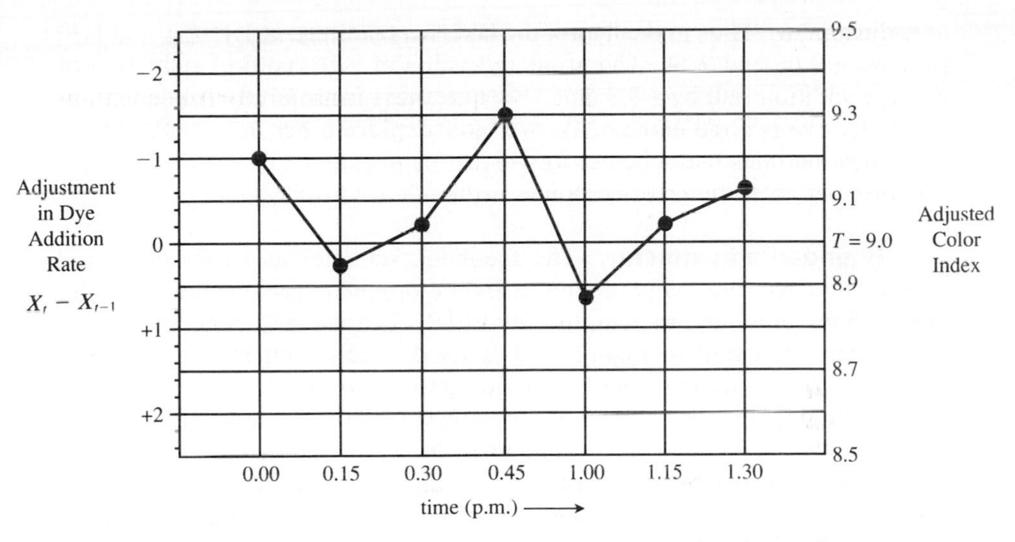

Figure 13.7 Manual adjustment chart for discrete integral control.

The corresponding adjustment equation is

$$x_t = -5(1 + 0.25\nabla)\varepsilon_t \tag{13.2.26}$$

To put this into effect manually, the chart in Figure 13.8 may be employed with the vertical dashed lines placed at a fraction $P = k_P/k_I = 0.25$ within each sampling interval. At each step the operator extrapolates the line through the last two points to the next dashed line and reads off the appropri-

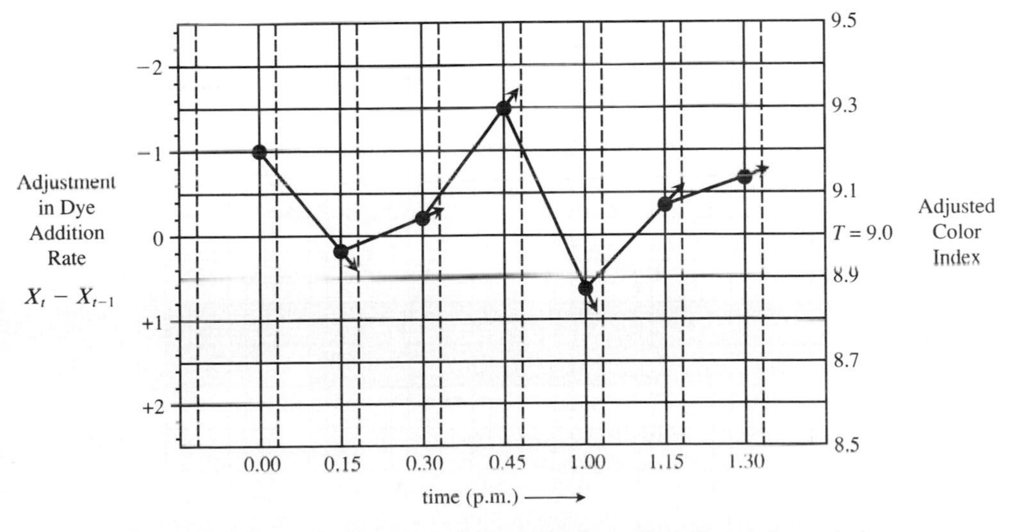

Figure 13.8 Manual adjustment chart putting into effect discrete integral plus proportional control.

ate adjustment. Thus in the figure, the last two readings, at 1:15 P.M. and 1:30 P.M., were 9.06 and 9.14. The projected value of 9.16 requires reduction of the dye addition rate by −0.8 unit. No exactness is required. A line extrapolated by eye is good enough. As we later explore other uses of PI charts, we will sometimes use schemes in which P is negative. This calls for *interpolation* between the last two points rather than for extrapolation.

Rounded adjustment. The feedback schemes as so far discussed require that we take *some* action at every opportunity—in this example, every 15 minutes. In practice, usually little is lost if the "rounded" adjustment chart indicated in Figure 13.9 is used. Such a chart is easily constructed from the original chart by dividing the action scale into bands. The adjustment made when an observation falls within the band is that appropriate to the middle point of the band on an ordinary chart. Figure 13.9 shows a rounded chart in which possible action is limited to −2-, −1-, 0-, 1-, or 2-unit catalyst formulation changes. The increase in mean square error (usually small), which results from using the rounded scheme, is often outweighed by the convenience of working with a small number of standard adjustments. A convenient width for the rounded bands is about one standard deviation σ_ε or a little less. Justification for the use of such charts was provided by and Box and Jenkins [48, Sec.13.1], where consideration is given to the effects of errors in the adjustment x_t. Note that the use of all these manual adjustment charts requires no calculation—they are simple and entirely graphical.

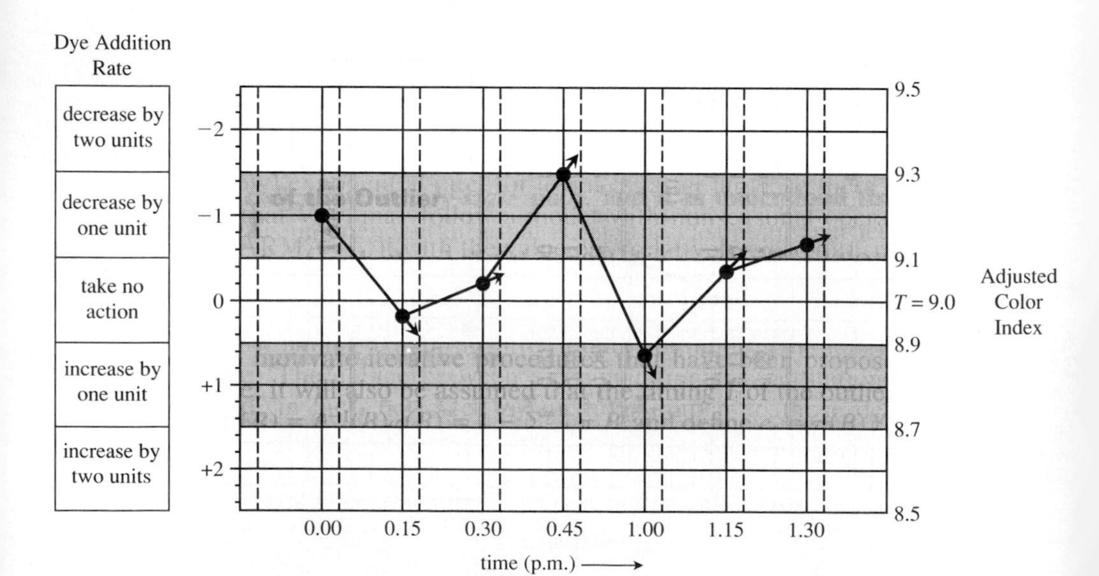

Figure 13.9 Rounded adjustment chart for proportional–integral control.

13.2.6 Complementary Roles of Monitoring and Adjustment

It is sometimes complained that feedback control can conceal the nature of a compensated disturbance which otherwise might be eliminated. However, when combined with appropriate monitoring, this need not happen. Adjustment schemes and monitoring schemes are complementary and should be used in consort. Figure 13.10 illustrates the point. This shows the behavior of a simulated feedback scheme in which the disturbance is an

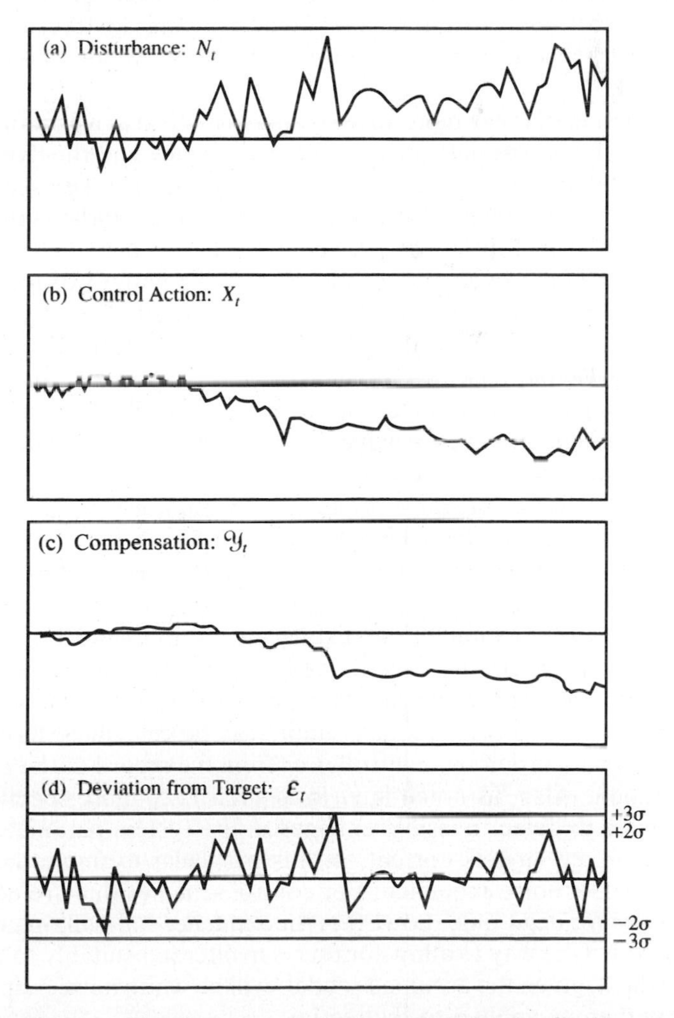

Figure 13.10 (a) Disturbance N_t; (b) feedback control action X_t; (c) compensation of the disturbance $\mathcal{Y}_t$, (d) resulting deviation ε_t from the target value.,,

IMA (0, 1, 1) process with $\lambda = 0.2$ and the process dynamics are represented by a first-order system (13.2.13) with $\delta = 0.5$ and $g = 1.0$. The calculations were made assuming that the system is controlled by the PI controller,

$$-X_t = \text{constant} + 0.20\varepsilon_t + 0.20 \sum_{i=1}^{t} \varepsilon_i \tag{13.2.27}$$

which, for these stated parameter values, produces MMSE. Although this is not usually done, the control action X_t in Figure 13.10(b), as well as the deviation from target $\{\varepsilon_t\}$ in Figure 13.10(d), can be charted (or better still, displayed on the screen of a process computer). Assuming the dynamics known, the exact compensation $\mathcal{Y}_t$ shown in Figure 13.10(c) can also be computed and hence the original disturbance N_t of Figure 13.10(a) can be reconstructed.

Examination of these monitoring displays motivates a generalized concept of common and special causes. The disturbance and the dynamic system together define the *common-cause* system, which is taken account of in the design of the controller. But management action could change the system and hence the appropriate form of control. For example, suppose it was discovered that in the operation of the system the pattern of the feedback control action X_t shown in Figure 13.10(b) mirrored that of a particular impurity in the feedstock. If this correlation checked out as a causative relation, management might decide to change the control system either by removing the impurity from the feedstock before it reached the process, or if that were impossible or too expensive, by measuring it and compensating for it by appropriate feedforward control.

In addition, a *special cause* producing a temporary deviation from the underlying system model, induced perhaps by misoperation of the controller or a mistake by the operator, can be evidenced in the residual sequence $\{\varepsilon_t\}$ leading to remedial action. To illustrate this we have added a deviation of size $3\sigma_a$ to the thirtieth value of the disturbance N_t in Figure 13.10(a). After the disturbance has been subjected to feedback control, this outlier is clearly visible in the record of the deviations ε_t from target plotted as a Shewhart chart in Figure 13.10(d). The control limits can be calculated directly from the models used to design the controller or from the record of the ε_t's during stable operation. Also, as noted later in Section 13.5, more specific checks may be applied to detect possible changes in the system parameters.

Assuming the models correct, in this particular example the residual ε_t's will be a white noise sequence. For control schemes that are not MMSE or which allow for dead time, however, the sequence $\{\varepsilon_t\}$ will, in general, be autocorrelated. One way to allow for this is to filter $\{\varepsilon_t\}$ suitably to produce a sequence which, given the assumed model, will be white noise. Appropriate checks may then be applied to that series.

13.3 EXCESSIVE ADJUSTMENT SOMETIMES REQUIRED BY MMSE CONTROL

One rationalization for the use of integral control and proportional–integral control is that for perhaps the simplest models for disturbance [equation (13.2.10)] and dynamics [equations (13.2.12) and (13.2.13)] which approximate reality, these forms of feedback adjustment can produce minimum mean square error.* Unfortunately, MMSE control sometimes requires unacceptably large manipulations of the compensating variable X. For illustration, consider again the situation where to an adequate approximation the disturbance model is the IMA (0, 1, 1) model of equation (13.2.10) with parameter θ and the dynamic model is the first-order difference equation (13.2.13) with parameters δ and g. Then the MMSE feedback control adjustment scheme can be written [see (13.2.14)]

$$x_t = -\frac{\lambda}{g}\frac{1 - \delta B}{1 - \delta}\varepsilon_t \tag{13.3.1}$$

where $\lambda = 1 - \theta$ and $\varepsilon_t = a_t$. If δ is negligibly small, MMSE control will be obtained with $x_t = -(\lambda/g)\varepsilon_t$ and let us then write

$$\sigma_x^2 = \text{var}[x_t] = \frac{\lambda^2}{g^2}\sigma_a^2 = k \tag{13.3.2}$$

But then, when δ is *not* negligible, $\sigma_x^2 = k[(1 + \delta^2)/(1 - \delta)^2]$. Thus if δ were near its upper limit of unity, σ_x^2 could become very large. For example, with $\delta = 0.9$ (so that only one tenth of the eventual change produced by a step input is experienced in the first interval), $\sigma_x^2 = 181k$. In fact, as δ approaches unity, the control action

$$x_t = -\frac{\lambda}{g(1 - \delta)}(\varepsilon_t - \delta\varepsilon_{t-1}) \tag{13.3.3}$$

takes on more and more of an "alternating" character,† the adjustment made at time t reversing a substantial portion of the adjustment made at time

* This theoretical formulation, which results in a discrete PI controller yielding MMSE, is, however, not unique. For example, a PI controller giving MMSE can be obtained from the models $\mathcal{Y}_t = gBX_t$ and $N_t = (1 - \theta_1 B - \theta_2 B^2)a_t$, as well as the dynamics model (13.2.13) with IMA(0, 1, 1) noise model (13.2.10).

† A value of $\delta = 0.9$ corresponds to a time constant for the system of over nine sampling intervals. The occurrence of such a value would immediately raise the question as to whether the sampling interval being taken was too short; whether in fact, the inertia of the process was so large that little would be lost by less frequent surveillance. Now (see Appendix A13.2) the question of the choice of sampling interval must depend on the nature of the noise that infects the system. Because the properties of the noise usually reflect system inertia as well, in many cases it would be concluded that the sampling interval should be increased.

$t-1$. The reason for such alternating and variable adjustment can also be understood from the consideration that with $\delta = 0.9$, the constant $P = \xi = 9$, of the manual adjustment chart for MMSE control would call for *extrapolation* of the line joining ε_{t-1} and ε_t *by nine sampling intervals!* In practice, constrained schemes can be used which at the expense of rather small increases in MSE at the output require much less compensatory manipulation.

13.3.1 Constrained Control

When the adjustments x_t form a stationary time series, such constrained schemes can be obtained by finding an unconstrained minimum of the expression

$$\sigma_\varepsilon^2 + \alpha\sigma_x^2 \tag{13.3.4}$$

where α can be regarded as an undetermined multiplier that allocates the relative *quadratic costs* of variations of ε_t and x_t. Such a scheme will be called a constrained MMSE scheme or CMMSE scheme. In particular, we have seen that for an IMA(0, 1, 1) disturbance and first-order dynamics, the *unconstrained* MMSE scheme calls for an adjustment of

$$x_t = -\frac{\lambda}{g}(1 + \xi\nabla)\varepsilon_t = -\frac{\lambda(1-\delta B)}{g(1-\delta)}\varepsilon_t \tag{13.3.5}$$

It is shown in Appendix A13.1 [in equation (A13.1.27)] that the corresponding CMMSE is of the form

$$x_t = [k_1 + (1-\lambda)k_0]x_{t-1} - (1-\lambda)k_1 x_{t-2} - \frac{\lambda(1-k_0)(1-\delta B)}{g(1-\delta)}\varepsilon_t \tag{13.3.6}$$

where k_0 and k_1 are fairly complicated functions of the parameters g, λ, δ, and α. A table for applying such control is also given in Appendix A13.1.

For illustration suppose that $\lambda = 0.6$, $\delta = 0.5$, and $g = 1$; then the optimal *unconstrained* MMSE scheme is

$$x_t = -1.2(1 - 0.5B)\varepsilon_t \tag{13.3.7}$$

with $\sigma_x^2 = 1.80\sigma_a^2$ and $\sigma_\varepsilon^2 = \sigma_a^2$. Suppose that this amount of variation in the adjustment x_t produced difficulties in process operation and it was desired to reduce it so that σ_x^2 was about $0.50\sigma_a^2$. Use of Table A13.2 shows that this can be achieved with the scheme

$$x_t = 0.32x_{t-1} - 0.06x_{t-2} - (0.57 \times 1.2)(1 - 0.5B)\varepsilon_t \tag{13.3.8}$$

which reduces σ_x^2 to $0.47\sigma_a^2$ with $\sigma_\varepsilon^2 = 1.07\sigma_a^2$. Thus an almost fourfold reduction in σ_x^2 is produced for an increase of only 7% in the output variance. Such optimal constrained schemes are extremely attractive since they often produce a very large reduction in σ_x^2 for only a small increase in σ_ε^2. See, for example, Whittle [198], Wilson [203], [204], MacGregor [141], Box and Jenkins [48], Harris, MacGregor, and Wright [104], Aström and Wittenmark [16], Rivera, Morari, and Skogestad [169], and Bergh and MacGregor [30]. Unfortunately, such schemes can become complicated.

In practice, however, exact "optimality" is to some extent an illusion because assumptions are never true. It turns out that a form of constrained control, which is almost as good as CMMSE control, can often be obtained using an *appropriately tuned* PI controller. Such a controller has the advantage that it is simple and, in particular, is easily adapted to manual control. The following example shows how suitably tuned PI controllers can do almost as well as optimal constrained schemes in producing great reductions in the variance σ_x^2 of the adjustment for only modest increases in the output variance σ_ε^2.

As an illustration, consider once again the situation where the process disturbance is represented by an IMA(0, 1, 1) process of (13.2.10) and the process dynamics by the first-order system (13.2.13) and suppose that $\lambda = 0.4$, $\sigma_a^2 = 1$, $g = 0.4$, and $\delta = 0.5$, so that $\xi = 1$. Then minimum mean square error control is achieved by the PI scheme (a) shown in Table 13.1, yielding an output variance σ_ε^2 of 1.00 with $\sigma_x^2 = 5$. Using the optimal constrained control equation (b) in Table 13.1, it is possible to achieve a 20-fold reduction in σ_x^2 (to 0.25) at the expense of a 20% increase in σ_ε^2 to 1.20. But almost nothing is lost by, instead, using the much simpler optimal constrained PI controller (c) in Table 13.1 for which, to two-decimal accuracy, the same result is obtained. Notice that if we use a manual adjustment chart for the MMSE PI scheme (a), it would be necessary to extrapolate one whole time period ahead from the current time t. However, for the constrained PI control (c), we must *interpolate* a quarter of a period back from the current time t. This accounts for the much greater stability of the latter scheme. A fuller discussion of this topic can be found in Box and Lucéno [53].

TABLE 13.1 Illustrative Results Comparing Different Control Schemes for Model (13.2.13) and (13.2.10), with $g = 0.4$, $\delta = 0.5$, $\lambda = 0.4$, and $\sigma_a^2 = 1$

		σ_ε^2	σ_x^2
(a) MMSE control	$-x_t = \{1 + \nabla\}\varepsilon_t$	1	5
(b) Optimal constrained control	$-x_t = -0.82x_{t-1} - 0.21x_{t-2} - 0.39\varepsilon_t + 0.19\varepsilon_{t-1}$	1.20	0.25
(c) Optimal constrained PI control	$-x_t = 0.52\{1 - 0.25\nabla\}\varepsilon_t$	1.20	0.25

13.4 MINIMUM COST CONTROL WITH FIXED COSTS OF ADJUSTMENT AND MONITORING

From the point of view of cost we can summarize the discussion so far as follows. If we assume that the *only* control cost we need consider is that of being off target and that this cost is proportional to the square of the deviation from target, unconstrained minimum mean square error control implies minimization of the total cost of the scheme. Suppose, however, that there is an additional quadratic loss associated with the size of the adjustment x, and that α is some measure of the *relative* cost of being off target and of making adjustments. Then $\sigma_\varepsilon^2 + \alpha\sigma_x^2$ can be a measure of the overall cost of the scheme, and minimization of this quantity can produce a control scheme yielding minimum cost, and as we have seen, suitably chosen PI schemes can often do almost as well. In either case, in practice it is rarely easy to gauge α in terms of relative costs. Instead, choice of a suitable scheme can be made by empirical judgment of what constitutes a satisfactory reduction of σ_x^2 in exchange for an acceptable increase in σ_ε^2. The same kinds of considerations apply to systems for which there are fixed adjustment and monitoring costs.

13.4.1 Bounded Adjustment Scheme for Fixed Adjustment Cost

Especially in the "parts" industries, situations occur where an adjustment often has immediate effect but entails a *fixed* cost incurred, for example, by stopping a machine or changing a tool.

Bounded adjustment charts. It was shown by Box and Jenkins [42] that in the latter case, on the assumption of quadratic off-target loss and an IMA disturbance, the minimum-cost feedback control is *not* achieved by repeated adjustment after each observation. Instead, it requires that an adjustment be made only when an exponentially weighted average $\hat{\varepsilon}_t(1)$ of the deviations from target falls outside some fixed limits, $\pm L$, say. We call this *bounded adjustment*. The adjustment that should then be made is that which will produce a change $-\hat{\varepsilon}_t(1)$ at the output. Such an adjustment can be put into effect manually using a "bounded adjustment chart" such as that discussed below, or automatically.

A bounded adjustment chart such as that shown in Figure 13.11 is superficially similar to that proposed for process monitoring by Roberts [170]. However, its purpose and design are different. The purpose is to decide when, and by how much, to *adjust* the process. The boundary lines are designed to minimize the overall cost, taking into account both the cost of making adjustments and the cost of being off target. Their purpose is *not* to discover statistically significant deviations from target. As the cost of

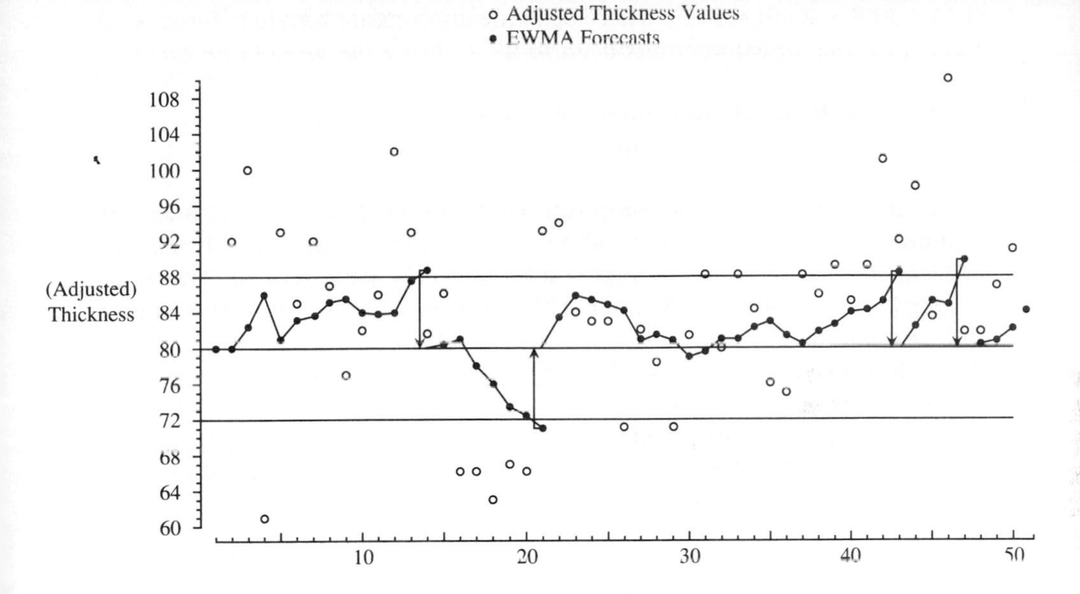

Figure 13.11 Bounded adjustment chart: the open circles are the thickness deviations ε_t (after adjustment), the closed circles are their EWMA forecasts $\hat{\varepsilon}_{t-1}(1)$ of these deviations.

adjustment approaches zero, the lines come closer together, converging on the target value when the cost of adjustment is zero and so yielding the "repeated adjustment" MMSE scheme.

Figure 13.11 shows an example of such a chart for the metallic thickness control problem that would be appropriate if there had been a fixed cost for changing the deposition rate X. As before, $\lambda = 0.2$, $g = 1.2$, and $\sigma_a = 11$. At time t an open circle represents the deviation from target ε_t obtained after periodically changing the deposition rate X_t as required by the chart. A filled circle represents an appropriate exponentially weighted moving average forecast. This is conveniently updated using the formula

$$\hat{\varepsilon}_t(1) = \lambda\varepsilon_t + \theta\hat{\varepsilon}_{t-1}(1)$$

The particular chart shown has boundary lines at 80 ± 8, that is, at $T + 0.72\sigma_a$. We discuss the rationale for this choice below. To understand how the chart operates, suppose initially that the deposition rate is some value X_0. This will remain unchanged until time $t = 13$, when the forecasted value 88.7 [i.e., $\hat{\varepsilon}_t(1) = 8.7$] falls outside the upper limit and the chart signals that a change is needed in the deposition rate which will reduce the thickness by -8.7. An adjustment of $X_{13} - X_0 = -8.7/1.2$ is now made in the deposition rate. Notice that such an adjustment does not upset the calculation of the next EWMA. For example, the forecasted thickness at time $t = 14$ is $(0.2 \times$

81.3) + (0.8 × 80.0) = 80.3, where 80 is the appropriate previous forecasted value *after the adjustment has been made to bring the process on target.*

13.4.2 Indirect Approach for Obtaining a Bounded Adjustment Scheme

Tables for calculating the positions of the appropriate limit lines for minimum cost schemes in terms of the *cost of being off target* and the *cost of adjustment* were provided by Box and Jenkins [42], Box, Jenkins, and MacGregor [50], and Box and Kramer [52]. However, as we said earlier, these costs are not always easy to assess and it seems more practical to use these results to provide an envelope of minimum cost schemes and then to choose among them empirically by considering the increased standard deviation at the output obtained in exchange for a longer interval between making adjustments. This approach was illustrated by Box [37]. Table 13.2 shows theo-

TABLE 13.2 Average Adjustment Interval (AAI) and Percent Increase in Standard Deviation of the Output (ISD) for Various Choices of L/σ_a Where the Limit Lines Are at $T \pm L$

λ	L/σ_a	*AAI*	Percent Increase in Standard Deviation ISD
0.1	0.5	32	2.4
	1.0	112	9
	1.5	243	18
	2.0	423	30
0.2	0.5	10	2.6
	1.0	32	9
	1.5	66	20
	2.0	112	32
0.3	0.5	5	2.6
	1.0	16	10
	1.5	32	20
	2.0	52	33
0.4	0.5	4	2.6
	1.0	10	10
	1.5	19	21
	2.0	32	34
0.5	0.5	3	2.5
	1.0	7	10
	1.5	13	21
	2.0	21	35

Source: Ref. [37].

deviation (ISD) of the adjusted process for various values of λ and L/σ_a, where limit lines of the bounded adjustment scheme are at $T \pm L$.

For illustration, consider again the thickness adjustment example. Entering Table 13.2 with $\lambda = 0.2$ shows how much inflation in the error standard deviation would occur for a bounded scheme for various choices of L/σ_a. Thus if L/σ_a were set equal to 0.5, a 2.6% increase in the standard deviation would occur, but on the average, adjustments would be needed only every 10 intervals. If L/σ_a were set equal to 1.0, a 9% increase in standard deviation would result but the average adjustment interval (AAI) would be 32. The scheme depicted in Figure 13.11 is a compromise in which L/σ_a was set equal to 0.72, which rough interpolation shows would give a 5% increase in the standard deviation with an AAI of about 20. To achieve this L was set equal to $8 \approx 0.72 \times 11$. A Monte Carlo study using the 100 observations of metallic thickness graphed in Figure 13.2 shows an actual inflation of the standard deviation of 8.5% for this example with an AAI of 14. In view of the rather limited sample size, the agreement must be considered quite good.

Interpolation chart. Any degree of technological sophistication can be used in applying these ideas: anything from transducers taking actions calculated by computers to operators taking actions based on a simple interpolation chart such as that shown in Figure 13.12, which uses a pushpin and a piece of thread to indicate the appropriate *manual* adjustment. In the situation depicted, a previous forecast made at time $t - 1$ was 86 and the observation, which has just been made at time t, is 66. Just before the current time t, therefore, the location of the pushpin on the current forecast scale would be at 86 with the thread hanging down from the pin. As soon as the actual value 66 became available, the thread would be pulled tightly to join the point 66 on the right-hand scale. The updated forecast of 82 would then be read off on the intermediate scale. This value lies within the boundaries, so the pushpin would be moved down to this new current forecast value with the thread hanging loose again until the next observation became available to produce a new updated forecast. As soon as an updated forecast fell outside either boundary, the appropriate adjustment in deposition rate to cancel out the forecasted deviation would be made, and the pushpin would then be *placed on the target value* ready for the next interpolation.

13.4.3 Inclusion of the Cost of Monitoring

It was shown by Box and Kramer [52] how these results could be extended to the case where the cost of monitoring the process had also to be taken into account. They considered the possibility of further reducing cost

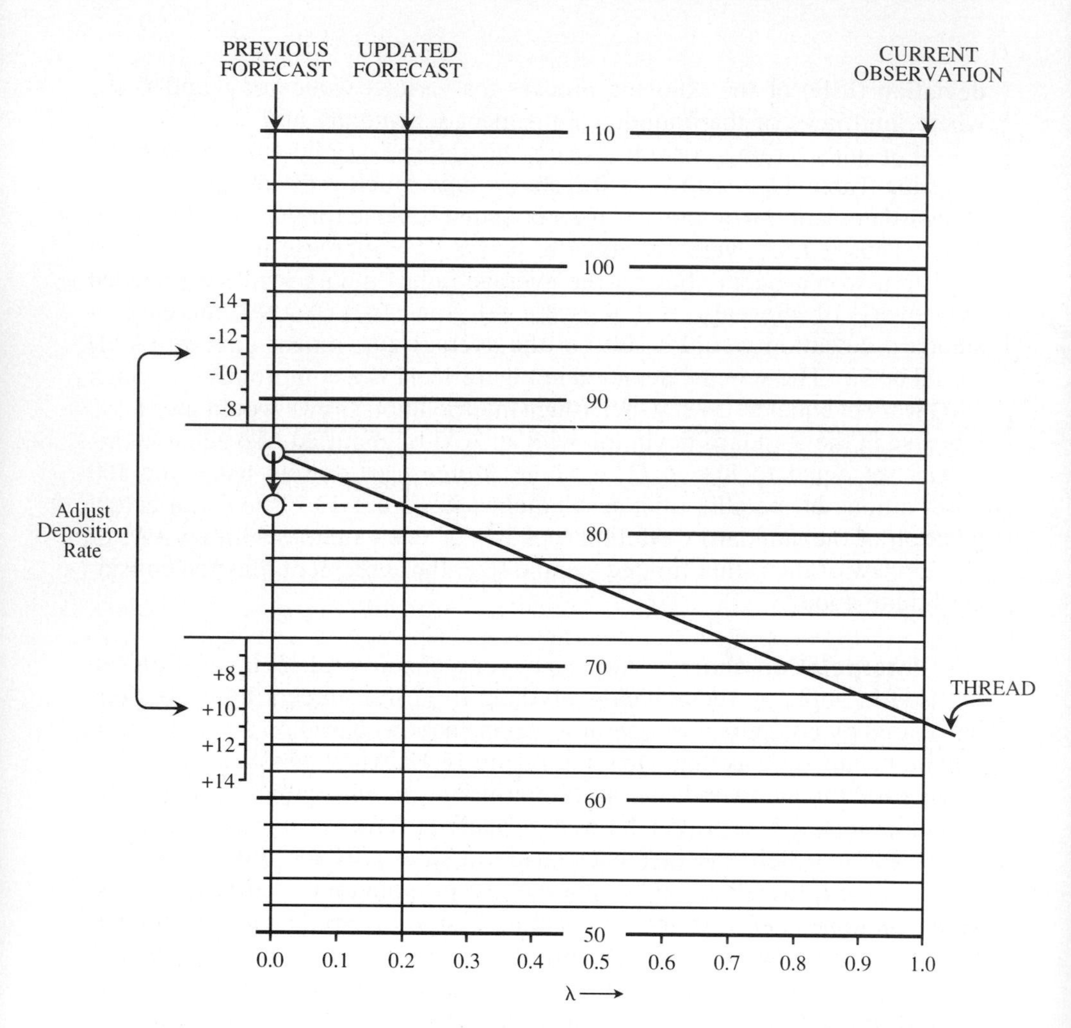

Figure 13.12 Interpolation chart to update the forecasted value of thickness and to indicate when and by how much the deposition rate should be adjusted.

by less frequent monitoring at an interval m instead of at a unit interval. They provided charts for obtaining minimum cost schemes given that in addition to σ_a and λ (estimated from plant data), three cost constants were known: (1) the (assumed quadratic) cost of being off target, (2) the fixed cost of making a change, and (3) the fixed monitoring cost of taking an observation. Given this information, the corresponding values of L/σ_a and of m yielding minimum cost could be read off their charts.

Again these three individual costs may not be easy to determine, and Box and Lucéno [53] used their results to allow the choice of scheme to be

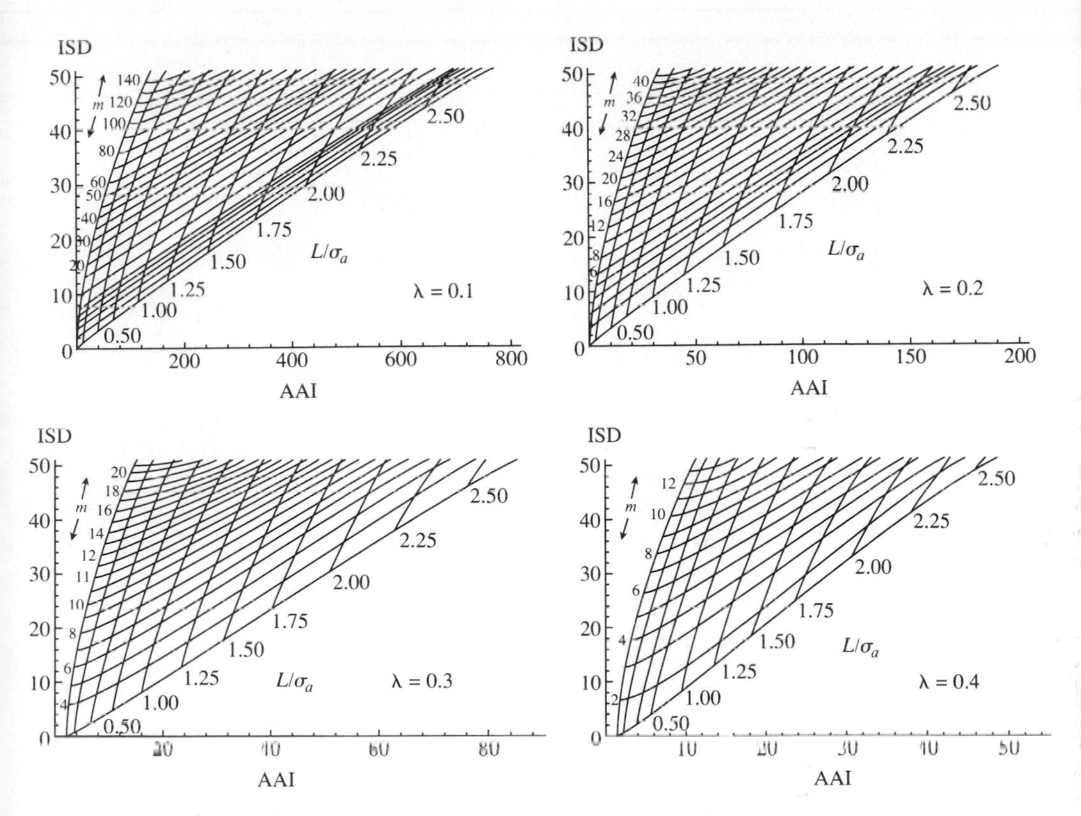

Figure 13.13 Charts for $\lambda = 0.1, 0.2, 0.3, 0.4, 0.5, 0.6, 0.8$, and 1.0 showing AAIs and ISDs obtained from various choices of L/σ_a and m.

based on empirical judgment. The charts shown in Figure 13.13 give the values of the average adjustment interval (AAI) and the percent increase in standard deviation (ISD) with respect to σ_a corresponding to value of the nonstationarity measure $\lambda = 0.1(0.1)\ 0.6,\ 0.8$, and 1.0, the standardized action limit $L/\sigma_a = 0.0(0.25)\ 2.5$, and the monitoring interval $m = 1, 2, 3, \ldots$. The charts cover small to moderate increases in the output standard deviation such as might be needed in practice. Thus the larger values of m appear only with smaller values of λ.

For example, we saw earlier that by using a bounded adjustment chart with $L/\sigma_a = 0.72$ instead of a continuous scheme, the average adjustment interval could be increased to about 20 at the cost of an increase of 5% in the standard deviation. This is confirmed by the chart of Figure 13.13 for $\lambda = 0.2$, which also shows, for example, that if we monitor the process half as frequently ($m = 2$) and we again set $L/\sigma_a = 0.72$ we could obtain about the same average adjustment interval (20) but with an 8% increase in the standard deviation.

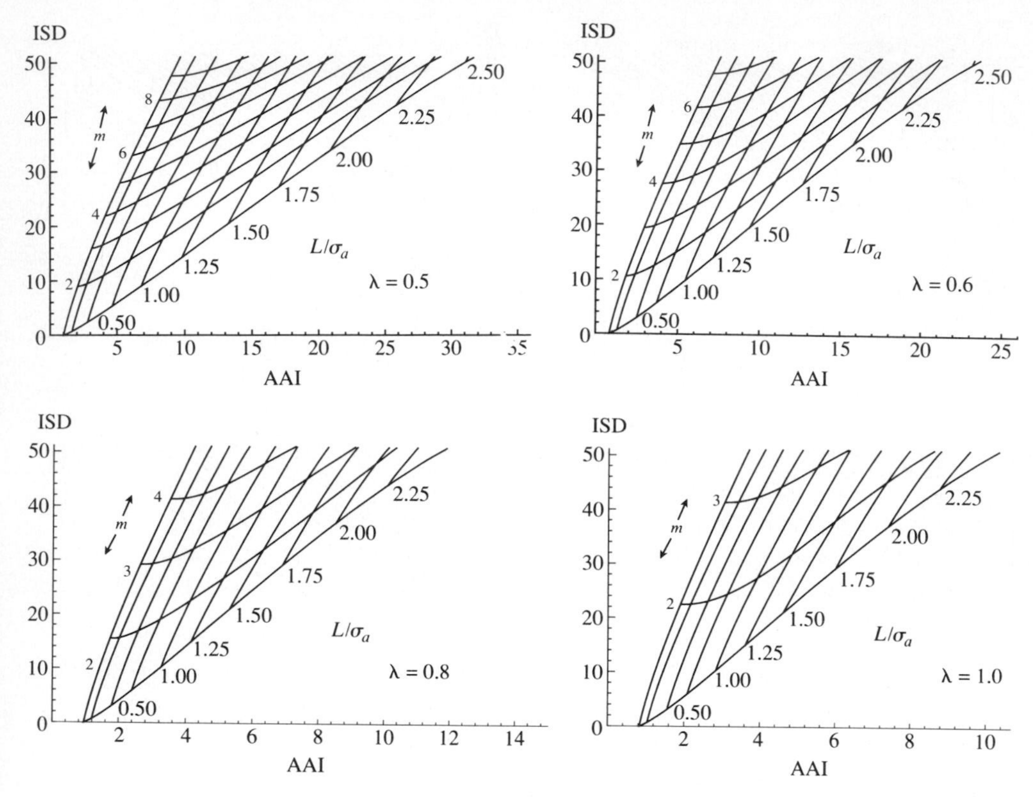

Figure 13.13 (Cont.)

13.5 MONITORING VALUES OF PARAMETERS OF FORECASTING AND FEEDBACK ADJUSTMENT SCHEMES

Earlier we mentioned the complementary roles of process adjustment and process monitoring. This symbiosis is further illustrated if we again consider the need to monitor the adjustment scheme itself. It has often been proposed that the series of residual deviations from target from such schemes (and similarly, the errors from forecasting schemes) should be studied and that a Shewhart chart or more generally a cumulative sum or other monitoring chart should be run on the residual errors to warn of changes. The cumulative sum is, of course, appropriate to look for small changes in mean level but often other kinds of discrepancies may be feared. A general theory of sequential directional monitoring based on a cumulative Fisher score statistic (Cuscore) was proposed by Box and Ramîrez [57] (see, also, Bagshaw and Johnson [19]).

Suppose that a model can be written in the form

$$e_t = e_t(\theta) \tag{13.5.1}$$

and *that if the correct value of the parameter* $\theta = \theta_0$ is employed in the model, $\{e_t\} = \{a_t\}$ is a sequence of NIID random variables. Then the cumulative score statistic appropriate to detect a departure from the value θ_0 may be written

$$Q = \sum e_t r_t \tag{13.5.2}$$

where $r_t = -(de_t/d\theta)|_{\theta=\theta_0}$ may be called the detector signal.

For example, suppose that we wished to detect a shift in a mean from a value θ_0 for the model $y_t = \theta + e_t$. We can write

$$e_t = e_t(\theta) = y_t - \theta \qquad a_t = y_t - \theta_0 \tag{13.5.3}$$

Then, in this example, the detector signal is $r_t = 1$ and $Q = \Sigma\, e_t$, the well-known *cumulative sum* statistic.

In general, for some value of θ close to θ_0, since e_t may be approximated by $e_t = a_t - (\theta - \theta_0)r_t$, the cumulative product in (13.5.2) will contain a part

$$-(\theta - \theta_0) \sum r_t^2 \tag{13.5.4}$$

which systematically increases in magnitude with sample size n when θ differs from θ_0. For illustration, consider the possibility that in the feedback control scheme for metallic thickness of Section 13.2.1, the value of λ (estimated as 0.2) may have changed during the period $t = 1$ to $t = 100$. For this example

$$e_t = e_t(\theta) = \frac{1 - B}{1 - \theta B} N_t \tag{13.5.5}$$

Thus

$$r_t = -\frac{1 - B}{(1 - \theta B)^2} N_{t-1} = -\frac{e_{t-1}}{1 - \theta B} = -\frac{\hat{e}_{t-1}(1)}{\lambda} \tag{13.5.6}$$

where $\hat{e}_{t-1}(1) = \lambda(1 - \theta B)^{-1}e_{t-1}$ is an EWMA of past e_t's. The cumulative score (Cuscore) statistic for detecting this departure is, therefore,

$$Q = -\frac{1}{\lambda} \sum e_t \hat{e}_{t-1}(1) \tag{13.5.7}$$

where the detector signal $\hat{e}_{t-1}(1)$ is, in this case, the EWMA of past values of the *residuals*. These residuals are the deviations from target plotted on the feedback adjustment chart of Figure 13.3. The criterion agrees with the commonsense idea that if the model is true, then $e_t = a_t$ and e_t is not predictable from previous values. The Cuscore chart shown in Figure 13.14

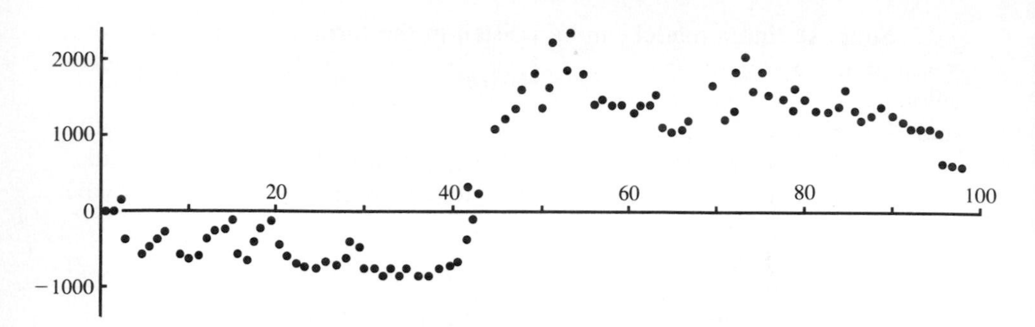

Figure 13.14 Cuscore monitoring for detecting a change in the parameter θ used in conjunction with the adjustment chart of Figure 13.3.

suggests that a change in parameter may have occurred at about $t = 40$. However, we see from the original data of Figure 13.2 that this is very close to the point at which the level of the original series appears to have changed and further data and analysis would be needed to confirm this finding.

The important point is that this example shows the *partnership* of two types of control (adjustment and monitoring) and the corresponding two types of statistical inference (estimation and criticism). A further development is to feed back the filtered Cuscore statistic to "self-tune" the control equation, but we shall not pursue this further here.

APPENDIX A13.1 FEEDBACK CONTROL SCHEMES WHERE THE ADJUSTMENT VARIANCE IS RESTRICTED

Consider now the feedback control situation where the models for the noise and system dynamics are again given by (13.2.10) and (13.2.13), so that $\varepsilon_t = \mathcal{Y}_t + N_t$ with

$$(1 - B)N_t = (1 - \theta B)a_t \quad \text{and} \quad (1 - \delta B)\mathcal{Y}_t = (1 - \delta)gX_{t-1+}$$

but some restriction of the input variance var$[x_t]$ is necessary, where $x_t = (1 - B)X_t$. The unrestricted optimal scheme has the property that the errors in the output $\varepsilon_t, \varepsilon_{t-1}, \varepsilon_{t-2}, \ldots$ are the uncorrelated random variables $a_t, a_{t-1}, a_{t-2}, \ldots$ and the variance of the output σ_ε^2 has the minimum possible value σ_a^2. With the restricted schemes, the variance σ_ε^2 will necessarily be greater than σ_a^2, and the errors $\varepsilon_t, \varepsilon_{t-1}, \varepsilon_{t-2}, \ldots$ at the output will be correlated.

We shall pose our problem in the following form: Given that σ_ε^2 be allowed to increase to some value $\sigma_\varepsilon^2 = (1 + c)\sigma_a^2$, where c is a positive constant, to find that control scheme which produces the minimum value for

$\sigma_x^2 = \text{var}[x_t]$. Equivalently, the problem is to find an (unconstrained) minimum of the expression $\sigma_\varepsilon^2 + \alpha\sigma_x^2$, where α is some specified multiplier which allocates the relative costs of variations in ε_t and x_t.

A13.1.1 Derivation of Optimal Adjustment

Let the optimal adjustment, *expressed in terms of the* a_t's, be

$$x_t = -\frac{1}{g}L(B)a_t \tag{A13.1.1}$$

where

$$L(B) = l_0 + l_1B + l_2B^2 + \cdots$$

Then we see that the error ε_t at the output is given by

$$\begin{aligned}\varepsilon_t &= \frac{(1-\delta)g}{1-\delta B}X_{t-1+} + N_t \\ &= -\frac{1-\delta}{1-\delta B}(1-B)^{-1}L(B)a_t + (1-B)^{-1}(1-\theta B)a_t \\ &= a_t + \left[\lambda - \frac{L(B)(1-\delta)}{1-\delta B}\right]Sa_{t-1}\end{aligned} \tag{A13.1.2}$$

where $S = (1-B)^{-1}$. The coefficient of a_t in this expression is unity, so that we can write

$$\varepsilon_t = [1 + B\mu(B)]a_t \tag{A13.1.3}$$

where

$$\mu(B) = \mu_1 + \mu_2B + \mu_3B^2 + \cdots$$

Furthermore, in practice, control would need to be exerted in terms of the observed output errors ε_t rather than in terms of the a_t's, so that the control equation actually used would be of the form

$$x_t = -\frac{1}{g}\frac{L(B)}{1 + B\mu(B)}\varepsilon_t \tag{A13.1.4}$$

Equating (A13.1.2) and (A13.1.3), we obtain

$$(1-\delta)L(B) = [\lambda - (1-B)\mu(B)](1-\delta B) \tag{A13.1.5}$$

Since δ, g, and σ_a^2 are constants, we can proceed conveniently by finding an unrestricted minimum of

$$C = \frac{(1-\delta)^2g^2V[x_t] + \nu V[\varepsilon_t]}{\sigma_a^2} \tag{A13.1.6}$$

where, for example,

$$V[x_t] = \text{var}[x_t]$$

and $\nu = (1 - \delta)^2 g^2/\alpha$. Now, from (A13.1.3), $V[\varepsilon_t]/\sigma_a^2 = 1 + \sum_{j=1}^{\infty} \mu_j^2$, while from (A13.1.1), $(1 - \delta)gx_t = -(1 - \delta)L(B)a_t = -\tau(B)a_t$, so that

$$\frac{(1 - \delta)^2 g^2 V[x_t]}{\sigma_a^2} = \sum_{j=0}^{\infty} \tau_j^2$$

where

$$\tau(B) = \sum_{j=0}^{\infty} \tau_j B^j = (1 - \delta)L(B) = [\lambda - (1 - B)\mu(B)](1 - \delta B)$$

from (A13.1.5). The coefficients $\{\tau_i\}$ are thus seen to be functionally related to the μ_i by the difference equation

$$\mu_i - (1 + \delta)\mu_{i-1} + \delta\mu_{i-2} = -\tau_{i-1} \qquad \text{for } i > 2 \qquad \text{(A13.1.7)}$$

with $\tau_0 = -(\mu_1 - \lambda)$, $\tau_1 = -[\mu_2 - (1 + \delta)\mu_1 + \lambda\delta]$. Hence we require an unrestricted minimum, with respect to the μ_i, of the expression

$$C = \sum_{j=0}^{\infty} \tau_j^2 + \nu\left(1 + \sum_{j=1}^{\infty} \mu_j^2\right) \qquad \text{(A13.1.8)}$$

This can be obtained by differentiating C with respect to each μ_i ($i = 1, 2, \ldots$), equating these derivatives to zero and solving the resulting equations. Now, a given μ_i only influences the values τ_{i+1}, τ_i, and τ_{i-1} through (A13.1.7), and we see that

$$\frac{\partial \tau_j}{\partial \mu_i} = \begin{cases} -1 & j = i - 1 \\ 1 + \delta & j = i \\ -\delta & j = i + 1 \\ 0 & \text{otherwise} \end{cases} \qquad \text{(A13.1.9)}$$

Therefore, from (A13.1.8) and (A13.1.9), we obtain

$$\begin{aligned} \frac{\partial}{\partial \mu_i} C &= 2\left(\tau_{i+1}\frac{\partial \tau_{i+1}}{\partial \mu_i} + \tau_i\frac{\partial \tau_i}{\partial \mu_i} + \tau_{i-1}\frac{\partial \tau_{i-1}}{\partial \mu_i} + \nu\mu_i\right) \\ &= 2[-\delta\tau_{i+1} + (1 + \delta)\tau_i - \tau_{i-1} + \nu\mu_i] \qquad \text{for } i = 1, 2, \ldots \end{aligned} \qquad \text{(A13.1.10)}$$

Then, after substituting the expressions for the τ_j in terms of the μ_i from equation (A13.1.7) in (A13.1.10) and setting each of these equal to zero, we obtain the following equations:

$$i = 1: \quad -\lambda(1 + \delta + \delta^2) + 2(1 + \delta + \delta^2)\mu_1 - (1 + \delta)^2\mu_2 + \delta\mu_3 + \nu\mu_1 = 0 \qquad \text{(A13.1.11)}$$

$$i = 2: \quad \lambda\delta - (1+\delta)^2\mu_1 + 2(1+\delta+\delta^2)\mu_2 - (1+\delta)^2\mu_3 + \delta\mu_4 + \nu\mu_2 = 0 \quad \text{(A13.1.12)}$$

$$i > 2: \quad [\delta B^2 - (1+\delta)^2 B + 2(1+\delta+\delta^2) - (1+\delta)^2 F + \delta F^2 + \nu]\mu_i = 0 \quad \text{(A13.1.13)}$$

Case where δ is negligible. Consider first the simpler case where δ is negligibly small and can be set equal to zero. Then the equations above can be written

$$i = 1: \quad -(\lambda - \mu_1) + (\mu_1 - \mu_2) + \nu\mu_1 = 0 \quad \text{(A13.1.14)}$$

$$i > 1: \quad [B - (2 + \nu) + F]\mu_i = 0 \quad \text{(A13.1.15)}$$

These difference equations have a solution of the form

$$\mu_i = A_1\kappa_1^i + A_2\kappa_2^i$$

where κ_1 and κ_2 are the roots of the characteristic equation

$$B^2 - (2+\nu)B + 1 = 0 \quad \text{(A13.1.16)}$$

that is, of

$$B + B^{-1} = 2 + \nu$$

Evidently, if κ is a root, so is κ^{-1}. Thus the solution is of the form $\mu_i = A_1\kappa^i + A_2\kappa^{-i}$. Now if κ has modulus less than or equal to 1, κ^{-1} has modulus greater than or equal to 1, and since $\varepsilon_t = [1 + B\mu(B)]a_t$ must have finite variance, A_2 must be zero with $|\kappa| < 1$. By substituting the solution $\mu_i = A_1\kappa^i$ in (A13.1.14), we find that $A_1 = \lambda$.

Finally, then, $\mu_i = \lambda\kappa^i$ and since μ_i and λ must be real, so must the root κ. Hence

$$\mu(B) = \frac{\lambda\kappa}{1 - \kappa B} \qquad 0 < \kappa < 1 \quad \text{(A13.1.17)}$$

$$1 + B\mu(B) = 1 + \frac{\lambda\kappa B}{1 - \kappa B} = \frac{1 - \theta\kappa B}{1 - \kappa B} \quad \text{(A13.1.18)}$$

where $\theta = 1 - \lambda$. Thus

$$\varepsilon_t = \frac{1 - \theta\kappa B}{1 - \kappa B} a_t$$

so that

$$\frac{V[\varepsilon_t]}{\sigma_a^2} = 1 + \frac{\lambda^2\kappa^2}{1 - \kappa^2} \quad \text{(A13.1.19)}$$

Also, using (A13.1.5) with $\delta = 0$,

$$L(B) = \lambda - \frac{(1 - B)\lambda\kappa}{1 - \kappa B} = \frac{\lambda(1 - \kappa)}{1 - \kappa B} \quad \text{(A13.1.20)}$$

Thus

$$x_t = -\frac{\lambda}{g}\frac{(1-\kappa)}{1-\kappa B}a_t$$

and

$$\frac{V[x_t]}{\sigma_a^2} = \frac{\lambda^2}{g^2}\frac{(1-\kappa)^2}{1-\kappa^2} = \frac{\lambda^2}{g^2}\frac{1-\kappa}{1+\kappa} \tag{A13.1.21}$$

Using (A13.1.4) with (A13.1.18) and (A13.1.20), we now find that the optimal control action, in terms of the observed output error ε_t, is

$$x_t = -\frac{1}{g}\frac{\lambda(1-\kappa)}{1-\theta\kappa B}\varepsilon_t$$

that is,

$$x_t = (1-\lambda)\kappa x_{t-1} - \frac{1}{g}\lambda(1-\kappa)\varepsilon_t \tag{A13.1.22}$$

Note that the constrained control equation differs from the unconstrained one in two respects:

1. A new factor $(1-\lambda)\kappa x_{t-1}$ is introduced, thus making present action depend partly on previous action.
2. The constant determining the amount of integral control is reduced by a factor $1-\kappa$.

We have supposed that the output variance is allowed to increase to some value $\sigma_a^2(1+c)$. It follows from (A13.1.19) that

$$c = \frac{\lambda^2\kappa^2}{1-\kappa^2}$$

that is,

$$\kappa = \sqrt{\frac{c}{\lambda^2 + c}}$$

where the positive square root is to be taken. It is convenient to write $Q = c/\lambda^2$. Then $Q = \kappa^2/(1-\kappa^2)$ and $\kappa^2 = Q/(1+Q)$ and the output variance becomes $\sigma_a^2(1+\lambda^2 Q)$.

In summary, suppose that we are prepared to tolerate an increase in variance in the output to some value $\sigma_a^2(1+\lambda^2 Q)$; then:

1. We compute $\kappa = \sqrt{Q/(1+Q)}$.
2. Optimal control will be achieved by taking action

$$x_t = (1-\lambda)\kappa x_{t-1} - \frac{1}{g}\lambda(1-\kappa)\varepsilon_t$$

3. The variance of the input will be reduced to

$$V[x_t] = \frac{\lambda^2}{g^2}\frac{1-\kappa}{1+\kappa}\sigma_a^2$$

that is, it will reduce to a value that is $W\%$ of that for the unconstrained scheme, where

$$W = 100\left(\frac{1-\kappa}{1+\kappa}\right)$$

Table A13.1 shows κ and W for values of Q between 0.1 and 1.0. For illustration, suppose that $\lambda = 0.4$. Then the optimal unconstrained scheme will employ the control action

$$x_t = -\frac{0.4}{g}\varepsilon_t$$

with $\varepsilon_t = a_t$. The variance of x_t would be $V[x_t] = (\sigma_a^2/g^2)0.16$. Suppose that it was desired to reduce this by a factor of 4, to the value $(\sigma_a^2/g^2)0.04$. Thus we require W to be 25%. Table A13.1 shows that a reduction of the input variance to 24% of its unconstrained value is possible with $Q = 0.60$ and $\kappa = 0.612$. If we use this scheme, the output variance will be

$$\sigma_\varepsilon^2 = \sigma_a^2(1 + 0.16 \times 0.60) = 1.10\sigma_a^2$$

Thus, by the use of the control action

$$x_t = 0.37x_{t-1} - \frac{1}{g}0.16\varepsilon_t$$

instead of

$$x_t = -\frac{0.4}{g}\varepsilon_t$$

the variance of the input is reduced to about $\frac{1}{4}$ of its previous value, while the variance of the output is increased by only 10%.

Case where δ is not negligible. Consider now the more general situation where δ is not negligible and the system dynamics must be taken

TABLE A13.1 Values of Parameters for a Simple Constrained Control Scheme

$c/\lambda^2 = Q$	κ	W	$c/\lambda^2 = Q$	κ	W
0.10	0.302	53.7	0.60	0.612	24.0
0.20	0.408	42.0	0.70	0.641	21.9
0.30	0.480	35.1	0.80	0.667	20.0
0.40	0.535	30.3	0.90	0.688	18.5
0.50	0.577	26.8	1.00	0.707	17.2

account of. The difference equation (A13.1.13) is of the form

$$(\alpha B^{-2} + \beta B^{-1} + \gamma + \beta B + \alpha B^2)\mu_i = 0$$

and if κ is a root of the characteristic equation, so is κ^{-1}. Suppose that the roots are $\kappa_1, \kappa_2, \kappa_1^{-1}, \kappa_2^{-1}$ and that κ_1 and κ_2 are a pair of roots with modulus <1. Then in the solution

$$\mu_i = A_1\kappa_1^i + A_2\kappa_2^i + A_3\kappa_1^{-i} + A_4\kappa_2^{-i}$$

A_3 and A_4 must be zero, because ε_t is required to have a finite variance.

Hence the solution is of the form

$$\mu_i = A_1\kappa_1^i + A_2\kappa_2^i \qquad |\kappa_1| < 1 \quad |\kappa_2| < 1$$

The A's satisfying the initial conditions, defined by (A13.1.11) and (A13.1.12), are obtained by substitution to give

$$A_1 = \frac{\lambda\kappa_1(1 - \kappa_2)}{\kappa_1 - \kappa_2} \qquad A_2 = -\frac{\lambda\kappa_2(1 - \kappa_1)}{\kappa_1 - \kappa_2}$$

If we write $k_0 = \kappa_1 + \kappa_2 - \kappa_1\kappa_2$, $k_1 = \kappa_1\kappa_2$, then

$$\mu(B) = \lambda\left[\frac{k_0 - k_1B}{1 - (k_0 + k_1)B + k_1B^2}\right] \tag{A13.1.23}$$

and

$$1 + B\mu(B) = \frac{1 - k_1B - (1 - \lambda)(k_0B - k_1B^2)}{1 - (k_0 + k_1)B + k_1B^2} \tag{A13.1.24}$$

Now substituting (A13.1.23) in (A13.1.5),

$$L(B) = \frac{\lambda(1 - \delta B)(1 - k_0)}{(1 - \delta)[1 - (k_0 + k_1)B + k_1B^2]} \tag{A13.1.25}$$

and

$$\frac{L(B)}{1 + B\mu(B)} = \frac{\lambda(1 - \delta B)(1 - k_0)}{(1 - \delta)[1 - k_1B - (1 - \lambda)(k_0B - k_1B^2)]}$$

Therefore, using (A13.1.4) we find that the optimal control action in terms of the error ε_t is

$$x_t = -\frac{\lambda}{g}\frac{(1 - \delta B)(1 - k_0)}{(1 - \delta)[1 - k_1B - (1 - \lambda)(k_0B - k_1B^2)]}\varepsilon_t \tag{A13.1.26}$$

or

$$x_t = [k_1 + (1 - \lambda)k_0]x_{t-1} - (1 - \lambda)k_1x_{t-2} - \frac{\lambda(1 - k_0)(1 - \delta B)}{g(1 - \delta)}\varepsilon_t \tag{A13.1.27}$$

Thus the modified control scheme makes x_t depend on both x_{t-1} and x_{t-2} (only on x_{t-1} if $\lambda = 1$) and reduces the standard integral and proportional action by a factor $1 - k_0$.

Variances of output and input. The actual variances for the output and input are readily found since

$$\varepsilon_t = a_t + \lambda \left[\frac{k_0 - k_1 B}{1 - (k_0 + k_1)B + k_1 B^2} \right] a_{t-1}$$

The second term on the right defines a mixed autoregressive–moving average process of order (2, 0, 1), the variance for which is readily obtained to give

$$\frac{V[\varepsilon_t]}{\sigma_a^2} = 1 + \lambda^2 \left\{ \frac{(k_0 + k_1)^2(1 - k_1) - 2k_1(k_0 - k_1^2)}{(1 - k_1)[(1 + k_1)^2 - (k_0 + k_1)^2]} \right\} = 1 + \lambda^2 Q \tag{A13.1.28}$$

Also,

$$\frac{V[x_t]}{\sigma_a^2} = \frac{\lambda^2}{g^2(1 - \delta)^2} \frac{(1 - k_0)[(1 + \delta^2)(1 + k_1) - 2\delta(k_0 + k_1)]}{(1 + k_0 + 2k_1)(1 - k_1)} \tag{A13.1.29}$$

Computation of k_0 and k_1. Returning to the difference equations (A13.1.13), the characteristic equation may be written

$$B^4 - MB^3 + NB^2 - MB + 1 = 0$$

where $M = (1 + \delta)^2/\delta$ and $N = [(1 + \delta)^2 + (1 + \delta^2) + \nu]/\delta$. It may also be written in the form

$$(B^2 - TB + P)(B^2 - P^{-1}TB + P^{-1}) = 0$$

where

$$T = \kappa_1 + \kappa_2 \quad \text{and} \quad P = \kappa_1 \kappa_2$$

Equating coefficients of B gives us

$$T + P^{-1}T = M$$

that is, $T - PM/(1 + P)$, and

$$P + P^{-1} + P^{-1}T^2 = N$$

Thus $P + P^{-1} + PM^2/(1 + P)^2 = N$, that is,

$$(P + 2 + P^{-1})(P + P^{-1}) + M^2 = N(P + 2 + P^{-1})$$

or

$$(P + P^{-1})^2 + (2 - N)(P + P^{-1}) + M^2 - 2N = 0$$

For suitable values of ν, this quadratic equation will have two real roots

$$u_1 = \kappa_1\kappa_2 + \kappa_1^{-1}\kappa_2^{-1} \qquad u_2 = \kappa_1\kappa_2^{-1} + \kappa_1^{-1}\kappa_2$$

the root u_1 being the larger. The required quantity P is now the smaller root of the quadratic equation

$$P^2 - u_1P + 1 = 0$$

and T is given by

$$T = [P(u_2 + 2)]^{1/2}$$

Table of optimal values for constrained schemes

Construction of the Table. Table A13.2 is provided to facilitate the selection of an optimal control scheme. The tabled values were obtained as follows for each chosen value of the parameter δ in the transfer function model:

1. Compute $M = \dfrac{(1+\delta)^2}{\delta}$ and $N = \dfrac{(1+\delta)^2 + (1+\delta^2) + \nu}{\delta}$ for a series of values of ν chosen to provide a suitable range for Q.
2. Compute $u_1 = \dfrac{1}{2}(N-2) + \left[\left(\dfrac{N-2}{2}\right)^2 + 2N - M^2\right]^{1/2}$

   and $u_2 = \dfrac{1}{2}(N-2) - \left[\left(\dfrac{N-2}{2}\right)^2 + 2N - M^2\right]^{1/2}$.
3. Compute $k_1 = P = \dfrac{1}{2}u_1 - \left[\left(\dfrac{1}{2}u_1\right)^2 - 1\right]^{1/2}$

   and $k_0 = T - P = [k_1(u_2 + 2)]^{1/2} - k_1$.
4. Compute $Q = \dfrac{(k_0 + k_1)^2(1 - k_1) - 2k_1(k_0 - k_1^2)}{(1 - k_1)[(1 + k_1)^2 - (k_0 + k_1)^2]}$.
5. Compute $W = \dfrac{(1 - k_0)[(1 + \delta^2)(1 + k_1) - 2\delta(k_0 + k_1)]}{(1 + k_0 + 2k_1)(1 - k_1)(1 + \delta^2)}$.
6. Interpolate among the W, k_0, k_1 values at convenient values of Q.

Use of the Table. Table A13.2 may be used as follows. The value of δ is entered in the vertical margin. Using the fact that $V[\varepsilon_t] = (1 + \lambda^2 Q)\sigma_a^2$, the percentage increase in output variance is $100Q\lambda^2$. A suitable value of Q is entered in the horizontal margin. The entries in the table are then (1) $100W$, the percentage reduction in the variance of x_t, (2) k_0, and (3) k_1.

For illustration, suppose that $\lambda = 0.6$, $\delta = 0.5$, and $g = 1$. The optimal unconstrained control equation is then

$$x_t = -1.2(1 - 0.5B)\varepsilon_t = -1.2(1 - 0.5B)a_t$$

TABLE A13.2 Table to Facilitate the Calculation of Optimal Constrained Control Schemes

		100Q				
δ		20	40	60	80	100
0.9	100W	21.7	11.3	6.7	4.5	3.1
	k_0	0.44	0.585	0.68	0.74	0.78
	k_1	0.18	0.27	0.34	0.39	0.44
0.8	100W	22.0	11.7	7.2	4.8	3.4
	k_0	0.44	0.585	0.68	0.74	0.78
	k_1	0.18	0.27	0.33	0.38	0.43
0.7	100W	22.7	12.4	8.0	5.6	4.1
	k_0	0.44	0.585	0.68	0.74	0.78
	k_1	0.17	0.25	0.32	0.36	0.40
0.6	100W	24.1	13.6	9.0	6.6	5.0
	k_0	0.44	0.58	0.67	0.73	0.78
	k_1	0.16	0.24	0.29	0.33	0.365
0.5	100W	26.5	15.5	10.5	7.9	6.2
	k_0	0.43	0.58	0.67	0.72	0.77
	k_1	0.15	0.21	0.26	0.29	0.32
0.4	100W	28.5	17.7	12.7	9.8	7.9
	k_0	0.43	0.57	0.66	0.72	0.76
	k_1	0.13	0.18	0.22	0.245	0.265
0.3	100W	31.5	20.5	15.2	12.0	9.9
	k_0	0.43	0.57	0.65	0.71	0.75
	k_1	0.105	0.145	0.17	0.19	0.20
0.2	100W	34.8	23.6	18.0	14.5	12.2
	k_0	0.42	0.56	0.64	0.69	0.73
	k_1	0.07	0.10	0.12	0.13	0.14
0.1	100W	38.2	26.7	21.0	17.3	14.6
	k_0	0.42	0.55	0.63	0.68	0.72
	k_1	0.04	0.05	0.06	0.065	0.07

and $\text{var}[x_t] = 1.80\sigma_a^2$. Suppose that this amount of variation in the input variable produces difficulties in process operation and it is desired to cut $\text{var}[x_t]$ to about $0.50\sigma_a^2$, that is, to about 28% of the value for the unconstrained scheme. Inspection of Table A13.2 in the row labeled $\delta = 0.5$ shows that a reduction to 26.5% can be achieved by using a control scheme with constants $k_0 = 0.43$, $k_1 = 0.15$, that is, by employing the control equation (A13.1.27) to give

$$x_t = 0.32x_{t-1} - 0.06x_{t-2} - (0.57 \times 1.2)(1 - 0.5B)\varepsilon_t$$

This solution corresponds to a value $Q = 0.20$. Therefore, the variance at the output will be increased by a factor of $1 + \lambda^2 Q = 1 + 0.6^2(0.2) = 1.072$, that is, by about 7%.

APPENDIX A13.2 CHOICE OF THE SAMPLING INTERVAL

By comparison with continuous systems, discrete systems of control, such as are discussed here, can be very efficient provided that the sampling interval is suitably chosen. Roughly speaking, we want the interval to be such that not too much change can occur during the sampling interval. Usually, the behavior of the disturbance that has to pass through all or part of the system reflects the inertia or dynamic properties of the system, so that the sampling interval will often be chosen tacitly or explicitly to be proportional to the time constant or constants of the system. In chemical processes involving reaction and mixing of liquids, rather infrequent sampling, say at hourly intervals and possibly with operator surveillance and manual adjustment, will be sufficient. By contrast, where reactions between gases are involved, a suitable sampling interval may be measured in seconds and automatic monitoring and adjustment may be essential.

In some cases, experimentation may be needed to arrive at a satisfactory sampling interval, and in others rather simple calculations will show how the choice of sampling interval will affect the degree of control that is possible.

A13.2.1 Illustration of the Effect of Reducing Sampling Frequency

To illustrate the kind of calculation that is helpful, suppose again that we have a simple system in which, using a particular sampling interval, the noise is represented by a (0, 1, 1) process $\nabla N_t = (1 - \theta B)a_t$ and the transfer function model by the first-order system $(1 - \delta B)\mathcal{Y}_t = g(1 - \delta)x_{t-1}$. In this case, if we employ the MMSE adjustment

$$x_t = -\frac{1 - \theta}{g(1 - \delta)}(1 - \delta B)\varepsilon_t \tag{A13.2.1}$$

then the deviation from target is $\varepsilon_t = a_t$ and has variance $\sigma_a^2 = \sigma_1^2$, say.

In practice, the question has often arisen: How much worse off would we be if we took samples less frequently? To answer this question, we consider the effect of sampling the stochastic process involved.

A13.2.2 Sampling an IMA(0, 1, 1) Process

Suppose that with observations being made at some "unit" interval, we have a noise model

$$\nabla N_t = (1 - \theta_1 B)a_t$$

with $\text{var}[a_t] = \sigma_a^2 = \sigma_1^2$, where the subscript 1 is used in this context to denote the choice of sampling interval. Then, for the differences ∇N_t, the autocovariances γ_k are given by

$$\begin{aligned}\gamma_0 &= (1+\theta_1^2)\sigma_1^2 \\ \gamma_1 &= -\theta_1\sigma_1^2 \\ \gamma_j &= 0 \qquad j \geqslant 2\end{aligned} \tag{A13.2.2}$$

Writing $\zeta = (\gamma_0 + 2\gamma_1)/\gamma_1$, we obtain

$$\zeta = -\frac{(1-\theta_1)^2}{\theta_1}$$

so that, given γ_0 and γ_1, the parameter λ of the IMA process may be obtained by solving the quadratic equation

$$(1-\theta_1)^2 - \zeta(1-\theta_1) + \zeta = 0$$

selecting that root for which $-1 < \theta_1 < 1$. Also,

$$\sigma_1^2 = -\frac{\gamma_1}{\theta_1} \tag{A13.2.3}$$

Suppose now that the process N_t is observed at intervals of h units (where h is a positive integer) and the resulting process is denoted by M_t. Then

$$\begin{aligned}\nabla M_t = N_t - N_{t-h} = (a_t + a_{t-1} + \cdots + a_{t-h+1}) \\ - \theta_1(a_{t-1} + a_{t-2} + \cdots + a_{t-h})\end{aligned}$$

$$\begin{aligned}\nabla M_{t-h} = N_{t-h} - N_{t-2h} = (a_{t-h} + a_{t-h-1} + \cdots + a_{t-2h+1}) \\ - \theta_1(a_{t-h-1} + \cdots + a_{t-2h})\end{aligned}$$

and so on. Then, for the differences ∇M_t, the autocovariances $\gamma_k(h)$ are

$$\begin{aligned}\gamma_0(h) &= [(1+\theta_1^2) + (h-1)(1-\theta_1)^2]\sigma_1^2 \\ \gamma_1(h) &= -\theta_1\sigma_1^2 \\ \gamma_j(h) &= 0 \qquad j \geqslant 2\end{aligned} \tag{A13.2.4}$$

It follows that the process M_t is also an IMA process of order (0, 1, 1),

$$\nabla M_t = (1 - \theta_h B)e_t$$

where e_t is a white noise process with variance σ_h^2. Now

$$\frac{\gamma_0(h) + 2\gamma_1(h)}{\gamma_1(h)} = -\frac{h(1-\theta_1)^2}{\theta_1}$$

so that

$$\frac{h(1-\theta_1)^2}{\theta_1} = \frac{(1-\theta_h)^2}{\theta_h} \tag{A13.2.5}$$

Also, since $\gamma_1(h) = -\theta_h\sigma_h^2 = -\theta_1\sigma_1^2$, it follows that

$$\frac{\sigma_h^2}{\sigma_1^2} = \frac{\theta_1}{\theta_h} \tag{A13.2.6}$$

Therefore, we have shown that the sampling of an IMA process of order (0, 1, 1) at interval h produces another IMA process of order (0, 1, 1). From (A13.2.5), we can obtain the value of the parameter θ_h for the sampled process, and from (A13.2.6) we can obtain the variance σ_h^2 of that process in terms of the parameters θ_1 and σ_1^2 of the original process.

In Figure A13.1, θ_h is plotted against log h, a scale of h being appended. The graph enables one to find the effect of increasing the sampling interval of a (0, 1, 1) process by any given multiple. For illustration, suppose that we have a process for which $\theta_1 = 0.5$ and $\sigma_1^2 = 1$. Let us use the graph to find the values of the corresponding parameters θ_2, θ_4, σ_2^2, σ_4^2 when the sampling interval is (a) doubled, (b) quadrupled. Marking on the edge of a piece of paper the points $h = 1$, $h = 2$, $h = 4$ from the scale on the graph, we set the paper horizontally so that $h = 1$ corresponds to the point on the curve for which $\theta_1 = 0.5$. We then read off the ordinates for θ_2 and θ_4 corresponding to $h = 2$ and $h = 4$. We find that

$$\theta_1 = 0.5 \qquad \theta_2 = 0.38 \qquad \theta_4 = 0.27$$

Using (A13.2.6), the variances are in inverse proportion to the values of θ, so that

$$\sigma_1^2 = 1.00 \qquad \sigma_2^2 = 1.32 \qquad \sigma_4^2 = 2.17$$

Suppose now that for the original scheme with unit interval, the dynamic constant was δ_1 (again we shall use the subscript to denote the sampling interval). Then, since in real time the same fixed time constant $T = -h/\ln(\delta)$ applies to all the schemes, we have

$$\delta_2 = \delta_1^2 \qquad \delta_4 = \delta_1^4$$

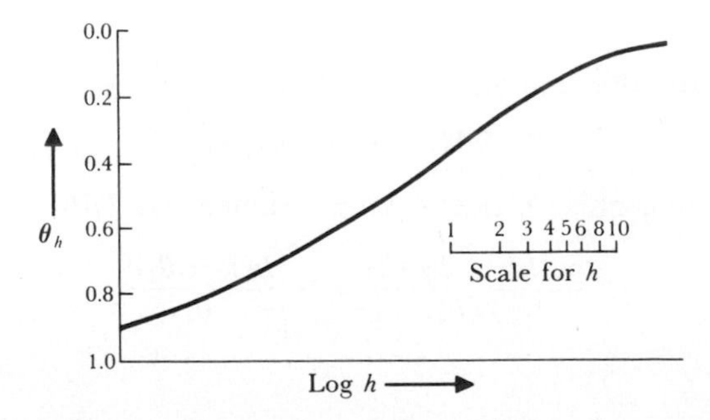

Figure A13.1 Sampling of IMA(0, 1, 1) process; parameter θ_h plotted against log h.

The scheme giving minimum mean square error for a *particular* sampling interval h would be

$$x_t(h) = -\frac{1-\theta_h}{g(1-\delta_1^h)}(1-\delta_1^h B)\varepsilon_t(h)$$

or

$$x_t(h) = -\frac{1-\theta_h}{g}\left(1+\frac{\delta_1^h}{1-\delta_1^h}\nabla\right)\varepsilon_t(h) \qquad \text{(A13.2.7)}$$

Suppose, for example, with $\theta_1 = 0.5$ as above, $\delta_1 = 0.8$, so tht $\delta_2 = 0.64$, $\delta_4 = 0.41$. Then the optimal schemes would be

$$h = 1: \quad x_t(1) = -\frac{0.5}{g}(1 + 4\nabla)\varepsilon_t(1) \qquad \sigma_\varepsilon^2 = 1.00 \qquad g^2\sigma_x^2 = 10.25$$

$$h = 2: \quad x_t(2) = -\frac{0.62}{g}(1 + 1.78\nabla)\varepsilon_t(2) \qquad \sigma_\varepsilon^2 = 1.32 \qquad g^2\sigma_x^2 = 5.50$$

$$h = 4: \quad x_t(4) = -\frac{0.73}{g}(1 + 0.69\nabla)\varepsilon_t(4) \qquad \sigma_\varepsilon^2 = 2.17 \qquad g^2\sigma_x^2 = 3.84$$

In accordance with expectation, as the sampling interval is increased and the dynamics of the system have relatively less importance, the amount of "integral" control is increased and the ratio of proportional to integral control is markedly reduced. We noted earlier that an excessively large adjustment variance σ_x^2 would usually be a disadvantage. The values of $g^2\sigma_x^2$ are indicated to show how the schemes differ in this respect. The smaller value for σ_x^2 would not of itself, of course, justify the choice $h = 4$. Using an optimal constrained scheme, such as is described in Appendix A13.1, with $h = 1$ a very large reduction in σ_x would be produced with only a small increase in the output variance. For example, entering Table A13.2 with $\delta = 0.8$, $100Q = 20$, we find that for a 5% increase of output variance to the value $(1 + \lambda^2 Q)\sigma_1^2 = 1.05\sigma_1^2$, the input variance for the scheme with $h = 1$ could be reduced to 22% of its unconstrained value, so that $g^2\sigma_x^2 = 10.25 \times 0.22 = 2.26$. Using (A13.1.27), we obtain for the constrained scheme with $h = 1$,

$$x_t = 0.40x_{t-1} - 0.09x_{t-2} - 0.56\left[\frac{0.5}{g}(1 + 4\nabla)\right]\varepsilon_t(1)$$

$$\sigma_\varepsilon^2 = 1.05 \qquad g^2\sigma_x^2 = 2.26$$

In practice, various alternative schemes could be set out with their accompanying characteristics and an economic choice made to suit the particular problem. In general, the increase in output variance which comes with the larger interval would have to be balanced off against the economic advantage, if any, of less frequent surveillance.

Part Five

CHARTS AND TABLES

This part of the book is a collection of auxiliary material useful in the analysis of time series. This includes tables and charts for obtaining preliminary estimates of autoregressive–moving average models, together with the usual tail-area tables of the Normal, χ^2, and t distributions. This is followed by a complete listing of all the time series analyzed in the book, as well as some additional time series that are discussed in the exercises and problems in Part Six.

COLLECTION OF TABLES AND CHARTS

Charts B, C, and D are adapted and reproduced by permission of the author from [183]. Tables E, F, and G are condensed and adapted from *Biometrika Tables for Statisticians,* Volume I, with permission from the trustees of Biometrika.

TABLE A TABLE RELATING ρ TO θ FOR A FIRST-ORDER MOVING AVERAGE PROCESS

θ	ρ_1	θ	ρ_1
0.00	0.000	0.00	0.000
0.05	−0.050	−0.05	0.050
0.10	−0.099	−0.10	0.099
0.15	−0.147	−0.15	0.147
0.20	−0.192	−0.20	0.192
0.25	−0.235	−0.25	0.235
0.30	−0.275	−0.30	0.275
0.35	−0.315	−0.35	0.315
0.40	−0.349	−0.40	0.349
0.45	−0.374	−0.45	0.374
0.50	−0.400	−0.50	0.400
0.55	−0.422	−0.55	0.422
0.60	−0.441	−0.60	0.441

TABLE A (*continued*)

θ	ρ_1	θ	ρ_1
0.65	−0.457	−0.65	0.457
0.70	−0.468	−0.70	0.468
0.75	−0.480	−0.75	0.480
0.80	−0.488	−0.80	0.488
0.85	−0.493	−0.85	0.493
0.90	−0.497	−0.90	0.497
0.95	−0.499	−0.95	0.499
1.00	−0.500	−1.00	0.500

The table may be used to obtain first estimates of the parameters in the (0, d, 1) process $w_t = (1 - \theta B)a_t$, where $w_t = \nabla^d z_t$, by substituting $r_1(w)$ for ρ_1.

CHART B CHART RELATING ρ_1 AND ρ_2 TO ϕ_1 AND ϕ_2 FOR A SECOND-ORDER AUTOREGRESSIVE PROCESS

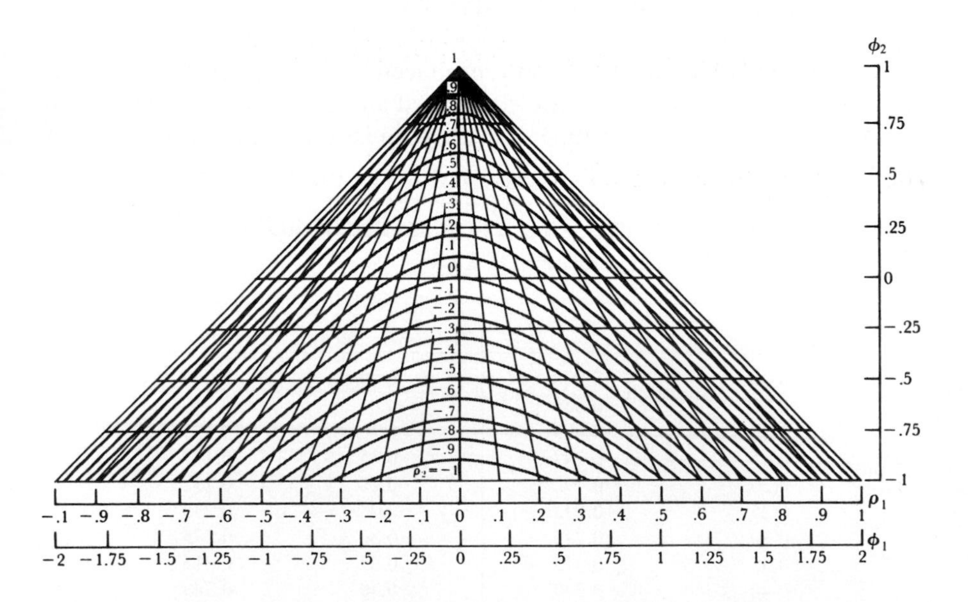

The chart may be used to obtain estimates of the parameters in the (2, d, 0) process: $(1 - \phi_1 B - \phi_2 B^2)w_t = a_t$, where $w_t = \nabla^d z_t$, by substituting $r_1(w)$ and $r_2(w)$ for ρ_1 and ρ_2.

CHART C CHART RELATING ρ_1 AND ρ_2 TO θ_1 AND θ_2 FOR A SECOND-ORDER MOVING AVERAGE PROCESS

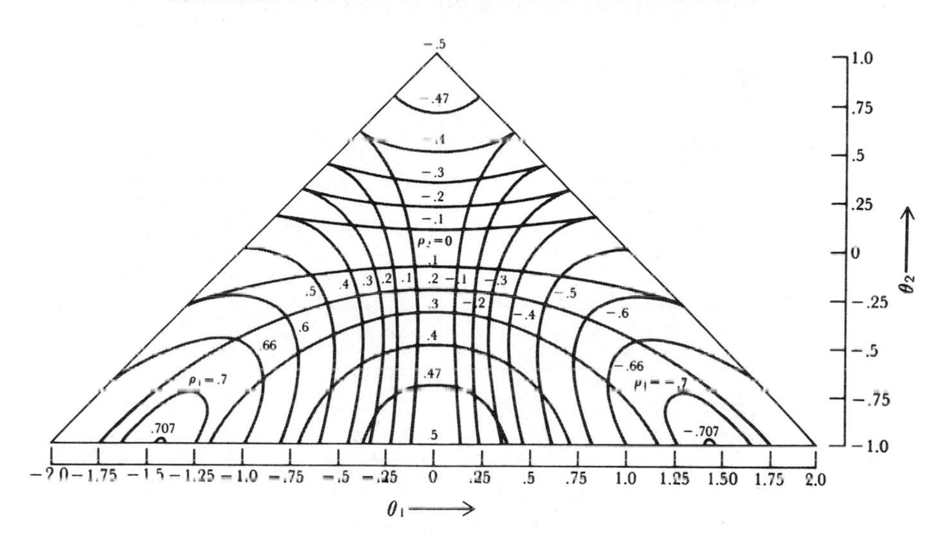

The chart may be used to obtain estimates of the parameters in the (0, d, 2) process $w_t = (1 - \theta_1 B - \theta_2 B^2)a_t$, where $w_t = \nabla^d z_t$, by substituting $r_1(w)$ and $r_2(w)$ for ρ_1 and ρ_2.

CHART D CHART RELATING ρ_1 AND ρ_2 TO ϕ AND θ FOR A MIXED FIRST-ORDER AUTOREGRESSIVE–MOVING AVERAGE PROCESS

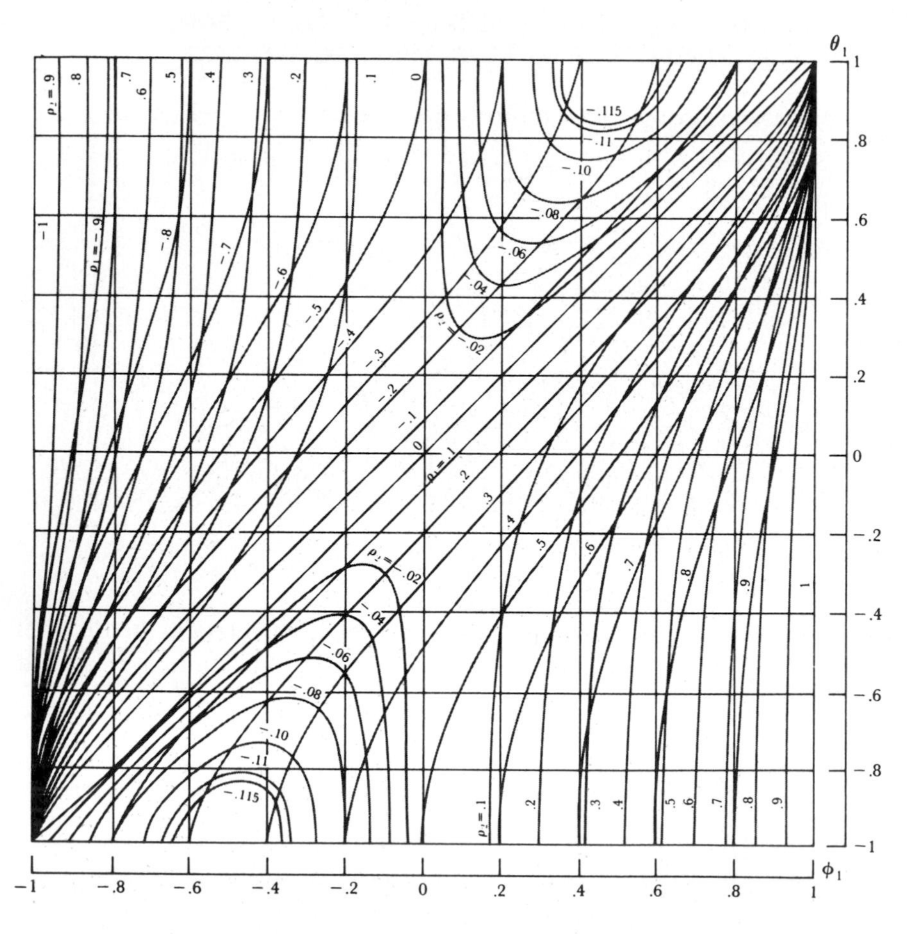

The chart may be used to obtain first estimates of the parameters in the $(1, d, 1)$ process $(1 - \phi B)w_t = (1 - \theta B)a_t$, where $w_t = \nabla^d z_t$, by substituting $r_1(w)$ and $r_2(w)$ for ρ_1 and ρ_2.

TABLE E TAIL AREAS AND ORDINATES OF UNIT NORMAL DISTRIBUTION

u_ε	ε	$p(u_\varepsilon)$	u_ε	ε	$p(u_\varepsilon)$
0.0	0.500	0.3989	1.6	0.055	0.1109
0.1	0.460	0.3969	1.7	0.045	0.0940
0.2	0.421	0.3910	1.8	0.036	0.0790
0.3	0.382	0.3814	1.9	0.029	0.0656
0.4	0.345	0.3683	2.0	0.023	0.0540
0.5	0.309	0.3521	2.1	0.018	0.0440
0.6	0.274	0.3322	2.2	0.014	0.0355
0.7	0.242	0.3123	2.3	0.011	0.0283
0.8	0.212	0.2897	2.4	0.008	0.0224
0.9	0.184	0.2661	2.5	0.006	0.0175
1.0	0.159	0.2420	2.6	0.005	0.0136
1.1	0.136	0.2179	2.7	0.003	0.0104
1.2	0.115	0.1942	2.8	0.003	0.0079
1.3	0.097	0.1714	2.9	0.002	0.0059
1.4	0.081	0.1497	3.0	0.001	0.0044
1.5	0.067	0.1295			

Note: Shown are the values of the unit Normal deviate u_ε such that $\Pr\{u > u_\varepsilon\} = \varepsilon$; also shown are the ordinates $p(u = u_\varepsilon)$.

TABLE F TAIL AREAS OF THE CHI-SQUARE DISTRIBUTION

	ε														
m	0.995	0.99	0.975	0.95	0.9	0.75	0.5	0.25	0.1	0.05	0.025	0.01	0.005	0.001	m
1	—	—	—	—	0.016	0.102	0.455	1.32	2.71	3.84	5.02	6.63	7.88	10.8	1
2	0.010	0.020	0.051	0.103	0.211	0.575	1.39	2.77	4.61	5.99	7.38	9.21	10.6	13.8	2
3	0.072	0.115	0.216	0.352	0.584	1.21	2.37	4.11	6.25	7.81	9.35	11.3	12.8	16.3	3
4	0.207	0.297	0.484	0.711	1.06	1.92	3.36	5.39	7.78	9.49	11.1	13.3	14.9	18.5	4
5	0.412	0.554	0.831	1.15	1.61	2.67	4.35	6.63	9.24	11.1	12.8	15.1	16.7	20.5	5
6	0.676	0.872	1.24	1.64	2.20	3.45	5.35	7.84	10.6	12.6	14.4	16.8	18.5	22.5	6
7	0.989	1.24	1.69	2.17	2.83	4.25	6.35	9.04	12.0	14.1	16.0	18.5	20.3	24.3	7
8	1.34	1.65	2.18	2.73	3.49	5.07	7.34	10.2	13.4	15.5	17.5	20.1	22.0	26.1	8
9	1.73	2.09	2.70	3.33	4.17	5.90	8.34	11.4	14.7	16.9	19.0	21.7	23.6	27.9	9
10	2.16	2.56	3.25	3.94	4.87	6.74	9.34	12.5	16.0	18.3	20.5	23.2	25.2	29.6	10
11	2.60	3.05	3.82	4.57	5.58	7.58	10.3	13.7	17.3	19.7	21.9	24.7	26.8	31.3	11
12	3.07	3.57	4.40	5.23	6.30	8.44	11.3	14.8	18.5	21.0	23.3	26.2	28.3	32.9	12
13	3.57	4.11	5.01	5.89	7.04	9.30	12.3	16.0	19.8	22.4	24.7	27.7	29.8	34.5	13
14	4.07	4.66	5.63	6.57	7.79	10.2	13.3	17.1	21.1	23.7	26.1	29.1	31.3	36.1	14
15	4.60	5.23	6.26	7.26	8.55	11.0	14.3	18.2	22.3	25.0	27.5	30.6	32.8	37.7	15
16	5.14	5.81	6.91	7.96	9.31	11.9	15.3	19.4	23.5	26.3	28.8	32.0	34.3	39.3	16
17	5.70	6.41	7.56	8.67	10.1	12.8	16.3	20.5	24.8	27.6	30.2	33.4	35.7	40.8	17
18	6.26	7.01	8.23	9.39	10.9	13.7	17.3	21.6	26.0	28.9	31.5	34.8	37.2	42.3	18
19	6.84	7.63	8.91	10.1	11.7	14.6	18.3	22.7	27.2	30.1	32.9	36.2	38.6	43.8	19
20	7.43	8.26	9.59	10.9	12.4	15.5	19.3	23.8	28.4	31.4	34.2	37.6	40.0	45.3	20
21	8.03	8.90	10.3	11.6	13.2	16.3	20.3	24.9	29.6	32.7	35.5	38.9	41.4	46.8	21
22	8.64	9.54	11.0	12.3	14.0	17.2	21.3	26.0	30.8	33.9	36.8	40.3	42.8	48.3	22
23	9.26	10.2	11.7	13.1	14.8	18.1	22.3	27.1	32.0	35.2	38.1	41.6	44.2	49.7	23
24	9.89	10.9	12.4	13.8	15.7	19.0	23.3	28.2	33.2	36.4	39.4	43.0	45.6	51.2	24
25	10.5	11.5	13.1	14.6	16.5	19.9	24.3	29.3	34.4	37.7	40.6	44.3	46.9	52.6	25
26	11.2	12.2	13.8	15.4	17.3	20.8	25.3	30.4	35.6	38.9	41.9	45.6	48.3	54.1	26
27	11.8	12.9	14.6	16.2	18.1	21.7	26.3	31.5	36.7	40.1	43.2	47.0	49.6	55.5	27
28	12.5	13.6	15.3	16.9	18.9	22.7	27.3	32.6	37.9	41.3	44.5	48.3	51.0	56.9	28
29	13.1	14.3	16.0	17.7	19.8	23.6	28.3	33.7	39.1	42.6	45.7	49.6	52.3	58.3	29
30	13.8	15.0	16.8	18.5	20.6	24.5	29.3	34.8	40.3	43.8	47.0	50.9	53.7	59.7	30

Note: Shown are the values of $\chi^2_\varepsilon(m)$ such that $\Pr\{\chi^2(m) > \chi^2_\varepsilon(m)\} = \varepsilon$, where m is the number of degrees of freedom.

TABLE G TAIL AREAS OF THE t-DISTRIBUTION

	ε					
ν	0.25	0.10	0.05	0.025	0.01	0.005
1	1.00	3.08	6.31	12.71	31.82	63.66
2	0.82	1.89	2.92	4.30	6.96	9.92
3	0.76	1.64	2.35	3.18	4.54	5.84
4	0.74	1.53	2.13	2.78	3.75	4.60
5	0.73	1.48	2.02	2.57	3.36	4.03
6	0.72	1.44	1.94	2.45	3.14	3.71
7	0.71	1.42	1.90	2.36	3.00	3.50
8	0.71	1.40	1.86	2.31	2.90	3.36
9	0.70	1.38	1.83	2.26	2.82	3.25
10	0.70	1.37	1.81	2.23	2.76	3.17
11	0.70	1.36	1.80	2.20	2.72	3.11
12	0.70	1.36	1.78	2.18	2.68	3.06
13	0.69	1.35	1.77	2.16	2.65	3.01
14	0.69	1.34	1.76	2.14	2.62	3.00
15	0.69	1.34	1.75	2.13	2.60	2.95
16	0.69	1.34	1.75	2.12	2.58	2.92
17	0.69	1.33	1.74	2.11	2.57	2.90
18	0.69	1.33	1.73	2.10	2.55	2.88
19	0.69	1.33	1.73	2.09	2.54	2.86
20	0.69	1.33	1.72	2.09	2.53	2.84
30	0.68	1.31	1.70	2.04	2.46	2.75
40	0.68	1.30	1.68	2.02	2.42	2.70
60	0.68	1.30	1.67	2.00	2.39	2.66
120	0.68	1.29	1.66	1.98	2.36	2.62
∞	0.67	1.28	1.64	1.96	2.33	2.58

Note: Shown are the values of $t_\varepsilon(\nu)$ such that $\Pr\{t(\nu) > t_\varepsilon(\nu)\} = \varepsilon$, where ν is the number of degrees of freedom.

COLLECTION OF TIME SERIES USED FOR EXAMPLES IN THE TEXT AND IN EXERCISES

Series A	Chemical process concentration readings: every 2 hours
Series B	IBM common stock closing prices: daily, May 17, 1961–November 2, 1962
Series B′	IBM common stock closing prices: daily, June 29, 1959–June 30, 1960
Series C	Chemical process temperature readings: every minute
Series D	Chemical process viscosity readings: every hour
Series E	Wölfer sunspot numbers: yearly
Series F	Yields from a batch chemical process: consecutive
Series G	International airline passengers: monthly totals (thousands of passengers) January 1949–December 1960
Series J	Gas furnace data
Series K	Simulated dynamic data with two inputs
Series L	Pilot scheme data
Series M	Sales data with leading indicator
Series N	Mink fur sales data of the Hudson's Bay Company: annual for 1850–1911
Series P	Unemployment and GDP data in UK: quarterly for 1955–1969
Series Q	U.S. hog price data: annual for 1867–1948
Series R	Monthly averages of hourly readings of ozone in downtown Los Angeles

SERIES A CHEMICAL PROCESS CONCENTRATION READINGS: EVERY 2 HOURS[a]

1	17.0	41	17.6	81	16.8	121	16.9	161	17.1
2	16.6	42	17.5	82	16.7	122	17.1	162	17.1
3	16.3	43	16.5	83	16.4	123	16.8	163	17.1
4	16.1	44	17.8	84	16.5	124	17.0	164	17.4
5	17.1	45	17.3	85	16.4	125	17.2	165	17.2
6	16.9	46	17.3	86	16.6	126	17.3	166	16.9
7	16.8	47	17.1	87	16.5	127	17.2	167	16.9
8	17.4	48	17.4	88	16.7	128	17.3	168	17.0
9	17.1	49	16.9	89	16.4	129	17.2	169	16.7
10	17.0	50	17.3	90	16.4	130	17.2	170	16.9
11	16.7	51	17.6	91	16.2	131	17.5	171	17.3
12	17.4	52	16.9	92	16.4	132	16.9	172	17.8
13	17.2	53	16.7	93	16.3	133	16.9	173	17.8
14	17.4	54	16.8	94	16.4	134	16.9	174	17.6
15	17.4	55	16.8	95	17.0	135	17.0	175	17.5
16	17.0	56	17.2	96	16.9	136	16.5	176	17.0
17	17.3	57	16.8	97	17.1	137	16.7	177	16.9
18	17.2	58	17.6	98	17.1	138	16.8	178	17.1
19	17.4	59	17.2	99	16.7	139	16.7	179	17.2
20	16.8	60	16.6	100	16.9	140	16.7	180	17.4
21	17.1	61	17.1	101	16.5	141	16.6	181	17.5
22	17.4	62	16.9	102	17.2	142	16.5	182	17.9
23	17.4	63	16.6	103	16.4	143	17.0	183	17.0
24	17.5	64	18.0	104	17.0	144	16.7	184	17.0
25	17.4	65	17.2	105	17.0	145	16.7	185	17.0
26	17.6	66	17.3	106	16.7	146	16.9	186	17.2
27	17.4	67	17.0	107	16.2	147	17.4	187	17.3
28	17.3	68	16.9	108	16.6	148	17.1	188	17.4
29	17.0	69	17.3	109	16.9	149	17.0	189	17.4
30	17.8	70	16.8	110	16.5	150	16.8	190	17.0
31	17.5	71	17.3	111	16.6	151	17.2	191	18.0
32	18.1	72	17.4	112	16.6	152	17.2	192	18.2
33	17.5	73	17.7	113	17.0	153	17.4	193	17.6
34	17.4	74	16.8	114	17.1	154	17.2	194	17.8
35	17.4	75	16.9	115	17.1	155	16.9	195	17.7
36	17.1	76	17.0	116	16.7	156	16.8	196	17.2
37	17.6	77	16.9	117	16.8	157	17.0	197	17.4
38	17.7	78	17.0	118	16.3	158	17.4		
39	17.4	79	16.6	119	16.6	159	17.2		
40	17.8	80	16.7	120	16.8	160	17.2		

[a]197 Observations.

SERIES B IBM COMMON STOCK CLOSING PRICES: DAILY, MAY 17, 1961–NOVEMBER 2, 1962[a]

460	471	527	580	551	523	333	394	330
457	467	540	579	551	516	330	393	340
452	473	542	584	552	511	336	409	339
459	481	538	581	553	518	328	411	331
462	488	541	581	557	517	316	409	345
459	490	541	577	557	520	320	408	352
463	489	547	577	548	519	332	393	346
479	489	553	578	547	519	320	391	352
493	485	559	580	545	519	333	388	357
490	491	557	586	545	518	344	396	
492	492	557	583	539	513	339	387	
498	494	560	581	539	499	350	383	
499	499	571	576	535	485	351	388	
497	498	571	571	537	454	350	382	
496	500	569	575	535	462	345	384	
490	497	575	575	536	473	350	382	
489	494	580	573	537	482	359	383	
478	495	584	577	543	486	375	383	
487	500	585	582	548	475	379	388	
491	504	590	584	546	459	376	395	
487	513	599	579	547	451	382	392	
482	511	603	572	548	453	370	386	
479	514	599	577	549	446	365	383	
478	510	596	571	553	455	367	377	
479	509	585	560	553	452	372	364	
477	515	587	549	552	457	373	369	
479	519	585	556	551	449	363	355	
475	523	581	557	550	450	371	350	
479	519	583	563	553	435	369	353	
476	523	592	564	554	415	376	340	
476	531	592	567	551	398	387	350	
478	547	596	561	551	399	387	349	
479	551	596	559	545	361	376	358	
477	547	595	553	547	383	385	360	
476	541	598	553	547	393	385	360	
475	545	598	553	537	385	380	366	
475	549	595	547	539	360	373	359	
473	545	595	550	538	364	382	356	
474	549	592	544	533	365	377	355	
474	547	588	541	525	370	376	367	
474	543	582	532	513	374	379	357	
465	540	576	525	510	359	386	361	
466	539	578	542	521	335	387	355	
467	532	589	555	521	323	386	348	
471	517	585	558	521	306	389	343	

[a]369 Observations (read down).

SERIES B′ IBM COMMON STOCK CLOSING PRICES: DAILY, JUNE 29, 1959–JUNE 30, 1960[a]

445	425	406	441	415	461
448	421	407	437	420	463
450	414	410	427	420	463
447	410	408	423	424	461
451	411	408	424	426	465
453	406	409	428	423	473
454	406	410	428	423	473
454	413	409	431	425	475
459	411	405	425	431	499
440	410	406	423	436	485
446	405	405	420	436	491
443	409	407	426	440	496
443	410	409	418	436	504
440	405	407	416	443	504
439	401	409	419	445	509
435	401	425	418	439	511
435	401	425	416	443	524
436	414	428	419	445	525
435	419	436	425	450	541
435	425	442	421	461	531
435	423	442	422	471	529
433	411	433	422	467	530
429	414	435	417	462	531
428	420	433	420	456	527
425	412	435	417	464	525
427	415	429	418	463	519
425	412	439	419	465	514
422	412	437	419	464	509
409	411	439	417	456	505
407	412	438	419	460	513
423	409	435	422	458	525
422	407	433	423	453	519
417	408	437	422	453	519
421	415	437	421	449	522
424	413	444	421	447	522
414	413	441	419	453	
419	410	440	418	450	
429	405	441	421	459	
426	410	439	420	457	
425	412	439	413	453	
424	413	438	413	455	
425	411	437	408	453	
425	411	441	409	450	
424	409	442	415	456	

[a]255 Observations (read down).

SERIES C CHEMICAL PROCESS TEMPERATURE READINGS: EVERY MINUTE[a]

26.6	19.6	24.4	21.1	24.4
27.0	19.6	24.4	20.9	24.2
27.1	19.6	24.4	20.8	24.2
27.1	19.6	24.4	20.8	24.1
27.1	19.6	24.5	20.8	24.1
27.1	19.7	24.5	20.8	24.0
26.9	19.9	24.4	20.9	24.0
26.8	20.0	24.3	20.8	24.0
26.7	20.1	24.2	20.8	23.9
26.4	20.2	24.2	20.7	23.8
26.0	20.3	24.0	20.7	23.8
25.8	20.6	23.9	20.8	23.7
25.6	21.6	23.7	20.9	23.7
25.2	21.9	23.6	21.2	23.6
25.0	21.7	23.5	21.4	23.7
24.6	21.3	23.5	21.7	23.6
24.2	21.2	23.5	21.8	23.6
24.0	21.4	23.5	21.9	23.6
23.7	21.7	23.5	22.2	23.5
23.4	22.2	23.7	22.5	23.5
23.1	23.0	23.8	22.8	23.4
22.9	23.8	23.8	23.1	23.3
22.8	24.6	23.9	23.4	23.3
22.7	25.1	23.9	23.8	23.3
22.6	25.6	23.8	24.1	23.4
22.4	25.8	23.7	24.6	23.4
22.2	26.1	23.6	24.9	23.3
22.0	26.3	23.4	24.9	23.2
21.8	26.3	23.2	25.1	23.3
21.4	26.2	23.0	25.0	23.3
20.9	26.0	22.8	25.0	23.2
20.3	25.8	22.6	25.0	23.1
19.7	25.6	22.4	25.0	22.9
19.4	25.4	22.0	24.9	22.8
19.3	25.2	21.6	24.8	22.6
19.2	24.9	21.3	24.7	22.4
19.1	24.7	21.2	24.6	22.2
19.0	24.5	21.2	24.5	21.8
18.9	24.4	21.1	24.5	21.3
18.9	24.4	21.0	24.5	20.8
19.2	24.4	20.9	24.5	20.2
19.3	24.4	21.0	24.5	19.7
19.3	24.4	21.0	24.5	19.3
19.4	24.3	21.1	24.5	19.1
19.5	24.4	21.2	24.4	19.0
				18.8

[a]226 Observations (read down).

SERIES D CHEMICAL PROCESS VISCOSITY READINGS: EVERY HOUR[a]

8.0	8.6	9.3	9.8	9.4	9.6	9.4
8.0	8.4	9.5	9.6	9.6	9.6	10.0
7.4	8.3	9.4	9.6	9.6	9.6	10.0
8.0	8.4	9.0	9.4	9.6	9.6	10.0
8.0	8.3	9.0	9.4	10.0	9.6	10.2
8.0	8.3	8.8	9.4	10.0	9.0	10.0
8.0	8.1	9.0	9.4	9.6	9.4	10.0
8.8	8.2	8.8	9.6	9.2	9.4	9.6
8.4	8.3	8.6	9.6	9.2	9.4	9.0
8.4	8.5	8.6	9.4	9.2	9.6	9.0
8.0	8.1	8.0	9.4	9.0	9.4	8.6
8.2	8.1	8.0	9.0	9.0	9.6	9.0
8.2	7.9	8.0	9.4	9.6	9.6	9.6
8.2	8.3	8.0	9.4	9.8	9.8	9.6
8.4	8.1	8.6	9.6	10.2	9.8	9.0
8.4	8.1	8.0	9.4	10.0	9.8	9.0
8.4	8.1	8.0	9.2	10.0	9.6	8.9
8.6	8.4	8.0	8.8	10.0	9.2	8.8
8.8	8.7	7.6	8.8	9.4	9.6	8.7
8.6	9.0	8.6	9.2	9.2	9.2	8.6
8.6	9.3	9.6	9.2	9.6	9.2	8.3
8.6	9.3	9.6	9.6	9.7	9.6	7.9
8.6	9.5	10.0	9.6	9.7	9.6	8.5
8.6	9.3	9.4	9.8	9.8	9.6	8.7
8.8	9.5	9.3	9.8	9.8	9.6	8.9
8.9	9.5	9.2	10.0	9.8	9.6	9.1
9.1	9.5	9.5	10.0	10.0	9.6	9.1
9.5	9.5	9.5	9.4	10.0	10.0	9.1
8.5	9.5	9.5	9.8	8.6	10.0	
8.4	9.5	9.9	8.8	9.0	10.4	
8.3	9.9	9.9	8.8	9.4	10.4	
8.2	9.5	9.5	8.8	9.4	9.8	
8.1	9.7	9.3	8.8	9.4	9.0	
8.3	9.1	9.5	9.6	9.4	9.6	
8.4	9.1	9.5	9.6	9.4	9.8	
8.7	8.9	9.1	9.6	9.6	9.6	
8.8	9.3	9.3	9.2	10.0	8.6	
8.8	9.1	9.5	9.2	10.0	8.0	
9.2	9.1	9.3	9.0	9.8	8.0	
9.6	9.3	9.1	9.0	9.8	8.0	
9.0	9.5	9.3	9.0	9.7	8.0	
8.8	9.3	9.1	9.4	9.6	8.4	
8.6	9.3	9.5	9.0	9.4	8.8	
8.6	9.3	9.4	9.0	9.2	8.4	
8.8	9.9	9.5	9.4	9.0	8.4	
8.8	9.7	9.6	9.4	9.4	9.0	
8.6	9.1	10.2	9.6	9.6	9.0	

[a] 310 Observations (read down).

SERIES E WÖLFER SUNSPOT NUMBERS: YEARLY[a]

1770	101	1795	21	1820	16	1845	40
1771	82	1796	16	1821	7	1846	62
1772	66	1797	6	1822	4	1847	98
1773	35	1798	4	1823	2	1848	124
1774	31	1799	7	1824	8	1849	96
1775	7	1800	14	1825	17	1850	66
1776	20	1801	34	1826	36	1851	64
1777	92	1802	45	1827	50	1852	54
1778	154	1803	43	1828	62	1853	39
1779	125	1804	48	1829	67	1854	21
1780	85	1805	42	1830	71	1855	7
1781	68	1806	28	1831	48	1856	4
1782	38	1807	10	1832	28	1857	23
1783	23	1808	8	1833	8	1858	55
1784	10	1809	2	1834	13	1859	94
1785	24	1810	0	1835	57	1860	96
1786	83	1811	1	1836	122	1861	77
1787	132	1812	5	1837	138	1862	59
1788	131	1813	12	1838	103	1863	44
1789	118	1814	14	1839	86	1864	47
1790	90	1815	35	1840	63	1865	30
1791	67	1816	46	1841	37	1866	16
1792	60	1817	41	1842	24	1867	7
1793	47	1818	30	1843	11	1868	37
1794	41	1819	24	1844	15	1869	74

[a]100 Observations.

SERIES F YIELDS FROM A BATCH CHEMICAL PROCESS[a]

47	44	50	62	68
64	80	71	44	38
23	55	56	64	50
71	37	74	43	60
38	74	50	52	39
64	51	58	38	59
55	57	45	59	40
41	50	54	55	57
59	60	36	41	54
48	45	54	53	23
71	57	48	49	
35	50	55	34	
57	45	45	35	
40	25	57	54	
58	59	50	45	

[a]70 Observations (read down). This series also appears in Table 2.1.

SERIES G INTERNATIONAL AIRLINE PASSENGERS: MONTHLY TOTALS (THOUSANDS OF PASSENGERS) JANUARY 1949–DECEMBER 1960[a]

	Jan.	Feb.	Mar.	Apr.	May	June	July	Aug.	Sept.	Oct.	Nov.	Dec.
1949	112	118	132	129	121	135	148	148	136	119	104	118
1950	115	126	141	135	125	149	170	170	158	133	114	140
1951	145	150	178	163	172	178	199	199	184	162	146	166
1952	171	180	193	181	183	218	230	242	209	191	172	194
1953	196	196	236	235	229	243	264	272	237	211	180	201
1954	204	188	235	227	234	264	302	293	259	229	203	229
1955	242	233	267	269	270	315	364	347	312	274	237	278
1956	284	277	317	313	318	374	413	405	355	306	271	306
1957	315	301	356	348	355	422	465	467	404	347	305	336
1958	340	318	362	348	363	435	491	505	404	359	310	337
1959	360	342	406	396	420	472	548	559	463	407	362	405
1960	417	391	419	461	472	535	622	606	508	461	390	432

[a]144 Observations.

SERIES J GAS FURNACE DATA[a]

t	X_t	Y_t	t	X_t	Y_t	t	X_t	Y_t
1	−0.109	53.8	51	1.608	46.9	101	−0.288	51.0
2	0.000	53.6	52	1.905	47.8	102	−0.153	51.8
3	0.178	53.5	53	2.023	48.2	103	−0.109	52.4
4	0.339	53.5	54	1.815	48.3	104	−0.187	53.0
5	0.373	53.4	55	0.535	47.9	105	−0.255	53.4
6	0.441	53.1	56	0.122	47.2	106	−0.229	53.6
7	0.461	52.7	57	0.009	47.2	107	−0.007	53.7
8	0.348	52.4	58	0.164	48.1	108	0.254	53.8
9	0.127	52.2	59	0.671	49.4	109	0.330	53.8
10	−0.180	52.0	60	1.019	50.6	110	0.102	53.8
11	−0.588	52.0	61	1.146	51.5	111	−0.423	53.3
12	−1.055	52.4	62	1.155	51.6	112	−1.139	53.0
13	−1.421	53.0	63	1.112	51.2	113	−2.275	52.9
14	−1.520	54.0	64	1.121	50.5	114	−2.594	53.4
15	−1.302	54.9	65	1.223	50.1	115	−2.716	54.6
16	−0.814	56.0	66	1.257	49.8	116	−2.510	56.4
17	−0.475	56.8	67	1.157	49.6	117	−1.790	58.0
18	−0.193	56.8	68	0.913	49.4	118	−1.346	59.4
19	0.088	56.4	69	0.620	49.3	119	−1.081	60.2
20	0.435	55.7	70	0.255	49.2	120	−0.910	60.0
21	0.771	55.0	71	−0.280	49.3	121	−0.876	59.4
22	0.866	54.3	72	−1.080	49.7	122	−0.885	58.4
23	0.875	53.2	73	−1.551	50.3	123	−0.800	57.6
24	0.891	52.3	74	−1.799	51.3	124	−0.544	56.9
25	0.987	51.6	75	−1.825	52.8	125	−0.416	56.4
26	1.263	51.2	76	−1.456	54.4	126	−0.271	56.0
27	1.775	50.8	77	−0.944	56.0	127	0.000	55.7
28	1.976	50.5	78	−0.570	56.9	128	0.403	55.3
29	1.934	50.0	79	−0.431	57.5	129	0.841	55.0
30	1.866	49.2	80	−0.577	57.3	130	1.285	54.4
31	1.832	48.4	81	−0.960	56.6	131	1.607	53.7
32	1.767	47.9	82	−1.616	56.0	132	1.746	52.8
33	1.608	47.6	83	−1.875	55.4	133	1.683	51.6
34	1.265	47.5	84	−1.891	55.4	134	1.485	50.6
35	0.790	47.5	85	−1.746	56.4	135	0.993	49.4
36	0.360	47.6	86	−1.474	57.2	136	0.648	48.8
37	0.115	48.1	87	−1.201	58.0	137	0.577	48.5
38	0.088	49.0	88	−0.927	58.4	138	0.577	48.7
39	0.331	50.0	89	−0.524	58.4	139	0.632	49.2
40	0.645	51.1	90	0.040	58.1	140	0.747	49.8
41	0.960	51.8	91	0.788	57.7	141	0.900	50.4
42	1.409	51.9	92	0.943	57.0	142	0.993	50.7
43	2.670	51.7	93	0.930	56.0	143	0.968	50.9
44	2.834	51.2	94	1.006	54.7	144	0.790	50.7
45	2.812	50.0	95	1.137	53.2	145	0.399	50.5
46	2.483	48.3	96	1.198	52.1	146	−0.161	50.4
47	1.929	47.0	97	1.054	51.6	147	−0.553	50.2
48	1.485	45.8	98	0.595	51.0	148	−0.603	50.4
49	1.214	45.6	99	−0.080	50.5	149	−0.424	51.2
50	1.239	46.0	100	−0.314	50.4	150	−0.194	52.3

SERIES J *(continued)*

t	X_t	Y_t	t	X_t	Y_t	t	X_t	Y_t
151	−0.049	53.2	201	−2.473	55.6	251	0.185	56.3
152	0.060	53.9	202	−2.330	58.0	252	0.662	56.4
153	0.161	54.1	203	−2.053	59.5	253	0.709	56.4
154	0.301	54.0	204	−1.739	60.0	254	0.605	56.0
155	0.517	53.6	205	−1.261	60.4	255	0.501	55.2
156	0.566	53.2	206	−0.569	60.5	256	0.603	54.0
157	0.560	53.0	207	−0.137	60.2	257	0.943	53.0
158	0.573	52.8	208	−0.024	59.7	258	1.223	52.0
159	0.592	52.3	209	−0.050	59.0	259	1.249	51.6
160	0.671	51.9	210	−0.135	57.6	260	0.824	51.6
161	0.933	51.6	211	−0.276	56.4	261	0.102	51.1
162	1.337	51.6	212	−0.534	55.2	262	0.025	50.4
163	1.460	51.4	213	−0.871	54.5	263	0.382	50.0
164	1.353	51.2	214	−1.243	54.1	264	0.922	50.0
165	0.772	50.7	215	−1.439	54.1	265	1.032	52.0
166	0.218	50.0	216	−1.422	54.4	266	0.866	54.0
167	−0.237	49.4	217	−1.175	55.5	267	0.527	55.1
168	−0.714	49.3	218	−0.813	56.2	268	0.093	54.5
169	−1.099	49.7	219	−0.634	57.0	269	−0.458	52.8
170	−1.269	50.6	220	−0.582	57.3	270	−0.748	51.4
171	−1.175	51.8	221	−0.625	57.4	271	−0.947	50.8
172	−0.676	53.0	222	−0.713	57.0	272	−1.029	51.2
173	0.033	54.0	223	−0.848	56.4	273	−0.928	52.0
174	0.556	55.3	224	−1.039	55.9	274	−0.645	52.8
175	0.643	55.9	225	−1.346	55.5	275	−0.424	53.8
176	0.484	55.9	226	−1.628	55.3	276	−0.276	54.5
177	0.109	54.6	227	−1.619	55.2	277	−0.158	54.9
178	−0.310	53.5	228	−1.149	55.4	278	−0.033	54.9
179	−0.697	52.4	229	−0.488	56.0	279	0.102	54.8
180	−1.047	52.1	230	−0.160	56.5	280	0.251	54.4
181	−1.218	52.3	231	−0.007	57.1	281	0.280	53.7
182	−1.183	53.0	232	−0.092	57.3	282	0.000	53.3
183	−0.873	53.8	233	−0.620	56.8	283	−0.493	52.8
184	−0.336	54.6	234	−1.086	55.6	284	−0.759	52.6
185	0.063	55.4	235	−1.525	55.0	285	−0.824	52.6
186	0.084	55.9	236	−1.858	54.1	286	−0.740	53.0
187	0.000	55.9	237	−2.029	54.3	287	−0.528	54.3
188	0.001	55.2	238	−2.024	55.3	288	−0.204	56.0
189	0.209	54.4	239	−1.961	56.4	289	0.034	57.0
190	0.556	53.7	240	−1.952	57.2	290	0.204	58.0
191	0.782	53.6	241	−1.794	57.8	291	0.253	58.6
192	0.858	53.6	242	−1.302	58.3	292	0.195	58.5
193	0.918	53.2	243	−1.030	58.6	293	0.131	58.3
194	0.862	52.5	244	−0.918	58.8	294	0.017	57.8
195	0.416	52.0	245	−0.798	58.8	295	−0.182	57.3
196	−0.336	51.4	246	−0.867	58.6	296	−0.262	57.0
197	−0.959	51.0	247	−1.047	58.0			
198	−1.813	50.9	248	−1.123	57.4			
199	−2.378	52.4	249	−0.876	57.0			
200	−2.499	53.5	250	−0.395	56.4			

[a]Sampling interval 9 seconds; observations for 296 pairs of data points. X, 0.60–0.04 (input gas rate in cubic feet per minute); Y, % CO_2 in outlet gas.

SERIES K SIMULATED DYNAMIC DATA WITH TWO INPUTS[a]

t	X_{1t}	X_{2t}	Y_t	t	X_{1t}	X_{2t}	Y_t
−2	0	0	58.3	30			65.8
−1			61.8	31			67.4
0			64.2	32	−1	−1	64.7
1			62.1	33			65.7
2	−1	1	55.1	34			67.5
3			50.6	35			58.2
4			47.8	36			57.0
5			49.7	37	−1	1	54.7
6			51.6	38			54.9
7	1	−1	58.5	39			48.4
8			61.5	40			49.7
9			63.3	41			53.1
10			65.9	42	1	−1	50.2
11			70.9	43			51.7
12	−1	−1	65.8	44			57.4
13			57.6	45			62.6
14			56.1	46			65.8
15			58.2	47	−1	−1	61.5
16			61.7	48			61.5
17	1	1	59.2	49			56.8
18			57.9	50			62.3
19			61.3	51			57.7
20			60.8	52	−1	1	54.0
21			63.6	53			45.2
22	1	−1	69.5	54			51.9
23			69.3	55			45.6
24			70.5	56			46.2
25			68.0	57	1	1	50.2
26			68.1	58			54.6
27	1	1	65.0	59			55.6
28			71.9	60	0	0	60.4
29			64.8	61			59.4

[a]64 Observations.

SERIES L PILOT SCHEME DATA[a]

t	x_t	ε_t	t	x_t	ε_t	t	x_t	ε_t
1	30	−4	53	−60	6	105	55	−4
2	0	−2	54	50	−2	106	0	2
3	−10	0	55	−10	0	107	−90	8
4	0	0	56	40	−4	108	40	0
5	−40	4	57	40	−6	109	0	0
6	0	2	58	−30	0	110	80	−8
7	−10	2	59	20	−2	111	−20	−2
8	10	0	60	−30	2	112	−10	0
9	20	−2	61	10	0	113	−70	6
10	50	−6	62	−20	2	114	−30	6
11	−10	−2	63	30	−2	115	−10	4
12	−55	4	64	−50	4	116	30	−1
13	0	2	65	10	−2	117	−5	0
14	10	0	66	10	−2	118	−60	6
15	0	−2	67	10	−2	119	70	−4
16	10	2	68	−30	0	120	40	−6
17	−70	6	69	0	0	121	10	−4
18	30	0	70	−10	2	122	20	−4
19	−20	2	71	−10	3	123	10	−3
20	10	0	72	15	0	124	0	−2
21	0	0	73	20	−2	125	−70	6
22	0	0	74	−50	4	126	50	−2
23	20	−2	75	20	0	127	30	−4
24	30	−4	76	0	0	128	0	−2
25	0	−2	77	0	0	129	−10	0
26	−10	0	78	0	0	130	0	0
27	−20	2	79	0	0	131	−40	4
28	−30	4	80	−40	4	132	0	2
29	0	2	81	−100	12	133	10	2
30	10	0	82	0	8	134	10	0
31	20	−2	83	0	−12	135	0	0
32	−10	0	84	50	−15	136	80	−8
33	0	0	85	85	−15	137	−80	4
34	20	−2	86	5	−12	138	20	4
35	10	−2	87	40	−14	139	20	0
36	−10	0	88	10	−8	140	−10	2
37	0	0	89	−60	2	141	10	0
38	0	0	90	−50	6	142	0	0
39	0	0	91	−50	8	143	−20	2
40	0	0	92	40	0	144	20	−1
41	0	0	93	0	0	145	55	−6
42	0	0	94	0	0	146	0	−3
43	20	−2	95	−20	2	147	25	−4
44	−50	4	96	−30	4	148	20	−4
45	20	0	97	−60	8	149	−60	4
46	0	0	98	−20	6	150	−40	6
47	0	0	99	−30	6	151	10	4
48	40	−4	100	30	0	152	20	0
49	0	−2	101	−40	4	153	60	−6
50	50	−6	102	80	−6	154	−50	2
51	−40	0	103	−40	0	155	−10	2
52	−50	3	104	−20	2	156	−30	4

SERIES L *(continued)*

t	x_t	ε_t	t	x_t	ε_t	t	x_t	ε_t
157	20	0	209	−40	4	261	−25	4
158	0	0	210	40	−2	262	35	−2
159	20	−2	211	−90	8	263	70	8
160	10	−2	212	40	0	264	−10	−5
161	10	−2	213	0	0	265	100	−20
162	10	−22	214	0	0	266	−20	−8
163	50	−6	215	0	0	267	−40	0
164	−30	0	216	20	−2	268	−20	2
165	−30	6	217	90	−10	269	10	0
166	−90	12	218	30	−8	270	0	0
167	60	0	219	20	−6	271	0	0
168	−40	4	220	30	−6	272	−20	2
169	20	0	221	30	−6	273	−50	6
170	0	0	222	30	−6	274	50	−2
171	20	−2	223	30	−6	275	30	−4
172	10	−2	224	−90	6	276	60	−8
173	−30	2	225	10	2	277	−40	0
174	−30	4	226	10	2	278	−20	2
175	0	2	227	−30	4	279	−10	2
176	50	−4	228	−20	4	280	10	0
177	−60	4	229	40	−2	281	−110	13
178	20	0	230	10	−2	282	15	4
179	0	0	231	10	−2	283	30	−2
180	40	−8	232	10	−2	284	0	−1
181	80	−12	233	−100	12	285	25	−3
182	20	−8	234	10	6	286	−5	−1
183	−100	6	235	45	−2	287	−15	1
184	−30	6	236	30	−4	288	45	−4
185	30	0	237	30	−5	289	40	−6
186	−20	2	238	−15	−1	290	−50	2
187	−30	4	239	−5	0	291	−10	2
188	20	0	240	10	−1	292	−50	6
189	60	−6	241	−85	8	293	20	1
190	−10	−2	242	0	4	294	5	0
191	30	−4	243	0	0	295	−40	4
192	−40	2	244	60	−4	296	0	6
193	30	−2	245	40	−6	297	−60	8
194	−20	1	246	−30	0	298	40	0
195	5	0	247	−40	4	299	−20	2
196	−20	2	248	−40	6	300	130	−12
197	−30	4	249	50	−2	301	−20	−4
198	20	0	250	10	−2	302	0	−2
199	10	−1	251	30	−4	303	30	−4
200	−15	1	252	−40	2	304	−20	0
201	−75	8	253	10	0	305	60	6
202	−40	8	254	−40	4	306	10	−4
203	−40	6	255	40	−2	307	−10	1
204	90	−6	256	−30	2	308	−25	2
205	90	−12	257	−50	6	309	0	1
206	80	−14	258	0	3	310	15	−1
207	−45	−2	259	−45	6	311	−5	0
208	−10	0	260	−20	5	312	0	0

[a]312 Observations.

SERIES M SALES DATA WITH LEADING INDICATOR[a]

t	Leading Indicator X_t	Sales Y_t	t	Leading Indicator X_t	Sales Y_t	t	Leading Indicator X_t	Sales Y_t
1	10.01	200.1	51	10.77	220.0	101	12.90	249.4
2	10.07	199.5	52	10.88	218.7	102	13.12	249.0
3	10.32	199.4	53	10.49	217.0	103	12.47	249.9
4	9.75	198.9	54	10.50	215.9	104	12.47	250.5
5	10.33	199.0	55	11.00	215.8	105	12.94	251.5
6	10.13	200.2	56	10.98	214.1	106	13.10	249.0
7	10.36	198.6	57	10.61	212.3	107	12.91	247.6
8	10.32	200.0	58	10.48	213.9	108	13.39	248.8
9	10.13	200.3	59	10.53	214.6	109	13.13	250.4
10	10.16	201.2	60	11.07	213.6	110	13.34	250.7
11	10.58	201.6	61	10.61	212.1	111	13.34	253.0
12	10.62	201.5	62	10.86	211.4	112	13.14	253.7
13	10.86	201.5	63	10.34	213.1	113	13.49	255.0
14	11.20	203.5	64	10.78	212.9	114	13.87	256.2
15	10.74	204.9	65	10.80	213.3	115	13.39	256.0
16	10.56	207.1	66	10.33	211.5	116	13.59	257.4
17	10.48	210.5	67	10.44	212.3	117	13.27	260.4
18	10.77	210.5	68	10.50	213.0	118	13.70	260.0
19	11.33	209.8	69	10.75	211.0	119	13.20	261.3
20	10.96	208.8	70	10.40	210.7	120	13.32	260.4
21	11.16	209.5	71	10.40	210.1	121	13.15	261.6
22	11.70	213.2	72	10.34	211.4	122	13.30	260.8
23	11.39	213.7	73	10.55	210.0	123	12.94	259.8
24	11.42	215.1	74	10.46	209.7	124	13.29	259.0
25	11.94	218.7	75	10.82	208.8	125	13.26	258.9
26	11.24	219.8	76	10.91	208.8	126	13.08	257.4
27	11.59	220.5	77	10.87	208.8	127	13.24	257.7
28	10.96	223.8	78	10.67	210.6	128	13.31	257.9
29	11.40	222.8	79	11.11	211.9	129	13.52	257.4
30	11.02	223.8	80	10.88	212.8	130	13.02	257.3
31	11.01	221.7	81	11.28	212.5	131	13.25	257.6
32	11.23	222.3	82	11.27	214.8	132	13.12	258.9
33	11.33	220.8	83	11.44	215.3	133	13.26	257.8
34	10.83	219.4	84	11.52	217.5	134	13.11	257.7
35	10.84	220.1	85	12.10	218.8	135	13.30	257.2
36	11.14	220.6	86	11.83	220.7	136	13.06	257.5
37	10.38	218.9	87	12.62	222.2	137	13.32	256.8
38	10.90	217.8	88	12.41	226.7	138	13.10	257.5
39	11.05	217.7	89	12.43	228.4	139	13.27	257.0
40	11.11	215.0	90	12.73	233.2	140	13.64	257.6
41	11.01	215.3	91	13.01	235.7	141	13.58	257.3
42	11.22	215.9	92	12.74	237.1	142	13.87	257.5
43	11.21	216.7	93	12.73	240.6	143	13.53	259.6
44	11.91	216.7	94	12.76	243.8	144	13.41	261.1
45	11.69	217.7	95	12.92	245.3	145	13.25	262.9
46	10.93	218.7	96	12.64	246.0	146	13.50	263.3
47	10.99	222.9	97	12.79	246.3	147	13.58	262.8
48	11.01	224.9	98	13.05	247.7	148	13.51	261.8
49	10.84	222.2	99	12.69	247.6	149	13.77	262.2
50	10.76	220.7	100	13.01	247.8	150	13.40	262.7

[a]150 Observations.

SERIES N MINK FUR SALES OF THE HUDSON'S BAY COMPANY: ANNUAL[a]

1850	29619	1866	51404	1882	45600	1897	76365
1851	21151	1867	58451	1883	47508	1898	70407
1852	24859	1868	73575	1884	52290	1899	41839
1853	25152	1869	74343	1885	110824	1900	45978
1854	42375	1870	27708	1886	76503	1901	47813
1855	50839	1871	31985	1887	64303	1902	57620
1856	61581	1872	39266	1888	83023	1903	66549
1857	61951	1873	44740	1889	40748	1904	54673
1858	76231	1874	60429	1890	35596	1905	55996
1859	63264	1875	72273	1891	29479	1906	60053
1860	44730	1876	79214	1892	42264	1907	39169
1861	31094	1877	79060	1893	58171	1908	21534
1862	49452	1878	84244	1894	50815	1909	17857
1863	43961	1879	62590	1895	51285	1910	21788
1864	61727	1880	35072	1896	70229	1911	33008
1865	60334	1881	36160				

[a]62 Observations.

SERIES P UNEMPLOYMENT AND GDP IN UK: QUARTERLY[a]

		UN	GDP			UN	GDP			UN	GDP
1955	1	225	81.37	1960	1	363	92.30	1965	1	306	108.07
	2	208	82.60		2	342	92.13		2	304	107.64
	3	201	82.30		3	325	93.17		3	321	108.87
	4	199	83.00		4	312	93.50		4	305	109.75
1956	1	207	82.87	1961	1	291	94.77	1966	1	279	110.20
	2	215	83.60		2	293	95.37		2	282	110.20
	3	240	83.33		3	304	95.03		3	318	110.90
	4	245	83.53		4	330	95.23		4	414	110.40
1957	1	295	84.27	1962	1	357	95.07	1967	1	463	111.00
	2	293	85.50		2	401	96.40		2	506	112.10
	3	279	84.33		3	447	96.97		3	538	112.50
	4	287	84.30		4	483	96.50		4	536	113.00
1958	1	331	85.07	1963	1	535	96.16	1968	1	544	114.30
	2	396	83.60		2	520	99.79		2	541	115.10
	3	432	84.37		3	489	101.14		3	547	116.40
	4	462	84.50		4	456	102.95		4	532	117.80
1959	1	454	85.20	1964	1	386	103.96	1969	1	532	116.80
	2	446	87.07		2	368	105.28		2	519	117.80
	3	426	88.40		3	358	105.81		3	547	119.00
	4	402	90.03		4	330	107.14		4	544	119.60

[a]60 pairs of data; data are seasonally adjusted; unemployment (UN) in thousands; gross domestic product (GDP) is composite estimate (1963 = 100). From reference [214].

SERIES Q LOGGED AND CODED U.S. HOG PRICE DATA: ANNUAL[a]

1867	597	1888	709	1909	810	1929	1112
1868	509	1889	763	1910	957	1930	1129
1869	663	1890	681	1911	970	1931	1055
1870	751	1891	627	1912	903	1932	787
1871	739	1892	667	1913	995	1933	624
1872	598	1893	804	1914	1022	1934	612
1873	556	1894	782	1915	998	1935	800
1874	594	1895	707	1916	928	1936	1104
1875	667	1896	653	1917	1073	1937	1075
1876	776	1897	639	1918	1294	1938	1052
1877	754	1898	672	1919	1346	1939	1048
1878	689	1899	669	1920	1301	1940	891
1879	498	1900	729	1921	1134	1941	921
1880	643	1901	784	1922	1024	1942	1193
1881	681	1902	842	1923	1090	1943	1352
1882	778	1903	886	1924	1013	1944	1243
1883	829	1904	784	1925	1119	1945	1314
1884	751	1905	770	1926	1195	1946	1380
1885	704	1906	783	1927	1235	1947	1556
1886	633	1907	877	1928	1120	1948	1632
1887	663	1908	777				

[a]82 observations; values are 1000 $\log_{10}(H_t)$, where H_t is the price, in dollars, per head on January 1 of the year. From reference [162].

SERIES R MONTHLY AVERAGES OF HOURLY READINGS OF OZONE IN DOWNTOWN LOS ANGELES[a]

	Jan.	Feb.	Mar.	Apr.	May	June	July	Aug.	Sept.	Oct.	Nov.	Dec.
1955	2.63	1.94	3.38	4.92	6.29	5.58	5.50	4.71	6.04	7.13	7.79	3.83
1956	3.83	4.25	5.29	3.75	4.67	5.42	6.04	5.71	8.13	4.88	5.42	5.50
1957	3.00	3.42	4.50	4.25	4.00	5.33	5.79	6.58	7.29	5.04	5.04	4.48
1958	3.33	2.88	2.50	3.83	4.17	4.42	4.25	4.08	4.88	4.54	4.25	4.21
1959	2.75	2.42	4.50	5.21	4.00	7.54	7.38	5.96	5.08	5.46	4.79	2.67
1960	1.71	1.92	3.38	3.98	4.63	4.88	5.17	4.83	5.29	3.71	2.46	2.17
1961	2.15	2.44	2.54	3.25	2.81	4.21	4.13	4.17	3.75	3.83	2.42	2.17
1962	2.33	2.00	2.13	4.46	3.17	3.25	4.08	5.42	4.50	4.88	2.83	2.75
1963	1.63	3.04	2.58	2.92	3.29	3.71	4.88	4.63	4.83	3.42	2.38	2.33
1964	1.50	2.25	2.63	2.96	3.46	4.33	5.42	4.79	4.38	4.54	2.04	1.33
1965	2.04	2.81	2.67	4.08	3.90	3.96	4.50	5.58	4.52	5.88	3.67	1.79
1966	1.71	1.92	3.58	4.40	3.79	5.52	5.50	5.00	5.48	4.81	2.42	1.46
1967	1.71	2.46	2.42	1.79	3.63	3.54	4.88	4.96	3.63	5.46	3.08	1.75
1968	2.13	2.58	2.75	3.15	3.46	3.33	4.67	4.13	4.73	3.42	3.08	1.79
1969	1.96	1.63	2.75	3.06	4.31	3.31	3.71	5.25	3.67	3.10	2.25	2.29
1970	1.25	2.25	2.67	3.23	3.58	3.04	3.75	4.54	4.46	2.83	1.63	1.17
1971	1.79	1.92	2.25	2.96	2.38	3.38	3.38	3.21	2.58	2.42	1.58	1.21
1972	1.42	1.96	3.04	2.92	3.58	3.33	4.04	3.92	3.08	2.00	1.58	1.21

[a]216 observations; values are in pphm.

REFERENCES

[1] Abraham, B., "Missing observations in time series," *Comm. Statist.,* **A10,** 1643–1653, 1981.

[2] Abraham, B., and G. E. P. Box, "Deterministic and forecast-adaptive time-dependent models," *Appl. Statist.,* **27,** 120–130, 1978.

[3] Akaike, H., "A new look at the statistical model identification," *IEEE Trans. Automatic Control,* **AC-19,** 716–723, 1974.

[4] Akaike, H., "Canonical correlation analysis of time series and the use of an information criterion," in *Systems Identification: Advances and Case Studies,* ed. R. K. Mehra and D. G. Lainiotis, Academic Press, New York, 1976, pp. 27–96.

[5] Anderson, B. D. O., and J. B. Moore, *Optimal Filtering,* Prentice Hall, Englewood Cliffs, N.J., 1979.

[6] Anderson, R. L., "Distribution of the serial correlation coefficient," *Ann. Math. Statist.,* **13,** 1–13, 1942.

[7] Anderson, T. W., *An Introduction to Multivariate Statistical Analysis,* 2nd ed., Wiley, New York, 1984.

[8] Anderson, T. W., *The Statistical Analysis of Time Series,* Wiley, New York, 1971.

[9] Anscombe, F. J., "Examination of residuals," *Proc. 4th Berkeley Symp.,* **1,** 1–36, 1961.

[10] Anscombe, F. J., and J. W. Tukey, "The examination and analysis of residuals," *Technometrics,* **5,** 141–160, 1963.

[11] Ansley, C. F., "An algorithm for the exact likelihood of a mixed autoregressive moving average process," *Biometrika,* **66,** 59–65, 1979.

[12] Ansley, C. F., and R. Kohn, "Exact likelihood of vector autoregressive–moving average process with missing or aggregated data," *Biometrika,* **70,** 275–278, 1983.

[13] Ansley, C. F., and R. Kohn, "Estimation, filtering, and smoothing in state space models with incompletely specified initial conditions," *Ann. Statist.,* **13,** 1286–1316, 1985.

[14] Ansley, C. F., and P. Newbold, "Finite sample properties of estimators for autoregressive moving average models," *J. Econometrics,* **13,** 159–183, 1980.

[15] Aström, K. J., and T. Bohlin, "Numerical identification of linear dynamic systems from normal operating records," in *Theory of Self-Adaptive Control Systems,* ed. P. H. Hammond, Plenum Press, New York, 1966, pp. 96–111.

[16] Aström, K. J., and B. Wittenmark, *Computer Controlled Systems,* Prentice Hall, Englewood Cliffs, N.J., 1984.

[17] Bachelier, L., "Théorie de la spéculation," *Ann. Sci. École Norm. Sup.,* Ser. 3, **17,** 21–86, 1900.

[18] Bacon, D. W., "Seasonal time series," Ph.D. thesis, University of Wisconsin–Madison, 1965.

[19] Bagshaw, M., and R. A. Johnson, "Sequential procedures for detecting parameter changes in a time-series model," *J. Amer. Statist. Assoc.,* **72,** 593–597, 1977.

[20] Baillie, R. T., "The asymptotic mean squared error of multistep prediction from the regression model with autoregressive errors," *J. Amer. Statist. Assoc.,* **74,** 175–184, 1979.

[21] Barnard, G. A., "Control charts and stochastic processes," *J. Roy. Statist. Soc.,* **B21,** 239–257, 1959.

[22] Barnard, G. A., "Statistical inference," *J. Roy. Statist. Soc.,* **B11,** 115–139, 1949.

[23] Barnard, G. A., "The logic of least squares," *J. Roy. Statist. Soc.,* **B25,** 124–127, 1963.

[24] Barnard, G. A., G. M. Jenkins, and C. B. Winsten, "Likelihood inference and time series," *J. Roy. Statist. Soc.,* **A125,** 321–352, 1962.

[25] Bartlett, M. S., "On the theoretical specification and sampling properties of autocorrelated time-series," *J. Roy. Statist. Soc.,* **B8,** 27–41, 1946.

[26] Bartlett, M. S., *Stochastic Processes,* Cambridge University Press, Cambridge, 1955.

[27] Bell, W., "A note on overdifferencing and the equivalence of seasonal time series models with monthly means and models with $(0, 1, 1)_{12}$ seasonal parts when $\Theta = 1$," *J. Business Econom. Statist.,* **5,** 383–387, 1987.

[28] Bell, W. R., and S. C. Hillmer, "Modeling time series with calendar variation," *J. Amer. Statist. Assoc.,* **78,** 526–534, 1983.

[29] Bell, W., and S. Hillmer, "Initializing the Kalman filter in the nonstationary case," *ASA Proc. Business and Economic Statistics Section,* 1987, pp. 693–697.

[30] Bergh, L. G., and J. F. MacGregor, "Constrained minimum variance controllers: internal model structure and robustness properties," *Ind. Eng. Chem. Res.,* **26,** 1558–1564, 1987.

[31] Bhattacharyya, M. N., and A. P. Layton, "Effectiveness of seat belt legislation on the Queensland road toll—an Australian case study in intervention analysis," *J. Amer. Statist. Assoc.,* **74,** 596–603, 1979.

[32] Birnbaum, A., "On the foundations of statistical inference," *J. Amer. Statist. Assoc.,* **57,** 269–306, 1962.

[33] Booth, G. W., and T. I. Peterson, "Non-linear estimation," *IBM Share Program Pa. No. 687 WL NLI,* 1958.

[34] Box, G. E. P., "Use and abuse of regression," *Technometrics,* **8,** 625–629, 1966.

[35] Box, G. E. P., "Sampling and Bayes' inference in scientific modelling and robustness," *J. Roy. Statist. Soc.,* **A143,** 383–430, 1980.

[36] Box, G. E. P., "Feedback control by manual adjustment," *Quality Engrg.,* **4**(1), 143–151, 1991.

[37] Box, G. E. P., "Bounded adjustment charts," *Quality Engrg.,* **4**(a), 331–338, 1991.

[38] Box, G. E. P., and D. R. Cox, "An analysis of transformations," *J. Roy. Statist. Soc.,* **B26,** 211–243, 1964.

[39] Box, G. E. P., and N. R. Draper, "The Bayesian estimation of common parameters from several responses," *Biometrika,* **52,** 355–365, 1965.

[40] Box, G. E. P., and W. G. Hunter, "The experimental study of physical mechanisms," *Technometrics,* **7,** 23–42, 1965.

[41] Box, G. E. P., W. G. Hunter, and J. S. Hunter, *Statistics for Experimenters,* Wiley, New York, 1978.

[42] Box, G. E. P., and G. M. Jenkins, "Further contributions to adaptive quality control: simultaneous estimation of dynamics: non-zero costs," *Bull. Internat. Statist. Inst., 34th Session,* Ottawa, Canada, 1963, pp. 943–974.

[43] Box, G. E. P., and G. M. Jenkins, "Discrete models for feedback and feedforward control," in *The Future of Statistics,* ed. D. G. Watts, Academic Press, New York, 1968, pp. 201–240.

[44] Box, G. E. P., and G. M. Jenkins, "Some recent advances in forecasting and control, I," *Appl. Statist.,* **17,** 91–109, 1968.

[45] Box, G. E. P., and G. M. Jenkins, "Discrete models for forecasting and control," in *Encyclopaedia of Linguistics, Information and Control,* ed. A. R. Meaham and R. A. Hudson, Pergamon Press, Elmsford, N.Y., 1969, p. 162.

[46] Box, G. E. P., and G. M. Jenkins, "Some statistical aspects of adaptive optimization and control," *J. Roy. Statist. Soc.,* **B24,** 297–331, 1962.

[47] Box, G. E. P., and G. M. Jenkins, "Mathematical models for adaptive control and optimization," *AIChE J. Chem. E Symp. Ser.,* **4,** 61, 1965.

[48] Box, G. E. P., and G. M. Jenkins, *Time Series Analysis: Forecasting and Control,* revised edition, Holden-Day, San Francisco, 1976.

[49] Box, G. E. P., G. M. Jenkins, and D. W. Bacon, "Models for forecasting seasonal and nonseasonal time series," in *Spectral Analysis of Time Series,* ed. B. Harris, Wiley, New York, 1967, pp. 271–311.

[50] Box, G. E. P., G. M. Jenkins, and J. F. MacGregor, "Some recent advances in forecasting and control, Part II," *Appl. Statist.,* **23,** 158–179, 1974.

[51] Box, G. E. P., G. M. Jenkins, and D. W. Wichern, "Least squares analysis with a dynamic model," *Technical Report 105,* Department of Statistics, University of Wisconsin–Madison, 1967.

[52] Box, G. E. P., and T. Kramer, "Statistical process monitoring and feedback adjustments—a discussion," *Technometrics,* **34,** 251–285, 1992.

[53] Box, G. E. P., and A. Lucéno, "Charts for optimal feedback control with recursive sampling and adjustment," *Report 89,* Center for Quality and Productivity Improvement, University of Wisconsin–Madison, 1993.

[54] Box, G. E. P., and J. F. MacGregor, "The analysis of closed-loop dynamic stochastic systems," *Technometrics,* **16,** 391–398, 1974.

[55] Box, G. E. P., and J. F. MacGregor, "Parameter estimation with closed-loop operating data," *Technometrics,* **18,** 371–380, 1976.

[56] Box, G. E. P., and D. A. Pierce, "Distribution of residual autocorrelations in autoregressive-integrated moving average time series models," *J. Amer. Statist. Assoc.,* **65,** 1509–1526, 1970.

[57] Box, G. E. P., and J. Ramírez, "Cumulative score charts," *Quality Reliability Engrg.,* **8,** 17–27, 1992.

[58] Box, G. E. P., and G. C. Tiao, *Bayesian Inference,* Addison-Wesley, Reading, Mass., 1973.

[59] Box, G. E. P., and G. C. Tiao, "Intervention analysis with applications to economic and environmental problems," *J. Amer. Statist. Assoc.,* **70,** 70–79, 1975.

[60] Box, G. E. P., and G. C. Tiao, "Comparison of forecast and actuality," *Appl. Statist.,* **25,** 195–200, 1976.

[61] Box, G. E. P., and G. C. Tiao, "Multiparameter problems from a Bayesian point of view," *Ann. Math. Statist.,* **36,** 1468–1482, 1965.

[62] Briggs, P. A. N., P. H. Hammond, M. T. G. Hughes, and G. O. Plumb, "Correlation analysis of process dynamics using pseudo-random binary test perturbations," *Inst. Mech. Eng., Advances in Automatic Control,* Paper 7, Nottingham, U.K., April 1965.

[63] Brockwell, P. J., and R. A. Davis, *Time Series: Theory and Methods,* Springer-Verlag, New York, 1987.

[64] Brown, R. G., *Smoothing, Forecasting and Prediction of Discrete Time Series,* Prentice Hall, Englewood Cliffs, N.J., 1962.

[65] Brown, R. G., and R. F. Meyer, "The fundamental theorem of exponential smoothing," *Operations Res.,* **9,** 673–685, 1961.

[66] Brubacher, S. R., and G. Tunnicliffe Wilson, "Interpolating time series with applications to the estimation of holiday effects on electricity demand," *Appl. Statist.,* **25,** 107–116, 1976.

[67] Bruce, A. G., and R. D. Martin, "Leave-*k*-out diagnostics for time series" (with discussion), *J. Roy. Statist. Soc.,* **B51,** 363–424, 1989.

[68] Chan, N. H., and C. Z. Wei, "Limiting distributions of least squares estimates of unstable autoregressive processes," *Ann. Statist.,* **16,** 367–401, 1988.

[69] Chang, I., G. C. Tiao, and C. Chen, "Estimation of time series parameters in the presence of outliers," *Technometrics,* **30,** 193–204, 1988.

[70] Chatfield, C., "Inverse autocorrelations," *J. Roy. Statist. Soc.,* **A142,** 363–377, 1979.

[71] Cleveland, W. S., "The inverse autocorrelations of a time series and their applications," *Technometrics,* **14,** 277–293, 1972.

[72] Cooper, D. M., and E. F. Wood, "Identifying multivariate time series models," *J. Time Ser. Anal.,* **3,** 153–164, 1982.

[73] Cornish, E. A., "The multivariate t-distribution associated with a set of normal sample deviates," *Austral. J. Phys.,* **7,** 531, 1954.

[74] Cox, D. R., "Prediction by exponentially weighted moving averages and related methods," *J. Roy. Statist. Soc.,* **B23,** 414–422, 1961.

[75] Damsleth, E., "Interpolating missing values in a time series," *Scand. J. Statist.,* **7,** 33–39, 1980.

[76] Daniel, C., "Use of half-normal plots in interpreting factorial two-level experiments," *Technometrics,* **1,** 311–341, 1959.

[77] Daniels, H. E., "The approximate distribution of serial correlation coefficients," *Biometrika,* **43,** 169–185, 1956.

[78] Davies, N., C. M. Triggs, and P. Newbold, "Significance levels of the Box–Pierce portmanteau statistic in finite samples," *Biometrika,* **64,** 517–522, 1977.

[79] Deming, W. E., *Out of the Crisis,* Center for Advanced Engineering Study, MIT, Cambridge, Mass., 1986.

[80] Dent, W., and A. S. Min, "A Monte Carlo study of autoregressive integrated moving average processes," *J. Econometrics,* **7,** 23–55, 1978.

[81] Dickey, D. A., W. R. Bell, and R. B. Miller, "Unit roots in time series models: tests and implications," *Amer. Statist.,* **40,** 12–26, 1986.

[82] Dickey, D. A., and W. A. Fuller, "Distribution of the estimates for autoregressive time series with a unit root," *J. Amer. Statist. Assoc.,* **74,** 427–431, 1979.

[83] Dickey, D. A., and S. G. Pantula, "Determining the order of differencing in autoregressive processes," *J. Business Econom. Statist.,* **5,** 455–461, 1987.

[84] Doob, J. L., *Stochastic Processes,* Wiley, New York, 1953.

[85] Dudding, B. P., and W. J. Jennet, "Quality control charts," *British Standard 600R*, 1942.

[86] Dunnett, C. W., and M. Sobel, "A bivariate generalization of Student's t-distribution, with tables for certain special cases," *Biometrika,* **41,** 153–169, 1954.

[87] Durbin, J., "The fitting of time-series models," *Rev. Internat. Statist. Inst.,* **28,** 233–244, 1960.

[88] Durbin, J., "Testing for serial correlation in least-squares regression when some of the regressors are lagged dependent variables," *Econometrica,* **38,** 410–421, 1970.

[89] Durbin, J., "An alternative to the bounds test for testing for serial correlation in least-squares regression," *Econometrica,* **38,** 422–429, 1970.

[90] Fearn, T., and P. I. Maris, "An application of Box–Jenkins methodology to the control of gluten addition in a flour mill," *Appl. Statist.,* **40,** 477–484, 1991.

[91] Fisher, R. A., *Statistical Methods and Scientific Inference,* Oliver & Boyd, Edinburgh, 1956.

[92] Fox, A. J., "Outliers in time series," *J. Roy. Statist. Soc.,* **B34,** 350–363, 1972.

[93] Fuller, W. A., *Introduction to Statistical Time Series,* Wiley, New York, 1976.

[94] Gardner, G., A. C. Harvey, and G. D. A. Phillips, "Algorithm AS 154: an algorithm for exact maximum likelihood estimation of autoregressive-moving

average models by means of Kalman filtering," *Appl. Statist.*, **29**, 311–322, 1980.

[95] Gersch, W., and G. Kitagawa, "The prediction of time series with trends and seasonalities," *J. Business Econom. Statist.*, **1**, 253–264, 1983.

[96] Godfrey, L. G., "Testing the adequacy of a time series model," *Biometrika*, **66**, 67–72, 1979.

[97] Gray, H. L., G. D. Kelley, and D. D. McIntire, "A new approach to ARMA modelling," *Comm. Statist.*, **B7**, 1–77, 1978.

[98] Grenander, U., and M. Rosenblatt, *Statistical Analysis of Stationary Time Series*, Wiley, New York, 1957.

[99] Hald, A., *Statistical Theory with Engineering Applications*, Wiley, New York, 1952.

[100] Hannan, E. J., *Time Series Analysis*, Methuen, London, 1960.

[101] Hannan, E. J., and M. Deistler, *The Statistical Theory of Linear Systems*, Wiley, New York, 1988.

[102] Hannan, E. J., W. M. Dunsmuir, and M. Deistler, "Estimation of vector ARMAX models," *J. Multivariate Anal.*, **10**, 275–295, 1979.

[103] Hannan, E. J., and J. Rissanen, "Recursive estimation of mixed autoregressive moving average order," *Biometrika*, **69**, 81–94, 1982.

[104] Harris, T. J., J. F. MacGregor, and J. D. Wright, "An overview of discrete stochastic controllers: generalized PID algorithms with dead-time compensation," *Canad. J. Chem. Engrg.*, **60**, 425–432, 1982.

[105] Harrison, P. J., "Short term sales forecasting," *Appl. Statist.*, **14**, 102–139, 1965.

[106] Harvey, A. C., "Finite sample prediction and overdifferencing," *J. Time Ser. Anal.*, **2**, 221–232, 1981.

[107] Harvey, A. C., and G. D. A. Phillips, "Maximum likelihood estimation of regression models with autoregressive-moving average disturbances," *Biometrika*, **66**, 49–58, 1979.

[108] Harvey, A. C., and R. G. Pierse, "Estimating missing observations in economic time series," *J. Amer. Statist. Assoc.*, **79**, 125–131, 1984.

[109] Harvey, A. C., and P. H. J. Todd, "Forecasting economic time series with structural and Box–Jenkins models: a case study," *J. Business Econom. Statist.*, **1**, 299–307, 1983.

[110] Hillmer, S. C., and G. C. Tiao, "Likelihood function of stationary multiple autoregressive moving average models," *J. Amer. Statist. Assoc.*, **74**, 652–660, 1979.

[111] Hillmer, S. C., and G. C. Tiao, "An ARIMA-model-based approach to seasonal adjustment," *J. Amer. Statist. Assoc.*, **77**, 63–70, 1982.

[112] Holt, C. C., "Forecasting trends and seasonals by exponentially weighted moving averages," *O.N.R. Memorandum 52*, Carnegie Institute of Technology, Pittsburgh, Pa., 1957.

[113] Holt, C. C., F. Modigliani, J. F. Muth, and H. A. Simon, *Planning Production, Inventories and Work Force*, Prentice Hall, Englewood Cliffs, N.J., 1963.

[114] Hougen, J. O., *Experience and Experiments with Process Dynamics,* Chemical Engineering Progress Monograph Series, **60**(4), 1964.

[115] Hutchinson, A. W., and R. J. Shelton, "Measurement of dynamic characteristics of full-scale plant using random perturbing signals: an application to a refinery distillation column," *Trans. Inst. Chem. Eng.,* **45,** 334–342, 1967.

[116] Ishikawa, K., *Guide to Quality Control,* Asian Productivity Organization, Tokyo, 1976.

[117] Jeffreys, H., *Theory of Probability,* 3rd ed., Clarendon Press, Oxford, 1961.

[118] Jenkins, G. M., "Tests of hypotheses in the linear autoregressive model, I," *Biometrika,* **41,** 405–419, 1954; II, *Biometrika,* **43,** 186–199, 1956.

[119] Jenkins, G. M., contribution to the discussion of the paper "Relationships between Bayesian and confidence limits for predictors," by A. R. Thatcher, *J. Roy. Statist. Soc.,* **B26,** 176–210, 1964.

[120] Jenkins, G. M., "The interaction between the muskrat and mink cycles in North Canada," *Proc. 8th International Biometric Conference,* Editura Academiei Republicii Socialiste Romania, Bucharest, 1975, pp. 55–71.

[121] Jenkins, G. M., *Practical Experiences with Modelling and Forecasting Time Series,* Gwilym Jenkins & Partners Ltd., Jersey, Channel Islands, 1979.

[122] Jenkins, G. M., and D. G. Watts, *Spectral Analysis and Its Applications,* Holden–Day, San Francisco, 1968.

[123] Jones, R. H., "Maximum likelihood fitting of ARMA models to time series with missing observations," *Technometrics,* **22,** 389–395, 1980.

[124] Kalman, R. E., "A new approach to linear filtering and prediction problems," *J. Basic Engrg.,* **82,** 35–45, 1960.

[125] Kalman, R. E., and R. S. Bucy, "New results in linear filtering and prediction theory," *J. Basic Engrg.,* **83,** 95–108, 1961.

[126] Kendall, M. G., "On the analysis of oscillatory time-series," *J. Roy. Statist. Soc.,* **A108,** 93–129, 1945.

[127] Kendall, M. G., and A. Stuart, *The Advanced Theory of Statistics,* Vol. 3, Charles Griffin, London, 1966.

[128] Kitagawa, G., and W. Gersch, "A smoothness priors-state space modeling of time series with trend and seasonality," *J. Amer. Statist. Assoc.,* **79,** 378–389, 1984.

[129] Kohn, R., and C. F. Ansley, "Estimation, prediction, and interpolation for ARIMA models with missing data," *J. Amer. Statist. Assoc.,* **81,** 751–761, 1986.

[130] Kolmogoroff, A., "Sur l'interpolation et l'extrapolation des suites stationnaires," *C.R. Acad. Sci. Paris,* **208,** 2043–2045, 1939.

[131] Kolmogoroff, A., "Stationary sequences in Hilbert space," *Bull. Math. Univ. Moscow,* **2**(6), 1–40, 1941.

[132] Kolmogoroff, A., "Interpolation und Extrapolation von stationären zufälligen folgen," *Bull. Acad. Sci. (Nauk) U.S.S.R., Ser. Math.,* **5,** 3–14, 1941.

[133] Koopmans, L. H., *The Spectral Analysis of Time Series,* Academic Press, New York, 1974.

[134] Koopmans, T., "Serial correlation and quadratic forms in normal variables," *Ann. Math. Statist.*, **13,** 14–33, 1942.

[135] Koopmans, T. C. (ed.), *Statistical Inference in Dynamic Economic Models,* Wiley, New York, 1950.

[136] Kotnour, K. D., G. E. P. Box, and R. J. Altpeter, "A discrete predictor-controller applied to sinusoidal perturbation adaptive optimization," *Inst. Soc. Amer. Trans.*, **5,** 255–262, 1966.

[137] Ljung, G. M., "Diagnostic testing of univariate time series models," *Biometrika,* **73,** 725–730, 1986.

[138] Ljung, G. M., and G. E. P. Box, "On a measure of lack of fit in time series models," *Biometrika,* **65,** 297–303, 1978.

[139] Ljung, G. M., and G. E. P. Box, "The likelihood function of stationary autoregressive-moving average models," *Biometrika,* **66,** 265–270, 1979.

[140] Loève, M., *Probability Theory I,* Springer-Verlag, New York, 1977.

[141] MacGregor, J. F., "Topics in the control of linear processes with stochastic disturbances," Ph.D. thesis, University of Wisconsin–Madison, 1972.

[142] Mann, H. B., and A. Wald, "On the statistical treatment of linear stochastic difference equations," *Econometrica,* **11,** 173–220, 1943.

[143] Marquardt, D. W., "An algorithm for least-squares estimation of nonlinear parameters," *J. Soc. Ind. Appl. Math.,* **11,** 431–441, 1963.

[144] Martin, R. D., and V. J. Yohai, "Influence functionals for time series" (with discussion), *Ann. Statist.,* **14,** 781–855, 1986.

[145] Montgomery, D. C., and G. Weatherby, "Modeling and forecasting time series using transfer function and intervention methods," *AIIE Trans.,* 289–307, 1980.

[146] Moran, P. A. P., "Some experiments on the prediction of sunspot numbers," *J. Roy. Statist. Soc.,* **B16,** 112–117, 1954.

[147] Muth, J. F., "Optimal properties of exponentially weighted forecasts," *J. Amer. Statist. Assoc.,* **55,** 299–306, 1960.

[148] Newbold, P., "Bayesian estimation of Box–Jenkins transfer function-noise models," *J. Roy. Statist. Soc.,* **B35,** 323–336, 1973.

[149] Newbold, P., "The exact likelihood function for a mixed autoregressive-moving average process," *Biometrika,* **61,** 423–426, 1974.

[150] Newbold, P., "The equivalence of two tests of time series model adequacy," *Biometrika,* **67,** 463–465, 1980.

[151] Noble, B., *Methods Based on the Wiener–Hopf Technique for the Solution of Partial Differential Equations,* Pergamon Press, Elmsford, N.Y., 1958.

[152] Osborn, D. R., "On the criteria functions used for the estimation of moving average processes," *J. Amer. Statist. Assoc.,* **77,** 388–392, 1982.

[153] Oughton, K. D., "Digital computer controls paper machine," *Ind. Electron.,* **3,** 358–362, 1965.

[154] Page, E. S., "On problems in which a change in a parameter occurs at an unknown point," *Biometrika,* **44,** 248–252, 1957.

[155] Page, E. S., "Cumulative sum charts," *Technometrics,* **3,** 1–9, 1961.

[156] Pierce, D. A., "Least squares estimation in dynamic-disturbance time series models," *Biometrika,* **59,** 73–78, 1972.

[157] Pierce, D. A., "Residual correlations and diagnostic checking in dynamic-disturbance time series models," *J. Amer. Statist. Assoc.,* **67,** 636–640, 1972.

[158] Plackett, R. L., *Principles of Regression Analysis,* Clarendon Press, Oxford, 1960.

[159] Poskitt, D. S., and A. R. Tremayne, "Testing the specification of a fitted autoregressive-moving average model," *Biometrika,* **67,** 359–363, 1980.

[160] Priestley, M. B., *Non-linear and Non-stationary Time Series Analysis,* Academic Press, London, 1988.

[161] Quenouille, M. H., "Approximate tests of correlation in time-series," *J. Roy. Statist. Soc.,* **B11,** 68–84, 1949.

[162] Quenouille, M. H., *Analysis of Multiple Time Series,* Hafner, New York, 1957.

[163] Quenouille, M. H., *Associated Measurements,* Butterworth, London, 1952.

[164] Rao, C. R., *Linear Statistical Inference and Its Applications,* Wiley, New York, 1965.

[165] Rao, J. N. K., and G. Tintner, "On the variate difference method," *Aust. J. Statist.,* **5,** 106–116, 1963.

[166] Reinsel, G., "Maximum likelihood estimation of stochastic linear difference equations with autoregressive moving average errors," *Econometrica,* **47,** 129–151, 1979.

[167] Reinsel, G. C., and G. C. Tiao, "Impact of chlorofluoromethanes on stratospheric ozone: a statistical analysis of ozone data for trends," *J. Amer. Statist. Assoc.,* **82,** 20–30, 1987.

[168] Reinsel, G. C., and M. A. Wincek, "Asymptotic distribution of parameter estimators for nonconsecutively observed time series," *Biometrika,* **74,** 115–124, 1987.

[169] Rivera, D. E., M. Morari, and S. Skogestad, "Internal model control. 4. PID controller design," *Ind. Eng. Chem. Process Des. Dev.,* **25,** 252–265, 1986.

[170] Roberts, S. W., "Control chart tests based on geometric moving averages," *Technometrics,* **1,** 239–250, 1959.

[171] Robinson, E. A., *Multichannel Time Series Analysis,* Holden-Day, San Francisco, 1967.

[172] Said, S. E., and D. A. Dickey, "Hypothesis testing in ARIMA(p, 1, q) models," *J. Amer. Statist. Assoc.,* **80,** 369–374, 1985.

[173] Savage, L. J., *The Foundations of Statistical Inference,* Methuen, London, 1962.

[174] Schuster, A., "On the investigation of hidden periodicities," *Terr. Mag. Atmos. Elect.,* **3,** 13–41, 1898.

[175] Schuster, A., "On the periodicities of sunspots," *Philos. Trans. Roy. Soc.,* **A206,** 69–100, 1906.

[176] Schwarz, G., "Estimating the dimension of a model," *Ann. Statist.,* **6,** 461–464, 1978.

[177] Shewhart, W. A., *The Economic Control of the Quality of Manufactured Product,* Macmillan, New York, 1931.

[178] *Short Term Forecasting,* ICI Monograph 2, Oliver & Boyd, Edinburgh, 1964.

[179] Silvey, S. D., "The Lagrangian multiplier test," *Ann. Math. Statist.,* **30,** 389–407, 1959.

[180] Slutsky, E., "The summation of random causes as the source of cyclic processes" (Russian), *Problems Econom. Conditions,* **3,** 1, 1927; English trans., *Econometrica,* **5,** 105–146, 1937.

[181] Solo, V., "The order of differencing in ARIMA models," *J. Amer. Statist. Assoc.,* **79,** 916–921, 1984.

[182] Stokes, G. G., "Note on searching for periodicities," *Proc. Roy. Soc.,* **29,** 122–123, 1879.

[183] Stralkowski, C. M., "Lower order autoregressive-moving average stochastic models and their use for the characterization of abrasive cutting tools," Ph.D. thesis, University of Wisconsin–Madison, 1968.

[184] Subba Rao, T., and M. M. Gabr, *An Introduction to Bispectral Analysis and Bilinear Time Series Models,* Springer-Verlag, Berlin, 1984.

[185] Thompson, H. E., and G. C. Tiao, "Analysis of telephone data: a case study of forecasting seasonal time series," *Bell J. Econom. Manage. Sci.,* **2,** 515–541, 1971.

[186] Tiao, G. C., G. E. P. Box, and W. J. Hamming, "Analysis of Los Angeles photochemical smog data: a statistical overview," *J. Air Pollut. Control Assoc.,* **25,** 260–268, 1975.

[187] Tintner, G., *The Variate Difference Method,* Principia Press, Bloomington, Ind., 1940.

[188] Tong, H., *Threshold Models in Non-linear Time Series Analysis,* Springer-Verlag, New York, 1983.

[189] Tong, H., and K. S. Lim, "Threshold autoregression, limit cycles and cyclical data," *J. Roy. Statist. Soc.,* **B42,** 245–292, 1980.

[190] Tsay, R. S., "Time series model specification in the presence of outliers," *J. Amer. Statist. Assoc.,* **81,** 132–141, 1986.

[191] Tsay, R. S., "Identifying multivariate time series models," *J. Time Ser. Anal.,* **10,** 357–372, 1989.

[192] Tsay, R. S., and G. C. Tiao, "Consistent estimates of autoregressive parameters and extended sample autocorrelation function for stationary and nonstationary ARMA models," *J. Amer. Statist. Assoc.,* **79,** 84–96, 1984.

[193] Tsay, R. S., and G. C. Tiao, "Use of canonical analysis in time series model identification," *Biometrika,* **72,** 299–315, 1985.

[194] Tukey, J. W., "Discussion, emphasizing the connection between analysis of variance and spectrum analysis," *Technometrics,* **3,** 191–219, 1961.

[195] Walker, G., "On periodicity in series of related terms," *Proc. Roy. Soc.,* **A131,** 518–532, 1931.

[196] Whittle, P., "Estimation and information in stationary time series," *Ark. Math.,* **2,** 423–434, 1953.

[197] Whittle, P., *Hypothesis Testing in Time Series Analysis,* University of Uppsala, Uppsala, Sweden, 1951.

[198] Whittle, P., *Prediction and Regulation by Linear Least-Squares Methods,* English Universities Press, London, 1963.

[199] Wichern, D. W., "The behaviour of the sample autocorrelation function for an integrated moving average process," *Biometrika,* **60,** 235–239, 1973.

[200] Wiener, N., *Extrapolation, Interpolation and Smoothing of Stationary Time Series,* Wiley, New York, 1949.

[201] Wilks, S. S., *Mathematical Statistics,* Wiley, New York, 1962.

[202] Wilson, G., "Factorization of the covariance generating function of a pure moving average process," *SIAM J. Numer. Anal.,* **6,** 1–7, 1969.

[203] Wilson, G. T., "Optimal control: a general method of obtaining the feedback scheme which minimizes the output variance, subject to a constraint on the variability of the control variable," *Technical Report 20,* Department of Systems Engineering, University of Lancaster, Lancaster, U.K., 1970.

[204] Wilson, G. T., Ph.D. thesis, University of Lancaster, U.K., 1970.

[205] Wincek, M. A., and G. C. Reinsel, "An exact maximum likelihood estimation procedure for regression-ARMA time series models with possibly nonconsecutive data," *J. Roy. Statist. Soc.,* **B48,** 303–313, 1986.

[206] Winters, P. R., "Forecasting sales by exponentially weighted moving averages," *Manage. Sci.,* **6,** 324–342, 1960.

[207] Wold, H. O., *A Study in the Analysis of Stationary Time Series,* Almqvist & Wiksell, Uppsala, Sweden, 1938 (2nd. ed. 1954).

[208] Woodward, W. A., and H. L. Gray, "On the relationship between the *S* array and the Box–Jenkins method of ARMA model identification," *J. Amer. Statist. Assoc.,* **76,** 579–587, 1981.

[209] Yaglom, A. M., "The correlation theory of processes whose *n*th difference constitute a stationary process," *Mat. Sb.,* **37**(79), 141, 1955.

[210] Yamamoto, T., "Asymptotic mean square prediction error for an autoregressive model with estimated coefficients," *Appl. Statist.,* **25,** 123–127, 1976.

[211] Young, A. J., *An Introduction to Process Control System Design,* Longmans, Green, New York, 1955.

[212] Yule, G. U., "On a method of investigating periodicities in disturbed series, with special reference to Wolfer's sunspot numbers," *Philos. Trans. Roy. Soc.,* **A226,** 267–298, 1927.

[213] Zadeh, L. A., and J. R. Ragazzini, "An extension of Wiener's theory of prediction," *J. Appl. Phys.,* **21,** 645, 1950.

[214] Bray, J., "Dynamic equations for economic forecasting with the G.D.P.–unemployment relation and the growth of G.D.P. in the United Kingdom as an example," *J. Roy. Statist. Soc.,* **A134,** 167–209, 1971.

[215] Hannan, E. J., *Multiple Time Series,* Wiley, New York, 1970.

[216] Haugh, L. D., and G. E. P. Box, "Identification of dynamic regression (distributed lag) models connecting two time series," *J. Amer. Statist. Assoc.,* **72,** 121–130, 1977.

[217] Liu, L.-M., and D. M. Hanssens, "Identification of multiple-input transfer function models," *Comm. Statist.*, **A11,** 297–314, 1982.

[218] Pankratz, A., *Forecasting with Dynamic Regression Models,* Wiley, New York, 1991.

[219] Poskitt, D. S., and A. R. Tremayne, "An approach to testing linear time series models," *Ann. Statist.*, **9,** 974–986, 1981.

[220] Priestley, M. B., *Spectral Analysis and Time Series,* Academic Press, New York, 1981.

[221] Walker, A. M., "Asymptotic properties of least squares estimates of parameters of the spectrum of a stationary non-deterministic time-series," *J. Aust. Math. Soc.*, **4,** 363–384, 1964.

Part Six

EXERCISES AND PROBLEMS

This part of the book is a collection of exercises and problems for the separate chapters. We hope that these will further enhance the value of the book when used as a course text and also assist private study. A number of examples point to extensions of the ideas and act as a first introduction to additional methods.

CHAPTER 2

2.1. The following are temperature measurements z made every minute on a chemical reactor:

200, 202, 208, 204, 204, 207, 207, 204, 202, 199, 201, 198, 200, 202, 203, 205, 207, 211, 204, 206, 203, 203, 201, 198, 200, 206, 207, 206, 200, 203, 203, 200, 200, 195, 202, 204.

(a) Plot the time series.
(b) Plot z_{t+1} versus z_t.
(c) Plot z_{t+2} versus z_t.
After inspecting the graphs, do you think that the series is autocorrelated?

2.2. State whether or not a stationary stochastic process can have the following values of autocorrelations.
(a) $\rho_1 = 0.80$, $\rho_2 = 0.55$, $\rho_k = 0$ for $k > 2$
(b) $\rho_1 = 0.80$, $\rho_2 = 0.28$, $\rho_k = 0$ for $k > 2$

2.3. Two stochastic processes z_{1t} and z_{2t} have the following autocovariance functions:

$$z_{1t}: \quad \gamma_0 = 0.5, \gamma_1 = 0.2, \gamma_j = 0 \ (j \geqslant 2)$$

$$z_{2t}: \quad \gamma_0 = 2.30, \gamma_1 = -1.43, \gamma_2 = 0.30, \gamma_j = 0 \ (j \geqslant 3)$$

Calculate the autocovariance function of the process $z_{3t} = z_{1t} + 2z_{2t}$ and verify that it is a valid stationary process.

2.4. Calculate c_0, c_1, c_2, r_1, r_2 for the series given in Exercise 2.1. Make a graph of r_k, $k = 0, 1, 2$.

2.5. On the supposition that $\rho_j = 0$ for $j > 2$,

(a) Obtain approximate standard errors for r_1, r_2, and $r_j, j > 2$.

(b) Obtain the approximate correlation between r_4 and r_5.

2.6. Using the data of Exercise 2.1, calculate the periodogram for periods 36, 18, 12, 9, 36/5, 6 and draw up an analysis of variance table showing the mean squares associated with these periods and the residual mean square.

2.7. A *circular* stochastic process with period N is defined by $z_t = z_{t+N}$.

(a) Show that (see, e.g., [63], [93], [122]) when $N = 2n$, the latent roots of the $N \times N$ autocorrelation matrix of z_t are

$$\lambda_k = 1 + 2 \sum_{i=1}^{n-1} \rho_i \cos\left(\frac{\pi i k}{n}\right) + \rho_n \cos(\pi k)$$

$k = 1, 2, \ldots, N$ and that the latent vectors corresponding to λ_k, λ_{N-k} (with $\lambda_k = \lambda_{N-k}$) are

$$l'_k = \left(\cos\left(\frac{\pi k}{n}\right), \cos\left(\frac{2\pi k}{n}\right), \ldots, \cos(2\pi k)\right)$$

$$l'_{N-k} = \left(\sin\left(\frac{\pi k}{n}\right), \sin\left(\frac{2\pi k}{n}\right), \ldots, \sin(2\pi k)\right)$$

(b) Verify that as N tends to infinity, with k/N fixed, λ_k tends to $g(k/N)/2$, where $g(f)$ is the spectral density function, showing that in the limit the latent roots of the autocorrelation matrix trace out the spectral curve.

CHAPTER 3

3.1. Write the following models in B notation.

(1) $\tilde{z}_t - 0.5\tilde{z}_{t-1} = a_t$

(2) $\tilde{z}_t = a_t - 1.3a_{t-1} + 0.4a_{t-2}$

(3) $\tilde{z}_t - 0.5\tilde{z}_{t-1} = a_t - 1.3a_{t-1} + 0.4a_{t-2}$

3.2. For each of the models in Exercise 3.1, obtain:

(a) The first four ψ weights

(b) The first four π weights

(c) The covariance generating function

(d) The first four autocorrelations

(e) The variance of $\tilde{z}_t$ assuming that $\sigma_a^2 = 1.0$

3.3. For each of the models of Exercise 3.1 and also for the following models, state whether it is **(a)** stationary; **(b)** invertible.

(4) $\tilde{z}_t - 1.5\tilde{z}_{t-1} + 0.6\tilde{z}_{t-2} = a_t$

(5) $\tilde{z}_t - \tilde{z}_{t-1} = a_t - 0.5a_{t-1}$

(6) $\tilde{z}_t - \tilde{z}_{t-1} = a_t - 1.3a_{t-1} + 0.3a_{t-2}$

3.4. Classify each of the models (1)–(4) in Exercises 3.1 and 3.3 as a member of the class of ARMA(p, q) processes.

3.5. **(a)** Write down the Yule–Walker equations for models (1) and (4).

(b) Solve these equations to obtain ρ_1 and ρ_2 for the two models.

(c) Obtain the partial autocorrelation function for the two models.

3.6. For the AR(2) process $\tilde{z}_t - 1.0\tilde{z}_{t-1} + 0.5\tilde{z}_{t-2} = a_t$,

(a) Calculate ρ_1.

(b) Using ρ_0 and ρ_1 as starting values and the difference equation form for the autocorrelation function, calculate the values of ρ_k for $k = 2, \ldots, 15$.

(c) Use the plotted function to estimate the period and damping factor of the autocorrelation function.

(d) Check the values in (c) by direct calculation using the values of ϕ_1 and ϕ_2.

3.7. **(a)** Plot the power spectrum $g(f)$ of the autoregressive process of Exercise 3.6 and show that it has a peak at a period which is close to the period in the autocorrelation function.

(b) Graphically, or otherwise, estimate the proportion of the variance of the series in the frequency band between $f = 0.0$ and $f = 0.2$ cycle per data interval.

3.8. **(a)** Why is it important to factorize the autoregressive and moving average operators after fitting a model to an observed series?

(b) It is shown in [120] that the number of mink skins z_t traded annually between 1848 and 1909 in North Canada is adequately represented by the AR(4) model

$$(1 - 0.82B + 0.22B^2 + 0.28B^4)[\ln(z_t) - \mu] = a_t$$

Factorize the autoregressive operator and explain what the factors reveal about the autocorrelation function and the underlying nature of the mink series. These data are listed as series N in the collection of time series in Part Five.

CHAPTER 4

4.1. For each of the models

(1) $(1 - B)z_t = (1 - 0.5B)a_t$

(2) $(1 - B)z_t = (1 - 0.2B)a_t$

(3) $(1 - 0.5B)(1 - B)z_t = a_t$

(4) $(1 - 0.2B)(1 - B)z_t = a_t$

(5) $(1 - 0.2B)(1 - B)z_t = (1 - 0.5B)a_t$

(a) Obtain the first seven ψ weights.

(b) Obtain the first seven π weights.

(c) Classify as a member of the class of ARIMA(p, d, q) process.

4.2. For the five models of Exercise 4.1, and using where appropriate the results there obtained,

(a) Write each model in random shock form.

(b) Write each model as a complementary function plus a particular integral in relation to an origin $k = t - 3$.

(c) Write each model in inverted form.

4.3. Given the following series of random shocks a_t: and given that $z_0 = 20$, $z_{-1} = 19$,

t	a_t	t	a_t	t	a_t
0	−0.3	5	−0.6	10	−0.4
1	0.6	6	1.7	11	0.9
2	0.9	7	−0.9	12	0.0
3	0.2	8	−1.3	13	−1.4
4	0.1	9	−0.6	14	−0.6

(a) Use the difference equation form of the model to obtain $z_1, z_2, \ldots, z_{14}$ for each of the five models in Exercise 4.1.

(b) Plot the derived series.

4.4. Using the inverted forms of each of the models in Exercise 4.1, obtain z_{12}, z_{13} and z_{14}, using only the values $z_1, z_2, \ldots, z_{11}$ derived in Exercise 4.3 and a_{12}, a_{13} and a_{14}. Confirm that the values agree with those obtained in Exercise 4.3.

4.5. If $\bar{z}_t = \sum_{j=1}^{\infty} \pi_j z_{t-j+1}$, then for the models (1) and (2) of Exercise 4.1, which are of the form $(1 - B)z_t = (1 - \theta B)a_t$, $\bar{z}_t$ is an exponentially weighted moving average. For these two models, by actual calculation confirm that $\bar{z}_{11}$, $\bar{z}_{12}$ and $\bar{z}_{13}$ satisfy the relations

$$z_t = \bar{z}_{t-1} + a_t \qquad \text{(see Exercise 4.4)}$$

$$\bar{z}_t = \bar{z}_{t-1} + (1 - \theta)a_t$$

$$\bar{z}_t = (1 - \theta)z_t + \theta\bar{z}_{t-1}$$

4.6. If $w_{1t} = (1 - \theta_1 B)a_{1t}$ and $w_{2t} = (1 - \theta_2 B)a_{2t}$, show that $w_{3t} = w_{1t} + w_{2t}$ may be written as $w_{3t} = (1 - \theta_3 B)a_{3t}$ and derive an expression for θ_3 and σ_{3a}^2 in terms of the parameters of the other two processes. State your assumptions.

4.7. Suppose that $Z_t = z_t + b_t$ where z_t is a first-order autoregressive process $(1 - \phi B)z_t = a_t$ and b_t is a white noise process with variance σ_b^2. What process does Z_t follow? State your assumptions.

CHAPTER 5

5.1. For the models

(1) $\tilde{z}_t - 0.5\tilde{z}_{t-1} = a_t$

(2) $\nabla z_t = a_t - 0.5a_{t-1}$

(3) $(1 - 0.6B)\nabla z_t = a_t$

express forecasts for lead times $l = 1$ and $l = 2$:

(a) From the difference equation
(b) In integrated form (using the ψ weights)
(c) As a weighted average of previous observations

5.2. The following observations represent the values $z_{91}, z_{92}, \ldots, z_{100}$ from a series fitted by the model $\nabla z_t = a_t - 1.1a_{t-1} + 0.28a_{t-2}$

166, 172, 172, 169, 164, 168, 171, 167, 168, 172.

(a) Generate the forecasts $\hat{z}_{100}(l)$ for $l = 1, 2, \ldots, 12$ and draw a graph of the series values and the forecasts (assume $a_{90} = 0$, $a_{91} = 0$).
(b) With $\hat{\sigma}_a^2 = 1.103$, calculate the estimated standard deviations $\hat{\sigma}(l)$ of the forecast errors and use them to calculate 80% probability limits for the forecasts. Insert these probability limits on the graph, on either side of the forecasts.

5.3. Suppose that the data of Exercise 5.2 represent monthly sales.
(a) Calculate the minimum mean square error forecasts for quarterly sales for 1, 2, 3, 4 *quarters* ahead, using the data up to $t = 100$.
(b) Calculate 80% probability limits for these forecasts.

5.4. Using the data and forecasts of Exercise 5.2, and given the further observation $z_{101} = 174$,
(a) Calculate the forecasts $\hat{z}_{101}(l)$ for $l = 1, 2, \ldots, 11$ using the updating formula

$$\hat{z}_{t+1}(l) = \hat{z}_t(l + 1) + \psi_l a_{t+1}$$

(b) Verify these forecasts using the difference equation directly.

5.5. For the model $\nabla z_t = a_t - 1.1a_{t-1} + 0.28a_{t-2}$ of Exercise 5.2,
(a) Write down expressions for the forecast errors $e_t(1), e_t(2), \ldots, e_t(6)$, from the same origin t.
(b) Calculate and plot the autocorrelations of the series of forecast errors $e_t(3)$.
(c) Calculate and plot the correlations between the forecast errors $e_t(2)$ and $e_t(j)$ for $j = 1, 2, \ldots, 6$.

5.6. Let the vector $e' = (e_1, e_2, \ldots, e_L)$ have for its elements the forecast errors made $1, 2, \ldots, L$ steps ahead, all from the same origin t. Then if $a' = (a_{t+1}, a_{t+2}, \ldots, a_{t+L})$ are the corresponding uncorrelated shocks

$$e = Ma \quad \text{where} \quad M = \begin{bmatrix} 1 & 0 & 0 & \cdots & 0 \\ \psi_1 & 1 & 0 & \cdots & 0 \\ \psi_2 & \psi_1 & 1 & \cdots & 0 \\ \cdot & \cdot & \cdot & \cdots & \cdot \\ \psi_{L-1} & \psi_{L-2} & \psi_{L-3} & \cdots & 1 \end{bmatrix}$$

Show (see, e.g., [60], [186]) Σ_e, the covariance matrix of the e's, is $\Sigma_e = MM'\sigma_a^2$ and hence that a test that a set of subsequently realized values $z_{t+1}, z_{t+2}, \ldots, z_{t+L}$ of the series do not jointly differ significantly from the forecasts made at origin t is obtained by referring

$$e' \Sigma_e^{-1} e = \frac{e'(MM')^{-1}e}{\sigma_a^2} = \frac{a'a}{\sigma_a^2} = \frac{1}{\sigma_a^2} \sum_{j=t+1}^{t+L} a_j^2$$

to a chi-square distribution with L degrees of freedom. [Note that a_{t+j} is the *one* step ahead forecast error calculated from $z_{t+j} - \hat{z}_{t+j-1}(1)$.]

5.7. It was found that a quarterly economic time series was well represented by the model

$$\nabla z_t = 0.5 + (1 - 1.0B + 0.5B^2)a_t$$

with $\sigma_a^2 = 0.04$.

(a) Given $z_{48} = 130$, $a_{47} = -0.3$, $a_{48} = 0.2$, calculate and plot the forecasts $\hat{z}_{48}(l)$ for $l = 1, 2, \ldots, 12$.

(b) Insert the 80% probability limits on the graph.

(c) Express the series and forecasts in integrated form.

CHAPTER 6

6.1. Given the identified models and the values of the estimated autocorrelations of $w_t = \nabla^d z_t$ in the following table:

	Identified Model			
	p	d	q	Estimated Autocorrelations
(1)	1	1	0	$r_1 = 0.72$
(2)	0	1	1	$r_1 = -0.41$
(3)	1	0	1	$r_1 = 0.40,\ r_2 = 0.32$
(4)	0	2	2	$r_1 = 0.62,\ r_2 = 0.13$
(5)	2	1	0	$r_1 = 0.93,\ r_2 = 0.81$

(a) Obtain preliminary estimates of the parameters analytically.

(b) Check these estimates using the charts and tables in Part Five of the book.

(c) Write down the identified models in backward shift operator notation with the preliminary estimates inserted.

6.2. For the (2, 1, 0) process considered in (5) of Exercise 6.1, the sample mean and variance of $w_t = \nabla z_t$ are $\overline{w} = 0.23$ and $s_w^2 = 0.25$. If the series contains $N = 101$ observations,

(a) Show that a constant term needs to be included in the model.

(b) Express the model in the form $w_t - \phi_1 w_{t-1} - \phi_2 w_{t-2} = \theta_0 + a_t$ with numerical values inserted for the parameters.

6.3. The following table shows the first 16 values of the ACF r_k and PACF $\hat{\phi}_{kk}$ for a series of 60 observations of logged quarterly unemployment in the United Kingdom. (These data, as given in [214], are listed as part of series P in the collection of time series in Part Five.)

k	r_k	$\hat{\phi}_{kk}$	k	r_k	$\hat{\phi}_{kk}$	k	r_k	$\hat{\phi}_{kk}$
1	0.93	0.93	7	0.03	0.05	13	−0.21	0.19
2	0.80	−0.41	8	−0.09	−0.07	14	−0.12	0.20
3	0.65	−0.14	9	−0.16	0.12	15	−0.01	0.03
4	0.49	−0.11	10	−0.22	−0.14	16	0.10	−0.11
5	0.32	−0.07	11	−0.25	0.03			
6	0.16	−0.10	12	−0.25	0.09			

(a) Draw graphs of the ACF and PACF.
(b) Identify a model for the series.
(c) Obtain preliminary estimates for the parameters and for their standard errors.
(d) Given $\bar{z} = 5.90$ and $s_z^2 = 0.08911$, obtain preliminary estimates for μ_z and σ_a^2.

6.4. The following table shows the first 10 values of the ACF and PACF of z_t and ∇z_t for a series of 60 observations consisting of quarterly measurements of gross domestic product (GDP) in the United Kingdom. (These data, as given in [214], are listed as part of series P in the collection of time series in Part Five.)

	z		∇z			z		∇z	
k	r_k	$\hat{\phi}_{kk}$	r_k	$\hat{\phi}_{kk}$	k	r_k	$\hat{\phi}_{kk}$	r_k	$\hat{\phi}_{kk}$
1	0.95	0.95	0.01	0.01	6	0.72	−0.01	0.00	−0.02
2	0.91	−0.01	0.06	0.06	7	0.68	−0.03	−0.26	−0.26
3	0.87	−0.03	0.13	0.13	8	0.63	−0.02	−0.19	−0.20
4	0.82	−0.02	0.03	0.02	9	0.59	0.01	0.03	0.06
5	0.77	−0.09	−0.04	−0.05	10	0.55	−0.02	−0.09	0.01

$$\bar{z} = 98.5;\ s_z^2 = 149.6;\ \bar{w} = 0.65;\ s_w^2 = 0.7335\ (w_t = \nabla z_t)$$

(a) Draw graphs of the ACF and PACF for z and ∇z.
(b) Identify a model for the series.
(c) Obtain preliminary estimates for the parameters.

6.5. The following table shows the first 10 values of the ACF of z_t and ∇z_t for a series defined by $z_t = 1000 \log_{10} (H_t)$, where H_t is the price of hogs recorded annually by the U.S. Census of Agriculture on January 1 for each of the 82 years 1867–1948.*

k	$r_k(z)$	$r_k(\nabla z)$	k	$r_k(z)$	$r_k(\nabla z)$
1	0.85	0.25	6	0.46	0.14
2	0.67	−0.25	7	0.42	0.18
3	0.55	−0.35	8	0.38	0.02
4	0.51	−0.13	9	0.35	−0.07
5	0.50	0.03	10	0.32	−0.10

* Further details of these data are given in [162].

Based on these autocorrelations, identify a model for the series and obtain preliminary estimates of the parameters. These data are listed as series Q in the collection of time series in Part Five.

CHAPTER 7

7.1. The following table shows calculations for an (unrealistically short) series z_t for which the (0, 1, 1) model $w_t = \nabla z_t = (1 - \theta B)a_t$ with $\theta = -0.5$ is being entertained with an unknown starting value a_0.

t	z_t	$w_t = \nabla z_t$	$a_t = w_t - 0.5a_{t-1}$
0	40		a_0
1	42	2	$2 - 0.50a_0$
2	47	5	$4 + 0.25a_0$
3	47	0	$-2 - 0.13a_0$
4	52	5	$6 + 0.06a_0$
5	51	-1	$-4 - 0.03a_0$
6	57	6	$8 + 0.02a_0$
7	59	2	$-2 - 0.01a_0$

(a) Confirm the entries in the table.
(b) Show that the conditional sum of squares

$$\sum_{t=1}^{7} (a_t|-0.5, a_0 = 0)^2 = S_*(-0.5|0) = 144.00$$

7.2. Using the data in Exercise 7.1:
(a) Show (using least squares) that the value $\hat{a}_0$ of a_0 which minimizes $S_*(-0.5|a_0)$ is

$$\hat{a}_0 = \frac{(2)(0.50) + (4)(-0.25) + \cdots + (-2)(0.01)}{1^2 + 0.5^2 + \cdots + 0.01^2} = \frac{-\Sigma_{t=0}^{n} \theta^t a_t^0}{\Sigma_{t=0}^{n} \theta^{2t}}$$

where $a_t^0 = (a_t|\theta, a_0 = 0)$ are the conditional values. Compare this expression for $\hat{a}_0$ with that for the exact back-forecast $[a_0]$ in the MA(1) model, where the expression for $[a_0]$ is given preceding equation (A7.3.9) in Appendix A7.3, and verify that the two expressions are identical.
(b) By first writing this model in the backward form $w_t = (1 - \theta F)e_t$ and recursively computing the e's, show that the value of a_0 obtained in (a) is the same as that obtained by the back-forecasting method.

7.3. Using the value of $\hat{a}_0$ calculated in Exercise 7.2,
(a) Show that the unconditional sum of squares $S(-0.5)$ is 143.4.
(b) Show that for the (0, 1, 1) model, for large n,

$$S(\theta) = S_*(\theta|0) - \frac{\hat{a}_0^2}{1 - \theta^2}$$

7.4. For the process $w_t - \mu = (1 - \theta B)a_t$ show that for long series the variance-covariance matrix of the maximum likelihood estimates $\hat{\mu}, \hat{\theta}$ is approximately

$$n^{-1}\begin{bmatrix} (1-\theta)^2\sigma_a^2 & 0 \\ 0 & 1-\theta^2 \end{bmatrix}$$

7.5. **(a)** Problems were experienced in obtaining a satisfactory fit to a series, the last 16 values of which were recorded as follows:

129, 135, 130, 130, 127, 126, 131, 152,
123, 124, 131, 132, 129, 127, 126, 124.

Plot the series and suggest where the difficulty might lie.

(b) In fitting a model of the form $(1 - \phi_1 B - \phi_2 B^2)z_t = (1 - \theta B)a_t$ to a set of data, convergence was slow and the coefficients in successive iterations oscillated wildly. Final estimates having large standard errors were obtained as follows: $\hat{\phi}_1 = 1.19$, $\hat{\phi}_2 = 0.34$, $\hat{\theta} = 0.52$. Can you suggest an explanation for the unstable behavior of the model? Why should preliminary identification have eliminated the problem?

(c) In fitting a model $\nabla^2 z_t = (1 - \theta_1 B - \theta_2 B^2)a_t$ convergence was not obtained. The last iteration yielded the values $\hat{\theta}_1 = 1.81$, $\hat{\theta}_2 = 0.52$. Can you explain the difficulty?

7.6. For the ARIMA(1, 1, 1) model $(1 - \phi B)w_t = (1 - \theta B)a_t$, where $w_t = \nabla z_t$,

(a) Write down the linearized form of the model.

(b) Set out how you would start off the calculation of the conditional nonlinear least squares algorithm with start values $\phi = 0.5$ and $\theta = 0.4$ for a series whose first nine values are shown below.

t	z_t	t	z_t
0	149	5	150
1	145	6	147
2	152	7	142
3	144	8	146
4	150		

7.7. **(a)** Show that the second-order autoregressive model $\tilde{z}_t = \phi_1 \tilde{z}_{t-1} + \phi_2 \tilde{z}_{t-2} + a_t$ may be written in orthogonal form as

$$\tilde{z}_t = \frac{\phi_1}{1-\phi_2}\tilde{z}_{t-1} + \phi_2\left(\tilde{z}_{t-2} - \frac{\phi_1}{1-\phi_2}\tilde{z}_{t-1}\right) + a_t$$

suggesting that the approximate estimates

$$r_1 \text{ of } \frac{\phi_1}{1-\phi_2} \quad \text{and} \quad \hat{\phi}_2 = \frac{r_2 - r_1^2}{1 - r_1^2} \text{ of } \phi_2$$

are uncorrelated for long series.

(b) Starting from the variance–covariance matrix of $\hat{\phi}_1$ and $\hat{\phi}_2$ or otherwise, show that the variance–covariance matrix of r_1 and $\hat{\phi}_2$ for long series is given approximately by

$$n^{-1}\begin{bmatrix}(1-\phi_2^2)(1-\rho_1^2) & 0\\ 0 & 1-\phi_2^2\end{bmatrix}$$

CHAPTER 8

8.1. The following are the first 30 residuals obtained when a tentative model was fitted to a time series:

t	Residuals					
1–6	0.78	0.91	0.45	−0.78	−1.90	−2.10
7–12	−0.54	−1.05	0.68	−3.77	−1.40	−1.77
13–18	1.18	0.02	1.29	−1.30	−6.20	−1.89
19–24	0.95	1.49	1.08	0.80	2.02	1.25
25–30	0.52	2.31	1.64	0.78	1.99	1.36

Plot the values and state any reservations you have concerning the adequacy of the model.

8.2. The residuals from a model $\nabla z_t = (1 - 0.6B)e_t$ fitted to a series of $N = 82$ observations yielded the following residual autocorrelations:

k	$r_k(\hat{e})$	k	$r_k(\hat{e})$
1	0.39	6	−0.13
2	0.20	7	−0.05
3	0.09	8	0.06
4	0.04	9	0.11
5	0.09	10	0.02

(a) Plot the residual ACF and determine whether there are any abnormal values.

(b) Calculate the chi-square statistic $\tilde{Q}$ and check whether the residual autocorrelation function as a whole is indicative of model inadequacy.

(c) What modified model would you now tentatively entertain, fit and check?

8.3. Suppose that a (0, 1, 1) model $\nabla z_t = (1 - \theta B)e_t$, corresponding to the use of an exponentially weighted moving average forecast, with θ arbitrarily chosen to be equal to 0.5, was used to forecast a series which was in fact well fitted by the (0, 1, 2) model $\nabla z_t = (1 - 0.9B + 0.2B^2)a_t$.

(a) Calculate the autocorrelation function of the lead 1 forecast errors e_t obtained from the (0, 1, 1) model.

(b) Show how this ACF could be used to identify a model for the e_t series, leading to the identification of a (0, 1, 2) model for the z_t series.

8.4. A long series containing $N = 326$ terms was split into two halves and a (1, 1, 0) model $(1 - \phi B)\nabla z_t = a_t$ identified, fitted and checked for each half. If the estimates of the parameter for the two halves are $\hat{\phi}^{(1)} = 0.5$ and $\hat{\phi}^{(2)} = 0.7$, is there any evidence that the parameter ϕ has changed?

8.5. **(a)** Show that the variance of the sample mean of n observations from a stationary AR(1) process $(1 - \phi B)\tilde{z}_t = a_t$ is given by

$$\text{var}[\bar{z}] \simeq \frac{\sigma_a^2}{n(1-\phi)^2}$$

(b) The yields from consecutive batches of a chemical process obtained under fairly uniform conditions of process control were shown to follow a stationary AR(1) process $(1 + 0.5B)\tilde{z}_t = a_t$. A technical innovation is made at a given point in time leading to 85 data points with mean $\bar{z}_1 = 41.0$ and residual variance $s_{1a}^2 = 0.1012$ before the innovation is made and 60 data points with $\bar{z}_2 = 43.5$, $s_{2a}^2 = 0.0895$ after the innovation. Is there any evidence that the innovation has improved the yield?

CHAPTER 9

9.1. Show that the seasonal difference operator $1 - B^{12}$, often useful in the analysis of monthly data, may be factorized as follows:

$$(1 - B^{12}) = (1 + B)(1 - \sqrt{3}B + B^2)(1 - B + B^2)(1 + B^2)(1 + B + B^2) \times (1 + \sqrt{3}B + B^2)(1 - B)$$

Plot the zeros of this expression in the unit circle and show by actual numerical calculation and plotting of the results that the factors in the order given above correspond to sinusoids with frequencies (in cycles per year) of 6, 5, 4, 3, 2, 1, together with a constant term. [For example, the difference equation $(1 - B + B^2)x_t = 0$ with arbitrary starting values $x_1 = 0$, $x_2 = 1$ yields $x_3 = 1$, $x_4 = 0$, $x_5 = -1$, etc., generating a sine wave of frequency 2 cycles per year.]

9.2. A method which has sometimes been used for "deseasonalizing" monthly time series employs an equally weighted 12-month moving average

$$\bar{z}_t = \tfrac{1}{12}(z_t + z_{t-1} + \cdots + z_{t-11})$$

(a) Using the decomposition $(1 - B^{12})/(1 - B) = 1 + B + B^2 + \cdots + B^{11}$, show that $12(\bar{z}_t - \bar{z}_{t-1}) = (1 - B^{12})z_t$.

(b) The exceedance for a given month over the previous moving average may be computed as $z_t - \bar{z}_{t-1}$. A quantity u_t may then be calculated which compares the current exceedance with the average of similar monthly exceedances experienced over the last k years. Show that u_t may be written as

$$u_t = \left(1 - \frac{B}{12}\frac{1 - B^{12}}{1 - B}\right)\left(1 - \frac{B^{12}}{k}\frac{1 - B^{12k}}{1 - B^{12}}\right) z_t$$

9.3. It has been shown [186] that monthly averages for the (smog-producing) oxidant level in Azusa, California, may be represented by the model

$$(1 - B^{12})z_t = (1 + 0.2B)(1 - 0.9B^{12})a_t, \quad \sigma_a^2 = 1.0$$

(a) Compute and plot the ψ weights.

(b) Compute and plot the π weights.

(c) Calculate the standard deviation of the forecast three months and 12 months ahead.

(d) Obtain the eventual forecast function.

9.4. The monthly oxidant averages in parts per hundred million in Azusa from January 1969–December 1972 were as follows:

	J	F	M	A	M	J	J	A	S	O	N	D
1969	2.1	2.6	4.1	3.9	6.7	5.1	7.8	9.3	7.5	4.1	2.9	2.6
1970	2.0	3.2	3.7	4.5	6.1	6.5	8.7	9.1	8.1	4.9	3.6	2.0
1971	2.4	3.3	3.3	4.0	3.6	6.2	7.7	6.8	5.8	4.1	3.0	1.6
1972	1.9	3.0	4.5	4.2	4.8	5.7	7.1	4.8	4.2	2.3	2.1	1.6

Using the model of Exercise 9.3, compute the forecasts for the next 24 months. (Approximate unknown a's by zeros.)

9.5. Thompson and Tiao [185] have shown that the outward station movements of telephones (logged data) in Wisconsin are well represented by the model

$$(1 - 0.5B^3)(1 - B^{12})z_t = (1 - 0.2B^9 - 0.3B^{12} - 0.2B^{13})a_t$$

Obtain and plot the autocorrelation function for $w_t = (1 - B^{12})z_t$ for lags $1, 2, \ldots, 24$.

9.6. The following table shows the first 12 autocorrelations of various differences of a series of $N = 41$ observations consisting of quarterly deposits in a bank.

	k													
	1	2	3	4	5	6	7	8	9	10	11	12	$\overline{w}$	s_w^2
$r_k(z)$	0.88	0.83	0.72	0.67	0.56	0.53	0.42	0.41	0.32	0.29	0.21	0.18	152.0	876.16
$r_k(\nabla z)$	−0.83	0.68	−0.70	0.70	−0.65	0.62	−0.70	0.69	−0.60	0.55	−0.60	0.62	2.46	97.12
$r_k(\nabla_4 z)$	0.21	0.06	0.15	−0.34	−0.10	0.03	−0.18	−0.13	0.02	−0.04	0.10	0.22	10.38	25.76
$r_k(\nabla\nabla_4 z)$	−0.33	−0.06	0.36	−0.45	0.08	0.22	−0.10	−0.05	0.14	−0.12	−0.05	0.28	0.42	38.19

Plot these autocorrelation functions and identify a model (or models) for the series. Calculate preliminary estimates for the parameters and for σ_a^2.

9.7. In the analysis of a series consisting of the logarithms of the sales of a seasonal product, it was found that differencing of the form $w_t = \nabla\nabla_{12} \ln(z_t)$ was needed to induce stationarity with respect to both monthly and seasonal variation. The following table shows the first 48 autocorrelations of the w_t series that contained $n = 102$ monthly observations.

Lag	Autocorrelations											
1–12	−0.39	−0.24	0.17	0.21	−0.27	−0.03	0.26	−0.10	−0.20	0.07	0.44	−0.58
13–24	0.09	0.17	0.01	−0.24	0.16	0.04	−0.12	−0.01	0.11	0.08	−0.33	0.28
25–36	0.01	−0.14	−0.02	0.18	−0.13	0.04	−0.01	0.10	−0.13	−0.09	0.27	−0.22
37–48	0.00	0.09	0.02	−0.18	0.17	−0.05	0.00	−0.06	0.06	0.06	−0.13	0.11

$$\overline{w} = 0.241 \qquad s_w^2 = 106.38$$

Identify a suitable model (or models) for the series and obtain preliminary estimates of the parameters.

CHAPTER 10

10.1. In the following transfer function models, X_t is the methane gas feedrate to a gas furnace, measured in cubic feet per minute, and Y_t the percent carbon dioxide in the outlet gas.

$$(1)\ Y_t = 10 + \frac{25}{1 - 0.7B} X_{t-1}$$

$$(2)\ Y_t = 10 + \frac{22 - 12.5B}{1 - 0.85B} X_{t-2}$$

$$(3)\ Y_t = 10 + \frac{20 - 8.5B}{1 - 1.2B + 0.4B^2} X_{t-3}$$

(a) Verify that the models are stable.
(b) Calculate the steady-state gain g, expressing it in the appropriate units.

10.2. For each of the models of Exercise 10.1, calculate from the difference equation and plot the responses to:
(a) A unit impulse $(0, 1, 0, 0, 0, 0, \ldots)$ applied at time $t = 0$
(b) A unit step $(0, 1, 1, 1, 1, 1, \ldots)$ applied at time $t = 0$
(c) A ramp input $(0, 1, 2, 3, 4, 5, \ldots)$ applied at time $t = 0$
(d) A periodic input $(0, 1, 0, -1, 0, 1, \ldots)$ applied at time $t = 0$
Estimate the period and damping factor of the step response to model (3).

10.3. Use equations (10.2.8) to obtain the impulse weights v_j for each of the models of Exercise 10.1, and check that they are the same as the impulse response obtained in Exercise 10.2(a).

10.4. Express the models of Exercise 10.1 in ∇ form.

10.5. **(a)** Calculate and plot the response of the two input system

$$Y_t = 10 + \frac{6}{1 - 0.7B} X_{1,t-1} + \frac{8}{1 - 0.5B} X_{2,t-2}$$

to the orthogonal and randomized input sequences shown below.

t	X_{1t}	X_{2t}	t	X_{1t}	X_{2t}
0	0	0	5	1	−1
1	−1	1	6	1	1
2	1	−1	7	−1	−1
3	−1	−1	8	−1	1
4	1	1			

(b) Calculate the gains g_1 and g_2 of Y with respect to X_1 and X_2, respectively, and express the model in ∇ form.

CHAPTER 11

11.1. If two series may be represented in ψ-weight form as

$$y_t = \psi_y(B)a_t \qquad x_t = \psi_x(B)a_t$$

(a) Show that the cross covariance generating function

$$\gamma^{xy}(B) = \sum_{k=-\infty}^{\infty} \gamma_{xy}(k)B^k$$

is given by $\psi_y(B)\psi_x(F)\sigma_a^2$

(b) Use the result above to obtain the cross covariance function between y_t and x_t when

$$y_t = (1 - \theta B)a_t \qquad x_t = (1 - \theta_1' B - \theta_2' B^2)a_t$$

11.2. After estimating a prewhitening transformation $\theta_x^{-1}(B)\phi_x(B)x_t = \alpha_t$ for an input series x_t and then computing the transformed output $\beta_t = \theta_x^{-1}(B)\phi_x(B)y_t$, cross correlations $r_{\alpha\beta}(k)$ were obtained as follows:

k	$r_{\alpha\beta}(k)$	k	$r_{\alpha\beta}(k)$
0	0.05	5	0.24
1	0.31	6	0.07
2	0.52	7	−0.03
3	0.43	8	0.10
4	0.29	9	0.07

with $\hat{\sigma}_\alpha = 1.26$, $\hat{\sigma}_\beta = 2.73$, $n = 187$.

(a) Obtain approximate standard errors for the cross correlations.

(b) Calculate rough estimates for the impulse response weights v_j.

(c) Suggest a model form for the transfer function with rough estimates of its parameters.

11.3. It is frequently the case that the user of an estimated transfer function model $y_t = \omega(B)\delta^{-1}(B)B^b x_t$ will want to establish whether the steady-state gain $g = \omega(1)\delta^{-1}(1)$ makes sense.

(a) For the first-order system

$$y_t = \frac{\omega_0}{1 - \delta B} x_{t-1}$$

show that an approximate standard error $\hat{\sigma}(\hat{g})$ of the estimate $\hat{g} = \hat{\omega}_0/(1 - \hat{\delta})$ is given by

$$\frac{\hat{\sigma}^2(\hat{g})}{\hat{g}^2} \simeq \frac{\text{var}[\hat{\omega}_0]}{\hat{\omega}_0^2} + \frac{\text{var}[\hat{\delta}]}{(1 - \hat{\delta})^2} + \frac{2\,\text{cov}[\hat{\omega}_0, \hat{\delta}]}{\hat{\omega}_0(1 - \hat{\delta})}$$

(b) Calculate $\hat{g}$ and an approximate value for $\hat{\sigma}(\hat{g})$ when $\hat{\omega}_0 = 5.2$, $\hat{\delta} = 0.65$, $\hat{\sigma}(\hat{\omega}_0) = 0.5$, $\hat{\sigma}(\hat{\delta}) = 0.1$, $\text{cov}[\hat{\omega}_0, \hat{\delta}] = 0.025$.

11.4. Consider the model

$$Y_t = \beta_1 X_{1,t} + \beta_2 X_{2,t} + N_t$$

where N_t is a nonstationary error term and $\nabla N_t = a_t - \theta a_{t-1}$. Show that the model may be rewritten in the form

$$Y_t - \overline{Y}_{t-1} = \beta_1(X_{1,t} - \overline{X}_{1,t-1}) + \beta_2(X_{2,t} - \overline{X}_{2,t-1}) + a_t$$

where $\overline{Y}_{t-1}$, $\overline{X}_{1,t-1}$, $\overline{X}_{2,t-1}$ are exponentially weighted moving averages so that for example:

$$\overline{Y}_{t-1} = (1 - \theta)(Y_{t-1} + \theta Y_{t-2} + \theta^2 Y_{t-3} + \cdots)$$

It will be seen that the fitting of this nonstationary model by maximum likelihood is equivalent to fitting the *deviations* of the independent and dependent variables from *local updated exponentially weighted moving averages* by ordinary least squares. (Refer to Section 9.4.2 for related ideas.)

11.5. Consider the unemployment and gross domestic product data from Exercises 6.3 and 6.4, respectively, which are listed as series P in the collection of time series in Part Five. Build (identify, estimate, and check) a transfer function-noise model that uses the GDP series X_t as input to help explain variations in the logged unemployment series Y_t.

CHAPTER 12

12.1. In an analysis [59] of monthly data y_t on smog-producing oxidant, allowance was made for two possible "interventions" I_1 and I_2 as follows:

I_1: In early 1960, diversion of traffic from the opening of the Golden State Freeway and the coming into effect of a law reducing reactive hydrocarbons in gasoline sold locally.

I_2: In 1966, the coming into effect of a law requiring all new cars to have modified engine design. In the case of this intervention, allowance was made for the well-known fact that the smog phenomenon is different in summer and winter months.

In a pilot analysis of the data the following intervention model was used:

$$y_t = \omega_1\xi_{1t} + \frac{\omega_2}{1 - B^{12}}\xi_{2t} + \frac{\omega_3}{1 - B^{12}}\xi_{3t} + \frac{(1 - \theta B)(1 - \Theta B^{12})}{1 - B^{12}}a_t$$

where

$$\xi_{1t} = \begin{cases} 0, & t < \text{Jan. } 1960 \\ 1, & t \geq \text{Jan. } 1960 \end{cases} \qquad \xi_{2t} = \begin{cases} 0, & t < \text{Jan. } 1966 \\ 1, & t \geq \text{Jan. } 1966 \end{cases} \qquad \xi_{3t} = \begin{cases} 0, & t < \text{Jan. } 1966 \\ 1, & t \geq \text{Jan. } 1966 \end{cases}$$

(summer months) (winter months)

(a) Show that the model allows for:

(1) A possible step change in January 1960 of size ω_1, possibly produced by I_1.

(2) A "staircase function" of annual step size ω_2 to allow for possible summer effect of cumulative influx of cars with new engine design.

(3) A "staircase function" of annual size ω_3 to allow for possible winter effect of cumulative influx of cars with new engine design.

(b) Describe what steps you would take to check the representational adequacy of the model.

(c) Assuming you were satisfied after (b), what conclusions would you draw from the following results? (Estimates are shown with their standard errors below in parentheses.)

$$\hat{\omega}_1 = -1.09, \quad \hat{\omega}_2 = -0.25, \quad \hat{\omega}_3 = -0.07, \quad \hat{\theta} = -0.24, \quad \hat{\Theta} = 0.55$$
$$(\pm 0.13) \qquad (\pm 0.07) \qquad (\pm 0.06) \qquad (\pm 0.03) \qquad (\pm 0.04)$$

(d) The data for this analysis are listed as series R in the collection of time series in Part Five. Use these data to perform your own intervention analysis.

12.2. A general transfer function model of the form

$$y_t = \sum_{j=1}^{k} \omega_j(B)\delta_j^{-1}(B)\xi_{jt} + \theta(B)\phi^{-1}(B)a_t$$

can include input variables ξ_j which are themselves time series and other inputs ξ_i which are indicator variables. The latter can estimate (and eliminate) the effects of interventions of the kind described in Exercise 12.1 and, in particular, are often useful in the analysis of sales data.

Let $\xi_t^{(T)}$ be an indicator variable which takes the form of a unit pulse at time T, that is

$$\xi_t^{(T)} = \begin{cases} 0 & t \neq T \\ 1 & t = T \end{cases}$$

For illustration, consider the models

(1) $y_t = \dfrac{\omega_1 B}{1 - \delta B}\xi_t^{(T)}$ (with $\omega_1 = 1.00$, $\delta = 0.50$)

(2) $y_t = \left(\dfrac{\omega_1 B}{1 - \delta B} + \dfrac{\omega_2 B}{1 - B}\right)\xi_t^{(T)}$ (with $\omega_1 = 1.00$, $\delta = 0.50$, $\omega_2 = 0.30$)

(3) $y_t = \left(\omega_0 + \dfrac{\omega_1 B}{1 - \delta B} + \dfrac{\omega_2 B}{1 - B}\right)\xi_t^{(T)}$ (with $\omega_0 = 1.50$, $\omega_1 = -1.00$, $\delta = 0.50$, $\omega_2 = -0.50$)

Compute recursively the response y_t for each of these models at times $t = T$, $T + 1$, $T + 2$, . . . and comment on their possible usefulness in the estimation and/or elimination of effects due to such phenomena as advertising campaigns, promotions and price changes.

12.3. For the tentatively identified model obtained for the price of hogs series in Exercise 6.5, fit the model by least squares to obtain final parameter estimates and perform model diagnostic checks on the residuals. In particular, entertain the possibility of outliers, or intervention events, for these data, during the periods 1932–1935 and 1942–1948. If appropriate, specify transfer function terms with indicator input variables to modify the basic ARIMA model and estimate the specified model.

CHAPTER 13

13.1. In a chemical process, 30 successive values of viscosity N_t which occurred during a period when the control variable (gas rate) X_t *was held fixed* at its standard reference origin were recorded as follows:

Time	Viscosities									
1–10	92	92	96	96	96	98	98	100	100	94
11–20	98	88	88	88	96	96	92	92	90	90
21–30	90	94	90	90	94	94	96	96	96	96

Reconstruct and plot the error sequence (deviations from target) ε_t and adjustments x_t which would have occurred if the optimal feedback control scheme

$$x_t = -10\varepsilon_t + 5\varepsilon_{t-1} \tag{1}$$

had been applied during this period. It is given that the dynamic model is

$$\mathcal{y}_t = 0.5\,\mathcal{y}_{t-1} + 0.10x_{t-1} \tag{2}$$

and that the error signal may be obtained from

$$\varepsilon_t = \varepsilon_{t-1} + \nabla N_t + \mathcal{y}_t \tag{3}$$

Your calculation sequence should proceed in the order (2), (3), and (1) and initially you should assume that $\varepsilon_1 = 0$, $\mathcal{y}_1 = 0$, $x_1 = 0$. Can you devise a more direct way to compute ε_t from N_t?

13.2. Given the following combinations of disturbance and transfer function models:

(1)
$$\nabla N_t = (1 - 0.7B)a_t$$
$$(1 - 0.4B)\mathcal{Y}_t = 5.0X_{t-1+}$$

(2)
$$\nabla N_t = (1 - 0.5B)a_t$$
$$(1 - 1.2B + 0.4B^2)\mathcal{Y}_t = (20 - 8.5B)X_{t-1+}$$

(3)
$$\nabla^2 N_t = (1 - 0.9B + 0.5B^2)a_t$$
$$(1 - 0.7B)\mathcal{Y}_t = 3X_{t-1+}$$

(4)
$$\nabla N_t = (1 - 0.7B)a_t$$
$$(1 - 0.4B)\mathcal{Y}_t = 5.0X_{t-2+}$$

(a) Design the minimum mean square error feedback control schemes associated with each combination of disturbance and transfer function model.

(b) For case (4), derive an expression for the error ε_t and for its variance in terms of σ_a^2.

(c) For case (4), design a nomogram suitable for carrying out the control action manually by a process operator.

13.3. In a treatment plant for industrial waste, the strength z_t of the influent is measured every 30 minutes and can be represented by the model $\nabla z_t = (1 - 0.5B)a_t$. In the absence of control the strength of the effluent Y_t is related to that of the influent z_t by

$$\tilde{Y}_t = \frac{0.3B}{1 - 0.2B}\tilde{z}_t$$

An increase in strength in the waste may be compensated by an increase in the flow u_t of a chemical to the plant, whose effect on Y_t is represented by the model

$$\tilde{Y}_t = \frac{21.6B^2}{1 - 0.7B}\tilde{u}_t$$

Show that minimum mean square error feedforward control is obtained with the control equation

$$\tilde{u}_t = -\frac{0.3}{21.6}\left[\frac{(0.7 - 0.2B)(1 - 0.7B)}{(1 - 0.2B)(1 - 0.5B)}\right]\tilde{z}_t$$

that is, $\tilde{u}_t = 0.7\tilde{u}_{t-1} - 0.1\tilde{u}_{t-2} - 0.0139(0.7\tilde{z}_t - 0.69\tilde{z}_{t-1} + 0.14\tilde{z}_{t-2})$.

13.4. A pilot feedback control scheme, based on the following disturbance and transfer function models:

$$\nabla N_t = a_t$$

$$(1 - \delta B)\mathcal{Y}_t = \omega_0 X_{t-1+} - \omega_1 X_{t-2+}$$

was operated, leading to a series of adjustments x_t and errors ε_t. It was believed that the noise model was reasonably accurate but that the parameters of the transfer function model were of questionable accuracy.

(a) Given the first 10 values of the x_t, ε_t series shown below:

t	x_t	ε_t	t	x_t	ε_t
1	25	−7	6	−30	1
2	42	−7	7	−25	3
3	3	−6	8	−25	4
4	20	−7	9	20	0
5	5	−4	10	40	−3

set out the calculation of the residuals $a_t (t = 2, 3, \ldots, 10)$ for $\delta = 0.5$, $\omega_0 = 0.3$, $\omega_1 = 0.2$ and for arbitrary starting values $\mathcal{Y}_1^0$ and x_0^0.

(b) Calculate the values $\hat{\mathcal{Y}}_1$, $\hat{x}_0$ of $\mathcal{Y}_1^0$ and x_0^0 that minimize the sum of squares $\sum_{t=2}^{10}(a_t|\delta = 0.5, \omega_0 = 0.3, \omega_1 = 0.2, \mathcal{Y}_1^0, x_0^0)^2$ and the value of this minimum sum of squares.

13.5. Consider [55] a system for which the process transfer function is gB and the noise model is $(1 - B)N_t = (1 - \theta B)a_t$ so that the error ε_t at the output satisfies

$$(1 - B)\varepsilon_t = g(1 - B)X_{t-1+} + (1 - \theta B)a_t$$

Suppose that the system is controlled by a known discrete 'integral' controller

$$(1 - B)X_{t+} = -c\varepsilon_t$$

(a) Show that the errors ε_t at the output will follow the ARMA(1, 1) process

$$(1 - \phi B)\varepsilon_t = (1 - \theta B)a_t \qquad \phi = 1 - gc$$

and hence that the problem of estimating g and θ using data from a pilot control scheme is equivalent to that of estimating the parameters in this model.

(b) Show also that the optimal scheme is such that $c = c_0 - (1 - \theta)/g$ and hence that if the pilot scheme used in collecting the data happens to be optimal already, then $1 - \theta$ and g cannot be separately estimated.

AUTHOR INDEX

A

B

C

D

F

G

H

W

Y

Z

SUBJECT INDEX

E

F

G

H

N

O

P

Q

R

S

T

U

V

W

Y